Thin Films: Stresses and Mechanical Properties VI

MATERIALS RESEARCH SOCIETY
SYMPOSIUM PROCEEDINGS VOLUME 436

Thin Films: Stresses and Mechanical Properties VI

Symposium held April 8-12, 1996, San Francisco, California, U.S.A.

EDITORS:

William W. Gerberich
University of Minnesota
Minneapolis, Minnesota, U.S.A.

Huajian Gao
Stanford University
Stanford, California, U.S.A.

Jan-Eric Sundgren
Linköping University
Linköping, Sweden

Shefford P. Baker
Max-Planck-Institut für Metallforschung
Stuttgart, Germany

MATERIALS RESEARCH SOCIETY

PITTSBURGH, PENNSYLVANIA

Single article reprints from this publication are available through
University Microfilms Inc., 300 North Zeeb Road, Ann Arbor, Michigan 48106

CODEN: MRSPDH

Published by:

Materials Research Society
9800 McKnight Road
Pittsburgh, Pennsylvania   15237
Telephone (412) 367-3003
Fax (412) 367-4373
Website: http://www.mrs.org/

Library of Congress Cataloging in Publication Data

Thin films: stresses and mechanical properties VI: symposium held April 8–12,
     1996, San Francisco, California, U.S.A. / editors, W.W. Gerberich, H. Gao,
     J-E. Sundgren, S.P. Baker
          p. cm—(Materials Research Society symposium proceedings ; v. 436)
          includes bibliographical references and index.
          ISBN: 1-55899-339-8
     I. Gerberich, W.W. II. Gao, H. III. Sundgren, J-E. IV. Baker, S.P.
     V. Series: Materials Research Society symposium proceedings; v. 436.

Manufactured in the United States of America

CONTENTS

PART I: MECHANICAL PROPERTIES OF FILMS AND MULTILAYERS

*Invited Paper

*Invited Paper

*Invited Paper

PART VI: <u>PROPERTIES OF POLYMER FILMS</u>

*Invited Paper

*Invited Paper

PREFACE

Interest in the mechanical properties of thin films remains high throughout the world, as evidenced by the large international contingent represented at this symposium. This volume contains papers presented at the symposium "Thin Films: Stresses and Mechanical Properties VI" at the 1996 MRS Spring Meeting. Regarding stresses, techniques are becoming quite varied and sophisticated in sorting out residual stress and strain states, including Raman scattering, nonlinear acoustic responses, and back-scattered electron imaging microscopies, along with more standard wafer-bending and X-ray techniques. Regarding mechanical behavior, the bourgeoning field of nanoprobe imaging, spectroscopy and indenting for the characterization of mechanical properties of thin films is highlighted.

The symposium was kicked off by a tutorial by Shefford Baker and Paul Townsend entitled *Mechanical Properties of Thin Films.* The following four days of papers, as represented in this volume, dealt with new issues. The first two parts of this volume deal with mechanical properties of films and multilayers, including their stress evolution and structure/property relations as related to strength, fracture and adhesion. With regards to the measurement and understanding of such structure/property relations, the next two sections deal with the nanoindentation of films and surfaces, as well as the mechanical property methods and modeling, the latter including atomistic and finite element approaches. With a major driving force being the protective aspects of softer substrates, the next two parts emphasize the tribological properties of thin films and properties of polymer films. In the former, the ability to measure various characteristics of diamond-like carbon films is at the leading edge. One of the driving forces here is in magnetic recording, where plans for recording densities in the range of $400\,Gbits/in^2$ by 2005 require quasi-contact recording. In the last two sections, stress effects in thin films and interconnects and epitaxy and strain relief mechanisms are emphasized. The joint session with Symposium L (on the stress effects) was particularly successful.

As with the previous proceedings from this symposium series, the papers contained in this volume represent a broad range of current research efforts in the mechanical properties of thin films. It is our expectation that further developments in this field will provide additional stimulating symposia in the future.

William W. Gerberich
Huajian Gao
Jan-Eric Sundgren
Shefford P. Baker

June 1996

ACKNOWLEDGMENTS

This work was supported by contributions from both Materials Test Systems, Incorporated and Hysitron, Incorporated, Minneapolis, Minnesota.

This work was also supported in part by the Department of Energy, Basic Energy Sciences, Division of Materials Research under Grant Number DE-AC04-94AL-85000. The United States Government has a royalty-free license throughout the world in all copyrightable materials contained herein.

MATERIALS RESEARCH SOCIETY SYMPOSIUM PROCEEDINGS

MATERIALS RESEARCH SOCIETY SYMPOSIUM PROCEEDINGS

Part I

Mechanical Properties of Films and Multilayers

Near-Plastic Threshold Indentation and the Residual Stress in Thin Films

J. E. Houston and T. A. Michalske
Sandia National Laboratories
Albuquerque, NM 87185-1413

Abstract

In recent studies, we used the Interfacial Force Microscope in a nanoindenter mode to survey the nanomechanical properties of Au films grown on various substrates. Quantitative tabulations of the indentation modulus and the maximum shear stress at the plastic threshold showed consistent values over individual samples but a wide variation from substrate to substrate. These values were compared with film properties such as the surface roughness, average grain size and interfacial adhesion and no correlation was found. However, in a subsequent analysis of the the results, we found consistencies which support the integrity of the data and point to the fact that the results are sensitive to some property of the various film/substrate combinations. In the present paper, we discuss these consistencies and show recent measurements which strongly suggest that the property that is being probed is the residual stress in the films caused by their interaction with the substrate surfaces.

Introduction

As the structure of advanced materials become ever smaller (the so called nanophase materials) and the sizes of electromechanical devices shrink (into the realm of nanofabrication), it becomes increasingly important to be able to determine material properties on the nanometer scale. In recent work, we have explored the use of the Interfacial Force Microscope (IFM) in studies of the nanomechanical properties of surfaces [1-3]. The IFM is a scanning force microscopy similar to the atomic force microscope but is distinguished by its use of a stable, self-balancing force sensor. Not only does this sensor eliminate the "snap to contact" so prevalent in adhesion, scanning probe and indentor studies but it also represents a zero compliance sensor, i.e., an applied force does not produce a sensor displacement and no sensor-stored energy results. This work involved a parabolic W tip indenting Au surfaces which had been passivated by a monolayer of alkanethiol self-assembling molecules (SAM). Under these conditions, the normally strong W-Au interfacial interaction was eliminated and a classic Hertzian contact mechanics was observed.

In order to investigate the efficacy of the IFM for nanomechanical studies, we surveyed a series of Au films grown under various conditions on various substrates. Values were tabulated for the indentation modulus and the maximum shear stress at the plastic threshold by averaging the measurements many times on each film/substrate combination. The results were very consistent for each of the individual samples but varied widely between the various sustrates. The source of the variation was not known and it was speculated to be the result of film properties such as grain size, surface roughness or film adhesion. In the present paper, we report the results of a more careful analysis of our earlier results in order to try and determine the origin of the wide variation in the observed nanomechanical properties for these films. In addition, we

Mat. Res. Soc. Symp. Proc. Vol. 436 © 1997 Materials Research Society

present the results of additional experiments which stongly suggest that the origin of the variation in the measured nanomechanical properties is the variation in the residual film stress for the various film/substrate combinations.

Results

In the earlier study, force profiles of the kind shown in Fig. 1 were analyzed to obtain the nanomechanical properties of each film/substrate combination. These data were obtained by ramping the W probe at a constant rate (20 nm/sec) into contact with the Au surface. The direction of the probe motion was reversed after a certain level of repulsive force was attained and the probe was removed from contact at the same rate. The films consisted of 200 nm thick Au deposited on cleaned surface of glass, mica and Si(001) with both 10 nm adhesion layers of Ti and Cr. The glass and mica depositions were done at 300 C followed by a 3 hr anneal at 275 C while the Au/Si depositions were done at room temperature. The films were then cleaned and a SAM (n-octadecanethiol) deposited. All measurements were made with the probe emersed in a drop of hexadecane to suppress the attractive van der Waal's interaction. All measurements were made with the same sensor and the same 250 nm radius tip.

Data like that in Fig. 1 were analyzed by standard contact-mechanics techniques [5]. The initial rise of the force with displacement is elastic and follows the Herzian relationship, while the deviation from this form signals the onset of plasticity. The sudden drops in force after the plastic threshold result from material relaxation (nano-quake events) and are only seen in this form with a zero-compliance sensor.

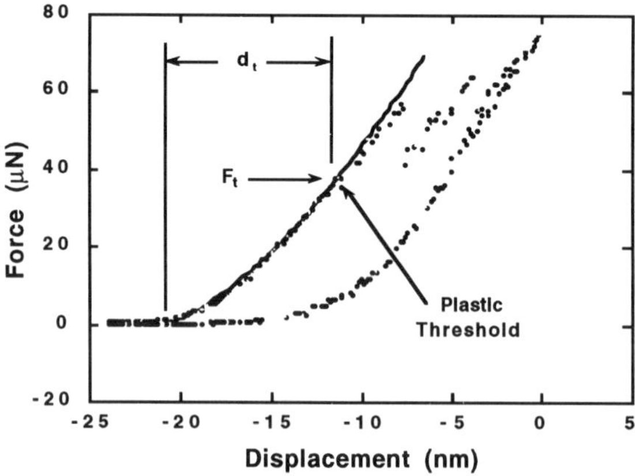

Fig. 1

Figure 1 A typical loading curve for a 200 nm Au film on a glass substrate. The solid line illustrates the fit of the Hertzian relationship to the data, which permits an evaluation of the quantities F_t and d_t.

The initial portion of the loading curve after contact is characterized by the Hertzian relation,

$$F = \frac{4}{3}\sqrt{R} \cdot E \cdot d^{3/2} \tag{1}$$

where F is the applied probe force, R is the radius of the tip, E is the indentation modulus and d is the deformation. From the fit to the data of this expression (shown as the solid curve in Fig. 1) and a knowledge of the tip radius, one can calculate the modulus value. The force and deformation at the point where the loading curve deviates from the Hertzian behavior signals the onset of plasticity. The maximum shear stress at this point can be calculated from the expression [5],

$$S_m = 0.47 \cdot \frac{F_t}{\pi \cdot \sqrt{R\, d_t}} \tag{2}$$

where S_m is the maximum shear stress and F_t and d_t are the probe force and deformation values at the plastic threshold.

Table I shows the results of our survey tabulated in ascending order of the modulus values. The error figures represent the statistical variation over 15-20 individual loading curves for each film/substrate combination and the value for single-crystal Au(111) has been included for comparison. Also shown are the values for the average grain sizes and the mean surface roughness of the surface for the various film/substrate combinations, as well as a column representing recent qualitative measurements of the film/substrate adhesion obtained from simple adhesive-tape stripping and stylus scratch tests.

Table I

Summary of the findings in the study of nanomechanical properties of 200 nm Au films grown on various substrates and for single-crystal Au(111) [4].

Sample	Elastic Modulus E (Gpa)	Max. Shear S_m (Gpa)	Grain Diameter (nm)	Roughness (nm)	Substrate Adhesion
Au/Mica	36 ± 5	1.7 ± 0.2	250	5.4	Very Weak
Au/Ti/Si	48 ± 5	2.1 ± 0.3	60	1.8	Strong
Au/Glass	75 ± 15	2.7 ± 0.5	500	4.1	Weak
Au/Cr/Si	110 ± 19	4.5 ± 0.4	150	2.4	Very Strong
Au(111)	70 ± 6	2.9 ± 0.1			

Discussion and Conclusions

As can be seen from Table I, the E and S_m values vary by almost a factor of three when, in fact, one would expect them to remain constant. In addition, it is clear that there is no correlation of this variation with any of the other film parameters listed. However, a consistancy in the data, which was not noted in the earlier publication, concerns the fact that the E and S_m values vary in the same way as the substrates are changed. In fact, the ratio S_m/E is very close to being constant at 4%. The ratio of S_m/E is a common figure of merit in material science and, in fact, can be crudely approximated from the Frenkel approximation to be ~5.5% [6]. Furthermore, if Eq. (1) is substituted into Eq. (2), we can show that a constant S_m/E ratio implies that the deformation to the plastic threshold d_t should also be a constant. Table II tabulates the narrow range of variation of these two constants for the various film/substrate combinations.

Table II

Values for the ratio S_m/E and the measured
values of d_t for the various film/substrate combinations.

Sample	S_m/E (%)	d_t (nm)
Au/Mica	4.7	13 ± 1
Au/Ti/Si	4.4	15 ± 2
Au/Glass	3.6	9 ± 1
Au/Cr/Si	4.1	13 ± 2

The remarkable consistency in the two parameters shown in Table II, in spite of the wide variation in film properties for the various substrates, strongly suggests that the wide range in measured E and S_m values has its origin in some other film property. A possible clue to this property can be found in the recent work with the nanoindentor on Al sample under lateral stress. Pharr and coworkers discussed earlier indentation work on films under stress and shows data for Al samples under applied lateral stress using a four-fold diamond Berkovich probe well beyond the initial plastic threshold. The results show a dependance of the measured modulus and hardness values on the applied lateral stress in the sample such that the ratio of the values remained essentially constant [7].

While this is an intriguing possibility for explaining our earlier results, there are several features which are different. First, the affect was only found to be at the 10% level. Second, Pharr, et al. conclude that the variation in values results from a stress-induced "pileup" of material at the periphery of the contact. Compressive stress causes additional pileup while tensile stress reduces the affect. The pileup changes the contact area and affects the calculation of the mechanical properties, which were made with contact areas <u>calculated</u> from the loading curves. When the loading curves were corrected for the measured contact areas, the affect

disappeared. Still, varying residual stresses in thin films are known to exist and the affect on
near-threshold indentation results bears investigation.

As a preliminary look at the affect of residual film stress on near-threshold indentation
results, we measured the stresses in indentically prepared films for two of our four film/substrate
combinations, i.e., 200 nm Au films deposited at room temperature on Si(001) substrates with 10 nm Cr
and Ti adhesion layers. The stresses were measured using a wafer-curvature technique [8] which
calculates film stress by determining the change in wafer warp before and after the deposition of the film.
This change, along with the elastic properties and thickness of the wafer material are used to calculate the
value of film stress. The results of these measurements, along with the corresponding values from Table
I are tabulated in Table III.

Table III

The correlation between the E and S_m values for two of the film/
substrate combinations with the measured residual-film stress.

Sample	E (GPa)	S_m (GPa)	Film Stress (MPa)
Au/Ti/Si(001)	48 ± 5	2.1 ± 0.3	+140
Au/Cr/Si(001)	110 ± 19	4.5 ± 0.4	-325

It is clear from the values tabulated in Table III that there is correlation between the E and
S_m values and the residual-film stress. The values of both E and S_m for Au/Ti/Si(001) are below
those for Au(111) while Au/Cr/Si(001) shows values greater than Au(111). The corresponding
stress figures indicate a compressive stress for Au/Ti/Si(001) and tensile for Au/Cr/Si(001). In
fact, even the scaling of the magnitudes are correlated. Although these results are preliminary, it
seems clear that the large variation in the near-threshold indentation measurements of E and S_m
for the various film/substrate combinations results from residual-film stress resulting from the
interfacial mismatch in structural properties for Au with these substrates. What is not clear from
these results is the details of the mechanisms that are responsible for this correlation. Illucidating
these mechanisms will have to await more careful measurements with films whose lateral stress
can be accurately controlled over a broad range (in a manner such as that employed by Pharr and
coworkers [7]), as well as calculations modeling experiments on similar systems.

ACKNOWLEDGEMENTS

The authors wish to express their gratitude to C. M. Matzke for sample preparation and
performing the residual stress measurements. We would also like to acknowledge support from
the U. S. Department of Energy, Office of Basic Energy Sciences under Contract DE-AC04-
94AL85000.

REFERENCES

1. S. A. Joyce, R. C. Thomas, J. E. Houston, T. A. Michalske and R. M. Crooks, Phys. Rev. Lett. 68, 2790 (1992).
2. R. C. Thomas,. J. E. Houston,. T. A. Michalske and R. M. Crooks, Science 259, 1883 (1993).
3. P. Tangyunyong, R. C. Thomas, J. E. Houston, T. A. Michalske, R. M. Crooks and A. J. Howard, Phys. Rev. Lett. 71, 3319 (1993).
4. P. Tangyunyong, R. C. Thomas, J. E. Houston, T. A. Michalske, R. M. Crooks and A. J. Howard, J. Adhes. Sci. Technol. 8, 897 (1994).
5. S. P. Timoshenko and J. N. Goodier, *Theory of Elasticity*, McGraw-Hill, New York (1970), Chapt. 12.
6. See, for example: R. W. Hertzberg, *Deformation and Fracture Mechanics of Engineering Materials*, John Wiley and Sons, New York 1989, Chap. 2.
7. G. M. Pharr, T. Y. Tsui, A. Boshakov and W. C. Oliver, Mat. Res. Soc. Proc. **338**, 127 (1994).
8. Flexus, Inc., Sunnyvale, CA

STRUCTURAL TRANSFORMATIONS DURING GROWTH OF EPITAXIAL Fe(001) THIN FILMS ON Cu(001) AND Pt(001)

B.M. CLEMENS, T.C. HUFNAGEL, M.C. KAUTZKY, and J.-F. BOBO
Department of Materials Science and Engineering, Stanford University, Stanford, CA 94305-2205

ABSTRACT

We have used grazing incidence x-ray diffraction to observe the structural evolution during growth of sputter-deposited epitaxial Fe films on Cu(001) and Pt(001). We find that on Cu(001), Fe is fcc up to a thickness of 10-12 monolayers, whereupon bcc Fe is observed in first the Pitsch and then the Bain orientations. The fcc Fe shows some relaxation of the misfit from the Cu, as do the Pitsch orientation bcc, which is in tension, and the Bain orientation bcc, which is in compression. All three Fe variants exist in a 40 monolayer thick film. On Pt(001) the Fe grows as bcc with the Bain orientation. However, a thin (20 Å) bcc Fe film is transformed to fcc Fe with cube-on-cube orientation by subsequent deposition of Pt. This behavior is consistent with intermixing of Pt into the Fe layer, which lowers the mismatch and bulk chemical energies of the fcc phase relative to that of the bcc phase.

INTRODUCTION

One intriguing aspect of thin film growth is the possibility of forming phases which are not stable in the bulk but can be produced in a metastable state due to the interaction between film and substrate. Epitaxial growth is one common mechanism by which such metastable phases can be formed, since the large strains present in thin epitaxial films can have a dramatic effect on the thermodynamic factors influencing phase stability. For instance, if the bulk stable phase of the film has a poor crystallographic misfit with the substrate, the strain and interfacial energies may be so large that growth of a nonequilibrium phase with lower misfit (but larger volume free energy) is favored. In such a case, metastable growth will continue until the increase in volume free energy exceeds the reduction in strain and interface energies; the film may then transform to the equilibrium structure. As a result of the competition between volume and interfacial free energy terms, we expect the transformation from the metastable phase to the bulk stable phase to occur at some critical thickness.

This theme of structural rearrangement occurring at some critical thickness has been studied extensively in the Fe on Cu(001) system. Fe deposited onto Cu(001) at room temperature will, after an initial period of distorted growth, grow epitaxially with the fcc structure to a thickness of 10-12 monolayers (ML) (see Reference [1] and references therein). Above this critical thickness, the fcc Fe transforms to epitaxial bcc(110). These epitaxial orientations are illustrated in Figure 1. A simple thermodynamic explanation of this behavior based on strain is challenged by the myriad of often contradictory results which have appeared in the literature in recent years. Three aspects of these results are particularly

9

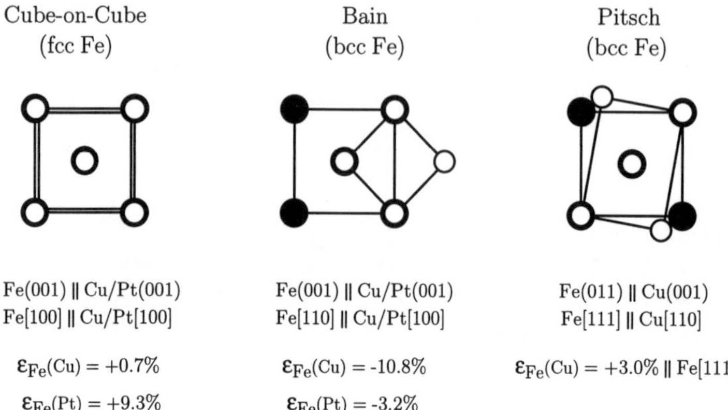

Figure 1: Common interface orientations for the growth of Fe on Cu(001) and Pt(001). Note that the Pitsch bcc orientation has four equivalent variants.

relevant. First, there is agreement neither on the type or magnitude of strain in the fcc layer before the transformation [2–5], nor on the precise dependence of the transformation on dislocation formation [6–8]. Secondly, some researchers report that the bcc Fe is fully unstrained immediately following the transformation [7,8], which is counter to expectation as bcc Fe has a non-zero misfit with both fcc Fe and Cu. Finally, most studies have focused only on coverage ranges in and around the fcc→bcc transition point; as a result, scarce data exists on the subsequent microstructural development of the bcc phase at larger coverages, even though changes in bcc orientation and grain structure have been observed [9]. Thus, several important questions concerning the structural evolution of Fe on Cu(001) remain open.

The growth of epitaxial Fe(001) on Pt(001) seems to be superficially similar to the case of Fe on Cu(001), based on observations of Fe(001)/Pt(001) multilayers [10–13]. For thin Fe layers(< 12 Å), the Fe is fcc and has a cube-on-cube orientation relationship with respect to the Pt underlayer (Fig. 1). For thicker Fe layers, however, the Fe is bcc and shows the Bain orientation with respect to the Pt. One might conclude from these observations that the initial growth of Fe on Pt is fcc, with a transformation to bcc at a critical thickness of around 12 Å. However, in contrast to the case of Fe on Cu, the epitaxial mismatch is actually *larger* for fcc Fe on Pt (9.3% tensile) than for bcc Fe on Pt (3.2% compressive). There is therefore no obvious thermodynamic reason why growth of fcc Fe on Pt should be favorable at any thickness, making the observation of fcc Fe in Fe/Pt multilayers somewhat surprising.

Our earlier studies of the strain in Fe(001)/Pt(001) epitaxial multilayers suggest that there is a significant amount of Pt intermixed with the Fe in the Fe layers [11]. This intermixing reduces both the volume and strain contributions to the free energy of the film.

We therefore suspected that the initial growth of Fe(001) on Pt(001) would actually be bcc, but that subsequent deposition of Pt on the Fe would result in the formation of an intermixed Fe/Pt alloy with a fcc structure. This would account for the observation of fcc Fe in Fe(001)/Pt(001) multilayers with thin Fe layers.

EXPERIMENTAL

All films in this study were deposited onto single-crystal MgO(001) substrates. For the Fe/Pt experiment a 5 Å Fe/400 Å Pt seed layer combination was grown at 600°C, after which the sample was cooled to room temperature for deposition of Fe. The use of several angstroms of Fe prior to Pt deposition has been found to improve the quality of epitaxial Pt films [14]. For the Fe/Cu experiment a seed layer of 50 Å Pd was deposited at 600°C, followed by a 400 Å layer of Cu at room temperature. For all depositions the sputtering gas was 3.0 mtorr Ar; the source-to-substrate distance in our chamber is approximately fifteen centimeters. The Fe deposition rate was approximately 0.25 Å/s for the Fe/Pt experiment and 0.05 Å/s for the Fe/Cu experiment. The base pressure of the chamber was between 5×10^{-9} and 5×10^{-8} torr during these experiments.

A specially designed sputter deposition chamber integrated with a z-axis diffractometer for performing grazing incidence x-ray scattering (GIXS) was used for these studies. The chamber is described in detail in Reference [15]. GIXS is uniquely capable of precise (0.001 Å resolution) in-plane lattice parameter measurements and can probe the structure of buried layers [16]. For all of the experiments described here the x-ray energy was 9695 eV.

RESULTS AND DISCUSSION

Fe on Cu(001)

A central question of this study was whether sputter deposition could be used to grow epitaxial fcc Fe on Cu(001). Figure 2 shows radial scans through the Cu(220) peak at several coverages below and above the reported transition thickness. The bare Cu peak occurs at a slightly lower angle than expected for bulk Cu (60.0 °2θ), corresponding to a 0.3% expansion of the Cu in-plane lattice parameter, and an increased misfit of +1.6% for fcc Fe. With Fe coverage, the peak both decreases in intensity and changes shape. Starting at the lowest coverage of 3 ML, an asymmetry gradually develops on the high 2θ side, corresponding to an increase in the population of smaller in-plane d-spacings. We attribute this change to the superposition of a fcc Fe(220) peak, since fcc Fe has a smaller lattice parameter than Cu (3.58 Å and 3.6148 Å respectively). The evolving peak is far from the position expected for bulk Fe (60.7 °2θ), indicating it is strained in tension to match the Cu. However, we can infer from the asymmetry of the combined peak that the Fe is not growing pseudomorphically, since the broadening does not occur on both sides of the Cu peak. Therefore, it appears that the Fe layer is at least partially relaxed from the full misfit strain during this initial growth phase. Finally, it is also clear from the shoulder evolution that above approximately 16 ML no new fcc Fe is growing. The continued asymmetry, which is still present even at 40 ML, nonetheless indicates that there is still fcc Fe at coverages well past the bcc transition point (as we shall show).

Another important question is to what thickness the fcc phase is stable. Figure 3 shows the evolution of radial scans through the Pitsch Fe(222) and Bain Fe(200) peaks. The first bcc Fe is detected in the Pitsch orientation at $\approx$10-12 ML, after which it grows stably out to 40 ML. Fe(110) is not the only stable bcc orientation, however, since epitaxial Bain-oriented bcc Fe appears later at $\approx$ 14 ML. This orientation also increases in intensity out to 40 ML, suggesting the two bcc orientations grow in tandem. This is verified in Figure 4a, which shows the volumetric intensity of the Pitsch (110) and Bain (200) reciprocal space spots versus coverage. The intensity was approximated as the product of the radial scan width (FWHM) and the rocking curve area. The intensity of both bcc phases scales approximately linearly with coverage until the thickness has become greater than the penetration depth of the GIXS evanescent wave. The absence of any abrupt change in the Pitsch intensity profile when Bain Fe appears confirms that the two orientations coexist and grow simultaneously. An extrapolation of the linear fits in the initial bcc growth regions to zero intensity yields the thicknesses at which each orientation appears. These values are 10.8 ML and 13.0 ML for the Pitsch and Bain Fe respectively, which agree well (within $\sim$1 ML) with the appearance of the respective diffraction peaks in the radial scans.

A final question in this experiment concerns the amount of strain in the film at various

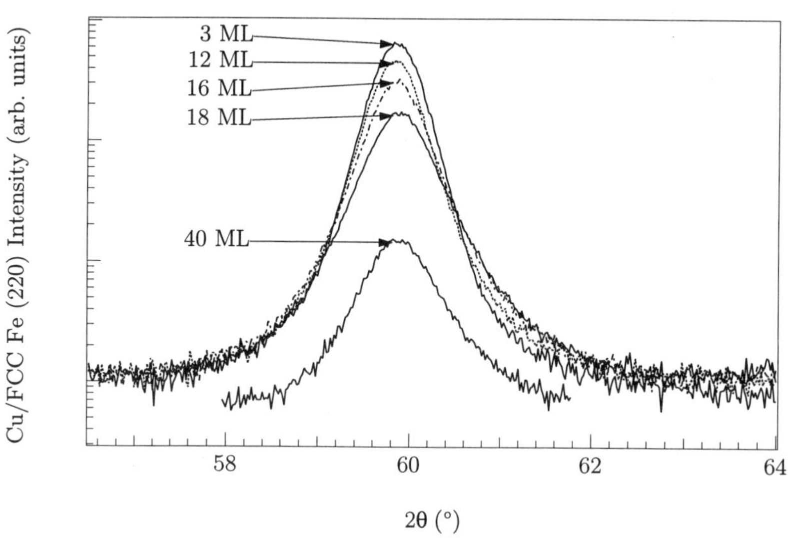

Figure 2: Radial scans of the Cu(220) peak during growth of Fe. The developing asymmetry on the high 2θ shoulder is due to superposition of a partially-relaxed fcc Fe(220) peak. For clarity only a portion of the offset 40 ML scan is shown.

stages of the transformation. Our data suggests that the fcc Fe is at least partially relaxed preceding the transformation to bcc, but still accommodates a sizable fraction of the total misfit strain. The notion is supported by preliminary fits to the combined Cu(220)/fcc Fe(220) peak at 40 ML, which indicate that the fcc portion still has a residual tensile strain of $\approx 1\%$ at this thickness. The case for strain in the bcc Fe, however, is much clearer. In Figure 3, the Pitsch (222) peak gradually moves to higher 2θ values (i.e. lower d-spacing), which is consistent with relaxation of a tensile in-plane misfit strain. The Bain (200) peak, on the other hand, moves to lower 2θ values, indicating this orientation is relaxing from a compressive misfit strain. Both observations are consistent with the misfits of bcc Fe on either Cu or fcc Fe. We also observe that the strain relaxation in both orientations is anisotropic. Figure 4b shows the Fe lattice parameter, as measured along orthogonal directions, for each bcc orientation. The Pitsch Fe relaxes quickly along <110> (ϵ=0.2% at 20 ML), but supports large tensile strains along <200> (ϵ=2.3% at 20 ML). The strain anisotropy in the Bain Fe, by comparison, is much smaller, though the absolute values of strain are large ($> -2\%$) at the beginning of growth. Both bcc orientations converge towards the bulk Fe lattice parameter as expected at large coverages.

From preliminary analysis of these results we can make some conclusions and compar-

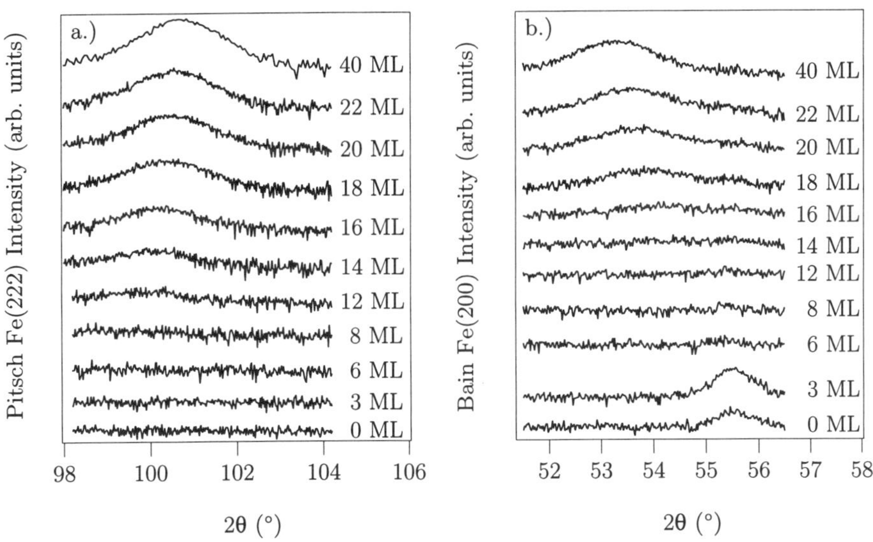

Figure 3: (a.) Evolution of the (222) peak for Pitsch-oriented bcc Fe on Cu. The shift to higher angles above 12 ML indicates relaxation of an in-plane tensile strain. (b.) Evolution of the (200) peak in Bain-oriented Fe on Cu. The shift to lower angles above 14 ML is due to relaxation of an in-plane compressive strain. The extra peak at 0 and 3 ML is from the Pd seed layer.

isons with previous studies. We find that fcc Fe is stable until a coverage of ≈10-12 ML, which is in good agreement with the results of other studies of evaporated Fe on Cu single crystals [1, 6, 17]. Our result that the fcc layer is strained, however, contradicts reports that fcc Fe has is to the transformation [3, 7, 8]. This discrepancy may be due to the different growth conditions (sputtering instead of evaporation) and substrate (Cu seed layer instead of Cu single crystal) in our experiment.

Our observation that fcc Fe persists in films as thick as 40 ML also disagrees with reports that the fcc Fe transforms completely to bcc [18]. This may be due to the extreme surface sensitivity of LEED and STM, which were used in the earlier study, which may make it impossible to detect fcc Fe remaining at the Cu/Fe interface. If fcc Fe is present at the Cu/Fe interface for thicknesses > 12 ML, then the reduction in interfacial energy between Cu and fcc Fe must be large enough to more than offset the energy increase required to create a new fcc/bcc Fe interface. This suggests that the *onset* of the fcc-to-bcc transformation is controlled by the gradual increase of the fcc volume free energy with coverage, but that the *extent* of the transformation into the film is largely controlled by the Cu/fcc Fe interfacial energy.

Our observation of the appearance of bcc Fe in the Bain orientation at about 14 ML is interesting because the misfit between Bain-oriented bcc Fe and either Cu or fcc Fe is

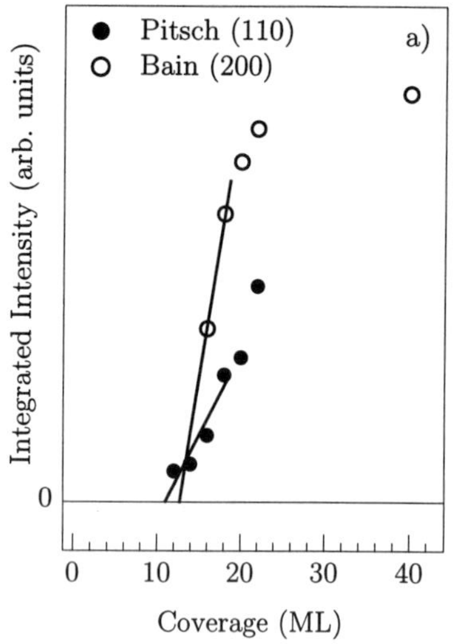

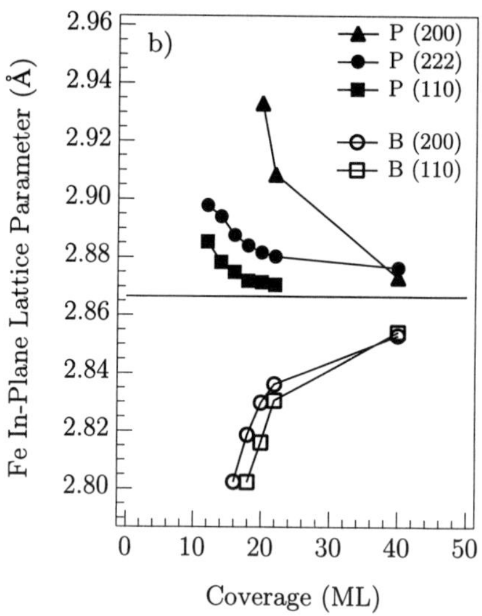

Figure 4: (a.) Integrated spot intensity versus coverage for Pitsch and Bain bcc Fe on Cu. The intensity was approximated as the radial scan FWHM times the rocking curve area. (b.) In-plane lattice parameter versus coverage for principal directions in bcc Fe. P and B denote Fe in the Pitsch and Bain orientations respectively. The straight line is the bulk Fe lattice parameter.

much larger than that for Pitsch-oriented bcc Fe. Also, (110) growth is typically favored in bcc metals due to the low surface energy of this plane [19]. We postulate that, since the misfit and strain in the Bain orientation is opposite that of the Pitsch orientation, the Bain orientation may be favored by global strain energy considerations.

Finally, we observe a gradual relaxation of the strain in the Fe layer during growth. Other studies indicate that the Pitsch-oriented bcc Fe is fully unstrained immediately following the fcc→bcc transition [7,8]. It may be that a relatively high dislocation density in the Cu seed layer affects the transformation by dislocation insertion into the Fe layer. This may occur in conjunction with, or even dominate, the martensitic shear transformation that is observed in films on single crystal substrates.

<u>Fe(001) on Pt(001)</u>

The geometry for the Fe/Pt system is considerably simpler than that for Fe/Cu, since the Pitsch orientation is not observed. Because the fcc Fe lattice parameter is smaller than the Pt lattice parameter (3.924Å), we would expect to see the bulk Fe(220) peak at a larger value of 2θ than the Pt(220) peak. For the bcc Fe in the Bain orientation, the bulk bcc Fe(200) peak appears at a smaller value of 2θ, since the bcc Fe(200) d-spacing ($d_{220} = \frac{\sqrt{2}}{4} a_{Fe}$) is larger than the Pt(220) d-spacing.

The evolution of the film structure as Fe is deposited on Pt(001) is shown in Figure 5. The large peak near 55 °2θ represents the Pt(220) d-spacing. The initial 10 Å of Fe is nearly lattice-matched with the Pt and therefore no separate Fe peak appears. The Pt(220) peak is, however, slightly asymmetric on the low 2θ side, indicating the presence of strained bcc Fe. As additional Fe is deposited, this asymmetry in the Pt(220) peak becomes more pronounced, with increasing intensity on the low 2θ side of the peak, until the peak ultimately splits into separate Pt and bcc Fe peaks. Notice that there is no evidence for the development of fcc Fe, which would be indicated by the presence of a peak or shoulder on the high 2θ side of the Pt(220) peak. These observations indicate that even very thin Fe layers have a bcc structure, strained in-plane towards the Pt in-plane spacing.

Figure 6 shows how the Fe(200) d-spacing changes with the deposition of Fe. Initially the 10 Å thick Fe layer is strained nearly 3% (with respect to bulk bcc Fe), to match the Pt spacing. Deposition of a second 10 Å of Fe actually increases this strain slightly, improving the lattice match with the underlying Pt. This effect has also been seen for growth of Cu on Fe [20] and Fe on MgO [21]. It may be associated with the agglomeration of Fe islands into a continuous film. Further deposition of Fe (to 50 Å total) results in a gradual relaxation of the strain in the Fe layer towards the bulk bcc Fe(200) d-spacing. Notice that deposition of up to 20 Å Pt on top of this 50 Å thick Fe layer results in continued expansion of the Fe lattice. This is due to intermixing of Pt into the Fe layer, as we shall see later. Continued deposition of Pt results in the formation of an epitaxial Pt surface layer, which moves the observed peak towards the bulk Pt(220) d-spacing.

To investigate the possibility that thin (~20 Å) bcc Fe layers are transformed to fcc by intermixing of subsequently deposited Pt, we deposited a 20 Å Fe layer on a Pt(001) surface, as before. As Figure 7 shows, an asymmetry on the low 2θ side of the Pt(220) peak

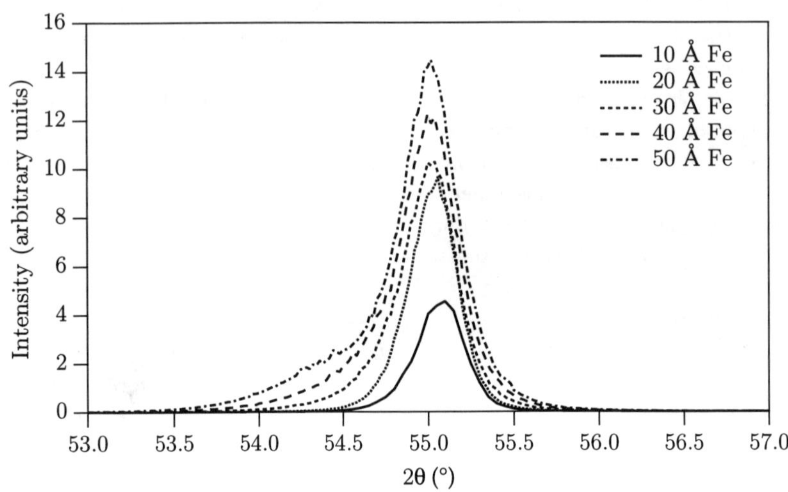

Figure 5: Evolution of Pt(220)/bcc Fe(200) peak during deposition of Fe.

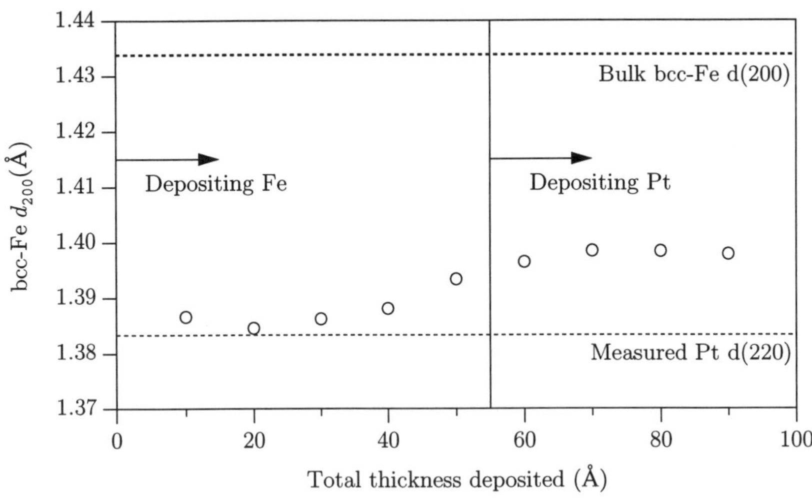

Figure 6: Evolution of Fe in-plane d-spacing during growth of 50 Å Fe followed by 50 Å Pt.

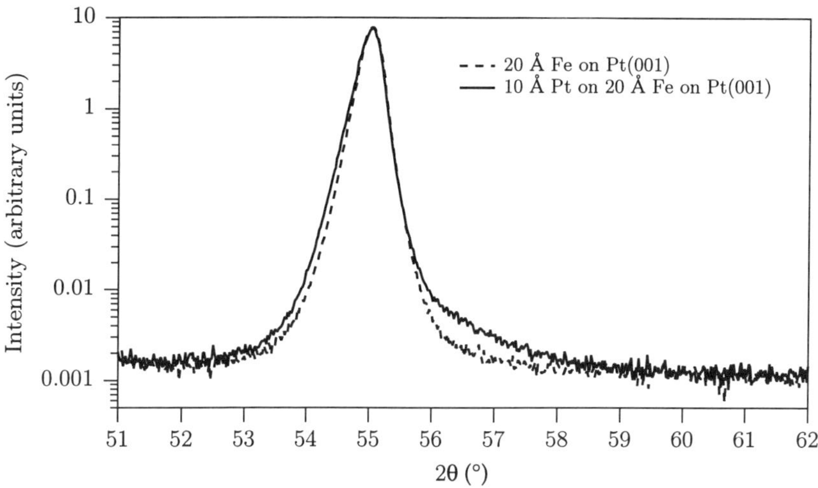

Figure 7: Evolution of Pt(220)/fcc Fe(220) peak during deposition of Pt on 20 Å Fe.

develops, and we conclude that the Fe layer has a strained bcc structure. When we deposit 10 Å of Pt on the Fe, we see the appearance of a new peak on the *high* 2θ side of the Pt(220) peak. This new peak cannot be due to strain of the bcc Fe layer, because introducing Pt into bcc Fe would increase the Fe lattice parameter and drive the bcc Fe peak to *lower* 2θ. Therefore, the new peak is most likely due to the formation of a layer of strained fcc Fe-Pt alloy on the surface of the sample.

We can understand the structural behavior of Fe in Pt/Fe/Pt sandwich structures with a simple thermodynamic model which incorporates the various contributions to the energy of a growing Fe layer, together with the intermixing which occurs during the growth of the Pt overlayer. The two most significant contributions to the free energy of the Fe layer are strain energy (because the elastic strain is quite large) and a volume energy term (due to the effect of intermixing Fe and Pt). Both of these contributions favor the fcc structure as the Pt content of the Fe layer increases.

In previous work on sputtered epitaxial Fe/Pt (001) multilayers, Daniels *et al.* reported that the unstrained lattice parameter of the bcc Fe layers was inversely proportional to their thickness [11]. This result and the results reported here are consistent with intermixing of Pt into the Fe layer as the Pt is deposited on top. Our previous results indicate that a constant amount of Pt (corresponding to about 2 monolayers of Pt) diffuses into the Fe. This intermixing will change the size misfit of the Fe layer with the underlying Pt substrate, and hence affect the misfit energy contributions for the bcc and fcc phases. Starting with well-established concepts and expressions for the misfit energy as a sum of misfit dislocation and strain energies [22], one can show that the equilibrium misfit energy is given by:

$$E_M^0 = \begin{cases} hM\epsilon_m^2 & \epsilon_m < \epsilon_{mc} \\ hM\epsilon_{mc}(2\epsilon_m - \epsilon_{mc}) & \epsilon_m > \epsilon_{mc} \end{cases}$$

where ϵ_{mc} is the critical misfit value which is given by:

$$\epsilon_{mc} = \frac{\mu b}{4\pi h M S(1-\nu)} \ln\left(\frac{\beta h}{b}\right)$$

where M is the biaxial modulus, h is the film thickness, b is the dislocation Burger's vector, S is the dislocation spacing, μ is the shear modulus, ν is Poisson's ratio, and β is the dislocation core parameter (which is of order unity).

If the misfit strain is less than ϵ_{mc}, the film will be coherent with the substrate so that the strain in the film will be equal to the total misfit strain and the misfit energy will be proportional to the square of this strain. If the misfit is greater than ϵ_{mc} there will be dislocations at the interface allowing some relaxation of the misfit strain, and the misfit energy will vary linearly with the misfit strain. In this case the misfit energy density will also vary approximately linearly with $1/h$. The misfit for either fcc Fe or bcc Fe on Pt is greater than the critical misfit value for film thicknesses greater than a few monolayers, so for the films studied here, the misfit energy will vary linearly with the size of the misfit.

The misfit with Pt of fcc and bcc Fe as a function of Pt content is shown in Fig. 8a, where we have used data from King [23] and Pearson [24] to calculate the lattice parameter of Fe. This figure shows that the magnitude of the misfit for the fcc Fe is decreased by incorporation of the larger Pt atoms. In contrast, the misfit of bcc Fe on Pt is *increased* by the introduction of Pt.

The effect of this misfit on the misfit energy is shown in Fig. 8b, where we see that the misfit energy for the fcc Fe is decreased by increasing Pt content, while that of the bcc Fe is increased, so that they cross at about $x_{Pt} = 0.15$. Since the average Pt concentration of a Fe layer increases with decreasing thickness, the misfit energy will tend to stabilize the fcc structure of thin Fe layers. If we estimate that the Pt content of the film corresponds to 2 monolayers of Pt, we find that the crossover thickness of 15% Pt is reached at a Fe thickness of 15-20 Å. Therefore, we expect that Fe layers less than 15-20 Å thick will be fcc when Pt is deposited on them.

The energy difference between fcc and bcc layers also includes a volume term associated with the chemical composition of the Fe layer. In the binary Fe-Pt phase diagram, the single-phase bcc region extends to about 5% Pt, and the fcc structure and its ordered variants are stable over the composition range for $x_{Pt} > 15\%$. Hence, with increasing Pt content the volume free energy of a the bcc phase increases relative to that of the fcc phase. As the thickness of the Fe layer decreases and the Pt content increases, therefore, the volume term also favors the fcc phase. From the phase diagram behavior we might estimate that the bulk free energies of the fcc and bcc phases cross at about 10% Pt, which for our estimate of intermixed Pt corresponds to a Fe layer thickness of about 24-30 Å.

These simple energy arguments predict that, with a constant intermixing of about 2 monolayers of Pt into an Fe layer, the misfit and bulk energies should favor the fcc phase for

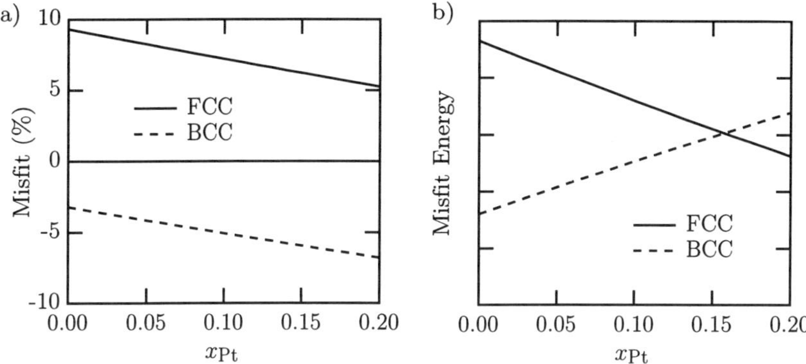

Figure 8: a) Misfit of Fe with Pt, as a function of atomic fraction Pt, for bcc and fcc Fe. b) Misfit energy for fcc and bcc Fe on Pt.

thicknesses below about 15-30 Å. The misfit energy should scale roughly with the inverse Fe layer thickness, and so might become the dominant term for small thicknesses.

CONCLUSIONS

The results of these studies demonstrate two very different ways in which the growth of metastable phases can occur in metal films . In the case of Fe on Cu(001), metastable fcc Fe develops during the initial stages of growth by virtue of its lower interfacial and strain energy relative to the bcc phase. We have observed that strained fcc growth occurs prior to the bcc transformation, and that the fcc phase exists, probably at the Fe/Cu interface, long after the transformation to bcc has taken place. These results suggest that the interfacial energy may be particularly significant in controlling the extent of the fcc→bcc transformation. With respect to the bcc phase, we have observed that bcc Fe(110) appears first at 11 ML, followed shortly thereafter by bcc Fe(001), with both orientations growing in tandem out to large thicknesses. Both bcc orientations also exhibited large strain relaxation with continued coverage. The thermodynamic basis of this strained tandem growth is not yet clear, but we postulate that it may be related to a more rapid reduction in the global film strain.

In contrast to this single-layer metastable growth, our results from the Fe on Pt experiment show that metastable phases can also be formed after deposition of bulk stable films by interaction with subsequent layers. We have observed that Fe deposited onto Pt grows as a strained bcc layer at all thicknesses, but that the deposition of Pt onto a thin Fe layer (20 Å) will cause a transformation to the fcc structure. We have explained this behavior in terms of a simple thermodynamic model based on intermixing of Pt in the Fe layer. This model predicts that for Fe layers thinner than 15-30 Å, approximately 2 ML Pt intermixed with the Fe layer will cause the misfit and volume free energies to favor fcc Fe.

We gratefully acknowledge the assistance of S. Brennan, C.T. Wang, K.P. Fahey, and A. Munkholm in setting up the experiment and collecting the data. SSRL is supported by the U.S. Department of Energy, Office of Basic Energy Sciences under Contract No. DE-AC03-76SF00515. This work was funded by the National Science Foundation under contracts NSF DNR-9408552 and NSF DNR-9400372.

REFERENCES

1. P. Schmailzl, K. Schmidt, P. Bayer, R. Döll, and K. Heinz. *Surface Science*, 312:73, 1994.

2. M. Wuttig and J. Thomassen. *Surface Science*, 282:237, 1993.

3. P. Bayer, S. Müller, P. Schmailzl, and K. Heinz. *Phys. Rev. B*, 48:17611, 1993.

4. H. Magnan, D.D. Chandesris, B. Vilette, O. Heckmann, and J. Lecante. *Phys. Rev. Lett.*, 67:859, 1991.

5. A. Clarke, P.J. Rous, M. Arnott, H. Jennings, and R.F. Willis. *Surface Science*, 292:L843, 1987.

6. M. Wuttig, B. Feldmann, J. Thomassen, F. May, H. Zillgen, A. Brodde, H. Hannemann, and H. Neddermeyer. *Surface Science*, 291:14, 1993.

7. S. Müller, P. Bayer, A. Kinne, P. Schmailzl, and K. Heinz. *Surface Science*, 322:21, 1995.

8. J.V. Barth and D.E. Fowler. *Phys. Rev. B*, 52:11432, 1995.

9. J. Koike and M. Nastasi. *Mat. Res. Soc. Symp. Proc.*, 202:13, 1991.

10. B.J. Daniels. Ph.D. thesis, Stanford University, 1995.

11. B.J. Daniels, W.D. Nix, and B.M. Clemens. *Mater. Res. Soc. Symp. Proc.*, 356:373, 1995.

12. M. Sakurai. *J. Appl. Phys.*, 76:7272, 1994.

13. M.R. Visokay, B.M. Lairson, B.M. Clemens, and R. Sinclair. *J. Magn. Magn. Mat.*, 126:136, 1993.

14. B.M. Lairson, M.R. Visokay, R. Sinclair, S. Hagstrom, and B.M. Clemens. *Appl. Phys. Lett.*, 61:1390, 1992.

15. A.P. Payne, B.M. Clemens, and S. Brennan. *Rev. Sci. Instr.*, 63:1147, 1992.

16. P.H. Fuoss and S. Brennan. *Ann. Rev. Mat. Sci.*, 20:365, 1990.

17. W. Jesser and J. Matthews. *Phil. Mag.*, 15:1097, 1967.

18. J. Giergiel, J. Shen, J. Woltersdorf, A. Kirilyuk, and J. Kirschner. *Phys. Rev. B*, 52:8528, 1995.

19. D.L. Smith. *Thin Film Deposition: Principles and Practice*. McGraw-Hill, Inc., New York, New York, 1995.

20. A.P. Payne, B.M. Lairson, S. Brennan, B.J. Daniels, N.M. Rensing, and B.M. Clemens. *Phys. Rev. B*, 47:16064, 1993.

21. B.M. Lairson, A.P. Payne, S. Brennan, N.M. Rensing, B.J. Daniels, and B.M. Clemens. *J. Appl. Phys.*, 78:4449, 1995.

22. W.D. Nix. *Met. Trans. A*, 20A:2217, 1989.

23. H.W. King. *J. Mat. Sci.*, 1:79, 1966.

24. P. Villars and L.D. Calvert. *Pearson's Handbook of Crystallographic Data for Dntermetallic Phases*. American Society for Metals, Metals Parks, OH, 1985.

MICROSTRUCTURE AND MECHANICAL PROPERTIES
OF THIN Al-Si-Ge FILMS

S. KIRCHNER, O. KRAFT, S. P. BAKER, and E. ARZT
Max-Planck-Institut für Metallforschung, and Institut für Metallkunde, University of Stuttgart,
Seestr. 71, D70174 Stuttgart, Germany

ABSTRACT

The mechanical properties are thought to play an important role in the performance of metallization materials for very large scale integration (VLSI) applications. From recent investigations on bulk materials it is known that Al-Si-Ge alloys can be very efficiently strengthened with only a small amount of the alloying elements. These alloys are potential candidates for future metallizations both because Si and Ge are compatible with the existing semiconductor technology, and because the resistivity is expected to be low.

We present the first results of detailed characterizations of Al-Si-Ge thin films as a function of sputter conditions and heat treatments. The microstructure was characterized using x-ray diffraction and transmission electron microscopy. The kinetics of precipitation were studied using resistance measurements. Room temperature hardness was investigated using nanoindentation, and the mechanical properties at temperatures up to 240°C were examined using a substrate curvature method. The correlation between precipitate structure and film properties is discussed.

INTRODUCTION

Aluminum alloys are widely used as metallization materials in VLSI applications. Due to high current densities, conductor line materials have to be made resistant against the damaging effects of electromigration [1]. In electromigration, material is transported along the line as a result of the "electron wind". In areas where material is depleted, tensile stresses arise which can lead to formation of voids. Material accumulation leads to compressive stresses which can further lead to formation of hillocks. In a classical study of the drift of metallic stripes, I. Blech [2] concluded that the stress gradient can drive backdiffusion which opposes electromigration. It was subsequently argued [3, 4] that a material with a high yield strength should therefore also have a high resistance to electromigration.

Recent investigations by Hornbogen and coworkers [5-7] on bulk materials have shown that aluminum can be strengthened with very small amounts of Si and Ge together as alloying elements. It is known that the binary alloys Al-Si and Al-Ge show no or very little precipitation hardening [8]. The Si or Ge diamond phase forms directly from Si or Ge-rich clusters in the fcc solid solution forming no metastable phases. Tab. 1 shows that there is a substantial size misfit between the individual alloying elements and Al. Due to the resulting strain energy, only a small fraction of clusters exceed the critical size and coarse precipitates are formed.

	atomic radius [nm]	misfit strain to Al
Al	0.286	—
Si	0.278	- 2.7 %
Ge	0.292	+ 2.1 %

Tab. 1 Atomic radii for fcc structure and misfit strain compared to Al [6]

21

Hornbogen and coworkers demonstrated that by combining the larger Ge and the smaller Si atoms as alloying elements, the strain energy can be reduced and the number of precipitates per unit volume in the ternary alloys can be increased by an order of magnitude compared with the binary alloys. Depending on the microstructure and the amount of alloying elements, they expect a strengthening of 100 to 400 MPa.

This hardening potential makes Al-Si-Ge an interesting candidate for metallizations. Its resistivity is not expected to be increased much in view of the low alloy content, and the elements Si and Ge are compatible with existing semiconductor technology. The purpose of this work is therefore to investigate the strengthening effects of Si and Ge as alloying elements in Al thin films.

SAMPLE PREPARATION

<u>Target Preparation:</u> Ternary alloy targets with a composition of Al-0.6at%Si-0.6at%Ge were produced by induction melting under high vacuum ($p \approx 10^{-5}$ mbar) in a MgO crucible from highly pure Al (99.995 %), Si, and Ge (both 99.999 %). The melt was cast under vacuum in a Cu mold (90×95×12 mm). The finished bars were homogenized at 500°C for 2 days and quenched. To avoid contamination during rolling, the slag was removed by milling. The alloys were cold-rolled about 25 % and subsequently recrystallized at 500°C for 2 h. To stabilize the microstructure of the targets, they were aged at 200°C for 2 days. The 3" targets (diameter 75.8 mm, thickness 6.3 mm) were manufactured out of these bars. In addition to the targets, bulk samples (9×9×5 mm) were produced for subsequent testing.

<u>Thin Film Preparation:</u> Film deposition was carried out in an UHV sputtering system with 3 torus magnetrons for 3" targets using Ar (99.999 %) as the sputtering gas. The base pressure was about 5×10^{-9} Torr. The substrate-target distance was about 20 cm. Al-0.6Si-0.6Ge films were produced using the target having the same composition. Pure Al films were sputtered using a highly pure Al (99.999 %) target. The substrates were (100) silicon wafers (4") with 100 nm of thermal oxide. They were cleaned prior to sputtering using 500 eV Ar ions for 60 s at 0.3 mTorr. Typical sputter conditions were 300 W dc power, 4 mTorr pressure, 100°C substrate temperature, and - 50 V bias voltage. The thickness of the films was typically about 500 nm and the sputtering rate was about 0.15 nm/s. Patterned thin film structures for electrical resistivity measurements were made by sputtering through a mask (length: 34 mm, width 0.5 mm). The film thickness of these samples was about 250 nm.

EXPERIMENTAL

Texture analyses were performed using a 4-circle Siemens D5000 texture goniometer with Cu Kα radiation. To quantify the texture of the films, a (111) pole figure was recorded in reflection geometry at tilting angles between 0° and 85°. The measured intensities were corrected for background scattering, geometric defocusing and penetration depth as described in [9].

For transmission electron microscopy (TEM) a JEOL200CX at a voltage of 200 kV was used. The TEM plan view samples were conventionally prepared by grinding, dimpling and ion milling until electron transmissibility was achieved.

Isothermal resistivity measurements were carried out using a standard 4 point method. The probe current was 1.7 mA, thermovoltages were eliminated by making measurements with the current direction reversed.

Macrohardness measurements of the bulk target material were carried out using a Vickers indenter with a 3 kg load. At least 12 indents were made in each sample. The samples were homogenized for 2 h at 500°C, quenched in water and then aged for different times at 200°C.

Nanohardness of the films was measured using a Nanoindenter II (Nano Instruments). The tests were conducted and analyzed using procedures similiar to those described by Oliver and Pharr [10]. At least 6 indents were made at every indentation depth.

The biaxial stresses as a function of temperature in the Al-Si-Ge films were determined from substrate curvature measurements using a laser-based optical lever device. The stresses were calculated using a data analysis method similiar to that in reference [11]. The samples were cycled in a nitrogen atmosphere several times between RT and 240°C with a constant heating and cooling rate of 10 K/min.

RESULTS AND DISCUSSION

Microstructure

Figs. 1a and b show plan view TEM micrographs of a sputtered Al-0.6Si-0.6Ge film which was aged in vacuum for 1 h at 200°C. The overview in Fig. 1a shows that the grain sizes vary from 0.3 to 1.1 µm, the average grain size is 0.7 µm. The precipitates are homogeneously distributed. Fig. 1b shows a close up view of the precipitates inside a single grain. Precipitate sizes vary from 10 to 50 nm, with an average of about 25 nm. A dislocation line which is bowed between the particles can be seen.

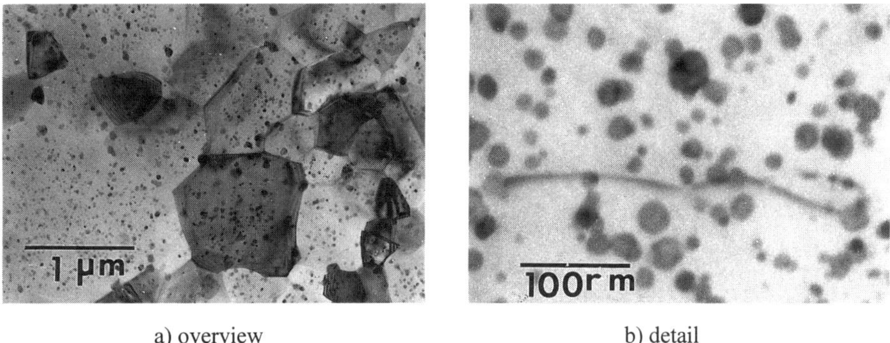

a) overview   b) detail

Fig. 1   TEM image of an Al-0.6Si-0.6Gc film aged for 1 h at 200°C

All films showed a fiber texture independent of the sputtering conditions. The pure Al-films exhibited the (111) texture common to Al films. The Al-Si-Ge-films, however, reveal a quite uncommon (110) texture even in the as-deposited state (see Tab. 2). During annealing, the texture sharpens in both film types. The tendency to form the (110) texture in the Al-Si-Ge films is very strong; changing the sputtering bias and pressure did not affect the (110) texture. This texture is assumed to arise from an interface effect because the Al-Si-Ge films which are sputtered on uncleaned substrates or on TiN have a mixed (111) and (100) texture. From literature it is known that Al-Si films normally reveal a (111) texture [12], therefore Ge must play an important role in the unusual texture development.

film type	heat treatment	Texture fraction (%)		
		(111)	(110)	random
Al	as-deposited	34	–	66
	10 min/ 475°C	61	–	39
Al-0.6Si-0.6Ge	as-deposited	–	86	14
	10 min/475°C	–	93	7

Tab. 2 Texture of as-deposited and annealed Al- and Al-Si-Ge-films prepared
under the same conditions

The isothermal kinetics of precipitation were investigated for temperatures between 126 and 225°C via the electrical resistivity (Fig. 2). The samples were homogenized at 500°C for 1 h, quenched and subsequently measured. The precipitation and the coarsening of the precipitates leads to a decrease in electrical resistivity. When the resistivity saturates, the microstructure is assumed to have stabilized. From Fig. 2 it can be seen that only a slight change of temperature changes the precipitation kinetics enormously. This indicates that at temperatures greater than 200°C the material ages very rapidly.

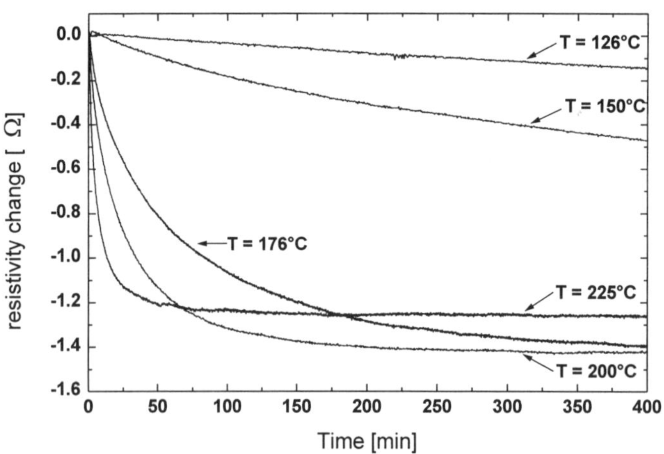

Fig.2   Isothermal resistivity measurements as function of time for homogenized samples

Mechanical Properties

Vickers hardness measurements of the target bulk material were carried out to examine its hardenability. Fig. 3 shows Vickers hardness as a function of ageing time at 200°C. The hardness rises from 29 HV for homogenized material to a peak of 50 HV at 1 h, an increase of 72%, which indicates very good hardenability.

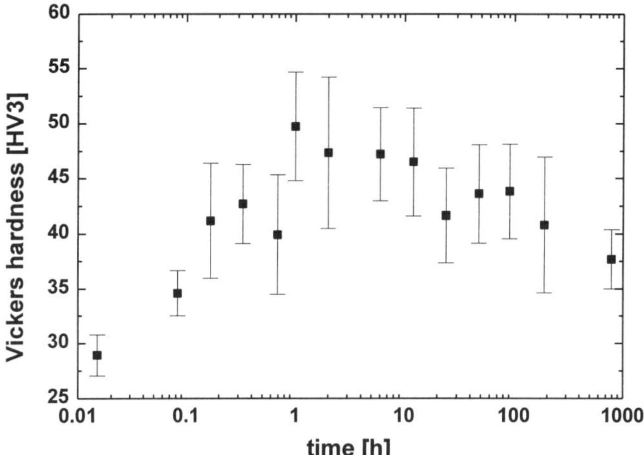

Fig. 3   Vickers hardness *vs.* ageing time at 200°C for the bulk target material

The hardnesses of an Al-Si-Ge film and of a pure Al-film were measured using nanoindentation. The Al-Si-Ge was aged for 1h at 200°C. The Al film was twice as thick as the Al-Si-Ge film. To compare the films, the hardness values at indentation depths of 70 % of the film thickness are shown in Tab. 3. It can be seen that the hardness of the Al-Si-Ge film is twice that of the Al film. As in the bulk material, this can be explained by the fine precipitate structure shown in Fig. 1 in addition to a possible thickness effect.

type of film	film thickness [nm]	indentation depth [nm]	hardness [GPa]
Al-0.6Si-0.6Ge	250	169	2.13 ± 0.25
Al	500	331	0.95 ± 0.07

Tab. 3   Comparison of the hardness at indentation depths of 70 % of the film thickness in an aged Al-Si-Ge film and an unaged Al film. Ageing conditions: 1 h at 200°C

Fig. 4 shows the first and fourth cycles of substrate curvature measurements of an as-deposited Al-Si-Ge film. During the first cycle, the typical compressive stress maximum occurs which is usually attributed to microstructural changes such as grain growth. The second and third cycles are similiar to the fourth, indicating that no age hardening is taking place. From the results in Fig. 2 we know that the alloy tends to overage quickly and it is possible that the film is already overaged after the first heating cycle.

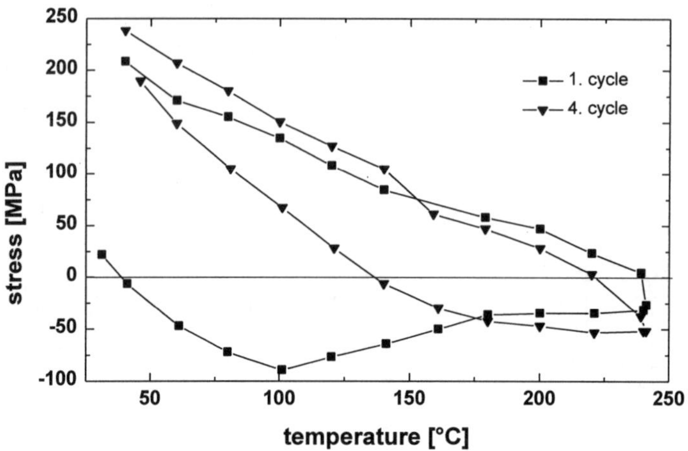

Fig.4   Substrate curvature measurements of an as-deposited Al-Si-Ge film.

CONCLUSIONS

Our results show that Al-Si-Ge films as well as bulk alloys can be effectively age hardened. However, the resistivity and the substrate curvature measurements indicate that the films overage more quickly than the bulk material of the same composition. The texture analysis shows an uncommon (110) texture for the Al-Si-Ge films in both the as-deposited and annealed states. Further work is necessary to fully understand the special behaviour of these films and to judge their potential for use in thin film metallizations.

ACKNOWLEDGEMENT

This work is financially supported by the Deutsche Forschungsgemeinschaft (project "Mikrowerkstoffe" number Ar 201/5-1).

REFERENCES

[1]   J. R. Black, IEEE Trans. Electr. Dev. **16**, 338 (1969)
[2]   I. A. Blech, E. Kinsbron, Thin Solid Films, **25**, 327 (1975)
[3]   E. Arzt and W. D. Nix, J. Mater. Res. **6**, 4113 (1991)
[4]   E. Arzt, O. Kraft, J. Sanchez, S. Bader, and W. D. Nix, Mat. Res. Soc. Proc. **239**, 677 (1991)
[5]   E. Hornbogen, A. K. Mukhopadhyay, and E. A. Starke Jr., Z. Metallkd. **83**, 577 (1992) 8
[6]   E. Hornbogen, A. K. Mukhopadhyay, and E. A. Starke Jr., Scr. Met. et Mat. **27**, 733 (1992)
[7]   E. Hornbogen, A. K. Mukhopadhyay, and E. A. Starke Jr., J. Mat. Sci. **28**, 3670 (1993)
[8]   L. F. Mondolfo, Aluminum Alloys, Butterworth, London 1979, pp. 292 - 368
[9]   D. B. Knorr, Mat. Res. Soc. Proc. **309**, 75 (1993)
[10]   W. C. Oliver, and G. M. Pharr, J. Mat. Res. **7**, 1564 (1992) 6
[11]   P. A. Flinn, J. Mater. Res. **6**, 1498 (1991)
[12]   A. N. Campbell, R. E. Mikawa, and D. B. Knorr, J. Elec. Mat. **22**, 589-596 (1993)

THE INFLUENCE OF STRENGTHENING MECHANISMS ON STRESS RELAXATION IN THIN ALUMINUM METALLIZATION

Jonathan Gorrell*, Paul Holloway*, and Hal Jerman**
*Dept. of Materials Science & Engineering, University of Florida 32611-6400
**EG&G IC Sensors, 1701 McCarthy Blvd., Milpitas, CA 95035.

ABSTRACT

With the development of microelectromechanical systems there is a need for stronger aluminum thin films that resist stress relaxation. A number of strengthening mechanisms are used extensively for bulk aluminum alloys, but very few have been used to improve the performance of thin films. Pure aluminum, standard microelectronics metallization (Al-.04Cu-.017Si), alloy T201 (Al-.046Cu-.006Ag-.004Mn-.003Mg-.003Ti), and alloy 2090 (Al-.026Cu-.021Li-.001Zr) were electron beam evaporated or sputter deposited onto (100) silicon substrates.. Stress versus temperature and stress relaxation were measured in the films. Pure aluminum and AlSiCu alloy films exhibited plastic deformation at low stresses and low temperatures. The T201 and 2090 films exhibited residual elastic stresses at room temperature of 350 MPa and 500 MPa, and did not plastically deform until 240°C at 100 MPa stress, or 270°C at 200 MPa stress, respectively. The T201 film also showed a low stress relaxation rate. We speculate that solid solution strengthening caused the increase in strength of the T201 film, and that age hardening caused the increase in strength of the 2090 film.

INTRODUCTION

The objective of this research is to identify ways of strengthening aluminum thin films for use in bimetallic actuators. We are interested in bimetallic actuation of microelectromechanical device (MEMS)[1]. MEMS are the broad classification of micro devices covering micromachines, micro valves, and micro sensors. Bimetallic actuators are used in MEMS since they produce a relatively constant force over a large actuation distance and are easy to fabricate. However, bimetallic actuators develop high internal stresses to produce a displacement. If there is any relaxation of these stresses, there is a corresponding reduction in displacement of the bimetallic actuators. Thus, aluminum alloy thin films with higher yield strength and minimal stress relaxation are desired.

The active component of a bimetallic actuator is a bimetallic strip, which is made by bonding together two materials with different coefficients of thermal expansion (CTE). Upon heating or cooling the bimetallic strip bends due to the difference in expansion or contraction between the two materials of the strip. Bimetallic strips are very sensitive to temperature changes and this has lead to there extensive use as thermostats[2]. The amount of movement in a bimetal strip for a given temperature change is proportional to the difference in the CTE of the two materials in the strip as shown by equation 1[3]

Mat. Res. Soc. Symp. Proc. Vol. 436 © 1997 Materials Research Society

$$w := \frac{3HL^2 \cdot (\alpha_1 - \alpha_2) \cdot \Delta T}{\left(H_1\right)^2 + \left(H_2\right)^2 + 3H^2 + \left[\dfrac{E_2 \cdot (H_2)^3 \cdot \left[1 - (v_1)^2\right]}{E_1 \cdot H_1 \cdot \left[1 - (v_2)^2\right]}\right] + \left[\dfrac{E_1 \cdot (H_1)^3 \cdot \left[1 - (v_2)^2\right]}{E_2 \cdot H_2 \cdot \left[1 - (v_1)^2\right]}\right]} \tag{1}$$

where H is the thickness of the bimetal strip, L is its length, α is the CTE, ΔT is the change in temperature, v is Poison's ratio, and w is the displacement of the bimetallic strip. Thus, the use of a stronger metal such as nickel[4] could eliminate the stress relaxation problem. It would also reduce the deflection of a bimtallic actuator for a given temperature change. Therefore strengthening of aluminum alloys is an effective way to increase the performance of a bimetallic actuator.

Bimetallic actuators are currently being used in microvalves[5]. Figure 1 shows a cross sectional diagram of a normally closed bimetallic actuated microvalve. Upon heating, the aluminum thin film expands more that the silicon substrate and causes the valve to open. A resistive heater is built into the silicon under the aluminum. The valve is closed by cooling the aluminum/silicon strip, which is done by conduction of the heat to the value body. However, plastic deformation and stress relaxation of the aluminum will also allow the valve to close. Plastic deformation of this aluminum film is likely as the aluminum film is under a high shear stress at an elevated temperature. Stress relaxation in aluminum and Al-Si-Cu films is known to occur at temperatures as low as room temperature [6,7]. Development of an aluminum alloy with a high CTE, high yield strength and resistance to creep would be desirable for this application.

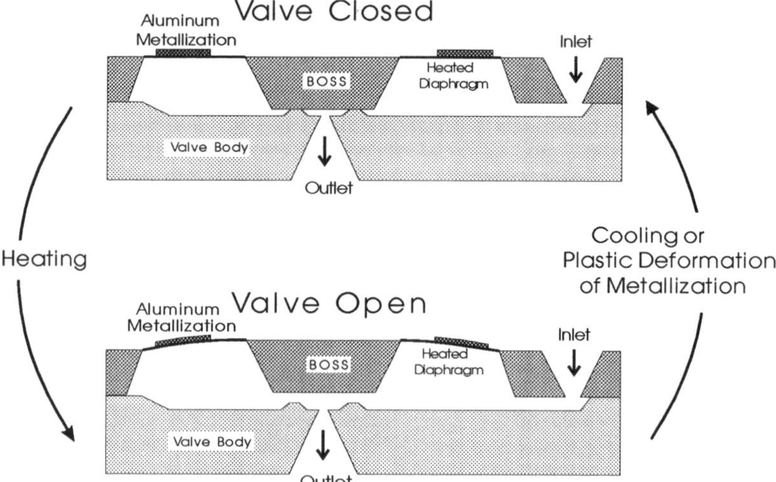

Figure 1: Cross sectional diagram of bimetallic actuated microvalve

Over the past century, a number of strengthening mechanisms have been developed that are now used extensively to strengthen bulk aluminum alloys. The five strengthening mechanisms frequintly used are[8]: solid solution strengthening, age hardening , multiphase strengthening, work hardening, and grain size reduction. In all cases, strengthening is achieved by making it more difficult for dislocations to move through the aluminum. In solid solution strengthening[8], the addition of another soluble element into the aluminum matrix causes strains in the matrix, due to the different atomic radii. A strained matrix provides greater resistance to dislocation movement, and thus increased strength. In age hardening[8], an alloying element is added to concentration greater than their solubility limit in the aluminum matrix and initially supersaturates the solution. Over time, generally at elevated temperatures, the alloying element precipitates out as a second phase in the aluminum matrix with a coherent interface to the aluminum matrix[8]. The precipitates blocks the movement of dislocations. In addition, the coherent interface can cause strains in the aluminum matrix, making slip difficult. Age hardening is the single most effective way to increase the strength of bulk aluminum alloys. Multiphase strengthening is similar to age hardening in that two or more phases are formed. However, in multiphase strengthening[3], the second phase does not have a coherent interface to the aluminum matrix. Thus, in multiphase strengthen, the precipitates may block the movement of dislocations, but do not strain the aluminum matrix. It should be noted that the coherent precipitates of many age hardened alloys become incoherent precipitates over time, at room or at elevated temperatures. The loss of the coherent interface results in a reduction in strength of the aluminum alloy. Work hardening[8] increases the dislocation density by 10^4-10^6 times and therefore strengthens the metal. However recrystallization at modest temperatures negate these effects[8]. Lastly, it is difficult for a dislocation to cross a grain boundary. Thus, reducing the grain size will increase the grain boundary area and impedes the movement of dislocations with concomitant strengthening. Of the five strengthening mechanism listed, this research focuses on using the first three, to improve bimetallic actuated MEMS devices.

In general, several strengthening mechanisms may be used simultaneously. For MEMS applications, the use of multiple strengthening mechanisms would be acceptable and beneficial. For microelectronics applications, all of these strengthening mechanisms will increase the electrical resistivity[9,10,11]. Thus, additional research would be needed to determine how best to strengthen aluminum with a minimum increase in resistivity for microelectronics application.

EXPERIMENT

Two commercially available aluminum alloys, T201 and 2090 (see Table 1), were deposited on (100) silicon substrates. Both, pure aluminum (99.998 % pure) and a standard microelectronic metallization of Al-0.04Cu-0.017Si were also deposited to bench mark the properties of the commercial alloys. The film thickness and deposition technique used for each film tested are shown in Table 1.

Table 1
List of Films Examined

FILM	COMPOSITION	THICKNESS µm	DEPOSITION METHOD - SUBSTRATE TEMP. °C
Al	99.998 Pure	1.43	EB - 25
AlSiCu	Al-.04Cu-.017Si	1.28	SP - 25
T201	Al-.046Cu-.006Ag-.004Mn-.003Mg-.003Ti	1.30	SP - 100
2090	Al-.026Cu-.021Li-.001Zr	0.80	SP - 100

EB - Electron beam evaporated
SP- Sputter deposited

Stress in the deposited films was determined by measuring the change in the substrate curvature using a Tencor FLX 2320. Stoney's equation[12] was used to calculate the film stress. After the initial stress was determined, the samples were heated to allow an equilibrium microstructure to develop and to allow the maximum film tensile stress to develop. Stress versus temperature was measured and plotted. These data show the temperature and stress at which the films begin to plastically deform. Lastly stress was measured as a function of time at a given temperature. These data provide a measure of the creep rate of the film at the test temperature.

RESULTS

The results of thermal cycles to 250°C (figure 2) show that pure aluminum can only withstand a tensile stress 150 MPa at room temperature before plastically deforming, and aluminum will plastically deform at 150°C under a compressive stress of 50 MPa. The AlSiCu film is slightly stronger, withstanding a tensile stress of 200 MPa at room temperature and plastically deforming at 175°C under a compressive stress of 50 MPa. The T201 alloy film is significantly stronger than the pure aluminum or the AlSiCu alloy. The T201 film withstood a tensile stress of 350 MPa at room temperature and showed no plastic deformation up to 250°C. The 2090 aluminum film was even stronger since it withstood a tensile stress of 500 MPa at room temperature and shows no plastic deformation up to 250°C.

Over a larger temperature range (figure 3) the T201 film plastically deformed above 240°C under a compressive stress of 100 MPa. The 2090 film plastically deformed above 270°C under a compressive stress of 200 Mpa.

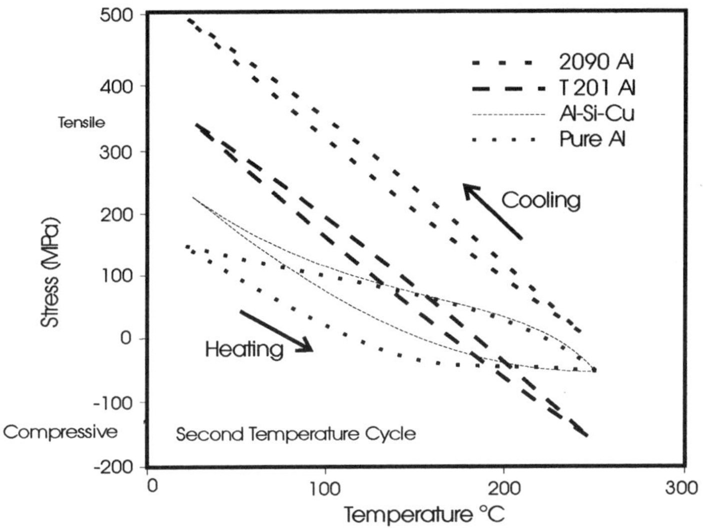

Figure 2: Stress versus temperature for aluminum alloy films to 250°C.

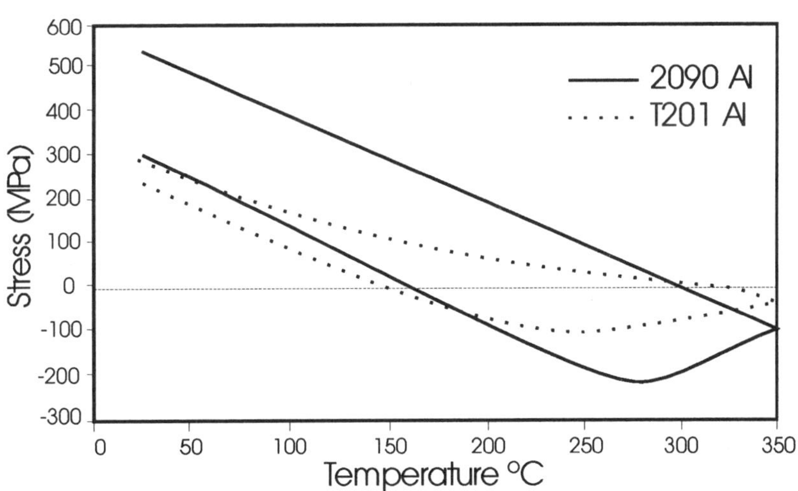

Figure 3: Stress versus temperature for aluminum alloy films to 350°C.

31

Stress relaxation at 50°C is shown in figure 4. The pure aluminum and AlSiCu films showed a 17% reduction in stress over 1000 minutes (17 hours), whereas the T201 alloy film show only a 2% reduction in stress over the same time. The smaller stress relaxation for the T201 film is even more impressive when it is recalled that at 50°C, the pure aluminum, AlSiCu and T201 films are stressed to levels of 75, 200 and 300 MPa tensile stress. Not only is the T201 film relaxing at a much slower rate than the other two films, but it is also under a significantly greater stress.

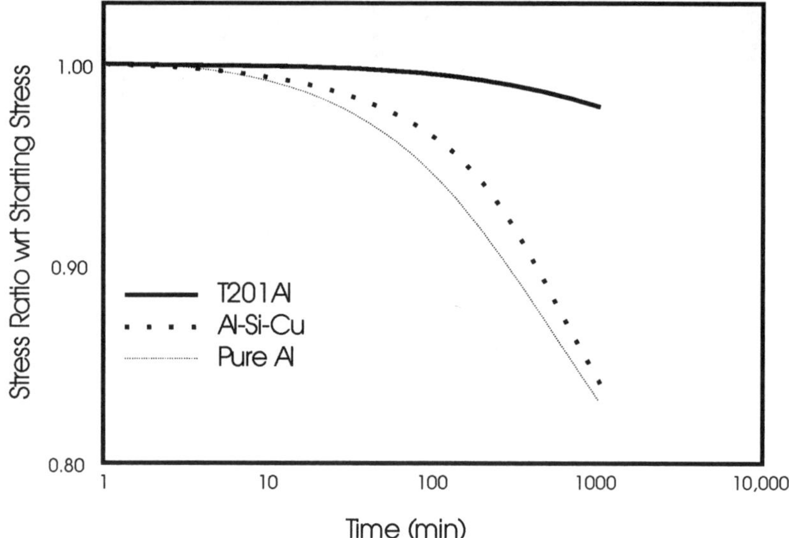

Figure 4: Stress relaxation of aluminum films at 50 °C.

DISCUSSION

The increased strength of the AlSiCu alloy versus pure aluminum can be attributed to a combination of three possible strengthening mechanism. First, supersaturated solid solutions of silicon and copper in the aluminum would result in some solid solution strengthening. Additionally, the $CuAl_2$ precipitates could provide some age hardening or multiphase strengthening. Given the thermal history of these samples over aging of the $CuAl_2$ precipitates is expected. Therefore, we believe that the $CuAl_2$ precipitates are acting as a multiphase strengthening agent in this application, and there is little or no age hardening of the alloy.

For the T201 film we attribute the increase in strength to solid solution strengthening by the manganese and magnesium alloying elements[13]. The silver is known to improve the age hardening affects, but age hardening is not believed to cause significant strengthening in this application. Titanium is used as a grain refiner and may maintain a small grain size even for thin film applications.

For 2090 aluminum films, we attribute most of the increase in strength to the formation of metastable, coherent Al_3Li precipitates[14], in age hardening. It is assumed that copper is precipitated as large, incoherent $CuAl_2$ particles and therefore have a minimal strengthening effect. Zirconium, like titanium in T210, is a grain refiner and may provide some positive strengthening effects.

These data are consistent with the fact that some of the strengthening mechanisms used in bulk alloys can be effective in improving the yield strength and creep resistance of thin film aluminum metallization. There is still a significant amount of research required to determine the most effective strengthening mechanisms, and which are compatible with the resistivity and defect/impurity properties of microelectronics application.

CONCLUSIONS

In this study, films of pure Al, AlSiCu, T201, and 2090 alloys were deposited by electron beam evaporation or sputter deposition. Stress versus temperature and stress relaxation were measured in the films. Pure aluminum and AlSiCu alloy films exhibited plastic deformation at low stresses and low temperatures. The T201 and 2090 films exhibited residual elastic stresses at room temperature of 350 MPa and 500 MPa, and did not plastically deform until 240°C at 100 MPa stress, or 270°C at 200 MPa stress, respectively. The T201 film also showed a low stress relaxation rate. We speculate that solid solution strengthening caused the increase in strength of the T201 film, and that age hardening caused the increase in strength of the 2090 film.

ACKNOWLEDGMENTS

This work was supported by DARPA Grant DAR5C679/8. The aluminum lithium 2090 alloy was provided by Alcoa.

REFERENCES

1.	Werner Riethmüller and Wolfgang Benecker, IEEE Transactions on Electron Devices, VOL. 35, NO. 6 (June 1988).

2.	S. Timoshenko, J.Opt. Soc. Am. and R.S.I., VOL. 11, (Sept. 1925), pp. 233-55.

3.	C. D. Pionke and G. Wempner, Journal of Applied Mechanics, Transactions of the ASME, VOL. 58, (Dec. 1991), pp.1015-20

4.	P. W. Barth, Solid State Sensor and Actuator Workshop, Hilton Head, June 1994, pp. 248-50

5.	Hal Jerman, J. Micromech. Microeng., 4, p. 210-6, (1994).

6.   Korhonen, Scripta Metallurgica, VOL. 24, (1990), pp. 2297-2302.

7.   B. L. Draper and T. A. Hill, J. Vac. Sci. Tech., 139 (4), (July 1991), pp. 1956-62.

8.   Charlie R. Brooks, Heat Treatment, Structure and Properties of Nonferrous Alloys, American Society for Metals, Metals Park, Ohio, 1982, pp. 1-137.

9.   Daniel D. Pollock, Electrical Conduction In Solids An Introduction, American Society for Metals, Metals Park, Ohio, 1995, pp. 125-40.

10.   Rolf E. Hummel, Electronic Properties of Materials, 2nd Edition, Springer-Verlag, New York, 1992, pp. 84-8.

11.   Charlie R. Brooks, Heat Treatment, Structure and Properties of Nonferrous Alloys, American Society for Metals, Metals Park, Ohio, 1982, pp. 275-85.

12.   Milton Ohring, The Materials Science of Thin Films, Academic Press, Inc., New York, 1992, pp. 415-8.

13.   Metals Handbook, VOL.2, Properties and Selection: Nonferrous Alloys and Special Purpose Materials, Tenth Edition, pp.152-3.

14.   Metals Handbook, VOL.2, Properties and Selection: Nonferrous Alloys and Special Purpose Materials, Tenth Edition, pp.178-99.

FILM THICKNESS EFFECT ON TENSILE PROPERTIES AND MICROSTRUCTURES OF SUBMICRON ALUMINUM THIN FILMS ON POLYIMIDE

Y.S. KANG, P.S. HO, R. KNIPE*, AND J. TREGILGAS*
Center for Materials Science and Engineering, University of Texas at Austin, Austin, TX 78712
* Texas Instruments Inc., Dallas, TX 75265

ABSTRACT

The mechanical behavior of the metal film on a polymer substrate becomes an important issue in microelectronics metallization. The metal/polymer structure is also useful to investigate the deformation behavior of very thin free-standing metal film since the flexible polymer serves as a deformable substrate. The tensile force-elongation curves have been measured using a microtensile tester for aluminum thin films, deposited on a PMDA-ODA polyimide film, in the thickness range from 60 nm to 480 nm. The stress-strain curves for aluminum films were constructed by subtracting these curves with polyimide curves measured separately. Tensile strength increases linearly with decreasing film thickness from 196 MPa to 408 MPa within the film thickness range studied. This is in good agreement with the published data for free-standing aluminum films in the same thickness range. The measured Young's modulus is lower than the bulk modulus and exhibits no systematic dependence on the film thickness. The microstructures of aluminum films have been examined using a transmission electron microscope (TEM). These films posses the (111)-textured columnar grain structures. Grain sizes exhibit log-normal distributions and the mean grain size increases monotonically with the film thickness. An attempt is made to evaluate the effect of film thickness and grain size on the strength of aluminum thin film and the result is discussed.

INTRODUCTION

Thin film materials have been used in microelectronics, data storage, and micromechanical devices such as digital micro mirrors [1], primarily for their electronic, magnetic, and optical properties. However, reliability concern for the structural integrity of these devices has stimulated considerable interests in the study of mechanical properties of thin films. While mechanical properties have been studied for thin films in a free-standing state [2,3], most extensive works have been concentrated on thin films deposited on rigid substrates [4-7]. These results indicate that the mechanical behaviors of thin films are strongly influenced by the presence of rigid substrate. It is of interest to investigate the behavior of free-standing thin films from both fundamental and practical viewpoints in order to understand the large strain behavior and for free-stranding application.

Mechanical behaviors of thin films differ from that of their bulk counterparts due to characteristic microstructures. Thin films usually have strongly textured columnar grain structures. Grain size is typically limited by the film thickness, even in an annealed film, resulting in a smaller grain size. While dislocation glide is still a dominant deformation mechanism, large pile-ups of dislocations at grain boundaries have hardly been observed in thin films [5]. This leads to the debate on the validity of Hall-Petch relation in thin films. On the other hand, the constraint of dislocation motion by being pinned at substrate and natural surface oxide has often been observed [6, 7]. This indicates that the mechanical strength is also influenced by the film thickness, as suggested by the dislocation bowing model.

Mat. Res. Soc. Symp. Proc. Vol. 436 © 1997 Materials Research Society

The objective of this paper is to investigate the effect of film thickness on mechanical properties and microstructures of thin films. For this purpose, aluminum thin films, deposited on a PMDA-ODA polyimide film, were chosen in the thickness range from 60 nm to 480 nm. The aluminum/polyimide structure is useful for interconnection applications in microelectronics. In addition, this metal/polymer bilayer structure is also useful to evaluate the overlayer metal properties. The tensile mechanical properties were measured by a microtensile tester. Aluminum properties were determined by subtracting polymer properties, which were obtained in separate measurements, from aluminum/polymer composite properties. Microstructures, measured using a transmission electron microscope (TEM), were correlated with the tensile property change as a function of the film thickness. Finally, we will attempt to decouple the contributions of film thickness and grain size to the strengthening of aluminum thin film.

EXPERIMENTS

The films used in this study were pure aluminum (Al >99.99%) films of thickness 60, 120, 240, and 480 nm. These films were deposited as 2 mm wide stripes onto polyimide-coated Si wafers at room temperature using MRC model 903 DC magnetron sputtering system. Prior to aluminum deposition, PMDA-ODA precursor solution (DuPont Pyraline PI2540) was spun onto a 6" Si wafer covered with 1000 Å of SiO_2, and then cured at 300 °C under nitrogen flow. Thickness of polyimide was about 3.5 μm. As-deposited aluminum films were used throughout this study.

Mechanical properties were measured using a home-built microtensile tester. Samples were cut through the polyimide on the wafers into stripes of 3 mm in width and about 50 mm in length. Polyimide was very easy to peel: it was lifted off almost by itself from the Si wafer with one drop of methanol. Samples were carefully observed under an optical microscope before and during the experiments for handling damages, cracks, and delamination of aluminum films. Damaged samples were discarded before the measurements. Gauge length was 16 mm and strain rate was 1×10^{-3}/s.

Microstructures of aluminum thin films were examined by transmission electron microscopy (TEM) using a JEOL EX1200 microscope operated at 120 KV. Polyimides were removed from the aluminum films by reactive ion etching. The grain size was determined from an average of the largest and the smallest dimensions of each grain. Mean grain size of the film was obtained from the measurements of about 100 grains.

RESULTS AND DISCUSSIONS

Tensile mechanical properties of aluminum thin films were measured by the microtensile tester using metal/polymer bilayer structures. The configuration of the test sample is shown in the inset of Figure 1. Both polymer substrate and metal overlayer were clamped and elongated at the same time. The strains of metal and polymer are equal to the measured strain. In the deformation process, the force output is partitioned into three parts: from the polymer substrate, the overlayer metal film, and the interphase. Bartha *et al.* [8] reported that aluminum formed an interphase of about 50Å with PMDA-ODA polyimide. Since this amount is comparable with the thickness (30-50 Å) of the natural oxide on the free surface of aluminum, we neglect the interphase effect at this moment. Instead, the surface oxide effect will be discussed later in this section. Then, the portion from metal can be determined by subtracting substrate portion measured in a separate microtensile test for the bare polyimide film. An example of the force

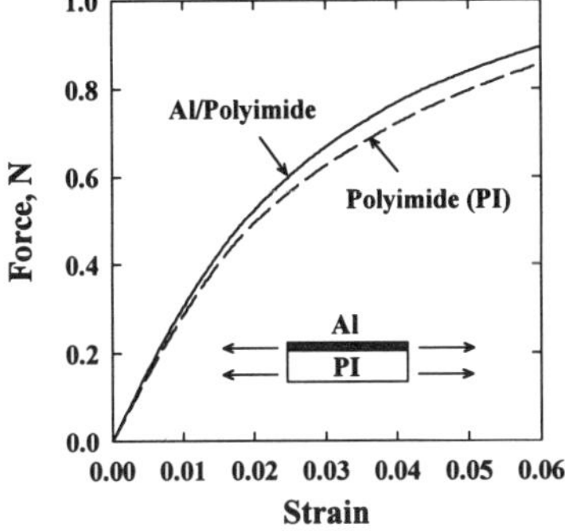

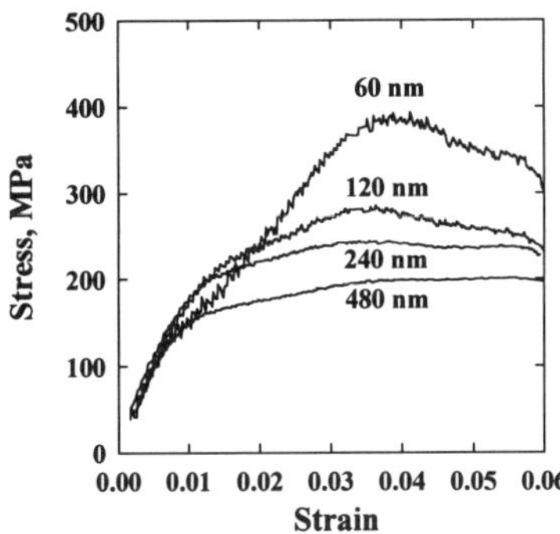

Figure 1. Force-strain curves for Al/polyimide and polyimide.

Figure 2. Stress-strain curves for Al thin films at different film thicknesses.

versus strain plot for a 60 nm Al/polyimide sample is shown in Figure 1, together with that for polyimide alone. The deformation of polyimide substrate is typical for this polymer, undergoing a smooth and gradual transition from the linear elastic to the plastic regime. The Al/polyimide bilayer deforms in a manner similar to that of the polyimide substrate, except that the addition of aluminum overlayer requires more force for the same deformation. The aluminum films showed no evidence of crack or delamination under an optical microscope up to a measured strain of 20% which is far beyond the range of interest. Therefore the additional force is applied only to deform the aluminum film. The amount of this additional force is determined directly from the force difference between the two curves. The division of the force difference by the cross-sectional area of aluminum film is the stress value of the aluminum overlayer at a given strain. Stress-strain curves for 60, 120, 240, and 480 nm aluminum thin films are plotted together in Figure 2. Stress increases polynomially with strain up to about 1% strain and then reaches a plateau with little hardening for all thicknesses except 60 nm. It is worth mentioning that this behavior is very similar to the tensile stress-strain behavior for free standing aluminum films in the thickness range of 110-160 nm [3].

To measure the elastic modulus, the sample was stretched up to 1% and then unloaded. Both the loading portion and the unloading portion of the curve were linear and their slopes were same. Therefore, the elastic modulus was determined from the initial slope of tensile stress-strain curves and shown in Figure 3 as a function of film thickness. No thickness dependence of modulus is observed. Overall modulus values are much lower than the bulk modulus, which has been reported by many other researchers [3]. It is believed that an artifact occurs upon initial loading in association with sample mounting. For example, in our case, the lifted sample has a small residual stress so that it curls down at both edges across the film width. Upon initial loading, part of the force input is consumed to straighten up the sample resulting in a smaller initial slope. Nevertheless, the sample becomes flat after few tenths of percentage of strain. Tensile strength was measured from the maximum stress value on the curve and plotted as a function of film thickness in Figure 4. This plot includes the tensile strength of free-standing aluminum films reported by Mearini and Hoffman [2]. The two sets of data are in a good

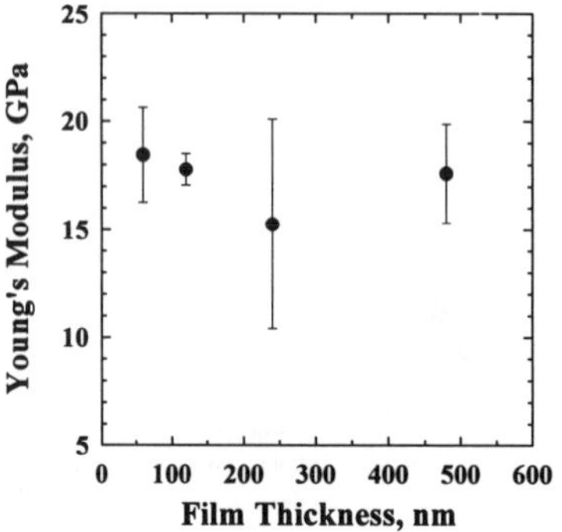

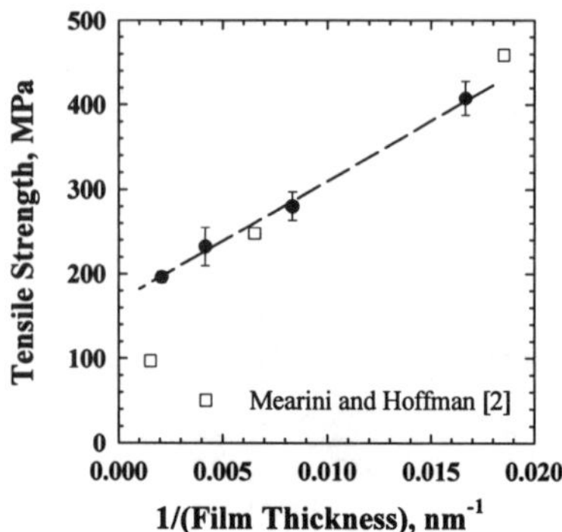

Figure 3. Young's modulus as a function of the film thickness for Al thin films.

Figure 4. Tensile strength vs. the inverse of film thickness plot for Al thin films.

agreement. It should be noticed that the tensile strength of aluminum thin film is much higher than the bulk strength. In addition, it decreases with increasing film thickness. Since the aluminum film has a 30-50 Å thick native oxide on its surface, its contribution to tensile strength was estimated using the published data [2]. Native oxide effect on tensile strength was about 5% for 60 nm film and smaller for thicker films. This rather small effect is due to the fact that an oxide is very brittle and cracks at small strain before it reaches a tensile strain. This result suggests that the variation of tensile strength is primarily due to film thickness change. However, it is well known that the microstructure of thin films such as grain size is also influenced by the film thickness. Therefore, the observed tensile strength behavior is the result of the superposition of film thickness effect and microstructural change.

Aluminum thin films were observed under transmission electron microscope (TEM) for their microstructure characterization. In all thicknesses, aluminum films exhibit a columnar grain structure in which grain boundaries are perpendicular to the film plane. However, some cases of two grains stacked along the film thickness were observed in the 480 nm film. These films have primarily (111) textures and show log-normal grain size distribution. Figure 5 shows the mean grain size versus film thickness plot. Grain size increases asymptotically with film thickness. This is rather unusual because the grain size of aluminum films deposited on rigid substrates often increases linearly with film thickness. The dotted curve represents the result of simple mathematical fitting of the data, which will be used later in the strengthening model.

The strength of thin film is influenced by many parameters such as film thickness, grain size, the presence of alloying elements, and texture. In a given material system, the thickness of film and grain size have been recognized as main factors which govern the mechanical strength of thin films. The thickness effect is predicted by a dislocation bowing mechanism, where the stress required to bow out a dislocation loop in the film is inversely proportional to the film thickness. The grain size effect on the strength of bulk materials has often been expressed by the Hall-Petch relation. This relation is based on the assumption of large pile-ups of dislocations at grain boundaries. In thin films, however, such large pile-ups have seldom been reported. On the other hand, it has been proposed that the flow stress increases inversely proportional to the grain size ($\sim g^{-1}$) rather than to the square root of grain size ($\sim g^{-1/2}$) [5, 9, 10]. In the following, the contribution of film thickness and grain size to the strengthening of aluminum is estimated. It is assumed that the tensile strength is composed of two components: first, the strengthening by film thickness which is inversely proportional to the film thickness and second, a grain boundary strengthening which follows a g^{-1} model. Then the tensile strength, σ_{TS}, is expressed as

$$\sigma_{TS} = \sigma_{TS,th} + \sigma_{TS,gb}$$
$$\sigma_{TS} = \frac{m}{t} + \frac{k}{g} \qquad (1)$$

where t is the film thickness, g is the grain size, and m and k are the coefficient of film thickness strengthening and grain boundary strengthening, respectively. As mentioned earlier, the tensile strengths measured for different thicknesses of films are the results of the combination of both film thickness and grain size effect. However, if the g parameter in equation (1) is replaced with the grain size as a function of the film thickness obtained in Figure 5, both strengthening components can be separated by fitting the plot in Figure 4 with this equation. In this way, the strengthening coefficients m and k are found to be 1466 (MPa·nm) and 17564 (MPa·nm), respectively. This is plotted in Figure 6 and the measured tensile strength is shown together. First of all, the level of strength component for grain size is much higher than that for the film thickness over the entire thickness range. This suggests that a high tensile strength of aluminum thin film is mainly due to a smaller grain size. While the film thickness component becomes zero at infinite thickness, the tensile strength due to the grain size effect is 155 MPa. This is because the grain size does not increase infinitely but asymptotically with the film thickness as shown in Figure 5. However, it should be pointed out that this result is based on the assumption of a g^{-1} model for grain boundary strengthening. In order to verify this assumption, further study is underway to examine the grain size effect on the tensile strength of aluminum thin film at a constant film thickness.

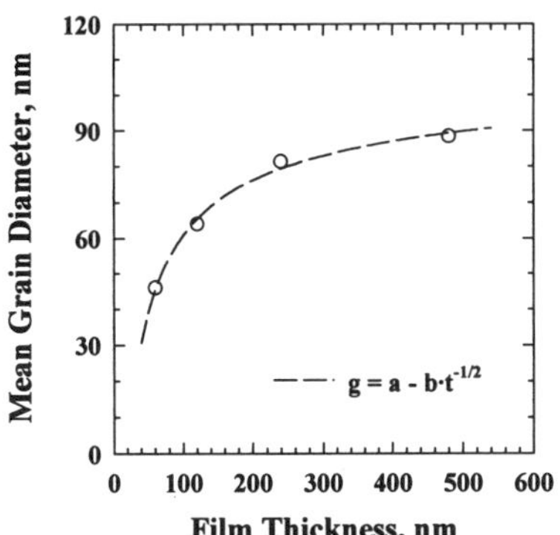

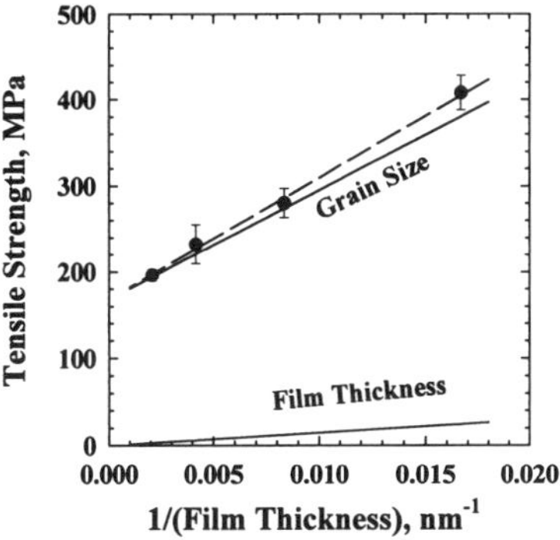

Figure 5. Mean grain size as a function of the film thickness for Al thin films.

Figure 6. Plot of the strengthening due to the film thickness and grain size strengthening component for Al thin films.

CONCLUSIONS

Aluminum films deposited on a PMDA-ODA polyimide film have been investigated for mechanical properties and microstructures in the thickness range from 60 nm to 480 nm. The tensile stress-strain curves of aluminum films were constructed from the measurements of force-elongation on Al/polyimide and polyimide using a microtensile tester. The result is in good agreement with the published data for free-standing aluminum films in a similar thickness range. The Young's modulus was lower than the bulk modulus and shown to be independent of the film thickness. The tensile strength was considerably higher than the bulk strength and enhanced from 196 MPa to 408 MPa as the film thickness decreased from 480 nm to 60 nm. The microstructure of aluminum films was examined using TEM. At all thicknesses, the grains had (111)-textured columnar structures and showed a log-normal distribution. The average grain size increased asymptotically with the film thickness.

The contributions of the film thickness and grain size to the strengthening were evaluated assuming that tensile strength constitutes from these two components. The high strength of aluminum thin film was found to be largely due to the grain size strengthening. However, further study is required to verify the assumption on g^{-1} model for the grain size strengthening.

ACKNOWLEDGMENT

We gratefully acknowledge the financial support of this work by Texas Instruments, Inc.

REFERENCES

[1]   L.J. Hornbeck, *SPIE* Critical Reviews **1150**, 86 (1989)
[2]   G.T. Mearini and R.W. Hoffman, J. Electron. Mat. **22**, 623 (1993)
[3]   R.W. Hoffman, Mat. Res. Soc. Symp. Proc. **130**, 295 (1989)
[4]   C.A. Volkert, C.F. Alofs, and J.R. Liefting, J. Mat. Res. **9**, 1147 (1994)
[5]   R. Venkatraman and J.C. Bravman, J. Mat. Res. **7**, 2040 (1992)
[6]   R. Venkatraman, J.C. Bravman, W.D. Nix, P.W. Davies, P.A. Flinn, and D.B. Fraser, J. Electron. Mat. **19**, 1231 (1990)
[7]   T.S. Kuan and M. Murakami, Metall. Trans. A **13A**, 383 (1982)
[8]   J.W. Bartha, P.O. Hahn, F. LeGoues, and P.S. Ho, J. Vac. Sci. Technol. A **3**, 1390 (1985)
[9]   P. Chaudhari, Phil. Mag. **39**, 507 (1979)
[10] C.V. Thompson, J. Mat. Res. **8**, 237 (1993)

MICROHARDNESS STUDY OF CATHODE–MOUNTED AMORPHOUS HYDROGENATED BORON CARBIDE

SHU–HAN LIN,* DONG LI** AND BERNARD J. FELDMAN*
*Department of Physics and Center for Molecular Electronics, University of Missouri, St. Louis, MO 63121
**Department of Materials Science and Engineering and Center for Engineering Tribology, Northwestern University, Evanston, IL 60208

ABSTRACT

We have grown cathode-mounted amorphous hydrogenated boron carbide thin films by rf plasma decomposition of diborane and methane. The chemical composition, infrared absorption, optical absorption, microhardness and adhesion of these thin films were measured. As a function of increasing diborane concentration in the feedstock, we observe increasing boron and decreasing hydrogen concentrations, increasing infrared absorption at 1300 cm^{-1} due to boron icosahedra, increasing optical band gaps, dramatically increased microhardness, and increased adhesion to the underlying substrates of these thin films. These results provide evidence that the presence of boron icosahedra increases microhardness, adhesion, and optical band gaps.

INTRODUCTION

Amorphous hydrogenated boron carbide (a–B:C:H) was first grown as a p-type doped material of amorphous hydrogenated carbon (a–C:H) [1,2]. Since then, it has been characterized [3,4] and used as an inner wall coating for fusion reactors [5], as a wear-resistant coating for mechanical systems [6], and as a diode material [7]. We previously investigated the transition from doping to alloying in the electrical properties of a–B:C:H [8].

Amorphous hydrogenated carbon has become an important industrial wear-resistant coating, presently being widely used to protect computer hard disks from damage by crashing magnetic heads. Because of its hardness and wear-resistance, a–C:H has been commonly called diamond-like carbon. Its physical, structural, optical, tribological and compositional properties have been extensively studied, and References 9 and 10 are excellent review articles.

Koidl and co-workers first measured the hardness of a–C:H and observed that its hardness decreased with increasing hydrogen concentrations [10]. Last year, we reported the microhardness of anode-mounted a–B:C:H, noting that its hardness was very small due to its large hydrogen concentrations, but possibly larger than that of a–C:H with the same hydrogen concentration [11]. In this report we extend our microhardness studies to cathode-mounted a–B:C:H which should have higher boron and lower hydrogen concentrations. We hope to elucidate the role of boron and hydrogen in the microhardness of these films.

Mat. Res. Soc. Symp. Proc. Vol. 436 © 1997 Materials Research Society

EXPERIMENTAL PROCEDURES

We grew both a-C:H and a-B:C:H from a feedstock of methane (CH_4), diborane (B_2H_6), and hydrogen (H_2) in a capacitively coupled rf plasma reactor with the conditions described in Reference 11. We varied the boron concentration in our films by varying the fraction of diborane in the feedstock, as listed in Table 1. The thin films were grown on the cathode, using quartz, aluminum foil, and silicon substrates. The silicon substrate samples were mounted in an ultramicro-indentation system (UMIS-2000). Multiple indentations were made at different locations on the film surface at different loads. At each load, the load versus displacement curve was recorded, from which the effective modulus and hardness can be calculated using standard formulae [12]. The films grown on aluminum foil were immersed in dilute HCl, dissolving the aluminum; the free standing films were sent to Galbraith Laboratories for chemical analysis. The films grown on silicon substrates were also used for infrared absorption measurements using a Perkin Elmer Model 1610 FTIR spectrophotometer. The films grown on quartz substrates were used for visible and ultraviolet absorption measurements on a Hitachi Model U3100 spectrophotometer; the optical band gaps were determined by fitting the optical absorption to the Tauc equation. The adhesion of the film to the underlying substrate was qualitatively determined by the scotch tape and immersion in liquid nitrogen tests.

TABLE I.

Sample No.	Cathode(C) or Anode(A)	CH_4 (mTorr)	5% B_2H_6 in H_2(mTorr)	Power (W)	Self Bias (Volts)
1	A	120	0	22	400
2	A	60	60	22	400
3	A	20	100	22	400
4	A	10	110	22	400
5	C	120	0	22	400
6	C	60	60	22	400
7	C	20	100	22	400
8	C	10	110	22	400

RESULTS

The chemical compositions, microhardness, and optical band gaps of our various cathode-mounted a-B:C:H samples are listed in Table 2. For comparison, we have also included the results on anode-mounted a-B:C:H and anode- and cathode-mounted a-C:H samples. Figure 1a shows the infrared absorption spectra of cathode-mounted a-B:C:H (sample 8); again, for comparison, Figure 1b shows the infrared absorption spectra of anode-mounted a-B:C:H (sample 3). Figure 2 is a plot of the microhardness of various amorphous samples as a function of their optical band gaps; the solid and open squares are from anode- and cathode-mounted a-B:C:H respectively, and the solid and open circles, from anode- and cathode-mounted a-C:H films, respectively. In general, the a-B:C:H films passed both adhesion tests, whereas the a-C:H films failed both.

TABLE II.

Sample No.	Anode(A) or Cathode (C)	B_2H_6/CH_4	Boron (at.%)	Carbon (at.%)	Hydrogen (at.%)	Hardness (GPa)	Band Gap (eV)
1	A	0	0	47	53	0.77	2.1
2	A	1/1	1.3	40	58.7	0.69	2.5
3	A	5/1	16	23	61	0.58	3.6
4	A	10/1	--	--	--	0.77	4.2
5	C	0	0	60	40	5.29	0.7
6	C	1/1	--	--	--	7.18	2.5
7	C	5/1	38	19.4	42.4	8.48	3.0
8	C	10/1	42	10	48	8.63	3.0

DISCUSSION

The following conclusions can be drawn from the above data: (1) The cathode-mounted a-B:C:H films have higher boron concentrations and lower hydrogen concentrations than the anode-mounted a-B:C:H films. Because the cathode is negatively charged, positive ions in the plasma are accelerated into the cathode and the growing film surface, breaking bonds and sputtering atoms. The C-H and B-H bonds are usually weaker than the B-B, B-C, and C-C bonds, depleting the film of hydrogen and enriching it with boron.

(2) The infrared absorption spectra of the cathode-mounted and anode-mounted a-B:C:H films are very similar, dominated by the broad peak around 1300 cm⁻¹. This peak is most likely due to boron icosahedra, which have an extremely large oscillator strength because of its large size and tight binding [11]. This peak in the cathode-mounted film is slightly red-shifted from that in the anode-mounted film -- from 1333 cm⁻¹ to 1295 cm⁻¹ -- probably due to the lower carbon concentration in the cathode-mounted films. It has been previously demonstrated in boron carbide alloys, that the presence of carbon in the boron icosahedra shifts the infrared peak to the blue [13].

(3) The microhardness of the cathode-mounted a-B:C:H films are significantly larger than the anode-mounted a-B:C:H films. Because both cathode- and anode-mounted films have very similar infrared

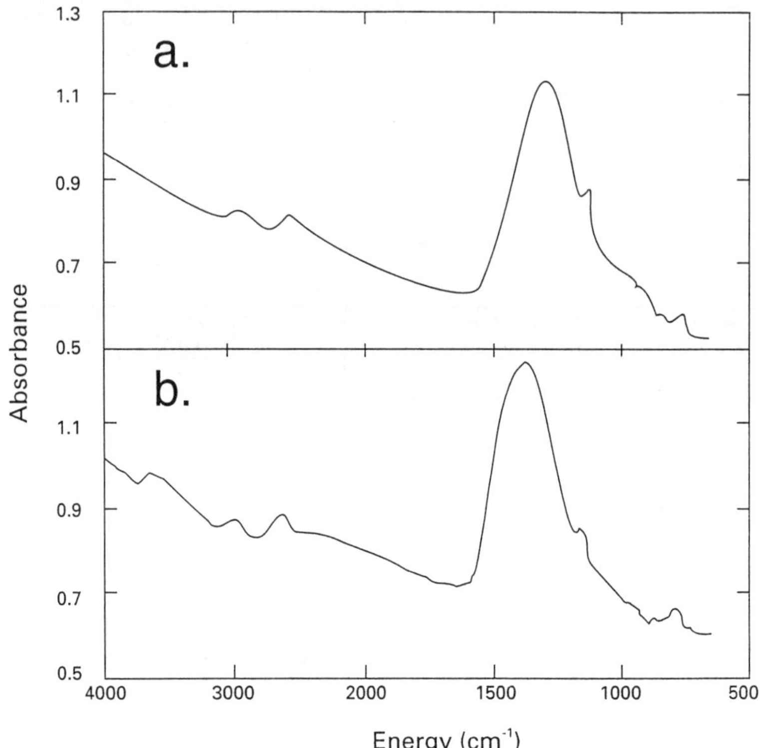

Figure 1.   The infrared spectra of (a) cathode-mounted a-B:C:H (Sample 8) and (b) anode-mounted a-B:C:H (Sample 3).

absorption spectra, all these films are structurally very similar -- boron icosahedra imbedded in an amorphous matrix.   Thus, the microhardness differences are not due to structural differences, but compositional differences.   Hydrogen has already been demonstrated to decrease microhardness [10,11].   This is due to the fact that hydrogen does not cross-link; instead, it terminates bonds, many times as methyl groups that are free to rotate.[14]   The fact that cathode-mounted a-B:C:H films are harder than cathode-mounted a-C:H films, even though they contain larger hydrogen concentrations, provides strong evidence that boron icosahedra contribute to the films' hardness.   This conclusion is supported by our previous work on anode-mounted a-B:C:H films [11].   So the major reasons for the cathode-mounted a-B:C:H films' increased microhardness are both its increased boron concentrations and decreased hydrogen concentrations.

(4) The optical band gap of our cathode-mounted a-B:C:H films increases with increasing boron concentrations.   It is well esta-

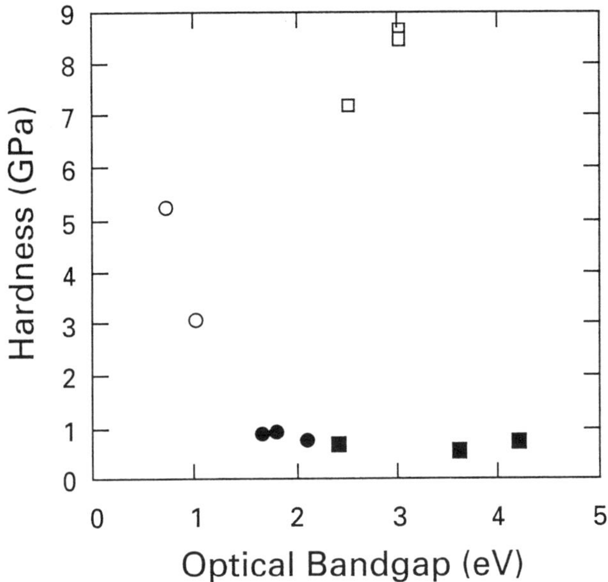

Figure 2. The microhardnesses of cathode- and anode-mounted a-B:C:H (open and solid squares), and cathode- and anode-mounted a-C:H (open and solid circles) thin films as a function of their optical band gaps.

blished that increasing hydrogen concentrations increases the optical band gap [10]. We have demonstrated that boron plays a similar role, even though its bonding in the films is very different.

(5) The adhesion of our a-B:C:H films is superior to our a-C:H films. Clearly, our understanding of adhesion at the atomic level is minimal, but boron icosahedra must be able to form strong bonds between the film and the underlying substrate.

CONCLUSIONS

We have grown a-B:C:H films with dramatically improved mechanical and optical properties. These films have greatly increased microhardness and adhesion and larger optical band gaps, all achieved with the addition of boron to the film and its growth on the cathode.

ACKNOWLEDGEMENTS

This research is partly supported by a University of Missouri Center for Molecular Electronics grant and the National Science

Foundation Surface Engineering and Tribology Program Contract No. MSS-92032239.

REFERENCES

1. D. I. Jones and A. D. Stewart, Phil. Mag. **46**, 423 (1982).

2. B. Meyerson and F. W. Smith, Solid State Commun. **41**, 23 (1982).

3. S. Lee, J. Mazurowski, G. Ramsmeyer and P. A. Dowben, J. Appl. Phys. **72**, 4925 (1992).

4. B. M. Way, J. R. Dahn, T. Tiedje, K. Myrtle and M. Kasrai, Phys. Rev. B **46**, 1697 (1992).

5. S. Veprek, S. Rambert, M. Heintze, F. Mattenberger, M. Jurick-Rajman and W. Portmann, J. Nuclear Materials **162-164**, 724 (1989).

6. J. Onate, A. Garcia, V. Bellido and J. Viviente, Surface and Coating Tech. **49**, 548 (1991).

7. S. Lee, Appl. Phys. A **A58**, 223 (1994).

8. B. Sylvester, S.-H. Lin and B. J. Feldman, Solid State Commun. **93**, 969 (1995).

9. J. C. Angus and C. C. Hayman, Science **24**, 913 (1988).

10. P. Koidl, Ch. Wild, B. Dischler, J. Wagner and M. Ramsteiner, Mat. Sci. Forum **52 & 53**, 41 (1989).

11. S.-H. Lin, D. Li and B. J. Feldman in Mechanical Behavior of Diamond and Other Forms of Carbon, edited by M. D. Drory, D. B. Bogy, M. S. Donley, and J. E. Field (Mater. Res. Soc. Proc. 383, Pittsburgh, PA, 1995) pp. 127-132.

12. M. F. Doermer and W. D. Nix, J. Mater. Res. **1**, 70 (1986).

13. H. Stein, T. Aselage and D. Emin in Boron-Rich Solids, (Amer. Inst. Phys. Conf. Proc. **231**, New York, NY, 1991) pp. 322-325.

14. S.-H. Lin, B. J. Feldman, J. R. Bodart, V. P. Bork, M. J. Kerman, P. A. Fedders, and R. E. Norberg in Novel Forms of Carbon II, edited by C. L. Renschler, D. M. Cox, J. J. Pouch and Y. Achiba (Mater. Res. Soc. Proc. **349**, Pittsburgh, PA, 1994) pp. 489-494.

MECHANICAL PROPERTIES OF CuNi FILMS

W. BRÜCKNER, F. MACIONCZYK, G. REISS
Institute for Solid State and Materials Research Dresden, Institute for Solid State Research,
D-01171 Dresden, Germany, brueckner@ifw-dresden.de

ABSTRACT

The mechanical properties of resistive CuNi films sputtered on both silicon wafers and flexible
Kapton foils are important for reliability and lifetime of resistors. Studies of stress-temperature
dependence and stress relaxation, stress-strain measurements and scanning scratch tests were
performed to investigate film stresses, elastic properties, plastic flow, stress for crack initiation
and film adhesion. The growth and annealing stresses were found to be tensile and can be seen as
the reason for resistance degradation by stress relaxation due to plastic flow. The strains for crack
initiation of 0.2 % to 0.7 % depending on film thickness and annealing restrict the application in
resistors on flexible substrates. The film adhesion can be improved by a NiCr base-layer.

INTRODUCTION

CuNi alloys are widely used for precision resistors and thermocouples. Recently, they have
been applied to low resistance thin film components. They are usually used with an addition of Mn
and in combination with a NiCr base and cover-layer [1,2]. The NiCr provides protection against
oxidation [3] and is important for the tuning of the temperature coefficient of resistance [4]. An
improvement of adhesion is attributed to a NiCr base-layer but up to now this has not been clearly
demonstrated. The application of low resistance films requires films with thicknesses $t_f \approx 1$ μm.
Silicon wafers and flexible polyimide foils are of interest as substrates.

Knowledge of the mechanical properties of CuNi(Mn) films is of considerable importance for
reliability and lifetime of resistors. Plastic deformation can result in resistance (aging) drift, bad
adhesion and cracking ends in breakdown. High mechanical stability (strength against crack
initiation) is desireable for films on flexible substrates.

The present paper discusses the following problems:
(1) The mechanical stress of films on silicon substrates (growth stress, stress evolution during
    annealing, thermal stress).
(2) The stress-strain behavior of the films on flexible polyimide foils (elastic properties, plastic
    flow, crack formation).
(3) The adhesion of the CuNi(Mn) films on oxidized silicon wafers and its improvement by both
    sputter etching of the substrate and a NiCr base-layer.

EXPERIMENT

The films were deposited by d.c. magnetron sputtering at room temperature. The 6 inch melt-
metallurgical target had a composition of 57 at.% Cu, 42 at.% Ni and 1 at.% Mn. The substrates
were oxidized 3 inch silicon wafers (oxide thickness 1.2 μm) with (100) orientation and polyimide
foils (Kapton of DuPont, thickness 8 μm) spread on alumina ceramics. The sputter conditions
were: Starting pressure 10^{-3} Pa, working pressure 0.2 Pa Ar, target-substrate distance 55 mm,
sputtering power 4 kW, and sputtering rate 100 nm/min. For the Ni_xCr_{1-x} base-layer a target
with x = 37 at.% was used and the sputter parameters were 0.7 kW and 16 nm/min. The film
thicknesses were 1000 and 1500 nm CuNi(Mn) for topics (1) and (2). 400 nm CuNi(Mn) with and

Mat. Res. Soc. Symp. Proc. Vol. 436 © 1997 Materials Research Society

without sputter etching before deposition and in some cases a base-layer of 65 nm NiCr were used for problem (3). The film thicknesses were measured by a DEKTAK stylus profiler. More details concerning the preparation and chemical and structural properties of the films can be found in previous publications on the electrical behavior of these CuNi(Mn) films [3-5].

The mechanical stress and its temperature dependence and relaxation were determined by conventional laser-optical wafer curvature measurements using a TENCOR FLX-2410 measuring system [6]. The heating and cooling rate amounted to 4 K/min in a flowing N_2/H_2(5 vol.%) gas mixture. The cooling ramp was interrupted by an isothermal annealing at 200 °C for 10 h.

The stress-strain measurements were done with coated and blank strips of Kapton foil of 50 mm (gauge length) × 5 mm using a ZWICK 1446 tensile testing machine. The strain rate amounted to 3.3×10^{-5}/s. As-deposited films as well as samples which had been annealed at 350 °C for 1 h in a N_2/H_2 gas mixture were tested. The stress-strain curves of the films were evaluated by the difference of the curves for the coated and blank foils. The same thermal treatment of the substrates with and without films was performed. In order to get information about cracking, resistance-strain measurements with Kelvin connections at the clamping jaws and microscopic video recording were performed.

The film adhesion was measured with a SHIMADZU Scanning Scratch Tester SST-101. The measuring parameters were: Tip radius of stylus 15 µm, scanning amplitude 100 µm, scratch speed 10 µm/s, and load rate 5 mN/s. The assessment of the adhesion was done by determination of the critical load L_c and by analysis of micrographs and depth profiles of the scratches.

RESULTS

<u>Mechanical Stress</u>

The growth stresses of the CuNi(Mn) films with thicknesses of 980 nm and 1460 nm are 900 - 1000 MPa in tension (Fig 1a). This was measured by the substrate curvature before and after deposition as well as checked in some cases by removing the film afterwards. The temperature during deposition increased to about T = 300 - 350 °C.

The thermal stress σ_{th} caused by elastic deformation due to the different thermal expansion of film and substrate results in a slope of the $\sigma(T)$ curve of

$$\frac{d\sigma_{th}}{dT} = \frac{E_f}{1-\nu_f} \cdot (\alpha_s - \alpha_f) \tag{1}$$

where $E_f/(1-\nu_f)$ is the biaxial modulus of the film, α_f and α_s are the coefficients of thermal expansion of film and substrate respectively [6]. The measured slope amounts to -2.7 MPa/K and -2.8 MPa/K between room temperature and 200 °C for the films with t_f = 980 nm and 1460 nm respectively, both for the heating and cooling curves. Defects in the as-deposited films and plastic flow during the cooling ramp slightly affect the slope. A value of -3.1 MPa/K was measured with 400 nm thick CuNi(Mn) films on silicon [7].

The increase of the mechanical stress above 350 °C (Fig 1a) is driven by the temperature against the tensile stress and can be ascribed to defect annealing, grain boundary relaxation and, above all, grain growth [6].

The isothermal annealing at 200 °C over 10 h during the cooling ramp results in a reduction of the tensile stress. The corresponding stress-time curves $\sigma(t)$ (Fig 1b) in the range of constant temperature between 1 h and 10 h were described by a simple relaxation model

$$\frac{\sigma(t)-\sigma_0}{\sigma_i-\sigma_0} = \exp(-t/\tau) \tag{2}$$

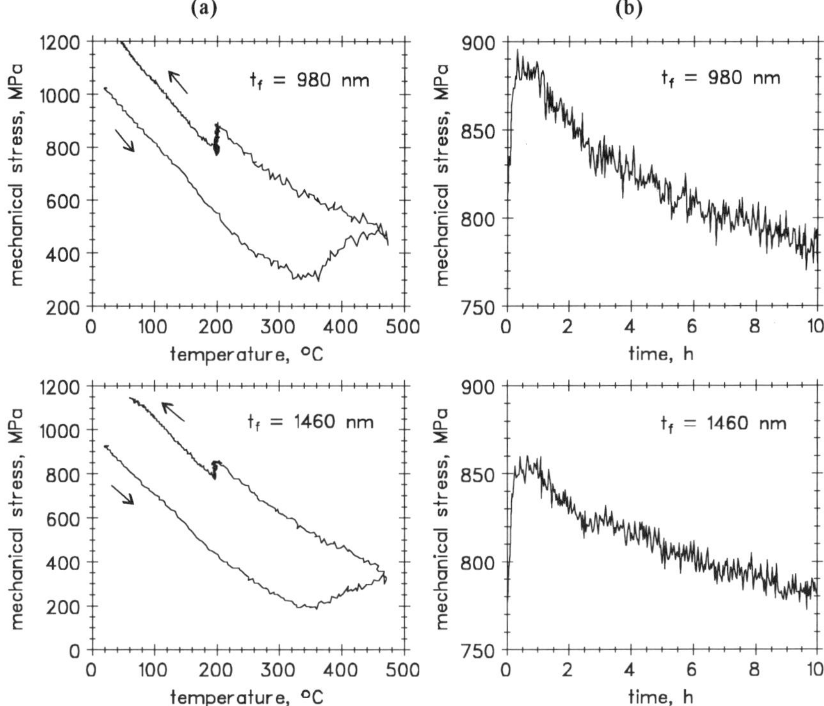

Figure 1. Mechanical stress in CuNi(Mn) films of different thicknesses t_f.
(a) Stress-temperature curves with 4 K/min and annealing at 200 °C during cooling.
(b) Stress relaxation at 200 °C in the cooling ramp.

where σ_0 is the remnant stress and σ_i the initial stress. The relaxation times τ were obtained to (4.5 ± 0.3) h and (7.4 ± 0.9) h for the 980 nm and 1460 nm CuNi(Mn) films respectively. The stress relaxes for the thinner film faster than for the thicker one. To understand details of plastic deformation further investigations are in progress. The results, however, already show that flow occurs in the CuNi alloy films. Therefore this deformation is assumed to cause the resistance increase (aging drift) during annealing of resistive films on silicon wafers. This effect is particularly large with films on alumina ceramic substrates due to structural inhomogeneities [8].

<u>Stress-Strain Diagrams</u>

Stress-strain diagrams $\sigma(\varepsilon)$ were obtained by tensile testing of coated foils and subtracting the data from blank foils (Fig 2a). Because the results at the beginning of the test may be influenced by clamping effects, the Young´s moduli were determined at $\varepsilon \approx 0.3$ % by partial relief of the load and reloading (Fig 2b). Figures 2c and 2d give the results of stress-strain and resistance-strain measurements respectively. The electrical resistance is related to the value without strain, R_0.

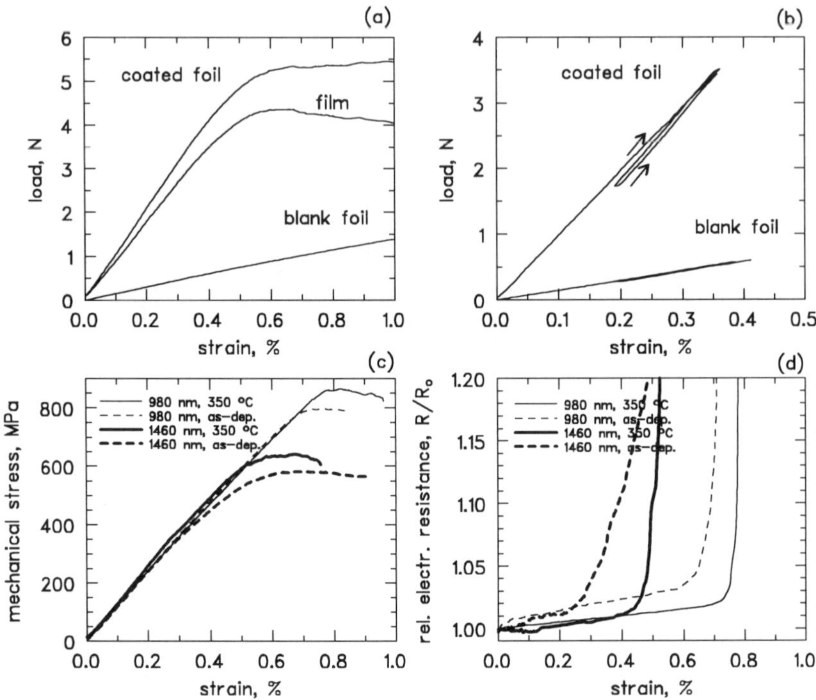

Figure 2. Stress-strain measurements of CuNi(Mn) films on 8 μm Kapton foils.
   (a)   The load-strain curve of the film is the difference between the results of coated and blank foils.
   (b)   The Young´s modulus is obtained by load relief at a strain of ≈ 0.3 %.
   (c, d) Typical stress-strain and resistance-strain curves for as-deposited and annealed (350 °C, 1 h) films with the given thicknesses t_f.

The following Young´s moduli E were obtained:

Film Thickness	As-Deposited Films	Annealed Films (350 °C, 1 h)
980 nm	E = (136 ± 3) GPa	E = (137 ± 6) GPa
1460 nm	E = (125 ± 5) GPa	E = (130 ± 5) GPa

Comparing the results with the bulk value $E_b = 163$ GPa [9], one should consider the preferred (100) texture of the films [5]. Anisotropic data, however, are not available.

The resistance-strain curves show a bending at $\varepsilon = 0.6 - 0.7$ % for the films with $t_f = 980$ nm and $\varepsilon = 0.2 - 0.4$ % for $t_f = 1460$ nm. Up to this bending the relative resistance change is proportional to the strain with a strain coefficient of electrical resistance of about 2 - 3. This value is typical for metallic films. Bending and steep resistance increase are caused by crack formation. This is also confirmed by the microscopic video recording. The cracks propagate from the edge through the sample and their density increases with strain.

The strain at the R(T) bending is the strain for crack initiation. The stress-strain curves obtained for the films can be considered as material properties only up to this crack initiation. Beyond this range they reflect the properties of the film-substrate compound.

The stress for crack initiation of the films is several hundred MPa. This stress decreases with film thickness and increases by annealing. This can be qualitatively explained by a decrease of the number of defects by annealing and an increase of the roughness (micro-flaws!) with the film thickness.

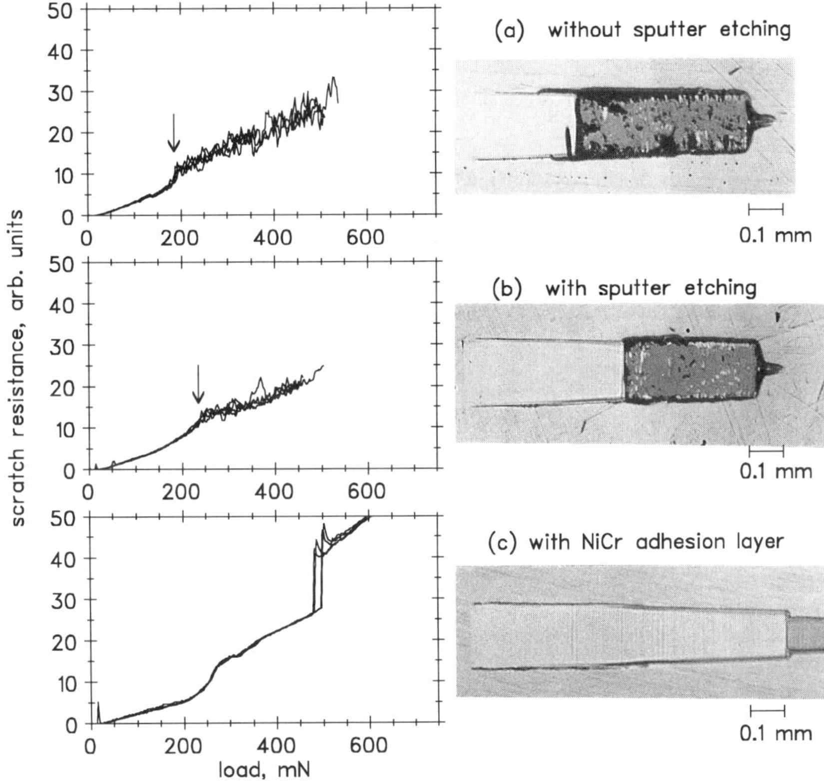

Figure 3. Characterization of the adhesion of 400 nm thick CuNi(Mn) films in scanning scratch tests by the load curves and micrographs.
(a - c) are scratches for different substrate treatments.
The critical loads are marked by arrows.

The deviations from the elastic behavior in the stress-strain diagrams is due to crack formation and only a little plastic flow exists.

<u>Film Adhesion</u>

Results of scanning scratch tests for characterization of the effect of the sputter etching and a NiCr base-layer on the adhesion are given in Fig 3. The critical load L_c is that load at which the film delaminates. The films really delaminate only for the cases without a NiCr adhesion layer. With the NiCr base-layer the CuNi(Mn) is gradually removed. The jump in the scratch curve happens with the transition into NiCr. Analysing the cases without NiCr the average values of the critical loads are (174 ± 8) mN and (214 ± 18) mN for CuNi(Mn) films without and with sputter etching before deposition respectively. These results show that the adhesion can be only moderately improved by sputter etching. In contrast to this, the NiCr base-layer increases the adhesion considerably. Thus, the film adhesion is higher than the toughness of both film and substrate in the scratch experiment in this case.

CONCLUSIONS

The growth stresses of our sputtered CuNi(Mn) films with thicknesses ≥ 1 µm amount to 900 - 1000 MPa tensile. The mechanical stress increases due to annealing by about 400 MPa. It results in cracking and delamination.

The films show stress relaxation by plastic flow. This is assumed to be the reason for aging drifts of resistance during annealing.

The films on Kapton foils crack at strains of 0.2 % - 0.7 % depending on film thickness and annealing. The crack formation restricts the application of such resistive films on flexible substrates.

Good adhesion of CuNi(Mn) on oxidized silicon can be attained by using a base-layer of NiCr.

ACKNOWLEDGEMENTS

The authors are indebted to Th. Knuth with microtech Teltow for the film deposition, to Dr. M. Heilmaier for valuable discussions and to R. Vogel for technical assistance.

REFERENCES

1. I. Nishino, Y. Ichinose, Y. Sorimachi, and I. Tsubata, Int. J. Hybrid Microelectronics 6, 18 (1985).
2. W. Brückner, J. Schumann, H. Grießmann, phys. stat. sol. (a) 140, K21 (1993).
3. W. Brückner, St. Baunack, G. Reiss, G. Leitner, Th. Knuth, Thin Solid Films 258, 252 (1995).
4. W. Brückner, St. Baunack, D. Elefant, G. Reiss, J. Appl. Phys., in press.
5. W. Brückner, J. Schumann, St. Baunack, W. Pitschke, Th. Knuth, Thin Solid Films 258, 236 (1995).
6. M. F. Doerner, W. D. Nix, CRC Critical Reviews in Solid State and Materials Sciences 14, 225 (1988).
7. W. Brückner, phys. stat. sol. (a) 148, K89 (1995).
8. W. Brückner, J. Edelmann, H. Vinzelberg, G. Reiss, Th. Knuth, Thin Solid Films, in press.
9. F. Kohlrausch, <u>Praktische Physik</u>, B. G. Teubner, Stuttgart, 1968, p. 22.

INVESTIGATION OF THE ELASTIC MODULUS OF SOL-GEL DERIVED TITANIA USING THREE-POINT BENDING TESTS

Å. K. JÄMTING[*], J. M. BELL[*,**] AND M. V. SWAIN[***]
[*] University of Technology, Sydney, Department of Applied Physics, PO Box 123, Broadway NSW 2007, Australia
[**] Queensland University of Technology, Department of Mechanical, Manufacturing and Medical Engineering, PO Box 2434, Brisbane QLD 4001, Australia
[***] CSIRO, Division of Applied Physics, PO Box 218, Lindfield NSW 2070, Australia

ABSTRACT

There is increasing interest in the use of sol-gel derived films in tribological applications, and this necessitates an understanding of the mechanical properties of these films. Few investigations into the mechanical properties of sol-gel films have been undertaken, and in this study we have concentrated on measurement of the elastic modulus of sol-gel derived titania films as a preliminary stage in a full investigation of stress in sol-gel deposited thin films. Sol-gel films are often very thin and in order to understand the influence of the substrate on the measured elastic modulus, we have used a multiple coating technique to deposit titania films of increasing thickness on various substrates. A three point bending apparatus is used to measure the elastic modulus. The three-point bending apparatus has very low load and displacement measuring capabilities as is required for the very thin sol-gel films. Measurements of the compositional uniformity of the films have been performed using RBS, and this has been combined with film thickness measurements to determine the film porosity. This information ensures that the measured properties relate to intrinsic film properties. The results of all these measurements will be presented.

INTRODUCTION

Sol-gel films are used in a wide range of applications, ranging from porous aero-gels[1] to densified protective coatings[2]. As the interest for use of sol-gel deposition of thin films increases, it is becoming more important to understand the mechanical properties of the films and how these properties can be controlled. Compared to several other techniques, sol-gel deposition provides a cost effective technique with the possibility to coat substrates of almost all shapes and sizes. A major drawback with sol-gel coatings is the limitation in film thickness[3], but using a multiple coating technique films up to several microns thick can be deposited[4]. In this study, we concentrate on the characterisation of the mechanical performance of titania films, and in particular the measurement of the elastic modulus of these films. During the last few years the development of highly sensitive micro indentation techniques[5] enables the monitoring of the response of very thin films during nano-scale indentation. The elastic response of different coatings during indentation have been presented in several recent reports[6,7] and recently Field et al.[8] have developed a model to establish the elastic modulus using spherical indenters. The use of spherical indenters also enables the transition from elastic to plastic penetration to be studied. The use of indentation techniques has proven to be quite successful, but when investigating very thin films there are limitations in the accuracy due to the limited film thickness. There are different criteria used to decide when the influence of the substrate material starts affecting the measured properties. These vary between a depth of penetration from 25 %[9] of the overall thickness to as little as 10%[10]. When indenting films of thicknesses down to part of micrometers, as in the case

Mat. Res. Soc. Symp. Proc. Vol. 436 © 1997 Materials Research Society

of sol-gel films, this constraint can severely limit the range of experimental parameters. This is the main reason for our interest in different methods for determination of elastic modulus of thin films. Recent work by Rouzaud et al[11]. describes an approach to determine the elastic modulus using a three point bending apparatus and very low loads and another report by Mencik et al.[12] incorporates beam bending techniques being applied to coated systems. In the latter report, the authors also analyse extensively the problems and errors that may appear.

In this paper we report of the preparation and characterisation of multilayered sol-gel titania and of the results from three point bending tests of various substrates, uncoated as well as coated.

THEORY

The deflection of a beam in bending can be expressed as:

$$w = \frac{Pl^3}{cS} \tag{1}$$

where w is the beam deflection, P is the load, l is the distance between the bars in the test set-up (see also figure 2 below), c is a constant (c = 48 for three-point bending) and S is the bending stiffness of the beam. For a rectangular beam cross section the stiffness S can be expressed as

$$S = \varphi EI \tag{2}$$

where φ is a correction factor for the width of the beam ($\varphi = 1/(1-v^2)$; v being the Poisson's ratio), E is the elastic modulus and I is the moment of inertia:

$$I = \frac{wt^3}{12} \tag{3}$$

For a coated specimen there are two different expressions for the stiffness, depending on whether the specimen is coated on both sides or on one side only[12]:

$$S = \frac{\varphi E_s w_s t_s^3}{12}\left\{1 + \frac{E_f t_f}{E_s t_s}\left(3\left(1+(\frac{t_f}{t_s})\right) + (\frac{t_f}{t_s})^2\right)\right\} \tag{4}$$ coating on **both** sides

$$S = \frac{\varphi E_s w_s t_s^3}{12}\left\{1 + \frac{E_f t_f}{E_s t_s}\left(3\left(\frac{1+(\frac{t_f}{t_s})^2}{1+\frac{E_f t_f}{E_s t_s}}\right) + (\frac{t_f}{t_s})^2\right)\right\} \tag{5}$$ coating on **one** side.

Mencik and Quandt[12] have addressed a number of the errors that are associated with bending of a thin beam, including errors in measurement of the specimen, and in particular the width and the thickness, and errors arising from the simplified deformation model. The deformation model neglects deformation of the beam due to shear and torsion, deformation of the specimen supports, local deformation at the loading point and the stiffening effect of the beam width. These errors were all found to affect the results somewhat and need to be taken into consideration when estimating the total deflection. To ensure that the method used is applicable in the case of very thin films, the response of very thin films on various substrates was modelled. Using measured values of the elastic modulus for the substrate and assuming various values for the film modulus, the theoretical deflections can be predicted.

FILM DEPOSITION

The titania films investigated in this work were deposited from a solution consisting of titanium propoxide (TiOH) in propanol. For the work described here the dip coating process was chosen, since it enables the deposition of uniform films on both sides of the substrate material. The substrates are placed in a dry box with a dry N_2 atmosphere and coated by dipping of the substrate into the solution with a constant dipping and withdrawal speed. As noted above, the dipping speed is of the most important factors which influences the film thickness. In the present study, the best results were found for a dipping speed of about 17 cm/min. After dipping, the samples were allowed to dry for 15 minutes in the dry box and then hydrolysed in air for 60 minutes before firing to final temperature. The choice of firing cycle was based on DTA/TGA data obtained using the precursor solution. This data indicates that hydrocarbon removal occurs at approximately 400°C and so the final temperature was chosen to be between 425 and 500°C.

FILM THICKNESS

A wet etching technique was used together with microlithographic masking to enable the exact measurement of the film thickness. This was used to create a step to be measured, both with profilometry and atomic force microscopy (AFM), to establish the film thickness accurately. Titania coated glass specimens were coated with a thin chromium film, a pattern was etched in the Cr film and the samples were then wet etched in hot sulfuric acid, to create a step in the TiO_2. The step was subsequently measured and the results from these measurements are presented in Table I.

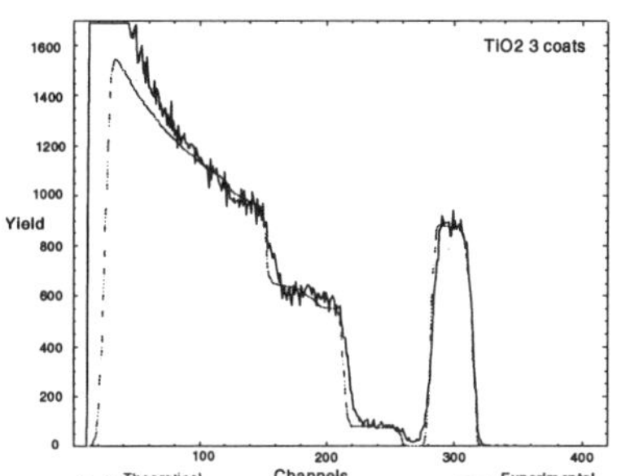

Figure 1. *RBS spectra of a specimen with three coats of TiO_2.*

Table I. Results from AFM measurements of two different thicknesses of TiO_2.

# coats	thickness (nm)
1	85
3	220

FILM COMPOSITION

To ensure that the deposited films have the appropriate composition and structure, these were studied using Rutherford Backscattering (RBS) and grazing incident angle x-ray diffraction (GIXD) respectively. GIXD analysis showed that the film fired to 500° C consist of anatase, which agrees with other studies of sol-gel titania[13]. A typical RBS spectrum from a 3-coat film is shown in figure 1, and a simulation of the spectrum is also shown. This indicates a reasonably uniform film, with a slightly less dense outer layer (~400Å) probably containing additional water. However, the films stoichiometry is very close to TiO_2.

From these measurements and the physical film thickness determined above, the film porosity could be determined. The results are shown in Table II below.

Table II. RBS measurement of thickness and corresponding porosity.

Number of coats	Thickness [nm]	Average porosity [%]
1 coat	85	38
2 coats	155	37
3 coats	226	37
4 coats	286	37

TEST SET-UP

A UMIS 2000 micro indentation system, with capabilities described elsewhere[14] was used for the bending tests. A spherical diamond indenter of radius 150 μm was used as the loading element and the test set-up used in these tests is shown in figure 2. The distance l between the rollers is 10.175 mm and the length of the rollers allows measurement of specimens with varying width.

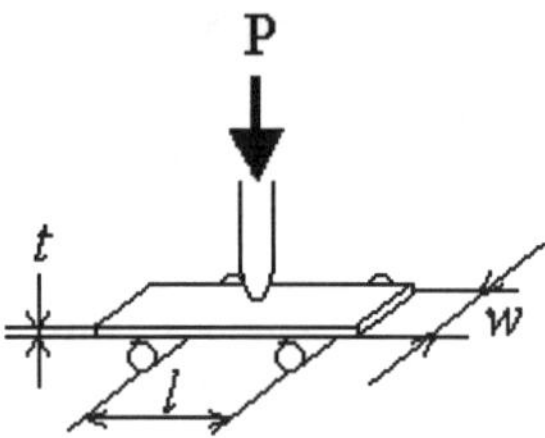

Figure 2. The test set-up

RESULTS AND DISCUSSION

Three-point bend tests were performed on the uncoated substrates prior to film deposition and initial tests on stainless steel beams (grade 304, thickness 0.128 mm) proved successful. These tests gave reproducible results and the data from the bending tests gave a corresponding elastic modulus of about 180 GPa. Tests were performed on stainless steel, silicon wafer strips (thickness 0.312 mm) and glass slides (borosilicate cover slip strips, thickness 0.216 mm) with varying width. The specimens were inserted into the set-up described in figure 2 and loads ranging from 1 mN up to 20 mN were applied as appropriate. For the stainless steel specimens 5, 10, 15 and 20 mN loads were chosen and cycles of typically 5 tests at each load were performed. The load was applied in 20 steps and a typical response is shown in figure 3 which shows the linear and very reproducible behaviour of the beam during bending.

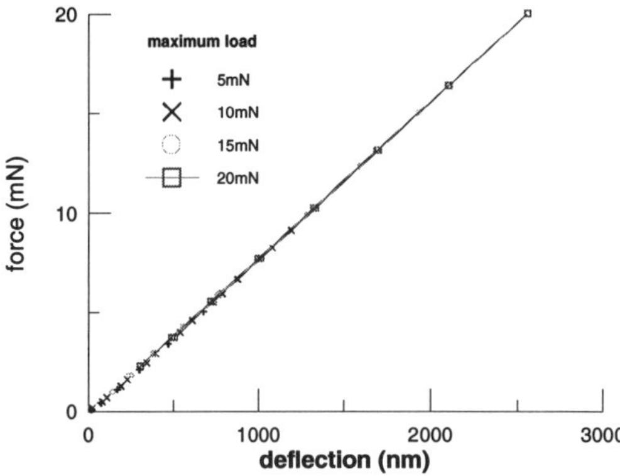

Figure 3. The load-unload data for deflection from the bending of a 5 mm wide stainless steel specimen to 4 different maximum load. The elastic modulus derived from this data is 180 GPa, about 10% lower than anticipated.

Glass and steel substrates showed the same linear behaviour, but there was some non-linearity of the results for the silicon, so these were not considered further. The bending of coated specimen were carried out for films deposited on glass substrates, since these substrates are less sensitive to any prior deformation. The results were adjusted according to the methods of Mencik and Quandt[12] when modelling the data.

The response of coated specimens was also corrected in the same way. Data for a four-coat TiO_2 deposited on glass is shown in figure 4.

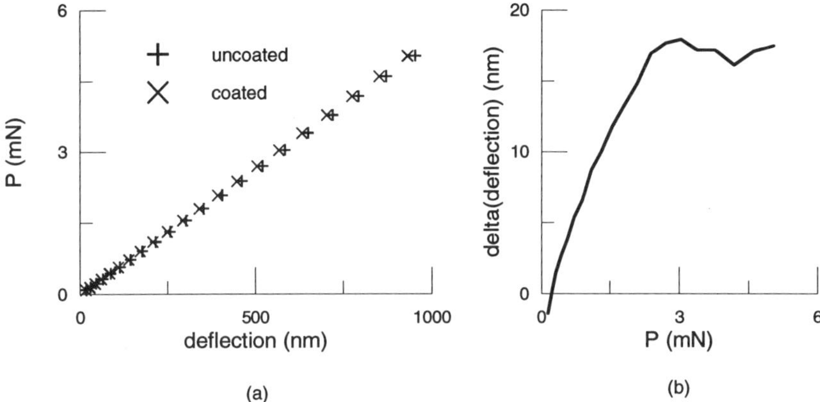

Figure 4. (a) The load deflection data for a coated and an uncoated glass substrate, and (b) the difference in deflection between the coated and uncoated samples

Despite the small change in deflection which is observed, there is a clear difference between the samples, which increases with load. The change in behaviour at a load of approximately 3 mN is not yet understood. Below this value, the change in deflection observed corresponds to an elastic modulus of the film of about 80 GPa. This is a rather low value for TiO_2, but the value obtained for the glass from this experiment was also somewhat low. Further work is in progress to calibrate this system to ensure that it is giving accurate load readings. In addition the titania is quite porous, and this will also account for a significant reduction in the measured elastic modulus.

CONCLUSION

We were able to deposit sol-gel titania films in a multi-layer structure to achieve sufficient thickness to enable the measurements of elastic modulus using a three point bending technique. The films were of anatase structure and we used a wet etching technique to create a step for the measurement of film thickness. The films were also analysed using RBS, and from these measurements the film composition and porosity were determined. It showed that films fired to 500 °C had an average porosity of about 40% which could account for the low values of elastic modulus. The three point bending tests are very promising and have sufficient resolution to measure the elastic modulus of thin ceramic films. Tests on uncoated substrates give elastic moduli for stainless steel and glass which are both a little lower than the expected values, and this suggests that there is either an undetermined source of error, or a calibration error in the load or deflection measurement. The tests on coated substrates showed that the set-up is capable of detecting the response of very thin films and this promises to be a useful method in the characterisation of sol-gel films.

ACKNOWLEDGMENTS

This project was sponsored by Australian Research Council and Silicon Technologies Australia. The authors would also like to thank Dr. Leszek Wielunski, DAP, CSIRO who performed the RBS tests and contributed to the evaluation of the results from these tests as well as M. Anast for the GIXD tests.

REFERENCES

[1] J. Fricke, J. Non-Cryst. Solids, **147-148** (1992) p. 356
[2] M. Shane and M. L. Mecartney, J. Mater. Sci. **25** (1990) p.1537
[3] A. Atkinson and R. M. Guppy, J. Mater. Sci., **26** (1991) p. 3869
[4] D. P. Partlow and T. W. O'Keeffe, Appl. Optics, **29** (1990) p. 1526
[5] W.C. Oliver and G. M. Pharr, J. Mater. Res. **7** (1992) p. 1564
[6] R. B. King, Int. J. Sol. & Struct., **23** (1987) p. 1657
[7] M. F. Doerner and W. D. Nix, J. Mater. Res. **1** (1986) p. 601
[8] J. S. Field and M. V. Swain, J. Mater. Res. **8** (1993) p. 297
[9] M. F. Doerner, D. S. Gardner and W. D. Nix, J. Mater. Res., **1** (1987) p. 845
[10] H. M. Pollock, D. Maugis and M. Barquins, <u>Microindentation techniques in materials science (ASTM STP 899)</u>, American Society for Testing and Materials, Philadelphia, PA (1986) p. 47
[11] A. Rouzaud, E. Barbier, J. Ernoult and E. Quesnel, Thin Solid Films, **270** (1995) p. 270
[12] J. Mencik and E. Quandt, submitted to J. Mater.Res.
[13] B. E. Yoldas, J. Mater. Sci. **21** (1986) p.1087
[14] T. J. Bell, A. Bendeli, J. S. Field, M. V. Swain and E. G. Twaite, Meterologia, **28** (1991/92) p. 463

PLASTIC DEFORMATION IN THIN COPPER FILMS DETERMINED BY X-RAY MICRO-TENSILE TESTS

A. KRETSCHMANN, W.-M. KUSCHKE, S. P. BAKER, and E. ARZT

Max-Planck-Institut für Metallforschung, Seestr. 92, 70174 Stuttgart, Germany

ABSTRACT

Plastic deformation in thin copper films has been studied at room temperature. Copper films having a thickness of 1 µm were made by sputtering onto nickel substrates with a Si_3N_4 underlayer and with or without a Si_3N_4 caplayer. Deformation experiments were conducted using a special micro-tensile tester built into a θ-θ diffractometer. The problems normally associated with tension tests of free-standing films were avoided by deforming the substrate and film together. *In-situ* x-ray measurements of the lattice spacings and lattice spacing distributions were used to determine both elastic and plastic strains. The effects of caplayer and annealing temperature on mechanical properties are reported.

INTRODUCTION

Copper is an interesting alternative material for device interconnections in microelectronic circuits because of the lower resistivity and better resistance to electromigration damage in comparison with aluminum. Unlike Al, Cu does not form a natural passivating oxide, and so a protective caplayer is necessary. Previous investigations [1] have shown that such a caplayer can strongly influence the mechanical properties of the film. Thermally-induced stresses in Cu films on Si substrates were measured using both a substrate curvature technique and x-ray diffraction. The magnitude as well as the characteristics of the stress evolution were seen to be strongly affected by the caplayer, and the differences were attributed to a Bauschinger-like dislocation effect.

In order to further study the effect of the capping layer on plastic deformation in Cu films, we have developed, following previous examples [2-8], a special micro-tensile tester. With this instrument, a sample consisting of a thin film on a substrate is pulled in tension while the strains in both components are measured, *in-situ*, using x-ray diffraction. The primary advantages of this method are that one can apply and measure strain without the effects of changing temperature, can investigate deformation in grains having a particular crystallographic orientation and can measure deformation of the film while it remains attached to its substrate.

The stresses and strains in the sample during tensile deformation result from a superposition of biaxial (due to interaction between the film and substrate) and uniaxial (imposed) strain states. We have developed a model for analysis of the x-ray data which accounts for this strain state. In this paper, we present the machine, the analysis model, and first results from Cu films deposited on Ni substrates with and without a Si_3N_4 caplayer. These results support the existence of a Bauschinger-like dislocation effect in capped Cu films.

EXPERIMENT

Micro-tensile Tester

A schematic of the micro-tensile tester is shown in Figure 1. Samples are clamped between grips. One grip is fixed to a load cell and the other is driven along track rods by a screw. The total strain is measured by an extensometer which is glued to the back of each sample. The micro-tensile tester is built on a special base plate which can be mounted on a Siemens D5000 Θ-Θ goniometer. During a test, the total load on the sample and the total strain are measured continuously by the load cell and the extensometer. The screw is stopped periodically (typically every 30 or 40 seconds) and an x-ray strain measurement is made. The micro-tensile tester allows tests with strain rates between 2×10^{-5} and 1×10^{-1} s^{-1} with a strain resolution of $\pm\,0.0125\%$. The load cell works in the range between 0 and 2 kN with an error of $\pm\,0.25\%$.

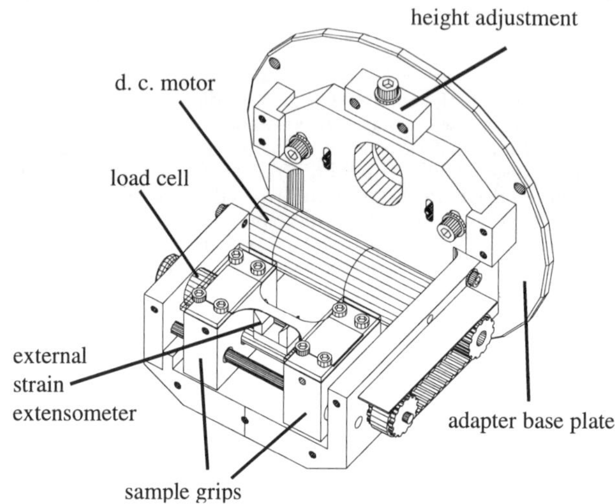

Figure 1: Schematic of the micro-tensile tester.

<u>Sample Preparation</u>

Flat, double-tapered substrates (shown in Figure 2) were spark eroded from a rolled 250 µm thick Ni sheet (approximately 99.7% pure). Afterwards they were mechanically polished on one side with 1200 grit paper, followed by 6, 3 and 1 micron diamond supensions to a mirror finish. The textures of the polycrystalline nickel substrates were determined using a separate four-circle x-ray diffractometer.

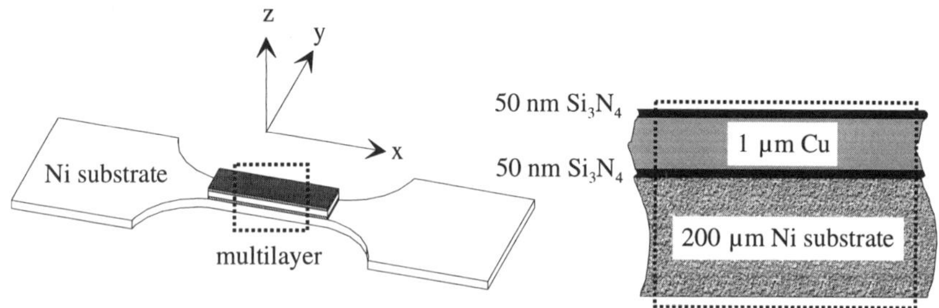

Figure 2:   Schematic of the sample configuration.

Each substrate was mounted on a plate, and a mask with a window in it was used to cover the substrate so that the subsequent films were only deposited on the narrow gage section. The gage section dimensions were 20 mm × 6 mm. Before starting the sputtering process, the surface of each substrate was cleaned by sputtering for 5 minutes with Ar-ions. Next, a 50 nm thick amorphous Si_3N_4 film was sputtered onto each substrate followed by deposition of a copper film to a thickness of 1 µm. Two identical sets of specimens were made; one with and the other without a Si_3N_4 caplayer sputtered on top of the Cu film. Sputter parameters are listed in Table 1.

The Si_3N_4 intermediate layer is necessary in order to isolate the Cu film from the Ni substrate. Without this barrier layer, the Cu film adopts both the crystallography and the deformation characteristics of the Ni substrate. In addition, the Si_3N_4 layer functions as a diffusion barrier between Cu and Ni. A further advantage of Si_3N_4 is that good adhesion with these metals is expected.

sputter parameters	Si_3N_4	Cu
target	B-doped Si	Cu
target diameter	2 inch	3 inch
substrate-target distance	12 cm	6 cm
pressure (total)	3.3×10^{-3} mbar	6×10^{-3} mbar
pressure (base)	3×10^{-6} mbar	3×10^{-6} mbar
pressure (N_2)	6×10^{-4} mbar	-
substrate temperature	room temperature	room temperature
deposition rate	5 nm / min	100 nm / min
power	100 W	150 W
bias	-	-50 V; 40 mA

Table 1: Sputter parameters for Si_3N_4 and Cu

The goal of these experiments was to investigate the plastic deformation characteristics of the Cu thin films. Therefore, it is important that the Si_3N_4 films deform only elastically up to the maximum strain in these tests. That is, no cracks or delamination may occur. In order to determine the fracture threshold for the intermediate Si_3N_4 layer, a Ni substrate having only the first Si_3N_4 layer was strained in the micro-tensile tester to various degrees and was observed after each step using a scanning electron microscope. These 50 nm thick amorphous Si_3N_4 films were found to be stable and free of cracks up to strains of 0.6%.

To investigate the influence of microstructure on the strain/stress behavior of the Cu films, the finished samples were annealed under vacuum for 1h at either 200°C or 400°C before micro-tensile testing.

X-ray Stress Analysis

X-ray diffraction was performed in order to measure the interplanar spacings, d, of the 420 (Ni) and 331 (Cu) planes as a function of the inclination angle Ψ between the scattering vector and the sample normal.

We define an orthogonal coordinate system (sample system) with x along the tension axis and z perpendicular to the plane of the film. The stress state in Cu and Ni during a tensile test is the superposition of an equal-biaxial stress, σ_b ($\sigma_x = \sigma_y = \sigma_b$; $\sigma_z = 0$), which arises due to the interaction of the Cu film with the Ni substrate, and the stresses σ_u imposed by the uniaxial tension test. The equal-biaxial stress arises from differential thermal expansion and also from differential microstructural evolution (e.g. grain growth) in Cu and Ni. The tension test imposes equal strains in Ni and Cu in the x and y directions. If we assume that the transverse (y) contraction is just that of Ni, the stress states of Cu and Ni can be written as

$(\sigma_b^{Cu} + \sigma_{x,u}^{Cu}; \ \sigma_b^{Cu} + \sigma_{y,u}^{Cu}; \ 0)$ and $(\sigma_b^{Ni} + \sigma_{x,u}^{Ni}; \ \sigma_b^{Ni}; \ 0)$.

The strain as a function of Ψ, is given by $\varepsilon(\psi) = \varepsilon_x \sin^2 \psi + \varepsilon_z (1 - \sin^2 \psi)$ which, when combined with the general form of Hooke's law, $\varepsilon_{ij} = S_{ijkl}\, \sigma_{kl}$, leads to an equation of the form $d(\Psi) = d_0 (b + m \sin^2 \Psi)$.

This equation can be solved for $\sigma_{x,u}^{Cu}$ and $\sigma_{x,u}^{Ni}$ in terms of m, b, the unstressed lattice spacing d_0, the equal-biaxial stress, and the elastic constants C_{ij}. The texture measurements showed the Ni substrates to be textured such that isotropic (polycristalline) elastic constants can be used. One then finds,

$$\sigma_{x,u}^{Ni} = \frac{E\,m}{(1+v)\,d_0} - \sigma_b^{Ni} \tag{1}$$

where E is the elastic modulus and v the Poisson's ratio of nickel.

In the case of the Cu films, quantitative analyses of the x-ray texture data following ref. [10] revealed a strong 111-fiber texture. In order to account for this, the stiffness matrix was transformed to the 111-fiber texture sample coordinate system, yielding stiffness coefficients C'_{ij} that are defined in the sample system and are functions of the stiffness coefficients C_{11}, C_{12} and C_{44} in the crystallographic system. One then finds

$$\sigma_{x,u}^{Cu} = \frac{r(b\,k - d_0\,k + 2\,C'_{13}\,d_0\,\sigma_b^{Cu})}{C'_{13}\,k\,d_0} + \frac{(C'_{12} - C'_{11})(2\,C'_{13}\,d_0\,\sigma_b^{Cu} + C'_{33}\,d_0\,\sigma - k\,m)}{k\,d_0} \tag{2}$$

with $r = C'_{12}\,C'_{13} - C'_{11}\,C'_{13} + C'^2_{13} - C'_{11}\,C'_{33}$ and $k = C'_{11}\,C'_{33} + C'_{12}\,C'_{33} - 2\,C'^2_{13}$.

The unstressed lattice spacing d_0 and the equal-biaxial stress σ_b were measured in each sample using a standard biaxial $d(\Psi)$ *vs.* $\sin^2 \Psi$ method prior to tensile testing. The uniaxial stress components were determined by plotting the measured values of $d(\Psi)$ as a function of $\sin^2 \Psi$, fitting a line to the data to determine b and m and using these values in Equations (1) and (2).

RESULTS AND DISCUSSION

The uniaxial stresses determined during loading and subsequent unloading for four samples are shown in Figure 3. The samples vary in annealing temperature and the presence or absence of a Si_3N_4 caplayer. The samples which were annealed at 200°C have about the same yield stress. However, the film with the caplayer shows a great deal of strain hardening while the film without a caplayer shows none. For the films which were annealed at 400°C, the yield stresses are lower and the strain hardening behavior is reversed. That is, the film with a caplayer shows no strain hardening and the film without the caplayer strengthens significantly during straining. This behavior is repeatable. For example, the two films which were annealed at 200°C and tested in tension were subsequently reannealed at 400°C and tested again. These films showed the same stress/strain behavior as the films which were first annealed at 400°C and then tested.

These results show that the caplayer has a great influence on the mechanical behavior of the Cu film, even in the absence of temperature-driven deformation mechanisms. This behavior is difficult to explain if only the typical effects of annealing on the microstructure of the film are taken into account. However, it is possible to understand these results if a Bauschinger-like dislocation effect similar to that proposed earlier [1] is considered.

A Bauschinger effect arises when dislocations are driven up against obstacles by plastic deformation in one direction, storing strain energy in the material. If the direction of the external loading is reversed, this strain energy acts in the same direction as the applied load and reduces the apparent yield stress of the material. The results in Fig. 3 are consistent with the existence of such an effect in these films.

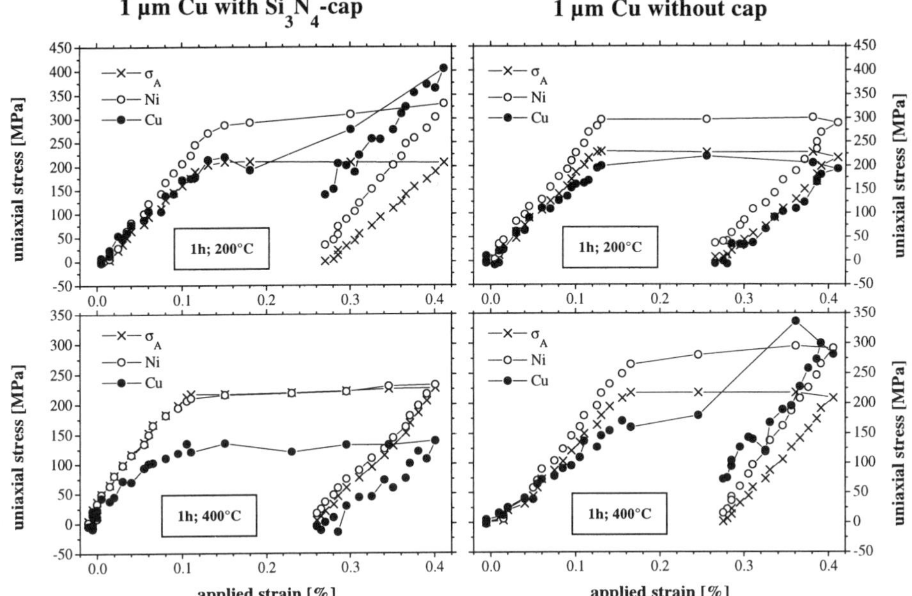

Figure 3:   Tensile test data from sputtered Cu films on Ni substrates *versus* the total applied strain. The variations of the applied stress from the load cell, σ_A, along with the stresses calculated from x-ray data from the Ni substrate and the Cu film are plotted against the strain gage reading from the back of the sample. The error for the Ni stress values are lower than 1% and for the Cu are about 10%.

The Cu in the sample with a caplayer which was annealed at 200°C shows a strong work hardening during plastic deformation. The biaxial stress in this sample was 200 MPa. Calculated from the differences in the thermal expansion coefficients of Cu and Ni, this stress drops to about zero during the annealing process. Thus, this sample is not plastically deformed before the start of the tensile test and the work hardening can be explained by an accumulation of dislocations (*i.e.* normal strain hardening) in the Cu film. In contrast, for the capped film which was annealed at 400°C, there is no work hardening effect. The biaxial stresses in this film were about 250 MPa. During annealing these stresses should have been reduced by more than 400 MPa, therefore, in comparison with the stress levels reported in [1] at this temperature, we expect that this film was deformed plastically in *compression* during heating. If it deforms only elastically in tension, as would be expected given the yield stress behavior near room temperature [1], then the dislocations necessary for a Bauschinger-like dislocation effect are present in the film. If this is the case, work hardening effects in this film could be reduced or eliminated because the dislocations which were driven through the film in compression can now move in the opposite direction in response to the applied tensile stress. The fact that the yield stress is lower in the film annealed at 400°C supports this argument as well.

The sample without a caplayer which was annealed at 200°C shows no work hardening during plastic deformation. Although this sample showed the same initial stress state as the sample with the caplayer, there is no barrier on the Cu surface. Consequently, dislocations can escape from the film and strain hardening would be expected to be much less effective. In contrast again, the sample without a caplayer which was annealed at 400°C does show strain hardening. Upon examination of the previous results [1], we expect the compressive stresses at the annealing temperature to relax to near zero (diffusion is much more effective as a relaxation mechanism in an uncapped film). Therefore it is likely that the film deforms plastically in *tension* rather significantly upon cooling. These dislocations will be driven further into the film during tensile straining and a work hardening effect is much more likely.

These initial microtensile test results are consistent with Bauschinger-like dislocation effects arising as a result of the caplayer. A better understanding of the deformation mechanisms in thin Cu films could be obtained through further comparisons of stress/temperature- and stress/strain-analyses. Further support for the proposed deformation mechanisms must be obtained from careful investigations of the microstructure by transmission electron microscopy.

CONCLUSIONS

These initial results indicate that x-ray micro-tensile testing can be a valuable analysis method to determine stress/strain behavior in thin Cu films. With the technique and experiments presented here, it is possible to show that great differences in mechanical properties exist between Cu thin films with and without a Si_3N_4 caplayer. These differences can be attributed to a Bauschinger-like dislocation effect and could arise in the present films as a result of the dislocation structure which is created in the film during of the pre-testing annealing treatment.

ACKNOWLEDGEMENTS

This work was supported by the Deutsche Forschungsgemeinschaft;
Project "Mikrowerkstoffe" AR 201/5-1

REFERENCES

1. R.-M. Keller, W.-M. Kuschke, A. Kretschmann, S. Bader, R.P. Vinci and E. Arzt in Materials Reliability in Microelectronics V, edited by A. S. Oates, W. F. Filter, R. Rosenberg, K. Gadepally, L. Greer (Mater. Res. Soc. Symp. Proc. **391**, Pittsburgh, PA 1995), p. 309-314.
2. R.H. Marion and J.B. Cohen in Advances in X-Ray Analysis, edited by H. F. McMurdie et.al. (**Vol. 20**, Plenum Press, New York, 1977), p. 355-367
3. K. Perry, I.C. Noyan, P.J. Rudnik and J.B. Cohen in Advances in X-Ray Analysis edited by J.B. Cohen et. al. (**Vol. 27**, Plenum Press, New York (1984), p. 159-170.
4. A. J. Perry and L. Chollet, J. Vac. Sci. Technol. **A 4**, p. 2801-2808 (1986.
5. G. Sheikh and I.C. Noyan in Advances in X-Ray Analysis, edited by C. S. Barrett (**Vol. 33**, Plenum Press, New York (1990), p.161-169.
6. H. Behnken and V. Hauk, Thin Solid Films **193/194**, p. 333-341 (1990).
7. I.C. Noyan and G. Sheikh, IBM Research Report **16927**, p. 1-11 (1991).
8. L.S. Schadler and I.C. Noyan in Thin Films: Stresses and Mechanical Properties III, edited by W.D. Nix, J.C. Bravman, E. Arzt and L.B. Freund (Mater. Res. Soc. Symp. Proc. **239**, Pittsburgh, PA 1992), p. 151-156.
9. I.C. Noyan and G. Sheikh in Thin Films: Stresses and Mechanical Properties IV, edited by P.H. Townsend, T.P. Weihs, J.E. Sanchez, Jr. and P. Børgesen (Mater. Res. Soc. Symp. Proc. **308**, Pittsburgh, PA 1993), p. 3-13.
10. D.P. Tracy and D.B. Knorr, J. Electron. Mater. **22**, p. 611-616 (1993).

GROWTH, STRUCTURE, AND MECHANICAL PROPERTIES OF PULSED LASER DEPOSITED TiN/TiB$_2$ MICROLAMINATES

ASHOK KUMAR*, R. B. INTURI**, U. EKANAYAKE*, H. L. CHAN*, and J. A. BARNARD**
*Department of Electrical Engineering, University of South Alabama, Mobile, AL 36608,
**Department of Metallurgical and Materials Engineering, University of Alabama Tuscaloosa, AL 35487

ABSTRACT

Hard coatings of TiN and TiB$_2$ have many interesting properties such as high thermal and electrical conductivity, high melting point, good thermodynamic stability and combination of these properties make them an interesting prospect for a wide range of tribological and electronic applications. It is understood that artificial multilayer structures have shown anamolously high hardness and modulii making them likely candidate for future protective coatings. Single layer of TiN, TiB$_2$, and TiB$_2$/TiN microlaminates coatings with varying thickness have been deposited on Si (100) and oxidized Si(111) substrates by in-situ pulsed laser deposition method. These films are deposited at 10 Hz repetition rate of excimer laser (λ = 248 nm). Our preliminary results show that elastic modulii and hardness values of multilayered coatings are superior than monolithic coatings of either of the two constituent materials. The coatings have been characterized by X-ray diffractometer and AFM techniques. Detailed results have been presented to correlate the effect of microlaminate thickness on the mechanical properties.

INTRODUCTION

A variety of hard nitride and boride coatings have attractive properties for a wide range of tribological electronic applications. Continuously increasing requirements on coating performance and reliability necessitates the development of modern deposition processes, novel coating compositions, and newer materials geometries. In this regard, pulsed laser deposition has been found to produce high quality TiN [1] and diamond-like carbon films [2] exhibiting excellent mechanical properties. Also, the use of multilayer and gradient coating geometries have been found to enhance the mechanical resistance, adhesion and functional properties of hard coatings. Examples of this category consisting of technologically important materials are TiC/TiN/Al$_2$O$_3$ [3], (TiC/TiB$_2$)$_n$ [4], TiC/Ti(C,N)/TiN [5], TiC/Ti(C,B)/TiB$_2$ [5]. Earlier studies on thin film microlaminates [6, 7] indicates that the deformation behavior and the thin film stress in Al/TiB$_2$[6], Al/SiC [7], Ti/SiO$_2$[6] microlaminates are strongly dependent on the thickness of soft and hard layers. Al/SiC microlaminates containing thin Al layers showed only elastic deformation. As the thickness of Al layer is increased from 240 Å to 1920 Å, Al layers underwent substantial plastic deformation both in tension and compression regimes. Stress-temperature plots determined for Ti/SiO$_2$ microlaminates indicated the plastic deformation of Ti in tension. On the other hand, Al/TiB$_2$ microlaminates containing very thin Al layers showed a purely elastic behavior. It is interesting and also highly beneficial to find out whether a

Mat. Res. Soc. Symp. Proc. Vol. 436 © 1997 Materials Research Society

supermodulus effect exists in TiN/TiB$_2$ microlaminates. In this investigation, a microlaminates system consisting of alternate TiN and TiB$_2$ layers is prepared by pulsed laser ablation technique. The mechanical properties such as elastic modulus and hardness were studied as a function of the thickness of TiN and TiB$_2$ layers, and the substrate material.

EXPERIMENTAL

The deposition of the films was carried out in the "Laser Processing of Materials" research laboratory at the University of South Alabama. This PLD system is capable of depositing multi-layer structures composed up to five different materials, in a vacuum up to 10^{-7} Torr, at temperatures up to 800^0C, and with a background atmosphere of nitrogen, argon or oxygen. A single layer of TiN and TiB$_2$, bilayer of (TiN/TiB$_2$) and (TiB$_2$/TiN) and microlaminates of (TiN/TiB$_2$ .... TiN/TiB$_2$.)$_n$, (TiB$_2$/TiN........ TiB$_2$/TiN)$_n$ films were deposited on Si (100) and oxidized Si (111) substrates at 500^0C in high vacuum. The repetition rate of the excimer laser was 10 Hz for the deposition of all films. The thickness of the single layer films of TiN and TiB$_2$ was approximately 300 nm. The bilayer thickness of TiN/TiB$_2$ or TiB$_2$/TiN was approximately 200 nm and 300 nm, respectively. In the case of multilayer films, the thickness of TiN or TiB$_2$ was increased to 50 nm in each case for the respective microlaminates.   The deposited film were characterized by XRD and AFM techniques. The mechanical properties of single, bilayer and microlaminates deposited on Si(100) and oxidized Si (111) substrates were determined by Nanoindenter II (supplied by Nanoinstruments Inc., Oak ridge, TN). Elastic modulus and hardness of the samples were evaluated from six indentation experiments. Different maximum depths ranging from 20-100 nm were used for each sample. Hardness and elastic modulii of the samples at each maximum indentation depth were evaluated using the formulations developed by Oliver and Pharr [8] and they are given in Table 1.

RESULTS

(a) **Single Layer of TiN and TiB$_2$ Films** - The Young's modulus and the hardness of a single layer of TiN and TiB$_2$ are shown in Figure 1 and Figure 2. Young's modulus and the hardness are almost the same for all indentation displacement. However, the Young's modulus and the hardness values of the single layer films of TiN and TiB$_2$ are comparable to the bulk material values.

(b) **Bilayer of TiN/TiB$_2$ and TiB$_2$/TiN Films** - Nanoindentation properties of the top TiB$_2$ and TiN layer were measured to find out if the inherent properties of top layer were dependent on the presence of TiN and TiB$_2$ in the underlayer. Young's modulus and the hardness values of the bilayers films on Si (100) are higher than that of the single layers. Hardness and the Young's modulus versus indenter displacement are shown in Figures 1 [200/300 nm, TiB$_2$/TiN/Si (100)] and 2 [200/300 nm, TiB$_2$/TiN/Si (100)].

(c) **Microlaminates** - With increasing thickness of the TiN (50 nm) the hardness of the microlaminates shows the highest hardness value of 21 GPa, which is higher than that of the single layer and the bilayer films. As the thickness of TiN increases (200 nm and 250 nm), the hardness and the Young's modulus values of the microlaminates [(TiB$_2$/TiN..... TiB$_2$/TiN)$_n$ 100/200 nm and 100/250 nm,)] films are comparable with the values of the single and bilayer films of TiB_2. In the case of film deposited on SiO_2 substrates, the hardness and Young's modulus values are in the same range of single layer films.

Table 1: Summary of Mechanical Properties of Single, Bilayer and Microlaminates Films
l = 50 nm, n = 3 (number of layers)

Film Number	Structure of Films	Young's Modulus (GPa)	Hardness (GPa)	Thickness (nm)
95-143	TiN/Si(100)	190	15	300
95-144	TiB_2/Si(100)	190	14	300
95-145	TiB_2/TiN/Si(100)	220	19	200/300
95-146	TiN/TiB_2/Si(100)	212	17	300/200
95-147	(2l TiB_2/ 1 TiN)$_n$ / Si(100)	195	16.5	(100/50)
95-151	(2l TiB_2/ 2l TiN)$_n$/ Si(100)	190	12.5	(100/100)
95-153	(2l TiB_2/ 3l TiN)$_n$/ Si(100)	215	19	(100/150)
95-156	(2l TiB_2/ 4l TiN)$_n$/ Si(100)	180	16.5	(100/200)
95-157	(2l TiB_2/ 5l TiN)$_n$/ Si(100)	195	14	(100/250)
95-158	(2l TiN/ 1 TiB_2)$_n$/Si(100)	185	13.5	(100/50)
95-159	(2l TiN/ 2l TiB_2)$_n$/ Si(100)	205	17.5	(100/100)
95-162	(2l TiN/ 3l TiB_2)$_n$/ Si(100)	185	13.5	(100/150)
95-160	(2l TiN/ 4lTiB_2)$_n$/ Si(100)	185	13	(100/200)
95-143	TiN/SiO_2	100	12.5	300
95-144	TiB_2/SiO_2	90	16.5	300
95-145	TiB_2/TiN/SiO_2	115	15.5	200/300
95-146	TiN/TiB_2/SiO_2	115	11	300/200
95-147	(2l TiB_2/ 1 TiN)$_n$/SiO_2	90	11	(100/50)
95-151	(2l TiB_2/ 2l TiN)$_n$/SiO_2	80	7	(100/100)
95-153	(2l TiB_2/ 3l TiN)$_n$/ SiO_2	90	11	(100/150)
95-156	(2l TiB_2/ 4l TiN)$_n$/SiO_2	90	11	(100/200)
95-157	(2l TiB_2/ 5l TiN)$_n$/SiO_2	95	11	(100/250)
95-158	(2l TiN/ 1 TiB_2)$_n$/ SiO_2	75	8.5	(100/50)
95-159	(2l TiN/ 2l TiB_2)$_n$/SiO_2	85	10.5	(100/100)
95-162	(2l TiN/ 3l TiB_2)$_n$/SiO_2	85	10	(100/150)
95-160	(2l TiN/ 4lTiB_2)$_n$/SiO_2	80	10	(100/200)

Young's modulus and hardness values of the (TiN/TiB_2....... TiN/TiB_2)$_n$ system are shown in Figures 3 and 4 with the increasing 50 nm thickness of TiB_2 films in each case. With the increasing thickness of TiB_2 films hardness and Young's modulus values increase with indentation displacement and then decreases with the displacement distance. The values of the Young's modulus and hardness are in same range of single layer films. For the SiO_2, when the thickness of TiN and TiB_2 increases, the values of the hardness and Young's modulus are lower than the single and the bilayers films.

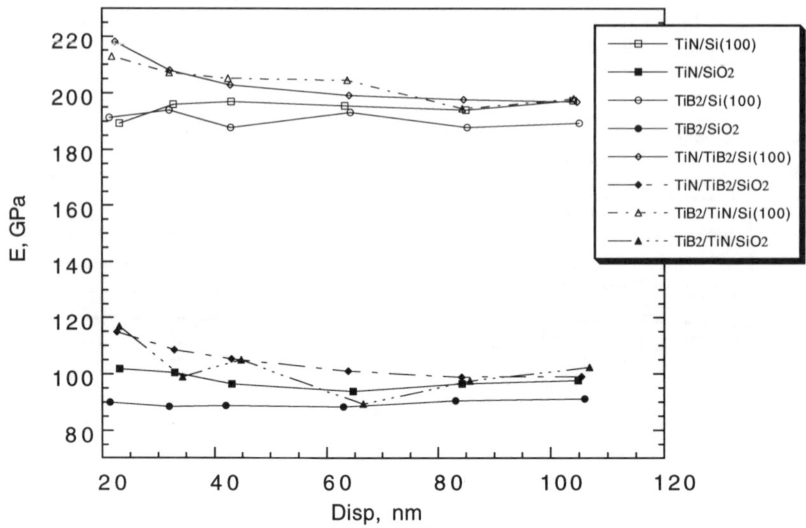

Figure 1: Young's Modulus of Microlaminates on Si (100) and SiO$_2$ Substrates as function of Penetration depth

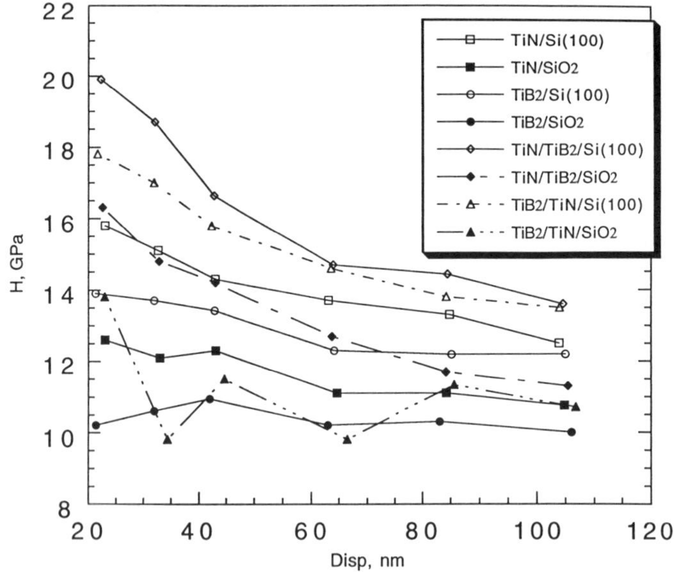

Figure 2: Hardness of Microlaminates on Si (100) and SiO$_2$ Substrates as function of Penetration depth

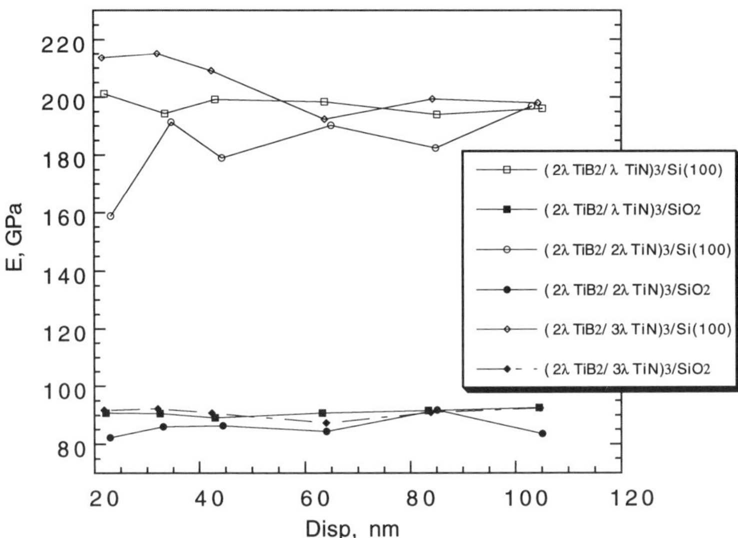

Figure 3: Young's Modulus of Microlaminates on Si (100) and SiO$_2$ Substrates as function of Penetration depth

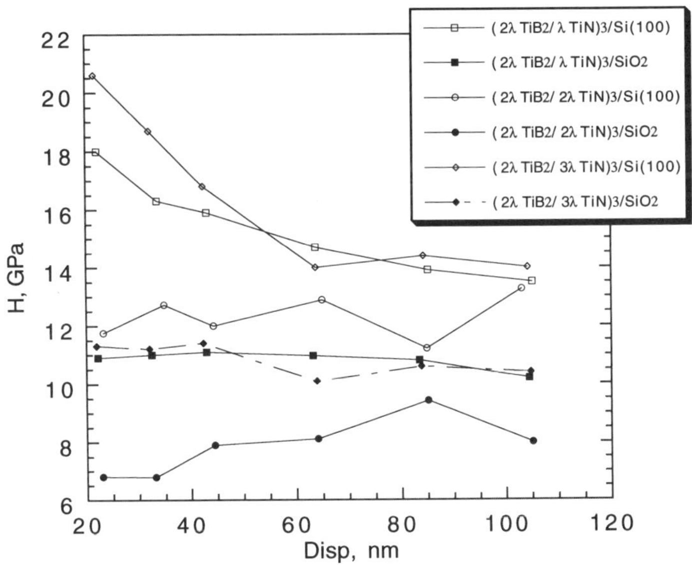

Figure 4: Hardness of Microlaminates on Si (100) and SiO$_2$ Substrates as function of Penetration depth

CONCLUSIONS

The single layer of TiN and TiB_2, bilayer of TiN/TiB_2 and TiB_2/TiN and microlaminates of (TiN/TiB_2.....TiN/TiB_2)$_n$ and (TiB_2/TiN.....TiB_2/TiN)$_n$ films were deposited on Si(100) and oxidized Si(111) substrates at 500^0C in high vacuum. Films deposited on Si(100) substrates show higher hardness and Young's modulus values compared to the films deposited on SiO_2 substrates. Bilayer films have always shown the higher modulus values compared to the single layer or the multilayer films on both substrates, but some of the multilayer films exceed or equal the hardness and modulus values compared to bilayer and single layer films.

ACKNOWLEDGEMENT

This research was supported by Alabama NASA EPSCoR program. AK also acknowledges the partial support from DOE EPSCoR Young Investigator Award.

REFERENCES

[1] R. B. Inturi, Ashok Kumar, U. Ekanayake, N. Shu and J. A. Barnard "MRS Fall Meeting" Boston, MA (1995) in press
[2] Ashok Kumar, R. B. Inturi, Y. Vohra, U. Ekanayake, N. Shu, D. Kjendal, G. Wattuhewa and J. A. Barnard "MRS Fall Meeting" Boston, MA (1995) in press
[3] I. Lhermitte-Sibirre, R. Colmet, R. Naslain, J. Desmaison and G. Gladel "Thin Solid Films", 138 (1986) p. 221
[4] H. Holleck and H. Schultz, "Surface Coating and Technology", 36 (1988) p. 707
[5] H. Holleck and H. Scultz, "Thin Solid Films" (1987) p. 11
[6] R. B. Inturi, J. Chen and J. A. Barnard "Proceedings of the Tenth Internation Conference on Composite Materials", Vol. IV (1995) p. 697
[7] R. B. Inturi, J. A. Barnard, M. Chinnmulgund and J. D. Jarratt "MRS Proceeding", Vol. 308 (1993) p. 719
[8] W. C. Oliver and G. M. Pharr, "J. Mater. Res.", 7 (1992) p. 1564

DEPENDENCE OF HARDNESS AND STIFFNESS ON DENSITY OF Ta₂O₅ AND TiO₂ LAYERS

S.P. Baker[*], C.R. Ottermann[†], M. Laube[‡], F. Rauch[‡], and K. Bange[†]
* Max-Planck-Institut für Metallforschung, Seestr. 92, D-70174 Stuttgart, Germany
† Schott Glaswerke R&D, PO Box 24 80, D-55014 Mainz, Germany
‡ Universität Frankfurt, Institut für Kernphysik, D-60486, Frankfurt, Germany

ABSTRACT

Highly refractive amorphous TiO_2 and Ta_2O_5 films with thicknesses between 270 and 514 nm were deposited on fused silica glass substrates by reactive evaporation and reactive ion plating. Density, hardness, and stiffness were investigated as a function of deposition process. The films were examined using Rutherford backscattering spectroscopy and were found to have densities between 72 and 100% of those of the corresponding bulk oxides. Nanoindentation studies indicated a strong correlation between density and both hardness and elastic stiffness of the oxide film materials. Hardness and modulus both varied by more than 40% over this density range.

INTRODUCTION

TiO_2 films are commonly used as high-refractive-index components in optical coatings and multilayers. Ta_2O_5 is a possible replacement for TiO_2 in certain applications. Although its index of refraction is 10% lower, Ta_2O_5 is stable in amorphous form to higher temperatures and is more transparent to ultraviolet wavelengths.

For production, one would like to deposit such coatings as quickly as possible. However, processes having higher deposition rates tend to produce films of lower density. This, in turn, can lead to high stresses, changes in elastic modulus and hardness, adhesion problems, and reduced corrosion resistance. Thus, to understand the performance of optical coatings, information regarding the relationships between deposition processes, density, and mechanical properties is needed. Amorphous oxide films are, as well, good model systems for investigations of these relationships since the density can be continuously varied over a wide range without other dramatic changes in microstructure (*e.g.* phase changes).

In this paper, we report on our investigations of the density and indentation behavior of TiO_2 and Ta_2O_5 films produced by two different methods. The hardness, modulus and strain rate sensitivity were found to depend strongly on the density of the film.

EXPERIMENTAL

TiO_2 and Ta_2O_5 thin films were deposited onto fused silica substrates by reactive electron beam evaporation (RE) and reactive ion plating (IP). The substrates and TiO_2 films were prepared as described previously [1,2]. Ta_2O_5 films were deposited onto the same substrates using the same equipment and similar conditions. The sample preparation conditions for the samples reported here are listed in Table 1.

The film thicknesses were determined using optical spectrometry. In this method the optical transmittance and reflectance of a film are measured as a function of frequency. A fit to the resulting data [3] gives thickness values to ± 1 nm, as well as the index of refraction and extinction coefficient. The areal density of each film was determined using Rutherford backscattering spectroscopy (RBS). A 15 nA beam of 2.8 MeV ^{4}He was used in a 165° scattering geometry to measure from a 2×2 mm^2 spot. The RUMP computer program [4] was used to simulate the data to determine the composition and areal density of the films. The thickness, composition and areal density can be combined to determine the volume density of the films [5]. Film thicknesses and densities are also reported in Table 1

Nanoindentation experiments were conducted using a commercially available instrument [6] with a diamond Berkovich indenter. Six indentations were made to each of 6 programmed maximum loads, P_{max} = 10, 5.0, 2.0, 0.9, 0.6 and 0.4 mN, in each sample. For each indentation, contact with the surface was first established at a very low load (< 1 μN). The load was then

Mat. Res. Soc. Symp. Proc. Vol. 436 ©1997 Materials Research Society

increased linearly to P_{max} in 10 seconds and was held constant for 30 seconds. The load was then reduced to $0.2P_{max}$, increased to P_{max}, and reduced again to $0.2P_{max}$, all at the initial loading rate. The load was then held at 20% of P_{max} for 30 seconds and the indenter was finally unloaded. This test was run in all the films shown in Table 1, as well as in an uncoated substrate.

A least-squares regression of a power law to the unloading data between 40 and 95% of P_{max} was performed for both unloading curves from each indentation. Following Oliver and Pharr [7], the indentation plastic depth, h_p, was taken to be $h_p = h_{max} - 0.75(h_{max} - h_i)$ where h_{max} is the maximum depth and h_i the intercept of the tangent to the fitted curve at P_{max} with the depth axis. The contact stiffness is the slope of this tangent.

Let A be the projected area of an indentation. The hardness can be simply calculated as P_{max}/A if the shape of the tip, $A = f(h_p)$, is known. An absolute measure of the tip shape is very difficult to obtain and an iterative method [7] is commonly used to obtain an approximate shape using test standards. For the experiments reported here, a special tip shape determination was performed using this method and the fused silica substrate. According to a standard model [7], the indentation contact compliance, C (reciprocal of the contact stiffness), can be written as

$$C = C_m + \frac{\sqrt{\pi}}{2}\frac{1}{E_r}\frac{1}{\sqrt{A}} \quad (1), \qquad \text{where} \qquad \frac{1}{E_r} = \frac{1-v_s^2}{E_s} + \frac{1-v_i^2}{E_i} \quad (2)$$

and C_m is the compliance of the testing machine, E is the Young modulus and v the Poisson ratio, and the subscripts s and i refer to the sample and indenter, respectively. For this tip shape calibration, the measured machine compliance and an assumed value of the indentation modulus of fused silica of 74 GPa were used. For each indentation in the fused silica, the measured contact compliance was used to determine the contact area. The resulting tip shape function was subsequently used to determine the contact areas for the indentations in the films.

From Eq's. 1 and 2, it can be seen that the indentation modulus, $E/(1-v^2)$, of a material can be obtained from the slope of the C vs. $1/\sqrt{A}$ data as long as the stiffness of the indenter is known. This may be done in two ways: 1) One may fit a straight line to C vs. $1/\sqrt{A}$ data to find the slope. This has the advantage that errors in C_m do not affect the data. However, for a thin film on a substrate, this approach can produce substantial errors. Consider a stiffer film on a more compliant substrate. For large indentations, the data should follow a line of greater slope determined by the substrate elastic properties. For small indentations, the data should follow a line of lower slope determined by the stiffer film. In between, the data must cross over on a path, the slope of which is even lower than that of the film, returning an indentation modulus which is even higher than that of the stiffer film. 2) If C_m is known with confidence, the indentation modulus may be simply determined from the slope between zero and each C vs. $1/\sqrt{A}$ datum and is thus constrained to lie between the film and substrate values. The second method was used for the results presented here.

Table 1. Deposition parameters and properties of the oxide materials investigated in this study. T_{sub}, P, $Rate$, and I_{arc} are the substrate temperature, total gas pressure, deposition rate and arc current, respectively, during deposition. t_f, ρ/ρ_o, and H are the measured film thickness, relative density, and hardness, respectively.

Sample	Deposition Parameters					Properties		
	process	T_{sub} (°C)	P (mPa)	Rate (nm/s)	I_{arc} (A)	t_f (nm)	ρ/ρ_o (%)	H (GPa)
Ta_2O_5	IP	190	100 (Ar,O$_2$)	0.3	55	338	98	7.3
	RE	210	16 (O$_2$)	0.3	-	380	72	4.3
TiO_2	IP	150	80 (Ar,O$_2$)	0.3	40	270	100	7.2
	RE	190	14 (O$_2$)	0.3	-	285	78	-
	RE	195	14 (O$_2$)	0.3	-	514	78	3.9
Substrate	-	-	-	-	-	-	-	8.2

RESULTS AND DISCUSSION

All films were of very high quality; flat, smooth (roughness < 2 nm as measured by AFM) and uniformly thick. As might be expected, the RE films are significantly less dense than the IP films. This has been attributed to lower adatom mobility in RE films [2,8].

Both TiO_2 and Ta_2O_5 films made by reactive evaporation are softer, more compliant and exhibit more time-dependent plastic deformation than films made by ion-plating. These characteristics are evident in the load-displacement curves for IP and RE TiO_2 films shown in Figure 1. The lower hardness and modulus of the RE film are indicated by the greater penetration depth and the lower slope of the unloading data, respectively. The additional creep in the RE film can be seen in the greater displacement at maximum load. Additional evidence of time-dependent deformation in these films is provided by the hysteresis loop which occurs during the second load/unload cycle. During the hold at a constant load of $0.2P_{max}$, the displacement is very small, indicating that both time-dependent plastic processes and thermal drift are not significant during this part of the test.

The hardness of the uncoated substrate is shown in Figure 2. In this and other plots, each value and error bar reported is the mean and standard deviation, respectively, of the values from the 6 indentations made at each given load. The hardness is 8.2 GPa at an indentation depth of about 200 nm and is nearly constant with load. The hardnesses of the IP and RE Ta_2O_5 oxide films are also shown in Figure 2. For the IP film, there is a pronounced variation of hardness with depth. The RE film is much softer, and the hardness values are relatively independent of depth. The hardnesses of the IP and RE TiO_2 oxide films are shown in Figure 3. Again, the IP film is much harder than the RE films. The variation of hardness with depth is small for both the IP and the thicker RE film, but the hardness of the thinner RE film increases markedly with increasing indentation depth.

In order to interpret these hardness data, one must understand the variations in hardness with depth. An indentation size effect, *i.e.* increasing hardness with decreasing depth, is commonly observed in glasses [9], and in particular in fused silica [10], but has not been adequately understood to date. Typical explanations, *e.g.* [10], do not require a gradient in mechanical properties near the surface of the material and indeed, the samples which are reported here are not expected to display any such gradient. Only the IP Ta_2O_5 film reveals a pronounced indentation size effect. Since this film is clearly softer than the substrate, this cannot be attributed to the hardness of the substrate.

As several studies have shown, *e.g.* [11], the depth at which the substrate influences the measurement depends heavily on the properties of the film and the substrate. The interface also plays a role, but these films have been shown to have excellent adhesion [12], so we neglect this factor. The RE films are substantially softer than the substrate. Thus, as the indentation depth increases, the hardness is expected to increase towards the substrate value. This is clearly indicated in the data from the thinner RE TiO_2 film. For the thicker TiO_2 film, as well as the RE

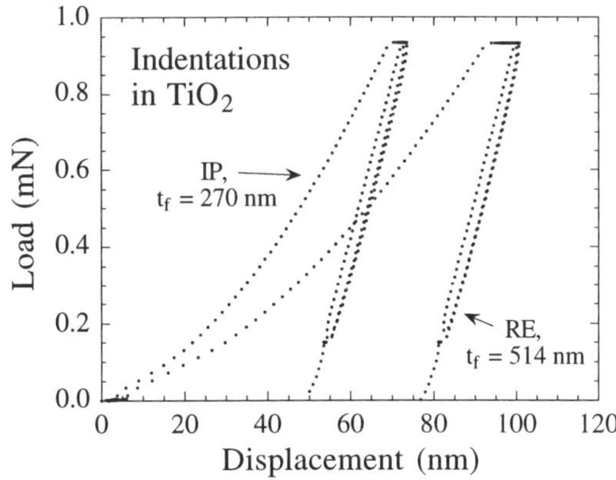

Figure 1. Indentation load-displacement curves for IP and RE TiO_2 films with film thicknesses of 270 and 514 nm, respectively.

Ta$_2$O$_5$ film, the hardness begins to increase slowly for plastic depths greater than about half of the film thickness. The hardnesses of both IP films are only slightly lower than those of the substrate and the differences reach a maximum at the maximum plastic depth. Thus, the substrate is not thought to affect these values appreciably.

The hardness values reported here for the substrate agree with nanoindentation results reported previously [7] and with Vickers indentation results [10] at very low loads. Thus, the tip shape calibration seems to be valid and the fused silica values may serve as a reference for the others. It is not normally possible to avoid indentation size effects and substrate effects for such a thin film. However, since the substrate is harder than all of the tested films, the minimum hardness can be taken as an upper bound for the hardness of each film.

As can be clearly seen in Figures 2 and 3, the deposition process, and thus the density, has a strong influence on the hardness. This is presumably because a less dense film densifies more and accommodates more indentation deformation at a given load than a denser film. For the Ta$_2$O$_5$ (TiO$_2$) films, reductions of density of roughly 26% (22%) cause the hardness to drop by about 40% (45%). A comparable relationship between density and hardness was found by Martin *et al.* [8] in considerably thicker Ta$_2$O$_5$ films.

Figures 4 and 5 show the indentation modulus as a function of indentation depth for the Ta$_2$O$_5$ and TiO$_2$ film/substrate systems, respectively. It is evident that the RE films are a great deal more compliant than the IP films. Again, the influence of the substrate must be considered. Note that the data from the substrate returns a uniform value of 74 GPa by definition from the tip shape calibration. The data from the RE TiO$_2$ films and the RE Ta$_2$O$_5$ film do not vary

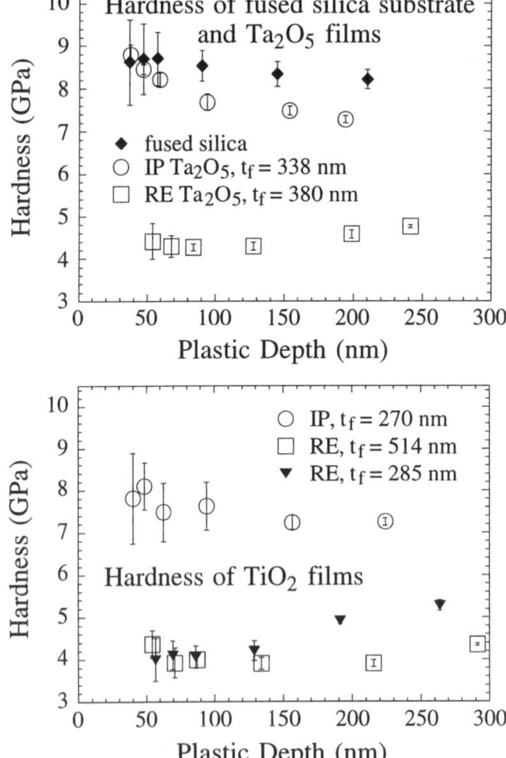

Figure 2. Hardness of the fused silica substrate, and RE and IP Ta$_2$O$_5$ films as a function of depth (t$_f$ = film thickness).

Figure 3. Hardness of RE and IP TiO$_2$ films as a function of depth (t$_f$ = film thickness).

dramatically with depth. This is at least in part due to the fact that these films are compliant, as well as soft, so that elastic deformation can be accommodated by the film. The indentation modulus of the thicker RE TiO_2 film is 105 GPa from the smallest indentations. For the RE Ta_2O_5 film, this value is 103 GPa. Moduli calculated from the slope of a least squares fit of a straight line to the C *vs.* $1/\sqrt{A}$ data from the smaller indentations return similar values indicating that the C_m value selected was good.

The behavior of the IP films is different. The indentation modulus increases strongly with decreasing indentation depth. This results from the fact that the larger indentations are sampling the more compliant substrate material more and more. These stiffer films are more efficient at transferring elastic displacements to the substrate, thus the substrate effect is more prominent than in RE films. Presumably, if it weren't for machine resolution limits and surface topography, smaller indentations would approach the film modulus value. As is indicated by the error bars, much smaller indentations would not be particularly helpful Thus, the highest values shown may be considered to be lower limits to the indentation moduli of these films.

In any case, it is quite clear that the IP films are much stiffer than the RE films and that all films are stiffer than the substrate. Similar large differences between the Young moduli of the IP and RE TiO_2 films have also been measured using a surface acoustic wave method [13]. These differences in the stiffness of these oxide films are strongly correlated with the film density.

Finally, examination of the load-time data taken during the hold at maximum load indicates that the less dense material is also more strain rate sensitive. This behavior is also to be expected

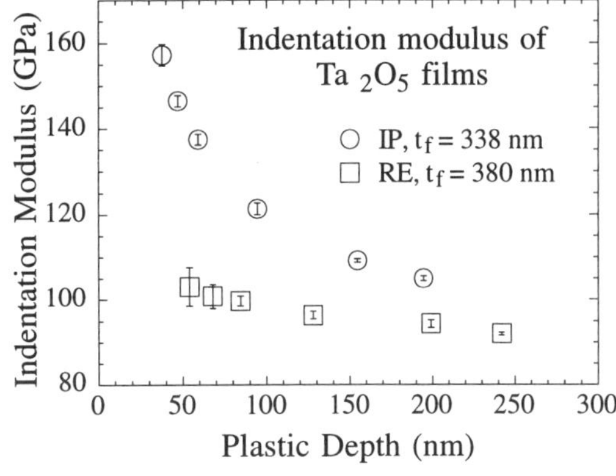

Figure 4. Indentation modulus of RE and IP Ta_2O_5 films as a function of depth (t_f = film thickness).

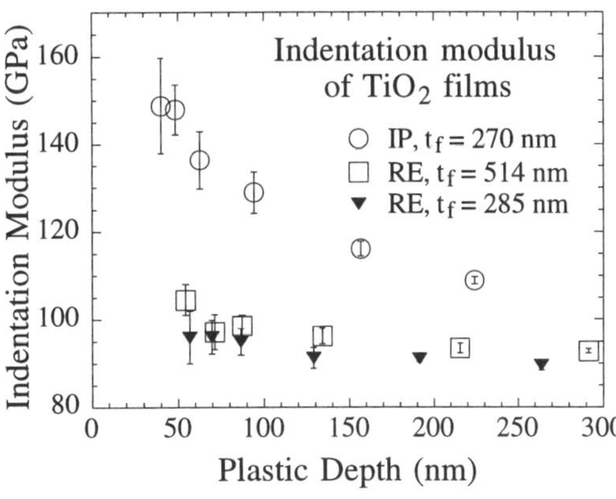

Figure 5. Indentation modulus of RE and IP TiO_2 films as a function of depth (t_f = film thickness).

as any densification processes taking place due to indentation deformation must be, at least in part, controlled by diffusion processes. This is an area requiring further study.

CONCLUSIONS

It has been shown using nanoindentation that the hardness and indentation modulus of thin high-refractive-index optical coatings are very sensitive to density and thus to the deposition process which is used. Denser IP films are harder, stiffer and less rate-sensitive than less-dense RE films. Indentation size effects and substrate effects place limits on the accuracy of the hardness and modulus values which can be obtained, but careful analysis returns upper bound hardness values which are thought to be close to the film values. The substrate effect has a much more significant impact on the modulus data due to the fact that the elastic displacement fields are much larger than the plastically-deformed zones. Lower bound values for the indentation modulus of each film were determined although these could be significantly below the value for the film material alone. Nanoindentation can be used to investigate thin optical coatings, but more work is required to separate film and substrate properties and to relate these properties to the density and to the densification processes which transpire during the indentation process.

ACKNOWLEDGMENTS

The authors would like to thank U. Jeschkowski and M. Plößer for the preparation of the coatings and U. Martens and H. Schwarz for technical assistance.

REFERENCES

1. C. Ottermann, J. Otto, U. Jeschkowski, O. Anderson, M. Heming and K. Bange, in *Thin Films: Stresses and Mechanical Properties IV*, (Edited by P. H. Townsend, T. P. Weihs, J. E. S. Jr. and P. Børgesen). Proc. Materials Research Society, Vol. 308, (1993) p. 69.

2. C.R. Ottermann, M. Heming and K. Bange, in *Thin Films: Stresses and Mechanical Properties V*, (Edited by S. P. Baker, C. A. Ross, P. H. Townsend, C. A. Volkert and P. Børgesen). Proc. Materials Research Society, Vol. 356, (1995) p. 187.

3. C. Ottermann, A. Temmink and K. Bange, *Proc. SPIE* **1272**, 111 (1990).

4. L.R. Doolittle, *Nucl. Instrum. Methods B* **9**, 344 (1985).

5. M. Laube, F. Rauch, C. Ottermann, O. Anderson and K. Bange, *Nucl. Instrum. Methods B* in press.

6. Nano Instruments Inc., "Nanoindenter II," Oak Ridge, TN

7. W.C. Oliver and G.M. Pharr, *J. Mater. Res.* **7**, 1564 (1992).

8. P.J. Martin, A. Benavid, M. Swain, R.P. Netterfield, T.J. Kinder, W.G. Sainty, D. Drage and L. Wielunsky, *Thin Solid Films* **239**, 181 (1993).

9. H. Scholze, *Glass: nature, structure and properties* , Springer-Verlag, New York (1991).

10. H. Li and R.C. Bradt, *Journal of Non-Crystalline Solids* **146**, 197 (1992).

11. T.A. Laursen and J.C. Simo, *J. Mater. Res.* **7**, 618 (1992).

12. C.R. Ottermann, K. Bange, A. Braband, H. Haefke and W. Gutmannsbauer, elsewhere in these Proceedings

13. C.R. Ottermann, R. Kuschnereit, O. Anderson, P. Hess and K. Bange, elsewhere in these Proceedings

Part II

Fracture and Adhesion

FRACTURE OF THIN SYNTHETIC DIAMOND FILMS

M. D. Drory
Crystallume, 3506 Bassett Street, Santa Clara, CA 95054, mddrory@aip.org

ABSTRACT

Large residual stresses in diamond coatings may result in film failure through splitting, delamination and substrate failure. In addition, the CVD diamond growth environment may degrade the substrate mechanical properties. These issues are examined for diamond-coating of a tool steel alloy. Diamond growth was achieved on the steel substrate with the use of a titanium interlayer. Embrittlement of the Ti interlayer was not evident, however the substrate hardness was severely degraded.

INTRODUCTION

Diamond coatings provide the opportunity for improving the wear and corrosion resistance of materials by virtue of diamond's high hardness and chemical inertness (at room temperature). In addition, diamond has the highest value of Young's modulus, a low coefficient of friction (in air), and a relatively low thermal expansion coefficient (Table 1). This combination of properties can be obtained in coating of complex shapes on a variety of ceramics and metals with CVD processes [1]. For a number of applications, attractive mechanical properties are sought with high thermal conductivity and the ability to tailor the electrical resistivity. CVD diamond may possess high thermal conductivity and a broad range of electrical resistivity achieved through doping.

A major challenge for obtaining useful diamond coatings is avoiding failure of the coating through the large residual film stresses arising on cooling from the deposition temperature, ΔT (~700 to 950°C) with typical differences in thermal expansion coefficients of the film and substrate, $\Delta\alpha$. As a consequence of the large value of the Young's modulus, E, enormous stresses, σ ($\sim E\Delta\alpha\Delta T$) can develop in the coating resulting in film splitting (i.e. mud-cracking), delamination, and failure of the substrate [2]. For example, failure of the film in tension occurs for diamond-coating of silica substrates (Fig. 1), while failure by large compressive film stresses is the result of diamond deposited on copper (Fig. 2). As a result of the low thermal expansion coefficient of diamond, the compressive failure mode is typical of most diamond-coated materials. Attempts to minimize the propensity for film failure rely on reducing the film stress through lower diamond growth temperatures. Significant modifications of the total film stress have not been achieved through changes in (CVD diamond) process chemistry.

Adherent diamond coatings can be obtained at present deposition temperatures for a number of metals and ceramics substrates. For example, diamond deposited on a titanium alloy possesses very large compressive film

Mat. Res. Soc. Symp. Proc. Vol. 436 © 1997 Materials Research Society

stresses ~7GPa, with an interface toughness of about 50J/m^2 [3]. In this system of materials, the film stress was found to be accounted for by the residual thermal stress, suggesting a negligible contribution of intrinsic or growth stresses in the diamond coating [4]. As a consequence of the simple (thermal elastic) film stress and high hardness, diamond coatings are a model material for investigating the reliability of brittle coatings.

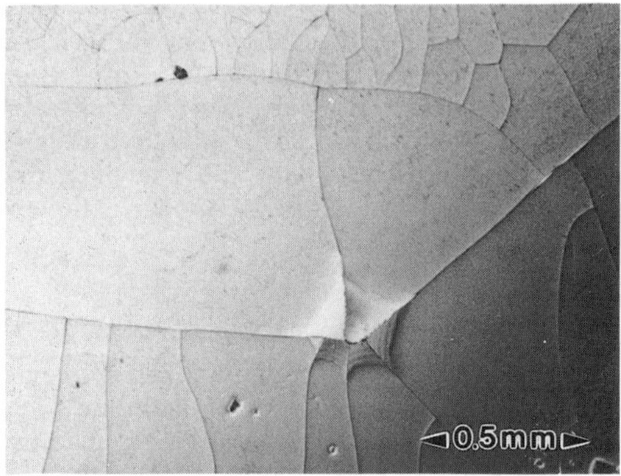

Fig. 1 - Low-magnification optical micrograph of diamond-coated silica substrates indicating "mud-cracking" and delamination.

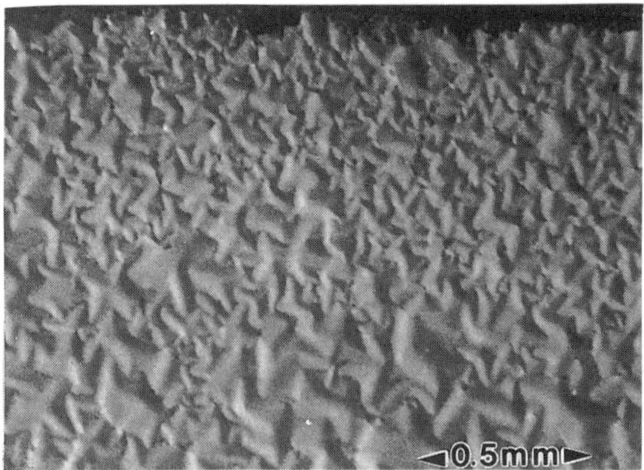

Fig. 2 - Large compression stresses result in film buckling for a diamond coated copper substrate as revealed by optical microscopy with side illumination.

In addition to coating reliability, an important ancillary problem is the effect of the diamond deposition process on the substrate mechanical properties. In particular, the relatively high temperatures and (several hours) duration of the diamond growth process may cause loss of hardness and dimensional changes in steel. The effect of the CVD diamond deposition environment on the hardness of a tool steel alloy is examined in the present work.

Table 1 - Selected Properties of Diamond

Property	Value	Reference
Density	$3.52g/cm^3$	[5]
Vickers Hardness	100GPa	[6]
Young's Modulus	1050GPa	[5]
Poisson's Ratio	0.07	[5]
Fracture Toughness	$5MPa\sqrt{m}$	[6]
Thermal Exp. Coefficient (20°C)	$0.8 \times 10^{-6}°C^{-1}$	[5]
Friction Coefficient (in air)	0.05-0.15	[5]
Thermal Conductivity(20°C)	2000W/m°K	[7]

EXPERIMENT

Diamond was deposited on a tool steel alloy to assess the effects of the diamond growth process on the substrate mechanical properties. A tool steel alloy, *Rex-20*, was selected for this purpose because it displays high values of Rockwell C hardness (HRC) and is currently used for bearing applications operating at elevated temperatures. Rex-20 is a commercial alloy (Crucible Materials Corporation) produced by powder metallurgical methods with a nominal composition of 1.30%C, 3.75%Cr, 6.25%W, 10.50%Mo, 2.00%V, (balance)Fe [8]. The Rockwell hardness varies from 68 HRC at room temperature to 48 HRC at 650°C.

Substrates were prepared by sectioning and grinding disks of 25mm diameter and 3.2mm thickness. A titanium interlayer of 2000A thickness was sputtered on the surface of the specimen to be exposed to the diamond growth process. An interlayer is needed for depositing diamond on steel because of the tendency to produce non-diamond carbon (e.g. graphite) in CVD diamond growth processes [9]. Samples were prepared for diamond deposition by scratching with a fine diamond powder to enhance nucleation followed by rinsing in solvents. Diamond growth was achieved in a 1% concentration of methane in hydrogen with a total flow rate of 200 std. cm^3/min at a pressure of 50 torr. The sample temperature was monitored by a thermocouple and a 2-color optical pyrometer. The temperature was measured as approximately 800°C, however there is considerable uncertainty in the temperature measurement because of local heating in the microwave plasma environment.

Diamond coated samples were characterized by optical and SEM microscopy and Raman spectroscopy. The latter is a non-destructive method for assessing diamond quality where a characteristic peak should be indicated at about 1332 cm^{-1}, however this value can be significantly shifted by residual film stresses [4]. Finally, the Rockwell C hardness was measured on the back surface of the substrate before and after diamond deposition with a minimum of three measurements for each condition.

RESULTS

Deposition experiments on the Rex-20 alloy with the titanium interlayer resulted in a continuous diamond coating as evidenced by the Raman spectrum (Fig. 3), but is accompanied by film buckling and delamination (Fig. 4). Film failure is expected from the enormous film compression arising from the large difference in thermal expansion coefficients, $\Delta\alpha \sim 1\mathrm{x}10^{-5}°C^{-1}$ between diamond and steel, the elevated processing temperature, and the extreme value of the Young's modulus of diamond (Table 1). A portion of the diamond film was readily removed from one sample in order to examine the delaminated interface. The titanium remained adherent to the steel substrate as evidenced by X-ray (EDAX) mapping of the delaminated (diamond) region. In addition, Rockwell C hardness impressions of the exposed area did not produce film delamination or other evidence of brittle failure of the Ti interlayer [10]. This suggests that the titanium layer was not significantly embrittled in the hydrogen-rich diamond deposition environment, thereby confirming the concept of a titanium interlayer on metal substrates providing a diffusion barrier to achieve diamond growth [9].

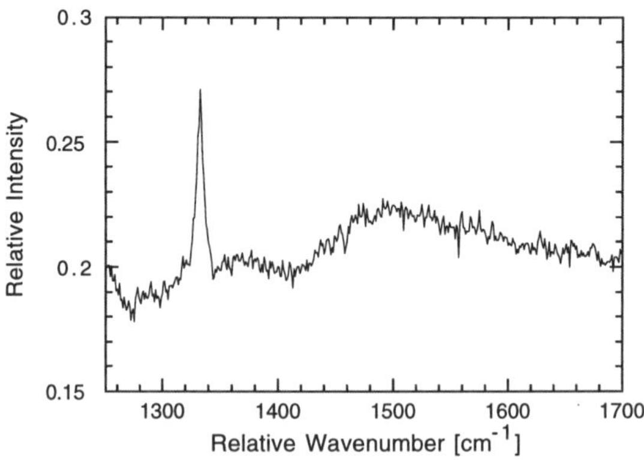

Fig. 3 -   Raman spectrum of diamond coated steel specimen.

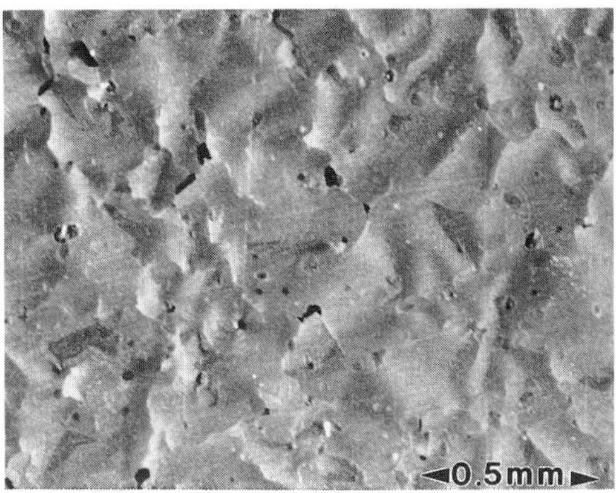

Fig. 4 - Optical micrograph of diamond-coated steel sample revealing film buckling.

Rockwell C hardness measurements of substrates before and after diamond deposition reveal significant reduction in hardness of the steel alloy. Relatively small changes in deposition temperature effect the hardness as indicated in the consistent difference between HRC values at the center and the half-distance to the edge of the disk. Temperature measurements during deposition were not sufficiently accurate to determine the variation in temperature with position or deposition experiment. Finally, it remains as future work to determine if the reduction in substrate hardness is deleterious to contact loading resistance with the presence of a continuous (and defect-free) diamond layer.

Table 2 - Rockwell C Hardness Measurement Data (Average Values)

Sample	Before Deposition	After Deposition: Center	Radius/2
I	67.3	49.8	52.1
II	65.5	36.6	39.1
III	69.5	47.8	48.6

CONCLUSIONS

The residual film stress in diamond coating a tool steel resulted in film buckling and delamination. In addition, a significant loss of hardness was measured in the steel substrate following diamond deposition. Further work is needed to achieve diamond growth at lower temperatures which will diminish the tendency for film failure and softening of the substrate.

REFERENCES

1. J. C. Angus and C. C. Hayman, Science **241**, p. 913 (1988).

2. A. G. Evans, M. D. Drory and M. S. Hu, J. Mat. Res. **3**, p. 1043 (1988).

3. M. D. Drory and J. W. Hutchinson, Science **263**, p. 1753 (1994).

4. J. W. Ager and M. D. Drory, Phys. Rev. B **48**, p. 2601 (1993).

5. J. E. Field in <u>The Properties of Natural and Synthesis Diamond</u>, edited by J. E. Field (Academic Press, London 1992), p. 668-698.

6. M. D. Drory, C. F. Gardinier and J. S. Speck, J. Am. Ceram. Soc. **12**, p. 3148 (1991).

7. M. N. Yoder in <u>Diamond Films and Coatings</u>, edited by R. F. Davis, (Noyes Publications, Park Ridge, NJ 1993), p. 1-30.

8. <u>Materials Handbook</u>, Vol. 1, 10th ed., ASM International, Materials Park, Ohio (1990).

9. M. D. Drory, Appl. Diam. Films & Related Mater.: 3rd Int. Conf., NIST Spec. Publ. 885, p. 313 (1995).

10. M. D. Drory and J. W. Hutchinson in <u>Mechanical Behavior of Diamond and Other Forms of Carbon</u>, edited by M. D. Drory, D. B. Bogy, M. S. Donley, and J.E. Field, (Mat. Res. Soc. Symp. Proc. 383, Pittsburgh, PA 1995), p. 173-182.

EFFECT OF IN AND OUT OF PLANE STRESSES DURING INDENTATION OF DIAMOND FILMS ON METAL SUBSTRATES

D.F. Bahr and W.W. Gerberich
Chemical Engineering and Materials Science, University of Minnesota, Minneapolis, MN 55455

ABSTRACT

Diamond films grown on molybdenum substrates are indented with conical diamond indenters to depths more than 5 times the thickness of the film. This causes delamination and spalling of the film. The effects of variations in indenter angle and film thickness are demonstrated. Using a radius of the delaminated area to the residual contact radius, sharper indenters are shown to cause a 10 to 20% change in that ratio. For films greater than 5 μm thick, there is no apparent film thickness effect. Possibilities for this observation are cracking due to bending stresses or the interfacial crack kinking out of the interface due to a reduction in compressive stresses, leading to mixed mode loading. The strain energy release rate of the interface for an 800 nm diamond film on molybdenum is between 2.4 and 7 J/m^2.

INTRODUCTION

Diamond film growth is presently done using a variety of techniques on different substrates. Growth methods ranging from microwave plasma assisted chemical vapor deposition (CVD) to DC Jet thermal plasma CVD produce growth rates which can vary from 1 to 100 μm per hour[1]. In addition, films are grown on substrates ranging from silicon to molybdenum to sintered tungsten carbide[2]. The variability of these conditions has led to difficulty in assessing the adhesion of diamond films.

Indentation using a standard Brale indenter, a conical diamond indenter with a 120° included angle, has become a common test for evaluating the adhesion of diamond films[3,4]. Indentations are made using a standard hardness tester through the diamond film and into the substrate. The deformation of the substrate causes delamination of the diamond film. The delaminated radius is then measured after the indentation.

Recently, a model has been developed by Drory and Hutchinson[5] which related the delaminated radius to the strain energy release rate of the interface. In this model, only uniform compressive stresses applied in the plane of the film via radial deformation of the substrate are considered. This method has been shown to provide values of interfacial toughness of diamond on a titanium alloy of approximately 45 J/m^2.

The present study will expand on this theory by applying different amounts of out of plane bending stresses to diamond films grown on molybdenum substrates. Since indentations made with conical indenters between 105° and 137° provide similar radial deformation while changing the amount of pile up around the indenter[6,7], applying the same load to different indenters will allow the effects of out of plane stresses on the delamination of diamond films on metal substrates to be evaluated.

EXPERIMENTAL PROCEDURE

Diamond films were grown using a DC Jet Triple Torch Reactor, which is described in detail elsewhere[8]. The reactor uses three concentric DC torches to create a plasma plume. Methane and hydrogen were fed into the plasma plume as reactant gasses, and a molybdenum substrate was raised into the plasma on a water cooled substrate holder. The chamber is kept at a pressure of 270 torr, and substrate temperatures are kept between 1000 and 1200°C.

The substrates used in this study are molybdenum discs 25.4 mm in diameter and 1 mm thick. Two substrate pre treatments were used for film growth. One set of molybdenum substrates was polished with 240 grit SiC paper, ultrasonically rinsed with methanol, and then placed in the reactor. The second method used was to polish the molybdenum with 240 grit SiC paper, rinse it with methanol, rub the surface of the substrate with 0.25 μm diamond powder,

Mat. Res. Soc. Symp. Proc. Vol. 436 © 1997 Materials Research Society

ultrasonically rinse it in methanol for 8 minutes, remove the sample, and ultrasonically rinsed in clean methanol. After placing the samples in the reactor, the reactor is pumped down, flushed with argon, and then the plasma started and film growth proceeds. By varying the treatment, the temperature, and the time of film growth, films between 800 nm thick and 21 μm thick were grown

After film growth, the samples are tested using conical diamond indenter tips of 105°, 120° and 137° included angles. The load is applied using a standard Rockwell hardness tester, which applies a minor load of 10 kg and the a major load which is selected by the user. The major load is applied for 15 seconds, and then removed automatically. The minor load is then removed manually. The indentation radius and delaminated area are measured using a mixture of optical microscopy, scanning electron microscopy using either a JEOL 840 SEM or Hitachi S800 field emission gun SEM. In cases where delaminations were not circular, an effective delamination radius was determined by measuring the delaminated region and determining the radius assuming a circular delamination of the same area. Profilometry was done with a Dektak surface profilometer.

RESULTS AND DISCUSSION

The films tested in this study fall into two categories. Films grown on substrates which were polished with diamond powder were approximately 800 nm thick and made of mircocrystalline diamond clusters. Films grown without rubbing with diamond powder began growing by nucleating 5 μm diamond crystals, which grew into a complete columnar film over 10 μm thick. These thick films show a high degree of faceting, with predominately (100) and (111) facets on the top of the film.

One concern is that the diamond film is not truly on molybdenum, but instead grows on a layer of molybdenum carbide[9]. Delamination occurs only between the diamond and carbide. Since this carbide is present on all the molybdenum substrates used in this study, all discussion in this paper will only refer to the diamond - molybdenum system and interface.

Previous observations of the indentations using a long focal length microscope have shown that the film cracks and spalls during loading. Thick films have diamond particles left in the cavity created by the indenter. This reinforces the idea that the indenter penetrates the film, travels into the substrate, and deforms the substrate before film delamination. Two cases can be identified, one where the entire cavity is covered in diamond, and one where diamond is only partially in the cavity, as shown in Figure 1a and 1b. If residual diamond is left in only part of the cavity, then the film must have delaminated and spalled past the point of the maximum contact radius before the maximum contact radius was reached. The unresolved issue is whether or not the film spalls all at once, or if it delaminates gradually while the load is applied. If it is assumed that the film breaks all at once, then the residual diamond contact area should be used for calculations involving the contact radius. However, if the film delaminates gradually as the load is applied then the maximum contact radius should be used. For the purposes of this paper, it will be assumed that delamination occurs gradually while the load is applied, and therefore the maximum contact radius will be used.

Previous studies have shown that the plastic zone radius around an indentation is relatively independent of indenter angle, but that the extent of pile up at the edge of the indentation can be greater by a factor of 3 for 105° indenters than for 137° indentations [8]. As can be seen from Figure 2, the sharper indenters cause larger delaminations of the diamond film. Due to a lack of space on each film, not all films were tested with all three angle indenters. It is assumed that the adhesion of each film should be relatively constant over the regions tested.

The deformation of the substrate provides the strain added to the film required for interfacial failure. Rather than measure delamination radius as a function of applied load, the normalizing factor of maximum contact radius will be used. Since the deformation around and indenter scales with the only the contact radius, this will allow substrates of different hardnesses to be compared. The hardness of a given substrate material could change given different heating conditions and amounts of carburization. Figure 3 shows the ratio of the delaminated radius, R, to the contact radius, a; for films of various thicknesses, t, and included indenter angle.

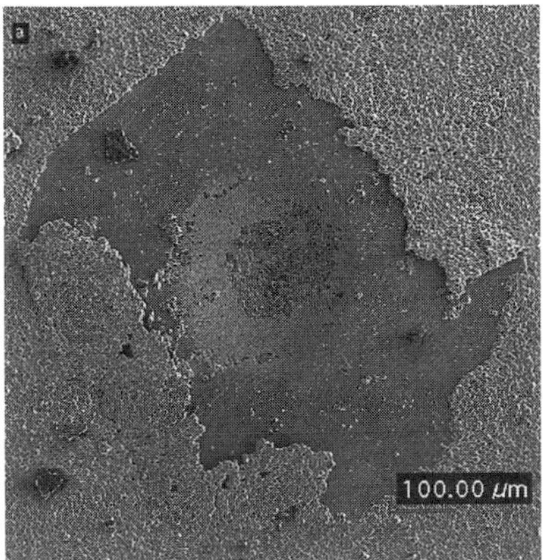

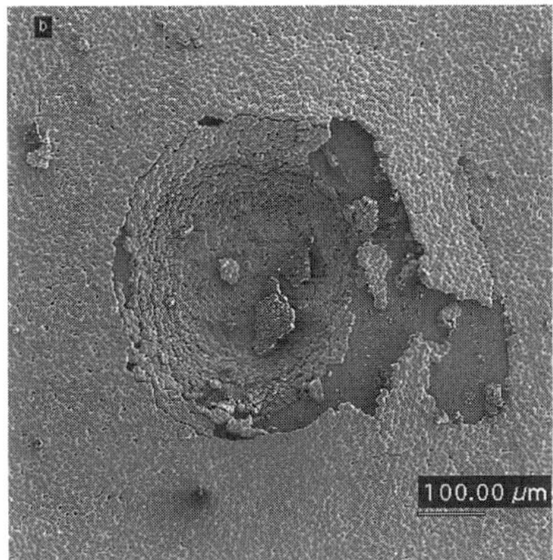

Figure 1: SEM images of indentations into diamond film on molybdenum showing residual diamond in cavity: a) 14 µm thick , b) 8 µm thick

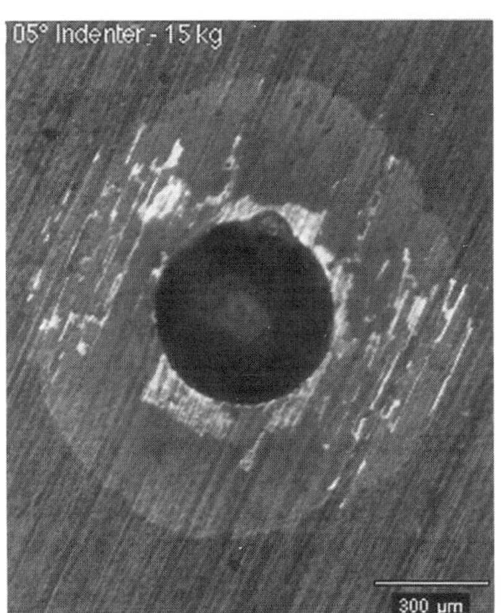

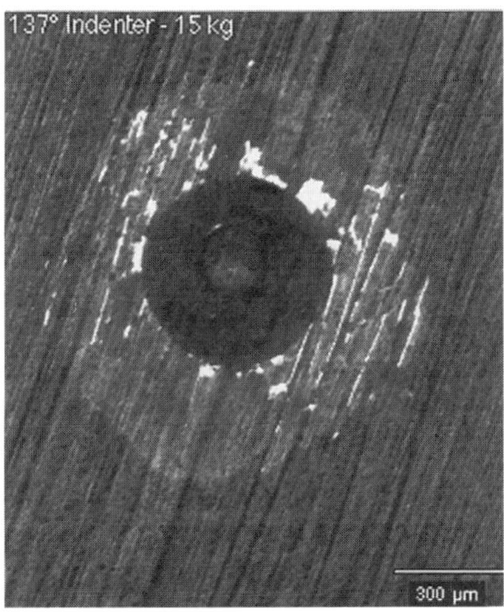

Figure 2: Optical images of indentations with conical diamond indenter into 800 nm thick diamond film on molybdenum; a) 105° included angle, b) 137° included angle

The model has been further developed by Drory and Hutchinson[10] to allow the strain energy release rate of the interface, G_i, to be determined by utilizing G_0, the initial strain energy present in the film due to residual stresses in the film from growth and strains due to indentation. G_0 is given by

$$G_0 = \frac{(1-v^2)t}{2E}\sigma_0^2 \qquad (1)$$

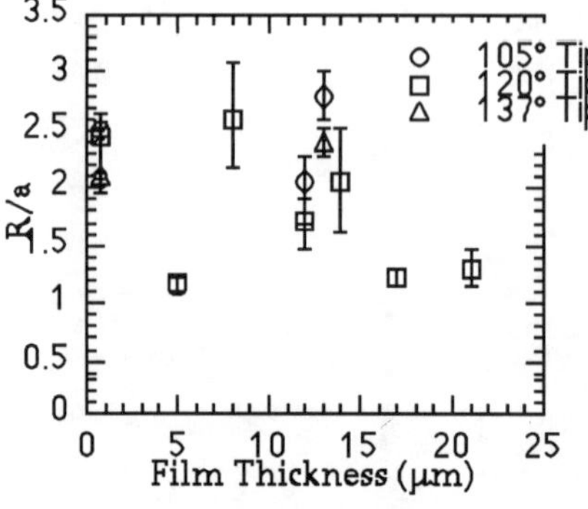
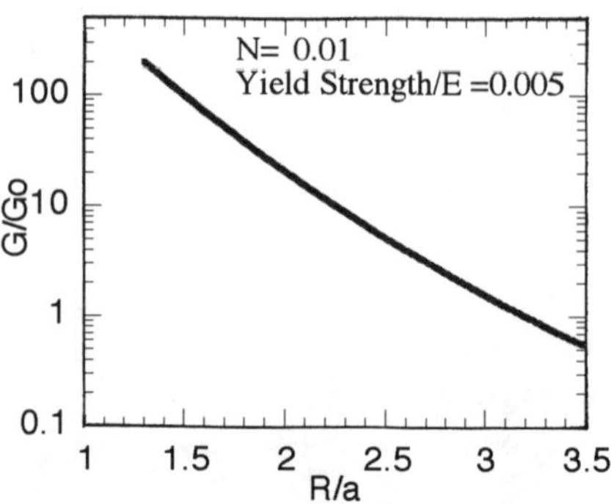

Figure 3: R/a ratio versus film thickness at various indenter angles.

Figure 4: G/G$_0$ ratio with properties similar to material used.

where t is the film thickness; ν the Poisson's ratio of the film; E the Young's modulus of the film; and σ_0 the initial stress in the film. When strains are added into the film through indentation and an interfacial crack grows, the total strain energy release rate, G, can be normalized by the initial strain energy release rate by

$$\frac{G}{G_0} = \frac{\left(\varepsilon_r(r) + \nu\varepsilon_\theta(r)\right)^2}{(1+\nu)^2 \varepsilon_0^2} \tag{2}$$

where $\varepsilon_r(r)$ is the radial strain due to indentation at a position r; $\varepsilon_\theta(R)$ is the circumferential strain due to indentation at a position R; and ε_0 is the initial strain in the film. Using an approximation of the finite element model of the strain field around an indentation, a relationship between G/G$_0$ and R/a can be made. The results and conditions for this study are presented in Figure 4 for the R/a ratios of interest in Figure 2, using a yield strength to Young's modulus ratio of 0.005, ε_0=0.001 (a residual stress of 1.1 GPa), and a strain hardening exponent, N, of the substrate of 0.01. The yield strength to Young's modulus ratio of the actual material is 0.0018, and N is approximately 0.05. By using the values available [10], it is hoped that a reasonable value of G$_i$ can be determined. By determining G at the radius R where the interfacial crack arrests, the strain energy release rate of the interface, G$_i$, can be determined. This combines the theory shown in Figure 4 with the data from Figure 2, as shown in Figure 5. The values of the G$_i$ are significantly higher than expected, which suggests that there is a limitation in using the radial compression model of Drory and Hutchinson[10] for these particular diamond films thicker than 1 µm.

There appears to be no relationship between film thickness and the ratio R/a. All the films grown to thicknesses greater than 1 µm were only polished with SiC, and not rubbed with diamond powder. The resulting film tends to nucleate as individual diamond crystals which grow together into a film. Upon delamination, all of the films exhibit the same type of structure at the interface, shown in Figure 6. The gaps in the film and the initial structure appear to be independent of film thickness. Therefore, one possible controlling mechanism may be the flaws in the film, instead of the strength of the interface.

Figure 7 shows the profile of the film bending and pile up which accompanies indentations. This bending is a source of stress in the film. Assuming a clamped plate being loaded at a center hole, the maximum stress in the plate is given by[11]

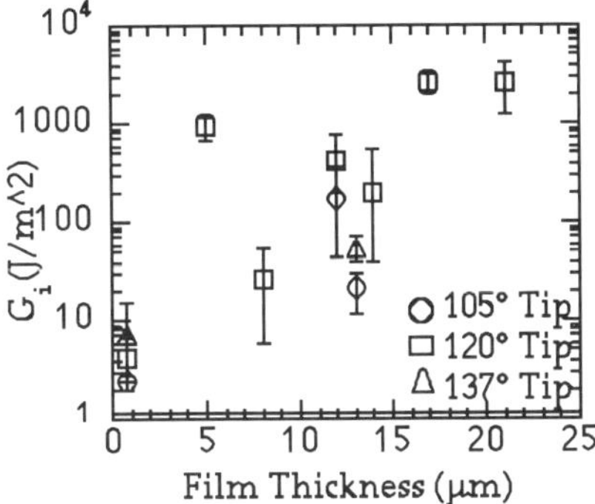

Figure 5: Strain energy release rate of
diamond on molybdenum.

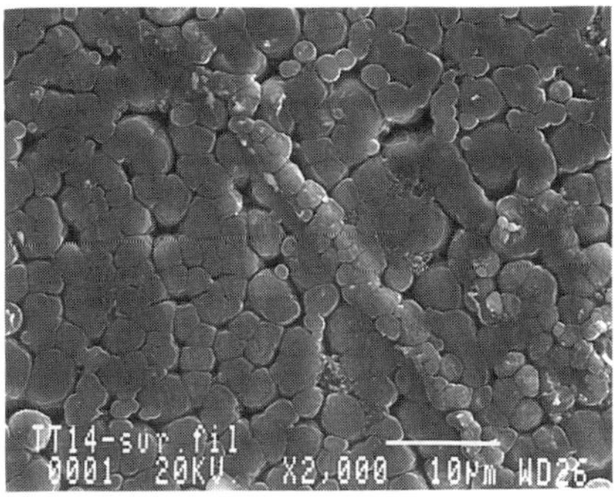

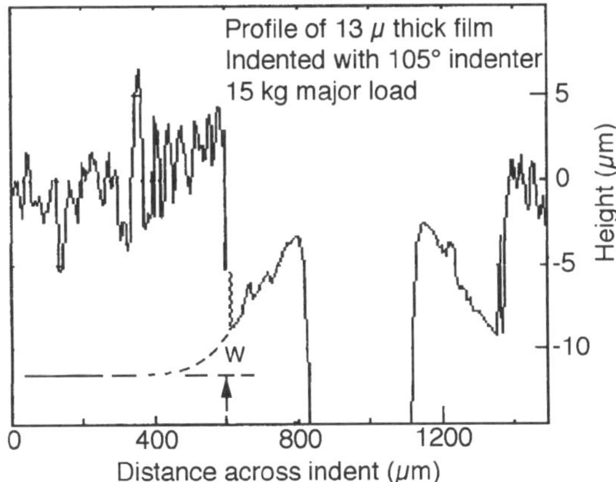

Figure 6: Morphology of delaminated
side of a diamond film.

$$\sigma_{max} = \frac{k}{k_1} Et \frac{w}{R^2} \qquad (3)$$

where σ_{max} is the maximum stress in the film; k and k_1 are constants determined by the ratio of the center hole radius to clamped radius; E is the Young's modulus of the film; t the film thickness; w the maximum deflection of the plate out the plane of the film (here, the pile up height at some distance from the indenter); and R is the clamped radius, which will be assumed in this case to be the same as the delaminated radius.

Figure 7: Pile up around indentation through diamond film on molybdenum substrate.
Note the amount of out of plane bending the film experiences.

The height, w, at some distance from the indenter can be related to the contact radius by assuming that the ratio of maximum pile up to contact radius is constant for a given indenter. as was previously found for indentations into film-free molybdenum[7]. In this case, w is approximately 1.1% of the contact radius at a distance of 2a from the center of the indenter, and approximately 3% of the contact radius at a distance of 1.5a from the center of the indenter. For the length scales of interest, k/k_1 is about 13.3. Therefore the maximum stress in the plane of the film due to bending for a contact radius of 183 μm, a delaminated radius of 551 μm, and a film thickness of 13 μm is 1.3 GPa. With a fracture toughness of 5.6 MPa$\sqrt{\text{m}}$[12], a critical flaw size of 5.7 μm through the film thickness would be needed to cause failure. Based on the shape of the growth surface of the diamond film, that is not an unreasonably sized flaw in these films.

This bending stress neglects residual stress effects, but thick diamond films grown on molybdenum substrates have been shown[13] to have steep stress gradients, with high compressive stresses at the interface, and small compressive stresses near the surface. The highest tensile stress will be applied where the highest residual stress is present. Even if the bending stress alone is not high enough to cause film fracture, it may still allow kinking from the interface into the diamond film due to mixed mode loading, leading to spalling of the film.

CONCLUSIONS

Diamond films grown on molybdenum substrates have been indented with conical indenters of included angles of 105°, 120°, and 137°. Indentations through the film thickness and into the substrate tend to cause delamination of the diamond film around the indentations. Since delamination of diamond films is controlled by deformation of the substrate, the extent of interfacial cracking is normalized by the contact radius of the indentation, R/a. This allows substrates of different hardnesses to be compared.

Larger included angle indentations appear to cause a lower R/a ratio than the sharper angle indentations. However, these changes are relatively small compared to the overall ratio, suggesting that the changes induced by changing the indenter angle in out of plane bending provides a small effect on the extent of delamination.

Film thickness appears to have no effect on the ratio of delaminated film radius to indenter contact radius. Therefore, it is proposed that for thick diamond films grown on metal substrates the extent of delamination is controlled by flaws in the films due to a relatively low nucleation density. These flaws, when subjected to small bending deflections, are placed under stresses large enough to initiate cracking through the film thickness, allowing the interfacial crack to kink through the film thickness and spall a portion of the film.

Indentations into 800 nm diamond films on molybdenum substrates show average interfacial strain energy release rates between 2.4 and 7 J/m^2, depending on the angle of the indenter being used. Due to the low bending moments in this film, this should be taken as a reliable value for the adhesion of a diamond film on molybdenum.

REFERENCES

1.   W.A. Yarbrough, , J. Am. Ceram. Soc., **75**, p. 3179 (1992).
2.   C.P. Sung, H.C. Shih, J. Mater. Res., **7**, p. 105 (1992).
3.   C.T. Kuo, T.Y. Yen, T.H. Huang, S.E. Hsu, J. Mater. Res., **5**, p. 2515 (1990).
4.   R.C. McCune, R.E. Chase, E.L. Cartwright, Surf. Coat. Tech., **53**, p. 189 (1992).
5.   M.D. Drory, J.W. Huthcinson, Science, **263**, p. 1753 (1994).
6.   D.S. Dugdale, J. Mech. Phys. Solids, **2**, p. 265 (1954).
7.   D.F. Bahr, W.W. Gerberich, *submitted to Met. Trans. A. (1995).*
8.   Z.P. Lu, J. Heberlein, E. Pfender, *Plasma Chem. Plasma Process.*, **12**, p.35 (1992)
9.   J.W. Hoehn, D.F. Bahr, J. Heberlein, E. Pfender, W.W. Gerberich, in <u>Mechanical Behavior of Diamond and Other Forms of Carbon,</u> ed. M.D. Drory, D.B. Bogy,   M.S. Donley, J.E. Field (Mater. Res. Soc. Proc. 383, Pittsburgh, PA 1996).
10.   M.D. Drory, J.W. Hutchinson, "Measurment of the adheison of a brittle film on a ductile substrate by indentation", *submitted to Proc. Roy. Soc. (1995).*
11.   W.A. Nash, in <u>ASME Handbook ,Metals Engineering Design</u>, ed. O.J. Horger, Mc Graw Hill, New York, 1951, pp. 174-177.
12.   M.D. Drory, R.H. Dauskardt, A. Kant, R.O. Ritchie, J. Appl. Phys., **78**, p. 3083   (1995).
13.   D.F. Bahr, D.V. Bucci, L.S. Schadler, J.A. Last, J. Heberlein, E. Pfender, W.W. Gerberich, *submitted to Dia. Rel. Mater., (1996).*

A Model of Cleavage Fracture along Metal/Ceramic Interfaces

D. M. LIPKIN*, G. E. BELTZ†, and D. R. CLARKE*
*Materials Department, †Department of Mechanical & Environmental Engineering, University of California, Santa Barbara, CA 93106

ABSTRACT

We propose a mechanism by which cleavage-type crack growth along a metal/ceramic interface could proceed concomitantly with significant plastic dissipation in the metal. The large strain gradients in the immediate vicinity of the crack tip are postulated to lead to extensive local hardening. A simple, continuum-based model is used to identify a characteristic length scale ahead of the crack tip, within which the material can not plastically deform subject to the crack-tip stress field. An analytical expression is derived for the crack-tip shielding afforded by the plastic zone. The coupling between the plastic dissipation and the ideal work of fracture is found to be synergistic, with slight variations in Griffith energy affecting order-of-magnitude changes in toughness. These results suggest a possible mechanism for a number of interfacial fracture phenomena observed in thick- and thin-film metal/ceramic systems, including segregation-induced interfacial embrittlement, the ductile-to-brittle transition, and stress corrosion cracking.

INTRODUCTION

A number of fracture processes exhibit superficial characteristics of cleavage fracture, while the corresponding fracture energy greatly exceeds the work of adhesion, implying extensive plastic dissipation. Early fracture models drew a sharp distinction between "brittle" and "ductile" fracture. Ideally brittle (or Griffith) fracture is identified with the energy to create fresh fracture surfaces [1], whereas fully plastic fracture is typically characterized by crack-tip blunting and hole growth [2,3]. Orowan's modification of the Griffith criterion to include the plastic work dissipated in the crack-tip process zone [4] was the first attempt to bridge the gap between ductile and brittle fracture models. However, Rice recognized that the plastic work dissipated during fracture has a strong functional dependence on the Griffith energy itself (the "valve" effect) [5]. The major subtlety of the valve effect is that the fracture events occur on vastly differing length scales: micron-scale plastic deformation at stresses near the yield strength and atomic-scale crack-tip decohesion at stresses approaching the cohesive strength of the solid.

Segregation-induced interfacial embrittlement in metal-ceramic couples [6-8] is a prime example of the strong role played by plastic dissipation in an otherwise cleavage-type fracture. Previous attempts at quantifying the valve effect, notably those by Thomson [9], Suo *et al.* [10] and Jokl *et al.* [11], have confirmed the inherently non-linear coupling between the crack-tip decohesion and the surrounding plastic deformation. However, these models fall short of a self-consistent description of the fracture process in terms of measurable material parameters. Recently, we presented an analysis for cleavage-type fracture in homogeneous materials [12], proposing a new methodology for describing the fracture process at all length scales. Presently, we extend this analysis to interfacial fracture in bimaterial systems. Employing concepts from strain-gradient plasticity and continuum asymptotic crack-tip field solutions, we obtain an inherent material length scale within which dislocation plasticity is inhibited. We ultimately arrive at the shielding ratio in terms of the interfacial work of adhesion, macroscopic toughness, yield strength, and work-hardening exponent.

THE MODEL

i. The Plastic Zone

Figure 1 illustrates the physical basis for the following discussion. A bimaterial couple containing a sharp interfacial crack is loaded in opening by remote tractions. For simplicity, we assume plane strain conditions. While the material in the lower half-plane is specified to be elastic, the upper half-plane is allowed to plastically deform and strain harden, such that a plastic zone develops about the crack tip. Following the small-strain limit of the generalized Ramberg-Osgood constitutive law:

Mat. Res. Soc. Symp. Proc. Vol. 436 © 1997 Materials Research Society

$$\overline{\sigma} = \sigma_o \left(\frac{E \overline{\varepsilon}_p}{\beta \sigma_o} \right)^n, \tag{1}$$

where $\overline{\sigma}$ is the effective stress ($\overline{\sigma}^2 = 3/2\, s_{ij} s_{ij}$, where s_{ij} is the stress deviator), σ_o is the uniaxial yield stress, $\overline{\varepsilon}_p$ is the effective plastic strain ($\overline{\varepsilon}_p^2 = 2/3\, \varepsilon_{ij} \varepsilon_{ij}$), E is Young's modulus, n is the work-hardening exponent, and β is an empirical prefactor of order unity (β=3/7 in the original Ramberg-Osgood formulation [13]). The asymptotic stress field in the vicinity of the crack tip is taken from a modified HRR [14,15] solution presented by Wang [16], henceforth referred to as WHRR. To maintain continuity of tractions across the interface, Wang noted that the stresses over the entire region must have the same singularity. Taking the stress fields in both solids to be separable, the effective stress about the crack tip can therefore be expressed as:

$$\overline{\sigma} = \sigma_o \left(\frac{\xi}{\beta I_n} \frac{K_{I,\infty}^2}{r \sigma_o^2} \right)^{n/(1+n)} \tilde{\sigma}_e(\theta, n), \tag{2}$$

where r is the distance ahead of the crack tip, $\xi = 1 - \nu^2$, ν is Poisson's ratio, and $K_{I,\infty}$ is the applied, or far-field, stress intensity factor that characterizes the elastic field well beyond the plastic zone. (We assume small-scale yielding, wherein the plastic zone remains small in relation to other length scales in the problem.) The factors $\tilde{\sigma}_e$ and I_n are weak functions of the work-hardening exponent, and are determined numerically in the WHRR formulation.[†] The requirement that the stress field be separable fixes the crack-tip mode mixity parameter, M_p, which is defined as $2/\pi \tan^{-1}(\tilde{\sigma}_{\theta\theta}(0)/\tilde{\sigma}_{r\theta}(0))$. The value of M_p is determined in the analysis. Figure 2 illustrates representative stress, strain, and displacement field solutions for n=0.3. Note that, aside from the discontinuity in the in-plane stresses across the interface, the fields in the elastic-plastic solid are quite similar to those of the homogeneous HRR solution [14].

ii. Crack-Tip Strain-Gradient Plasticity (HRR-type Field)

As for any asymptotic continuum formulation, the WHRR field does not extend to the very crack tip, being superseded by highly nonlinear behavior at the atomic length scale. As we have proposed in [12], however, the validity of conventionally accepted continuum asymptotic fields can break down at substantially larger-than-atomic length scales. The conjecture was based on recent work of Fleck et al. [17], who have shown that in the presence of large plastic strain gradients, conventional plasticity formulations must be severely modified at length scales far exceeding atomic dimensions (up to several microns). (More detailed discussion of the theoretical and experimental basis of strain-gradient plasticity is found in [17] and references therein.) On this basis, we have calculated an inherent distance within which the flow strength of the material exceeds the magnitude of the singular stress field of a crack. This parameter, which has hitherto been ill-defined in the application of strain-gradient plasticity theory to cracks, suggests a new approach for incorporating vastly different length scales into a unified fracture model.

Conventionally, the local flow strength of a plastically deforming material is related to the local dislocation density, ρ_T, through a modified Orowan-Taylor relation [18]:

$$\overline{\sigma}_{\text{flow}} = \alpha E b \sqrt{\rho_T}, \tag{3}$$

where α is a constant prefactor of order unity and b is the Burgers vector. In conventional crystal hardening theory, all of the dislocations are "statistically stored," whether they existed prior to deformation or arose during deformation via various dislocation-generating sources. However, as argued by Ashby [19], an additional distribution of dislocations is "geometrically necessary" to maintain displacement compatibility in spatially gradient strain fields. The net dislocation density is therefore the sum of the statistically stored and geometrically necessary

---

[†] The fourth-order o.d.e. was solved on a Silicon Graphics workstation using a shooting method similar to that described in References 14 and 15.

densities ($\rho_T = \rho_G + \rho_S$). Where local strain gradients become large -- as occurs near crack tips -- the geometrically necessary dislocation density can far exceed the statistically stored contribution ($\rho_G \gg \rho_S$), whereupon the flow strength in such regions can be approximated by substituting ρ_G into Eqn.3. For isotropic plasticity following J_2-deformation theory, the density of geometrically necessary dislocations is directly proportional to the effective curvature, χ_e:

$$\rho_G \approx \frac{\chi_e}{b},\tag{4}$$

where $\chi_e = \sqrt{\frac{2}{3}\chi_{ni}^{pl}\chi_{ni}^{pl}}$ and the curvature tensor, χ^{pl}, is related to the plastic strain, ε^{pl}, through:[§]

$$\chi_{ni}^{pl} = e_{nkj}\,\varepsilon_{ij,k}^{pl},\tag{5}$$

where e_{nkj} is the permutation tensor and $\varepsilon_{ij,k}^{pl} \equiv \partial\varepsilon_{ij}^{pl}/\partial x_k$.

The asymptotic strain field for a crack exhibiting a WHRR singularity is:

$$\varepsilon_{ij}^{pl} = \frac{\beta\sigma_o}{E}\left(\frac{\xi}{\beta I_n}\frac{K_{I,\infty}^2}{r\sigma_o^2}\right)^{1/(1+n)}\tilde{\varepsilon}_{ij}(\theta,n),\tag{6}$$

from which the effective curvature can be calculated [12]:

$$\chi_e = \sqrt{\frac{2}{3}}\frac{\beta\sigma_o}{E}\left(\frac{\xi}{\beta I_n}\frac{K_{I,\infty}^2}{r^{2+n}\sigma_o^2}\right)^{1/(1+n)}\tilde{V}_e(\theta,n),\tag{7}$$

where $\tilde{V}_e^2 = (\frac{1}{1+n}\tilde{\varepsilon}_{r\theta} + \tilde{\varepsilon}_{rr,\theta})^2 + (\frac{1}{1+n}\tilde{\varepsilon}_{\theta\theta} + \tilde{\varepsilon}_{r\theta,\theta})^2$. (For clarity, the designation 'pl' has been omitted; henceforth, it is implicitly assumed that both the curvature tensor and strain components are associated with plastic deformation.) Combining Eqns.3, 4, and 7, we can identify the asymptotic strain-gradient-induced hardening distribution about the crack tip:

$$\overline{\sigma}_{flow} = \left(\tfrac{2}{3}\right)^{1/4}\alpha\sqrt{\beta\tilde{V}_e\,b\sigma_o E}\left(\frac{\xi}{\beta I_n}\frac{K_{I,\infty}^2}{r^{2+n}\sigma_o^2}\right)^{1/2(1+n)}.\tag{8}$$

iii. The Core

From Eqn.8, we find that the rate of hardening due to geometrically necessary dislocations diverges as the crack tip is approached. For realistic values of the work-hardening exponent ($0<n<0.5$), the flow strength attains an approximately inverse-r singularity. Meanwhile, the WHRR asymptotic solution predicts that the stress field has *at most* an inverse-$\sqrt{r}$ singularity (LEFM), with no singularity whatsoever in the limit of perfect plasticity (Eqn.2). Thus, approaching the crack tip, the rate of increase of the flow strength exceeds the divergence of the stress field. *As a consequence, a point is reached at which material sufficiently close to the crack tip can no longer plastically deform under the prevailing crack-tip stress field.* The characteristic dimension thus defined is referred to as the crack-tip core.

In the absence of dislocation emission, the stress field inside the core retains an elastic singularity described by a crack-tip stress intensity factor, K_{tip}. (Note that the assumptions just made rely on the additional condition that the elastic core remains sufficiently larger than atomic dimensions for continuum elasticity theory to be valid.) The effective stress in the immediate vicinity of the crack tip can therefore be evaluated from linear-elastic fracture mechanics (LEFM):

93

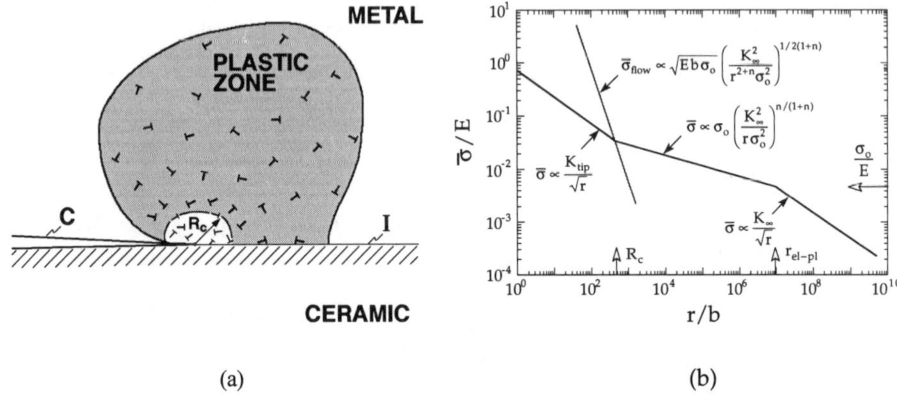

(a)   (b)

Fig. 1 Illustration of the fracture model. (a) Plastic zone about a sharp interfacial crack (I=interface, C=crack). (b) Idealized stress distribution ahead of the crack tip. Approaching the crack tip, one moves from the far-field K-dominant region to the plastically deforming zone and finally to the core region. Stress continuity defines the location of the core boundary.

$$\bar{\sigma} = \lambda(\theta)\frac{K_{I,tip}}{\sqrt{r}},\tag{9}$$

where the function λ conveys the angular dependence of the stress field $(4\pi\lambda^2 = [8v^2 - 8v + 5 - 3\cos(\theta)]\cos^2(\theta/2))$.

iv. Core Size and Shielding Ratio

It remains to establish the location of the boundary across which the scaling of the stress field switches from WHRR (Eqn.2) to LEFM (Eqn.9), and subsequently to evaluate the degree of crack-tip shielding afforded by the plastic zone.

At the point of incipient fracture, the crack-tip stress intensity must attain its critical value, K_c. Using the Irwin relation, we can restate the fracture criterion in terms of the critical energy release rate, $\mathcal{G}_c^{tip}$:

$$K_{I,tip} \rightarrow \sqrt{\tfrac{1}{\xi}\mathcal{G}_c^{tip}E},\tag{10}$$

where $\mathcal{G}_c^{tip}$ is the Griffith toughness. The value of $\mathcal{G}_c^{tip}$ is simply the work of adhesion, W_{ad}, of the solid-solid interface -- a material parameter that can be calculated from atomistic models or estimated using various experimental techniques.

As seen from Fig.1b, the location of the interface corresponds to the crossover point at which the flow strength (Eqn.8) exceeds the stress attainable in a WHRR crack-tip field. (This empirical construction does not in itself guarantee compatibility, but has been used in limited circumstances to provide boundary conditions at elastic-plastic interfaces.) At distances closer to the crack tip than the core length, the stress is insufficient to induce plastic flow and the stress field is described by Eqn.9. Equations 2, 8 and 9 represent a system of three equations in three unknowns. Enforcing continuity of effective stress across the core boundary, $r = R_c$, along a fixed direction (say $\theta = 0$), we find that:

$$\frac{R_c}{b} = \left(\frac{g_n}{\tilde{\sigma}_e^2}\frac{E}{\sigma_o}\right)\left(\frac{\lambda^2}{g_n\xi}\frac{W_{ad}}{b\sigma_o}\right)^{1-2n}\tag{11}$$

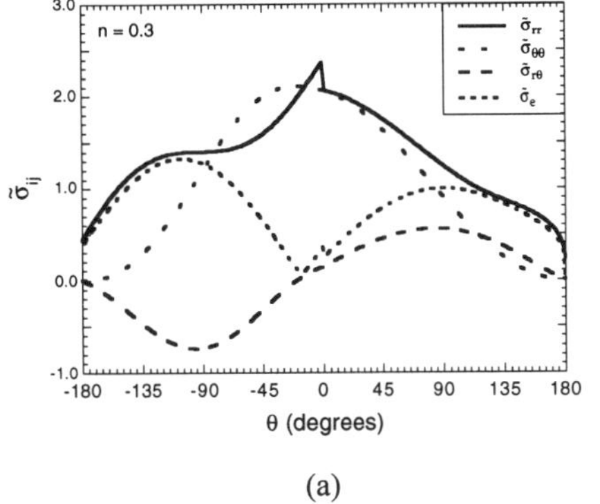

Table I. Selected values of plane strain WHRR parameters for $\theta = 0$.

n	M_p	$\tilde{\sigma}_e$	I_n	$\tilde{\nabla}_e$
0.1	0.92	0.63	2.76	0.055
0.2	0.94	0.40	2.80	0.074
0.3	0.96	0.25	2.92	0.095
0.4	0.97	0.17	3.06	0.136
0.5	0.98	0.10	3.22	0.174

(a)

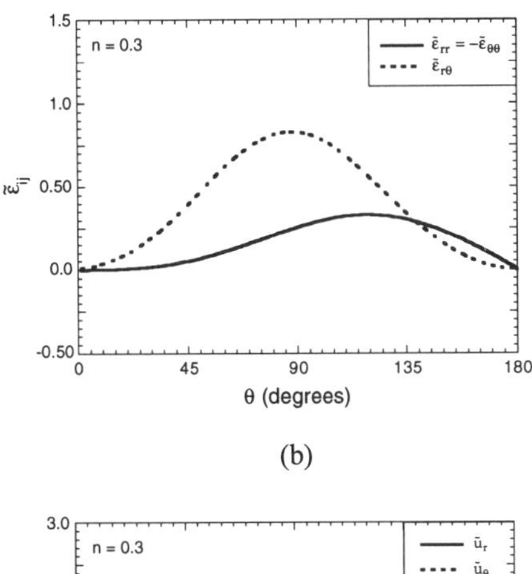

(b)

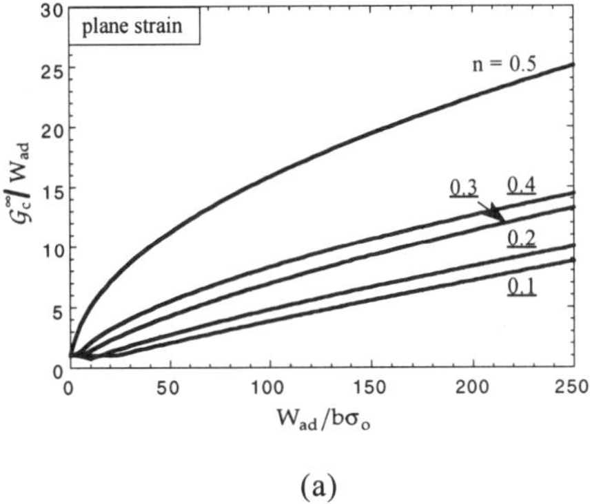

(a)

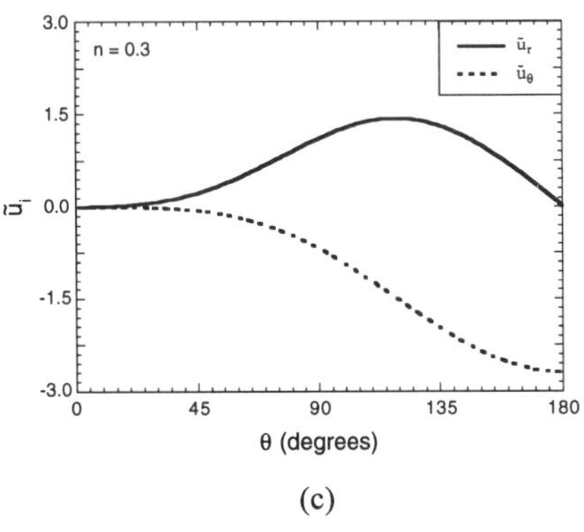

(c)

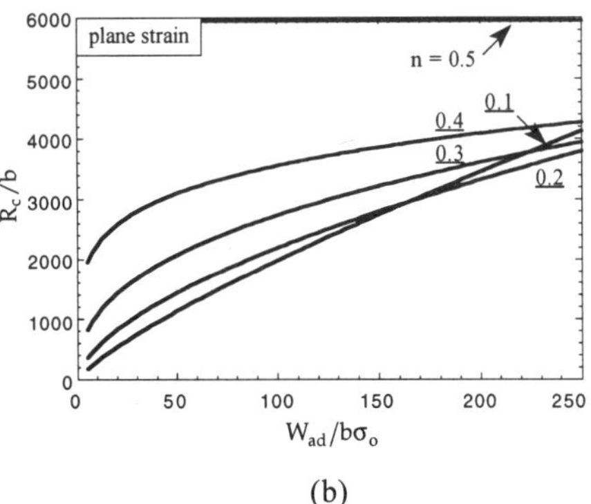

(b)

Fig. 2 The θ-variation of (a) stresses, (b) strains, and (c) displacements at the tip of a mode-I interfacial crack for n=0.3.

Fig. 3 Variation of (a) shielding ratio and (b) core size with the work of adhesion for a range of work-hardening exponents. α=1, β=3/7, ν=1/3, and E/σ_o=1000.

and

$$\frac{\mathcal{G}_c^\infty}{W_{ad}} = \left(\frac{\beta I_n}{\tilde{\sigma}_e^2}\frac{\lambda^2}{\xi}\right)\left(\frac{\lambda^2}{g_n\xi}\frac{W_{ad}}{b\sigma_o}\right)^{1-n},\tag{12}$$

where $\mathcal{G}_c^\infty/W_{ad}$ defines the crack-tip shielding ratio and $g_n \equiv \sqrt{\tfrac{2}{3}}\alpha^2\beta\tilde{V}_e$.

DISCUSSION

Using appropriate values for the proportionality constants in Eqns.11 and 12 (Table I), the core size and shielding ratio are plotted in Figs.3a and b, respectively, for a range of work-hardening exponents. The core size is generally much larger than the Burgers vector, suggesting that the use of continuum approximations of the prevailing fields is reasonable. The mode mixity parameter is nearly unity over the entire range of exponents considered, such that the loading is well approximated as mode I. The positive slopes of the shielding ratio curves in Fig.3a indicate that the strain-gradient model does predict a synergistic coupling between the macroscopic toughness (attributed to plastic deformation) and the inherent material toughness (represented by the ideal cleavage energy, W_{ad}). Such behavior is in qualitative agreement with observations of segregation-induced embrittlement [6-7], whence a slight reduction in the work of adhesion leads to a dramatic decrease in the macroscopic toughness. The magnitude of the shielding ratios appears to be approximately half of the respective values for the homogeneous fracture model [12]. This implies that, within the framework of the present model, the primary influence of the interface is to reduce the total volume of the plastic zone. The dimensionless material parameter, $W_{ad}/b\sigma_o$, can be utilized as a performance index for classifying the propensity for interfacial debonding in metal/ceramic systems.

REFERENCES

1. A.A.Griffith, *Phil.Trans.Roy.Soc.Lond. A* **221**, 163 (1920).
2. G.Hahn, M.Kanninen, and A.Rosenfeld, *Ann.Rev.Mater.Sci.* **2**, 381 (1970)
3. R.M.McMeeking, *J.Mech.Phys.Solids* **25**, 357 (1977).
4. E.Orowan, *Trans.Inst.Engrs.Shipbuilders Scot.* **89**, 165 (1945).
5. J.R.Rice, in *Proceedings of the 1st International Conference on Fracture*, (ed. by T.Yokobori, T.Kawasaki, and J.L.Swedlow), p.309. Sendai, Japan (1966).
6. D.Korn, G.Elssner, H.F.Fischmeister, and M.Rühle, *Acta metall.mater.* **40**, s355 (1992).
7. D.M.Lipkin, D.R.Clarke, and A.G.Evans, "Effects of Interfacial Segregation on Fracture in the Gold/Sapphire System," work in progress (1996).
8. J.R.Rice, Z.Suo, and J.-S.Wang, in *Metal-Ceramic Interfaces*, (ed. by M.Ruhle, A.G.Evans, M.F.Ashby, and J.P.Hirth), p.269. Pergamon Press, NY (1990).
9. R.Thomson, *J.Mater.Sci.* **13**, 128 (1978).
10. Z.Suo, C.F.Shih, and A.G.Varias, *Acta metall.mater.* **41**, 1551 (1993).
11. M.L.Jokl, V.Vitek, and C.J.McMahon Jr., *Acta metall.mater.* **28**, 1479 (1980).
12. D.M.Lipkin, D.R.Clarke, and G.E.Beltz, *Acta mater.*, in press (1996).
13. W.Ramberg and W.R.Osgood, NACA TN 902 (1943).
14. J.W.Hutchinson, *J.Mech.Phys.Solids* **16**, 13 & 337 (1968).
15. J.R.Rice and G.F.Rosengren, *J.Mech.Phys.Solids* **16**, 1 (1968).
16. T.C.Wang, *Engineering Fracture Mechanics* **37**[3], 527 (1990).
17. N.A.Fleck *et al.*, *Acta metall.mater.* **42**, 475 (1994).
18. M. Döner, H. Chang, H. Conrad, *JMPS* **22**, 555 (1974)
19. M.F.Ashby *et al.*, *Phil.Mag.* **21**, 413 (1970).

FRACTURE OF THIN TANTALUM NITRIDE FILMS ON AlN SUBSTRATES

N. R. MOODY, D. MEDLIN, D. P. NORWOOD*
Sandia National Laboratories, Livermore, CA 94551-0969
*Sandia National Laboratories, Albuquerque, NM, 87185

ABSTRACT

Nanoindentation, continuous nanoscratch testing, and transmission electron microscopy were used in this study to determine the structure-property relationships of thin tantalum nitride resistor films on aluminum nitride substrates. The films were sputter-deposited to a nominal thickness of 200 nm during one production run and then tested at room temperature. Most films were uniform in structure and thickness, consisting of fine equiaxed crystallites along the film-substrate interfaces and long columnar grains further away from the interface. However, one film varied greatly in thickness across the substrate. It had large crystallites along the film-substrate interface and clusters of columnar grains. Most importantly, it contained a far greater amount of porosity than the other films. The high porosity content led to significantly lower elastic moduli and hardness values than the low porosity films and a much greater susceptibility to fracture.

INTRODUCTION

Thin tantalum nitride films are used extensively in microelectronic applications such as resistors and capacitors because of their long term stability and low thermal coefficients of resistance. [1,2] Although these films are strong heat generators, they exhibit no changes in structure or composition of the interface with aluminum oxide substrates that degrade performance or reliability. However, the use of high power density components is driving a move to replace aluminum oxide with aluminum nitride. [3,4] This creates concern as elevated temperature exposure and long-term operation can induce detrimental changes along the thin film interface not observed in aluminum oxide devices. We therefore combined nanoindentation [5-9] and continuous nanoscratch testing [10-12] with transmission electron microscopy to determine the relationship between the structure, properties, and fracture resistance of thin tantalum nitride films on aluminum nitride substrates.

MATERIALS AND PROCEDURE

In this study, thin tantalum nitride films were reactively sputtered onto polished polycrystalline aluminum nitride substrates to a nominal thickness of 200 nm using a d. c. magnetron sputtering unit. The substrates were prepared by cleaning in detergent, soaking in trichloroethane, and rinsing in methanol. They were then heated to 850°C in air to produce a surface oxide. After air-heating, they were transfered to a deposition chamber and heated to 170°C in vacuum for two hours to drive off moisture. This was followed by an RF backsputter for 15 seconds to remove contaminants and expose fresh material. With a vacuum maintained to less than 1.3×10^{-5} Pa (10^{-7} torr), the films were deposited on the substrates using a tantalum target, argon as a carrier gas, and controlled additions of nitrogen to form Ta_2N. Deposition was at a rate of 0.3 nm/s onto the surfaces of the aluminum nitride substrates. The films were deposited sequentially in one production run. Profilometry was then used to measure film thickness.

The elastic moduli of the as-sputtered films were determined from a series of indentations employing an indentation test system developed by Nano Instruments, Inc. [5-7] The tests were conducted at a constant loading rate of 300 μN/s to depths ranging from 20 to 600 nm using a Berkovich indenter. Indentation loads and corresponding depths were recorded continuously throughout each test. The contact areas for one-half of the indentations on one film were measured directly in a JEOL 840 SEM. These areas were used to determine the elastic modulus and hardness as well as to verify the applicability of the indenter shape function established from earlier tests on single crystal aluminum, fused silica, and thin tantalum nitride and aluminum films. From this relationship the contact areas for all other tests were determined indirectly. These areas were then

Mat. Res. Soc. Symp. Proc. Vol. 436 © 1997 Materials Research Society

used with the slopes from the unloading curves to determine stiffness and elastic modulus as a function of indentation depth following the method of Oliver and Pharr. [7]

The resistance to interfacial fracture was determined using the same system configured to control normal loads and lateral indenter displacements. These tests employed a conical diamond indenter with a nominal 1μm tip radius and a 90° included angle that was simultaneously driven into the films at a rate of 500 μN/s and across the films at a rate of 0.5 μm/s until a portion of the film spalled from the substrate. During each test, the normal and tangential loads, the indenter depth, and the lateral position were continuously monitored. The fracture energy and interfacial fracture toughness values then were determined from the data using the elastic approach of Venkataraman et al. [11,12]

RESULTS AND DISCUSSION

The microstructure of both the polycrystalline aluminum nitride substrates and the tantalum nitride films were examined using SEM and TEM techniques. Both techniques detected yttrium rich phases embedded in the aluminum nitride substrates. This was expected as yttrium oxide is commonly added to aluminum nitride to improve both processing and heat transfer properties. The yttrium oxide can react to form several different yttrium aluminates. Our analyses have identified the presence of at least $Y_2O_3 \cdot Al_2O_3$ (yttrium aluminum Perovskite or YAP). Our initial analysis has detected no significant effect of these second phases on the structure of the tantalum nitride films.

Tantalum nitride can exist in several compositionally and structurally distinct forms. [13] Electron diffraction measurements identified the films in this study as Ta_2N in the trigonal form (space group P$\bar{3}$1m; lattice parameters: a=0.5255 nm and c=0.4919 nm). This structure consists of hexagonally close-packed tantalum atoms with interstitially distributed nitrogen. The films were uniform in thickness with fine, nanometer-sized crystallites forming near the interface followed by 20-to-40-nm wide columnar grains, corresponding to individual crystallites, extending to the surface of the film. (Figure 1a) The columnar grains formed with a preferential [0002] out-of-plane crystal orientation as determined from electron diffraction analysis. Some fine-scale porosity was also identified along the columnar grain boundaries.

One 200-nm-thick film exhibited a markedly different structure than the other films (Figure 1b) even though it was processed following the same cleaning and deposition procedures. Rather than forming in a narrow columnar morphology, as in Figure 1a, the film grew with clusters of

Figure 1. Most films (a) were uniform in thickness with fine nanometer-sized crystallites along the film substrate interface and 20-to-40-nm-wide columnar grains extending through the film thickness. One film (b) formed multi-grain clusters of irregular width and height with a much greater amount of porosity than observed in the other films.

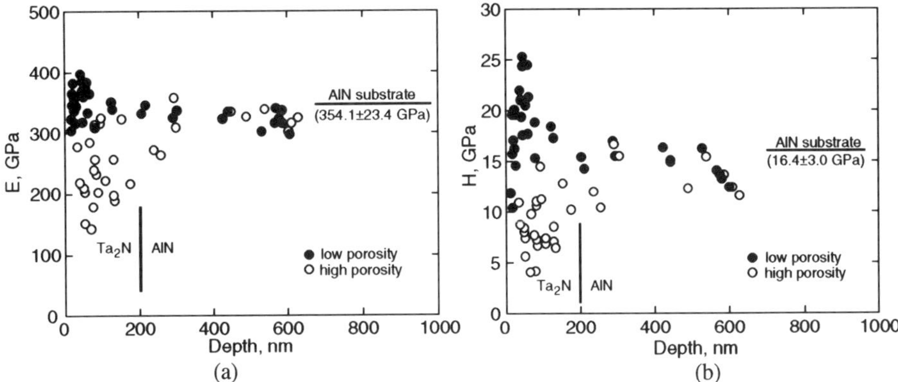

Figure 2. In low porosity films, the measured (a) elastic moduli of film and substrate were equal while (b) hardness decreased with indenter depth. The corresponding values measured in high porosity films were lower with a greater amount of scatter.

irregular width and height, giving an almost 'popcorn-like' appearance to the film. Darkfield TEM studies showed that these large clusters are composed of agglomerations of small individual crystallites. Furthermore, the porosity along the cluster boundaries and film-substrate interface was of a much larger scale and was more prevalent than observed in other films. Although not apparent in Figure 1, the large-scale interface roughness due to grain-to-grain substrate surface relief is the same in both films.

Nanoindentation of the low porosity films showed that the measured elastic moduli were independent of indenter depth and equal to the value measured for the aluminum nitride substrates as shown in Figure 2a. In all these films, the measured values exhibited significant scatter within 50 nm of the film surface due in large part to variations in surface topology. Beyond 50 nm in depth the values exhibited little scatter. In contrast with the elastic property behavior, measured hardness values varied with indenter depth. This effect is shown in Figure 2b where the measured hardness values for low porosity 200-nm-thick films decreased from a near surface value near 25 GPa to the alumnium nitride substrate value of 16 GPa at the film-substrate interface. As in the modulus measurements, there was a significant amount of scatter within 50 nm of the surface.

The measured properties for the high porosity film varied markedly from the values measured in all other films. This difference is shown in Figure 2a, where the elastic moduli measured in the high porosity film are lower and exhibit significantly more scatter than values measured in low porosity films. The measured hardness values in the high porosity film are also lower and exhibit more scatter than values measured in the low porosity film as shown in Figure 2b. This behavior is a direct result of the higher volume fraction of pores, larger pores, and more irregularity in film thickness compared with low porosity films. The sampling of pores leads to substantial scatter in the measurements and lower elastic moduli and hardness values. Of particular note, the measured elastic moduli and hardness values in the low porosity films increased to substrate values with indenter depth.

A series of ten microscratch tests were run on each film to determine resistance to fracture. These tests showed that the critical loads for fracture decreased with decreasing thickness and increasing porosity. Fracture of the low porosity films occurred consistently at loads near 15 mN and was characterized by an abrupt change in tangential load as shown in Fig. 3a. The fractures were always characterized by well-defined spalls along the film and substrate interface (Figure 3b), often occurring sequentially during each scratch test. In all tests, the first spall provided the data needed to estimate elastic fracture energy and toughness. Compositional scans across the fracture surface showed no evidence of tantalum from the film suggesting that fracture occurred along the interface or just within the aluminum nitride substrate. The measurement of normal and tangential

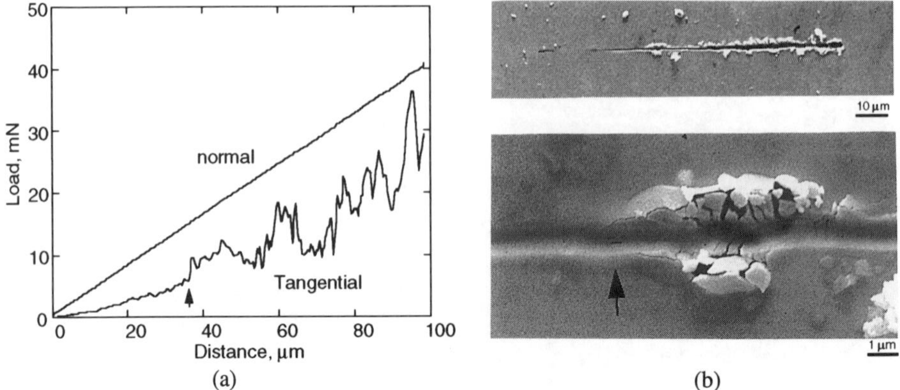

Figure 3. Tantalum nitride films on aluminum nitride substrates (a) consistently failed at loads near 15 mN and were accompanied by (b) well-defined spalls along the film and substrate interface during the scratch tests.

forces provided a coefficient of friction equal to 0.28, which is significantly lower than the value of 0.63 for ductile grooving. [11]

Fracture in the low porosity films occurred consistently near 16 mN by well-defined spall formation. The first spall along each scratch trace was a characterized by a continuous delamination. (Figure 4a) It was similar in appearance to spalls in thicker films on aluminum nitride and aluminum oxide substrates. [14,15] In contrast, fracture in the high porosity film occurred consistently at loads near 8 mN. All loads for fracture in the high porosity film were less than any load for fracture in the low porosity films. The fractures were also smaller and less defined, being characterized by multiple fractures along the cluster grain boundaries in the first full delamination as well as along the film-substrate interface. (Figure 4b)

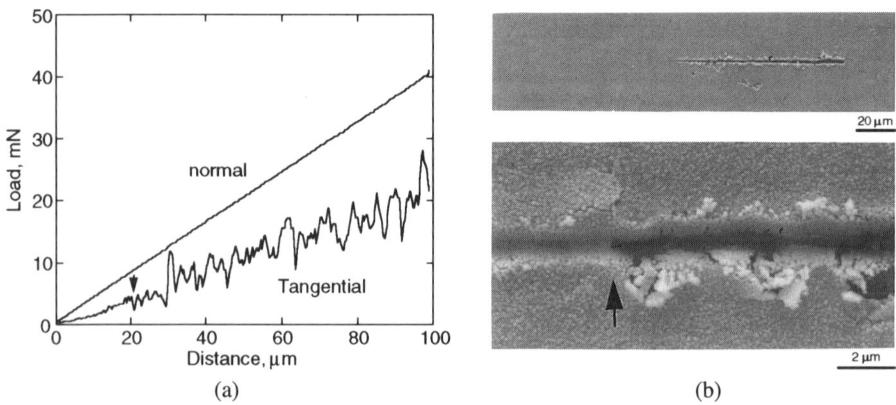

Figure 4. Failures in high porosity tantalum nitride films (a) occurred consistently at loads near 8 mN with (b) spalls that were smaller and less defined than found in low porosity films. The spalls also contained multiple fractures that occurred along the cluster boundaries.

Table I. Elastic fracture energy and interfacial fracture toughness values along with thickness, critical loads, average areas of spallation, and elastic moduli for thin tantalum nitride films on polished polycrystal aluminum nitride substrates.

Film/Substrate	t (nm)	A (μm^2)	P_{cr} (mN)	L_{cr} (mN)	E (GPa)	Γ_i (J/m^2)	K_i (MPa-m$^{1/2}$)
Low Porosity Ta$_2$N/AlN	160	8.1	15.8	5.8	350	0.4±0.2	0.2±0.1
High Porosity Ta$_2$N/AlN	180	5.1	8.0	3.1	350	0.6±0.3	0.3±0.1
					150	1.1±0.5	0.3±0.1

The average elastic energies for fracture, Γ_i, and fracture toughness values, K_i, along with average areas of spallation, A, critical normal and tangential loads at failure, P_{cr} and L_{cr}, film thickness, t, and elastic modulus, E, are given in Table I. Two elastic modulus values are used to calculate the energy for fracture of the high porosity 200-nm-thick film and these represent measured bounds to the high porosity data. Table I shows that the average elastic energy for fracture of the low porosity films is on the order of van der Waals forces, as expected for a brittle elastic film on an essentially rigid elastic substrate. The average fracture energy is also similar to values measured on sapphire substrates even though the near-interface microstructure of films on aluminum nitride and sapphire substrates differ markedly. [14,16,17] This reinforces the conclusion that bonding along the interface controls resistance to fracture of these brittle films. [16] In contrast, the fracture energies for the high porosity film at both bounds of measured elastic modulus are greater than the energies of the low porosity films. This anomaly arises from a limitation of the continuum approach, which does not account for the energy dissipation effect of multiple grain boundary fractures in estimating interfacial fracture energy.

SUMMARY

Nanoindentation and continuous microscratch testing were used in this study to determine the relationship between microstructure and properties for thin tantalum nitride films on polished polycrystalline aluminum nitride substrates. These films were sputter-deposited at room temperature to a nominal thickness of 200 nm. The structure of these films consisted of fine equiaxed crystallites along the film-substrate interfaces and long columnar grains extending to the film surface with only fine scale porosity along the columnar grain boundaries. The elastic moduli were independent of thickness and equal to the modulus of the aluminum nitride substrate. In contrast, measured hardness decreased from relatively high values at the surface to aluminum nitride values at the substrate. All films failed along the film-substrate interface at fracture energies on the order of van der Waals forces. One film had larger crystallites along the film-substrate interface than low porosity films and columnar grains that clustered. Most importantly, it contained a far greater amount of porosity than the other films. The elastic moduli and harness values were smaller than the low porosity film values. Moreover, the critical loads to fracture were much less reflecting a much greater susceptibility to fracture than low porosity films.

ACKNOWLEDGMENTS

The authors would like to thank C. Rood, M. Clift, and D. Boehme from Sandia National Laboratories, Livermore, CA for their technical support. The authors also acknowledge the support of the U.S. DOE through Contract DE-AC04-94AL85000.

REFERENCES

1. J. R. Adams and D. K. Kramer, Surface Science, **56**, 486 (1976).
2. C. L. Au, W. A. Anderson, D. A. Schmitz, J. C. Flassayer, and F. M. Collins, J. Mater. Res., **5** , 1224 (1990).

3. P. Garrou, Advancing Microelectronics, **21**, 6 (1994).
4. R. H. Palmer, Advancing Microelectronics, **21**, 11 (1994).
5. M. F. Doerner and W. D. Nix, J. Mater. Res., **1**, 601 (1986).
6. W. D. Nix, Metall. Trans. A, **20A**, 2217 (1989).
7. W. C. Oliver and G. M. Pharr, J. Mater. Res., **7**, 1564 (1992).
8. S. K. Venkataraman, D. L. Kohlstedt, and W. W. Gerberich, J. Mater. Res., **8**, 685 (1993).
9. T. W. Wu, C. Hwang, J. Lo, and P. S. Alexopoulos, Thin Solid Films, **166**, 299 (1988).
10. T. W. Wu, J. Mater. Res, **6**, 407, (1991).
11. S. K. Venkataraman, D. L. Kohstedt, and W. W. Gerberich, J. Mater. Res., **7**, 1126 (1992).
12. S. K. Venkataraman, D. L. Kohlstedt, and W. W. Gerberich, Thin Solid Films, **223**, 269 (1993).
13. Binary Alloy Phase Diagrams, 2nd Edition, edited by T. B. Massalski (ASM International, Metals Park, OH, 1990).
14. N. R. Moody, S. Venkataraman, J. Nelson, W. Worobey, and W. W. Gerberich, in Covalent Ceramics II, edited by A. R. Barron, G. S. Fischman, M. A. Fury, and A. F. Hepp (MRS, Pittsburgh, PA, vol. **327**, 1994) pp. 337-342.
15. N. R. Moody, unpublished research, Sandia National Laboratories, Livermore, CA, 94551-0969, 1996.
16. N. R. Moody, S. V. Venkataraman, J. C. Nelson, W. Worobey, and W. W. Gerberich, in Polycrystalline Thin Films: Structure, Texture, Properties, and Applications, edited by K. Barmak, M. A. Parker, J. A. Floro, R. Sinclair, and D. A. Smith (MRS, Pittsburgh, PA, vol. **343**, 1994) pp. 603-608.
17. J. E. Angelo, N. R. Moody, S. K. Venkataraman, and W. W. Gerberich, in Structure and Properties of Defects in Ceramics, edited by D. A. Bonell, M. Ruhle and U. Chowdhry (Mater. Res. Soc. Proc. **357**, Pittsburgh, PA, 1995) pp. 195-200.

Mechanics Of Interfacial Crack Propagation In Microscratching

Maarten P. de Boer*, John C. Nelson and William W. Gerberich
Dept. of Chemical Engineering and Materials Science, University of Minnesota, Minneapolis, MN
55455. *Present address - Sandia National Laboratories, Albuquerque, NM 87185 MS 1413

Abstract

A new probing technique has been developed to test thin film mechanical properties. In the Microwedge Scratch Test (MWST), a wedge shaped diamond indenter tip is drawn along a fine line, while simultaneously being driven into the line. We compare microwedge scratching of Zone 1 and Zone T thin film specimens of sputtered W on SiO_2. Symptomatic of its poor mechanical properties, the Zone 1 film displays three separate crack systems. Because of its superior grain boundary strength, the Zone T film displayed only one of these - an interfacial crack system. Using bimaterial linear elastic fracture mechanics, governing equations are developed for propagating interfacial cracks, including expressions for strain energy release rate, bending strain, and mode mixity. Grain boundary fracture strength information may be deduced from the Zone 1 films, while adhesion may be inferred from the Zone T films.

Introduction

There are two main issues regarding thin film debonding. The first is the *nucleation* of interfacial cracks, while the second is the *propagation* of cracks. From a mechanical testing point of view, scratch testing [1-3] primarily serves to address the former issue, while indentation testing [4, 5] is a method of addressing the latter. To date, the results of the two tests cannot be quantitatively compared with good confidence. This is for two reasons. First, the scratch test geometry is three dimensional, while indentation test mechanics analysis is based on an axisymmetric geometry. Second, in the scratch test a critical load drop is measured which may be associated with a complicated elastic-plastic stress distribution near the indenter tip, while the indentation analysis applies to a crack tip significantly removed from the indenter tip so that linear elastic conditions apply. The objective of this work is to relate the work of adhesion, Γ_i, in the scratch test to Γ_i in the indentation test. This is done by considering *crack propagation* in the two cases. Both types of tests have been placed in a plane strain geometry to simplify the comparisons. The sample is a thin film fine line, rather than a planar thin film. The mechanical probe is a microwedge, rather than a conical tip. The associated mechanics and experimental results [6, 7] for the Microwedge Indentation Test (MWIT) have been reported. Here, we investigate the Microwedge Scratch Test (MWST), compare it to the MWIT, show how the mechanics are related and that similar values for adhesion arise.

The important variable in this study is that one film has a Zone 1 microstructure (voided grain boundaries), while the other has a Zone T microstructure (metallurgical grain boundaries) [8]. Sputtered W on SiO_2 is used as the experimental system of interest, as it is common in the microelectronics industry, can exhibit poor adhesion especially when subject to high tensile stresses of chemical vapor deposition, and exhibits brittle failure so that bimaterial linear elastic fracture mechanics [9] is applicable.

Experimental

Silicon wafers were thermally oxidized to 1 µm thickness. Zone 1 and Zone T films were obtained by rf-sputtering at 20 and 13 mtorr respectively. The residual stress of these films measured by wafer curvature was 15 and 100 MPa, small compared to stresses induced by the microwedge. Zone 1 grains were about 20 nm in diameter, while Zone T film grains were bimodal in size distribution, with large elongated grains, in agreement with ref. [10]. Photolithography and wet etching was employed to define long lines of 10 µm width. A 20 µm wide diamond 90° included angle microwedge with a flatted surface of 0.7 µm was used for scratching. The microwedge was mounted to a continuous microindenter system [11] with normal load resolution of 16 µN and depth resolution of 0.5 nm. A spring loaded tangential load cell was constructed, and its resolution is 50 µN. Three of the four outputs (scratch distance SD, normal load P_{norm}, depth D, and tangential load P_{tang}) could be monitored during any given test. The

Mat. Res. Soc. Symp. Proc. Vol. 436 © 1997 Materials Research Society

coefficient of friction $cof = P_{tang} / P_{norm}$. At the end of the scratch test, the wedge tip was removed vertically from the substrate. Results of the mechanical testing were observed ex-situ by SEM on a JEOL 840 instrument with a LaB_6 filament.

In the first series of tests, the scratch speed was 0.5 µm/sec. At moderate loads as denoted by region "A" in Fig. 1a, tensile cracking behind the indenter tip occurs due to voided grain boundaries characteristic of Zone 1 films. These lead to the "delamination crack" seen in Fig. 1b.

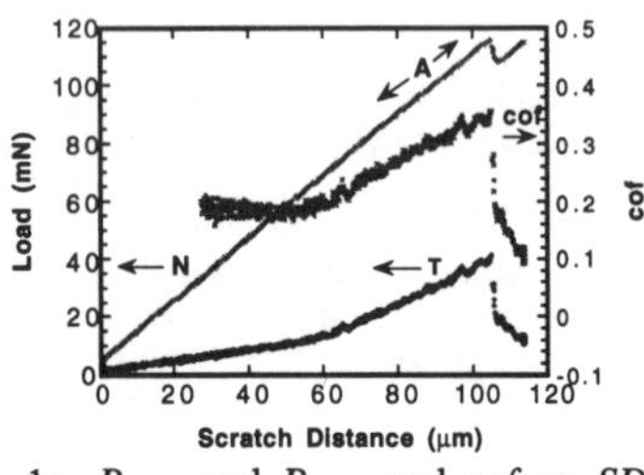

Fig. 1a P_{norm} and P_{tang} and cof vs. SD

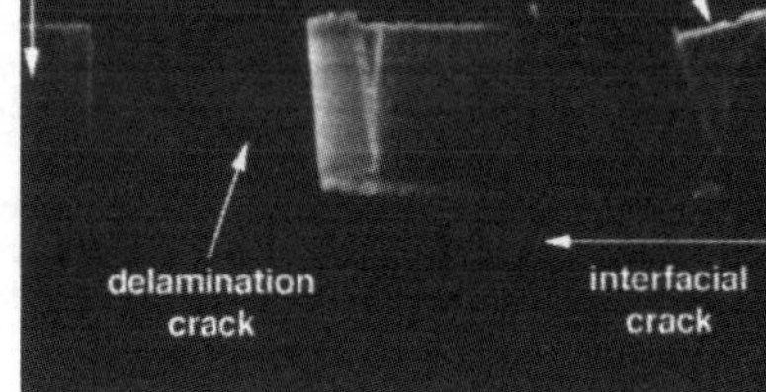

Fig. 1b Associated SEM fractograph

As P_{norm} increases above 60 mN in Fig. 1a, pileup of the tungsten thin film in front of the indenter tip causes the cof to rise, due to grooving in the film. At a critical load $P_{norm,c}$, a load drop occurs. This is associated with an "interfacial crack" system in front of the indenter tip, as labeled in Fig. 1b. The final interfacial crack length is determined by a spallation event, as will be shown. A value of cof =0.35 at $P_{norm,c}$ will be used for the Zone 1 film. Also seen in Fig. 1b is a short strip in front of the microwedge which has been pushed ahead by a scratch distance approximately equal to the travel after the load drop. As the indenter tip continues its motion after the load drop, the short strip in front of it is broken off, and the tip strikes the substrate, where P_{norm} begins to rise again. This is called the "short strip" crack system. A magnified view of P_{norm} and P_{tang} vs. SD near $P_{norm,c}$ revealed a sudden load drop of P_{norm} and P_{tang} after $P_{norm,c}$, suggesting a significant change in compliance of the sample.

In order to obtain greater insight into the sequence of events, an investigation of the P_{norm} curve near $P_{norm,c}$ was undertaken. The scratch speed was reduced to 0.1 µm/sec, the minimum available on the system. Data collection rate was increased to 12 per channel per second. The wedge motion was stopped with 0.1 µm accuracy at various points of interest on the P_{norm} vs. SD curve, as monitored on a strip chart. No change in $P_{norm,c}$ was observed, although the delamination and interfacial crack systems were now spaced more closely, as would be expected. The depth reading from the strip chart gave the greatest in-situ signal with regard to critical events. SEM micrographs of samples from which the wedge was removed just before the load drop indicated no interfacial cracking. Had an interfacial crack initiated at this point, it would remain open due to induced residual stress. Just after the load drop, interfacial cracking and spallation had both already taken place. It was not possible to obtain a stable crack for the Zone 1 films.

The 0.5 µm thick Zone T film was tested next. As seen in Fig. 2a, a buckle shape leading to a stable crack of 26 µm length now forms. Scratching was stopped about 0.15 µm after $P_{norm,c}$ was reached for this sample. $P_{norm,c}$ is 85 mN, lower than for the Zone 1 film, in part because it is thinner. A close-up of the area near the indenter tip reveals an initiation of film fracture near the indenter tip. There is only minor tensile cracking behind the indenter tip, and no evidence of a delamination crack. This may be attributed to the greater Zone T grain boundary strength. For the Zone T films, there is no apparent "short strip" pushed out in front of the indenter tip.

A good correlation existed for length of the crack vs. scratch distance after the load drop. However, if scratching continued for too long, no crack at all was observed. Fig. 2b shows a line in which scratching continued for 2 µm after the load drop. Rather than exhibiting an even longer interfacial crack, the line overlays the original line near the indenter tip. Apparently, after the indenter tip was removed, the line recoiled, and snapped over the bonded part of the line behind the indenter tip. Given that the line was originally in slight tension, it is surprising that the freed film has increased in length. It is believed that large residual compression has been induced into the line

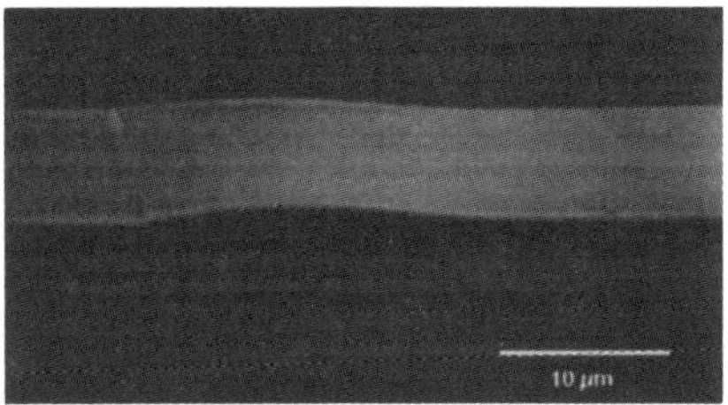
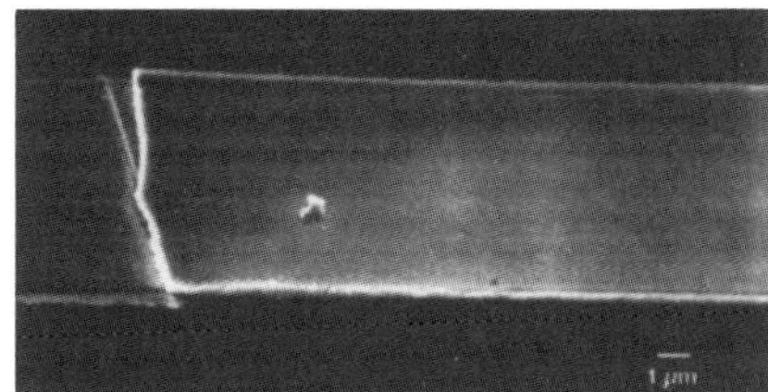

Fig.2a *SD*=0.15 µm *after* $P_{norm,c}$ - buckle shape. Fig.2b *SD*=2.0 µm *after* $P_{norm,c}$ -film detaches

both behind and ahead of the indenter tip. When the tip is removed the compression relaxes, so that the freed film actually increases in length in its unstressed state. Likely, the crack length was much greater than 50 µm, but once the driving force was removed, it lay back down, and the interfacial crack is imperceptible.

Although the Zone T film did not spall, a load drop, while smaller, was detectable. Hence, for the Zone T film the load drop can be associated with the large decrease in stiffness of a buckled film in comparison with an unbuckled film. As for the Zone 1 film, a larger change in P_{tang} than P_{norm} accompanied the load drop. For the Zone T films, a significantly smaller *cof* =0.1 was measured. It is believed that this is due to less plastic deformation, resulting in less plowing in front of the tip, and perhaps less adhesion between the indenter tip and the film.

Expression for G

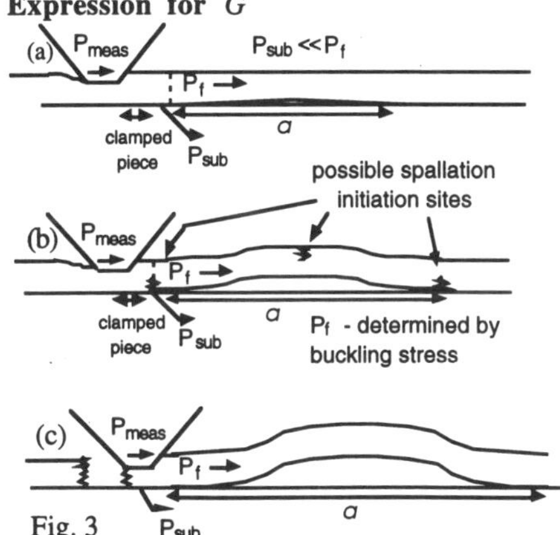

An overall schematic of the sequence of events is constructed, as seen in Fig. 3. Once the interfacial crack nucleates, the tangential load is substantially higher than that required for propagation, and the crack rapidly extends as in Fig. 3a. Buckling ensues as in Fig. 3b. If the bending strains are survived, crack extension continues with further motion of the indenter tip (Fig. 3c). For the interfacial crack system, an approach well founded in fracture mechanics regarding adhesion and spallation is to find the strain energy release as a function of crack length, $G \equiv \pm \partial U / \partial A$ at the advancing end of the crack front (U is the system strain energy, and A is the surface area created). A combination of the idealized thin film post-buckling theory and elastic half-space line loading of the substrate are used to find G. While the crack propagates, the tangential forces are not carried by the film alone. If P_f is the tangential load (per unit width) carried by the thin film, and P_s the tangential load (per unit width of the thin film) carried by the substrate,

$$P_f = \xi\, P_{meas}, \qquad (1a) \qquad \text{and} \qquad P_s = [1 - \xi(a)]\, P_{meas} \qquad (1b)$$

where ξ is a fraction between 0 and 1 best determined by detailed elastic-plastic considerations. In Fig. 3, $P_{meas} = P_{tang} / b$ is the measured tangential load per unit width of the line. ξ is a weak function of crack length a *before buckling* because in the solution for tangential line loading of an elastic half space [12], shear stresses near the surface approach 0. The substrate strain energy must be considered because of the long range contribution from the elastic field of the tangential contact transferred by friction to the substrate. Near the surface in the tangential direction stresses from P_{norm} are small [12]. Therefore the interfacial crack propagation driven by the linear elastic stress field outside the indenter tip plastic zone is dominated by P_{meas}. A fixed load condition is assumed in taking $G = G_f + G_s$. Given the weak dependence of ξ on a, $G_s \approx 0$, before buckling it may be shown that

$$G \approx G_f = \frac{\xi^2(a)P_{meas}^2}{2hE_f'}.$$ (2)

The buckling condition is satisfied when induced stress σ_e equals the Euler buckling stress σ_c:

$$\sigma_e = \frac{\xi P_{meas}}{h} = \sigma_c = \frac{\pi^2 E_f'}{3}\left(\frac{h}{a_c}\right)^2, \quad (3a) \qquad \text{or} \quad a_c = \frac{\pi}{\sqrt{3}}h\left[\frac{E_f'}{(\xi P_{meas}/h)}\right]^{1/2}. \quad (3b)$$

where a_c is the critical crack length for buckling. After buckling, the load carried by the film is determined by the buckling stress, while $P_s = P_{meas} - \sigma_c h$. where P_s is the tangential load per unit width transferred to the substrate through the film which remains under the indenter tip. At the buckling condition where $a = a_c$, P_f may be associated with an effective plastic displacement

$$P_f = (\Delta_{eff}/a_c)hE_f'.$$ (4)

Equivalently, this may be associated with an effective plastic volume at the indenter tip $V_{eff} = \Delta_{eff}\,bh$.

The source of buckling amplitude is lateral displacement of the film near the indenter tip. Because of the spring loading of the tangential load cell, the indenter tip surges forward as the tangential load drops, corresponding to a volume V_{ld}, or a displacement $\Delta_{ld} = V_{ld}/bh$. Crack extension continues resulting in one of two possible events. The first possibility is spallation due to large bending strains, as in Fig. 4b. The second possibility is crack arrest, which will occur if the initial surge in V_{ld} is survived. If crack stability is reached, continued forward displacement, Δ_{for}, of the indenter tip creates further displacement, such that

$$\Delta_{elf} = \Delta_{eff} + \Delta_{ld} + \Delta_{for}.$$ (5)

A total volume $V_{elf} = \Delta_{elf}\,bh$ is swept out by this process. As V_{elf} increases, crack extension continues. The standard expression $G_f = \sigma^2 h/2 E_f'$ inherent in (2) must be modified to account for buckling. Furthermore, the substrate strain energy now changes with crack length, and can no longer be ignored. From a fixed grip analysis, the resulting expression is [13]

$$G = G_f + G_s = \frac{(\sigma_{elf})^2 h}{2E_f'}\left[4\left(\frac{\sigma_c}{\sigma_{elf}}\right) - 3\left(\frac{\sigma_c}{\sigma_{elf}}\right)^2\right] - \frac{(\sigma_e)^2 h}{E_s'}\left(\frac{4\eta h}{ba}\right)\left[\left(\frac{\sigma_c}{\sigma_e}\right) - \left(\frac{\sigma_c}{\sigma_e}\right)^2\right],$$ (6)

where $\sigma_{elf} = E_f'\Delta_{elf}/a$ is the stress that would be induced in the film were it not for buckling. $\eta = 3.45 \times 10^{-5}$ m is a coefficient for substrate strain energy, and derives from the elastic tangential line loading solution [12]. The first term in (6) represents G_f accounting for buckling, while the second is due to the increasing tangential load in the substrate as the crack extends, which reduces the effective stress intensity. Buckling and transfer of load to the substrate result in crack stability, but only if spallation does not occur first. The ultimate crack length depends on when the scratch test is stopped, because V_{elf} will increase V_{for} with as the scratch test proceeds.

An analytical form for the total strain ε_t vs. a can be determined from one-dimensional buckling theory by superimposing bending and compressive strains [13]. Likewise an analytical expression for the phase angle Ψ vs. a describing the mode mixity has been developed [13] using bimaterial linear elastic fracture mechanics [9]. Ψ decreases from 53° before buckling continuously to -37° after buckling as Δ_{for} increases.

Assessment of G, ε and Ψ

The measured data are given in Table 1. In this assessment of the model, it is assumed that $\xi=1$. This is reasonable given that the film is much stiffer than the substrate (also P_{norm} will increase P_f). The value for $E_s^{'}$ is taken to be 150 GPa, since the substrate elastic strain energy is dominated by silicon under the silica. As reported above, the *cof* was found to be 0.35 and 0.1 for the Zone 1 and Zone T films respectively. For the Zone 1 film, $E_f^{'}$=400 GPa, while $E_f^{'}$=445 GPa for the Zone T, with $\upsilon_f = 0.28$. The value for $\Delta_{\mathrm{eff}} = \left(\pi / \sqrt{3}\right)\left(P_{\mathrm{meas}}h / E_f^{'}\right)^{1/2}$ is calculated consistent with (3a). In the assessment of Δ_{for} for the Zone T films it is assumed that slippage between the indenter tip in the film is negligible after the load drop, and that the crack does not relax after the indenter tip is removed. For the Zone T data, the film remained attached near the indenter tip, so that crack relaxation should have been small.

Table 1 - Measured data

Zone	h (μm)	a (μm)	$b\,P_{\mathrm{tang}}$ (mN)	b (μm)	$\Delta_{\mathrm{eff}}(\mu)$	Δ_{ld} (μm)	Δ_{for} (μm)
1	1.0	20	48.0	9.0	0.209	(0.067)	-
1	0.8	14	42.0	8.7	0.178	(0.083)	-
T	0.45	26	8.5	8.5	0.059	-	0.15
T	0.45	40	9.0	8.5	0.058	-	0.4
T	0.45	49	8.75	8.5	0.058	-	0.6

Table 2 - Calculated Data

Zone	$\Delta_c(a)$ (μ)	a_c (μm)	G_f (J/m^2)	G_s (J/m^2)	G (J/m^2)	ε (%)	Ψ (°)
1	0.1645	15.7	47.2	-16.7	**30.5**	**0.75**	**7**
(1)	0.1645	15.7	74.1	-16.7	57.5	1.59	-6
1	0.1504	11.8	48.2	-18.0	**30.2**	**0.52**	**12**
(1)	0.1504	11.8	72.8	-18.0	54.9	2.12	-7
T	0.026	11.6	3.3	-0.7	**2.7**	**0.88**	**-27**
T	0.0167	11.4	2.4	-0.2	**2.2**	**0.78**	**-32**
T	0.0136	11.5	1.5	-0.1	**1.3**	**0.63**	**-33**

Tables 1 and 2

The Δ_{ld} column in Table 1 was calculated for the Zone 1 films by assuming that the tangential load drop was 3 times the normal load drop, i.e. 2.4 mN (as calculated from the high data rate test.) Then Δ_{ld}=2.4mN/k, where k=1.63X10^4 N/m is the tangential load cell spring constant. The data has been corrected for compression of the substrate as calculated from the tangential line loading solution in elasticity theory [14] (otherwise Δ_{ld} would be still larger). For the Zone T films, Δ_{ld} is neglected because it is small, and the small tension in the films approximately compensates it.

The values for G, ε and Ψ are calculated at the measured a's and reported in Table 2. For the spalled Zone 1 films, G is very high compared with the known $\Gamma_i \approx 0.5$ J/m^2 from the MWIT [7] Therefore, the crack cannot yet be stable. Furthermore, values for ε and Ψ are close to those which accounted for spallation in the MWIT [7] ($\varepsilon_{\mathrm{max}}$=0.4% and Ψ=7°). Note the distinction made between calculations made for Δ_{ld}=0 and $\Delta_{ld}\neq 0$. If Δ_{ld} is used, then the values for G and ε are even larger, and the above statements still hold. *This corroborates the experimental inference that spallation occurs before the next data point can be taken for the Zone 1 films.* For the Zone T films, the values for G are reasonably close to those measured more accurately [15] where $\Gamma_i \approx 0.7$ J/m^2 was found, as are the values of strain where a strain of 0.5 % was survived. Stable cracks grow according to (6), and they do not spall because the critical strain (fracture toughness) of the Zone T film is not exceeded. As expected by the strong dependency of Ψ on Δ_{for}, the phase angle is near the minimum of -37° for these films.

The phenomena are explained well by the theory presented. Better numerical accuracy can be achieved with more precise evaluation of ξ ($\xi=1$ overestimates the true value, making the calculated values somewhat high), improved modeling of the buckling phenomena, and in-situ evaluation of Δ_{for}. Also, the effect of P_{norm} on ξ, at present ignored, requires further consideration. A primary reason why the Zone 1 films spall is simply that the measured tangential force at spallation is greater. In effect, the Zone 1 is subject to a double jeopardy. First, nucleation of the crack is influenced primarily by the normal load, which is much larger in magnitude than the tangential load, and dominates shear at the interface near the indenter tip. The *cof* for the Zone 1 film is higher. Therefore, once it is nucleated, a much higher driving force induces larger bending strains. Second, the Zone 1 film has weaker grain boundaries.

A major drawback in the MWIT [7] was that substrate cracking rendered the calculated effective stresses invalid. In spite of the need to resort to effective displacements in the present model (inherently less accurate than load measurement for scratch testing), the results are quite good. This may be traced to the <u>absence of substrate cracking in the MWST</u>. As Δ_{for} increases, the first term of (6) becomes the major contribution to G. Also, the effect of P_{norm} becomes minimal because the post-buckling behavior for eccentrically and perfectly aligned beams merges at large a. Therefore, although $\xi=1$ is too high, the choice of its value becomes less important and the $\Gamma_i=1.3$ J/m^2 at $a=49$ μm (bottom row of Table 2) becomes quite close to the known value of $\Gamma_i \approx 0.7$ J/m^2 It is believed that the above results are directly applicable to the conical tip scratch test of a planar film. Although then a 3-D stress problem must be solved, mixed mode I, II and III crack resistance in a basis 3 space can conceivably result.

Summary and conclusions

In part due to the bluntness of the wedge (as required to survive scratching of the hard W film), there is some difficulty in nucleating an interfacial crack. Once it nucleates, the crack is overdriven, growing unstably until buckling. In the case of a Zone 1 film, the bending strains result in spallation. For the Zone T film, a stable interfacial crack is observed *if scratch motion is ended soon after the load drop*. The buckling and transfer of tangential load to the substrate act in concert to bring about a stable crack. <u>Quantitative</u> results from the model developed are in good agreement with this scenario. For stable cracks (Zone T films), the somewhat high values of Γ_i may be explained because in reality $\xi<1$. The present article shows the relationship between the indentation and scratch testing for the case of plane strain.

Acknowledgments

This work was supported by ONR under grant No. N00014-89-J-1726. M. P. de Boer acknowledges the University of Minnesota Graduate School for a doctoral dissertation fellowship.

References

1. P. Benjamin and C. Weaver, Proc. Roy. Soc. **A254**, 163 (1960).
2. M. T. Laugier, Thin Solid Films **117**, 243 (1984).
3. P. J. Burnett and D. S. Rickerby, Thin Solid Films **157**, 233 (1988).
4. D. B. Marshall and A. G. Evans, J. Appl. Phys. **56** (10), 2632 (1984).
5. L. G. Rosenfeld, J. E. Ritter, T. J. Lardner and M. R. Lin, J. Appl. Phys. **67** (7), 3291 (1990).
6. M. P. de Boer and W. W. Gerberich, Acta Metall. Mater., in press (1996).
7. M. P. de Boer and W. W. Gerberich, Acta Metall. Mater., in press (1996).
8. J. A. Thornton, Ann. Rev. Mater. Sci. **7**, 239 (1977).
9. Z. Suo and J. W. Hutchinson, Int. J. Frac. **43**, 1 (1990).
10. A. M. Haghiri-Gosnet, F. R. Ladan, C. Mayeux, H. Launois and M. C. Joncour, J. Vac. Sci. Tech. **A7** (4), 2663 (1989).
11. T. W. Wu, J. Mater. Res. **6**, 407 (1991).
12. K. L. Johnson, *Contact Mechanics*, (Cambridge, Malta, 1985).
13. M. P. de Boer, J. C. Nelson and W. W. Gerberich, Proc. Roy. Soc. **A**, submitted (1996).
14. S. P. Timoshenko and J. N. Goodier, *Theory of Elasticity*, (Mc-Graw-Hill, New York, 1970).
15. M. P. de Boer and W. W. Gerberich, J. Mater. Res., submitted (1996).

MICROSCRATCH ANALYSIS OF THE ADHESION FAILURE ON OXIDE THIN FILMS WITH DIFFERENT THICKNESS

C.R. OTTERMANN[1], K. BANGE[1], A.BRABAND[2], H. HAEFKE[2], and
W. GUTMANNSBAUER[3]
[1]SCHOTT GLASWERKE, R&D, P.O. Box 24 80, D-55014 Mainz, Germany
[2]CSEM Swiss Center for Electronics and Microtechnology, Inc., P.O. Box 41,
CH-2007 Neuchâtel, Switzerland
[3]University of Basel, Institute of Physics, Klingelbergstrasse 82, CH-4056 Basel, Switzerland

ABSTRACT

Adhesion failures of TiO_2 and Ta_2O_5 thin films deposited by reactive evaporation (RE), reactive ion plating (IP) and plasma impulse chemical vapour deposition (PICVD) on fused silica, AF 45, TEMPAX and soda-lime glass substrates are investigated by means of a micro-scratch tester. The oxide films possess thickness between 60 and 500 nm and show different mass densities depending on the deposition conditions. Scratch testing exhibits well pronounced detachment for thicker films on hard substrates. The clearance of the scratch signal is reduced with decreasing layer thickness or for softer substrate materials. The test results are also influenced by the various substrates and different chemical and mechanical properties of the films due to the alternate deposition techniques.

INTRODUCTION

Mechanical properties of coatings are still one of the primary challenges in thin film technologies. Coatings and thin films on glass are widely used to tailor optical performances, to increase wear resistance, or to get better stability against a chemical attack of surrounding aggressive media. The complexity of the coatings has increased in the past decade due to the increased extension of applications and the necessity to fulfill tougher specifications on coated glass products. It is obvious, that good adhesion is one of the most important demands for coatings and films on glass. Therefore, knowledge of chemical and physical key factors, influencing the adhesion behavior is of main interest. Understanding of processes reducing the adhesion strength is needed to optimize and guarantee the reliability of coatings on a long-term scale.

The influence of different factors, related to the used adhesion testing method and its specific performance parameters, and those related to film and substrate properties have been earlier extensively discussed [1-5]. However, most of this work was done on hard coatings or films on metal substrates. The information on adhesion properties and respective key factors is rather limited for oxide coatings on glass [6-8]. The aim of this work is to provide correlations between adhesion strength and film and substrate properties for TiO_2 and Ta_2O_5 films on different glasses as a model system.

EXPERIMENTAL DETAILS

The oxide films were deposited simultaneously on four different glass materials. These glass substrates differ in indentation modulus, plastic hardness, and chemical composition, especially

Mat. Res. Soc. Symp. Proc. Vol. 436 © 1997 Materials Research Society

in the alkali content (cf. Table I). Before coating, all substrates were cleaned by a standard cleaning procedure [9] and treated by a glow discharge plasma.

The TiO_2 and Ta_2O_5 films were deposited with thicknesses between 60 and 500 nm by different deposition techniques: reactive evaporation (RE), reactive ion plating (IP), and plasma impulse chemical vapour deposition (PICVD). Only a limited set of process parameters was varied to alter the film properties. For RE, the substrate temperature was chosen between 180 and 320°C, the arc current to create the plasma of the IP process was set to 40 and 70 A, and the microwave power and precursor concentration were varied for PICVD films. Film thickness was chosen due to different deposition times and controlled by optical spectroscopy. Further information on the deposition methods is given elsewhere [8-10].

The layers were characterized by a multi-method approach to obtain information on morphology, stoichiometry, optical properties, film density, hardness, Young's modulus and film stress. A comprehensive description of the used methods and respective results has been published elsewhere [9-13].

Adhesion properties of the films were investigated by scratch testing using CSEM's Micro-scratch Tester (MST). A description of the instrument has been reported in ref. [14]. For the dynamical indentations, diamond indenters (type Rockwell C) were used with tip radii between 20 and 300 μm. The maximal load was chosen between 3 and 30 N according to the respective tip curvature. Scratch speed and load rate were set to 16 mm/min and 3 N/min, respectively. The acoustic emission signal and the tangential (friction) force were monitored as a function of the progressively increased load. In addition, the scratch lines were also investigated by means of optical and electron microscopy to determine the critical load, L_c, at which a separation between film and substrate occurs. Atomic force microscopy (AFM) was employed to reveal adhesion failures on a submicrometer scale. The force microscope was operated in the contact mode, in air, and at room temperature.

RESULTS and DISCUSSION

All tantalum oxide and titanium dioxide thin films are amorphous, except RE titania films, deposited at substrate temperatures above 300°C, which have an anatase structure. The film densities vary between 70 and 100% of the respective value of bulk materials. The stress of TiO_2 films is tensile for RE and PICVD films and is compressive for IP ones [10]. First results on hardness and Young's modulus E indicate that RE TiO_2 films are softer (E ≈ 65 - 100 GPa) than corresponding IP films (E ≈ 150 GPa) [11, 13].

Fig. 1a shows an AFM image of a scratch line on a 285 nm thick RE TiO_2 film deposited on fused silica. The tip radius of the indenter is 20 μm and the scratch direction is from top to bottom of the image. In this figure, the derivatives of the AFM scan lines at the sample surface are visualized by a gray scale. The scratch line can be clearly recognized and shows the beginning of film detachment at the bottom of the image. This region is also depicted in Fig. 1b at higher magnification. The film is partially detached and it seem that the separation occurs in (or close to) the interface between film and substrate, as indicated by an SEM/EDX analysis. In addition, cracks are also observed in the glass surface. These cracks possess circular shapes with radii of typically 20 μm.

In Fig. 2 an AFM image of a scratch line is shown on a 284 nm thick IP TiO_2 film deposited on soda-lime glass. In the scratch line the film material is still remaining at the substrate as indicated by an SEM/EDX investigation. But, the scratch extends into the glass bulk by more than 1 μm deep, due to plastic deformation of the substrate material. Further increase of the load causes cohesive cracking of the substrate material and, finally, the film is removed due to a

Table I: Indentation modulus $E/(1-v^2)$, plasic hardness h_{plast}, and alkali content of different glasses used as substrates for TiO_2 and Ta_2O_5 films.

Substrate	$E/(1-v^2)$ [GPa]	h_{plast} [kN/mm^2]	alkali content
fused silica	75	12.6	free
AF 45	66	8.1	free
TEMPAX (TPX)	62	9.3	Na, K (3 %)
Soda-lime glass	75	8.2	Na (~10 %)

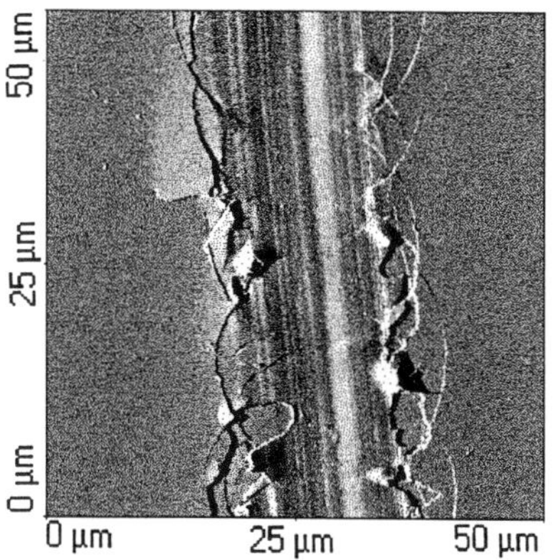

Fig. 1a: RE TiO_2 film (d = 285 nm) on fused silica. AFM image of a scratch line taken in the region of film detachment (indenter radius, R = 20 µm).

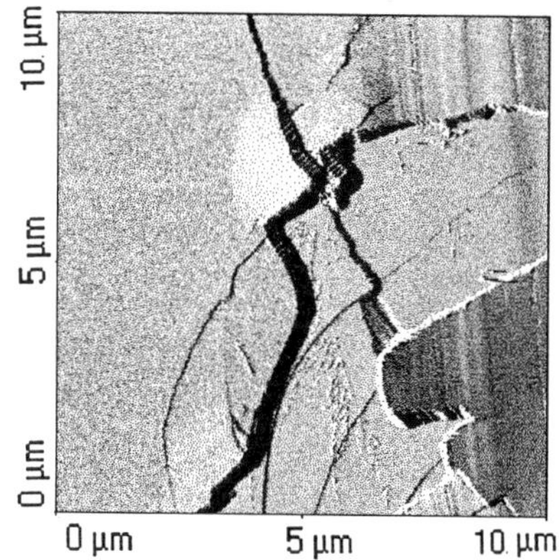

Fig. 1b: AFM image of Fig. 1a at higher magnification showing the untreated sample surface (left side) and a detail of the scratch line (right side).

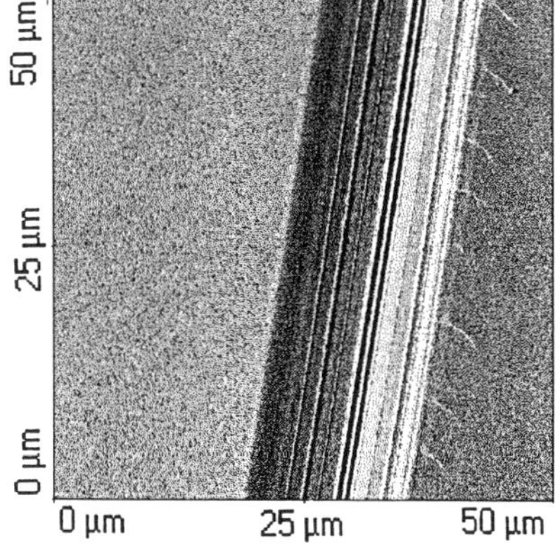

Fig. 2: AFM image of a scratch line obtained on a IP TiO_2 film (d = 285 nm) deposited on soda-lime glass (R = 20 µm).

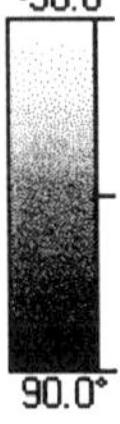

Gray scale for Fig. 1 and Fig 2, see text.

complete destruction of the substrate surface along the scratch line. In this case, no reliable value can be obtained for the critical load.

Fig 3 shows the influence of the film thickness on the adhesion properties of tantalum oxide films. Critical load, obtained by a diamond indenter with 300 μm tip radius, is depicted as a function of thickness for Ta_2O_5 thin layers deposited on AF 45 by reactive evaporation and reactive ion plating. In case of RE Ta_2O_5, the scratch investigations are reliable even for the thinnest films having a thickness of 66 nm. However, no clear detachment was observed for the thinnest IP films (d = 61 nm). The load at which film separation from substrate occurs is close to 25 N. The surface of uncoated AF 45 glass plates cracks also at this load, indicating that the observed removement of the thinnest IP films is maybe caused by a cohesive failure of the substrate material. The adhesion of the films produced by ion plating is better than for those deposited by reactive evaporation, in agreement with results given in ref. [6]. The critical load and, therefore, the film adhesion decrease with increasing thickness for both deposition methods. This finding is in contradiction to results published for hard coatings on soft substrates [2] and TiO_2 films on glasses [6], but is in agreement with observations obtained by a different scratch-test method on oxidic films and Cr thin layers on fused silica [8, 13]. Also from energy balance arguments a decrease of critical load is predicted for increasing film thickness.

Film stress is an other factor which influences adhesion properties of hard coatings [2]. Fig. 4 shows respective results obtained on TiO_2 films deposited by PICVD on fused silica substrates. Stress, σ, was controlled by the density, ρ, of these films according to the dependence of σ on ρ given in [10]. In this figure critical load is plotted as a function of stress for films with thicknesses of 145 and 250 nm. Adhesion decreases with increasing thickness also in this case. In addition, critical load is reduced for films which possess higher stress, and that is in agreement with the finding published in [2]. But, it has to be taken into account that different production condition are used to obtain different stress values. Therefore, other film properties, such as growth conditions, film density, and related to ρ hardness and Young's modulus of the films [12, 11], are also changed. It cannot be ruled out, that the observed differences in adhesion properties also may be influenced by the changes of those film properties.

In Fig. 5 critical load is depicted as a function of Young's modulus for 280 nm thick titania films on fused silica substrates. These films are obtained by reactive evaporation and ion plating, respectively. Critical load seems to increase with increasing modulus. This finding disagrees with predictions given by an energy-balance consideration [2]. But, the experimental situation is complex in this case, too. Different deposition conditions were used to obtain TiO_2 films with different Young's moduli [11]. This also has an impact on other film properties which may influence the scratch-testing results.

The chemical composition of substrate materials is expected to have a strong impact on film adhesion properties. In addition, the mechanical properties of substrates also influence significantly the results obtained by indentation measurement. This issue has been reported by different authors [2, 4, 6] and studied by finite-elements methods on various film/substrate combinations [15]. In Fig. 6 critical load is shown for RE TiO_2 films deposited simultaneously with different thicknesses on several glasses. The critical load decreases with increasing thickness for each individual substrate material. Two different groups of glasses which differ remarkably by their adhesion properties are observed. TiO_2 films on fused silica and alkali free AF 45 glass possess only a moderate adhesion strength. Whereas, critical loads obtained on the same films deposited on the alkali containing substrates TEMPAX and soda-lime glass are more than two times higher. It seems that the alkali content of the glasses is mostly responsible for this observation and the mechanical properties possess only a minor influence. Because, indentation modulus and plastic hardness of fused silica are significant higher than for AF 45 (cf. Table I), whereas the later one is quite similar for AF 45, TEMPAX and soda-lime glass, in disagreement with the cor-

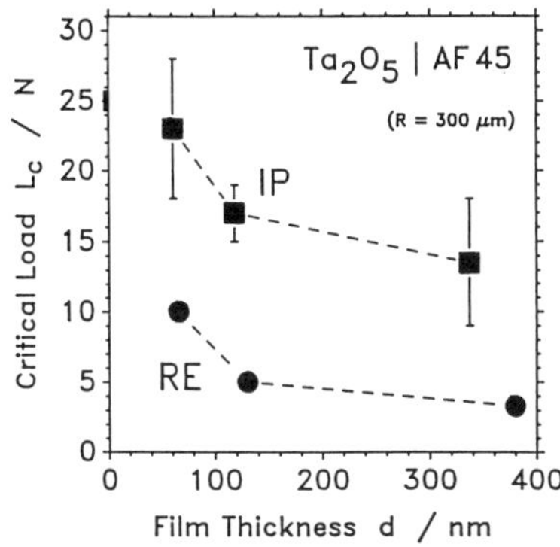

Fig. 3: Critical load L_c as a function of film thickness d for RE and IP Ta$_2$O$_5$ films deposited on AF 45.

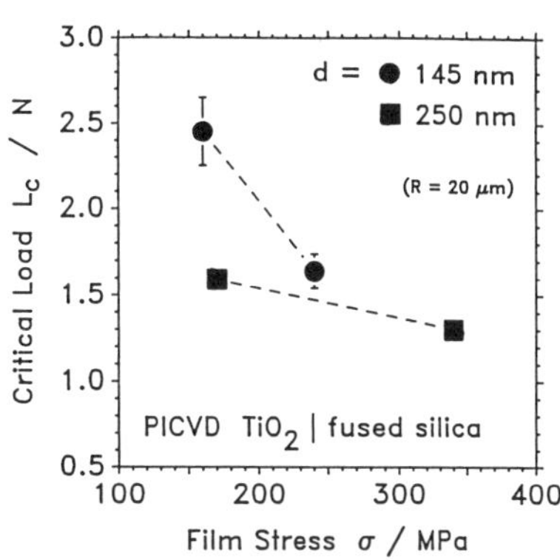

Fig. 4: Critical load L_c as a function of film stress for PICVD TiO$_2$ layers deposited with different thickness on fused silica.

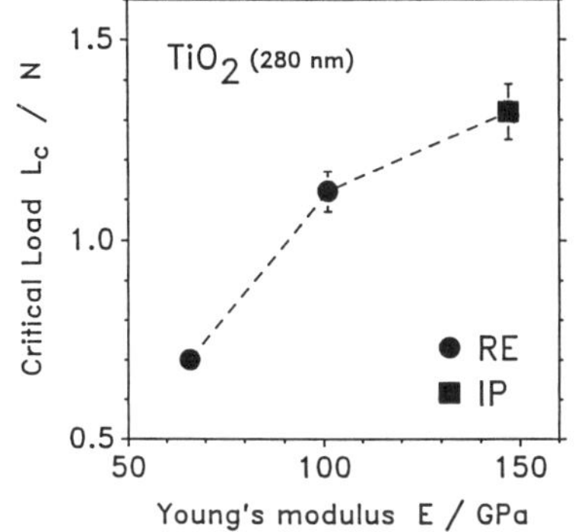

Fig. 5: Dependence of critical load on Young's modulus for TiO$_2$ films (d~280 nm) deposited by RE and IP.

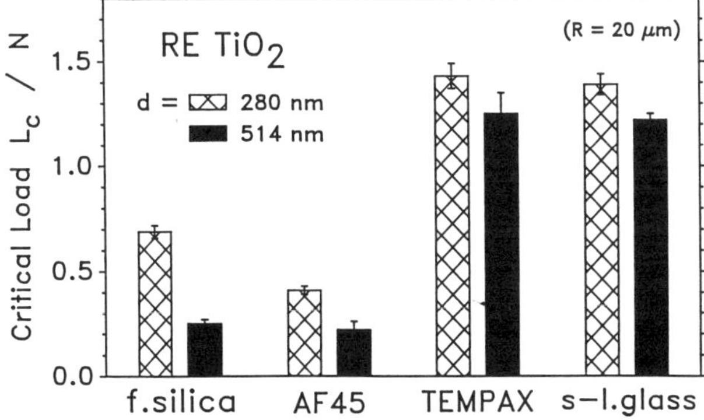

Fig. 6: Critical load of RE TiO$_2$ films deposited with thicknesses of 285 nm and 514 nm on fused silica, AF 45, TEMPAX, and soda-lime glass.

relation described before. An influence of alkali on properties of TiO_2 films is also reported in ref. [9]. Here, sodium diffusion into TiO_2 films was observed by means of SIMS investigations for different titania films deposited on Na containing substrates.

A comparison of the results obtained on fused silica and AF 45 samples should only reflect the influence of their different hardnesses on the adhesion properties. Here, critical load increases with increasing substrate hardness. This finding is in agreement with respective results obtained on different film/substrate combinations [3-4]. A similar observation is also published in ref. [6] for TiO_2 films on several glasses. However, according to the finding of this work, an additional influence of alkali content on the reported results cannot be ruled out.

SUMMARY

Several mechanical and chemical factors influencing the adhesion properties of TiO_2 and Ta_2O_5 thin films on glass are investigated in this work. It is found that critical load decreases with increasing film thickness and the adhesion of TiO_2 films is increase by an alkali content of the substrate materials. Other mechanical film properties, such as stress, hardness, and Young's modulus, also correlate to the critical loads. However, the variation of these properties implies also changes of additional film properties, and the analytical situation is complex.

ACKNOWLEDGMENT

We would like to thank U. Jeschkowski, M. Lohmeier and M. Plößer for the preparation of the films, C. Julia-Schmutz for helpful discussions, U. Martens, H. Koglin, and H. Schwartz for technical assistance, and M. Griepentrog for evaluating the mechanical properties of the substrates.

REFERENCES

[1] P.A. Steinmann, Y. Tardy, and H.E. Hintermann, Thin Solid Films 154, 333-349 (1987)

[2] P.J. Burnett and D.S. Rickerby, Thin Solid Films 154, 403-416 (1987)

[3] S.J. Bull and D.S. Rickerby, Surf. Coat. Technol. 42, 149-164 (1990)

[4] S.J. Bull, Surf. Coat. Technol. 50, 25-32 (1991)

[5] K.L. Mittal (ed.), *Adhesion Measurement of Films and Coatings* (VSP, Utrecht, The Netherlands, 1995)

[6] A.J. Perry and H.K. Pulker, Thin Solid Flms 124, 323-333 (1985)

[7] F. Kimura, S. Baba, A. Kikuchi, and A. Kinbara, Thin Solid Films 181, 435-441 (1989)

[8] C.R. Ottermann et al., Mat. Res. Soc. Symp. Proc. Vol. 308, 627 (1993); Vol. 356, 839 (1995)

[9] K. Bange et al., BMFT-Abschlußbericht FKZ 13 N 5476/6 (1991)

[10] C.R. Ottermann et al., Mat. Res. Soc. Symp. Proc. 308, 69 (1993); Vol. 356, 187 (1995)

[11] C.R. Ottermann, R. Kuschnereit, O. Anderson, P. Hess, and K. Bange, Mat. Res. Soc. Symp. Proc, in preparation

[12] C.R: Ottermann, S.P. Baker, M. Laube, K. Bange, and F. Rauch, Mat. Res. Soc. Symp. Proc, in preparation

[13] C.R. Ottermann et al., in preparation

[14] C. Julia-Schmutz and H.E. Hintermann, Surf. Coat. Techn. 48, 1-6 (1991)

[15] T.A. Laursen and J.C. Simo, J. Mater. Res. 7, 618 (1992)

MEASURING INTERFACIAL FRACTURE TOUGHNESS WITH THE BLISTER TEST

R. J. HOHLFELDER*, H. LUO**, J. J. VLASSAK*, C.E.D CHIDSEY**, W. D. NIX*
* Department of Materials Science & Engineering, Stanford University, Stanford, CA 94305
** Department of Chemistry, Stanford University, Stanford CA 94305

ABSTRACT

The adhesion of thin films to substrates can be quantified using the *blister test,* which measures the crack extension force (G) required to propagate a crack along the film/substrate interface. We summarize the derivation of crack extension force for the blister test, and discuss how blister tests can be conducted by measuring only the pressure and volume of liquid injected into the test system. We describe a way to calculate the velocity of the interface crack front.

Data from blister tests of acrylate films (14 μm thick) on nitride substrates are analyzed. The critical crack extension forces (G_C) measured were 25 - 34 J/m^2 for samples which had a commercial adhesion promoter at the interface, and 0.5 - 2.0 J/m^2 without the adhesion promoter. G_C was observed to increase with the velocity of the interface crack, and the dependence appears to obey a power-law.

INTRODUCTION

In the *blister test*, a free-standing window of a thin film, bonded to its substrate at the window edges, is pressurized by pumping fluid into the cavity below the film. As pressure increases, the film will either break or start to debond from the substrate, forming a circular "blister" which grows outward (Figure 1b). Debonding can be treated as a crack advancing along the film/substrate interface. The energy expended debonding the film per unit area (the critical crack extension force G_C) can be determined from the applied pressure and the height of the blister.[1-4]

The blister test is closely related to the *bulge test* (Figure 1a), in which a free-standing film window is pressurized without debonding the film from the substrate. The applied pressure and resulting deflection of the window are related to the film's residual stress and biaxial modulus.[3,5-7] An understanding of the bulge test is necessary to calculate the blister test energy release rate.

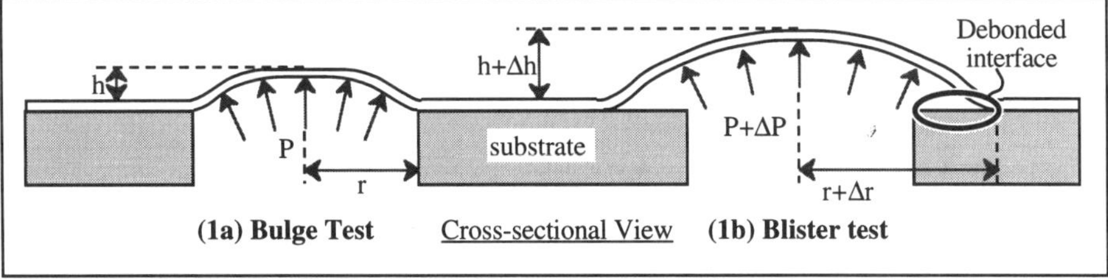

In this paper, we call attention to the fact that neither G_C nor the interface crack velocity (dr/dt) is necessarily constant during a blister test. We describe a way to determine G and (dr/dt) from volumetric measurements, and present data from blister tests of polymer films in which G_C displays dependence on crack velocity.

ANALYSIS

To understand the blister test, we must first consider the mechanics of a pressurized thin-film window, or "*bulge*". It has been shown[3,6-9] that for a film which deforms linear-elastically, the applied pressure (P) and deflection (h) of a window can be related by:

$$P = c_1 \sigma_0 t \frac{h}{r^2} + c_2 M t \frac{h^3}{r^4} \quad , \tag{1}$$

where r is the radius of the window, σ_0 is the film's residual stress, M is the film's biaxial modulus (equal to $E/(1-\nu)$ for an isotropic film), t is the film thickness, and c_1 and c_2 are geometric parameters which account for the shape (square, circular, etc.) of the window.

The volume enclosed underneath a bulge is approximately:

$$V = \kappa_v r^2 h \quad , \tag{2}$$

where the coefficient κ_v, which accounts for the shape of the bulge, has a value of roughly 1.62 for circular windows and 1.94 for square windows[10].

As the derivation of the crack extension force was described in a previous publication[11], we only outline it here. Equations (1) and (2) may be used to calculate the strain energy of a pressurized film, which is the amount of work required to inflate a window from an initially flat state to some final height. The assumption that the window is flat at zero pressure requires that the residual stress be tensile; the window will buckle otherwise. After pressurizing the window, we allow its radius to increase by a small increment dr, which is equivalent to increasing the length of the interface crack. The amount of work performed by the pressurizing liquid during this incremental growth is greater than the increase in the film's strain energy; the difference between the two is the crack extension force (G), the amount of energy available to propagate the interface crack. The end result of this analysis is the *blister equation*:

$$G = P h \frac{\kappa_v}{\pi} \left(\frac{4 + 5\psi}{4 + 4\psi} \right) \qquad \text{where } \psi = \frac{c_2 M}{c_1 \sigma_0} \left(\frac{h}{r} \right)^2$$

$$= P h \frac{\kappa_v}{\pi} g(\psi) \quad \left(\sim \tfrac{1}{2} P h \right) \tag{3}$$

The function $g(\psi)$ varies weakly with ψ and is in the range 1.00-1.25. We note that our previous publication[11] contained an error in the calculation of strain energy and of G; the result here is correct. Using equation (3), one can calculate the crack extension force at any point in a blister test.

<u>Blister testing using volumetric measurements</u>

To determine crack extension force during a blister test, one needs to measure pressure (P) and window deflection (h). It is convenient to determine the h from the volume of liquid injected into the system. If liquid is injected at constant rate, then only pressure and time must be recorded. Blister height can be calculated from pressure and volume data by rearranging (1) and (2) to solve for h:

$$h^5 + h^2 \left(\frac{c_1 \sigma_0 V}{c_2 M \kappa_v} \right) - P \left(\frac{V^2}{c_2 M t \kappa_v^2} \right) = 0 \quad . \tag{4}$$

Equation (4) can be solved numerically for each value of (P_i, V_i) as long as the modulus and residual stress of the film are known. While calculating the height from pressure and volume is much more convenient than measuring height directly, it does require one to first obtain reasonably accurate values for modulus and stress. The modulus and stress may be determined by fitting the initial loading segment of the data, when debonding has not yet occurred, to the bulge equation (1). Figure 2 shows a sketch of what blister test data should look like.

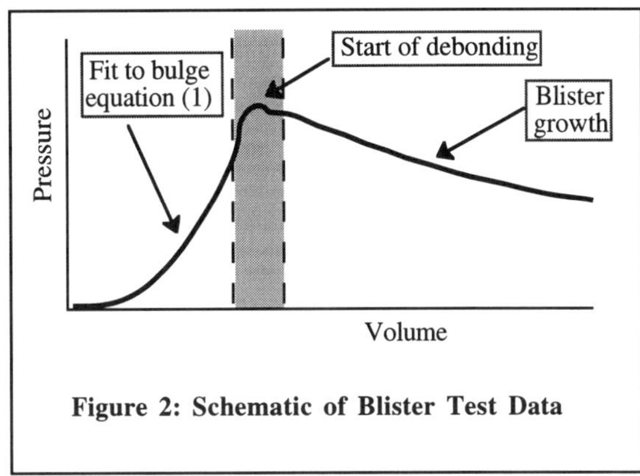

Figure 2: Schematic of Blister Test Data

The volume of liquid injected into the system will be greater than the actual volume of the blister; no test system is perfectly stiff, and a certain amount of liquid will be required to pressurize the system. For measurements of blister volume to be accurate, the stiffness of the test system must be calibrated and corrected for:

$$V_{Blister} = V_{Injected} - \frac{P}{(stiffness)} \quad . \tag{5}$$

The correction is most significant for the initial loading (bulge) portion of the data, where pressures are high and blister volumes are low. The stiffness can be measured by mounting a solid piece of substrate in place of a thin film window, injecting liquid, and finding the slope of the pressure vs. volume curve.

Once the heights have been calculated, the blister radius can be computed using (2), and G determined using (3). Hence, crack extension force can be computed at each point along the debonding curve.

<u>Interface Crack Velocity</u>

If a blister test is conducted at a constant pump rate (dV/dt), then the speed at which the interface crack advances (dr/dt) will continuously decrease as the blister grows. As the critical crack extension force for polymer films may vary with debonding rate[4,12], the variation in crack advance rate should not be neglected.

Multiplying the bulge equation (1) by h, we obtain:

$$Ph = c_1 \sigma_0 t \left(\frac{h}{r}\right)^2 + c_2 M t \left(\frac{h}{r}\right)^4 = function\left(\frac{h^2}{r^2}\right) \quad . \tag{6}$$

The crack extension force (G) is proportional to Ph, and is therefore a function of $(h/r)^2$. If G does not vary by much during a test then (h/r) will be roughly constant. Inserting $(h/r) = (constant)$ into equation (2) and differentiating, we obtain an estimate for (dr/dt):

$$\left(\frac{dr}{dt}\right) \approx \left(\frac{dV}{dt}\right)\Big/ 3 \cdot \sqrt{\kappa_v\, hV} \quad . \tag{7}$$

EXPERIMENTS

Sample Preparation

Free-standing thin film windows were fabricated using a simple micromachining process (Figure 3). First, PECVD SiN$_X$ was grown on both sides of a Si (100) substrate. The backside nitride was patterned and etched to form square windows roughly 3 mm wide. A KOH/methanol mixture was used to etch a hole through the Si, leaving free-standing nitride windows on the front side of the wafer.

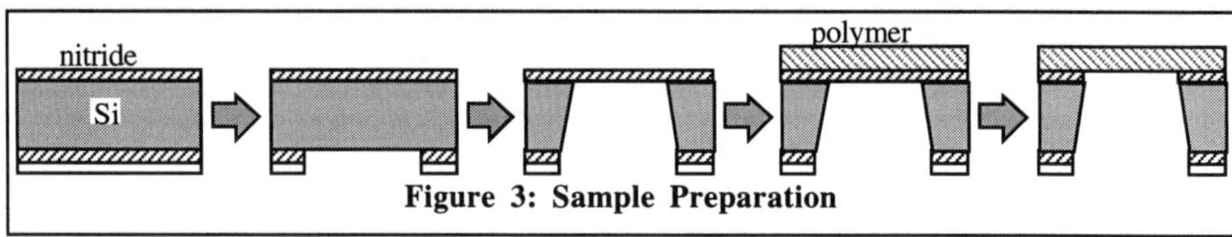

Figure 3: Sample Preparation

Some of the samples were then treated with a commercially available adhesion promoter (MPS- mercaptopropyltrimethoxysilane). The polymer film was spin-coated on top of the nitride windows to a thickness of 13.6 μm. The polymer was a highly cross-linked photopolymerized acrylate, composed of 52.5% oligomer (ALU-303, *Echo Resins and Laboratory*), 42.5% monomer (poly(propylene glycol) diacrylate, *Aldrich*) , and 5% photoinitiator (Igacure 184, *Ciba-Geigy*). To induce cross linking, the films were UV irradiated (360 nm) for 15 minutes in an Ar atmosphere and then baked for 20 minutes at 120°C. Finally, a plasma etch was used to remove the nitride from the undersides of the polymer windows, leaving free-standing polymer windows. The wafers were diced into squares, each square containing one window.

Test Apparatus

The blister test apparatus used for these experiments is depicted in Figure 4. It consists of a compact plexiglass cube to which the pressure transducer, syringe pump, and sample are

mounted. The sample is epoxied to the top of a bolt which is then screwed into the block. A hole drilled down the center of the bolt permits water to contact the sample.

The main concern in the design of the apparatus was the elimination of air bubbles, the main cause of low and inconsistent values of stiffness. The transparent cube makes it easy to see any trapped bubbles. Distilled water was used as the pressurizing liquid for all experiments. The measured stiffness of this apparatus is 328 ± 8 kPa/μl.

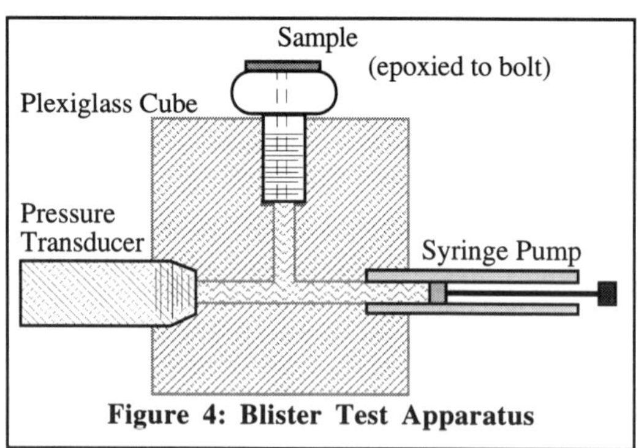

Figure 4: Blister Test Apparatus

<u>Results</u>

Figure 5a shows raw data from blister tests of three samples with the adhesion promoter and three samples without the promoter. A pump rate of 1.4 µl/sec was used in all tests. Pressure increased monotonically until the crack extension force reached a critical value, at which point the films began to debond. The blisters were grown to a final volume of about 15 µl. The initial loading segment of each plot was fitted to the bulge equation. The fit was good at low values of pressure (40 kPa and below), but the data deviated below the ideal linear-elastic behavior of the bulge equation at higher pressures, indicating the presence of plastic or time-dependent flow in the film at higher stresses.

In figure 5b, the calculated crack extension force (G) is plotted against volume. The most obvious feature of the data is that the adhesion promoter increased the critical crack extension force by approximately 1.5 orders of magnitude, from 0.5 - 2.0 J/m^2 for untreated samples to 25 - 34 J/m^2 for samples with the adhesion promoter. In each case, once debonding had initiated, G_C decreased as the test proceeded.

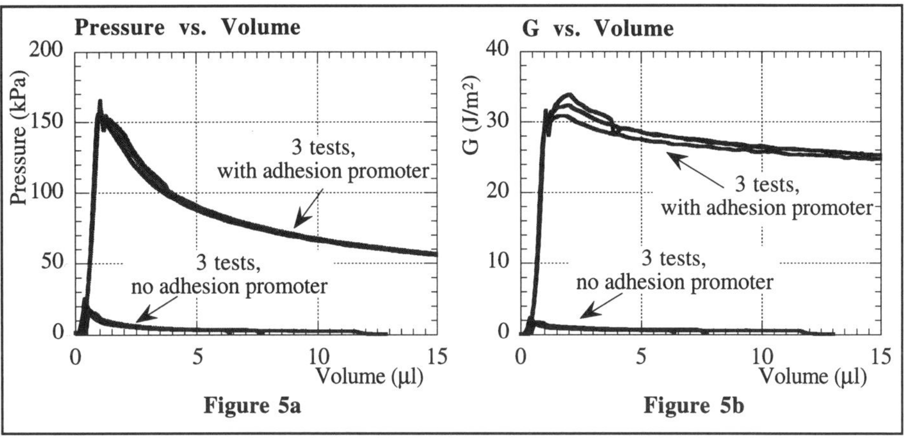

Figure 5a   **Figure 5b**

The interface crack velocity (dr/dt) was calculated using equation (7). Figures 6a and 6b show plots of $\log_{10}(G_C)$ vs. $\log_{10}(dr/dt)$ for the tests with and without adhesion promoter. The data can be fitted to a power-law:

$$G_C \propto \left(\frac{dr}{dt}\right)^m \quad , \qquad (8)$$

where m, the rate exponent, is the slope of a plot of $\log_{10}(G_C)$ vs. $\log_{10}(dr/dt)$.

Rate sensitivity is observed both at low pressures, where the polymer film is within the linear-elastic regime, and at high pressures, where the linear-elastic bulge model may not accurately describe the film. This indicates that the sensitivity is not just an artifact resulting from the idealized linear elastic model. The rate exponents obtained in the high-pressure tests were lower than those for the low-pressure tests (see Figures 6a and 6b).

Rate sensitivity could result from time-dependent strain relaxation of the polymer near the crack tip; at slow crack velocities, there is more time for relaxation to occur, and the energy required to propagate the crack would be lower. A power-law rate dependency with m=0.43 was observed by *Briscoe & Panesar* in blister tests of 4.5 mm thick polyurethane on a stainless steel substrate pressurized by nitrogen gas.[12]

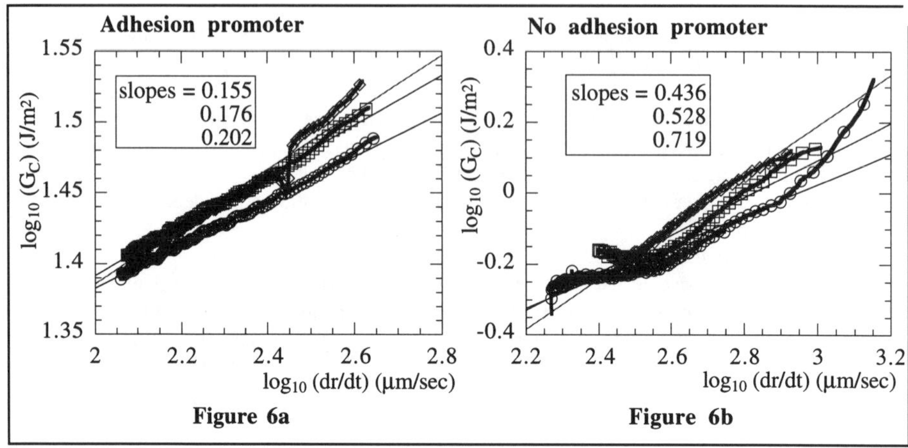

Figure 6a   Figure 6b

CONCLUSIONS

Blister tests can be performed using volumetric measurements as long as care is given to the design of a stiff test apparatus. The critical crack extension force and crack front velocity should be evaluated throughout a blister test, not assumed to be constant. In the acrylate films tested, G_C increased with crack growth rate.

ACKNOWLEDGMENTS

The authors gratefully acknowledge the support of the National Science Foundation (CHE-9412720-002), the Department of Energy (DE-FG03-89ER-45387), Failure Analysis Associates (the G. Marshall Pound Fellowship), the NSF-MRL Program through the Center for Materials Research at Stanford University, and the NSF-MRSEC Program through the Center on Polymer Interfaces and Molecular Assemblies.

REFERENCES

1. A. N. Gent, L. H. Lewandowski, *J. App. Polymer Sci.*, **33**, pp. 1567-1577 (1987).
2. H. M. Jensen, *Engineering Fracture Mechanics*, **40**, pp. 475-486 (1991).
3. M. G. Allen, S. D. Senturia, *J. Adhesion*, **25**, pp. 303-315 (1988).
4 Chu, Y. Z., H. S. Jeong, R. C. White, C. J. Durning, *Mater. Res. Soc. Symp. Proc.*, **276**, pp. 209-220 (1992).
5. J. W. Beams, *Structure and Properties of Thin Films*, C.A. Neugebauer, J. B. Newkirk, and D.A. Vermilyea, Eds., John Wiley and Sons, Inc., 1959, p. 183.
6. M. K. Small, W. D. Nix, *J. Mater. Res.*, Vol. 7, 1553-1563 (1992).
7. J. J. Vlassak, W. D. Nix, *J. Mater. Res.*, Vol. 7, 3242-3249 (1992).
8. M. K. Small, *Ph. D. thesis*, Stanford University, 1992.
9. J. J. Vlassak, *Ph. D. thesis*, Stanford University, 1994.
10. J. J. Vlassak, R. J. Hohlfelder, unpublished.
11. R.J. Hohlfelder, J. J. Vlassak, W. D. Nix, H. Luo, C.E.D. Chidsey, *Mat. Res. Soc. Symp. Proc.*, **356**, (1995).
12. B. J. Briscoe, S. S. Panesar, *Proc. R. Soc. Lond.*, A **433**, pp. 23-43 (1991).

EFFECTS OF MECHANICAL PROPERTIES OF METAL FILMS ON THE ADHESION STRENGTH OF Cr/POLYIMIDE INTERFACES

JIN WON CHOI, TAE SUNG OH
Department of Metallurgy and Materials Science, Hong Ik University, Seoul 121-791, Korea

ABSTRACT

Effects of mechanical properties of Cu/Cr metal films on the peel strength of Cr/PI interfaces have been studied. Cr and Cu thin films were successively sputter-deposited on *in-situ* RF plasma-treated polyimides, and 20 µm-thick Cu was electroplated. With increasing the yield strength of Cu/Cr films from 156 MPa to 325 MPa, peel strength of Cr/PMDA-ODA and Cr/BPDA-PDA were lowered from 75 g/mm to 57 g/mm and from 69 g/mm to 20 g/mm, respectively. With identical Cu/Cr metal films, lower peel strength was obtained on Cr/BPDA-PDA interfaces, compared to the values of Cr/PMDA-ODA. Peel strength was also decreased more pronouncedly on Cr/BPDA-PDA with increasing the yield strength of Cu/Cr metal films. With T/H (80℃/94% R.H.) exposure, however, peel strength was lowered much more pronouncedly on Cr/PMDA-ODA than on Cr/BPDA-PDA, especially for specimens with Cu/Cr metal films of lower yield strength.

INTRODUCTION

Thin film packaging is one of the key technologies to develop advanced electronic systems such as computers and telecommunication systems. As insulation layers in thin film packaging structures such as multilayer multichip modules, flexible circuit boards and TAB tapes, polyimides have been widely used due to their low dielectric constant, good planarization and easy processibility [1-8]. In thin film packaging structures, there are numerous metal/PI (metal on polyimide) and PI/metal (polyimide on metal) interfaces, and their mechanical integrity during fabrication and operation is one of the critical issues which determine the device performance and reliability.

In general, adhesion of metal/PI interfaces, where the metal film is deposited on fully cured polyimide, is much lower than that of PI/metal interfaces, where polyamic acid is spin-coated on metal films and cured to form the polyimide. Since the reliability of multilayer structures is governed by the most weakly bonded interface, improvement of metal/PI adhesion by enhancing the interfacial reactions has been extensively investigated using various surface treatments of polyimide prior to metal deposition [6,9,10]. Adhesion strength of metal/PI interfaces, which is usually measured using peel test, consists of the work done in creating new surfaces of metal and polyimide (interfacial fracture toughness of metal/PI) and plastic work done by deformation of metal films during peeling process [11-14]. Thus, peel strength of metal/PI interfaces should be dependent on mechanical properties of metal films as well as interfacial reactions. However, few reports can be found for systematic study on the variation of peel strength of metal/PI interfaces with mechanical properties of metal films.

In this study, effects of mechanical properties of Cu/Cr metal films on the peel strength of Cr/PI interfaces have been characterized. Adhesion degradation behavior of Cr/PI interfaces during exposure to T/H environment have also been investigated.

EXPERIMENTAL

Two types of cured polyimide films, 50 µm-thick PMDA-ODA (Kapton) and 25µm-thick BPDA-PDA (Upilex), are used as substrates to fabricate Cu/Cr/PI specimens. Polyimide surfaces were modified using RF plasma treatment with Ar at a power density of 0.14 W/cm² for 3 minutes prior to metal deposition. Cr and Cu films were successively sputter deposited on RF plasma-treated polyimide. Thickness of Cr and Cu films was 50 nm and 500 nm, respectively. After Cr and Cu deposition, 20 µm-thick Cu was electroplated in CuSO₄ solution at a current density of 15 mA/cm². In order to change the mechanical properties of metal films, various CuSO₄ solutions (Table 1) were used to electroplate Cu.

Mat. Res. Soc. Symp. Proc. Vol. 436 © 1997 Materials Research Society

Peel strength of Cr/PI interfaces was measured with 30 × 5 mm Cu/Cr/PI strips using T-peel test at a peeling rate of 5 mm/min (Fig. 1). During peel test, any instrument to maintain the peeling angle as **T** shape was not used. Microstructure of electroplated Cu films were observed using TEM. Deformation of metal films during peel test was characterized by FWHM increment of Cu (331) K_{d1} X-ray diffraction peaks of the unpeeled and peeled Cu/Cr metal strips. To verify the effects of mechanical properties of metal films on the peel strength of Cr/PI interfaces, specimens were annealed in N_2 atmosphere at 200℃ for 1 hour. To examine the hydrothermal stability of Cr/PI interfaces, specimens were exposed to high temperature/humidity environment (80℃/94% R.H.) up to 1000 hours, and adhesion degradation behavior was characterized using T-peel test.

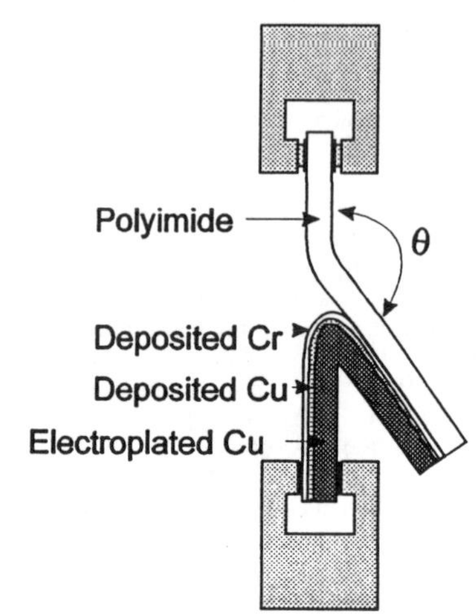

Fig. 1 Schematic illustration of T-peel test.

RESULTS and DISCUSSION

Mechanical properties of Cu/Cr metal films were measured using Cu/Cr/PI strips. Under tensile loading, strains of Cu/Cr metal films and polyimide should be identical, and the total stress applied to the specimen can be expressed as Eq. 1, where E_M is elastic modulus of Cr/Cu metal films, E_{PI} is elastic modulus of polyimide, V_f^M and A_f^M is the volume fraction and area fraction of metal films, respectively [15].

$$\sigma_{total} = E_M \varepsilon V_f^M + E_{PI}\varepsilon(1 - V_f^M) = E_M \varepsilon A_f^M + E_{PI}\varepsilon(1 - A_f^M) \qquad (Eq.\ 1)$$

0.2% offset yield strength of Cu/Cr metal film ($E_M\varepsilon$) was obtained from 0.2% offset yield strengths of Cu/Cr/PI multilayer structure (σ_{total}) and PMDA-ODA film ($E_{PI}\varepsilon$) measured on the stress-strain curves of Cu/Cr/PI and polyimide film, respectively. Yield strength of Cu/Cr metal films with different electroplated Cu films is listed in Table 1. Yield strength of Cu/Cr metal film could be increased by adding brightener, $SC(NH_2)_2$, into $CuSO_4$ electroplating solution due to grain refinement of the electroplated Cu films, which could be confirmed with TEM observation.

Table 1. Composition of Cu electroplating solutions and yield strength of Cu/Cr metal films.

Solution No.	$CuSO_4$ (g)	H_2SO_4 (g)	Brightener (g)	Yield strength (MPa)
E1	295	75	0	156
E2	160	225	0	174
E3	295	75	0.25	284
E4	295	75	6	325

* total 1 ℓ, water balance

Fig. 2 shows Cu (331) X-ray diffraction peaks obtained from the electroplated Cu side of unpeeled Cu/Cr metal strips. Broadening of Cu (331) K_α doublet for Cu films, electroplated in $CuSO_4$ solutions containing the brightener, clearly illustrates that increase of yield strength of Cu/Cr films is due to the grain refinement of electroplated Cu film.

As shown in Fig. 3, peel strength of Cr/PMDA-ODA and Cr/BPDA-PDA interfaces was greatly affected by the yield strength of Cu/Cr metal film. With increasing the yield strength of Cu/Cr metal film from 156 MPa to 325 MPa, peel strength of Cr/PMDA-ODA and Cr/BPDA-PDA was lowered from 75 g/mm to 57 g/mm and from 69 g/mm to 20 g/mm, respectively.

Peel strength of Cr/PI interfaces in Cu/Cr/PI structure can be expressed as Eq. 2, where P is the peel strength (peeling force per unit width), ξ is the interfacial fracture energy or decohesion energy, and ψ is the plastic work dissipated during plastic bending of the peeled strips [13].

$$P = \xi + \psi \qquad\qquad (Eq.\ 2)$$

Contrary to 90° peel test, plastic bending of the peeled polyimide film as well as Cu/Cr metal film occurs during T-peel test, as shown in Fig. 1. Comparing the yield strength of PMDA-ODA (~ 15 MPa), measured from the stress-strain curve, with yield strength of Cu/Cr metal films (Table 1), however, it may be assumed that the plastic work ψ for T-peel test is also associated mainly with plastic bending of the peeled metal strip.

Mechanical properties of Cu/Cr film were changed by electroplating 20 μm-thick Cu layer with different electroplating solutions (Table 1). However, R.F. pre-sputtering treatment of the polyimide surface prior to metal deposition was identical for all specimens. Then, the interfacial fracture energy ξ, which is associated with the reactions between polyimide and sputter-deposited Cr, should be the same for all specimens regardless of Cu electroplating conditions. Thus, lowering of the peel strength of Cr/PI interfaces with increasing the yield strength of Cu/Cr films is clearly due to less plastic bending of the peeled metal strip. This could be confirmed by observing the peeling angle θ (Fig. 1), which was decreased with increasing the yield strength of Cu/Cr film.

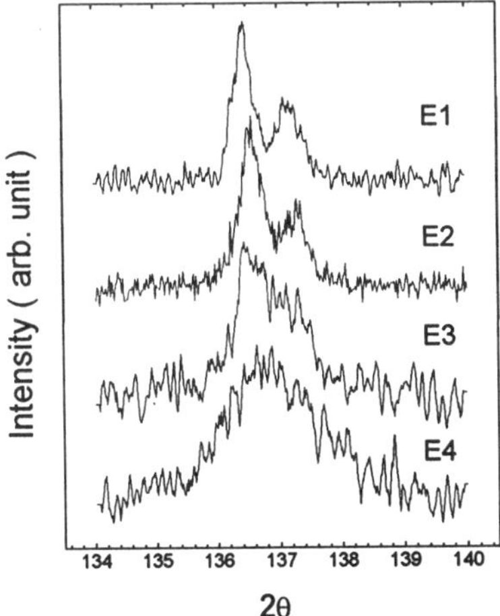

Fig. 2 Cu (331) XRD peak of the electroplated Cu films before peel test.

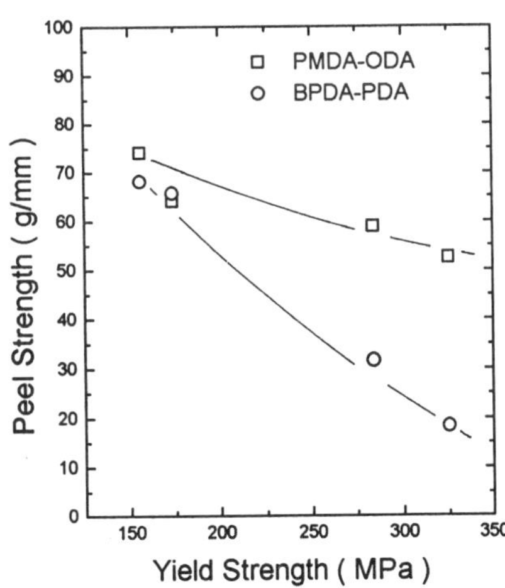

Fig. 3 Peel strength of Cr/PMDA-ODA and Cr/BPDA-PDA interfaces vs. yield strength of Cu/Cr film.

With identical Cu/Cr metal films, lower peel strength was obtained on Cr/BPDA-PDA interfaces, compared to the values of Cr/PMDA-ODA (Fig. 3). Adhesion enhancement of Cr/PI interfaces by RF sputtering of the polyimide surface occurs due to chemical modification of the polyimide surface facilitating chemical bonding with the deposited Cr [6,9]. BPDA-PDA is relatively more stable than PMDA-ODA due to hydrocarbon-like character of BPDA-PDA structure [9]. With identical RF sputtering treatment on PMDA-ODA and BPDA-PDA surfaces, interfacial reactions would be more active at Cr/PMDA-ODA than Cr/BPDA-PDA [9], resulting in higher interfacial fracture energy of Cr/PMDA-ODA. Peel strength is mainly determined by plastic deformation energy ψ of the peeled metal film. However, ψ is dependent on the interfacial fracture energy ζ [12,13]. In Fig. 3, peel strength was decreased more pronouncedly on Cr/BPDA-PDA with increasing the yield strength of Cu/Cr metal film. As this result, effects of the mechanical properties of metal films on the peel strength of Cr/PI interfaces would be enlarged with lower interfacial fracture energy ζ.

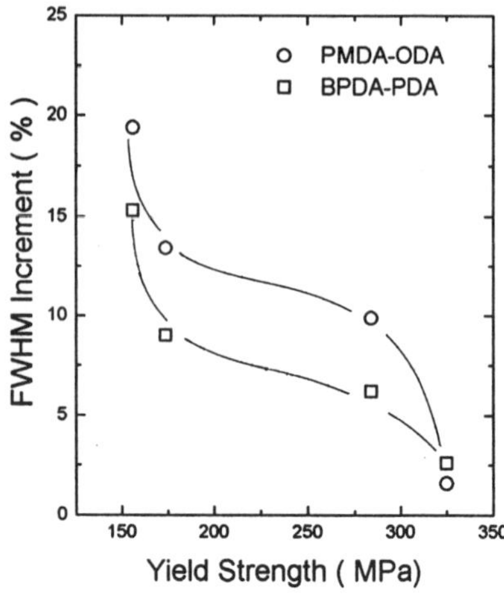

Fig. 4   FWHM increment of Cu (331) XRD peak after peel test, normalized with Cu (331) FWHM of the unpeeled Cu/Cr film.

Fig. 4 shows FWHM increment of Cu (331) XRD peak after peel test, normalized with Cu (331) FWHM values of the unpeeled Cu/Cr films. With increasing the yield strength of Cu/Cr films from 156 MPa (E1) to 325 MPa (E4), normalized FWHM increment of the peeled metal strip was decreased from 19% to 3% for Cr/PMDA-ODA and from 16% to 6% for Cr/BPDA-PDA interfaces, respectively.

To verify the effects of mechanical properties of metal films on the peel strength of Cr/PI interfaces, specimens were annealed in N_2 atmosphere at 200℃ for 1 hour in two different ways. In first annealing process, heat-treatment was performed after electroplating of 20 μm-thick Cu on the sputter-deposited Cu/Cr/PI. With this annealing process, yield strength of Cu/Cr films was lowered (156 MPa to 149 MPa for E1, 174 MPa to 159 MPa for E2, 284 MPa to 253 MPa for E3, and 325 MPa to 297 MPa for E4) due to the relaxation of residual stresses in the electroplated Cu films. TEM microscopy showed no evidence of grain growth of the electroplated Cu films during annealing process. Even annealed in N_2 atmosphere, it has been reported that interfacial fracture energy of Cr/PI was reduced due to the formation of Cr oxide at the Cr/PI interface and thermal expansion mismatch between polyimide and metal films [16-18]. Comparing to specimens which were not annealed, both interfacial fracture energy and mechanical properties of Cu/Cr films were changed with first annealing process. In second annealing process, only sputter-deposited Cu/Cr/PI was annealed and then 20 μm-thick Cu was electroplated. Comparing to specimens which were not annealed, only the interfacial properties could be changed. Specimens annealed by second method had the same interfacial properties as those of specimens annealed by first method. In Fig. 5, peel strengths of Cr/PMDA-ODA and Cr/BPDA-PDA were illustrated with yield strength of Cu/Cr films. With identical Cu/Cr metal films, peel strength of Cr/PI was lowered by heat treatment due to the degradation of interfacial fracture energy during heat treatment. With the same interfacial properties, however, lowering of the peel strength of Cr/PI with increasing the yield strength of Cu/Cr films was fitted nicely into one curve regardless of the annealing processes.

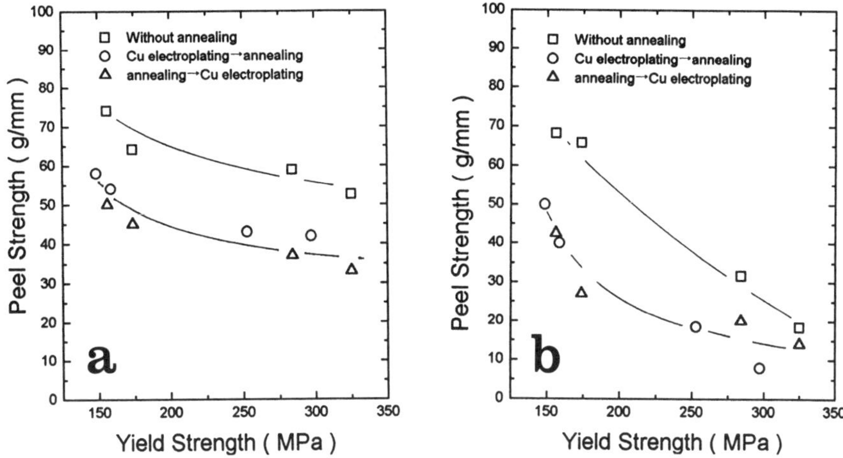

Fig. 5 Peel strength of (a) Cr/PMDA-ODA and (b) Cr/BPDA-PDA vs.
yield strength of Cu/Cr films with different annealing process.

As shown in Fig. 6, peel strength of Cr/PMDA-ODA and Cr/BPDA-PDA interfaces was decreased with increasing T/H (80℃/94% R.H.) exposure time. Comparing the adhesion degradation behavior of Cr/PMDA-ODA and Cr/BPDA-PDA, peel strength was lowered much more pronouncedly on Cr/PMDA-ODA, especially for specimens with Cu/Cr films of lower yield strength. This result implys that chemical bonding between Cr and PMDA-ODA is more susceptible to hydrothermal degradation than chemical bonding at Cr/BPDA-PDA interface.

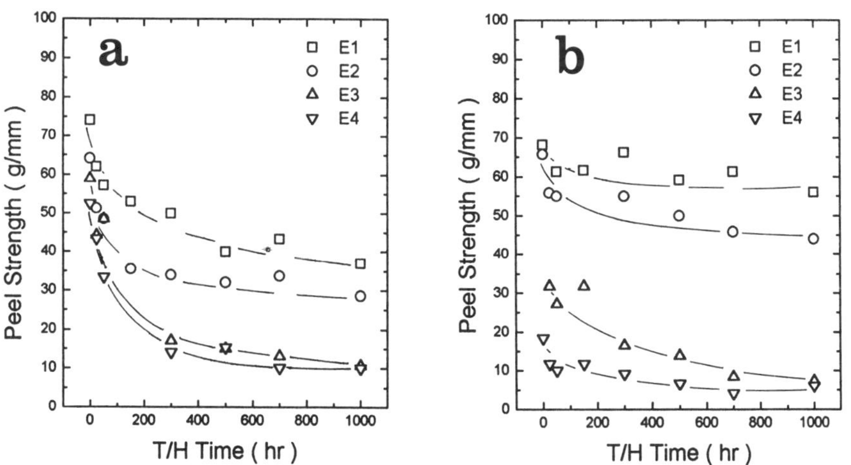

Fig. 6 Peel strength degradation of (a) Cr/PMDA-ODA and (b) Cr/BPDA-PDA
vs. T/H (80℃/94% R.H.) exposure time.

SUMMARY

Effects of mechanical properties of Cu/Cr metal films on the peel strength of Cr/PI interfaces have been investigated. Yield strength of Cu/Cr films in Cu/Cr/PI multilayer structures could be changed by electroplating 20 μm-thick Cu in various CuSO$_4$ solutions onto the sputter-deposited Cu/Cr/PI. With increasing the yield strength of Cu/Cr films from 156 MPa to 325 MPa, peel strength of Cr/PMDA-ODA and Cr/BPDA-PDA were lowered from 75 g/mm to 57 g/mm and from 69 g/mm to 20 g/mm due to less plastic bending of Cu/Cr films of higher yield strength during peel test. With identical Cu/Cr metal films, lower peel strength was obtained on Cr/BPDA-PDA interfaces, compared to the values of Cr/PMDA-ODA. This may imply that interfacial reactions would be more active at Cr/PMDA-ODA than Cr/BPDA-PDA, resulting in higher interfacial fracture energy of Cr/PMDA-ODA. Peel strength was decreased more pronouncedly on Cr/BPDA-PDA with increasing the yield strength of Cu/Cr metal films, indicating that effects of the mechanical properties of metal films on the peel strength of Cr/PI interfaces would be enlarged with lower interfacial fracture energy. Peel strength of Cr/PMDA-ODA and Cr/BPDA-PDA interfaces was decreased with increasing T/H (80℃/94% R.H.) exposure time. However, peel strength of Cr/PMDA-ODA was lowered much more pronouncedly with T/H exposure time than peel strength of Cr/BPDA-PDA, indicating that chemical bonding at Cr/PMDA-ODA interface is more susceptible to hydrothermal degradation than chemical bonding at Cr/BPDA-PDA.

ACKNOWLEDGMENT

This work was supported by the Korea Science and Engineering Foundation(95-0300-08-01-3).

REFERENCES

1. R. R. Tummala and E. J. Rymaszewski, *Microelectronic Packaging Handbook*, VSR, New York (1989).
2. L. J. Matienzo, F. Emmi, D. C. Vanhart, and J. C. Lo, *J. Vac. Sci. Technol.*, **A9**, 1278 (1991).
3. G. D. David, B. J. Rees, and D. L. Whisnant, *J. Vac. Sci. Technol.*, **A12**, 2378 (1994).
4. S. G. Anderson, J. Leu, B. D. Silverman, and P. S. Ho, *J. Vac. Sci. Technol.*, **A11**, 368 (1993).
5. K. S. Sengupta, and H. K. Birnbaum, *J. Vac. Sci. Technol.*, **A9**, 2928 (1991).
6. T. S. Oh, S. P. Kowalczyk, D. J. Hunt and J. Kim, *J. Adh. Sci. Technol.*, **4**, 119 (1990).
7. D-Y. Shin, N. Klymko, R. Flitsch, J. Paraszczac, and S. Nunes, *J. Vac. Sci. Technol.*, **A9**, 2963 (1991).
8. Y. Nakayama, P. Baltzer, B. Wannberg, and U. Gelius, *J. Vac. Sci. Technol.*, **A12**, 772 (1994).
9. D. L. Pappas, and Jerome J. Cuomo, *J. Vac. Sci. Technol.*, **A9**, 2704 (1991).
10. J. E. E. Baglin, *Nucl. Instr. and Meth.*, **B39**, 764 (1989)
11. H. S. Jeong and R. C. White, *J. Vac. Sci. Technol.*, **A11**, 1373 (1993).
12. J. Kim, K. S. Kim and Y. H. Kim, *J. Vac. Sci. Technol.*, **A3**, 175 (1989).
13. K. S. Kim, and J. Kim, *J. Eng. Master. Technol.*, **110**, 266 (1988).
14. A. N. Gent and G. R. Hamed, *J. Appl. Polymer Sci.*, **21**, 2717 (1977).
15. G. E. Dieter, *Mechanical Metallurgy*, p.222, McGraw Hill, London (1988).
16. B. K. Furman, K. D. Childs, H. Clearfield, R. David, S. Purushothaman, *J. Adh. Sci. Technol.*, **10**. 2913 (1992).
17. J. T. Pan and S. Poon, *Mat. Res. Sympo. Proc.*, **154**, 27 (1989).
18. C. H. Yang and P. C. Chen, *Mat. Res. Sympo. Proc.*, **154**, 335 (1989).

INFLUENCE OF Cu DEPOSITION CONDITIONS ON THE MICROSTRUCTURE AND ADHESION OF Cu/Cr THIN FILM TO POLYIMIDE FILM

CHEOL-HO JOH and YOUNG-HO KIM
Department of Materials Engineering, Hanyang University, Seoul, KOREA
kimyh@hyunp1.hanyang.ac.kr

ABSTRACTS

The microstructure of Cu thin films in various deposition conditions and the influence of their microstructure on the adhesion strength between Cu/Cr films and polyimides were studied. Cr films (50 nm thick) and Cu films (500 or 1000 nm thick) were deposited on polyimide films by DC magnetron sputtering. The Ar pressure during Cu deposition was controlled to 5, 50, and 100 mtorr. The microstructure was characterized using SEM and TEM. The adhesion strength between Cu/Cr and polyimide was measured using a modified T-peel test. Plastic deformation of the peeled metal was qualitatively measured using the XRD technique. The Cu film sputtered at 5 mtorr has a dense and uniform structure, while low-density regions or open boundaries between columns exist in the film deposited at higher pressure. As sputtering pressure increases, open boundaries are observed more frequently. The peel adhesion strength of Cr film to polyimide increases with Cu sputtering pressure. The adhesion change of Cu/Cr film can be interpreted as the difference in plastic deformation. Open boundaries in the Cu film seem to play an important role in increasing the amount of plastic deformation in the metal film during peeling.

INTRODUCTION

The peel test is a simple mechanical test which has been extensively used to measure the adhesion strength between thin metal film and substrate [1]. The peel adhesion strength represents the total works required for a thin film to be separated from the substrate [2,3]. The total work includes the work needed to break the interfacial bond and other work expended in the plastic deformation of the thin film. The contribution of the latter to the total works is important to control the peel strength. Several investigators showed that the peel strength may vary due to the differences in the amount of plastic deformation even though the interfacial bond strength is the same [3,4]. The plastic deformation depends on the mechanical properties of the film such as yield strength and ductility. The mechanical properties are well known to be sensitive to the microstructure.

Thornton [5] proposed a structure zone model, where the sputtered film structure is a function of Ar pressure and substrate temperature. The change of the sputtering conditions modifies the structure of thin film, which will affect the mechanical properties of the thin film. In this study, Cu was deposited on Cr/polyimide with different sputtering conditions, the microstructure of Cu thin films was characterized using SEM (scanning electron microscope) and TEM (transmission electron microscope), and the peel adhesion of Cu/Cr film onto polyimide was evaluated with Cu deposition conditions.

EXPERIMENT

The polyimide used in this study was 40 μm thick BPDA-PDA type film (Upilex-s). Cr and Cu films were deposited by using a DC magnetron sputtering unit. As an adhesion layer, 50 nm thick Cr was deposited at 5 mtorr on the water-cooled polyimide film, which had been RF plasma treated for 3 min. Then, 500 nm or 1000 nm (1 μm) thick Cu was deposited on Cr/polyimide. Sputtering pressure during Cu deposition was controlled to 5, 50, and 100 mtorr. Surface morphology and microstructure of Cu sputtered films were characterized by SEM and

TEM. Ion milling was conducted to prepare TEM specimens. The ion milling rate was carefully controlled to observe the region near the top surface of the sputtered Cu film.

For peel tests, 20 μm thick Cu was electroplated on Cu/Cr/Polyimide in a $CuSO_4$ solution. The adhesion between the metal film and polyimide was measured using a modified T-peel test [6]. The XRD (X-Ray Diffraction) technique was successfully used to measure qualitatively the amount of plastic deformation on peeled metal strips [4,6]. This technique was also applied here to compare the degree of plastic deformation in the metal films. The FWHM (Full Width at Half Maximum intensity) of the deconvoluted Cu (331) $K_{\alpha 1}$ peak was measured and compared with the peel adhesion strength.

RESULTS AND DISCUSSION

Surface Morphology and Microstructure in the Sputtered Cu Film

Fig. 1 consists of scanning electron micrographs which show the effects of sputtering pressure and thickness on the surface morphology in the Cu sputtered films. The Cu film deposited at 5 mtorr has a very dense structure. No distinct open boundaries are observed. Low density regions or open boundaries are formed in the film deposited at high sputtering pressure. As deposition pressure increases, open boundaries are more frequently found. Also as the film grows thicker, open boundaries become more distinct; open boundaries are wider and the density of open boundaries is higher. These topographies are similar to the previous results [5,7], which is called the columnar structure. Near surface regions in the same Cu thin films observed by TEM are shown in Fig. 2. Fig. 2 was taken at the same magnification as Fig. 1, so it is possible to compare each other. The dependence of open boundary size and distribution on sputtering pressure and Cu thickness observed in Fig. 2 reaches the same conclusions as in Fig. 1. Open boundaries are more clearly seen in TEM micrographs. Grains and twins are seen in all films. Columnar structure is evident specially in the 1 μm thick film which is sputter-deposited at 100 mtorr. The formation of open boundary and columnar structure is attributed to the shadowing effect of atomic flux [7,8]. The comparison between SEM and TEM indicates that each column contains a number of grains. The grain size measurement shows that the variation of grain size is not strongly related to the sputtering pressure.

Peel Adhesion

Peel adhesion strength was measured by separating the metal layers (plated Cu/Sputtered Cu/Sputtered Cr) from polyimide. Table 1 summarizes the peel test results and FWHMs of Cu peaks obtained using XRD technique. All peel strength values in Table 1 are over 50 g/mm, which indicates very strong adhesion between Cr and polyimide. Increasing sputtering pressure during Cu deposition generally increases the average peel strength value. The peel adhesion of the specimen in which Cu was deposited at 5 mtorr does not vary with Cu film thickness. Meanwhile, the peel adhesion of the specimen in which Cu was deposited at 100 mtorr increases with the Cu thin film thickness. Both results showing the influence of sputtering pressure and Cu thin film thickness are closely related to the size and the distribution of the open boundaries in the Cu thin film. When the size and the density of open boundary increase significantly, peel adhesion strength increases excessively. The 90° peel test results yield the same conclusions [9].

The relative amount of plastic deformation in the peeled metal layer was measured using XRD spectra [4,6]. Cu (331) spectra were smoothed and deconvoluted into $K_{\alpha 1}$ and $K_{\alpha 2}$. The line broadening, measured as $\Delta 2\theta$ at half maximum intensity is shown in Table 1. The larger $\Delta 2\theta$ means heavier deformation since non-uniform strain causes line broadening [10]. The trend of the relative amount of plastic deformation with sputtering pressure and Cu film thickness is similar to that of peel strength. The metal strip in which Cu was deposited at 5 mtorr is deformed least and the metal strip in which 1 μm thick Cu was deposited at 100 mtorr is deformed most. The peeled metal surfaces of both specimens are shown in Fig. 3. Both specimens show the surface cracks propagating into the Cu film from Cr layer. The cracks lie

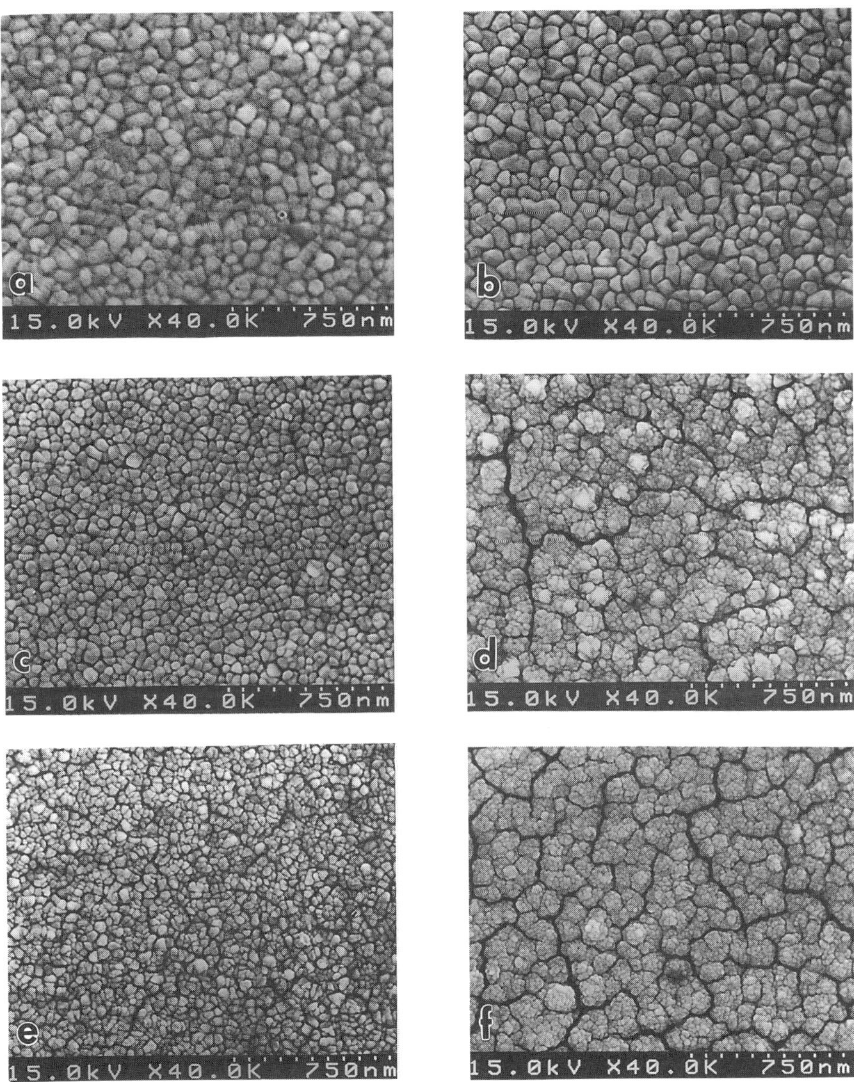

Fig. 1. SEM micrographs showing surface morphology as functions of sputtering pressure and Cu film thickness. (a) 5 mtorr, 500 nm (b) 50 mtorr, 500 nm (c) 100 mtorr, 500 nm (d) 5 mtorr, 1000 nm (e) 50 mtorr, 1000 nm (f) 100 mtorr, 1000 nm

perpendicular to the peeling direction. The 100 mtorr specimen has more surface cracks than 5 mtorr specimen. The cracks are larger and deeper in the 100 mtorr specimen. The crack size and distribution on Cr surface suggest that the 100 mtorr specimen be deformed more during peeling. This result is consistent with XRD result.

All procedures to make peel test specimens are identical except Cu sputter deposition method. Therefore, the interfacial bond strength between Cr and polyimide is the same in all specimens. Variation in 20 μm thick electroplated Cu layers due to the difference of the

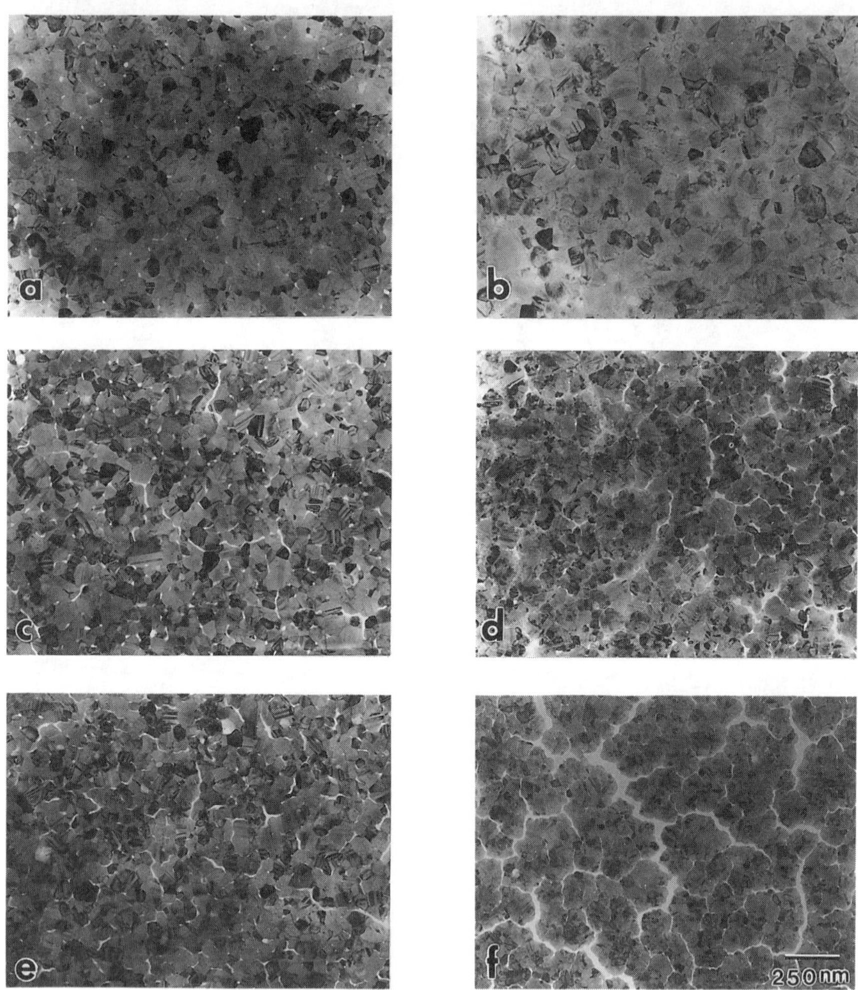

Fig. 2. TEM images showing the microstructure of near surface regions in Cu thin films.
(a) 5 mtorr, 500 nm (b) 50 mtorr, 500 nm (c) 100 mtorr, 500 nm
(d) 5 mtorr, 1000 nm (e) 50 mtorr, 1000 nm (f) 100 mtorr, 1000 nm

Table 1. Peel test and XRD results with different deposition conditions

Deposition Condition	Peel Strength	FWHM($\Delta 2\theta$)
Cu(5 mtorr, 500 nm)/Cr(5 mtorr, 50 nm)/Polyimide	52.1 g/mm	0.496
Cu(50 mtorr, 500 nm)/Cr(5 mtorr, 50 nm)/Polyimide	62.1 g/mm	0.630
Cu(100 mtorr, 500 nm)/Cr(5 mtorr, 50 nm)/Polyimide	62.2 g/mm	0.636
Cu(5 mtorr, 1000 nm)/Cr(5 mtorr, 50 nm)/Polyimide	51.6 g/mm	0.494
Cu(50 mtorr, 1000 nm)/Cr(5 mtorr, 50 nm)/Polyimide	60.8 g/mm	0.660
Cu(100 mtorr, 1000 nm)/Cr(5 mtorr, 50 nm)/Polyimide	73.2 g/mm	0.694

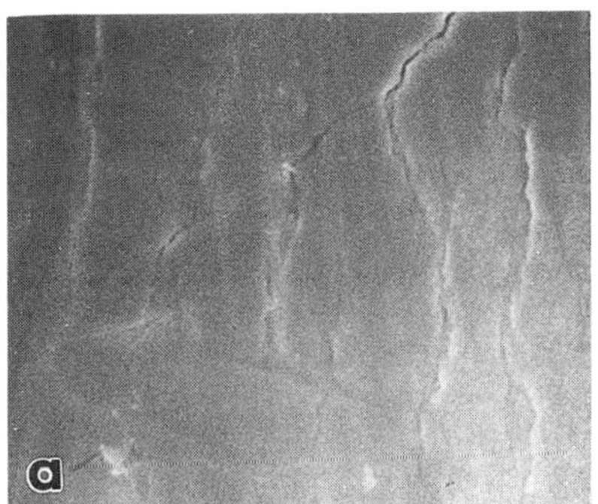
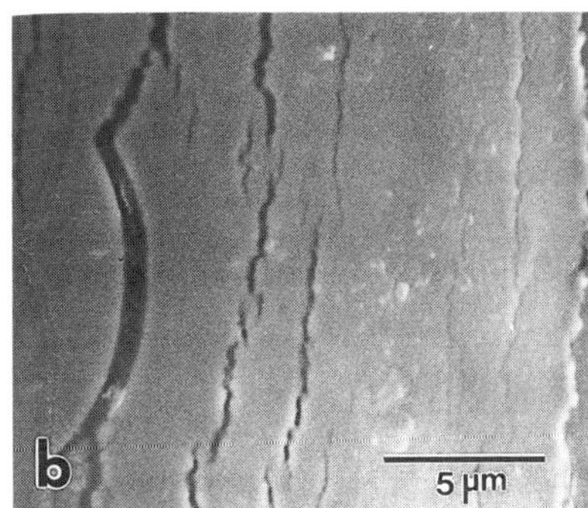

Fig. 3. SEM micrographs of peeled metal surface (Cr surface).
(a) 5 mtorr specimen (b) 100 mtorr specimen

morphology of Cu sputtered film was not noticed under the microscopic observation. The properties of electroplated layers are expected to be the same in all specimens. Consequently, the change in the peel strength values comes from the structural change of sputtered Cu films which are affected by sputtering pressure and thickness. Open boundaries formed during sputter deposition are sustained by the tensile stress during bending the metal strip for peel test. Open boundary will induce the concentrated force to the plated Cu layer so that the ductile electroplated Cu layer deformed locally. Increasing crack density and crack size increases the local deformation of electroplated Cu layer. If the film is hard and brittle, an open boundary will serve as a crack which provides an easy fracture path. However, the electroplated Cu layer is very ductile and easily deformed under local stress before the stress is released due to interface separation[2]. As a result, deposition at high sputtering pressure yields an increasing number of wide and large open boundaries in the Cu film, which assist the electroplated layer to be deformed easily during peeling. The thicker Cu film is deposited at high pressure, the more effective it is to increase the plastic deformation. Plastic deformation in the electroplated layer governs the peel strength for the metal film strongly adherent to polyimide [3,4]. Increasing the amount of plastic deformation in the plated layer is important to enhance the peel adhesion.

The existence of surface cracks in Fig. 3 implies that Cr layer is torn during peel test. The work used to tear Cr layer is also included in the total work which is expressed as peel strength. There are many surface cracks in 100 mtorr specimen, so they may contribute to increase the peel strength even in a small amount.

CONCLUSIONS

1. The surface morphology and microstructure in the sputtered Cu films are closely related to sputtering pressure and film thickness. The size and the density of open boundaries increase with sputtering pressure and film thickness. Columnar structure is formed in the Cu film deposited at high pressure, where one column contains several grains.

2. The peel adhesion strength depends on the microstructure of the film when the interfacial bond strength is the same. The peel strength of Cu/Cr film onto polyimide increases with Cu sputtering pressure and sputtered Cu film thickness. The metal film having higher peel strength is deformed more. Open boundaries seem to play an important role in increasing the amount of plastic deformation in the metal film during peeling.

ACKNOWLEDGMENT

This work was supposed by KOSEF under contract No. 95-0300-08-01-3.

REFERENCES

·1. P.A. Steinmann and H.E. Hintermann, J. Vac. Sci. Tchnol., **7**, 2276 (1989).
2. N. Aravas, K.S. Kim, and M.J. Loukis, Mater. Sci. Eng., **107**, 159 (1989).
3. J. Kim, K.S. Kim, and Y.-H. Kim, J. Adhesion Sci. Technol., **3**, 175 (1989).
4. J.H. Lim, T.G. Lee, and Y.-H. Kim, The Second Pacific Rim International Conference on Advanced Materials and Processing, **2**, 1127 (1995).
5. J.A. Thornton, J. Vac. Sci. Technol., **11**, 666 (1974).
6. J.H. Lim, Y.-H. Kim, and S.H. Han, J. Kor. Inst. Surf. Eng., **26**, 327 (1993).
7. S. Craig and G.L. Harding, J. Vac. Sci. Technol., **19**, 205 (1981).
8. M. Ohring, The Materials Science of Thin Films, (Academic Press, San Diego, 1991) p. 225.
9. C.-H. Joh, and Y.-H. Kim, to be published.
10. B.D. Cullity, Elements of X-ray Diffraction, 2nd ed., (Addison Wesley, Reading, 1978) p. 249.

PEEL STRENGTH IN THE Cu/Cr/POLYIMIDE SYSTEM

I. S. Park, Jin Yu and Y. B. Park
Department of Materials Science and Engineering, KAIST
Chongryang P.O.Box 201, Seoul, Korea

ABSTRACT

In order to understand the effects of **plastic deformation** and the **interfacial fracture energy** on the **peel strength, thickness** of the metal layer and the pretreatment conditions of polyimide were varied in the **Cu/Cr/polyimide** system. The work expenditure during the peel test was estimated using the stress strain curves of metal films, **X-ray measurements** of the plastic strain in the peeled films, and the elastoplastic beam analysis. Results indicate that the peel strength is strongly affected by the film thickness and the pretreatment condition in a synergistic way, and that the measured peel strength is more a measure of the plastic deformation during the peel test than a measure of the true interfacial energy.

1. INTRODUCTION

In the microelectronics industry, the adhesion strength of thin metal films to dielectric substrate is often measured by the peel test and the peel strength is directly related to the interfacial fracture resistance[1]. However, extensive plastic deformation occurs during the peel test, and the peel strength does not truly represent the interfacial fracture energy because the measured value includes not only the debonding force of the interface but also the force necessary for the plastic deformation of the film. Based on the pure bending of an elastoplastic beam and slender beam theory, Kim and coworkers [2-6] calculated the work expenditure rate caused by plastic deformation of the film, Ψ , and derived the interfacial fracture energy, γ. Their result showed that the plasticity effect can be very important and that the peel strength varies sensitively on the thickness and yield stress of the film, and the compliance of the substrate. A critical factor that determines Ψ is the maximum curvature, K_B, which is generally difficult to know.

In the present work, the plasticity effect in the peel test was studied for a Cu/Cr/polyimide system. Effects of the metal film thickness and the Cr/polyimide interface energy on the peel strength were studied by varying the film thickness and the pretreatment condition of the polyimide surface by rf plasma, respectively. Then, the amount of work expenditure rate during the plastic deformation was estimated from the X-ray peak broadening experiments of peeled metal films, and the elastic-perfectly plastic beam analysis. The plastic strain remaining in the film, ε^p, was found from the broadening of Cu(331) peak and a calibration plot based on the uniaxial tension data, which allowed us to estimate K_B, and ultimately Ψ.

2. EXPERIMENTAL PROCEDURE

The alumina substrate covered with 13 μm thick polyimide was prepared by spin coating polyamic acid(PAA), PIQ, over the Cr sputter coated alumina, and by following appropriate curing processes [7]. The polyimide surface was pretreated by Ar+ rf plasma for 300 seconds in 1×10^{-2} torr under the plasma density of 0.030, 0.036, 0.05 and $0.24W/cm^2$ before the following depositions of Cr and Cu onto polyimide. Here, Cr is a 20nm thin interlayer used to improve the adhesion between copper and polyimide, or polyimide and alumina [8]. The Cu layer was sputter deposited up to 480 nm, and thickened to 17, 34, 51, 85, 119, and 170μm respectively, by electroplating in the cupric sulfate solution [7]. Then, the peel strength was measured by conducting the $90°$ peel test at the crosshead speed of 2mm/min [7].

Stress-strain curves were obtained by conducting uniaxial tension tests using specimens made of Cu films according to ASTM E 345-87. The work expenditure during the plastic deformation was estimated from the X-ray peak broadening measurement of peeled metal films. The full width at half maximum, *FWHM*, of the Cu(331) peak was measured after separating

Mat. Res. Soc. Symp. Proc. Vol. 436 © 1997 Materials Research Society

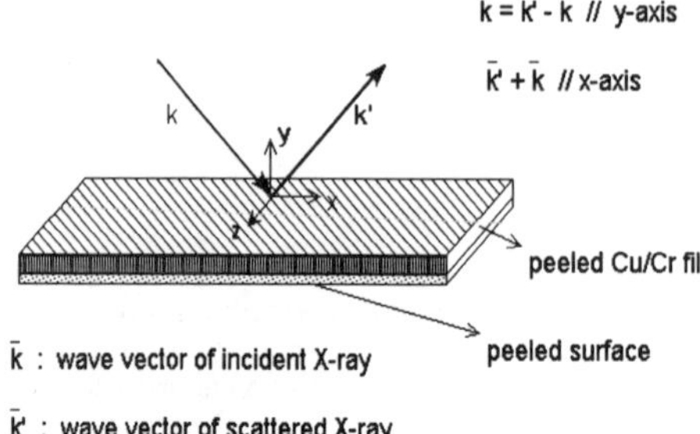

Fig. 1. Alignment of peeled Cu/Cr film with respect to X-ray beam.

the unresolved K_α doublet by the Rachinger method [9]. The specimen alignment with respect to the incident and diffracted X-ray beams is schematically described in Fig.1 where the coordinate center is located on the neutral plane of the metal film.

In order to eliminate the particle size or instrumental peak broadening, *FWHM* of untested Cu/polyimide specimens, *FWHM*°, was subtracted from *FWHM* of specimens after the peel or tension test to give the net increase of *FWHM*, $\triangle FWHM$, due to the plastic deformation. A calibration plot relating $\triangle FWHM$ to the plastic strain ε_{yy}^p was made using the uniaxial tension specimen which has the dimension of 3mm x 60mm x varying thickness. After unloading, ε_{xx}^p was found from the stress-strain curve by subtracting the elastic strain from the total strain, and related to ε_{yy}^p assuming the uniform deformations of the incompressible material ($\nu = \frac{1}{2}$). In the case of peeled metal film, the X-ray data was regarded to originate from the top surface($y = \frac{t}{2}$), and the ε_{yy}^p ($y = \frac{t}{2}$) read from the calibration plot was related to ε_{xx}^p ($y = -\frac{t}{2}$) assuming beam bending under the plane strain condition.

3. RESULTS

3.1 Uniaxial tension tests

The uniaxial stress-strain (σ-ε) curves of Cu thin films of varying thickness, t, are presented in Fig. 2. The stage I work hardening is virtually absent at all thickness levels, and the curve is well described as parabolic ($\sigma = A\varepsilon^N$) at large t, but better described as an elastic-perfectly plastic type at low t.

As mentioned in the previous section, a calibration plot relating $\triangle FWHM$ to ε_{yy}^p was obtained from the X-ray analysis of the uniaxial tension specimens based on the actual measurement of ε_{xx}^p , and Fig. 3 shows variations of $\triangle FWHM$ with ε_{yy}^p at the four specimen thickness levels. For all the specimens except the ones with the metal thickness of 17 μm, $\triangle FWHM$ is well described by a single parabolic curve;

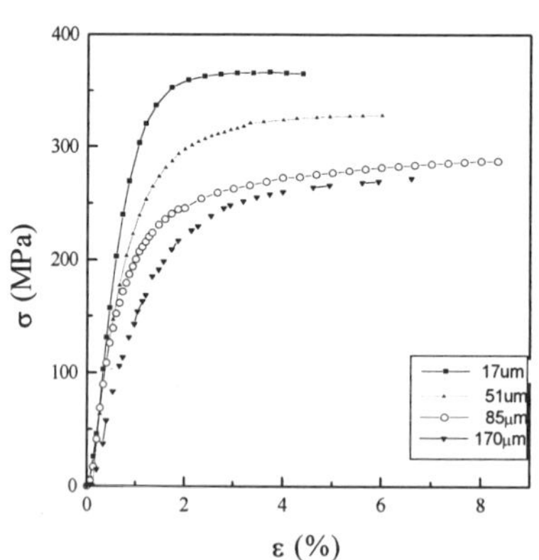

Fig. 2. The uniaxial stress-strain curves of Cu films of various thickness.

$$\Delta FWHM(\deg ree) = 2.521 \mid \varepsilon_{yy}^{p} \mid^{0.708} \text{----------} (1)$$

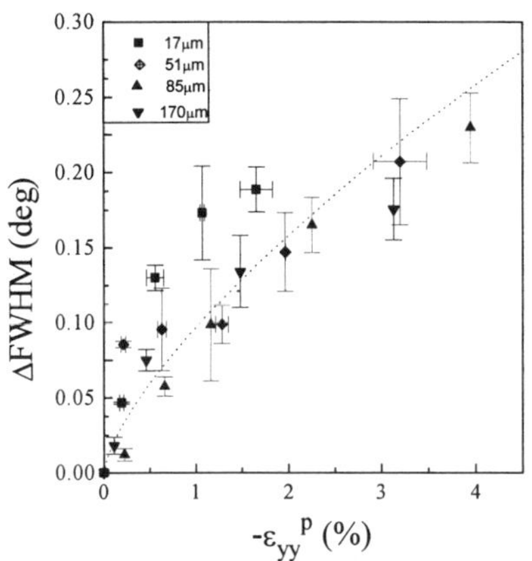

Fig. 3. A master curve relating ΔFWHM of Cu(331) peak to plastic strain of Cu films after the uniaxial tension test.

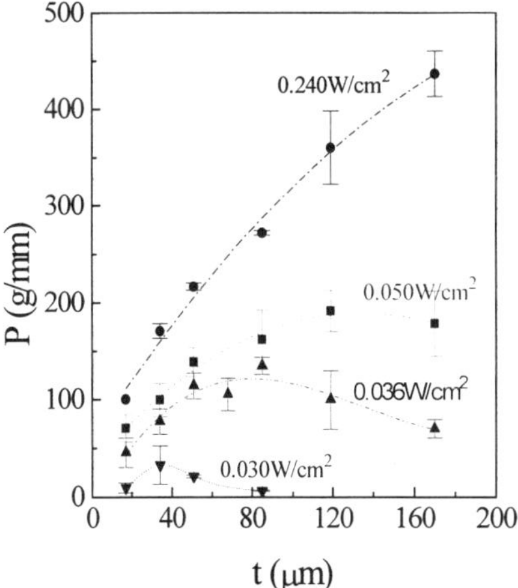

Fig. 4. The peel strength, P, as a function of the metal layer thickness for various pretreatment conditions. All the pretreatments were done for 5 minutes at the indicated plasma power density.

Later, this equation is used to find ε_{yy}^{p} in the peeled metal films, sometimes by extrapolation. It is not clear at the present moment why the specimens with the thickness of 17 μm showed higher $\Delta FWHM$ for given $\mid \varepsilon_{yy}^{p} \mid$.

3.2 Effects of the film thickness on the peel strength

The peel strength data of Cu/Cr/polyimide specimens under varying metal thickness and pretreatment conditions are presented Fig. 4. It becomes clear that the peel strength is very strongly affected by the metal thickness and the pretreatment conditions. The four pretreatment conditions were introduced to change the interfacial fracture energy, γ. Pretreating the polyimide surfaces by rf plasma before the sputter depositions of Cr is known to increase the peel strength by activating the C or O bonds of polyimide to form more carbide or oxide bonds with Cr[14]. However, we do not expect the presumed increase of γ can account for such a huge difference in the peel strength by itself. It appears that the amount of plasticity accompanying the peel test is strongly affected by γ.

The peel strength increases with metal thickness(t) at small t, reaches maximum at an intermediate t, t_{max}, and ultimately decreases at large t. The trend is quite clear for specimens that were treated under the rf plasma density of 0.03 and 0.036 W/cm^2 for 5min. , but not quite so at higher plasma density because the t_{max} becomes larger than the largest film thickness studied. Here, two completing factors seem to be at work. One is the crystal volume effect, and the other the thickness effect. Obviously the amount of plastic work should become very small as the film thickness approaches zero. However, the peel strength required for the plastic deformation of the film decreases with film thickness because it becomes more difficult to bend and thereby causing smaller amount of plastic bending deformation. Thus the crystal volume dominates the thickness effect for $t < t_{max}$, and the reverse is true for $t > t_{max}$. Softening of the matrix with the film thickness(cf. Fig. 2) tends to increase

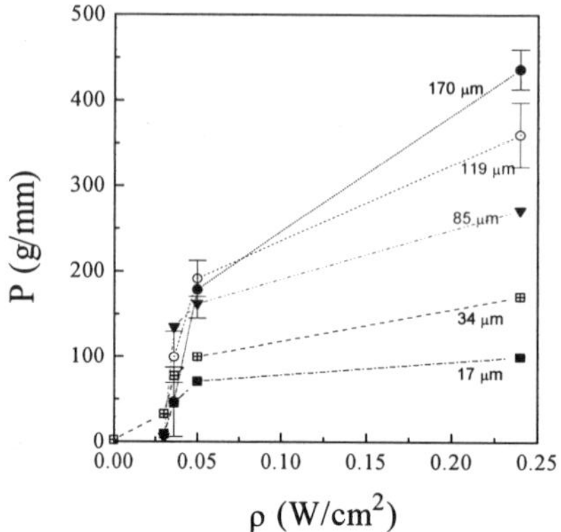

Fig. 5. The peel strength as a function of the rf plasma power density, ρ, for various metal layer thickness.

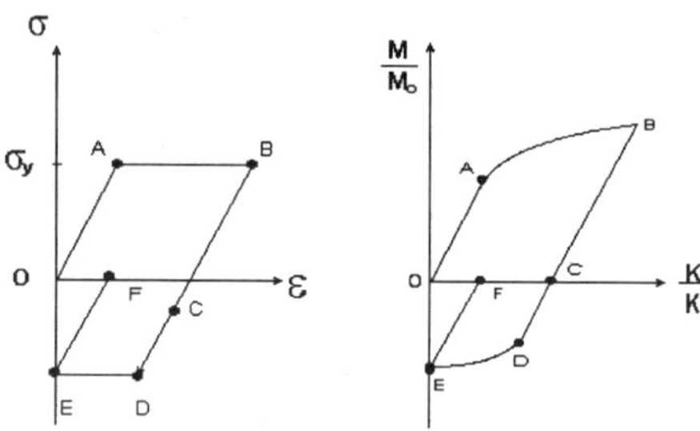

Fig. 6. Moment-curvature distribution and stress-strain curve for the steady state peeling of an elastic perfectly plastic material after Kim and Aravas.[2]

the peel strength, but seems to be dominated by other factors.

Variations of the peel strength with the rf plasma density, ρ, is shown is Fig. 5 at several film thickness levels. For $\rho > 0.05$ W/cm², the peel strength increases mildly with the power density with a rate that increases with the film thickness. Since the interfacial fracture energy is independent of the film thickness by nature, we suppose that γ increases very mildly with ρ and that is responsible for a large increase of plastic deformation and peel force.

3.3 An estimation of the work expenditure ψ.

We estimated the work expenditure by the following procedure ;
a) approximate the stress strain curve of Fig. 2 as elastic-perfectly plastic by taking the ultimate tensile stress in the figure as the yield stress σ_Y, b) regard the measured X-ray data to belong to the stage C of the M-K curve(cf. Fig.6), c) read $\varepsilon_{xx}^p(y=-\frac{t}{2})$ from Fig. 3, and find K_C from eq. (2), d) use $K_B = K_C + 1$, and to find ψ using eqs. (35)-(39) of the reference 3.

According to Kim and Aravas [2], beam bending is the dominant mode of deformation during peeling, and the work expenditure rate per unit advance of crack other than γ during the steady state peeling is the area surrounded by the loading path OABCDEO in the moment curvature distribution shown in Fig. 6.

Variations of $\Delta FWHM$ of the specimens that were pretreated at 0.036 W/cm² and 0.24 W/cm² are shown in Fig. 7. Note that $\Delta FWHM$ is smaller at the lower plasma power density suggesting that the amount of work expenditure was smaller there for a given film thickness. For the specimens with 0.24 W/cm², $\Delta FWHM$ was more or less constant suggesting that $\varepsilon_{yy}(\frac{t}{2})$ did not vary significantly with the film thickness. Since ε_{xx} is related to the film curvature K by

$$\varepsilon_{xx}(y) = -K\,y \quad \text{------------------------ (2)}$$

thinner specimens were bent more and expected to show higher peel strength. A comparison with the peel strength data of Fig. 4 suggests that the crystal volume effect dominates the thickness effect for these specimens. On the other hand, $\Delta FWHM$ of the specimens pretreated under 0.036 W/cm² was lowest for the thickest film, and the presence of a maximum is clear from this data. Here, we find stronger thickness effect than the previous one and the peel

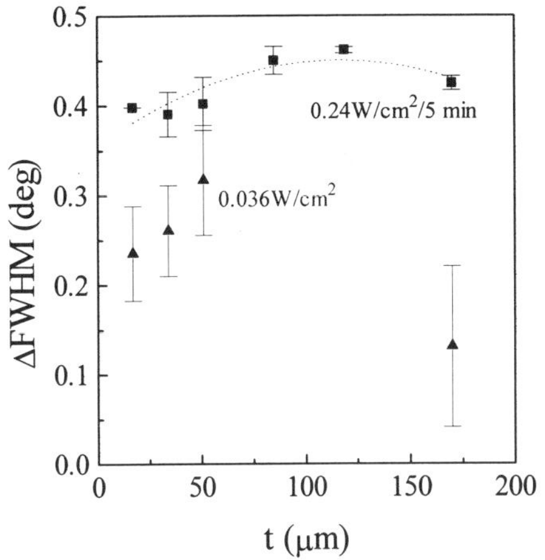

Fig. 7. ΔFWHM of peeled Cu/Cr/polyimide films pretreated under the rf plasma power density of 0.24W/cm2 for 5 minutes as a function of the metal layer thickness.

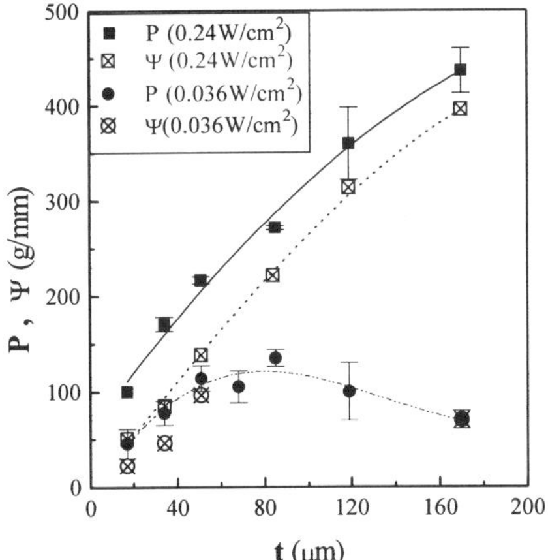

Fig. 8. Measured peel strength(P) and estimated work expenditure(Ψ) at the two pretreatment condition.

strength data with a maximum is what we expected.

Finally, we come up with the estimate work expenditure and compare it with the measure peel strength in Fig. 8. Compared to the approximation made in the estimation procedure, the agreement between the peel strength and ψ is good. Note that the estimated work expenditure is reasonably close to the peel strength data except the their specimens($t < 40$ μm). If the errors which involved in the interpretation of the X-ray data are corrected, the agreement can be improved. This results show that the measured peel strength is more a measure of the plastic deformation occurring during the peel test rather than a measure of the interfacial fracture energy.

4. CONCLUSIONS

In order to understand the effects of plastic deformation and interfacial fracture energy on the peel strength, thickness of the metal film and the pretreatment condition of polyimide surface were varied. The results indicate that ;

1. The peel strength increases with the film thickness, reaches a maximum at t_{max}, and decreases, which is clearer for specimens pretreated at lower plasma power density. This can be reasonably explained by the competition of the crystal volume effect and the thickness effect.

2. The peel strength is strongly affected by the pretreatment of polyimide, and presumably by the interfacial fracture energy. The pretreatment effect is magnified with the increase of the film thickness.

3. The work expenditure rate during the peel test was estimated using the stress-strain curves of Cu films at several thickness levels, the X-ray measurement of ε_{xx}^{p} in peeled metal films, and elastic-perfectly plastic beam analysis, which gives a reasonable estimate of the peel strength. This indicates that the peel strength is more a measure of the plastic deformation occurring in the peel test rather that a measure of the interfacial fracture energy.

REFERENCE

1. A. G. Evans, Harvard University, DAS report Mech-274, 1995.
2. K. S. Kim and N. Aravas, Inst. J. Solids Struct. 24, 417(1988)
3. K. S. Kim and J. Kim, Trans. ASME, J. Eng. Mater. Tech. 110, 266(1988)
4. N. Aravas, K. S. Kim and M. J. Loukis, Materials Sci. and Eng., A7, 159(1989)
5. K. S. Kim, Mat. Res. Soc. Symp. Proc., 119, 31(1988)
6. J. Kim, K. S. Kim and Y. H. Kim, J. Adhesion Sci. Technol., 3, 175(1989)
7. I. S. Park, Ph.D. Thesis, to be published.
8. T. G. Chung, Y. H. Kim and Jin Yu, J. Adhesion Sci. Technol., 8, 41,(1994)
9. Harold P. Klug and Leory E. Alexander, X-Ray Diffraction Procedure, 2nd ed.(JOHN WILEY & SONS, New York, 1974) p625
10. Leon I. Maissel and Reinhard Glang, Handbook of Thin Film Technology, (McGRAW HILL BOOK COMPANY, 1983), p.12-44.
11. J. S. Yang, Ph.D. Thesis, Seoul National University, 1995.
12. R. F. Bunshah, Deposition Technologies for Films and Coatings, (NOYES PUBLICATIONS,1982) p385
13. N. J. Chou and C. H. Tang, J. Vac. Sci. and Technol., A2, 751(1984)
14. J. L. Jordan, P. N. Sanda, J. F. Morar, C. A. Kovac, F. J. Himpsel and R. A. Pollak, J. Vac. Sci. and Technol., A4, 1046(1986)
15. C. J. McMahon and Jr, V. Vitek, Acta Met., 27,509(1979)
16. J. R. Rice and R. Thomson, Phil. Mag., 29, 73(1974)
17. I. S. Park and Jin Yu, to be published.

Part III

Nanoindentation of Films
and Surfaces

FINITE ELEMENT STUDIES OF THE INFLUENCE OF PILE-UP ON THE ANALYSIS OF NANOINDENTATION DATA

A. BOLSHAKOV*, W.C. OLIVER** and G.M. PHARR*
*Department of Materials Science, Rice University, 6100 Main St., Houston, TX 77005
**Nano Instruments, Inc., 1001 Larson Drive, Oak Ridge, TN 37830

ABSTRACT

Methods currently used for analyzing nanoindentation load-displacement data give good predictions of the contact area in the case of hard materials, but can underestimate the contact area by as much as 40% for soft materials which do not work harden. This underestimation results from the pile-up which forms around the hardness impression and leads to potentially significant errors in the measurement of hardness and elastic modulus. Finite element simulations of conical indentation for a wide range of elastic-plastic materials are presented which define the conditions under which pile-up is significant and determine the magnitude of the errors in hardness and modulus which may occur if pile-up is ignored. It is shown that the materials in which pile-up is not an important factor can be experimentally identified from the ratio of the final depth after unloading to the depth of the indentation at peak load, a parameter which also correlates with the hardness-to-modulus ratio.

INTRODUCTION

This paper deals with effects of pile-up on the measurements of contact area, hardness and effective modulus from nanoindentation load-displacement data using the Oliver-Pharr method [1]. The effects are investigated using finite element simulation of the indentation of an elastic-plastic half-space by a rigid conical indenter with a semi-vertical angle of 70.3°, which gives the same area to depth ratio as the Berkovich pyramidal indenter used frequently in nanoindentation experiments. The Oliver-Pharr method may be applied to the simulated indentation load-displacement data in the following manner. First, the unloading curve is fit to the power-law relation:

$$P = B\left(h - h_f\right)^m \tag{1}$$

where P is the load, h is the displacement, h_f is the final displacement after complete unloading, and B and m are power law fitting parameters. Next, the unloading stiffness, S, is found as the derivative of Eqn.1 at the maximum depth of indentation, $h = h_{max}$, using:

$$S = dP/dh(h = h_{max}) = Bm(h_{max} - h_f)^{m-1} \tag{2}$$

The projected contact area, A, hardness, H and effective modulus of the material, E_{eff}, (assuming that the indenter is rigid), are then determined using:

$$A = \pi\left(h_{max} - 0.75P_{max}/S\right)^2 \tan^2(\phi) \tag{3}$$

Mat. Res. Soc. Symp. Proc. Vol. 436 © 1997 Materials Research Society

$$H = \frac{P}{A} \qquad (4)$$

$$E_{eff} = \frac{E}{1 - \nu^2} = \frac{S\sqrt{\pi}}{2\sqrt{A}} \qquad (5)$$

where E and ν are the Young's modulus and Poisson's ratio of the material and ϕ is the semi-vertical angle of the cone [1].

The attractiveness of this approach is that no direct measurements of contact areas are needed, thus facilitating the measurement of mechanical properties from very small indentations. Clearly, however, the accuracy of the method depends on how well Eqns.3-5 describe the deformation behavior of the material. In this regard, it is important to note that Eqns.3-5 were obtained from a purely elastic contact solution developed by Sneddon [2], and thus may not work well for elastic/plastic indentation. One reason for this is that in Sneddon's elastic solution, the material around the indenter always sinks in, while for elastic/plastic indentation, there may be either sink-in or pile-up. Therefore, it is not surprising that method has been found to work well for hard materials, as shown in [1], but one may expect errors in the application of the method to softer materials (see [3], for example).

Here, we investigate by finite element methods the influences that pile-up has on the accuracy with which hardness and modulus can be measured by nanoindentation methods. The key parameter in the investigation is the contact area, which may be determined either from application of Eqn.3 to the simulated indentation load-displacement data (the Oliver-Pharr method), or by direct examination of the contact profiles in the finite element mesh. The two areas obtained from the finite element calculations can be substituted into Eqns.4-5 to obtain two separate estimations of the hardness and elastic modulus. If the Oliver-Pharr method outlined in Eqns.1-5 works well, then one should expect good agreement between the area and hardness obtained by the method and by finite element analysis. Also, the estimation of effective modulus obtained by Eqn.5 should match well with the value for effective modulus that was used as input in the finite element program. What will be shown is that the measurement of H and E using Eqns.3-5 works well when there is little pile-up, but when the pile-up is large, significant errors can result.

FINITE ELEMENT PROCEDURES

Elastic/plastic indentation was simulated using the axisymmetric capabilities of the ABAQUS finite element code. The indenter was modeled as a rigid cone with a semi-vertical angle of 70.3°. All simulations were performed to the same depth, $h_{max} = 500$ nm, using a finite element mesh similar to one used previously [3].

The material examined in the simulation had a Young's modulus $E=70$ GPa and Poisson's ratio of $\nu = 0.25$. The von Mises yield criterion with isotropic hardening was used to model the material behavior, and no friction was assumed between the punch and material. The yield stress input in the code was varied between $\sigma_y = 114.25$ MPa and $\sigma_y = 26.62$ GPa to simulate different materials. Two separate cases of strain hardening were considered, one with no strain hardening, that is, a work hardening rate $\eta = d\sigma/d\epsilon = 0$, (i.e., the material was assumed to be elastic-perfectly plastic) and the other with a work hardening rate of

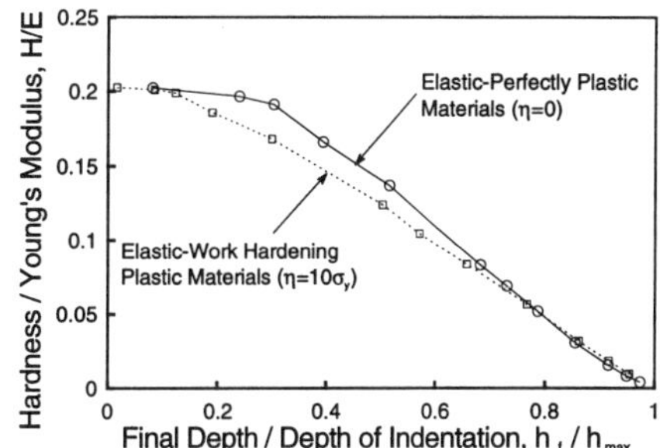

Fig.1. The dependence of H/E on h_f/h_{max} for the two different types of materials under consideration.

$\eta = 10\sigma_y$. These yield strengths and work hardening rates cover a wide range of metals, ceramics and polymers.

Note that in the finite element analysis, the value of the Poisson's ratio was kept constant at $\nu = 0.25$. The influence of the Poisson's ratio on conical indentations in elastic-plastic materials will be reported at a later date (see [6]).

In the course of our finite element study, we found a convenient, experimentally measurable parameter which can be used to identify the expected indentation behavior of a given material. This parameter is the ratio of final depth, h_f, (i.e. depth of the indentation impression after unloading) to the depth of the indentation at peak load, h_{max}. Note that this parameter, h_f/h_{max}, can be easily extracted from the unloading curve in a nanoindentation experiment. Furthermore, because the conical indenter has a self-similar geometry, h_f/h_{max} does not depend on the depth of indentation in a given material. The natural limits for this parameter are $0 \leq h_f/h_{max} \leq 1$. The lower limit corresponds to the fully elastic case, whereas $h_f/h_{max} = 1$ corresponds to the case of no elastic recovery after unloading. In the next section, a parametric study of the dependencies of different measured material properties on h_f/h_{max} will be presented which reveals the importance of this parameter for nanoindentation experiments.

FINITE ELEMENT RESULTS AND DISCUSSION

We start the discussion with Fig.1, which shows the dependence of the hardness to Young's modulus ratio, H/E, on h_f/h_{max}. In computing the data used for the plot, h_f and h_{max} were taken directly from the load-displacement curves generated in the finite element simulations, H was computed as the indentation load divided by the actual projected contact area at peak load determined by examination of the finite element mesh, and the value of E was that used as input for the finite element calculations. The H/E ratio is thus that of the simulated material rather than the value that would be derived from an analysis of the load-displacement data by the Oliver-Pharr method.

An examination of the curves in Fig.1 reveals that:

$$\lim_{h_f/h_{max} \to 1} H/E = 0 \tag{6}$$

$$\lim_{h_f/h_{max} \to 0} H/E = 0.207 \tag{7}$$

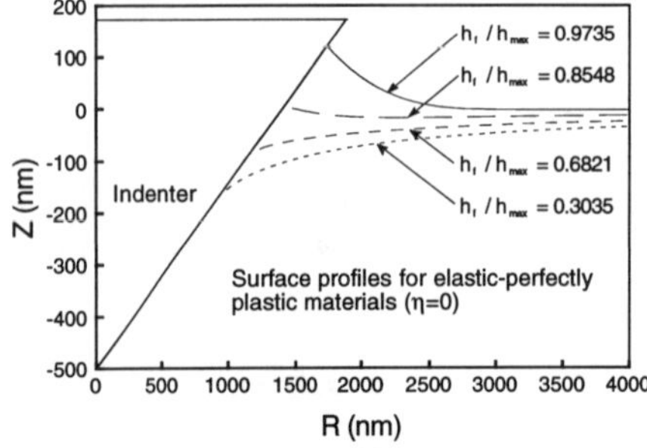
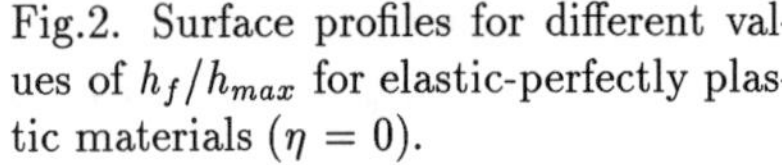

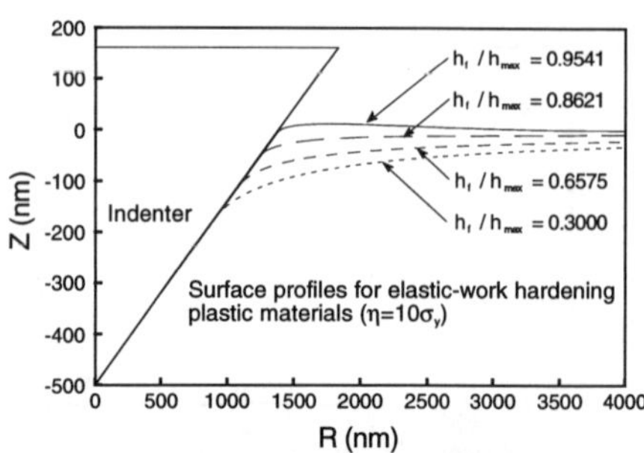

Fig.2. Surface profiles for different values of h_f/h_{max} for elastic-perfectly plastic materials ($\eta = 0$).

Fig.3. Surface profiles for different values of h_f/h_{max} for elastic-work hardening plastic materials ($\eta = 10\ \sigma_y$).

The first limit represents the case of rigid/plastic deformation, for which there is no elastic recovery during unloading and $h_f = h_{max}$. The second limit (at $h_f/h_{max} = 0$) represents purely elastic behavior. The limiting value $H/E = 0.207$ is a little surprising, since Sneddon's analysis for elastic indentation by a rigid cone gives $H/E = 0.191$. (Note that in deriving this result, it is assumed that the hardness is the indentation load divided by the projected contact area at peak load rather than the projected area of the residual hardness impression.) However, in a separate study [5], we have recently shown that the difference between Sneddon's solution and the finite element result is due to the fact that Sneddon's analysis applies to small deformations while the finite element code accounts for finite deformations. When the effects of finite deformations are included, the finite element calculations give the right limits at both ends of the curves.

Note that if for a given material the ratio h_f/h_{max} is known, then one can make predictions about the value of H/E based on the curves shown in Fig.1. Furthermore if it is assumed that two curves represent extreme cases and that data for most real materials lie between them, then a mid-line can be drawn between the two curves which can be used to estimate the H/E ratio from a known value of h_f/h_{max}. An error analysis showed that the use of such a curve for H/E determination will have less than a 15% error for $0.8 < h_f/h_{max} < 1$ and less than a 7% error for $0.0 < h_f/h_{max} < 0.8$.

Figures 2 and 3 show the surface profiles for the two different types of materials under consideration. Examination of these figures reveals that the amount of pile-up or sink-in depends on the amount of work hardening as well as on the value of h_f/h_{max}. More specifically, the amount of pile-up is large only when h_f/h_{max} is close to one and the amount of work hardening is small. It should also be noticed that for $h_f/h_{max} < 0.7$, very little pile-up is found no matter what the work hardening behavior of the material is.

These observations are particularly important when one considers the results of the contact area estimations shown in Fig.4. The contact area is normalized with respect to the shape function (or area function) of the indenter, that is, the projected area when no pile-up or sink-in occurs. From this figure one may see that since the Oliver-Pharr method is based on an elastic analysis, it fails to predict the correct contact areas for materials in which pile-up is important. Furthermore, when $h_f/h_{max} > 0.7$, the accuracy of the Oliver-Pharr method depends significantly on the amount of work hardening in the material. If

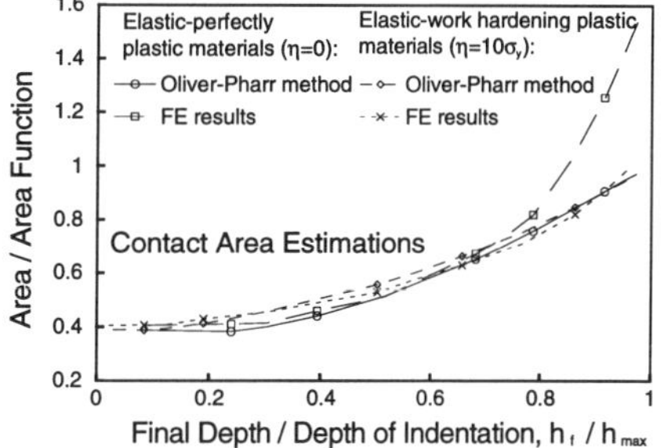

Fig.4. Contact areas as obtained by the Oliver-Pharr method and from finite element calculations.

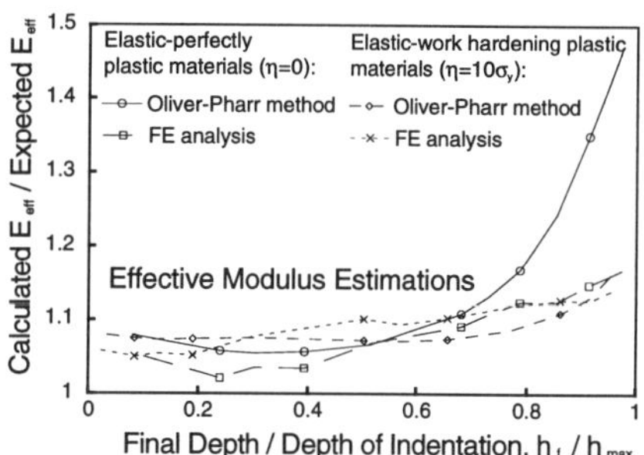

Fig.5. Estimations of effective modulus according to Eqn.5 with contact areas obtained by the Oliver-Pharr method and from finite element calculations.

the material is elastic-perfectly plastic, the Oliver-Pharr method underestimates the contact area by more then 50% at $h_f/h_{max} = 0.9735$. On the other hand, contact areas for materials with a large amount of work hardening are predicted very well by the method. Note that from the experimental point of view, it is not possible to predict if a material work hardens based solely on the load-displacement data. Therefore, in an indentation experiment, care needs to be exercised when $h_f/h_{max} > 0.7$, since use of the Oliver-Pharr method can lead to large errors in the contact area. On the other hand, when the pile-up is small, i.e., $h_f/h_{max} < 0.7$, then, according to Fig.4, the contact areas given by the Oliver-Pharr method match very well with the contact areas obtained from the finite element analysis, independent of the amount of work hardening. Since the hardness of the material is given by $H = P/A$, where P is measured during the experiment, the inaccuracies in the contact area will lead to similar inaccuracies in the hardness (in a reciprocal sence).

The same effects may be seen in effective modulus calculations as shown in Fig.5. The effective modulus was calculated using Eqn.5 with two different estimations for the contact area (the Oliver-Pharr method and the finite element mesh). These calculated effective moduli are normalized with respect to the effective modulus used as input in the finite element simulations (the expected E_{eff}). As in the case of the hardness, the effective modulus will be overestimated when pile-up is significant.

Another noteworthy feature of Fig.5 is that even when the amount of pile-up is negligible, that is, materials with a high yield stresses and/or high work hardening (characterized by $h_f/h_{max} <0.7$), the Oliver-Pharr method overestimates the effective modulus by 2 to 9%. As shown in Fig.5, this overestimation is predicted even at values of h_f/h_{max} close to 0. In fact the value of effective modulus at $h_f/h_{max} = 0$ is overestimated by 6 to 9%, depending on the method used in calculation. This problem is addressed in [5], in which it is shown that corrections to Sneddon's solution are needed in order to account for finite deformations and rotations that are automatically taken into account in the finite element calculations. The magnitude of the correction depends on the Poisson's ratio, ν, and the cone angle, ϕ. According to [5], the correction will reduce the estimated effective modulus by 8.53%, which is very close to what is needed to correct the moduli deduced from the finite element load-displacement data to the expected value. Thus, for $h_f/h_{max} < 0.7$, there is a simple explanation for the error in the modulus observed in the finite element results.

The same is not true, however, when $h_f/h_{max} > 0.7$. In this range, the effective modulus is overestimated by 10 to 16%, even if correct contact area is used in Eqn.5. This is due to the fact that the plastic properties of the material effect the unloading curves in such a way that the analysis can no longer be done by means of elastic solutions only.

CONCLUSIONS

The results of a finite element study have shown that pile-up can have important consequences for the measurement of contact area, hardness and effective modulus by nanoindentation methods. The parameter h_f/h_{max}, which can be measured experimentally, can be used as an indication of when pile-up may be an important factor. Pile-up is significant only when $h_f/h_{max} > 0.7$ <u>and</u> the material does not appreciably work harden. For such materials, failure to account for the pile-up can lead to an underestimation of the contact area deduced from nanoindentation load-displacement data by as much as 50%. This, in turn, results in an overestimation of the hardness and elastic modulus. When $h_f/h_{max} < 0.7$, or in all materials that strongly work harden, pile-up is not a significant issue, and currently existing nanoindentation data analysis procedures can be expected to give accurate results.

ACKNOWLEDGMENTS

This research was sponsored in part by the Division of Materials Science, U.S. Department of Energy, under contract DE-AC05-96OR22464 with Lockheed Martin Energy Research Corp., and through the SHaRE program under contract DE-AC05-76OR00033 with the Oak Ridge Associated Universities. One of the authors (GMP) is grateful for sabbatical support provided by the Oak Ridge National Laboratory.

REFERENCES

1. W.C. Oliver and G.M. Pharr, J. Mater. Res. **7**, p. 1564 (1992).
2. I.N. Sneddon, Int. J. Engng. Sci. **3**, p. 47 (1965).
3. A. Bolshakov, W.C. Oliver and G.M. Pharr, J. Mater. Res. **11**, p. 760 (1996).
4. D. Tabor, in <u>Microindentation Techniques in Materials Science and Engineering</u>, ASTM STP 889, P.J. Blau and B.R. Lawn, Eds., American Society for Testing and Materials, Philadelphia, 1986, pp. 129-159.
5. A. Bolshakov and G.M. Pharr, submitted to this conference proceedings.
6. A. Bolshakov, W.C. Oliver and G.M. Pharr, to be published.

INDENTER GEOMETRY EFFECTS ON THE MEASUREMENT OF MECHANICAL PROPERTIES BY NANOINDENTATION WITH SHARP INDENTERS

T.Y. TSUI*, W.C. OLIVER**, and G.M. PHARR*
* Department of Materials Science, Rice University, 6100 Main St., Houston, TX 77005
** Nanoinstruments Inc., 1001 Larson Drive, Oak Ridge, TN 37830

ABSTRACT

The measurement of mechanical properties by nanoindentation methods is most often conducted using indenters with the Berkovich geometry (a triangular pyramid) or with a sphere. These indenters provide a wealth of information, but there are certain circumstances in which it would be useful to make measurements with indenters of other geometries. We have recently explored how the measurement of hardness and elastic modulus can be achieved using sharp indenters other than the Berkovich. Systematic studies in several materials were conducted with a Vickers indenter, a conical indenter with a half-included tip angle of 70.3°, and the standard Berkovich indenter. All three indenters are geometrically similar and have nominally the same area-to-depth relationship, but there are distinct differences in the behavior of each. Here, we report on the application of these indenters in the measurement of hardness and elastic modulus by nanoindentation methods and some of the difficulties that occur.

INTRODUCTION

The measurement of mechanical properties such as hardness, H, and elastic modulus, E, by nanoindentation methods has been conducted largely with indenters having the Berkovich or spherical geometries [1-4]. The Berkovich indenter, a three-sided pyramid with the same area-to-depth ratio as the Vickers indenter commonly used in microindentation testing, is useful in experiments where full plasticity is needed at very small penetration depths, such as the measurement of hardness of very thin films and surface layers [1,2]. The spherical indenter, on the other hand, is useful when purely elastic contact or the transition from elastic to plastic contact is of interest [3,4].

Of the many sharp indenters used for microhardness testing, the Berkovich has proven the most useful in nanoindentation work. This is because the three sided pyramidal geometry of the Berkovich naturally terminates at a point, thus facilitating the grinding of diamonds which maintain their sharpness to very small scales. Berkovich tip defects, as characterized by the effective tip radius, are frequently less than 50 nm in many of the better diamonds. For the Vickers indenter, on the other hand, it is more difficult to maintain geometric similarity at such small scales because the square-based pyramidal geometry does not terminate at a point but rather at a "chisel" edge. The conical indenter is the most difficult to grind, and as a result, conical diamonds often have severe tip rounding.

Nevertheless, there are certain circumstances in which it may be useful to make nanoindentation property measurements with a Vickers or a conical indenter. The Vickers, for example, may be useful in measuring the properties of single crystals with 4-fold symmetry, or to directly compare hardnesses obtained in nanoindentation experiments with conventional Vickers microhardness results. In addition, the Vickers indenter is the primary indenter used in measuring the fracture toughness of brittle materials by the indentation cracking method [5]. The conical indenter has advantages when one wishes to avoid deformation phenomena caused by the sharp edges of the Berkovich and Vickers indenters, e.g., when cracking is to be avoided in brittle materials or when large strain gradients around the circumference of the contact complicate the understanding of indentation phenomena. The conical indenter is also more amenable to analysis than the Berkovich or Vickers; virtually all modelling of indentation contact by sharp indenters uses the conical geometry [6], and with very few exceptions [7], most finite element simulations of indentation by sharp indenters have used the conical geometry to take advantage of the axial symmetry [8-10].

Mat. Res. Soc. Symp. Proc. Vol. 436 © 1997 Materials Research Society

For these reasons, an elementary study to examine the problems that are encountered in using Vickers and conical indenters in nanoindentation testing was undertaken. The study focused on three sharp diamonds: a Berkovich, a Vickers, and a conical diamond with a half-included tip angle of 70.3°. All three have the same nominal area-to-depth relationship, thus removing this factor as a variable in the study and thereby simplifying the understanding of results. Here we present a select set of results for four materials chosen to represent the behavior of metals and ceramics. A much larger number of materials were studied, but space considerations prohibit us from including all of them. A more complete report is in preparation [11].

EXPERIMENTAL PROCEDURE

The four materials examined in the study were fused quartz, a well-annealed aluminum single crystal, a mechanically polished specimen of polycrystalline gold (large grains), and a single crystal of (001) sapphire. The first two materials are used as standards in nanoindentation testing for the determination of indenter area functions, and the latter two are included to illustrate specific points that help in understanding the indentation behavior of the three diamonds. Indentations were made in each of the materials at a variety of loads, and the indentation load-displacement behavior was analyzed to determine H and E using a variation on the method of Oliver and Pharr [2].

The variation was developed to account for recent observations that the machine compliance in a nanoindentation experiment may not be constant for a given machine, but rather may depend on how the specimen is mounted and the way in which the mount is secured to the nanoindentation system. As a result, a single, specimen-independent machine compliance cannot be used in precision work. This is not an important consideration for materials and experimental conditions for which the contact stiffness is small (i.e, contact stiffnesses much less than the machine stiffness), but can lead to significant errors in some experimental situations. The errors become particularly important when making large indentations (loads of 50 mN or more) in very soft materials like well-annealed aluminum. Since aluminum is a material we use as a standard in making area function determinations, we have altered the procedure by which the area function is measured to account for the influences of the mount on the machine compliance.

To briefly describe the procedure, a number of indentations are made to various peak loads up to 300 mN, and the areas, A, of the largest indentations are measured optically. The highest load indentations are large enough that their areas can be measured with accuracies of about ±5%. Using Young's modulus E=70 GPa and Poisson's ratio ν=0.34 (the values for polycrystalline aluminum), the machine compliance, C_m, is then found from :

$$C_m = C_T - \frac{\sqrt{\pi}}{2} \frac{(1-\nu^2)}{E} \frac{1}{\sqrt{A}} \quad , \tag{1}$$

where C_T is the total compliance measured in the experiment (see [2] for the origin of this equation). Using this value for C_m to evaluate the load-displacement data obtained at other loads, an area function is then established by solving Eqn.1 for A. Unfortunately, aluminum is not a good calibration material for indentation depths smaller than 500 nm because of scatter in the data, possibly caused by the oxide film. For this reason, the small-depth portion of the area function is established in separate experiments conducted on fused quartz, which can be conveniently indented to depths ranging from 5 to 1500 nm. The machine compliance for the fused quartz experiments is determined by a trial and error procedure in which a value of C_m is sought which, when applied to Eqn.1 and using E=72 GPa and ν=0.17 (the values for fused quartz), gives contact areas which match those of the aluminum data in the region of overlap (500 nm<h_c<1500 nm). After combining the two data sets, curve fitting procedures are used to produce an area function which applies over the range 20-7000 nm. This technique is used in preference to the procedure outlined in [2] because the latter involves an extrapolation which can lead to errors in C_m.

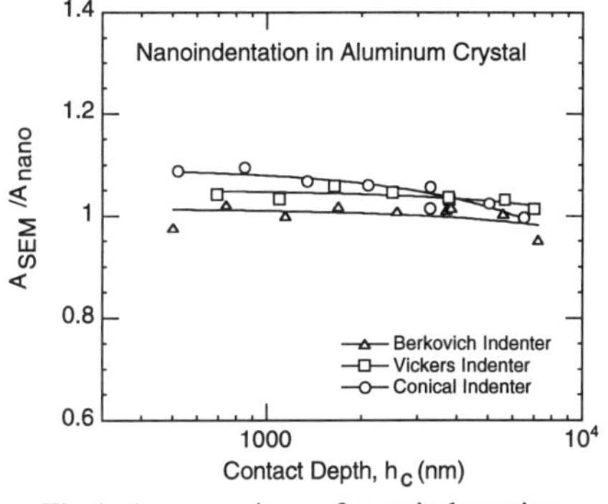
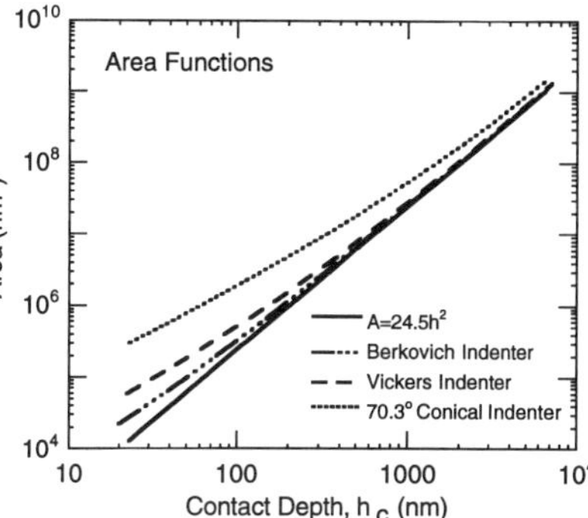

Fig.1. A comparison of nanoindentation areas with SEM measurements for Al.

Fig.2. Experimentally determined area functions for each of the indenters.

RESULTS AND DISCUSSION

The area function determination procedure just described was used in the current study to establish the area function of each of the three indenters. To check how well the procedure works, Figure 1 presents results for indentations in aluminum which compare the areas predicted from analysis of the nanoindentation load-displacement data , A_{nano}, to areas measured by direct examination of scanning electron microscopy (SEM) images, A_{SEM}. The measurements are limited to the indentation depths greater than 500 nm due to difficulties in obtaining clear, measurable images at smaller depths. Over the range of depths examined, the SEM areas for the Berkovich and Vickers indentations deviate from predictions of the area function by no more than 5%. For the conical indenter, the deviation is slightly larger, but only at small depths where it is difficult to make good measurements. Thus, the procedure for deriving the area functions appears to work reasonably well for each three indenters, at least over the range of indentation sizes which are convenient to check in the SEM.

The complete area functions are shown in Figure 2, where they have been evaluated over their range of applicability, i.e., h_c = 20 to 7000 nm. Also shown in the figure is the ideal area function, $A=24.5h^2$. The ideal area function is that which all three diamonds would exhibit if they were ground to perfect geometric form. The plot shows that while the geometry of the Berkovich indenter is in closest agreement with ideal geometry, the 70.3° cone shows significant deviations over most of the applicable contact range. The deviations are consistent with the notion that the 70.3° cone is very blunt and possibly more sphere-like than sharp at the scales of interest in this study. The Vickers indenter lies somewhere in between. As will be shown shortly, deviations from perfect geometry play an important role in understanding the results of the nanoindentation hardness measurements.

Measurements of elastic modulus and hardness for each of the four materials by each of the three diamond indenters are summarized in Figure 3. The elastic moduli are shown in the plots in the left-hand column. Not surprisingly, the moduli for aluminum and fused quartz are very close to the expected values, E = 70 GPa and 72 GPa, irrespective of indenter geometry. This, of course, must be the case, since these values were used as input in the procedure by which the area functions were established. The true test comes from the other two materials, gold and sapphire. For the gold data, all three indenters give approximately the same value for E. Furthermore, the measured modulus is independent of indentation depth and has a value which compares quite favorably with the known polycrystalline gold value, E=80 GPa [12]. Comparison of the sapphire data with actual values for E is a little more difficult because the modulus of sapphire varies depending on crystallographic orientation in the range 400-500 GPa [2]. However, inspection of the data in Figure 3 shows that all three indenters give depth-independent elastic moduli which fall generally in this range. Thus, based on these limited

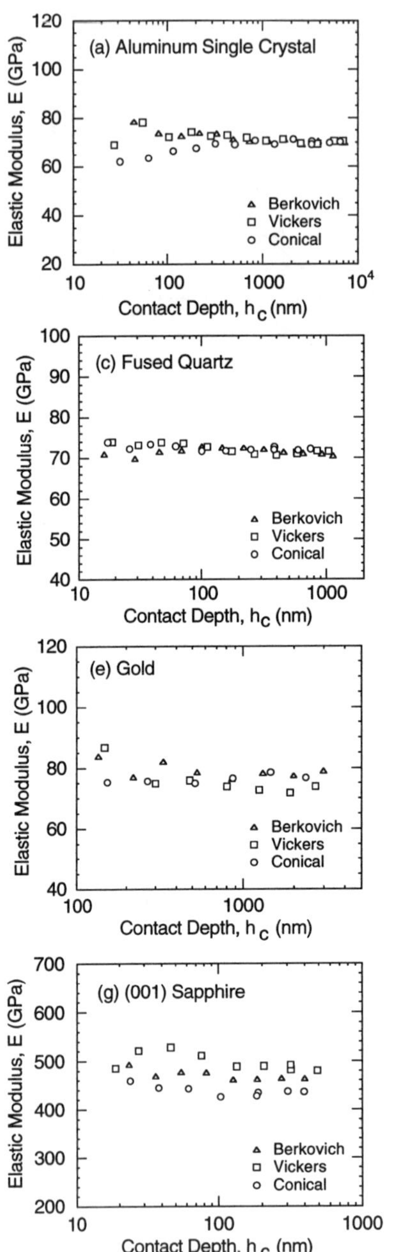
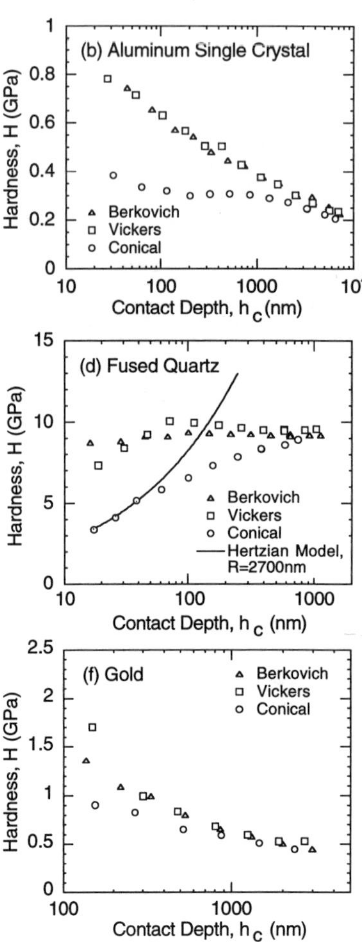

Fig.3. Nanoindentation measurements of elastic modulus and hardness for four materials using the Berkovich, Vickers, and 70.3° conical indenters.

observations, it appears that nanoindentation measurement of elastic modulus using sharp indenters other than the Berkovich is indeed possible. It should be noted, however, that accurate measurements can be achieved only when the deviations from the ideal indenter geometry are taken into account. Had the ideal area function been used, different results would have been obtained.

The hardness data shown in the right hand column of Figure 3 show some peculiar and interesting behaviors. For example, the aluminum data at the top of the column shows a significant indentation size effect (ISE) for the Berkovich and Vickers indenters, but not for the conical indenter. Based on SEM measurement of the actual contact areas (Fig.1), it is clear that the indentation size effect is real and not just an artifact of the data analysis procedures (at least for contact depths greater than 500 nm). Collectively, these data suggest that the ISE in aluminum may be caused by localized plastic deformation at the edges of the indenter. A model based on this concept has been developed by Ma and Clarke [13]. The gold hardness data exhibit behavior similar to that of the aluminum, with the exception that the hardnesses measured with the conical indenter are not quite as flat .

The hardness behavior of the fused quartz specimen is quite different. For this material, the Berkovich and Vickers indenters give very similar, depth-independent hardnesses, but there is a large increase in hardness with increasing depth for the data obtained with the conical indenter. Thus, it appears that there is a completely different indentation size effect in this material, one which runs contrary to most ISE observations reported in the literature (the usual observation is increasing hardness with decreasing depth), and one which is unique to the conical indenter. However, shortly we will show that this behavior is not real, but results from a transition from purely elastic to elastic/plastic deformation caused by the bluntness of the conical indenter.

The key to understanding the behavior of the fused quartz data was unveiled while testing the (001) sapphire. Hardness data for this material are presented in Figure 4, where the individual data points, rather than the averages plotted in previous graphs, are shown. A curious feature in the data is the division of the Vickers hardnesses at depths less than 100 nm into 2 distinct groups, one considerably higher than the other. Examination of the individual load-displacement curves revealed that the higher data points correspond to purely elastic indentation contact, as evidenced by loading and unloading curves which perfectly retrace themselves. For each of the lower data points, on the other hand, there was a plastic pop-in event similar to that reported by Page et al. [14]. Subsequent investigation of the data for the conical indenter revealed the same behavior, with the transition from purely elastic to elastic/plastic deformation occurring at slightly larger contact depths. Purely elastic indentation contact is promoted by tip blunting. The transition is not observed in the Berkovich data, presumably because the Berkovich is considerably sharper than the other two indenters.

The observation of purely elastic contact at small indentation depths for the Vickers and conical indenters has important consequences for the interpretation of hardnesses measured in nanoindentation experiments. Specifically, the hardness, H=P/A, obtained by applying the Oliver/Pharr analysis procedure to nanoindentation load-displacement data is not the traditional hardness based on the residual area of the hardness impression, but a different hardness based on the area of contact at peak load. In soft materials, these two hardnesses are essentially the same, but for hard materials they can be different, particularly when deformation during indentation is mostly elastic and recoverable. To amplify on this point, note that in the limit of purely elastic contact, the conventional hardness rises without bound, while that derived by the Oliver/Pharr procedure has a finite value. In fact, assuming that the blunting of indenters can be characterized by assigning a radius of curvature to the tip, one can compute from Hertzian contact theory what the Oliver/Pharr hardness should be for purely elastic contact. A simple analysis yields:

$$ H = \frac{4\sqrt{2}}{3\pi} \frac{E}{(1-v^2)} \left(\frac{h_c}{R}\right)^{1/2} \quad , \tag{2} $$

where R is the effective radius of curvature of the tip.

To show that this phenomenon is indeed the source of the decrease in hardness at low loads

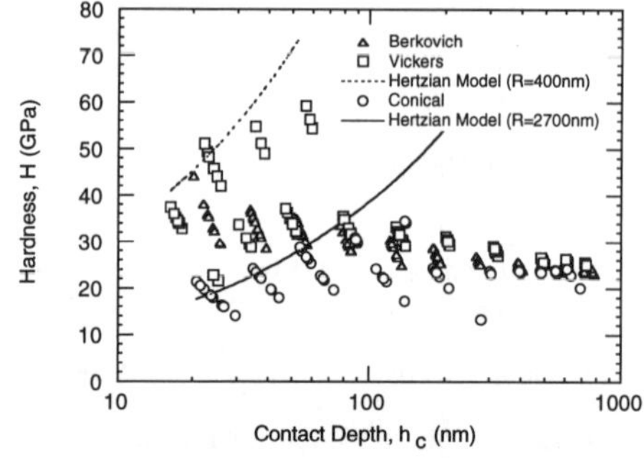

Fig.4. Nanoindentation hardness measurement of (001) sapphire using the Berkovich, Vickers , and 70.3° conical indenters.

in sapphire, Hertzian analysis was used to extract an effective tip radius from the low-load, reversible load-displacement data obtained for both the conical and Vickers indenters. The tip radii were found to be R=400 nm for the Vickers and R=2700 nm for the cone. Using these radii in Eqn. 2 along with reasonable estimates of the elastic constants, predictions based on the Oliver/Pharr definition of hardness and Hertzian contact theory are included in Figure 4, where it is seen that relatively good agreement is obtained with the experimental data points corresponding to purely elastic contact. Using R=2700 nm, the predictions of Eqn. 2 have also been computed for comparison to the conical indentation of fused quartz and are included in hardness data of Figure 3. Once again, there is relatively good agreement with the low-load data.

CONCLUSIONS

1. Using careful techniques to establish the area function of the indenter, it is possible to use conical and Vickers indenters in the measurement of elastic modulus by nanoindentation methods, even though these indenters suffer more from tip defects than the Berkovich indenter used more commonly in nanoindentation testing.

2. The tip defect confuses the interpretation of hardnesses measured by nanoindentation methods, since there is a change in the mode of deformation from purely elastic at low loads to elastic/plastic at high loads. This phenomena is more important in hard materials such as ceramics and glasses than it is in soft metals.

ACKNOWLEDGMENTS

This research was sponsored by the Division of Materials Sciences, U.S. Department of Energy, under contract DE-AC05-96OR22464 with Lockheed Martin Energy Research Corp., and through the SHaRE Program under contract DE-AC05-76OR00033 between the U.S. Department of Energy and Oak Ridge Associated Universities.

REFERENCES

1. G.M. Pharr and W.C. Oliver, MRS Bull. **XVII**, 28 (1992).
2. W.C. Oliver and G.M. Pharr, J. Mater. Res. **7**, 1564 (1992).
3. J.S. Field and M.V. Swain, J. Mater. Res. **8**, 297 (1993).
4. J.S. Field and M.V. Swain, J. Mater. Res. **10**, 101 (1995).
5. G.R. Anstis, P. Chantikul, B.R.Lawn, and D.B.Marshall, J. Am. Cer. Soc. **63**, 574 (1980).
6. K.L. Johnson, Contact Mechanics (Cambridge University Press, Cambridge, UK, 1985).
7. A.E. Giannakopoulos, P.L.Larsson & R.Vestergaard, Int. J. Solids Struct. **31**, 2679 (1994).
8. A.K. Bhattacharya and W.D. Nix, Int. J. Solids Structures **24**, 1287 (1988).
9. T.A. Larsen and J.C. Simo, J. Mater. Res. **7**, 618 (1992).
10. A. Bolshakov, W.C. Oliver, and G.M. Pharr, J. Mater. Res. **11**, 760 (1996).
11. T.Y. Tsui and G.M. Pharr, in preparation.
12. G. Simmons and H. Wang, Single Crystal Elastic Constants and Calculated Aggregate Properties: A Handbook, 2nd ed. (The MIT Press, Cambridge, MA, 1971).
13. Q. Ma and D.R. Clarke, J. Mater. Res. **10**, 853 (1995).
14. T.F. Page, W.C. Oliver and C.J. McHargue, J. Mater. Res. **7**, 450 (1992).

DISLOCATION NUCLEATION AND MULTIPLICATION DURING NANOINDENTATION TESTING

A.B.MANN AND J.B.PETHICA
Oxford University, Dept. of Materials, Parks Road, Oxford, OX1 3PH, UK.

ABSTRACT

Small volumes of crystalline material can exhibit near theoretical lattice strengths due to the absence of pre-existing defects. We show here that careful control of experimental parameters and surface preparation are important in observing these increased stresses. In particular, tip impact kinetic energies at the initial contact of only a few eV have been found to influence events occurring at a much later stage in the loading cycle. This can be understood in terms of the energy required to deform a small number of atomic scale asperities and thus release sufficient surface energy to permit dislocation nucleation at impact. If the K.E. is not sufficient to overcome the asperities, dislocations must be nucleated at a later stage in the loading cycle, giving a apparently different mechanical property. We outline the implications for thin film systems where a hard surface layer may play a vital role, firstly, in preventing dislocation nucleation on impact and, secondly, in impeding the propagation of any such dislocations into the bulk below.

INTRODUCTION

The processes involved in adhesion and tribology (friction and wear) when two bodies make contact are highly dependent on the initiation of permanent deformation. Recently high precision point-probe techniques, nanoindentation and Atomic Force Microscopy, have been used to examine the mechanical deformation occurring in single asperity size contacts, and hence, they have allowed the origin of inelastic deformation to be studied in volumes of crystalline material which are small enough that there are no pre-existing defects (under normal circumstances defects will tend to dominate the material's mechanical response). Nanoindentation testing has shown that such small volumes of material at a sample's surface can exhibit strengths approaching those predicted theoretically [1-5]. The mechanical response in these circumstances is initially elastic with the loading curve being almost totally reversible, however, at a critical load there is a sudden and pronounced increase in the depth of the indentation (called a "pop-in") which marks the point at which the material begins to deform inelastically.

In ceramics, Transmission Electron Microscopy has been used to examine indentations to peak loads both above and below that required for pop-in [1]. The results indicate that the pop-in is associated with the rapid generation of dislocations at or near to the sample's theoretical shear stress. A similar type of mechanical deformation has been observed in both FCC (e.g. Au, Ni) and BCC (e.g. Fe-3wt%Si, Mo, W) metals [3-7], but this seems to be heavily dependant on the presence of a hard surface film. This has been demonstrated by nanoindentation tests on Fe-3wt%Si (100) in which the surface oxide has been deliberately removed using aqueous HCl [3,4]; this leads to irreversible deformation for even the lightest of loads. Earlier work examining the effects of oxygen monolayers on Ni [6] and hydro-carbon contaminants on Au [7] have reinforced the view that even the thinnest of surface films can lead to a substantial increase in the load required to cause plastic yielding in metals.

In general, for any given experiment all of the studies described have shown a wide scatter of results. This is normally attributed to variations in surface properties resulting from sample preparation and the presence of surface films. In this paper we present results for GaAs (100) and (111) showing that if the kinetic energy (K.E.) of the tip when it first contacts the surface is reduced to less than 1 eV there is a dramatic reduction in this scatter and an overall increase in the pop-in load. We are able to explain this extreme sensitivity to the impact K.E. in terms of atomic-size asperities which for low energy impacts are able to support the indenter tip during the initial contact. However, impact K.E.s of just a few eV are sufficient to deform the smallest

153

asperities and, thus, allow surface forces to overcome the barrier presented by the asperities to intimate, roughness-free contact over a much larger area. In this way the surface forces, and the associated change in surface energy resulting from the contact, can cause the generation of defects on impact, and hence, permit the formation of dislocation loops either on contact or during the early stages of the indentation.

It is clear from this model that the mechanical strength of the surface asperities and the nature of the surface forces is of paramount importance in determining the measured nano-mechanical properties of a material. We are able to reinforce this with nanoindentation test results for W (100) indented, firstly, in air and, secondly, in aqueous HCl: in air the behaviour is directly analogous to that seen for GaAs with a clear dependence between the observed mechanical properties and the impact K.E.; in HCl the dependence on impact energy is not seen. This result implies that the surface oxide presents a barrier to defect generation on contact which must be overcome by the impact K.E. if defects (and dislocations) are to be generated on contact. If the impact energy is not sufficient dislocations must be nucleated at a much later stage in the indentation cycle.

EXPERIMENTAL RESULTS

To gain a better understanding of the mechanisms operating during nanoindentation testing we have undertaken a series of tests on crystalline GaAs (100), GaAs (111) and W (100). These have given a number of interesting results some of which have been presented elsewhere [8], here we emphasise two of them: the effect of the initial contact between tip and sample on the mechanical properties seen at a much later stage in the indentation; the role of the surface oxide in permitting near theoretical shear stresses to be reached during nanoindentation of tungsten.

Sample Preparation

The GaAs samples were prepared by cutting single crystals of GaAs with a diamond saw, then mechanically and chemically polishing them to give defect free (100) and (111) As-faced surfaces. The samples were then stored in a sealed container until required for nanoindentation testing. The W (100) sample was an annealed single-crystal which had been electro-polished using 5% NaOH in water at 2 V A.C..

Nanoindentation Testing of Gallium Arsenide

The nanoindentation testing of both the GaAs and W samples was performed using a Nanoindenter (Nanoinstruments Ltd., Knoxville, TN) the exact operation of which is documented elsewhere [e.g. 9]. During the testing the method used to make the indentations was varied by modifying the following: approach velocity; loading rate; unloading rate; hold periods; peak load (or depth); preparation of surface and tip; tip geometry; tip/sample orientation; ambient environment (by immersion in liquid or by varying the relative humidity in air). All of these parameters affected the shape of the indentation curves and the occurrence of pop-ins, but only one, the approach velocity, had a clear, well-defined and quantifiable effect.

Approach velocities of around 10 nm/s, equivalent to an impact K.E. of 2.3 eV, were found to give small pop-ins spread over a wide range of loads and occurring during both the loading and unloading cycles (see figures 1a,b & 2a). However, approach velocities of 4 nm/s or less, impact K.E.<0.4 eV, gave only a single, large and reproducible pop-in during the loading cycle (figures 1c & 2b). The reproducibility of this pop-in is highlighted by the statistics of table I.

Table I Pop-in statistics for GaAs (100) and (111) using an impact K.E.<0.4 eV. The statistics are based on 42 indentations for the (100) sample and 15 indentations for the (111) sample.

Sample	Pop-in Property: Mean Value (Standard Deviation)		
	Load/mN	Depth/nm	Size of Pop-in/nm
GaAs (100)	0.731 (0.059)	43.8 (3.5)	7.7 (1.1)
GaAs (111)	0.53 (0.14)	33.2 (8.6)	5.2 (1.5)

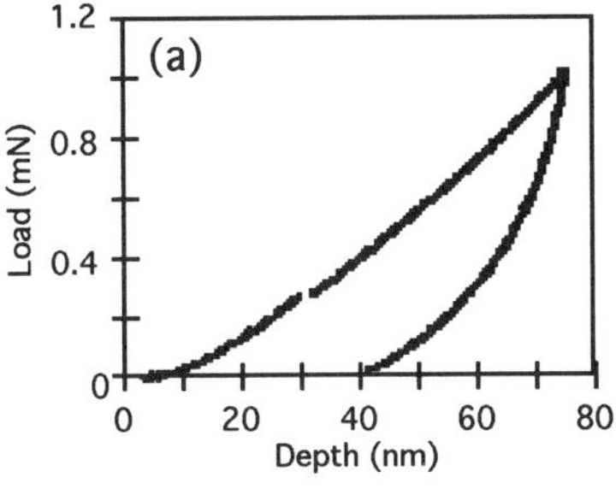

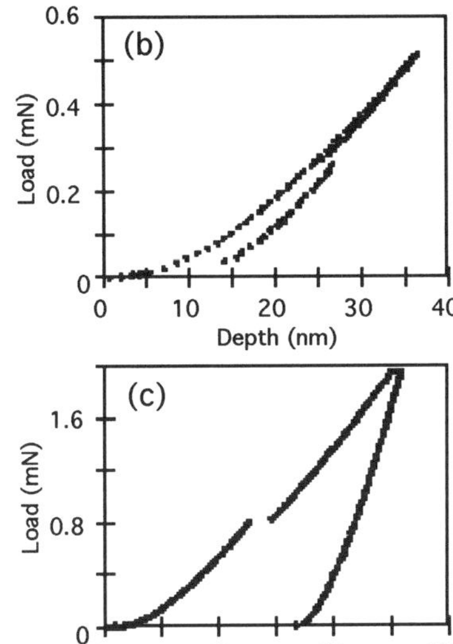

Figure 1 Nanoindentation curves for GaAs (100). The impact K.E.s used for (a) & (b) were 2.3 eV, while for (c) the impact K.E. was <0.4 eV. The higher energy impacts were found to cause small pop-ins during loading and occasionally unloading and, in general, to give permanent deformation at lower loads than those required for the lower energy impacts.

The pop-in statistics for the GaAs (111) sample have larger standard deviations than those of the GaAs (100) sample partly because of the number of indentations used in the analysis and partly because of the lower load and depth at which the pop-in occurs (this leads to a relatively larger error in the measured values of load and depth).

For both the GaAs samples the differences in the load/depth curves for the low and high K.E. impacts (figures 1 & 2) seem to imply that the deformation even at very lowest loads is not entirely elastic. In fact it appears that defects are generated on contact for the higher impact K.E.s. There is also evidence from the first few nms of the stiffness/depth and load/depth curves (figure 3) that the initial contact varies considerably with impact K.E., this is particularly pronounced for the contact stiffness at depths less than 10 nm. The differences in the initial contact might seem to imply there is a soft film present which is penetrated by the higher K.E./velocity impact, but this can be discounted as it would lead to an offset in the depth measurement which is not the case (for depths greater than 10 or 12 nm the load/depth and stiffness/depth curves converge for different approach velocities and only diverge again much later in the indentation).

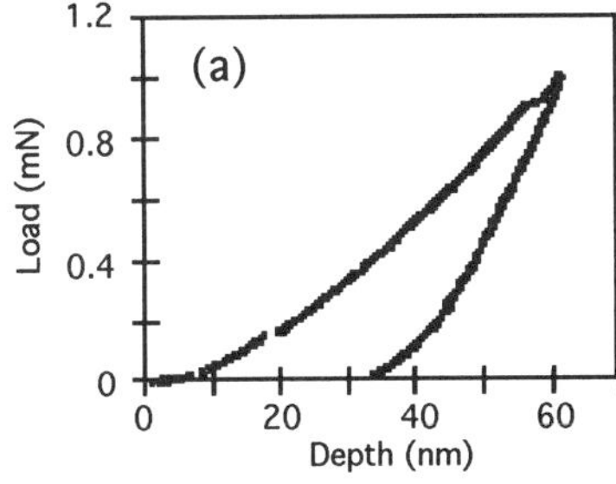

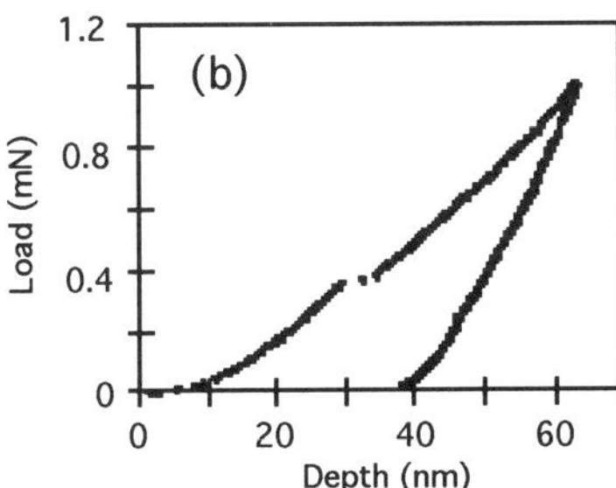

Figure 2 Nanoindentation curves for GaAs (111). For (a) the impact K.E. was.2.3 eV, while for (b) it was <0.4 eV. The curves show that a low K.E. impact gives a single large pop-in at a well defined load and depth, but higher K.E. impacts give several small pop-ins which occur over a wide range of loads and depths.

155

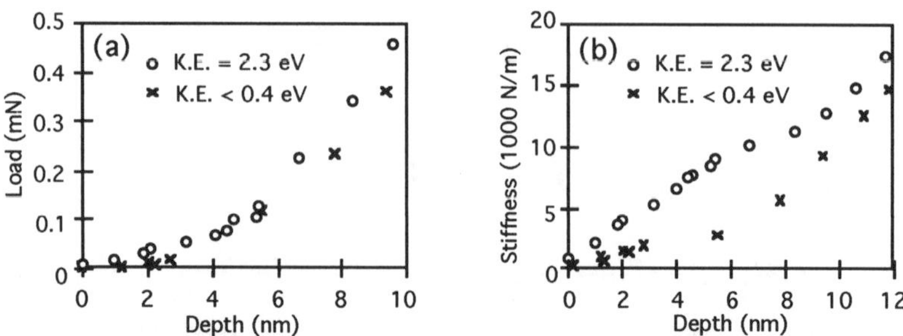

Figure 3 Load/depth and stiffness/depth curves for the initial contact on GaAs (111). For impact K.E.s <1 eV the first few nms of the loading curve is "flat" and the contact stiffness is extremely low. However, if impact energies of just 2 or 3 eV are used the initial load/depth and stiffness/depth curves look like an extension of those obtained for greater loads. The same variation in the initial contact with impact energy is seen for GaAs (100).

Nanoindentation Testing of Tungsten

The results for W (100) fall into two distinct categories: those performed in air, which show a clear dependence on impact velocity, and those performed in aqueous HCl, which do not show a dependence on impact velocity (the method used to perform nanoindentation testing in liquids is detailed elsewhere [8]). The indentation curves obtained in each case are illustrated by figure 4.

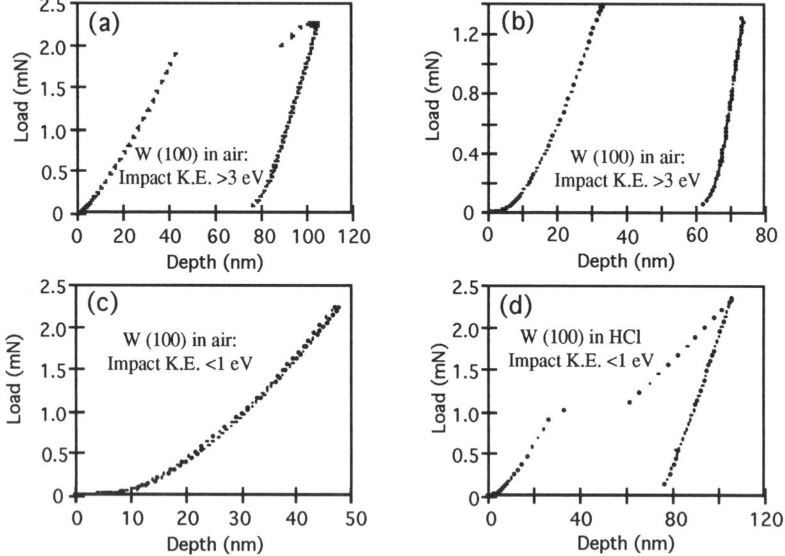

Figure 4 Load/depth curves for W (100). Though the curves for high energy impacts in air (a,b) show large pop-ins they are found to occur over a broad range of loads. In addition, pop-ins are often seen during unloading cycles and extended hold periods at loads which are only a small fraction of the peak value. By contrast, low energy impacts in air (c) give almost totally elastic indentation curves even when repeated loading and long hold periods are used. Immersing the sample in aqueous HCl to remove the surface oxide is found to lower the load required for permanent deformation to commence (d) and to alleviate the dependence on impact K.E., see figure 5.

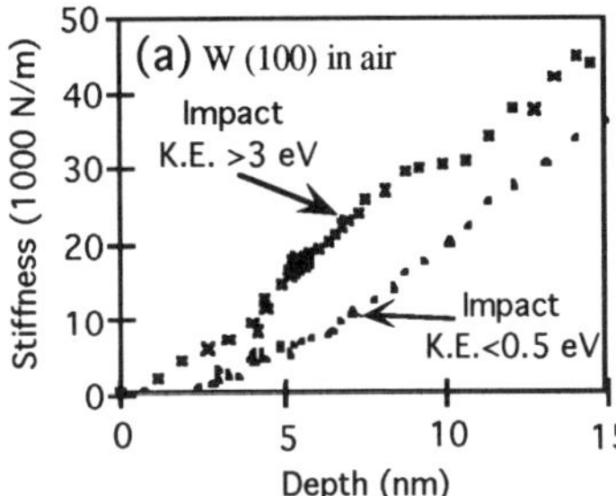

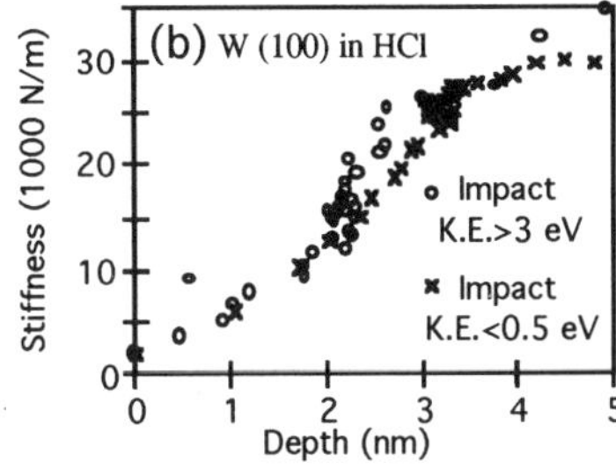

Figure 5 The initial stiffness/depth curves for indentations on W (100) in air (a) and aqueous HCl (b). For indentations performed in air there is a passivating, surface oxide layer present and the stiffness of the contact shows a dependence on impact K.E.. That is the stiffness is substantially lower for an impact K.E. <0.5 eV than the stiffness for an impact K.E. >3 eV. Removal of the oxide by immersion in HCl alleviates the dependence on impact K.E. and gives stiffness/depth curves which are indistinguishable for different impact energies even at the shallowest depths.

The main points which need to be highlighted about the load/depth curves of W (100) are that large pop-ins are seen in all cases (high or low impact K.E. and in air or HCl) as is a near elastic response for the lowest loads, but immersion in HCl or a high impact K.E. leads to pop-ins at lower loads with broad statistical spreads.

The differences in the load/depth curves for indentations performed in air and HCl are replicated in the initial portion of the stiffness/depth curves for the two cases (figure 5). Overall the implication is that the oxide layer (present when testing in air but not when testing in HCl) makes the tungsten more resilient to the generation of defects on contact.

MODEL FOR IMPACT VELOCITY/K.E. DEPENDENCE

The principal clue to the origin of the dependence on impact energy/velocity lies in the stiffness data for the first few nms of the indentation. Given the relationship between stiffness and contact area [5,10] this suggests that the initial contact area for a low impact K.E. is much smaller than that for an impact K.E. of 2 or 3 eV. In addition the data for tungsten shows that the response is totally different if there is no surface oxide present.

We are able to explain these phenomena by considering the effect on the initial contact of the sample being intrinsically rough on at least the near atomic scale (with the exception of certain cleaved crystals this is a valid assumption for all surfaces no matter how well prepared they may be). Using the model of Fuller and Tabor [11], which describes the contact between a flat and rough surface in terms of the JKR model of adhesive contact [12], we find that even the smallest asperities are sufficient to prevent contact over a large area if the surface energy of the sample is low (say 100 mJ/m^2) and the sample has a high elastic modulus (say 100 GPa). However, atomic-size asperities could be deformed by impact K.E.s of just a few eV, and hence, permit the surface forces acting in the area around the asperities to pull the tip and sample into contact over a much larger area.

Now applying the JKR model to the adhesive contact between a perfectly flat surface and a rounded tip (radius 500 nm, estimated using the method of Oliver and Pharr [9]) we are able to estimate the total change in surface energy as a result of the contact being formed. For the surface energy and modulus given above we find the potential change in surface energy is of the order of hundreds of eVs. Since the elastic strain energy of a dislocation is around 5 eV per atom plane threaded by the dislocation [13,14] it is clear that the surface forces and energy are potentially sufficient to generate short lengths of dislocation line on contact [15], but critically this will only be possible if the impact K.E. is able to overcome the barrier to a large contact area presented by surface nano-asperities.

DISCUSSION

The results presented highlight the extreme sensitivity of the measured nano-mechanical properties to the initial contact between tip and sample during nanoindentation testing. In particular impact energies of 2 or 3 eV are sufficient to permit defect generation on contact, and hence, alter the mechanical response of the material during the subsequent indentation. Impact K.E.s <1 eV are not sufficient to generate defects on contact and as a result dislocations must be nucleated at a much later stage in the loading cycle.

In metals, it appears that a hard surface film with a low surface energy (e.g. an oxide layer) can inhibit the generation of defects on impact. If the surface layer is removed the resilience of the surface to defect generation during the initial contact is lost. However, the results for W(100) imply that certain BCC metals can still behave pseudo-elastically and give large pop-ins if defects are present. This is likely to be due to the large shear stress required to cause cross-slip and multiplication of short lengths of dislocation line in these metals [13,14]. Further to this, the interface between the oxide and the metal may impede the propagation of any dislocations generated on impact into the substrate, and thus, prevent them from multiplying in the large shear stresses which occur some way below the surface during the loading cycle.

We conclude that surface properties (topography, chemistry and surface forces) play a vital role in the mechanical deformation of very small volumes of material at a surface. Specifically, surface coatings and films may provide some protection against the permanent damage which occurs during tribological processes by preventing dislocations, firstly, being generated on impact and, secondly, entering the bulk material.

ACKNOWLEDGEMENTS

We would like to thank our colleagues at Oxford for their assistance during this work. A.B.Mann was supported by the EPSRC through research studentship number 92312138 while conducting this research.

REFERENCES

[1] T.F.Page, W.C.Oliver, C.J.McHargue, J. Mater. Res. 7, 450, (1992).
[2] A.B.Mann, J.B.Pethica, W.D.Nix, S.Tomiya in Thin Films: Stresses & Mechanical Properties V, edited by S.P.Baker, P.Børgesen, P.H.Townsend, C.A.Ross, and C.A.Volkert (Mater. Res. Soc. Symp. Proc. 356, 1995) p. 271.
[3] S.K.Venkataraman, D.L.Kohlstedt, W.W.Gerberich, J. Mater. Res. 8, 685 (1993).
[4] W.W.Gerberich, S.K.Venkataraman, H.E.Huang, S.E.Harvey, D.L.Kohlstedt, Acta. Metall. Mater. 43, 1569, 1995.
[5] J.B.Pethica and W.C.Oliver in Thin Films: Stresses & Mechanical Properties, edited by J.C.Bravman, W.D.Nix, D.M.Barnett, and D.A.Smith (Mater. Res. Soc. Symp. Proc. 130, 1989) p. 13.
[6] J.B.Pethica, D.Tabor, Surf. Sci. 89, 182 (1979).
[7] N.Gane, F.P.Bowden, J. Appl. Phys. 39, 1432 (1968).
[8] A.B.Mann, J.B.Pethica, Langmuir Accepted 1996.
[9] W.C.Oliver, G.M.Pharr, J. Mater. Res. 7, 1564 (1992).
[10] J.B.Pethica, W.C.Oliver, Physica Scripta T19, 61 (1987).
[11] K.N.G.Fuller, D.Tabor, Proc. Roy. Soc. Lond. A 345, 367 (1974).
[12] K.L.Johnson, K.Kendall, A.D.Roberts, Proc. Roy. Soc. Lond. A 324, 301 (1971).
[13] J.P.Hirth, J.Lothe, Theory of Dislocations, 2nd Ed. (Wiley, 1982).
[14] D.Hull, D.J.Bacon, Introduction to Dislocations, 3rd Ed. (Pergamon, 1984).
[15] A.B.Mann, J.B.Pethica, Appl. Phys. Lett. Submitted 1996.

NANOINDENTATION STUDIES OF YIELD POINT PHENOMENA ON GOLD SINGLE CRYSTALS

S.G. CORCORAN*, R.J. COLTON*, E.T. LILLEODDEN**AND W.W. GERBERICH**
*Naval Research Laboratory, Code 6177, Washington, DC 20375-5342
**University of Minnesota, Department of Chemical Engineering and Materials Science, Minneapolis, MN 55455-0132

ABSTRACT

Nanoindentation studies often show an instantaneous displacement-excursion in the load-displacement curve. This anomaly is generally associated with a surface contamination effect, dislocation emission, a phase transition, or an oxide break-through event. The determination of which effect is operative is often difficult when investigating oxide covered surfaces. We have performed a detailed nanoindentation study on clean, flame annealed single-crystal Au thus eliminating the effects of a surface oxide or contamination layer. Multiple displacement excursions were observed exhibiting a new phenomenon of "staircase" yielding. Owing to the fact that our radius of contact is more than an order of magnitude smaller than the average dislocation spacing expected for well annealed Au, the excursions are explained in terms of multiple dislocation nucleation events on parallel slip bands. Indentation data were taken on Au (111), (110), and (100) single crystal surfaces.

INTRODUCTION

The recent advances in nanoindentation techniques has allowed for the detailed investigation of the contact of small volumes where the average dislocation spacing ($\sim 1\ \mu m$) for a well annealed crystal is greater than the contact radius of the indentation. Recently, displacement excursions during indentation have been observed on several metal surfaces[1-7] and attributed typically to oxide breakthrough by dislocations or a contamination effect. Gerberich, et. al. [6] demonstrated the existence of a 2nd displacement excursion at very low loads for the system Fe-3wt%Si which is an indication of dislocation emission below the oxide film prior to oxide breakthrough. Motivated by this finding and the yield instabilities orginally observed on Au by Gane and Bowden[8] and Pethica and Tabor[9] via TEM and resistivity techniques, we investigated the yield behavior of single crystal Au in order to eliminate the effect of a surface oxide. We find that the three low index faces of Au exhibit a reproducible displacement excursion in agreement with a dislocation nucleation event. In fact, a new phenomenon of "staircase" yielding was observed in which the plasticity was confined to a series of excursions separated by elastic deformation.

EXPERIMENT

A commercial Hysitron, Inc. Picoindenter® was used in conjunction with the Digital Instruments, Nanoscope™ III AFM/STM to perform indentations on Au single crystals oriented in the [111], [110], and [100] directions. The specially designed force transducer of Hysitron, Inc. allows for the direct imaging of the surface before and after the indentation. A Berkovitch diamond indenter was used for all indentations. The blunt end of the indenter has a spherical radius of 205 nm as determined by imaging the tip with an asperity on the Au surface[7].

Mat. Res. Soc. Symp. Proc. Vol. 436 © 1997 Materials Research Society

The gold single crystals[*] (99.99%) were prepared immediately before each series of indentations by electrochemical polishing at 2.5 Amps in a 1:2:1 electrolyte of ethylene glycol:ethanol:HCl at 55 °C. The surfaces were then flame annealed until glowing red hot under a H_2 flame. The samples were then quenched and stored under triply distilled water and transferred into a dry N_2-purged glove bag. One sample, Au (111), was deliberately left in the ambient overnight before placing in the glove bag. This proved to significantly affect the yielding and reproducibility of the indentation data.

All indentations were performed in a dry N_2-purged glove bag. The N_2 was flowed through a column with a mixture of $CaSO_4$ and 5 nm molecular sieve to eliminate H_2O and organic impurities.

STM images of the Au surfaces were obtained after the indentation tests to characterize the surface. The Au (110) surface had a typical RMS roughness of 1.0 nm. The Au (100) surface consisted of 200 - 300 nm wide atomically flat terraces separated by 2 nm high step bunches. Indentations were performed on these atomically flat regions. The Au (111) surface consisted of parallel atomic terraces typically 15 nm wide separated by single atomic steps 0.24 nm in height.

RESULTS AND DISCUSSION

Figure 1 (a)-(c) shows the typical load-displacement curves obtained for the Au <111>, <100>, and <110> orientations, respectively. These data were reproducible as long as the indentations were performed in a dry N_2 glove bag. Data demonstrating the reproducibility of the indentations as well as the effect of organic contamination can be found elsewhere[7].

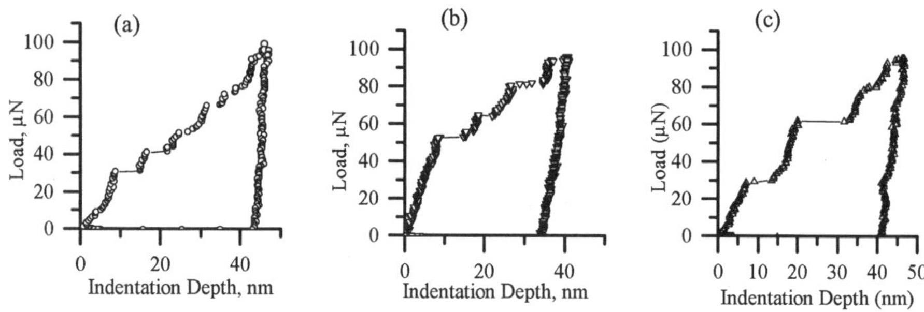

Figure 1: Typical load - displacement curves for Au single crystals (a) (111), (b) (100), and (c) (110).

We note that the number of displacement "steps" and their size varies between indentations but the sum total of all displacements is conserved, i.e., the average shape of the indentation curve is highly reproducible. As shown in Figure 1, a new phenomenon of "staircase" yielding is observed for Au single crystals. This phenomenon is unique because the plasticity appears to be confined to the instantaneous displacement excursions, i.e., the Au surface predominantly deforms elastically between each yield excursion. Figure 2 shows an example of a loading cycle in which the

---

[*] Monocrystals, Inc. Cleveland, OH

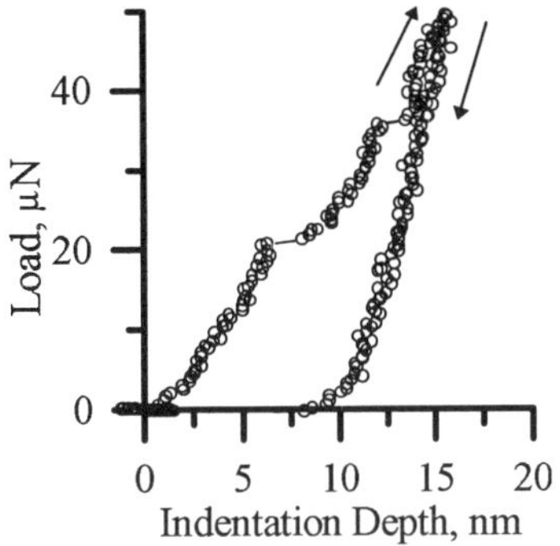
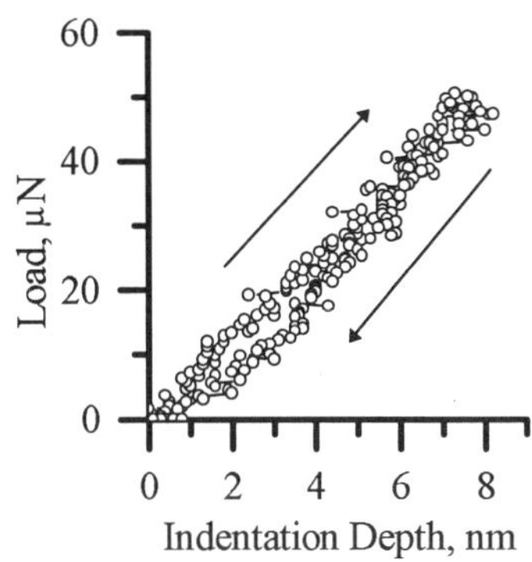

Figure 2: Load-displacement curve for Au (110).

Figure 3: Load-displacement curve for Au (100).

unloading portion of the curve occurs just before the next yield excursion event. The initial unloading approximately traces the upper portion of the loading curve indicating elastic behavior between the yield excursions. Figure 3 demonstrates the elastic response of Au (100) prior to the first yield excursion. The onset of staircase yielding (yield point) varies with the surface morphology. An atomically flat surface, Au(100) data plotted in Figure 1b, has a greater value of the yield point as compared to a surface with a high density of atomic steps, Au(111) Figure 1a. No orientational dependence on the yield point was observed owing to the apparent larger dependence on surface morphology.

Further evidence that the loading behavior between the displacement steps is elastic can be shown by assuming the first dislocation nucleated at the displacement excursion acts as a Frank-Read source. This source will operate until the back forces from the inverse pile-up is sufficient to sum the forces to zero. Equation 1 gives the relation between the applied load at which a yield excursion occurs and the extent of the displacement[6] expected before the Frank-Read source is deactivated:

$$P_{yp} = \frac{3\pi a}{2C(\alpha)}\left\{\frac{\mu_s b}{2}\left[1+\frac{N}{\pi(\alpha-1)}\right]\right\} \tag{1}$$

where N is the number of dislocations in the pile-up, b is the Burgers vector, a is the contact radius after the yield event, μ_s is the shear modulus, and $C(\alpha)$ is a geometric term equal to 0.119 for $\alpha = 4$ where α is the position of the super dislocation Nb in units of the contact radius, a. If all of the plasticity is occurring at the yield excursions equation 1 should apply to displacement excursion, i, if δ_{exc} is taken as the sum of all excursions up to load P_i. If we take $Nb \approx \delta_{exc}$, $\mu_s = 30$ GPa, $\alpha = 4$, $C(\alpha) = 0.119$ then:

$$\frac{P_{yp}}{a} = 174 \text{Nm}^{-1} + 64 \text{ GPa } \delta_{exc} \tag{2}$$

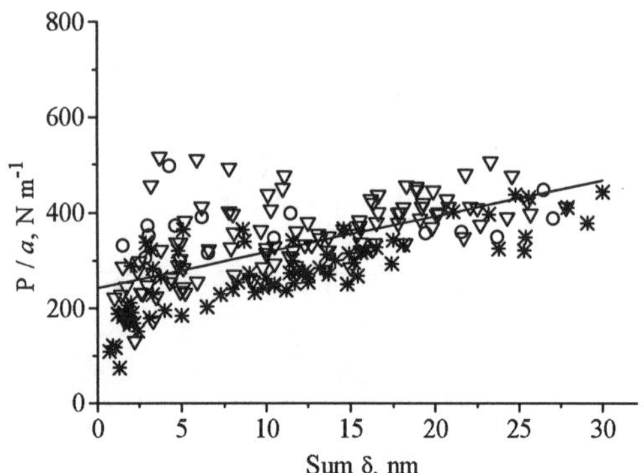

Figure 4: Normalized load, P_i / a, at excursion, i, versus the sum of all displacements from excursions 1 through i. (∗ Au(111), ∇ Au(100), and IAu(110))

Figure 4 is a plot of the load at excursion, i, normalized by the contact radius a versus the sum of all excursions 1 to i for Au (111), (100) and (110). Linear regression using all of the data yields the equation:

$$\frac{P_{yp}}{a} = 242 \mathrm{Nm}^{-1} + 7.5 \, \mathrm{GPa} \; \delta_{exc}$$

The constant in the above equation, which is directly related to the stress necessary to activate the Frank-Read source, is in fairly good agreement with the predicted value of equation 2. However, the experimentally determined slope is significantly less than the predicted value. This discrepancy may be due to the fact that the preliminary model was intended for one active slip band. For the "staircase" yielding behavior of Au, we believe that each excursion corresponds to the activation of a parallel slip band. Shielding effects between the dislocations in these parallel slip bands would be expected and may account for the overestimation of the applied load. That is, a parallel set of slip bands is not as effective at supporting the applied load as a single band.

Calculation of an elastic modulus from the unloading slopes of the indentation curves was not reliable owing to creep processes which often resulted in negative slopes on the initial unloading curves. However the initial loading slopes give a value of 77 ± 13 Gpa and 60 ± 13 GPa for the <100> and <110> orientations. These values are similar within the error. The (111) surface yields at too low of a load to accurately measure a loading slope. The lower value of the modulus calculated by the loading slopes may be a result of errors in estimating the contact radius of the indenter. The moduli were calculated from equation 3 which is derived from the analysis of Sneddon for indentation with a flat punch[10]:

$$\frac{E}{1-\nu^2} = \frac{\mathrm{slope}}{2\sqrt{R\delta_{elastic}}} \tag{3}$$

where R was taken as 205 nm and $\delta_{elastic}$ is 4, 8, 8 nm as in the previous calculation and ν is Poisson's ratio = 0.4. The uniaxial elastic moduli for Au is given in Table 1 as compared to the moduli calculated for indentation.

Surface	Uniaxial Modulus[11], GPa	Indentation Modulus[12], GPa
(111)	117	99
(110)	82	97
(100)	43	90

Table 1: The orientational dependence of the elastic modulus of Au as measured in uniaxial tension versus the elastic modulus of Au calculated for indentation.

As shown in Table 1, the elastic modulus on single crystals measured by an indentation technique averages over all orientations owing to the 3-dimensional elastic fields under the indenter tip. Within the error of the loading slopes we are unable to detect the difference in modulus between the (110) and (100) surfaces.

CONCLUSIONS

A new phenomenon of "staircase" yielding has been observed for the indentation of clean Au single crystal surfaces. The staircase is composed of multiple dislocation-nucleation events separated by elastic deformation and is explained in terms of dislocation nucleation on parallel slip bands as the indenter penetrates into the Au crystal. Qualitative agreement with a model of Gerberich et. al.[6] is shown for the relationship between the applied load and the sum total of all excursions. The elastic modulus as determined by indentation has very little dependence on crystallographic orientation in agreement with theoretical models[10]. Surface morphology at the atomic scale (atomic step density) has a large influence on the value of the yield point.

ACKNOWLEDGMENTS

We gratefully acknowledge Joel Hoehn, UMN, for providing the indentation moduli in Table 1. S.G.C. acknowledges the NRC for post-doctoral support. S.G.C. and R.J.C. acknowledge Hysitron, Inc. for their technical assistance. W.W.G. and E.T.L. acknowledge support by the Center for Interfacial Engineering under grant NSF / CDR-8721551.

REFERENCES

1. P. Tangyunyong, R.C. Thomas, J.E. Houston, T.A. Michalske, R.M. Crooks and A.J. Howard, *Phys. Rev. Lett.*, **71**, 3319 (1993).

2. W.C. Oliver and G.M. Pharr, *J. Mater. Res.*, **7**, 1564 (1992).

3. S.K. Venkataraman, D.L. Kohlstedt and W.W. Gerberich, *J. Mater. Res.*, **8**, 685 (1993).

4. W.W. Gerberich, S.K. Venkataraman, H. Huang, S.E. Harvey and D.L. Kohlstedt, *Acta Metall. Mater.* **43**(4), 1569 (1995).

5. E. Lilleodden, W. Bonin, J. Nelson and W.W. Gerberich, *J. Mater. Res. Commun.*, **10** (9), 2162 (1995).

6. W.W. Gerberich, J.C. Nelson, E.T. Lilleodden, P. Anderson, and J.T. Wyrobek, *Acta. Metall. Mater.*, in press.

7. S.G. Corcoran, R.J. Colton, E.T. Lilleodden, W.W. Gerberich, in preparation.

8. N. Gane and F.P. Bowden, J. Appl. Phys., **39**, 1432 (1968).

9. J.B. Pethica and D. Tabor, Surf. Sci., **89**, 182 (1979).

10. I.N. Sneddon, *Int. J. Engng. Sci*, **3**, 47 (1965).

11. An average of the values compiled in: <u>Single Crystal Elastic Constants and Calculated Aggregate Properties: A Handbook</u>, edited by Gene Simmons and Herbert Wang (MIT Press, Cambridge, MA, 1971).

12. The indentation-modulus values were calculated by Joel Hoehn, UMN, based on a model developed originally by Joost Vlassak and W. Nix, Stanford U.

EVOLUTION OF THE CONTACT AREA DURING AN INDENTATION PROCESS

Kangjie Li[*], T.W. Wu[**] and J.C.M. Li[*]
[*] Materials Science Program, Department of Mechanical Engineering, University of Rochester, Rochester, NY 14627
[**] IBM Research Division, Almaden Research Center, San Jose, CA 95120

ABSTRACT

The evolution of the contact area during an indentation process has been examined by an a.c. technique and also by finite element analysis on five different bulk materials. The consistency of the results obtained by the two approaches not only show the validity and advantage of the a.c. technique but also corroborate the applicability of the elastic contact equation.

INTRODUCTION

Great strides have been made in the development of microindentation and nanoindentation techniques over the last ten years. By using the depth sensing microindentation technique, not only the hardnesses but also the elastic moduli of materials can be obtained [1-3]. Measurement of elastic modulus is based on an assumption that the displacement upon unloading is mostly elastic, and the relationship between the contact stiffness (S) and the projected contact area can be derived from the elastic contact theory [4], i.e

$$S = \frac{dP}{dh} = \frac{2}{\sqrt{\pi}}\sqrt{A}E_r.$$

(1)

Where P and h are the applied load and indenter penetration depth respectively, A is the projected contact area. E_r is the reduced modulus defined as

$$\frac{1}{E_r} = \frac{1-v^2}{E} + \frac{1-v_i^2}{E_i},$$

(2)

in which E, v and E_i, v_i are Young's modulus and Poisson's ratio of the specimen and indenter respectively. Eq. 1 has been shown to be indenter geometry independent [5,6].

If the contact between the indenter and sample surface is elastic and the reduced modulus, E_r, is known in advance, Eq. 1 then allows us to calculate the projected contact area, A. However, during the unloading of an indentation process, the stress-free condition is not a flat surface, but a surface containing a hardness impression. Because of the complexity of the initial contact geometry, an analytic description of the evolution of the contact area during unloading is not yet available. However, some finite element simulation work has been implanted to investigate this issue [9].

In this study, a displacement-controlled a.c. indentation technique was applied on five mechanically different bulk materials, including Si, Al, Al_2O_3, glass and amorphous carbon. The purpose is to present a methodology which is capable of measuring the contact stiffness and, if E_r is known, the contact area throughout an entire indentation process. Finite element simulation has also been implanted to verify the experimental findings. The results not only illustrate the validity and advantage of the a.c. indentation technique, but also shed some light on the indenter/surface contact process.

A.C. INDENTATION TECHNIQUE

The ac-indentation technique was first introduced and also applied on the Nanoindenter by Pethica and Oliver [10]. In this study, a microindentation system has been upgraded to equip with

Mat. Res. Soc. Symp. Proc. Vol. 436 ©1997 Materials Research Society an a.c. measuring capability by superposing a small displacement modulation on an otherwise linear indenter motion [11,12]. The advantage of the a.c. technique is that it acquires not only the information as the d.c. method regularly does, but also the instantaneous contact stiffness throughout an indentation process. The instantaneous contact stiffness, $S_i(t)$, can be obtained by the formula [12]

$$S_i(t) = \frac{k_o \cdot \Delta LC}{\Delta IND(t) - \Delta LC(t)} \quad , \tag{3}$$

where k_o is the spring constant of the loadcell, $\Delta LC(t)$ is the peak-to-peak amplitude of the induced oscillation of the applied load and $\Delta IND(t)$ is the peak-to-peak amplitude of the indenter dithering motion.

If $S_i(t)$ is known the plastic depth, h_p, at any stage of an indentation process can be obtained through the simple flat punch theory [7], namely by

$$h_p(t) = h_t(t) - [P(t) / S_i(t)] \quad . \tag{4}$$

With a prior knowledge of the indenter shape function, the hardness as well as contact modulus depth profiles can then be calculated [12]. Furthermore, if the reduced modulus is known, the contact area during an indentation process can be obtained through Eq. 1.

EXPERIMENTAL

30μm thick dc-sputtered Al_2O_3 coating on Si<100> substrate, glass disk substrate, amorphous carbon, aluminum alloy 6061-T6 and Si<100> wafers were intentionally selected as the testing samples for their different mechanical behavior, i.e. hard/brittle (Si, Al_2O_3 and glass), soft/ductile (aluminum alloy) and rubber-like (amorphous carbon). The a.c. microindentations implemented under an identical testing condition except the maximum load. In all cases, the indenter speeds were set at 5.0 and 7.5 nm/sec for loading and unloading, respectively. A displacement dither with an amplitude of 6nm and modulated at a frequency of 15.3Hz was employed.

FINITE ELEMENT SIMULATION

The process of indenting materials by a Berkovich indenter is numerically simulated by pushing an axisymmetric rigid conical indenter into corresponding elastoplastic materials. The large elastoplastic feature of the ABAQUS finite element code was utilized. Along the contact interface, the interfacial elements IRS22As in the ABAQUS code were used to monitor the contact between the indenter and the sample surface. One advantage of the numerical simulation is that the indenter contact area at any point is known. To assure a good resolution of measuring the contact area, the interface layer has been refined to a width about 0.02μm.

In order to facilitate the comparison between the simulation and experiment, the rigid conical indenter used in the simulation was designed in such a way that closely matched the shape function of the Berkovich indenter, i.e. a 143.74° conical indenter with 0.239μm tip radius.

RESULTS

Fig.1(a) shows the experimental indentation loading curves of the five sample materials. The shapes of the loading curves clearly reflect their distinctive mechanical properties. The values of the measured hardnesses and Young's moduli, together with other parameters are listed in Table.1. The yield stresses were extracted from the measured hardness via the empirical relation $\sigma_y = H/3$. Except for the aluminum sample where a linear, isotropic hardening mechanism was assumed, all other materials were assumed to be elastic-perfectly plastic. Fig.1(b) shows the simulated loading curves for the five materials and a close similarity between the two sets of results is evident.

Table 1. Materials' Properties used for Simulation and Calculation

Material	Young's Modulus (GPa)	Hardness (GPa)	Poisson's Ratio	Yield Stress (GPa)	Hardening Modulus (GPa)
Silicon	170	12.6	0.23	4.2	0
Alumina	135	8.4	0.23	2.8	0
Glass	82	7.8	0.23	2.6	0
Carbon	35	5.1	0.23	1.7	0
Aluminum	79	1.0	0.33	0.33	0.146
Diamond	1000	----	0.10	----	----

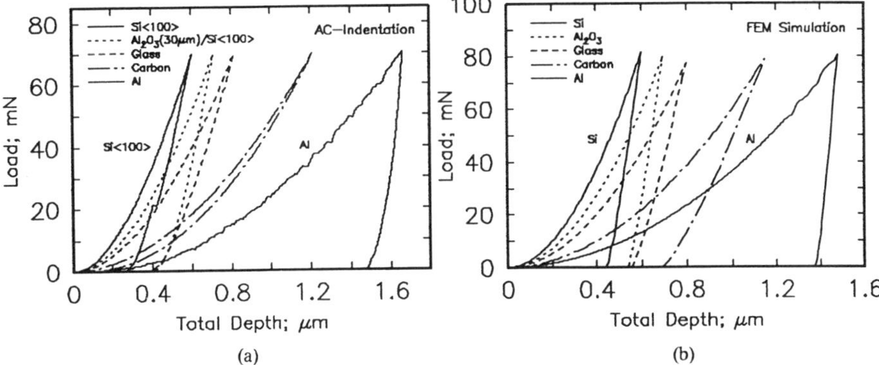

Fig.1 Indentation loading curves of Si, Al$_2$O$_3$, glass, Al and amorphous carbon. (a) The experimental measurements and (b) the FEM simulation.

Fig.2 shows a superposition of five S_i versus applied load curves acquired from the Al$_2$O$_3$ sample under different maximum loads. As stated by Eq. 1, the projected contact area is proportional to the square of the unloading slope if the contact modulus is constant. The latter condition is satisfied in this study. Fig.3(a) and (b) show the contact area evolution as function of the applied load. In order to avoid overlapping among the curves, offsets of 5μm^2 and 2μm^2 along the y-axis for glass and Al$_2$O$_3$, respectively, were introduced in Fig.3(a). In Fig.3(b), the contact area of the aluminum alloy was reduced by 75% to fit the scale. During loading, all the

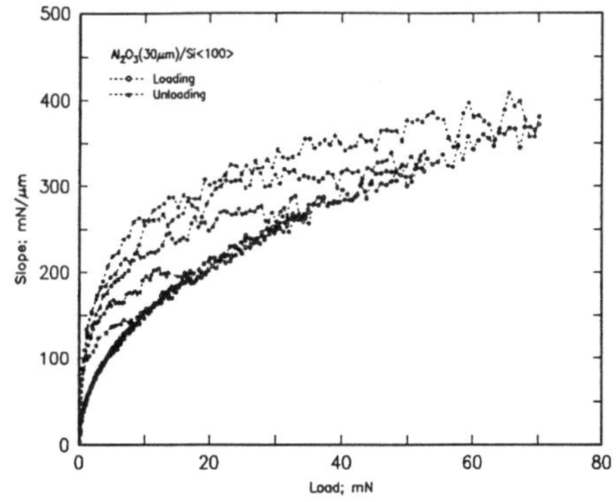

Fig.2 Instantaneous contact stiffness versus applied load under different peak loads for Al$_2$O$_3$.

materials showed a linear behavior and the differences in the slopes exactly reflected their differences in hardness. During unloading, however, their behavior were dramatically different. For silicon, Al$_2$O$_3$, glass and aluminum alloy, the contact areas remained unchanged as the unloading process began, then it decreased as the load dropped further down. On the contrary, the amorphous carbon showed no discernible constant regime and the unloading path almost

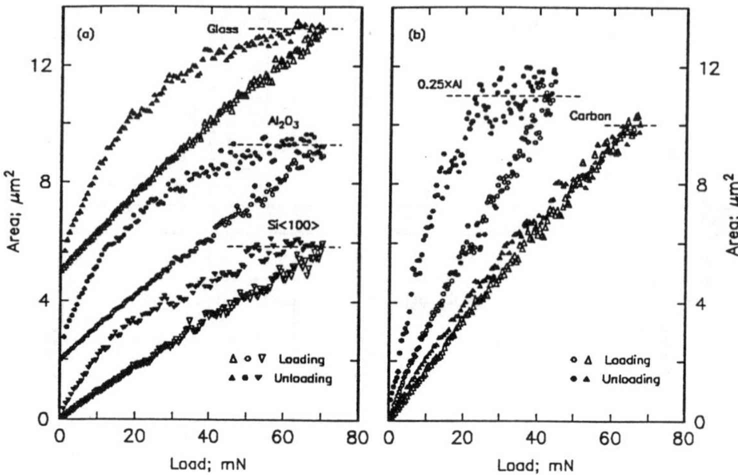

Fig.3 Contact areas evolution as functions of applied load for (a) Si, Al$_2$O$_3$ and glass; (b) Al and amorphous carbon.

coincided with the loading one. The constant area regime of the aluminum sample is clearly larger than those of silicon, Al$_2$O$_3$ and glass samples.

DISCUSSIONS

To study the relationship between the projected contact area (A) and the load (P) applied on the indenter, two normalized parameters, namely (A/A_{max}) and (S/S_{max}), were adopted to facilitate the analysis. Where A_{max} and S_{max} are the maximum values of A and S, respectively, occurred in an indentation cycle. Based on Eq. 1, the relation between the two normalized parameters is $(S/S_{max})^2 = A/A_{max}$, and which was used to convert the normalized stiffness into the normalized contact area.

Fig.4 shows the plot of normalized area versus normilized load for the Al$_2$O$_3$ sample. The curves denoted by symbols are the experimental measurements acquired under five different maximum loads. The solid curve is the result of the finite element simulation. The figure clearly shows that all the normalized curves coincide each other and form a master curve regardless of their maximum loads. Furthermore, the master curve is reasonably consistent with the simulation result. The match between the experimental and FEM simulation in Fig.4 further confirms the validity of applying Eq. 1 to the elastic contact situations starting with pre-indented surfaces.

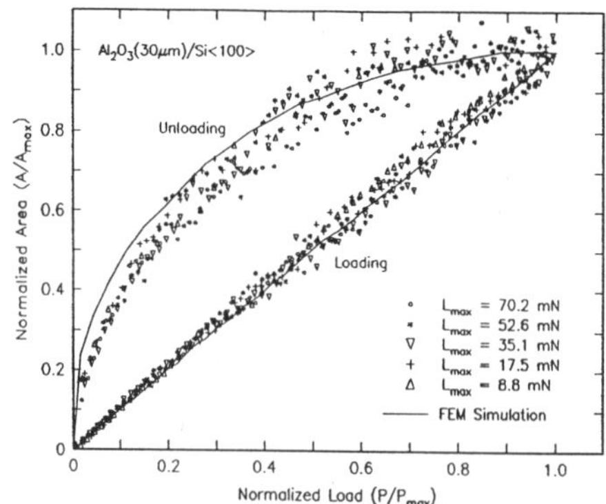

Fig.4 Normalized contact area versus normalized load for the Al$_2$O$_3$ sample. Symbol curves are the experimental measurements and the solid line is the FEM simulation.

Fig.5 shows the unloading portions of the aluminum alloy, Al_2O_3 and amorphous carbon samples. Since the curves of the silicon and glass samples are very close to that of the Al_2O_3, for the sake of clarity, they are not included in the figure. For the aluminum alloy and Al_2O_3, the contact areas started to decrease after the applied load had dropped 35% and 15%, respectively, from their maximum values. However, the amorphous carbon sample showed no discernible constant area regime and the contact area started to decrease immediately after unloading. Fig.6 shows the simulated unloading portions of the aluminum, glass and Al_2O_3 samples. It clearly shows that the aluminum has the widest constant area regime and the glass has the narrowest. Furthermore, the contact area of the aluminum dropped relatively faster than the other two samples once the indentation had left the constant area regime and generated a cross-over. Interestingly enough, the cross-over phenomenon was also observed experimentally (ref. Fig.5). A simulation result by Laursen and Simo [9] on an Al film on Si substrate was also included in Fig.6 and it shows a good agreement with our findings.

By using Sneddon's approach [4], the indentation load-displacement relationships can be expressed as $P = \alpha D^m$. Where P is the load, D is the elastic displacement of the indenter, α and m are constants, and m is 1 for flat end punches, 2 for cones and 1.5 for paraboloids. A rather simple relationship between the normalized contact area and normalized load can also be derived [4], it is namely,

$$A / A_{max} = \left(P / P_{max} \right)^k, \tag{5}$$

where $k=0$ for a flat end punch, $k=2/3$ for a parabolic indenter and $k=1$ for a conical indenter.

As shown by Oliver and Pharr [8], the unloading curves can be reasonably represented by a power law relation between load and displacement for a wide range of materials, and the value of m was between 1.25 and 1.5. Based on this, they suggested that the unloading behavior is close to elasticaly deforming a flat surface by a parabolic indenter. However, except for amorphous carbon, whose contact area has changed in a manner close to the conical indenter behavior (i.e. $k=1$ in Eq.5), neither the experimental nor the numerical result of the contact area is consistent with the parabolic or conical indenter behavior (ref. Fig.5, Fig.6 and Eq.5), even though a conical indenter was indeed employed in the simulation. Fig. 5 shows that the deformation of the amorphous carbon was almost purely elastic during entire indentation, and after unloading, the indented area was recovered to nearly its original flatness. If we take the amorphous carbon's

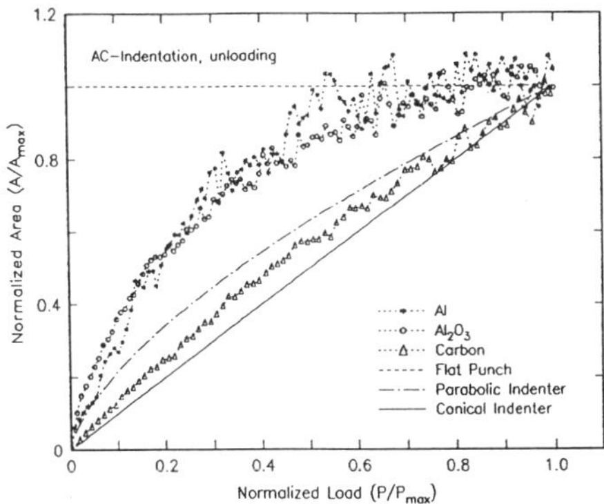

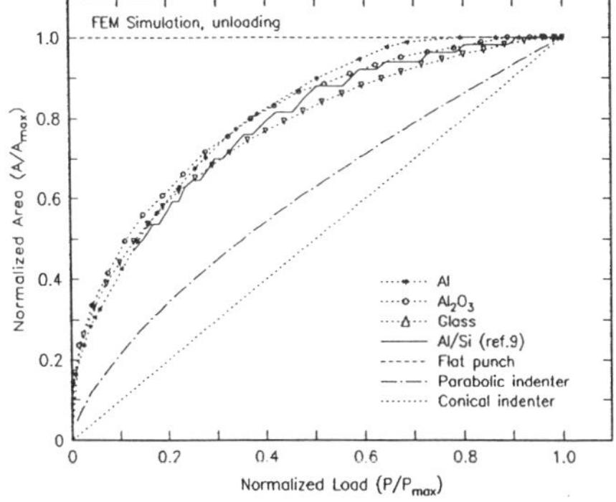

Fig.5 Experimental results of the aluminum alloy, Al_2O_3 and amorphous carbon samples during unloading.

Fig.6 FEM simulation results of the aluminum alloy, Al_2O_3 and glass samples during unloading.

mechanical behavior as an example of elastically deforming from a flat surface, then the large in the contact area behavior observed in other materials reflects the influences from existed hardness indents. Further work is under way to investigate whether the pre-indent is the only factor which cause the discrepancy. Plastic deformation during unloading can also contribute some differences. For other materials, the contact areas did remain constant during the initial unloading and that supports the flat punch assumption. This implies that although the unloading curve can be described by a power law relation, however, m doesn't maintain strictly constant all the way. In other words, due to the existence of an indent, there is a complicated stress redistribution over the contact area during the course of the unloading and the contact behavior during this stage is no longer properly described by the elastic contact formulation based on a single indenter.

CONCLUSIONS

The evolution of the contact area during an indentation cycle has been studied by the a.c. technique and also by finite element simulation.

Constant contact area regimes during the initial unloading were observed experimentally as well as numerically. The duration of that regime strongly depends on materials' mechanical behavior.

The normalized contact area versus normalized load curve has been utilized to facilitate data analysis. The formation of the master curve reflects the contact geometric similarity.

Good agreement established between the a.c. measurements and finite element simulations has confirmed the applicability of the local contact stiffness equation, $S \propto \sqrt{A} \cdot E_r$, even the elastic contact is started from a pre-indented surface.

The distinctive unloading behavior of the amorphous carbon suggested that a hardness indent can significantly affect the following unloading elastic characteristics, altoough it may not be the only factor.

ACKNOWLEDGMENTS

This work is a joint study between IBM and the University of Rochester. K. Li and J.C.M. Li would like to express their thanks to Dr. Jim Lyerla at IBM Almaden Research Center for his management support, and acknowledge the financial support from NSF through DMR 9221326 monitored by Dr. Bruce McDonald.

REFERENCES

1. F. Frolich, P. Grau and W. Grellmann, Phys. Stat. Sol.(a) **42**, 79-89 (1977).
2. D. Newey, M.A. Wilkins and H.M. Pollock, J. Phys. E: Sci. Instrum. **15**, 119-122 (1982).
3. J. Pethica, R. Hutchings and W.C. Oliver, Philos. Mag. A **48**, 593 (1983).
4. I.N. Sneddon, Int. J. Eng. Sci. **3**, 47-57 (1965).
5. G.M. Pharr, W.C. Oliver, and F.R. Brotzen, J. Mater. Res. **7**, 613-617 (1992).
6. R.B. King, Int. J. Solids Structures **3**, 1657 (1987).
7. M.F. Doerner and W.D. Nix, J. Mater. Res. **1**, 601-609 (1986).
8. W.C. Oliver and G.M. Pharr, J. Mater. Res. **7**, 1564-1583 (1992).
9. T.A. Laursen and J.C. Simo, J. Mater. Res. **7**, 618-626 (1992).
10. J.B. Pethica and W.C. Oliver, in <u>Thin Films: Stresses and Mechanical Properties</u>, edited by J.C. Bravman, W.D. Nix, D.M. Barnett, and D.A. Smith (Mater. Res. Soc. Symp. Proc. **130**, 13-23 (1989)).
11. T.W. Wu, J. Mater. Res. **6**, 407-426 (1991).
12. T.W. Wu, Mater. Chem. & Phys. **33**, 15-30 (1993).

MECHANICAL PROPERTY DATA FOR COATED SYSTEMS - THE PROSPECTS FOR MEASURING "COATING ONLY" PROPERTIES USING NANOINDENTATION

S.V. HAINSWORTH, T.F. PAGE
Materials Division, University of Newcastle upon Tyne, Newcastle upon Tyne, NE1 7RU, UK.

One of the critical aspects of assessing the properties of coated systems is the prospect of being able to measure the properties of the coating in isolation of the substrate. This has led to an increase in the use of continuously-recording indentation techniques (or nanoindentation techniques) for assessing the mechanical properties of coated systems since they can be used to measure the materials response to contact deformation at a scale relevant to the coating thickness. This paper presents the results of a new method for analysing the nanoindentation loading curves for coated systems. The analysis of the loading curve uses the relationship $P = K_m \delta^2$ which describes the indenter displacement, δ, in terms of the applied load P. For each material, K_m can be predicted from its modulus E and hardness H. One result is that if either of E or H is known, then the other may be calculated from the experimental loading curve. Further, the method has the potential to differentiate between the responses of the coating only, the coating and substrate in combination, and that dominated by the substrate once the load has become sufficient for cracking of the coating such that it no longer plays a significant role in supporting the applied load. In many cases, careful analysis of the loading curves allows the displacement ranges associated with these regimes to be identified. This is shown to be a powerful experimental tool for the interpretation of the mechanical properties of coated systems.

INTRODUCTION

Nanoindentation techniques are widely used for measuring the mechanical responses of coated systems due to their ability to probe depths considerably less than the coating thickness. These films may range in thickness from a few 100nm (e.g. thin, protective coatings) to 30-40µm (e.g. thick, wear-resistant coatings). In all of these systems, there is often a desire to measure the coating properties in isolation of the substrate either for property-modelling purposes or in order to best optimise the properties of the coating itself before attempting to match it to various substrates.

A further attraction of nanoindentation is that the load-displacement curve allows many parameters such as the hardness, Young's modulus and works of indentation to be calculated e.g. [1-5]. (For more details on the complete range of accessible parameters see e.g. [1]). However, an on-going contentious issue is to whether or not "coating only" properties can ever be measured by sufficiently shallow nanoindentations and, if so, how the experimental conditions for such measurements might be identified.

For most bulk solids, the hardness and Young's modulus are simply analysed from the unloading portion of the load-displacement curve using either a linear [3] or power-law fit[4]. However, in some cases, such as that shown in Figure 1 where the unloading segment is highly curved, then neither of these methods is particularly suitable for estimating the real contact area and thus mechanical properties of the material. Thus a recently-proposed alternative strategy is to analyse the mechanical properties of the system by modelling the loading curve [6]. In this method, the overall response of the material to the applied load is modelled as the sum of it's elastic and plastic components. This follows an earlier approach proposed by Loubet *et al* [7], but has two important differences to account for the indenter geometry.

Mat. Res. Soc. Symp. Proc. Vol. 436 © 1997 Materials Research Society

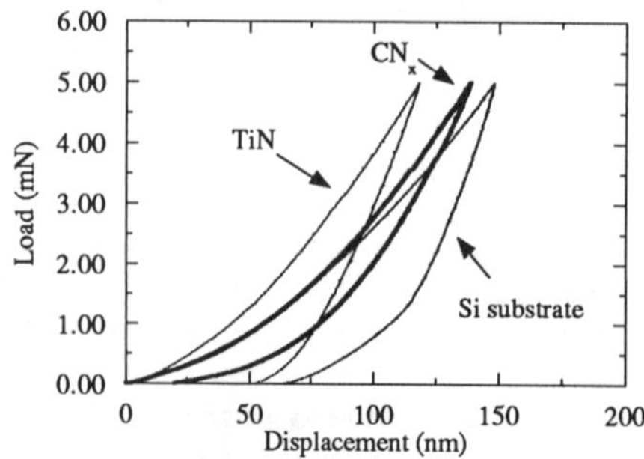

Figure 1: Load-displacement curves for a 300nm TiN coating on silicon, a 300nm CNx coating on silicon and the (111) silicon substrate alone. Note the unloading curve for the CN$_x$ system is highly curved and thus it is difficult to calculate parameters such as hardness and modulus for this system by conventional analysis of the unloading curves.

If bulk materials are considered then, for a pyramidal indenter, (e.g. Berkovich), the relationship $P \propto \delta^2$ should hold where P is the applied load and δ is the displacement. Thus, if the load is plotted against the square of the displacement for various materials, then

$$P = K_{exp} \delta^2$$

where K_{exp} is the experimentally-observed gradient of the resultant straight line plot.

In order to model the loading curve, the indenter diplacement is taken as the sum of elastic and plastic components. Thus, if the material being indented is rigid plastic, the contact radius, a, can be expressed as

$$a = \sqrt{\frac{P}{H}}$$

where P is the load and H the hardness (which is assumed constant with decreasing load).

For conical and pyramidal indenters, the plastic depth of indentation, δ_p, is directly related to the characteristic radius of the contact, a, by

$$\delta_p = \phi a$$

where ϕ is a constant which is dependent on the shape (i.e. the cross-section and angle) of the indenter. This leads to

$$\delta_p = \phi \sqrt{\frac{P}{H}}$$

In addition, for real materials, there is a further instantaneous elastic contribution, δ_e, which, by dimensional analysis, can only have the form

$$\delta_e = \psi \frac{P}{Ea}$$

where E is the Young's modulus and ψ is another empirical constant.

Summing these contributions - and with a little manipulation - this can be written as

$$P = E\left(\psi\sqrt{\frac{H}{E}} + \phi\sqrt{\frac{E}{H}}\right)^{-2}\delta^2 = K_m\delta^2$$

The constants ϕ and ψ - and thus K_m for any material of known E and H - can then be evaluated from experimental values of K_{exp} observed for a range of samples of well-characterised E & H values. For a Berkovich indenter, this analysis yields $\psi = 0.930$ and $\phi = 0.194$ [6].

APPLICATION OF THE MODEL TO COATED SYSTEMS

It is possible to predict the shape of the P vs δ^2 curve for coated systems. Initially, the indenter should just probe the properties of the coating alone and therefore the expected dependence is simply a straight line whose gradient would be that predicted from the above expression for K_m by inserting the Young's modulus and hardness values for the coating alone. There should then be a transition region where the gradient of the curve is more complex as the coating, the interfacial region and the substrate all deform together. In this region, the relationship P vs δ^2 will not necessarily display either the same slope or even be a straight line. Finally, at large displacements, the highly-fractured coating will no longer play any significant role in supporting the applied load and the predicted curve should again be a straight line with a gradient that can be predicted from the Young's modulus and hardness of the substrate alone. Thus the expected curve may be similar to that shown schematically in Figure 2. The extent of each of the three distinct regions will depend on the mechanical properties of the system's components and the thickness of the coating being tested.

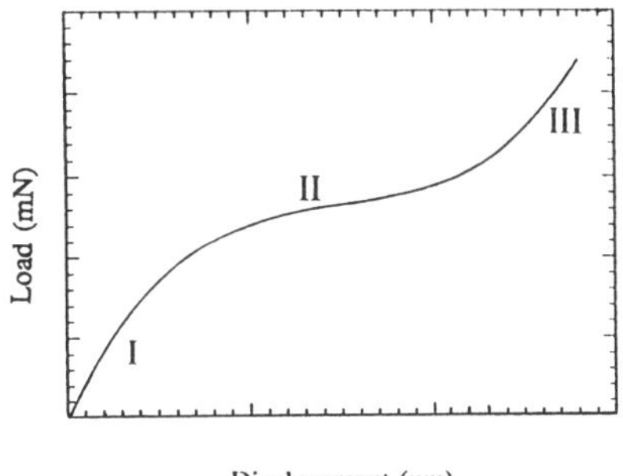

Figure 2: The predicted P vs δ^2 relationship for coated materials showing three distinct regions of behaviour. I = the coating itself deforms elastically & plastically while the substrate only supports some of the load elastically; II = the substrate deforms plastically causing both coating and substrate to deform significantly; III = substrate deformation dominates as the coating is increasingly cracked and pushed into it.

EXPERIMENTAL

In order to test whether the predicted behaviour described above correlates with the behaviour of real coated systems, a set of indentations were made on a series of experimental TiN_x coatings (where x~0.8) produced by magnetron sputtering with nitrogen pressure control. The coatings were all deposited on an M2 tool steel substrate which had been pre-polished to a colloidal silica finish to ensure that the surface roughness of the final coatings could be minimised. The coatings were produced to a range of well-controlled thicknesses which included 1,5,10,20 and 30 μm. As far as possible, the coatings were all believed to have the same properties, although the microstructures may have varied slightly as the thinner coatings will have different growth structures to the thickest. Nanoindentation experiments were performed using a Nano Indenter II™ (Nano Instruments Inc, Knoxville, TN) to peak loads of 5,10,50,100, 250 and 500mN using loading rates of Pmax/loading rate = 1000s and incorporated a hold-segment to allow the correction of any thermal drift. The data was processed using proprietary software to produce load-displacement curves while further analysis generated plots of P vs δ^2 for the loading segments.

RESULTS

The results of the nanoindentation experiments for indentations into the 5μm thickness TiN$_x$ coating are given in Figure 3. Figure 3a shows a typical load-displacement curve for a 500mN indentation. From this it can be seen that the maximum penetration depth at peak load is ~1500nm; therefore, even with nearly the maximum possible load applied by the nanoindenter we have not penetrated the coating sufficiently far to reach even one third of the coating thickness. Figure 3b shows the plot of P vs δ^2 for this same indentation. It can be seen that the curve departs from linearity at a load of ~ 80mN. Therefore, this can be attributed to the influences of both the interface and the substrate playing a role in the behaviour after this point. In order to verify that this behaviour was repeated at lower loads, Figure 3c and Figure 3d show the response at lower peak loads of 100mN and 5mN respectively. Figure 3c shows that for a 100mN peak load, the departure from linearity again occurs at ~80mN. For indentations below this critical load,

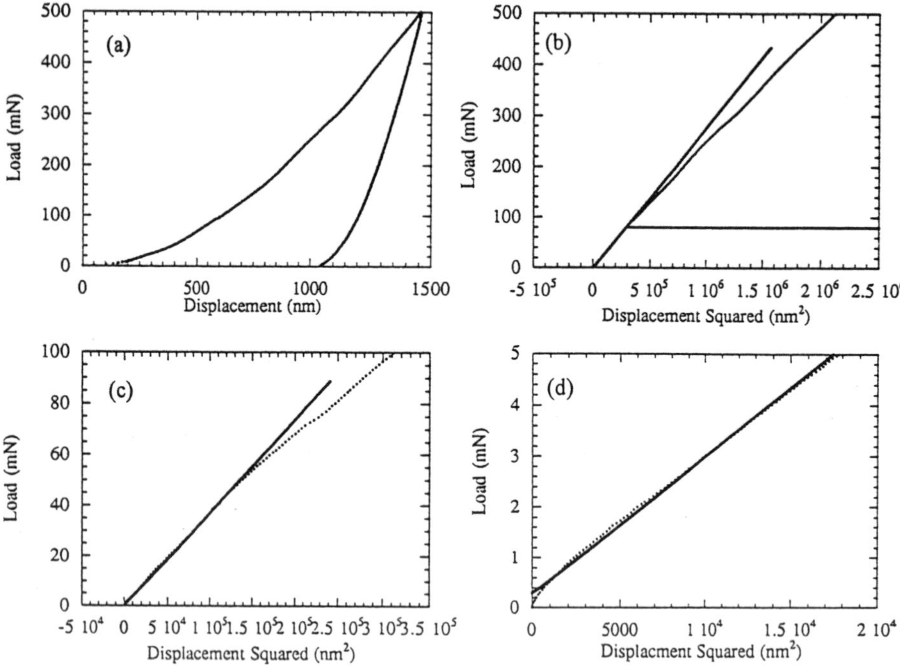

Figure 3 a) the load-displacement curve for a 500mN indentation into the 5μm-thick TiN$_x$ coating; b) the plot of P vs δ^2 for this indentation. c) and d) show P vs δ^2 plots for indentations to peak loads of 100mN and 5mN respectively (see text). (Key: • = data, — = straight line fit).

then the P vs δ^2 plot should be linear over the full extent of the data and this is indeed found to be the case as shown in Figure 3d. The small non-linear region at the very start of this plot is due to a combination of the detailed tip-end shape (i.e. the tip will not be ideally sharp at it's apex) together with any loading rate effects as the tip first makes contact with the surface etc. Thus plotting P vs δ^2 is an excellent test of "when is shallow, shallow enough" or, "when is the indentation response completely dominated by the coating alone". A further noteworthy point is that if the load of ~ 80mN where the P vs δ^2 relationship becomes non-linear is converted to a depth into the coating (by reading from figure 3a), it can be seen that the displacement is ~500nm, i.e. ~ t/10 where t is the coating thickness. This transition point has been evaluated for indentations into the 1 and 10μm coatings and the displacements at which non-linearity occurs for each of these coatings thicknesses were found to be ~175nm for the 1μm thick coating and ~700nm for the 10μm coating. For both the 20 and 30μm coatings, the P vs δ^2 relationship was found to be linear over the whole indentation depth, although some scatter was found in the data for these coatings which could be attributed to variations in the surface roughness.

The gradient of the straight line region in Figure 3b was calculated and this gives a value of K_{exp} = 3.03E-12Nm^{-2}. This can the be compared to the value of K_m (3.06E-12Nm^{-2}) which was obtained by substituting values of Hardness (H=18.1GPa) and Young's modulus (E = 408GPa) that were calculated using the Oliver & Pharr method [4]. Thus, in this case, there is excellent correlation between the loading curve model and unloading curve method. The added advantage of the loading curve method is that it is possible to detect whether - and over what range of displacement/load - coating only properties are being measured.

DISCUSSION

The transition between the P vs δ^2 being linear or non-linear appears to reproducibly occur at ~ t/10 for the range of coatings studied here (and also for a number of other coated systems studied [8]). This suggests that, up to this relative depth - which is often taken as a "rule of thumb" estimate [9] - both elastic and plastic deformation is concentrated within the coating with the substrate only providing elastic support of a small fraction of the load. For the elastic contribution, models such as that of Yu *et al* [10] would predict that for our indenter geometry and elastic modulus ratio, only ~ 10% of the elastic deformation occurs within the substrate.

However, a *relative plasticity criterion* is more important if the hardness of the coating is to be measured by indentation. If the sub-surface contact-induced shear stress distribution is broadly likened to that from a Hertzian contact, then this reaches a maximum of 0.3 x p_o (p_o = mean contact pressure) at a depth of $0.47a$ (where a is the contact radius). The shear stress then decays with a further depth (e.g. [11]). A *necessary criterion* for measuring the plastic properties of the coating *is that it yields before the substrate*. Thus, the maximum Hertz shear stress not only needs to lie in the coating but needs to exceed the shear stress of the coating (as the contact load is increased) before the shear stress experienced at depth ~ t exceeds the shear yield stress of the substrate. For a typical ratio of coating:substrate hardness of ~3:1, and with a typical Berkovich or Vickers indenter geometry, this gives a necessary indenter displacement of ~ t /10. Obviously, the use of soft substrates (e.g. ductile metals, polymers, etc) will always create the situation where the substrate yields first with the coating then being elastically bent and flexed into it followed by coating fracture (see, for example, the references cited in [1, 2]). Under such conditions, it will be impossible to measure the hardness of the coating itself by indentation tests, at least without changing to a harder substrate.

CONCLUSIONS

This paper has shown that a model previously developed to describe the nanoindentation loading curve for bulk materials can be successfully used to characterise the behaviour of coated systems. Not only does this P vs δ^2 analysis allow the differing regimes of behaviour of coated systems to be identified, but it also allows experimental verification of the load/displacement range over which "coating only" behaviour is being assessed. For the coatings used here, this depth appears as approximately one tenth of the coating thickness - agreeing with an often-used "rule of thumb" - and we have suggested a relative yield stress/Hertz shear stress criterion by which this depth fraction may be explained. A further necessary criterion for assessing "coating only" behavior is that the coating should yield plastically before the substrate.

ACKNOWLEDGEMENTS

The UK EPSRC are thanked for supporting the Newcastle NanoIndenter and for financial support (SVH). We thank Prof. H. W. Chandler (University of Aberdeen) for discussions on the modelling and Dr Steve Bull (AEA Technology, Harwell) for producing the coatings. Dr Hans Sjöström and Prof. Jan-Eric Sundgren (Linköping University) prompted this work with their investigations of CN_x and TiN films. TFP thanks the MRS for support to attend this meeting.

REFERENCES

1.   T.F. Page and S.V. Hainsworth, Surface Coatings and Technology **61** 201-208 (1993)

2.   T.F. Page, these proceedings, (1996)

3.   M.F. Doerner and W.D. Nix, J. Mater. Res. **1** 601-609 (1986)

4.   W.C. Oliver and G. M Pharr, J. Mater. Res. **7** 1564-1583 (1992)

5.   M.V. Swain and J. Mencík, Thin Solid Films **253** 204-211 (1994)

6.   S.V. Hainsworth, H.W. Chandler and T.F. Page, accepted for J. Mater. Res., 1996

7.   J.L. Loubet, J. M. Georges and J. Meille, in <u>Microindentation Techniques in Materials Science and Engineering</u> Eds. P.J. Blau and B.R. Lawn, American Society for Testing and Materials, Philadelphia, 1986, 72-89

8.   S.V. Hainsworth, unpublished work

9.   G.M.Pharr and W.C. Oliver, MRS Bulletin, **17** 28-33 (1992)

10.   H.Y. Yu, S.C. Sanday and B.B. Rath, J. Mech. Phys. Solids **38** 745-764 (1990)

11.   J.A.Williams, <u>Engineering Tribology</u>, Oxford University Press, 1994, p. 95

PROBING SILICATE/SAPPHIRE INTERFACES WITH AFM AND NANOINDENTATION

E.T. LILLEODDEN, A.V. ZAGREBELNY, S. RAMAMURTHY AND C.B. CARTER
Department of Chemical Engineering and Materials Science, University of Minnesota,
Minneapolis, MN 55455, carter@cems.umn.edu

Abstract

A nanoindentation study of a calcium-aluminosilicate/α-alumina system is described. Thin-films of $CaAl_2Si_2O_8$ glass were deposited on α-alumina single crystals of $\{11\bar{2}0\}$ and $\{1\bar{1}02\}$ orientations by pulsed-laser deposition. These samples were then subjected to a heat treatment causing the glass to crystallize and to dewet the sapphire surface. A nanoindentation device retrofitted on an atomic force microscope was employed to probe the mechanical properties of the samples. It uniquely enables the characterization of sub-micron features such as the dewet crystallites which necessitate small displacements and exact positioning of the indenter. Strong dependencies on the crystallographic anisotropy were observed in the morphology, but less so on the mechanical properties. In addition, displacement-excursions were observed and are explained in terms of yielding in the substrate.

Introduction

Glassy intergranular phases are present in bulk polycrystalline aluminas due to impurity elements in the starting materials and as a result of liquid-phase sintering. The presence of silicate glasses can often affect the mechanical properties of the bulk material. By the application of suitable heat treatments or under service conditions, the glassy phases may be crystallized. The mechanical integrity of bulk aluminas is controlled by the intergranular phase, as intergranular fracture is the most common mechanism of failure [1]. It has been shown that grain bridging acts as an effective toughening mechanism, and may be dependent on the adherence of the intergranular phase to the alumina grains [2]. It has been argued that the degree of crystallinity of the intergranular phase will therefore promote grain bridging [3]. In order to improve such properties of commercially relevant materials, a fundamental understanding of the interfacial properties is important. The use of conventional mechanical testing techniques is not suitable for probing specific interfacial properties due to many interdependent process variables present in such experiments. Therefore, a model sample geometry has been adopted and a unique nanoindentation device has been employed to probe the mechanical properties of the interface of both amorphous and crystallized anorthite ($CaAl_2Si_2O_8$) with α-alumina single crystals [4]. The overall objective of the present work is to study the interfacial properties, both structural and mechanical, and investigate any orientational or phase (amorphous versus crystalline) dependencies with the surface crystallography of sapphire.

Experimental

Two orientations of single-crystal α-alumina were employed in this study; namely $\{11\bar{2}0\}$ and $\{1\bar{1}02\}$. Square samples (2.5mm) were cut from as-received crystals obtained from Crystal Systems, Inc. (Salem, MA), and then annealed at 1400°C for 8 hours in air. A 100nm thick film of amorphous calcium aluminosilicate (CAS) was deposited using pulsed-laser deposition (PLD) as described elsewhere [5]. The films were deposited at 200°C in an oxygen partial pressure of 20mTorr. The laser energy was maintained at 100mJ/pulse and a pulse-repetition rate of 10Hz was used. The silicate/sapphire samples were then subjected to a heat treatment for 3 hours at 1200°C in air.

Atomic-force microscopy (AFM) was used to characterize the surface morphology of the samples (annealed, as-deposited and heat-treated). The mechanical properties were probed using a nanoindentation device [6] used in conjunction with an AFM. This device permits imaging of the pre- and post-indent surface, thereby allowing exact positioning of the indentation. The imaging capabilities eliminate ambiguities associated with quantifying the mechanical properties of inhomogeneous surfaces. The force is electrostatically generated with a three-plate capacitive

Mat. Res. Soc. Symp. Proc. Vol. 436 © 1997 Materials Research Society

structure. The load and resulting displacement are continuously measured during indentation. During imaging, the contact force is held constant; the deflection is mapped to generate the AFM image. An STM type diamond probe (straight shaft) is used as both the imaging and the indenting probe. This eliminates any time delay in imaging the indentation, and any problems with repositioning the imaging probe on the indented area. It should be noted that since the same tip is used for imaging and indenting, the character of the tip should be carefully addressed. A sample of fused quartz was used for calibrating the tip-response characteristics. A set of indentation experiments was accomplished in the quartz before and after each set of indentations into the CAS/sapphire samples.

Load-displacement curves were generated from indentations with different values of maximum load. The surfaces were imaged using AFM before and after each indentation. Three tips were used for the experiments. A large, conical indenter was used on all samples, as it is the most robust tip of the three. Two sharper pyramidal tips were also used on several of the samples, but the tips cracked during indentations deep into the substrate. The integrity of the tip is an important concern during indentation into hard materials. Often the tip cracks rather than wears which can be seen in the post indent images as the shape of the contact area shape changes dramatically.

Results and Discussion

The morphology of the dewetted, heat-treated CAS on sapphire is shown in Figure 1. It is found to be strongly dependent on the crystallographic orientation. AFM images of the CAS on α-alumina $\{11\bar{2}0\}$ shows (a) the directionality of the overall morphology and (b) the facetting of individual islands. The morphology of the heat-treated glass on the $\{1\bar{1}02\}$ alumina surface can be seen to be different in comparison to the islands on the $\{11\bar{2}0\}$ surface (Figure 1 c and d). The faceted morphology of the islands on both the surfaces of sapphire suggests crystallization of the glass film. Similar heat treatments to amorphous CAS films on (0001) α-Al_2O_3 have produced islands that epitactically crystallized on the sapphire surface [5].

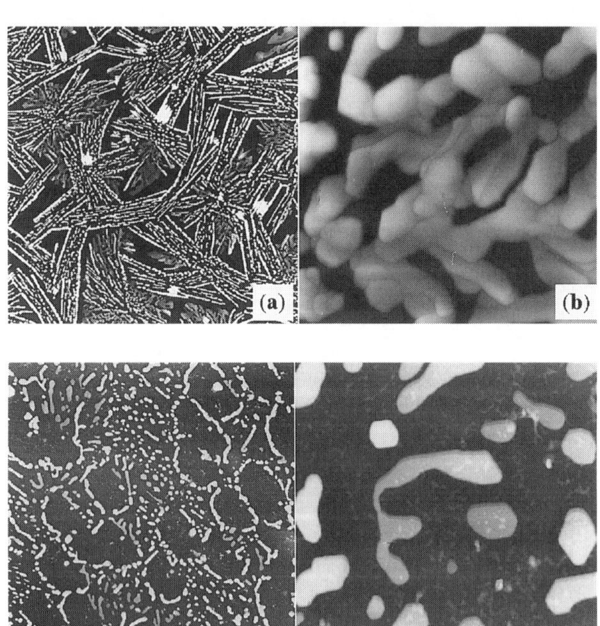

Figure 1. AFM images of the heat treated CAS/alumina samples. A 75μm square area of the $\{11\bar{2}0\}$ sample exhibits (a) directionality of the overall morphology, and a 5μm square area shows (b) faceting of the individual islands. (c) A 30μm square area and (d) 5μm square area of the dewetted CAS on $\{1\bar{1}02\}$ alumina exhibit a very different overall morphology, but show similar small scale faceting of the individual islands in comparison to the CAS/$\{11\bar{2}0\}$ alumina sample.

Modulus values, E*, were calculated using the Oliver and Pharr method [7] based on Sneddon's model for linear elastic unloading:

$$E* = \frac{dP}{dh}\frac{\sqrt{\pi}}{2\sqrt{A}} \tag{1}$$

where A is the projected contact area, as determined by the imaged contact area from post-indent AFM images. This is a reasonable approximation because the elastic pull-back is negligible for most metals and ceramics. However, for these relatively shallow depth indentations, the imaged area will underestimate the actual contact area, thereby tending to provide an upper bound value for E*. Table 1 summarizes the average values for each sample. Although the influence of the substrate on the elastic modulus of thin films may be assumed to be negligible for indentation depths less than 10 to 20% of the total film thickness [8], it is reasonable that the values are larger than that expected for a silicate glass because substrate effects are present. Similarly, contributions from the film will always be present, resulting in lower calculated values of modulus than expected for sapphire for indentations through the interface.

Table 1. Summary of modulus values calculated for CAS (δ<20nm), and for sapphire (through the interface), and a comparison of the experimentally observed critical load for dislocation emission with that predicted by a derivation of Johnson's axisymmetric indentation model. Also given are the radius of curvature for the indenter tips used in the critical load calculations, and the film thickness or crystallite height, h_{fo}.

	As-Deposited/ $\{11\bar{2}0\}$	As-Deposited/ $\{1\bar{1}02\}$	Heat-Treated/ $\{11\bar{2}0\}$	Heat-Treated/ $\{1\bar{1}02\}$
R (radius of curvature)	60nm	80nm	200nm	N/A
h_{fo} (film thickness)	100nm	100nm	200nm	N/A
E* (h<20nm)	115Gpa	126GPa	131GPa	145GPa
E* (through interface)	273Gpa	256GPa	253GPa	243GPa
P_{UYP} (experimental)	460μN	910μN	3550μN	N/A
P_{UYP} (model)	500μN	760μN	2700μN	N/A

It has been proposed that dislocation emission events can be observed in load-displacement curves generated in nanoindentation experiments, and are represented by displacement excursions, or "pop-ins" [e.g. 9, 10]. Indentation experiments on the annealed sapphire samples have verified that plasticity is initiated at the excursion; indentations below this point produced a load-displacement curve with superimposed loading and unloading segments which follow Hertzian contact theory. This has also been verified with immediate post-indent imaging. The critical point at which this pop-in occurs in the annealed, uncoated sapphire is dependent on the radius of curvature of the tip, R, the reduced modulus, E*, the critically resolved shear stress, τ_c, the resolved shear stress, τ_R, and maximum pressure under the indenter, p_o, as described by Johnson's axisymmetric indentation model [11]:

$$\frac{P_{UYP}}{E*R^2} = \frac{1}{6}\left[\frac{\pi\tau_c/E*}{\tau_R/p_o}\right]^3 \tag{2}$$

The predicted critical load is in good agreement with that observed experimentally.

Figure 2a shows superimposed load-displacement curves for indentations into the as-deposited CAS/ α-alumina $\{1\bar{1}02\}$ sample. Two peak-load indentations at two different regions on the same sample are shown. The reproducibility is shown by the perfect overlap of the loading part of the data. For the indentation up to a peak load of 500μN plasticity is observed below the excursion point, as verified by the unloaded residual displacement and the post-indent image shown in Figure 2b. This observation is expected because the deposited CAS film is an amorphous phase and therefore, any observation of pop-ins cannot be explained in terms of dislocation emission events. The sample was re-indented in a new region up to 1000μN and a pop-in was evident at 900μN. The enlarged residual contact area is shown in the post-indent image in Figure 2c. It is proposed that this signifies yielding in the substrate. A modification of equation (2) can be used in order to relate the film thickness to the critical load for substrate yielding, as follows [9]:

$$\frac{\tau_R}{p_o} = \frac{1}{\sqrt{18}}\left\{-2(1+\upsilon)\left[1-\frac{h_{fo}}{a}\tan^{-1}\left(\frac{a}{h_{fo}}\right)\right]+3\left(1+\frac{h_{fo}^2}{a^2}\right)^{-1}\right\} \qquad (3)$$

where h_{fo} is the film thickness and "a" is the contact radius.

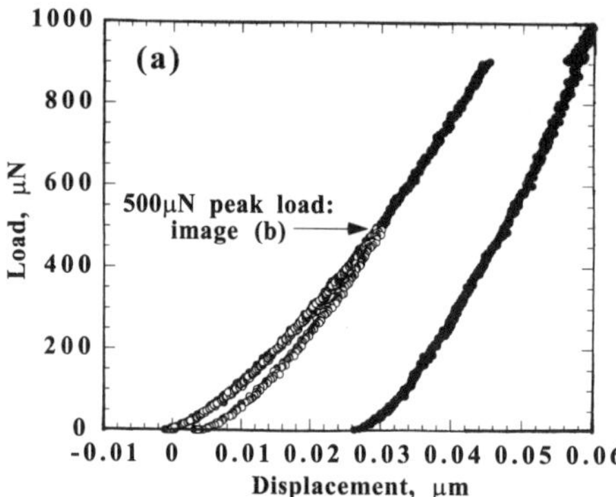

Figure 2. (a) The load-displacement curves for indentations up to 500μN and 1000μN maximum loads into the as-deposited CAS/α-alumina {1$\bar{1}$02} sample are superimposed. AFM images (500nm square) show of the post-indent topography of the as-deposited CAS on alumina {1$\bar{1}$02} (b) after a 500μN peak load (z range: 20nm), and (c) after a 1000μN (z range: 50nm) peak load indentation.

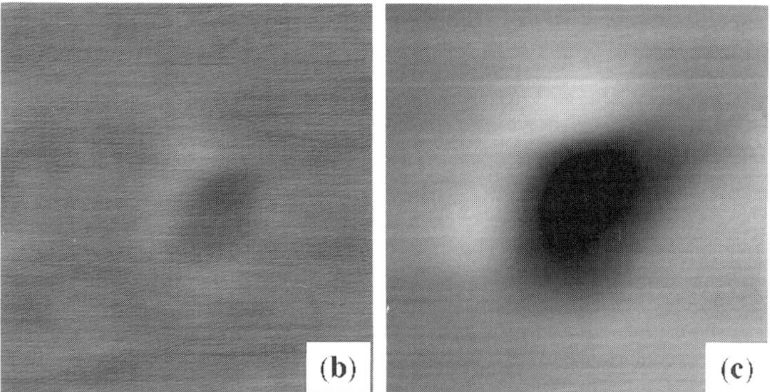

Since it has been proposed that dislocation emission events can be observed in load-displacement behavior, displacement excursions are expected for indentations into the heat-treated crystallites. Although anomalies were sometimes observed in the lower load regime they were not reproducible. Often the initial loading curve was characterized by sharp changes in slope, sometimes with increasing stiffness and sometimes with decreasing stiffness. The variations may be explained in terms of varying thickness and geometry of these islands, radial cracking, and/or interfacial delamination.

Figure 3 is an example of how *in situ* imaging is important in gaining a better understanding of the load-displacement behavior. Figure 3a shows a load-displacement curve of a 6mN peak-load indentation into the heat-treated CAS/alumina {11$\bar{2}$0} sample. Two pop-ins are observed; the first at 0.4mN and the second at 3.56mN load. These two distinct features in the load-displacement curve may correspond to dislocation-emission events in the crystallized CAS phase and in the substrate, respectively. However, upon examining the pre- and post-indent AFM images (Figures 3b and 3c) corresponding to this curve, it is observed that the indentation regime was coincident with a triple-junction in the crystallized phase. The grains have been displaced laterally from each other and radial cracks can be seen in the image. Therefore, it is likely that the first pop-in (which initiates at 115nm and terminates at the CAS/alumina interface) is not due to a dislocation emission event. The second pop-in occurs after an extended segment of continuous load-displacement data extending into the substrate, and at a critical load comparable to that for the annealed, uncoated sapphire experiments employing the same indenter tip.

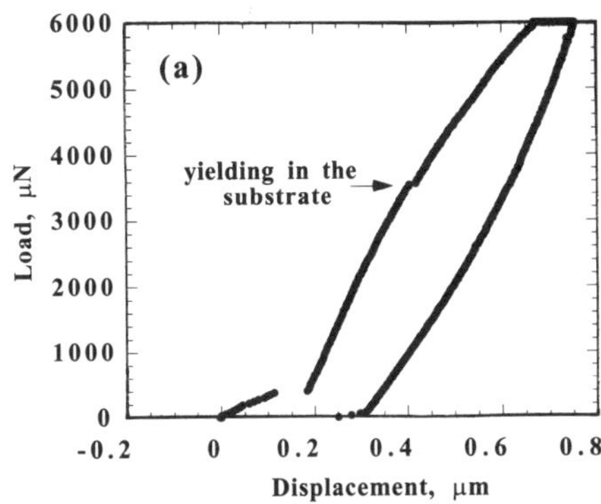

Figure 3. (a) Two discontinuities are observed in the load-displacement curve for a 6000μN maximum load indentation into the heat-treated CAS/α-alumina {11$\bar{2}$0} sample. AFM images show a 3μm square area corresponding to the (b) pre-indent and (c) post-indent surface morphology.

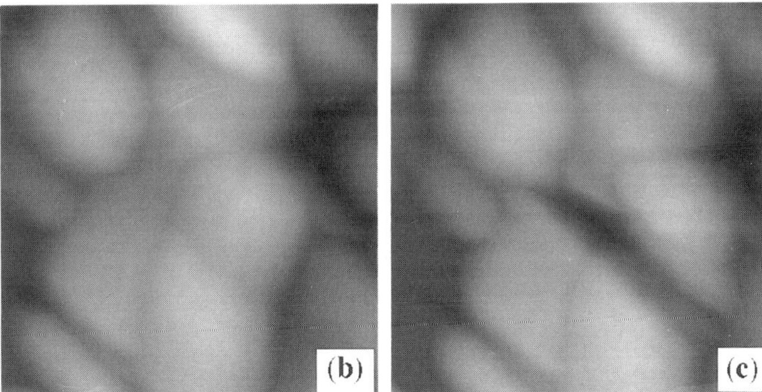

Equations (2) and (3) have been used to predict and compare the critical load for the upper excursion. It should be mentioned that although the film is not continuous with a thickness h_{fo}, 200nm (the thickness of the dewet island) has been used. Although the boundary conditions for small volumes is not addressed in this model, the model can be used for predicting the critical load to first approximation. The average thickness of the heat-treated CAS dewetted crystallites on alumina {1$\bar{1}$02} surface is 50nm which implies that the film has not completely dewet. Therefore this set of samples was not used in the predicted critical load calculations.

Summary

A model geometry has been used in this study for analyzing the mechanical properties of an amorphous phase and crystallized phase anorthite in contact with single-crystal α-alumina. A nanoindentation device with *in situ* imaging capabilities was employed to probe the submicron size island structures on the heat-treated samples. In addition to the advantage of exact positioning of the indenter tip, the imaging has allowed a better understanding of the load-displacement behavior. Pop-ins in the load-displacement curves for indentations into as-deposited CAS/sapphire samples have been observed, and correspond to substrate yielding. Similar load-displacement anomalies were also observed in the heat-treated CAS/sapphire samples. The critical load observed is in good agreement with that predicted by a derivation of Johnson's cavity model. However, the boundary conditions for small volumes must be addressed.

Acknowledgments

This research was supported the U.S. Department of Energy under Grant No. DE-FG02-92ER45465 (AVZ, SR and CBC) and by Hysitron, Inc. (ETL). Many helpful discussions with Prof. W.W. Gerberich are also acknowledged.

References
1. S.M. Weiderhorn, in <u>Ultra-Fine Grain Ceramics</u>, Edited by J.J. Burke, N.L. Reed and V. Weiss, Syracuse University Press, NY (1970).
2. H.L. O'Donnell, M.J. Readey and D. Kovar, *J. Am. Ceram. Soc.* **78**(4) 849 (1995).
3. N.P. Padture and H.M. Chan, *J. Mater. Res.,* **26** 2711-5 (1991).
4. A.V. Zagrebelny, E.T. Lilleodden, S. Ramamurthy and C.B. Carter, *Mater .Res.Soc. Symp. Proc.* **401** (1995).
5. M. P. Mallamaci, "Interfaces Between Alumina and Silicate-glass Films", Ph.D. Thesis, Cornell University, Ithaca, NY (1995).
6. Hysitron, Inc. (Minneapolis, MN).
7. W.C. Oliver and G.M. Pharr, *J. Mater. Res.* **7**(6) 1564 (1992).
8. W.D. Nix, *Met. Trans.* **20A** 2217 (1989).
9. W.W. Gerberich, J.C. Nelson, E.T. Lilleodden, P. Anderson and J.T. Wyrobek, *Acta Met.* In Press (1996).
10. T.F. Page, W.C. Oliver and C.J. McHargue, *J. Mater. Res.* **7**(2) 450 (1992).
11. K.L. Johnson, <u>Contact Mechanics</u>, Cambridge University Press, NY (1985).

Stress Reduction in PACVD amorphous Diamond Coatings by Boron Addition

V. Wagner[*], E.H.A. Dekempeneer[**], J. Geurts[+], L.J. van IJzendoorn[++], R.Sporken[*],
R.Caudano[*]

[*]LISE, Facultés Universitaires Notre-Dame de la Paix, Rue de Bruxelles 61, 5000 Namur,
Belgium, vwagner@fundp.ac.be
[**]Materials Division, Vlaams Instelling voor Technologisch Onderzoek, Boeretang 200, 2400
Mol, Belgium
[+]I. Physikalisches Institut, RWTH-Aachen, Sommerfeldstr. 28, 52056 Aachen, Germany
[++]Cyclotron Laboratory, Technical University of Eindhoven, 5600 MB Eindhoven, Netherlands

ABSTRACT

The applications of super hard coatings of diamond-like carbon (DLC) films are often limited by
adhesion problems caused by intrinsic stress in the deposited layer due to the PACVD growth
process. One approach to reduce the stress are less optimal process parameters. However, they
also affect the desired high hardness of the films. Our approach in this study is to add B_2H_6 to the
gas mixture, which shifts the composition towards another very hard compound (B_4C). A
sequence of amorphous B_xC_{1-x}:H films with a thickness of $0.5\mu m$ and a boron content between 0
and 50 at.-% were deposited on silicon substrates using a capacitively coupled r.f. PACVD
reactor. The tribological film properties and the internal stress were determined by depth-sensing
indentation and laser reflection measurements. For an addition of $x \geq 5$ at.-% boron the decrease
of the internal film stress is found to be clearly larger than the effect on the hardness value. For
boron contents $x > 18\%$ the internal stress is reduced by a factor of 5 while the reduction of
hardness is only a factor of 2.3. For microscopic structure analysis Raman and infrared
spectroscopy are applied. They reveal an increasing hydrogenation of the carbon network.
Therefore, the softening is attributed to a boron induced modification to a more polymeric-like
material.

INTRODUCTION

Diamond-like hydrogenated amorphous carbon (a-C:H) films synthesized by plasma
decomposition of hydrocarbons and by other ion-beam techniques generated much interest due to
their extreme mechanical, optical, and electrical properties. Earlier studies have revealed that the
a-C:H materials represent a broad range in structure. They are primarily amorphous with variable
proportion of sp2 and sp3 bonds; hydrogen concentration up to 60% were found in polymer-like
films [1]. However, films with high hardness values exhibit remarkably high values of internal
compressive stress due to the growth process [2-5]. This is often the limiting factor in practical
application due to stress induced adherence problems of the coating. In order to overcome this
problem either an interlayer is introduced to increase the adherence to the substrate, or less
optimal deposition conditions are used to decrease the internal stress value [6]. The first approach
leads, after finding an appropriate interlayer, to a more complicated growth sequence while the
latter approach reduces the desired high hardness values of the layers. In this paper we will follow
the second concept. However, instead of affecting the optimal deposition conditions a new
process parameter — a diborane (B_2H_6) gasflow — is introduced and varied systematically. The
addition of boron shifts the composition of the layers toward another very hard compound (B_4C).
This should make it possible to retain the desired high hardness values achieved for pure a-C:H
layers. In addition, boron doping of crystalline diamond films appears not to influence the crystal

quality [7] while reducing the internal layer stress [8]. Successful doping of Diamond to even several percent of boron already has been reported [9].

EXPERIMENT

Amorphous $C_{1-x}B_x$:H films were prepared in a capacitively coupled r.f. PACVD reactor using a gas mixture of CH_4 and B_2H_6 (10% in He) at a pressure of 0.015mbar. The diborane to methane gas flow ratio was varied up to 9:1. The films were deposited on silicon and glass substrates which were placed on the powered electrode (cathode). The negative bias voltage was fixed at -200V. The temperature stayed near 150°C during the deposition process. The carbon and boron contents of the films were measured by electron probe microanalysis (EPMA) while the hydrogen content was determined by elastic recoil detection analysis (ERDA). Film density values are also provided by the ERDA data. The film thicknesses were determined by locally etching away the films on the glass substrates in a CF_4-O_2 plasma and measuring the resulting steps by a stylus profilometer. Hardness and Young's modulus was measured with a depth-sensing indentation instrument (NanoTest 500) which provides continuous hysteresis plots of indentation depth vs. load [10]. The maximum indentation depth was limited to 100nm. The compressive film stress was deduced from the bending radius of the samples with silicon substrates. This bending was measured by a laser beam reflection technique. For this purpose the change of the reflection angle was monitored during a translation of the sample [5]. The microstructure of the films was investigated by infrared spectroscopy (double-beam spectrometer) in the 400-4000 cm^{-1} wavenumber range and by Raman spectroscopy (400-2000 cm^{-1}). The Raman spectra were excited by the 514.5nm line of an Ar^+ ion laser at 25mW beam power at a focus diameter of 100µm. The spectra were analyzed in an nonpolarized configuration by a double monochromator equipped with a multichannel detector.

RESULTS

<u>Sample preparation</u>
One film was deposited at optimized growth conditions without boron addition and serves as a reference a-C:H sample. Because of the high internal stress (see table I) the film thickness was

Table I: Composition, total hydrogen content y, tribological data, and internal stress values for the different samples with varying boron contents x. The stress and hardness values are plotted versus the boron content x in figure 1.

	composition $(C_{1-x}B_x)_{1-y}$:H_y		thickness	hardness	Young's modulus	int. stress
	x	y	(µm)	(GPa)	(GPa)	(GPa)
GLC811	-	0.33	0.37	23	172	2.6
GLC812	0.03	0.32	0.43	17.5	154	2.1
GLC813	0.05	0.32	0.47	15.6	137	1.7
GLC814	0.08	0.31	0.49	13	132	1.4
GLC815	0.10	0.32	0.50	14	128	1.2
GLC822	0.17	0.37	0.79	12.4		0.9
GLC823	0.23	0.37	0.64	10.3		0.6
GLC831	0.32	0.39	0.40	10.8	124	0.7
GLC832	0.46	0.41	0.27	9.1	109	0.5
typical error	<0.01	<0.01	0.02	8%	10	4%

chosen to be as low as 0.37µm to avoid adherence problems. The observed 33 atomic percent hydrogen is a typical value for hard hydrogenated amorphous diamond-like carbon layers [6]. For the other samples the EPMA composition results in table I prove a successful incorporation of boron in the layers. The total hydrogen content is found to decrease slightly until about 8 percent boron addition . Higher boron contents leads to re-increase of the total hydrogen portion up to 41% atomic percent. The gas flow ratio of diborane to methane necessary for the incorporation of approximately x=50% boron is remarkably high (9:1). This suggests either not optimal excitation (e.g. cracking) of diborane at the present plasma conditions or reduced acceptance of the boron species compared to carbon at the growing surface region. The decrease of the film density (not shown) implies that on average each boron atom introduces the volume occupied normally by approximately two carbon atoms.

Internal stress and tribological properties
The internal stress was determined from the measured sample curvature using the Stoney equation [5,6,11]. The successful reduction of the internal film stress by boron addition is a striking feature of the plotted values in figure 1. Up to 5% boron content the internal stress decreases almost linearly (-0.2 GPa / %-boron). Above 18% boron the stress remains constant at a residual value of 0.6 GPa. Unfortunately, the film hardness is also affected (figure 1). For small boron contents (x < 5%) the hardness is affected even proportionally. However, for x > 5% the stress reduction is considerably higher than the hardness decreases. Finally for a boron content above 18% the internal stress is reduced by a factor of 5 while the factor for the hardness reduction is only 2.3. These results have to be compared with a stress reduction of 30% accompanied by a 10% hardness reduction, as reported by Dekempeneer et al. [10] for pure a-C:H material at non-

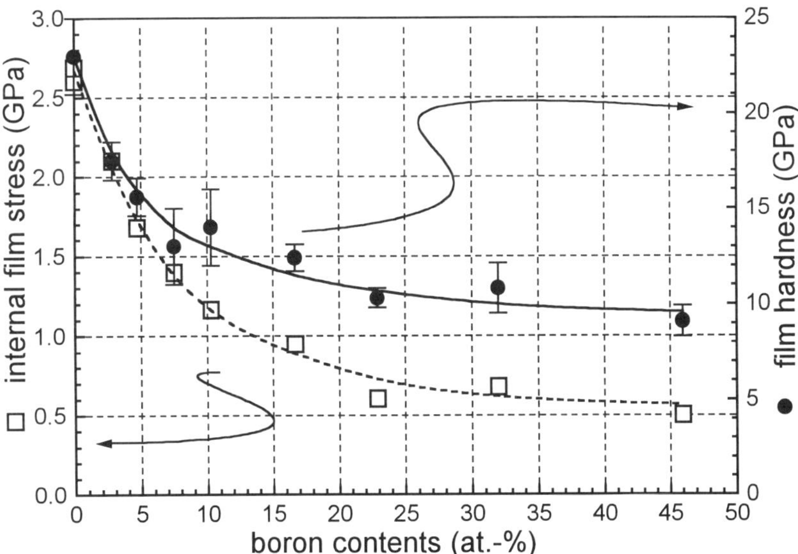

Figure 1: Internal stress and hardness versus the boron content x of the a-$C_{1-x}B_x$:H sample sequence. While for x < 5% there is an equal proportional decrease of stress and hardness, the stress reduction dominates at higher boron contents.

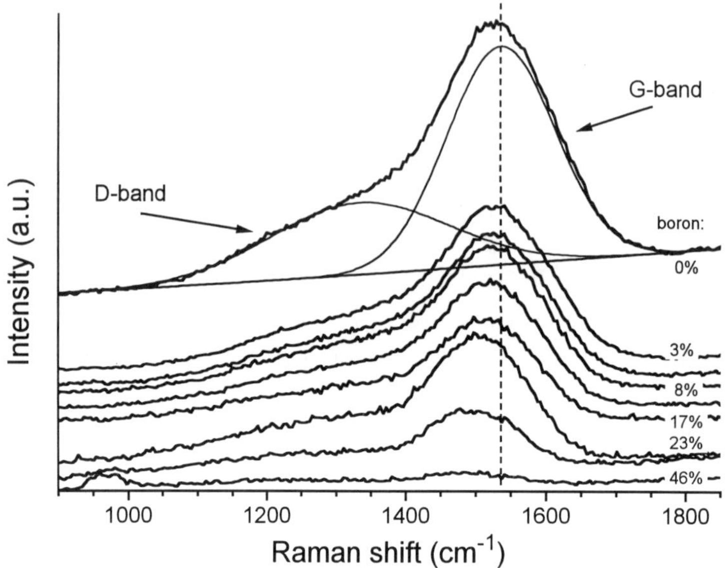

Figure 2: Raman spectra of a-$C_{1-x}B_x$:H films for different boron contents. In the x=0% spectra a fit of two gaussian peaks is incorporated to illustrate the position and shape of the carbon D- and G-band. With higher boron contents the G-band shifts to lower frequencies while the total intensity decreases.

optimized growth conditions (note: the hardness values reported in [6] have to be doubled to refer to the same 12GPa silicon reference value.). The hardness values presented in this paper are deduced from the indentation curves by the method of Oliver and Pharr [12]. However, since the exact treatment of the indentation curves is still controversial, all hardness values are normalized to the hardness obtained for a silicon(111) sample. The normalization factor was adjusted to obtain a stress value of 12 GPa for the silicon reference sample. The limit of 100nm for the indentation depth ensures an indentation depth to film thickness ratio below 1/2.7, which is well below the critical ratio of 1/1.5 given by Manika [13].

Assuming a Poissions ratio of $v=0.3$ the average film strain can be calculated by the measured Young's modulus E (which is almost constant) and the internal stress value σ:

$$\varepsilon_{xx} = (1-v)/E \cdot \sigma \quad , \quad \varepsilon_{zz} = -2v/E \cdot \sigma \tag{1}$$

Thus the compressive strain is ε_{xx}=1.1% for the a-C:H sample and reduces to ε_{xx}=0.3% for higher boron contents. The vertical elongation range is accordingly ε_{zz}=0.9% to 0.3%.

Microscopic film structure

Both, Raman and infrared spectra provide information about the intrinsic microscopic vibrational properties of the material. While infrared is very sensitive to polar vibration (e.g. C-H, C-B), Raman spectroscopy is able to detect also the non-polar vibrations (e.g. C-C, C=C). This has proved to provide important insight into the properties of hydrogenated and non-hydrogenated

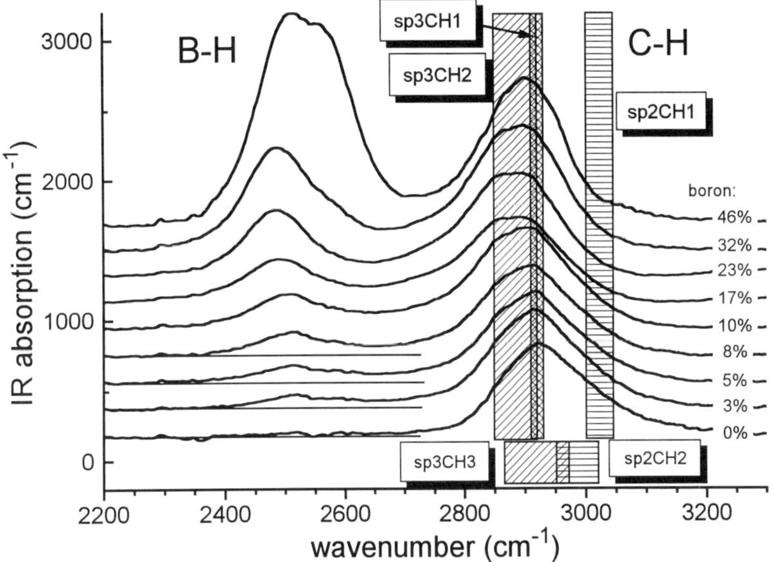

Figure 3: Infrared absorption spectra of the B-H and C-H stretching modes. The B-H feature is continuously increasing according to the boron contents while the C-H mode reveals an increasing hydrogenation (see text). Spectral positions of specific hydrogen-carbon coordination are indicated.

DLC films [14]. Raman spectra for the whole sample sequence are shown in figure 2. For the pure a-C:H sample a fit of the carbon D-band (1350cm^{-1}) and G-band (1580cm^{-1}) is included. With increasing boron content the G-band shifts to lower wavenumbers and the overall intensity decreases. For a quantitative comparison with the data reported by Ager et al. [14] all spectra were fitted by two gaussian peaks. Up to 10% boron the D- and G-band parameters obtained by the fitting procedure are consistent with pure a-C:H films which become more polymeric. This attribution is also consistent with the reduced scattering efficiency of less graphite-like clusters. Clear deviations of the D- and G-band parameters are observed for boron contents higher than 18%. This is explained by a larger amount of direct C-B bonds which excludes the description as a pure a-C:H-like material. The structure at 900 cm^{-1} for the 46% sample is a 2-phonon structure of the silicon substrate and is not related to the film. Infrared absorption spectra are shown in figure 3. They are restricted to the range of the B-H and C-H stretching modes. At lower wavenumbers (not shown) a continuously increasing absorption band at 1200 cm^{-1} occurs, indicating the formation of B-C bonds. The B-H absorption structure in figure 3 increases linearly with the boron contents, as may be expected. In contrast the intensity of the C-H mode stays almost constant. This behaviour suggests that mainly non-hydrogenated carbon is replaced by boron. Since the total hydrogen content stays almost constant up to 10% boron (table I), the increase of the total infrared absorption is explained by an increasing bonded to non-bonded hydrogen fraction in the layer (only bonded hydrogen is detected by infrared spectroscopy). Let us investigate the C-H structure in more detail. In the amorphous phase several C-H$_x$ bonding types are possible for sp2- or sp3-coordinated carbon atoms. The specific frequencies of the different configurations were measured by Dischler [15]. For hard a-C:H films the sp3CH$_1$, sp3CH$_2$ and sp2CH$_1$ configurations are dominating, as illustrated by the indicated frequency positions in figure 3. With increasing boron content the sp2CH$_1$ and sp3CH$_1$ components decrease until at 17% the C-H absorption structure can be explained by a single sp3CH$_2$ component. For the highest boron content a slight shift to higher wavenumbers might indicate even the appearance of the hydrogen-rich sp3CH$_3$-bonding configuration. Additionally a new B-H component appears at 2550cm^{-1} which is attributed to hydrogen in a boron-rich phase. This picture of boron-rich clusters in a polymeric carbon matrix agrees with the increase of the total amount of hydrogen observed for the highest boron contents.

CONCLUSIONS

It was demonstrated that boron addition can be used successfully to reduce the internal stress of PACVD-grown amorphous diamond-like carbon films. For small boron additions (x<5%) the desired high hardness value is observed to be reduced by the same ratio as the internal stress value. However, for larger boron contents (x>18%) the stress is reduced by a factor of 5 while the hardness is affected only by a factor of 2.3. An increasing hydrogenation of the carbon network induced by the boron addition is observed by infrared and Raman spectroscopy. Therefore the reduced hardness can be correlated to a softer and more polymeric-like material. This result suggests the possibility of a further optimization by a corresponding adjustment of the growth parameters.

ACKNOWLEDGEMENTS

The authors acknowledge the European Union for supporting this work via the Human Capital and Mobility Programme (networks), contract no. CHRX-CT94-0575, and Jan Meneve for carrying-out the nano-indentation measurements.

REFERENCES

1. J.C. Angus and F. Jansen, J. Vac. Sci. Technol. **A6**, 1778 (1988).
2. D.W.Hoffmann and J.A. Thornton, Thin Solid Films **40**, 355 (1977).
3. J.A.Thornton, J.Tabock and D.W.Hoffmann, Thin Solid Films **64**, 111 (1979).
4. H.Windischmann, J.Appl.Phys. **62**, 1800 (1987).
5. H.Windischmann, G.F.Epps, Y.Cong and R.W.Collins, J. Appl. Phys. **69**, 2231 (1991).
6. E.H.A. Dekempeneer, R.Jacobs, J.Smeets, J.Meneve, L.Eersels, B.Blanpain, J.Roos and D.J.Oostra, Thin Solid Films, **217**, 56-61 (1992).
7. K.Okano, H. Naruki, Y. Akiba, T. Kurosu, M.Iida, Y. Hirose and T.Nakamura, Jap. J. Appl. Phys. **28** (6), 1066 (1989).
8. K.Okano, T.Kurosu, M.Iida, T.Eickhoff, H.Wilhelm and D.R.T.Zahn in *Hard Materials in Optics*, (SPIE **1275** , 1990), pp.25-35.
9. M.W. Geis and J.C. Angus, Scientific American, 1992, 411.
10. E.H.A.Dekempeneer, J.Meneve, S.Kuypers, J.Smeets, Surf. Coat. Techn. **74-75**, 399 (1995).
11. J.W. Ager III, S. Andres, A. Andres, I.G. Brown, Appl.Phys.Lett. **66**(25), 3444 (1995).
12. W.C. Oliver and G.M. Pharr, J. Mater. Res. 7(6), 1564 (1992).
13. Manika et al., Thin Solid Films **208**, 223-227 (1992).
14. M.A. Tamor, W.C. Vassel, J. Appl. Phys. **76** (6), 3823-3830 (1994).
15. B. Dischler in *Amorphous Hydrogenated Carbon Films*, edited by P. Koidl and P. Oelhafen (European Mat. Res. Soc. Proc. **17**, Paris, France, 1987) p.189.

INACCURACIES IN SNEDDON'S SOLUTION FOR ELASTIC INDENTATION BY A RIGID CONE AND THEIR IMPLICATIONS FOR NANOINDENTATION DATA ANALYSIS

A. BOLSHAKOV* and G.M. PHARR*
* Department of Materials Science, Rice University, 6100 Main St., Houston, TX 77005

ABSTRACT

Methods currently used for analyzing nanoindentation load-displacement data to determine a material's hardness and elastic modulus are based on Sneddon's solution for the indentation of an elastic half-space by a rigid axisymmetric indenter. Although this solution is widely used, no attempts have been made to determine how well it works for conditions of finite deformation, as is the case in most nanoindentation experiments with sharp indenters. Analytical and finite element results are presented which show that corrections to Sneddon's solution are needed if it is to be accurately applied to the case of deformation by a rigid cone. Failure to make the corrections results in an underestimation of the load and contact stiffness and an overestimation of the elastic modulus, with the magnitude of the errors depending on the angle of the indenter and Poisson's ratio of the half-space. For a rigid conical indenter with a half-included tip angle of 70.3°, i.e., the angle giving the same area-to-depth ratio as the Berkovich indenter used commonly in nanoindentation experiments, the underestimation of the load and contact stiffness and overestimation of the elastic modulus may be as large as 13%. It is shown that a simple first order correction can be achieved by redefining the effective angle of the indenter in terms of the elastic constants. Implications for the interpretation of nanoindentation data are discussed.

INTRODUCTION

This paper deals with Sneddon's solution for the indentation of an elastic half-space by a rigid axisymmetric indenter [1,2]. Sneddon's solution is the cornerstone for determining elastic moduli from nanoindentation load-displacement data [3,4], and as such, the accuracy with which the elastic moduli can be measured depends on how well Sneddon's solution describes real material behavior.

The primary result of Sneddon's work of importance in this paper is the relation [3]:

$$E_{eff} = \frac{E}{1-v^2} = \frac{\sqrt{\pi}}{2} \frac{S}{\sqrt{A}} \tag{1}$$

Here, E_{eff} is the effective elastic modulus defined in terms of Young's modulus E and Poisson's ratio v, S is the contact stiffness, and A is the projected contact area. The equation applies to any rigid axisymmetric indenter, including singular indenters such as cones [3]. Since all of the quantities on the right hand side can be either measured or estimated from indentation load-displacement data, Eqn.1 forms the basis of elastic modulus measurement by load and displacement sensing indentation methods [3-5].

One potential problem in using Sneddon's solution is that it is derived for conditions of small deformation. These conditions are often well-satisfied for shallow indentations made by blunt indenters such as a sphere, but for sharp indenters like the Berkovich triangular pyramid used commonly in nanoindentation work, the small-deformation formulation may be inaccurate. An indication that there may be a problem for sharp indenters was recently encountered in finite element simulations of elastic/plastic deformation by a rigid cone using a finite element code which accounts for finite strain and rotation [6]. No matter how carefully the mesh size and shape were controlled, the elastic modulus derived from the simulated indentation load-displacement curves using Eqn.1 was always 10-20% larger than that used as input in the simulation. A similar result has been reported by Ritter et al [7]. Larsen and Simo [8], on the

189

Mat. Res. Soc. Symp. Proc. Vol. 436 © 1997 Materials Research Society

other hand, found relatively good agreement between the input and derived elastic modulus in their finite element simulations, but in carefully re-assessing their results, we have found that the procedure they used to measure the contact stiffness significantly underestimated its actual value. When corrected, the discrepancy between the input and derived elastic modulus in their finite element studies is even greater than in ours.

The discrepancies in the finite element results just described led us to undertake a careful study of the problems which can occur in applying Sneddon's equations to sharp indenters. Here, we report some observations for elastic indentation by a rigid cone and discuss their implications for the measurement of elastic modulus by nanoindentation methods. A more complete report, including implications for elastic/plastic deformation and the measurement of contact area and hardness, will be presented in a subsequent report [9].

INACCURRACIES IN SNEDDON'S SOLUTION FOR A RIGID CONE

The problem in applying Sneddon's solution to deformation by sharp indenters can be illustrated by considering his analysis of the indentation of an elastic half-space by a rigid cone. For a cylindrical coordinate system with a free surface initially at z=0, the boundary conditions used by Sneddon are:

$$\sigma_{zz}\big|_{z=0} = 0, \quad r > a \tag{2}$$

$$\sigma_{rz}\big|_{z=0} = 0, \quad r \geq 0 \tag{3}$$

$$w\big|_{z=0} = D - \cot(\phi)\, r, \quad r < a \quad . \tag{4}$$

Here, a is the contact radius, D is the depth of penetration, ϕ is the half-included angle of the cone, w is the vertical displacement of the surface, and the σ_{ij} are the components of the stress tensor. The first two boundary conditions specify the tractions on the z=0 surface (no friction is assumed), and the third condition forces the vertical displacements of the surface inside the circle of contact to be consistent with the geometry of the cone. Posed in this manner, the problem can be solved by the method of Hankel transforms [1,2]. The primary result of importance in this discussion is the expression for the radial displacements, u, of points inside the circle of contact, which is given by:

$$u\big|_{z=0}(r,\phi) = \frac{(1-2\nu)}{4(1-\nu)} \cot(\phi)\, r \left[\ln \frac{r/a}{1+\sqrt{1-(r/a)^2}} - \frac{1-\sqrt{1-(r/a)^2}}{(r/a)^2} \right], \quad r < a \quad . \tag{5}$$

Close examination of this equation shows that the surface radial displacements vanish only when Poisson's ratio ν is 0.5, i.e, the material is incompressible, or when ϕ=90°. Thus, during indentation by a sharp cone, the surface radial displacements for most materials of practical interest are finite. An important consequence is that when the radial displacements are taken into account, the deformed surface inside the area of contact is not conical but rather described by (see [9] for details):

$$r = \frac{z}{\cot(\phi)} + \frac{(1-2\nu)}{4(1-\nu)}\, z \left[\ln \frac{z/[a\cot(\phi)]}{1+\sqrt{1-z^2/[a\cot(\phi)]^2}} - \frac{1-\sqrt{1-z^2/[a\cot(\phi)]^2}}{z^2/[a\cot(\phi)]^2} \right] . \tag{6}$$

Based on the predictions of Eqn. 6, the shapes of surfaces deformed by a 70.3° cone for materials with various values of Poisson's ratio are shown in Figure 1. Note that only in the case of ν=0.5 is the deformed surface consistent with the 70.3° conical geometry; that is, for all other Poisson's ratios, the surface is displaced inward from the 70.3° cone and is slightly curved.

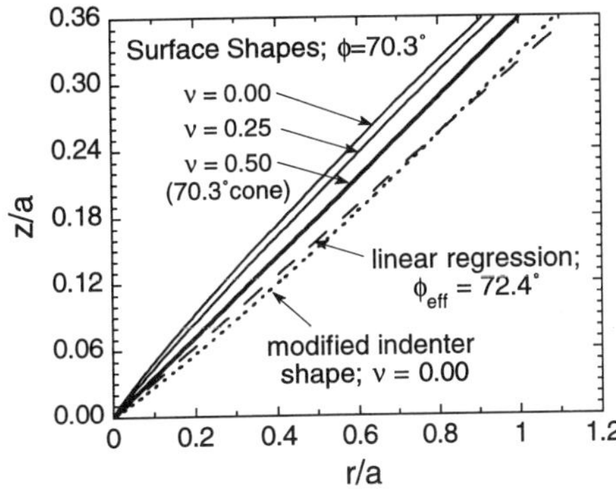

Fig. 1. The shapes of surfaces indented by a 70.3° cone when the radial displacements in Sneddon's solution are included. Also shown is the shape of the alternative indenter needed to account for the radial displacements for the case of $v=0$, and the linear approximation to the shape which defines the effective cone angle, ϕ_{eff}.

What this means is that in most cases of practical interest, Sneddon's solution applies not to a perfect cone, but to a cusped shaped indenter which approximates to a cone. As we will show in the next sections, this small deviation has important consequences for the measurement of elastic moduli by nanoindentation methods.

MODIFICATIONS TO THE SOLUTION

A solution which more accurately describes the indentation of an elastic material by a rigid cone rather than a cusp-shaped indenter can be achieved by modifying the boundary condition in Eqn. 4 to account for the finite radial displacements of points along the surface of contact. Under these circumstances, a more appropriate statement of the boundary condition is:

$$w \big|_{z=0} = D - \cot(\phi)\,[r + u(r)], \quad r < a \quad . \tag{7}$$

This boundary condition assures that the geometry of the deformed surface will be conical with a half-included angle ϕ which matches that of the indenter.

Two different approaches may be used to solve this problem - one exact and one approximate. The exact solution makes use of an indenter which deviates from the conical geometry, chosen so that the deformed surface will be a perfect cone of angle ϕ when the radial displacements are taken into account. The solution is obtained by treating the function u(r) in Eqn. 7 as an unknown and deriving an expression for u(r) using relations developed by Sneddon for the radial displacements of a general axisymmetric indenter [2]. Normalizing all length dimensions with respect to the contact radius, a, the approach leads to non-dimensional radial displacements given by:

$$\bar{u}(\rho) = \frac{(1-2v)}{2(1-v)}\frac{\cot(\phi)}{\pi}\left[\frac{\pi}{4}\rho\ln\frac{\rho}{1+\sqrt{1-\rho^2}} - \frac{\pi}{4}\frac{1-\sqrt{1-\rho^2}}{\rho} + \frac{1}{\rho}\int_0^1\frac{y\bar{u}(y)}{\sqrt{1-y^2}}dy - \right.$$

$$\left. \frac{1}{\rho}\int_\rho^1\frac{x^2}{\sqrt{x^2-\rho^2}}dx\int_0^x\frac{\bar{u}'(y)}{\sqrt{x^2-y^2}}dy\right], \tag{8}$$

where $\rho = r/a$ and $\bar{u}(\rho) = u(r)/a$. Numerical techniques can be used to solve for u(r), which can be used in combination with other results derived by Sneddon to provide an exact solution [2]. However, because the solution requires numerical evaluation, it is somewhat cumbersome and awkward. As a result, we limit the discussion here to the second solution technique which has the distinct advantage of having a very simple, closed form. The solution is only approximate, but may prove to be more practical in nanoindentation data analysis. The accuracy of the solution will be checked by comparison to finite element simulations.

In the approximate solution, the alternative indenter is constructed by taking the perfect conical geometry with half-included angle ϕ and increasing the radius at each point along its surface by an amount equal to the magnitude of the radial displacements in Eqn. 5. The shape of the indenter constructed in this manner for the case of $\nu=0$ is shown in Figure 1. The rationale is that since the radial displacements of the surface in Sneddon's solution are negative, increasing the radius of the perfect cone by an amount equal to the expected radial displacements should, to a first approximation, produce a deformed surface having a geometry close to the intended conical shape. A second assumption made in the analysis is that the curvature of the surface of the modified indenter is small enough to be ignored. In this case, the modified indenter can be modelled as a cone with an effective included angle, ϕ_{eff}, slightly greater than that of the ideal cone (see Figure 1). Linear regression of the shape of the modified indenter yields:

$$\cot(\phi_{eff}) = \frac{\cot(\phi)}{\beta} \tag{9}$$

where the parameter β is given by:

$$\beta = 1 + \frac{(1-2\nu)}{4(1-\nu)} \left(3 - \frac{\pi}{2} \right) \cot(\phi) \tag{10}$$

Eqn. 9 shows that the effective indenter angle, ϕ_{eff}, depends on the cone angle, ϕ, and the Poisson's ratio of the material. In the limit of incompressible deformation ($\nu=0.5$), ϕ_{eff} is exactly equal to ϕ, but when ν is small, the two angles can be significantly different. For $\nu=0$ and $\phi=70.3°$, i.e., for a cone with the same area-to-depth ratio as the Berkovich indenter, the effective cone angle is $\phi_{eff} = 72.4°$. The attractiveness of this approach is that all of the relations derived by Sneddon for the indentation of an elastic half-space by a rigid cone still apply provided the indenter angle ϕ is replaced by ϕ_{eff}. Thus, the load-displacement relation is:

$$P = \frac{2}{\pi} \frac{E}{1-\nu^2} D^2 \tan(\phi_{eff}) = \beta \frac{2}{\pi} \frac{E}{1-\nu^2} D^2 \tan(\phi) \tag{11}$$

and the modified form of Eqn. 1 is:

$$E_{eff} = \frac{E}{1-\nu^2} = \frac{1}{\beta} \frac{\sqrt{\pi}}{2} \frac{S}{\sqrt{A}} \tag{12}$$

The extent to which the modified relations deviate from Sneddon's original formulation depends on the magnitude of the parameter β, which in turn depends on both Poisson's ratio and the cone angle. Limiting the discussion here to $\phi=70.3°$, the largest deviations from Sneddon's original solutions occur when $\nu=0$, for which $\beta=1.13$. Thus, for materials with small Poisson's ratios, Sneddon's solution can underestimate the load and overestimate the elastic modulus by as much as 13%. As previously discussed, an overestimation of the modulus is exactly what has been observed in finite element simulations. For a more typical value of ν like 0.25, $\beta=1.09$, so a 9% overestimation is expected in the modulus. When $\nu=0.5$, $\beta=1.00$, and Sneddon's original equations are accurate.

COMPARISON TO FINITE ELEMENT SIMULATIONS

To check on the applicability and accuracy of the approximate solution, finite element simulations were conducted using the ABAQUS finite element code. The indenter was modeled as a rigid cone with a half-included angle of 70.3° and the material as isotropic elastic with Young's modulus E and Poisson's ratio ν. All simulations were performed to the same indentation depth, D=100 nm, using a finite element mesh similar to one used previously [6].

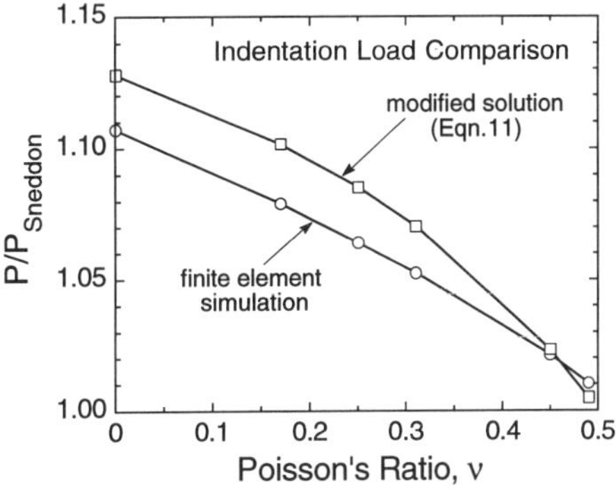

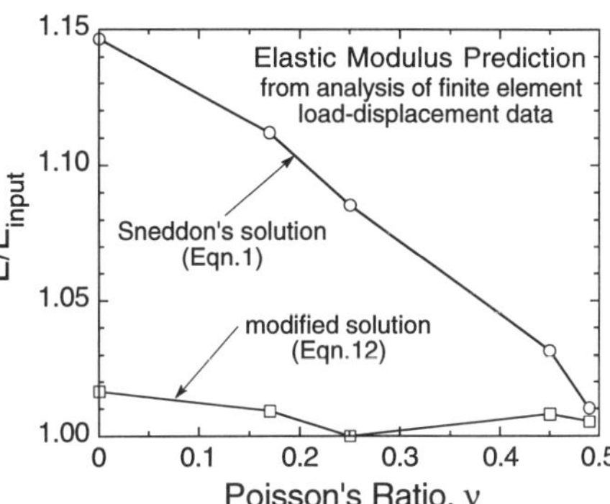

Fig. 2. A comparison of the indentation loads observed in the finite element calculations to Sneddon's solution and the approximate modified solution. Loads are normalized with respect to Sneddon's results and plotted as a function of Poisson's ratio.

Fig. 3. A comparison of the effective elastic moduli derived by applying Eqns. 1 and 12 to the finite element load-displacement data. The moduli are normalized with respect to the effective modulus input into the simulations and plotted as a function of Poisson's ratio.

The simulations were conducted for various ν's and E's but with the effective modulus $E_{eff} = E/(1-\nu^2)$ held constant at $E_{eff} = 90.85$ GPa. The reason for performing the simulations in this way was to explore an important difference between the dependencies on Poisson's ratio of Sneddon's original solution and the modified solution derived here. The difference may be seen by comparing Eqns. 1 and 12, which shows that E_{eff} computed from Sneddon's solution should be independent of Poisson's ratio, while E_{eff} derived from the modified solution is not because β is a function of ν. The same is true of the indentation load at a fixed depth.

Figures 2 and 3 present a comparison of the finite element results with the two solutions. Figure 2 shows the dependence of the predicted indentation loads on Poisson's ratio (the loads are normalized with respect to those predicted by Sneddon's solution), while in Figure 3, the variation of the elastic moduli predicted by Eqns. 1 and 12 are compared. In computing the elastic moduli, the contact areas were determined directly by examination of the finite element mesh rather than indirectly through an analysis of the load-displacement data, as would be done in nanoindentation experiments.

The results in Figure 2 show that the indentation loads in the finite element simulations are larger than those predicted in Sneddon's solution. For ν=0, the error in Sneddon's solution according to the finite element calculations is about 11%, but the error decreases essentially to zero at ν=0.5, consistent with the fact that the radial displacements of the surface vanish as ν approaches 0.5. On the other hand, the modified solution matches the finite element results reasonably well, thereby providing justification for the assumptions made in deriving it, as well as demonstrating its potential for application in nanoindentation analysis procedures. Figure 3 demonstrates the implications for Young's modulus measurement from nanoindentation load-displacement data. It shows that if Sneddon's results are used (Eqn. 1), errors as large as 14% may be obtained, but these errors can be largely avoided by using the modified solution (Eqn. 12). The results emphasize the need for modification of Eqn. 1 in the manner outlined here if accurate elastic moduli are to be derived from analyses of nanoindentation data.

CONCLUSIONS AND IMPLICATIONS FOR NANOINDENTATION DATA ANALYSIS

The analyses and finite element simulations presented in this paper show that corrections to Sneddon's solution for indentation of an elastic half-space by an axisymmetric indenter are needed if accurate measurements of elastic modulus are to be obtained in nanoindentation experiments with sharp indenters. Failure to make the corrections results in an underestimation of the indentation load and contact stiffness and an overestimation of the elastic modulus by as much as 13%. The amount of the correction depends on Poisson's ratio and can be estimated using procedures developed in the paper.

It is important to note that the contact area deduced from Sneddon's solution may also be affected by the problems identified here. Corrections to the contact area, if required, would have implications not only for the measurement of elastic modulus, but for the measurement of hardness as well. It is also important to note that the analyses presented in this paper apply strictly only to elastic deformation. For sharp contact involving both elastic and plastic deformation, the situation is considerably more complex. We are currently studying the role that these issues play in modifying the results [9].

ACKNOWLEDGMENTS

This research was sponsored by the Division of Materials Sciences, U.S. Department of Energy, under contract DE-AC05-96OR22464 with Lockheed Martin Energy Research Corp. One of the authors (GMP) is grateful for sabbatical support provided by the Oak Ridge National Laboratory.

REFERENCES

1. I.N. Sneddon, Int. J. Engng. Sci. **3**, 47 (1965).

2. I.N. Sneddon, Fourier Transforms (McGraw-Hill Book Company, Inc., 1951), pp. 450-467.

3. G.M. Pharr, W.C. Oliver, and F.R. Brotzen, J. Mater. Res. **7**, 613 (1992).

4. W.C. Oliver and G.M. Pharr, J. Mater. Res. **7**, 1564 (1992).

5. M.F Doerner and W.D. Nix, J. Mater. Res. **1**, 601 (1986).

6. A. Bolshakov, W.C. Oliver, and G.M. Pharr, J. Mater. Res. **11**, 760 (1996).

7. J.E. Ritter, T.J. Lardner, D.T. Madsen, and R.J. Gionazzo, in <u>Materials in Microelectronic and Optoelectronic Packaging, American Ceramic Society,</u> p. 379.

8. T.A. Larsen and J.C. Simo, J. Mater. Res. **7**, 618 (1992).

9. A. Bolshakov and G.M. Pharr, unpublished work.

THE EFFECT OF Ta AND N CONTENT ON MECHANICAL PROPERTIES OF DC MAGNETRON SPUTTERED FeN AND FeTaN THIN FILMS

Hong Deng, M. Kevin Minor and John A. Barnard

Department of Metallurgical and Materials Engineering and Center for Materials for Information Technology, The University of Alabama, Tuscaloosa, AL 35487-0202

ABSTRACT

This paper reports nanoindentation studies of the effect of Ta and N content on the mechanical properties of magnetically soft high moment FeN and FeTaN thin films prepared by dc magnetron sputtering. The FeTaN films were deposited on oxidized silicon (100) substrates with a series of FeTa targets in which the Ta content varies from 0 to 25wt%. The hardness (H) and Young's modulus (E) were measured by the Nano Indenter at nine indenter penetration depths: 20, 30, 40, 50, 60, 80, 100, 120 and 200 nm. The inherent hardness values of these films (no substrate effect) can be determined at penetration depths ranging from 20 to 60 nm for the 500 nm thick film used in the study. It was found that for the films deposited from the pure Fe target when the nitrogen flow rate increases from 0 to about 0.5 sccm the hardness of the film increases. However, a decreasing trend in hardness of these films was observed on further increasing the nitrogen flow rate. On the other hand, for the films prepared from the targets with the Ta content in the range of 5-15wt%, the hardness increases whenever Ta and N contents increase. These effects are clearly illustrated by 3-D and contour hardness and Young's modulus maps in this paper.

I. INTRODUCTION

Magnetically soft high moment FeXN (X=Ta, Al etc.) films have potential application in next-generation recording heads for magnetic storage systems. In addition to their excellent magnetic properties, superior to those of traditional head pole materials such as permalloy (NiFe), these films also possess mechanical and tribological properties better than permalloy, as demonstrated in our previous work [1]. It is known that mechanical properties such as hardness, elastic modulus and strain rate sensitivity are important for tribological performance. Recently, penetration depth sensing nanoindentation techniques such as the Nano Indenter (Nano Instruments, Inc. Oak Ridge, TN) and UMIS 2000 (CSIRO, Australia) have proved very effective, powerful and versatile for characterizing mechanical properties of thin films and hard coatings [2, 3]. Jones et al measured the hardness of Fe(N) films sputtered in an Ar-N_2 mixture by nanoindentation [4]. We have reported a series of studies on the mechanical and tribological properties of FeTaN head materials [1, 5, 6]. The objective of this study is to provide systematical mechanical property maps of sputtered FeN and FeTaN thin films which account for the effect of nitrogen and tantalum content. A series of FeN and FeTaN films were deposited by dc magnetron sputtering with a systematic variation of tantalum (Ta) (0-25wt%) and nitrogen (N) (0-3.5 sccm deposition flow rates) contents. The hardness (H) and Young's modulus (E) of these films were measured by nanoindentation. Based on these experimental data, three dimensional (3-D) and contour hardness and elastic modulus maps were generated so that the effects of tantalum and nitrogen content on these properties of FeN and FeTaN sputtered films can be clearly illustrated.

EXPERIMENTAL DETAILS

<u>Film Deposition</u>

The FeN and FeTaN films were deposited on oxidized silicon (100) substrates with a series of FeTa targets in which the Ta content varies (0, 5, 10, 15 and 25wt%) by dc magnetron sputtering

Mat. Res. Soc. Symp. Proc. Vol. 436 ® 1997 Materials Research Society
"

in an Ar-N$_2$ mixture using a Vac-Tec Model 250 Batch Side Sputtering System. The base pressure of the sputtering system is 2.4x10^{-7} mTorr. In order to examine the effects of tantalum and nitrogen contents, a fixed deposition power, total Ar pressure and deposition time were used. The sputtering deposition power and total pressure were fixed at 220 W and 3.0 mTorr respectively. Films were sputtered in a mixed gas atmosphere using separate bottles of pure Ar and mixed 90%Ar+10%N$_2$ gas in order to control nitrogen incorporation in the film. Nine flow rates (0, 0.3, 0.5, 0.7, 1.0, 1.5, 2.5, 3.0 and 3.5 sccm) of 90%Ar plus 10% N$_2$ mixture were employed. With an 8 minute deposition time the thicknesses of resultant films are about 500 nm. Because the fixed deposition power and total pressure were fixed a nearly constant deposition rate of 1 nm/s was achieved. The relationships between deposition rate, power and total pressure have been discussed previously [7]. The sputtering condition details are summarized in Table I. After deposition the specimens were used for the characterization of magnetic, structure, stress, mechanical and tribological properties. Magnetic and structural properties of these films were reported elsewhere [8].

Table I. Sputtering Deposition Conditions for FeN and FeTaN Magnetic Thin Films.

Target Ta content (wt%)	0, 5, 10, 15, 25	Deposition power(W)	220
Nitrogen flow rates (sccm)	0, 0.3, 0.5, 0.7, 1.0, 1.5, 2.5, 3.0, 3.5	Sputtering time (min)	8
Base pressure (mTorr)	2.4x10^{-7}	Film thickness (nm)	500
Total pressure (mTorr)	3.0	Deposition rates (nm/s)	1

Mechanical Property Characterization

The hardness (H) and Young's modulus (E) of the FeN and FeTaN films were measured by the Nano Indenter® II Mechanical Properties Microprobe. 6x4 arrays with 50 μm spacing were used for the indentations on each film specimen. Three experiments were conducted to obtain the information at nine indenter penetration depths: 20, 30, 40, 50, 60, 80, 100, 120 and 200 nm. In each experiment, three constant displacement loading rate load segments were employed with three pre- controlled depth limits. Loading rates were usually 10% of the total penetration depth, such as 5 nm/s for the depth of 50 nm penetration depth. A typical experiment procedure is (1) A: approach; (2) LD: load 30 nm depth/3rate; (3) H: hold 4 sec; (4) UL: unload 90%unload/100%rate; (5) LD: load 60 nm depth/6rate; (6) H: hold 4 sec; (7) UL: unload 90%unload/100%rate; (8) LD: load 120 nm depth/10rate; (9) H: hold 4 sec; (10) UL: unload 90%unload/100%rate; (11) H: hold 2 log/50 points; (12) UL: unload 100%unload/300%rate. In addition, eight indentations were made for each experimental condition in order to get good statistical results. For this purpose, an Excel program has been developed to perform the calculation. This program is very effective for post-experiment data processing, performing all the calculations and formatting data for plotting figures. Thus, the averaged hardness and Young's modulus values and standard deviation reported in Figure 1-4 are from 8 data points for each indentation depth and the values reported in Figure 5 are averaged from 32 experimental data points, i.e. from 30 to 60 nm indentation depths. Young's modulus values were obtained from the 8 averaged data points at 30 nm depth. All results are from as-deposited films.

III. RESULTS AND DISCUSSION

Typical nanoindentation hardness and Young's modulus plots at different penetration depths for as-deposited FeN films as a function of nitrogen content are shown in Figs. 1 and 2. These experimental results indicate that the hardness of the pure Fe film is about 13 GPa, which is similar to the results obtained by Jones et al [4]. After adding a small amount of nitrogen (0.5

sccm flow rate during the film deposition), the hardness of the film is improved to 14.5 GPa and elastic modulus is increased from 200 to 220 GPa. However, a further increase of the nitrogen content in the films results in a decrease in the hardness and elastic modulus of FeN films. These results differ from those reported in reference [4], in which the hardness of pure iron films initially increased and then stayed constant as the nitrogen content was increased in the films. The trend of increasing and then decreasing hardness upon introducing nitrogen was also observed in $TiB_2(N)$ films [9]. For the films with tantalum, the trend in hardness and elastic modulus with nitrogen content is completely different. Figures 3 and 4 are typical examples of nanoindentation hardness and Young's modulus at different penetration depths of as-deposited FeTa (Ta=15wt%) films as a function of nitrogen content. It can be seen that introducing nitrogen into films containing tantalum continuously increases the hardness of the film. This trend is similar to that in TaN films reported in reference [10]. Interestingly, this trend is observed in all the films deposited from FeTa (5 - 15wt%) targets in this work. The films with 25 wt% tantalum were unstable and all results such as magnetic, stress and mechanical properties for these films were abnormal (as shown in Figures 5 and 6 for mechanical properties). Further investigations are needed for these films. Moreover, the reason why introducing nitrogen into sputtered films improves hardness for some materials but deteriorates it for other materials is not clear and could be a very interesting subject for continuing studies. The variation of hardness and Young's modulus of the FeN and FeTaN films as a function of nitrogen content are summarized in Figures 5 and 6. In addition, the substrate effect on the measurement of thin film hardness by nanoindentation is also shown in Figures 1-4. Because the hardness of the oxidized silicon (100) substrate is about 11 GPa, the inherent hardness values of these films can be determined at penetration depths ranging from 30 to 60 nm. Also, it can be seen that the substrate effect on Young's modulus is of long range compared to its effect on film hardness. Even at very small indentation depth the substrate effect on Young's modulus can still be observed. It seems that if a film is not thick enough, the elastic modulus obtained by nanoindentation is going to be affected by a large substrate effect. Hence, this caution should be noted when the Young's modulus of a film is going to be determined by nanoindentation, though it is not emphasized in the literature.

Figure 7 is the plot of the square root of the (E/H) ratio for the films with 0-15 wt% tantalum as a function of nitrogen and tantalum content. According to Fabes and Oliver et al. [11,12], the square root of the (E/H) ratio is proportional to the plastic strain field under the indenter, hence it can be used as a measure of plastic deformation. Contact plastic deformation of a material is related to both hardness as well as elastic stiffness, i.e., elastic modulus of this material. The higher hardness a material has, the more difficult it is deform plastically, and vice versa. On the other hand, suppose two materials have the same yield strength and hardness (hardness can be related to the yield strength), obviously the material with the higher elastic modulus is more easily

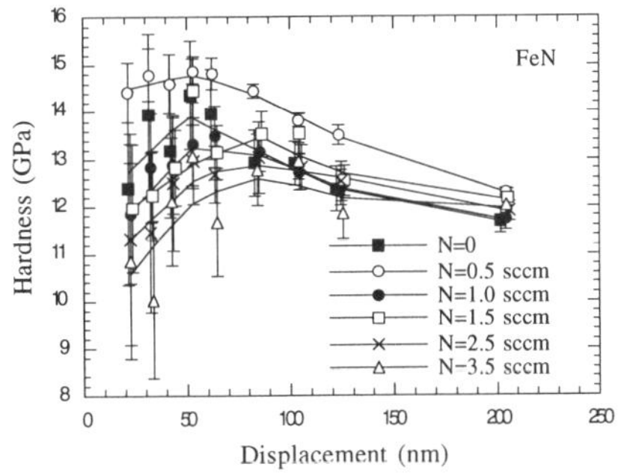

Fig. 1. Hardness of FeN films as a function of penetration depth and nitrogen content.

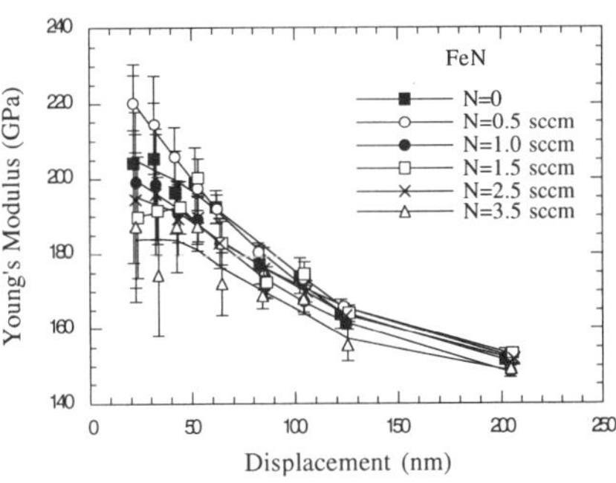

Fig. 2. Young's modulus of FeN films as a function of penetration depth and nitrogen content.

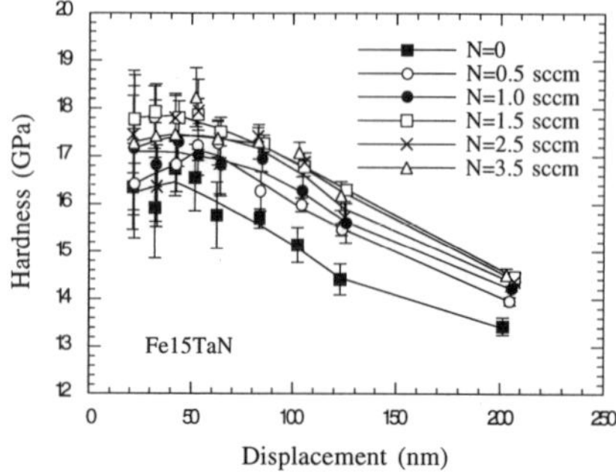

Fig. 3. Hardness of FeTaN (Ta=15wt%) films as a function of penetration depth and nitrogen content.

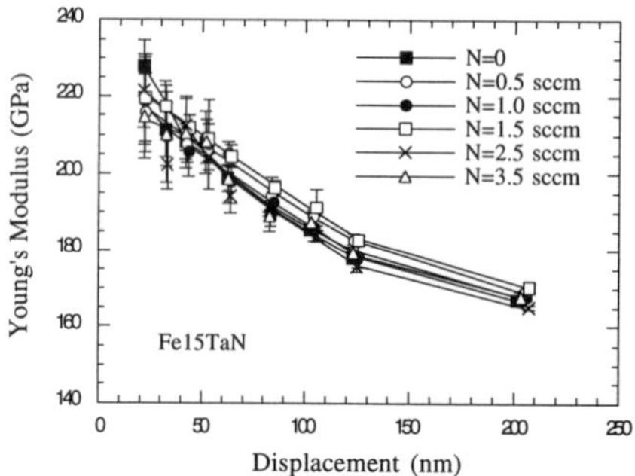

Fig. 4. Young's modulus of FeTaN (Ta=15wt%) films as a function of penetration depth and nitrogen content.

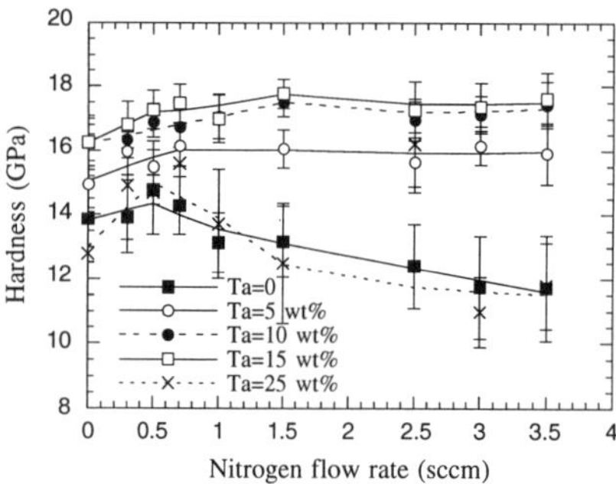

Fig. 5. Hardness of FeN and FeTaN (Ta=5-25wt%) films as a function of tantalum and nitrogen content.

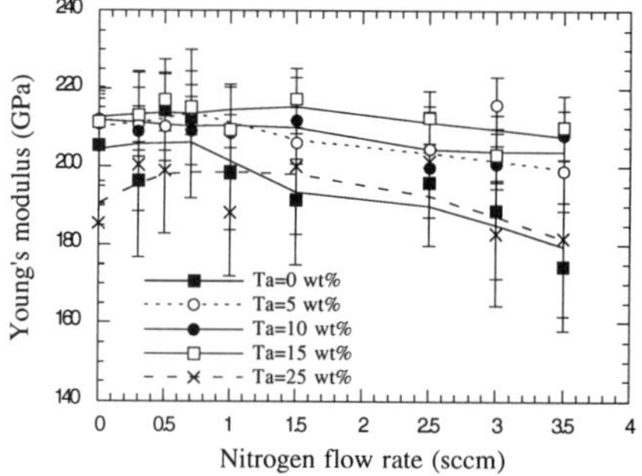

Fig. 6. Young's modulus of FeN and FeTaN (Ta=5-25wt%) films as a function of tantalum and nitrogen content.

subject to plastic deformation at lower strain. Therefore, the (E/H) ratio was used in this paper to evaluate the effect of Ta and N content on the plastic deformation of the films tested. It is obvious that the trends of the films with and without Ta are completely different. Tantalum generally makes the films more resistant to plastic deformation. Introducing nitrogen into simple Fe films results in more plastic deformation. By contrast, addition of N to films containing Ta increases their resistance to plastic deformation. Notice that in the calculation of the (E/H) ratio the hardness data is the averaged value of the hardness obtained at the penetration depth 20 - 60 nm, which are free from substrate effects, and elastic moduli used were taken at 30 nm depth for all the films, which have the minimum substrate

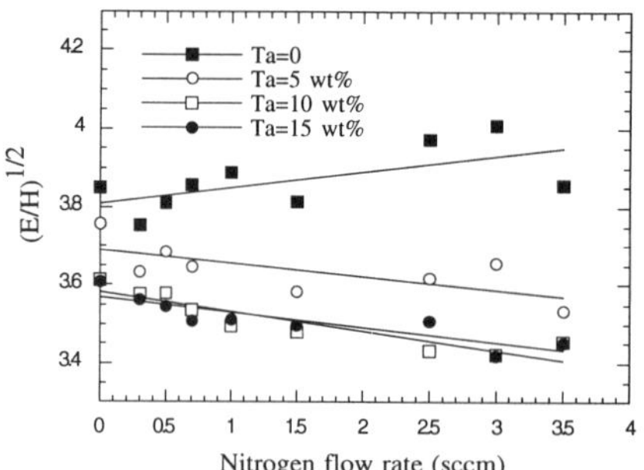

Fig. 7. The square root of (E/H) for the films with 0-15 wt% Ta as a function of nitrogen and tantalum content. This ratio is proportional to the plastic strain field (b=ac(E/H)$^{1/2}$) of the materials tested under the indenter, according to reference [11, 12]

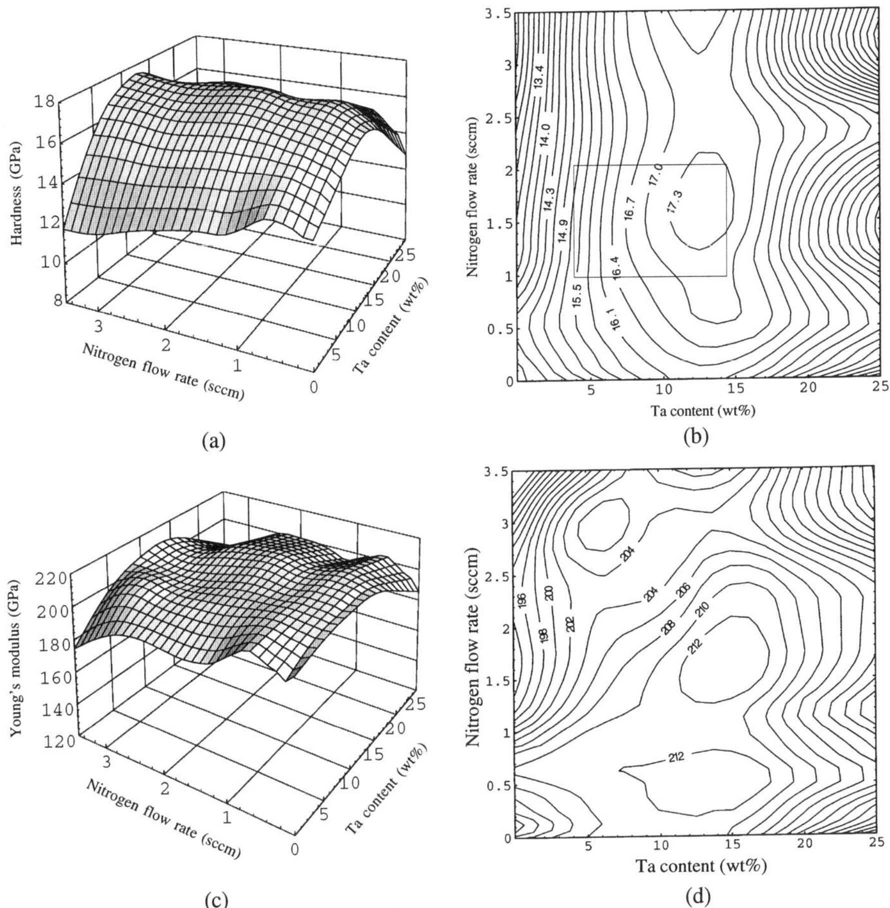

Fig. 8. Hardness and elastic modulus distributions of FeN and FeTaN films as a function of Ta content and N flow rates. (a) 3-D hardness map, (b) Hardness contour map (unit: GPa), (c) 3-D Young's modulus map, (d) Young's modulus contour map (unit: GPa). The selected area (a small rectangle in (b)) indicates the range in which excellent magnetic properties can be obtained with dc magnetron sputtering for these films, as reported in reference [8].

effects in the data collected. Therefore, the trend reported in Figure 7 reflects the intrinsic property responses from tested films with the minimum substrate effects.

Figure 8 shows the 3-D and contour maps for the hardness and Young's modulus of FeN and FeTaN sputtered films. From the 3-D and contour hardness maps, it is clear that for the films without tantalum the hardness increases as nitrogen flow rate increases up to 0.5 sccm, however, the hardness drops upon a further increase of the nitrogen content. Young's moduli of FeN films follow the same trend as nitrogen content increases. For films with tantalum in the range of 5 - 15wt% hardness increases and then trends to be constant and Young's modulus drops slightly as nitrogen increases. Furthermore, if the tantalum content is more than 20wt% the mechanical behavior of the films becomes unstable, as can be seen in the range of 20-25wt% tantalum. On the other hand, for a fixed nitrogen flow rate, both the hardness and Young's modulus of these films increases to a maximum point and then drops as tantalum content increases. The best mechanical properties (highest hardness and elastic modulus) of these films can be located in the range of 10-15wt% tantalum and 1-2 sccm nitrogen flow rate in the contour maps. The best magnetic properties (lowest coercivity and highest magnetization) of these films are indicated on the contour maps as shown by the small rectangle in Figure 8 (b). We can see that good magnetic and mechanical properties overlap. The magnetic properties of the similar films were reported in detail in reference [8].

IV. CONCLUSIONS

The mechanical properties of magnetically soft high moment FeN and FeTaN thin films were investigated by nanoindentation techniques. The films tested were deposited on oxidized silicon (100) substrates with a series of FeTa targets (Ta=0 to 25wt%) by dc magnetron sputtering in a argon plus nitrogen mixed atmosphere with nine nitrogen flow rates (0, 0.3, 0.5, 0.7, 1.0, 1.5, 2.5, 3.0 and 3.5 sccm). It was found that for the films deposited from the pure Fe target when the nitrogen flow rate increases from 0 to about 0.5 sccm the hardness of the film increases, however, a decreasing trend in hardness was observed on further increasing the nitrogen flow rate. On the other hand, for the films prepared from the targets with Ta content in the range of 5-15wt%, the hardness increases whenever Ta and N contents increase. The films prepared from the target with 25wt% tantalum did not show stable mechanical properties in this work. These effects are clearly illustrated by 3-D and contour hardness and Young's modulus maps in this paper.

ACKNOWLEDGMENT

This work was supported primarily by the MRSEC Program of the National Science Foundation under Award Number DMR - 9400399, by NSF-PYI DMR-9157402 and by the National Storage Industry Consortium-Advanced Technology Program funded through the Department of Commerce.

REFERENCES

[1]   H. Deng, M. K. Minor and J. A. Barnard, a paper presented at *the InterMag '96*, **EQ-11**, Seattle, WA, April 9-12, 1996.

[2]   B. Bhushan, in *Handbook of Micro/Nano Tribology*, p. 321, edit. B. Bhushan, CRC Press, (1995).

[3]   H. Fujimoto, T. Marieb, and B. Sun, *Conference Proceedings: International Workshop on Instrumented Indentation*, p. 13, San Diego, CA, April 22-23, 1995.

[4]   R. E. Jones, Jr., R. L. White, J. L. Williams and X-W. Qian, *IEEE Transactions on Magnetics*, **29**,3966, (1993).

[5]   H. Deng, V. .R. Inturi and J. A. Barnard, *IEEE Transactions on Mag.*, 31(6), 2697, (1995).

[6]   H. Deng, V. R. Inturi and J. A. Barnard, *Thin Films: Stress and Mechanical Properties V, MRS Symposium Proceedings*, **356**, 773, (1995).

[7]   H. Deng, J. Chen, R. B. Inturi and J. A. Barnard, *Thin Films: Stress and Mechanical Properties V, MRS Symposium Proceedings*, **356**, 181, (1995).

[8]   V. .R. Inturi and J. A. Barnard, *IEEE Transactions on Mag.*, **31**(6), 2660, (1995).

[9]   H. Deng, J. Chen, R. B. Inturi and J. A. Barnard, *Surface and Coatings Technology*, **76-77**, 609, (1995).

[10]   R. Saha, R. B. Inturi and J. A. Barnard, *Surface and Coatings Technology*, (in press), 1995.

[11]   B. D. Fabes, W. C. Oliver, R. A. McKee, and F. J. Walker, *Journal of Materials Research*, **7**(11), 3056, (1992).

[12]   B. D. Fabes, W. C. Oliver, *Thin Films: Stress and Mechanical Properties II, MRS Symposium Proceedings*, **188**, 127, (1990).

NANO SCALE CREEP AND THE ROLE OF DEFECTS

S.A.SYED ASIF & J.B.PETHICA
Department of materials, University of Oxford, Parks road, Oxford OX1 3PH, UK.

ABSTRACT

The modulating force method in nanoindentation gives a direct measure of contact stiffness, and being insensitive to drift, allows the accurate observation of creep in small indents to be carried out over long time periods. We present results for a range of metals at room temperature. Strain rate indices similar to those for macroscopic creep are found. Reverse creep occurs for step unloading greater than about half the starting load. For electropolished tungsten, we find quite different behaviour before and after the sudden pop-in. Afterwards, creep is as in other metals, but beforehand, it is essentially zero. The slight changes of stiffness observed at the very smallest loads are due to diffusion of adsorbed surface films into the contact zone. Our results show that the dislocations nucleated and multiplied at pop-in provide the mechanism of creep.

INTRODUCTION

The development of the depth sensing indentation technique has resulted in the ability to probe mechanical properties on the submicron scale. Commonly measured properties are hardness, young's modulus [1-2] and indentation creep [3]. An indentation creep test provides a simple method to investigate the time dependent flow properties of materials over a range of temperature [4]. In an indentation creep test, a constant load is applied to the indenter and the change in indentation size or depth is monitored as a function of time. Compared to conventional creep tests, indentation creep experiments are particularly useful when the sample volumes are very small. With the depth sensing indentation technique if the indentation depth is monitored as a function of time, the mean strain rate is commonly defined as

$$\dot{\varepsilon} = \frac{1}{h}\frac{dh}{dt} \qquad \text{--------(1)}$$

where h is the depth of indentation. However, with the depth sensing technique, as the scale of indentation decreases, the calculation of contact area from displacement become fairly inaccurate and the thermal drift can render displacement measurements useless for creep studies if the experiments exceed several seconds. To overcome this problem, Weihs and Pethica [5] used the AC modulation technique and measured the contact stiffness as a function of time. As stiffness is proportional to the area of contact [5], the effective strain rate can be defined as

$$\dot{\varepsilon} = \frac{1}{S}\frac{dS}{dt} \qquad \text{----------(2)}$$

where S is the contact stiffness. The mean stress or Hardness can be defined as

$$\sigma = \frac{F}{\pi}\frac{4}{S^2}E^{*2} \qquad \text{----------(3)}$$

where F is the load and E^{*} is the reduced modulus. With the AC modulation technique, since the resolution of the stiffness measurement can be 50 N/m or less the sensitivity is enough to monitor the motion of relatively few atoms.

In this work the AC modulation technique is used to study the nano scale creep of tungsten single crystal and polycrystalline copper samples. Tungsten has been shown to deform elastically at loads below ~1.5 mN [6]. When the load exceeds this value, plastic deformation of the sample occurs. In nanoindentation experiments, the transition between these two stages of deformation can be observed by a sudden increase in the penetration depth at constant load, and is known as " pop-in". The pop-in is interpreted [6] as due to the sudden multiplication of dislocations, producing a plastic zone beneath the indenter.This implies that the time dependent deformation response for electropolished single crystal tungsten will be very different before and after pop-in. The present paper will concentrate on time dependent deformation of tungsten both in the elastic regime and elastoplastic regime. We demonstrate that the sudden multiplication of mobile dislocation due to pop-in is the main source for creep and that before pop-in the creep is essentially zero.

EXPERIMENTAL PROCEDURE

The materials used for this study were tungsten and copper. The tungsten was a single crystal with [100] normal to the test surface. The tungsten samples were prepared in two sets. One set of sample surfaces was mechanically polished to a 0.25 μm finish, and the other one was electropolished. The polycrystalline copper sample was mechanically polished to a 0.25 μm finish.

The indentation experiments were carried out using a Nano Indenter [7] with a Berkovich diamond tip. To continuously measure the contact stiffness during an indentation, a small sinusoidal AC force ~16 μN at 40 Hz was added to the applied force, and the resulting oscillation in displacement was monitored using a lock-in amplifier. The amplitude and the phase difference were measured to calculate the contact stiffness as described by Pethica and Oliver [6]. Two types of experiments were carried out. First, the indenter was ramped to maximum load in 100 sec and then steadily unloaded without a hold period at maximum load. This was done to study the elastic and elastoplastic response of electropolished tungsten. The second set   of experiments were carried out to study the time dependent deformation response and consisted a sequence of eight segments. The details of each segment is given in table 1.

TABLE 1

Approach at 10 nm/s	Load 0.05, 0.5, 1, 2 mN in 100 sec	First Hold for 600 sec	Step Unload 10%, 30% 50%, 70% of Maximum load	Second Hold for 600 sec	Unload by 90 % of maximum load	Hold for 100 sec to find the drift	Unload 100%

After the tip contacted the specimen surface, the load was ramped to the desired maximum load in 100 sec. To monitor the time dependent deformation, the load was then held constant for 600 secs. After the first hold segment the indenter load was abruptly dropped to different percentages of maximum load as shown in the table 1. After this step unloading, the load was held constant again for 600 sec. All the tests were run at room temperature and care was taken to establish maximum thermal stability prior to testing. The thermal drift was less than 0.05 nm/s.

RESULTS AND DISCUSSION

The indentation load-displacement data for electropolished tungsten at 1mN maximum load is shown in Fig 1a. The elastic nature of deformation is clear from the reversibility on unloading. Fig1b shows the corresponding stiffness variation with time. The stiffness increases as the load is ramped up to 1mN. During the first hold segment the stiffness remains constant. For the second hold segment the load is held constant at 0.9 mN. It can be observed that, as in the first hold segment, the second hold segment does not show any change in stiffness as a function of time. Fig2c shows the mean stress (Hardness) calculated using equation (4) as a function of time. It can be seen that the mean stress remains constant in the first and second hold segments. This observation clearly indicates that time dependent deformation does not occur for electropolished tungsten at 1mN load. The indent volume remains elastic and no mobile dislocations were nucleated.

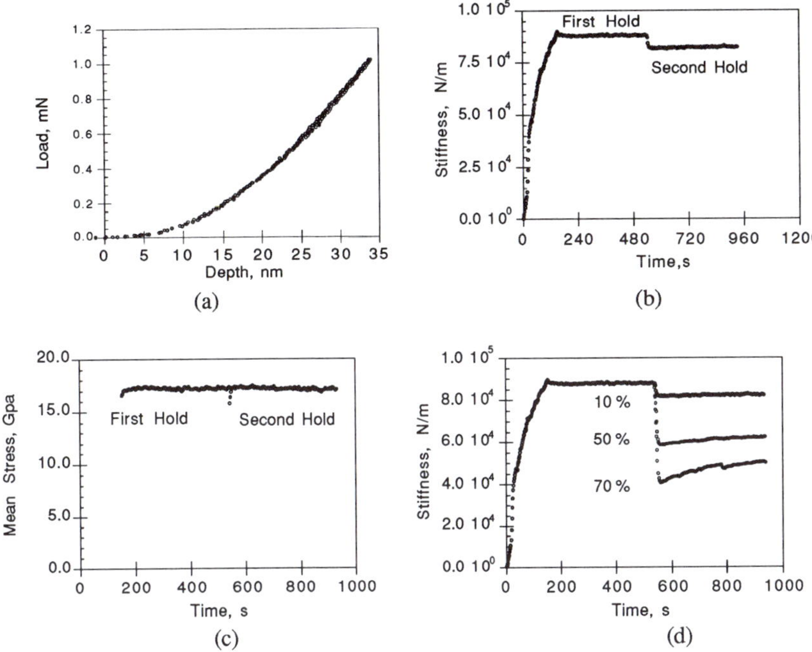

Figure 1: 1mN load indentation into electropolished tungsten showing elastic deformation (a), Zero creep in first and second hold(b), and constant mean stress(c). Unloading to different percentages show stiffness to increase with time in second hold(d).

When the indenter was unloaded by more than 10 percentage of maximum load, and the contact stiffness was reduced from 8×10^4 to 4×10^4 N/m there is a slight changes in stiffness as shown in Fig 1d. This could be due to diffusion of adsorbed surface films into the contact zone. A similar observation was made by Pethica and Weihs [5] and they attributed it to the surface diffusion of atoms from the specimen surface to the indenter tip.

Fig2a shows the load-displacement data for a maximum load of 2mN on electropolished tungsten. The corresponding stiffness-time data is shown in fig2b. From the load displacement graph it can be observed that the pop-in occurs around 1.5 mN. The corresponding jump in stiffness can be observed around 1×10^5N/m. It should be noted that, unlike in the case of 1mN load (Fig1b) which is below the pop-in load, the stiffness increases with time in the first hold segment after pop-in(Fig2b). This indicates that there is creep after pop-in. Fig2c shows the stiffness-time data for different percentages of unload from 2mN maximum load. The corresponding mean stress is shown in Fig 2d. It can be seen that the stress relaxes in the first hold segment. The dislocation network created by pop-in is therefore the source for stress relaxation.
When the load is dropped by 10 % of maximum load, positive creep occurs(Fig 2c). When it is dropped by 50% of maximum load, zero creep occurs and when it is dropped by 70% of maximum load, negative creep occurs(Fig 2c). The transition from positive creep to negative creep depends on the magnitude of the stress reduction. This indicates that the internal back stress evolved due to the dislocation interaction is the main reason for the positive, zero and negative creep. When the applied stress(mean stress) is greater than the internal back stress, positive creep occurs, when it is equal to internal back stress, zero creep occurs and when it is less than the internal back stress,

negative creep occurs. The effect of internal stress on the creep process is well known in bulk creep testing [8] and our nano scale tests show similar behaviour.

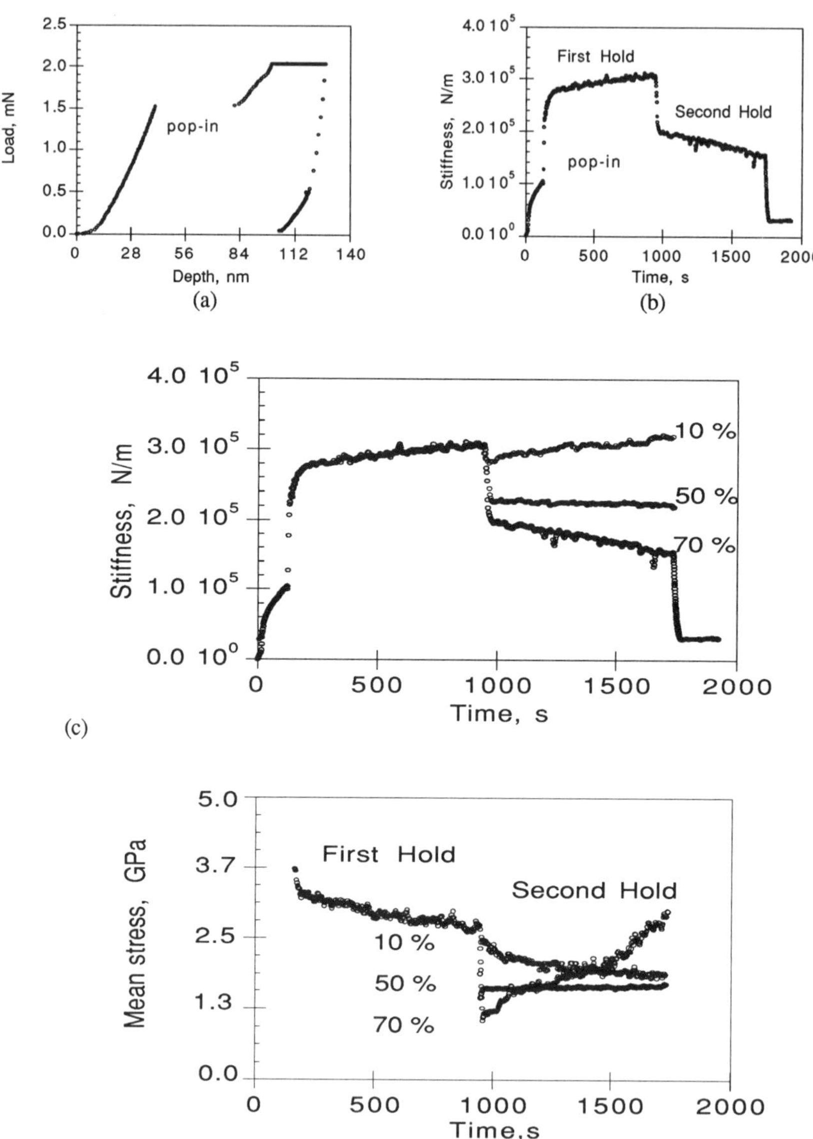

Figure 2: 2mN load indentation into electropolished tungsten showing sudden discontinuity or "pop-in" in load displacement (a) and stiffness(b) data. The stiffness time data at different percentages of unload showing positive, zero and negative creep in the second hold segment(c) and the corresponding stress relaxation and recovery(d).

The mechanical deformation of the material in creep testing is not driven by the full applied stress, but rather by a reduced effective stress[8]. This net or effective stress is given by the difference between the applied stress and the internal back stress, reflecting the resistance of the material to deformation.

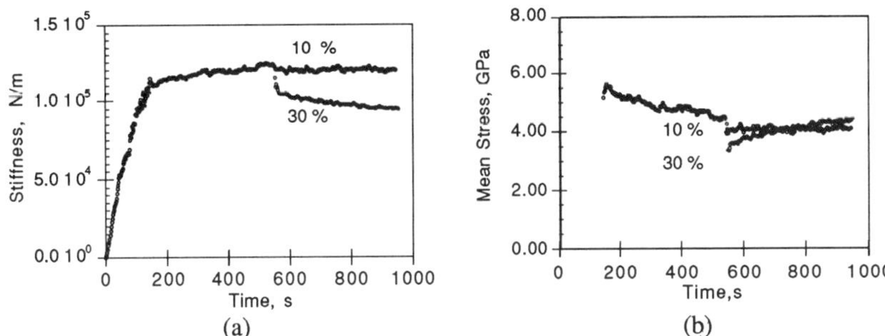

(a)   (b)

Figure 3: 500μN load indentation into mechanically polished tungsten showing positive creep in the first hold, zero and negative creep in the second hold segment(a), and the corresponding stress relaxation and recovery(b)

Fig3a shows the stiffness-time plot for mechanically polished single crystal tungsten at 500 μN load. The corresponding mean stress time plot is shown in Fig3b. It can be seen that, unlike the electropolished tungsten, creep occurs in mechanically polished tungsten even for 500μN load. Mechanical polishing leaves the surface with significant mobile dislocation density. It should be noted that the mean stress (Hardness) of mechanically polished tungsten (Fig3b) is higher than that of electropolished tungsten after pop-in (Fig2d) and the negative creep occurs for 30% unload from maximum load. This indicates higher mobile dislocation density and higher internal back stress for the mechanically polished tungsten compared to electropolished tungsten.

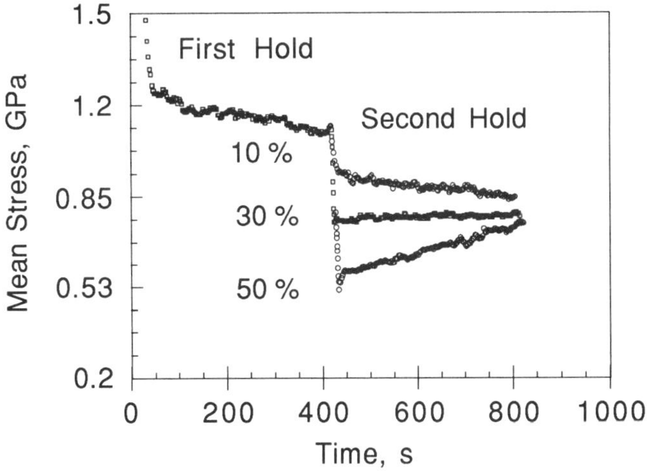

Figure 4: 500μN load indentation into mechanically polished copper polycrystal showing positive creep in the first hold segment, positive, zero and negative creep in the second hold segment .

Fig(4) shows mean stress time data for 500 µN load indentation into mechanically polished copper. It can be seen that, as the load is dropped and mean stress is reduced, a transition from positive to negative creep occurs in the second hold segment. When the percentage of unload is 30 %, zero creep occurs, indicating the internal back stress is just equal to applied mean stress. Similar experiments were performed for a range of metals[9] and in all the case we found similar transition from positive creep to negative creep depends on the magnitude of stress drop.

In bulk creep testing, the stress drop technique is mainly used to measure the internal back stress[8] at elevated temperature. Our experimental result suggests that nano indentation with AC modulation technique can be used to study the internal back stress evolved in an indentation experiment at room temperature. As the scale of indentation is very small involving very small volume of material nano indentation creep with stress drop technique can be used to study the internal stress and dislocation dynamics in thin film systems.

CONCLUSIONS

Using continuous stiffness measurement technique we have shown that below pop-in load, electropolished tungsten does not show any time dependent plastic deformation and the creep observed at very low load is mainly due to diffusion of adsorbed surface film. Above the pop-in load time dependent plastic deformation occurs. This clearly indicates pop-in is mainly due to the sudden multiplication of mobile dislocation. When the applied stress is dropped to different percentages of original stress, positive, zero and negative creep occurs depending up on the magnitude of the stress drop. Internal back stress due to dislocation interaction is the main reason for the transition from positive to negative creep.

ACKNOWLEDGEMENT

S.A. Syed Asif is grateful to the **INLAKS** foundation for the financial support.

REFERENCES

1) J.B.Pethica, R.Hutchings and W.C. Oliver, Philos. Mag. A 48,593,(1983)
2) W.C.Oliver and G.M.Phar, J.Mater.Res, 7 (6),1654, (1992)
3) A.G. Atkins, A.Silverio and D.Tabor, J.Inst.Metals 94, 369 (1966)
4) M.J.Mayo and W.D.Nix, Acta Met., 6,2183, (1988)
5) T.P.Weihs and J.B.Pethica , Mat.Res.Symp.Proc.108,325 (1992)
6) J.B.Pethica and W.C. Oliver, Mat, Res.Soc. Sym.Proc. 130,13 (1989)
7) Nano Instruments, Inc., P.O.Box 14211, Knoxville, TN 37914.
8)J.C.Gibeling and W.D. Nix Material Science and Engineering, 45, 123, (1980)
9) S.A.Syed Asif and J.B.Pethica, to be published.

NANOINDENTATION OF SOFT FILMS ON HARD SUBSTRATES: THE IMPORTANCE OF PILE-UP

T.Y. TSUI*, W.C. OLIVER**, and G.M. PHARR*
* Department of Materials Science, Rice University, 6100 Main St., Houston, TX 77005
** Nanoinstruments Inc., 1001 Larson Drive, Oak Ridge, TN 37830

ABSTRACT

Nanoindentation is a common technique for measuring the mechanical properties of thin films. Here, we address the potential measurement errors caused by pile-up when soft films deposited on hard substrates are tested by nanoindentation methods. Pile-up is exacerbated in soft film / hard substrate systems because of the constraint the substrate exerts on plastic deformation of the film. To experimentally examine pile-up effects, aluminum films with thicknesses of 240 and 1700 nm were deposited on hard glass substrates and tested by standard nanoindentation techniques. The aluminum/glass system is interesting because the film and substrate have similar elastic moduli; thus, any unusual behavior in the nanoindentation results may be attributed to differences in the plastic flow characteristics alone. A detailed scanning electron microscopy examination of nanoindentation hardness impressions in the film revealed that common methods for analyzing nanoindentation data underestimate the true contact areas by as much as 80%, which results in overestimations of the hardness and modulus by as much as 80% and 35%, respectively. The sources of these errors and their influence on the measurement of hardness and elastic modulus are discussed, and a simple model for the composite hardness of the film/substrate system is developed. The model could prove useful in measuring the hardness and elastic modulus of soft-film / hard substrate systems when it is not possible to make indentations shallow enough to avoid the substrate influences.

INTRODUCTION

Nanoindentation is a widely used technique for measuring mechanical properties on the micron and sub-micron scale [1,2]. The technique has proven particularly useful for measuring the properties of thin films, since measurements can often be made without having to remove the film from its substrate [2]. However, when the film is very thin, e.g., less than a micron, accurate measurements are sometimes difficult to obtain because the substrate can influence the indentation load-displacement behavior. As a result, numerous experimental and theoretical investigations have addressed the issue of substrate effects and what can be done in the analysis of nanoindentation data to obtain substrate independent measurements of mechanical properties [3-11]. Most work to date has focused on the measurement of hardness, H, and elastic modulus, E.

Many thin film systems are comprised of very soft films on very hard substrates. This is particularly common, for instance, in the semiconductor industry, where films of aluminum, gold and copper are often deposited on silicon (H = 12 GPa), germanium (H = 10 GPa), glass (H = 5-8 GPa) and ceramic substrates (H = 10-40 GPa). The hardness of these films is usually in the range 0.1-1 GPa, making them at least an order of magnitude softer than the materials on which they are deposited. One important consequence is that when a hardness impression is made, there is a tendency for material to pile-up around the hardness impression to a much greater degree than it would in a bulk material. This is due to the severe constraint imposed on plastic deformation in the film by the relatively undeformable substrate. Since current techniques for analyzing nanoindentation data do not make provisions for the extra contact area produced by the pile-up, the enhancement of pile-up in soft films on hard substrates has important consequences for the measurement of mechanical properties by nanoindentation methods. This may be seen, for example, by considering the equations used to extract H and E from nanoindentation load-displacement data. They are [1]:

Mat. Res. Soc. Symp. Proc. Vol. 436 ⓒ 1997 Materials Research Society

$$H = \frac{P}{A} \qquad (1)$$

and

$$E_{eff} = \frac{E}{1-v^2} = \frac{\sqrt{\pi}}{2}\frac{S}{\sqrt{A}} \quad , \qquad (2)$$

where P is the indentation load, A is the projected area of the contact, E_{eff} is the effective elastic modulus defined in terms of Young's modulus E and Poisson's ratio v, and S is the experimentally measured contact stiffness. The contact area in nanoindentation measurements is indirectly deduced from an analysis of unloading data which does not account for pile-up. Thus, in materials which are prone to pile-up, an error in the contact area introduces an error of similar magnitude into the hardness and and an error which scales as $A^{1/2}$ in the elastic modulus. As documented elsewhere [12,13], similar effects explain the apparent dependence of the nanoindentation hardness and elastic modulus on residual stress.

In order to gain a better understanding of how significant pile-up errors may be for soft films on hard substrates, we recently undertook an experimental investigation of the pile-up behavior of a model system. Here, we report some important observations of that study and discuss their implications for mechanical property measurement by nanoindentation methods.

EXPERIMENTAL PROCEDURE

The model system used in the investigation was high purity aluminum deposited on glass. In addition to a large difference in hardness (the hardness of the film is about 0.5 GPa while that for the substrate is 7.0 GPa), an equally important consideration in choosing this system was the similarity of the elastic moduli of the two components. The modulus of bulk aluminum is 70 GPa, while that for the glass used in the study was 57 GPa. The fact that the moduli are so similar minimizes the role that a film/substrate modulus difference would play in determining the indentation behavior, thus simplifying the interpretation of results.

The films were sputter-deposited to two different thicknesses, $t_f = 240$ nm and 1700 nm, and indented to various penetration depths, h_t, in the range 0.1-10 t_f with a sharp Berkovich indenter. The load-displacement data obtained at each depth were analyzed using the method of Oliver and Pharr [1] to determine the apparent contact area, hardness, and elastic modulus. Such measurements do not account for substrate influences. The quantities measured in this way will be referred to throughout this paper as A_{nano}, H_{nano}, and E_{nano}. Subsequently, the larger indentations were imaged in a high resolution scanning electron microscope (SEM) to determine their actual contact areas, A_{actual}. Care was taken in these measurements to include the pile-up in the contact area determination. This was accomplished by tracing a digital image of each indentation along the contact edge and computing the area enclosed inside the figure. The A_{actual} were then used in conjunction with Eqns. 1 and 2 to provide a second measurement of the hardness and modulus, H_{actual} and E_{actual} , based on the actual contact areas. Note that the hardness measured in this way is, by definition, the true hardness. Thus, any deviations of H_{nano} from H_{actual} must be attributed to inaccuracies in the procedure by which contact areas are deduced from the experimental load-displacement data.

RESULTS AND DISCUSSION

Figure 1 summarizes the results of the hardness and modulus measurements for the 1700 nm film. The thickness of this film makes it convenient for measurements at penetration depths less than film thickness ($h_t \leq t_f$). Later on, data for the 240 nm film will be presented to illustrate the behavior for $h_t \geq t_f$.

The filled symbols in Figs. 1a and 1b show the dependencies of H_{nano} and E_{nano} on the maximum penetration depth normalized with respect to the film thickness, i.e., h_t/t_f. The behavior of the composite hardness of the film/substrate system based on these data seems

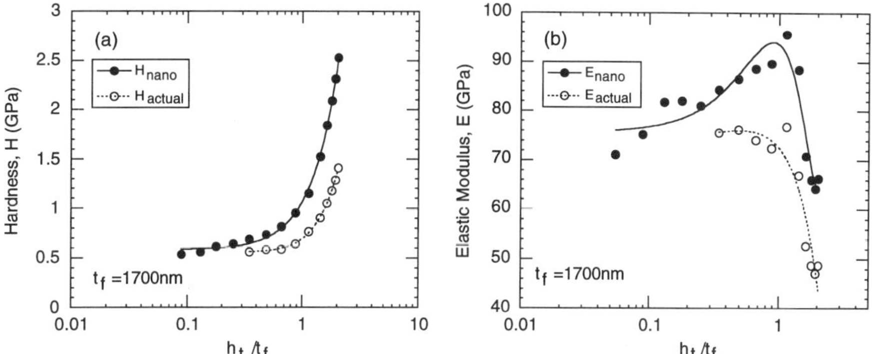

Fig. 1. Hardness and elastic modulus of 1700 nm aluminum films on glass substrates.

perfectly reasonable. At small depths, the hardness tends toward a limiting value of about 0.5 GPa, presumably the hardness of the film, but at large depths, the hardness increases markedly, consistent with a substrate hardness of 7.0 GPa. Note that most of the increase occurs at penetration depths close to the film thickness. The behavior of the elastic modulus, on the other hand, is quite different. At small depths, E_{nano} is close to the expected film modulus of 70 GPa, but as the depth increases, the modulus shows an unexpected increase into the 90-100 GPa range. The modulus then peaks at a penetration depth very close to the film thickness, followed by an abrupt decrease. The non-constancy of the modulus and increases above 70 GPa are especially perplexing given that modulus of the substrate, 57 GPa, is smaller that of the film.

The reason for the increase in the elastic modulus above that of the film and substrate is directly related to the pile-up behavior. Fig. 2 presents SEM images of indentations made at three different normalized depths: $h_{max}/t_f = 0.18$, 1.43, and 9.96. The first two indentations were made in the 1700 nm film, and the third in the 240 nm film. What is apparent from an examination of these micrographs is that the pile-up behavior depends on the depth of the indentation relative to the film thickness and that the amount of pile-up can be quite large under certain circumstances. At small depths (Fig. 2a), there is very little pile-up, consistent with the behavior of well-annealed, bulk aluminium. However, at depths close to the film thickness (Fig. 2b), the amount of pile-up is substantial, giving the indentation a circular appearance even though it was made with a triangular pyramid. For depths much greater than the film thickness (Fig. 2c), the amount of the pile-up is not as large because a greater portion of the deformed volume is in the hard glass substrate. Thus, it is apparent that the constraint that the hard

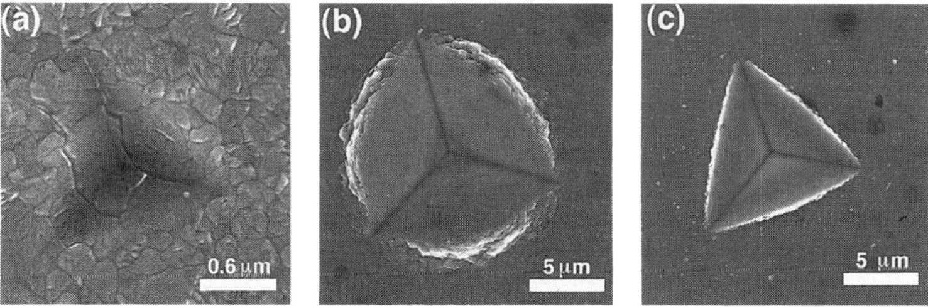

Fig. 2.   SEM images of indentations: (a) h_t/t_f=0.18; t_f=1700 nm; (b) h_t/t_f=1.43; t_f=1700 nm; and (c) h_t/t_f=9.96; t_f=240 nm.

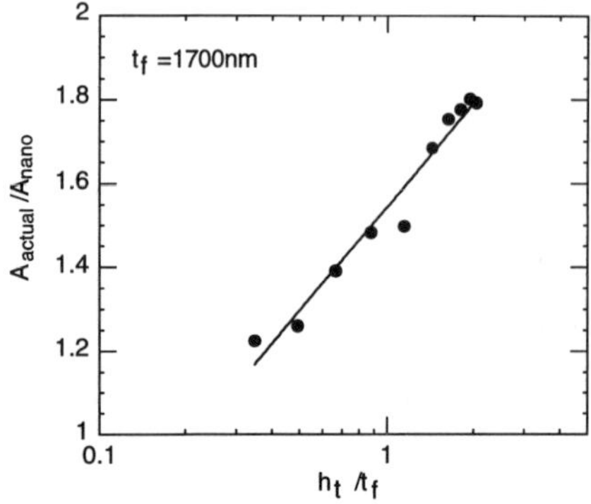 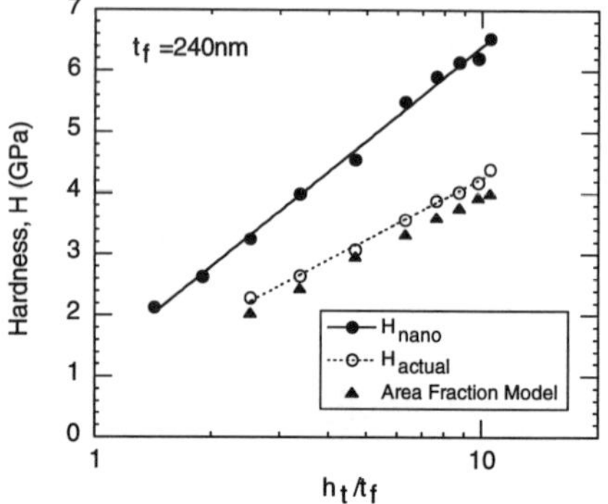

Fig. 3. Indentation depth dependence of A_{actual}/A_{nano} for the 1700 nm film.

Fig. 4. Indentation depth dependence of the hardness of the 240 nm film.

substrate imposes on the plastic deformation in the film enhances pile-up by an amount which is greatest at indentation depths around the film thickness.

To further quantify this behavior, the actual areas of the indentations made in the 1700 nm film measured from SEM images are plotted as A_{actual}/A_{nano} vs. h_t/t_f in Figure 3. The results show that there is indeed an increase in contact area due to the pile-up and that the nanoindentation analysis procedures underestimate the true contact area by as much as 80%. Clearly, this must be accounted for if accurate measurements of E and H are to be obtained; otherwise, errors as large as 80% in the hardness and 35% in the elastic modulus will result. For purposes of comparison, the hardnesses and elastic moduli computed from Eqns. 1 and 2 using the actual contact areas are plotted as the open circles in Fig. 1. It is seen that H_{actual} and E_{actual} are reduced by a considerable amount, with the reduction being most important for penetration depths close to the film thickness. Furthermore, the measured elastic modulus does not increase in the region $h_t < t_f$, but is rather constant at a value close 70 GPa, the modulus of aluminum. Thus, it can be concluded that the experimentally observed increase in E_{nano} with penetration depth is an artifact caused by not accounting for pileup in the measurement of contact area.

The decrease in elastic modulus at greater penetration depths has two separate origins. First, the modulus of the substrate, 57 GPa, is approximately 20% smaller than the film, so a small decrease is naturally expected. However, this by itself can explain neither the abruptness of the decrease nor the reduction in E_{nano} to values less than 57 GPa. Space considerations prohibit a complete explanation for this behavior, but in short, the lower than expected modulus is an artifact caused by the procedures used to estimate the unloading stiffness when the indenter penetrates through the film. During unloading, glass exhibits a much greater elastic recovery than aluminum. Thus, when the indenter penetrates the film into the glass, the displacements recovered during unloading are significantly different in the lower portion of the unloading curve, i.e., after contact with the aluminum is lost. This produces a bend in the curve and changes its shape in a way that the assumed power-law fitting relation does not fit well. The net effect is that unloading stiffness derived from the power-law fit is underestimated, which translates via Eqn. 2 to an underestimation of the elastic modulus. Indentations made in the 240 nm film were used to explore the film/substrate composite behavior when the indenter penetrates through the film. The composite elastic modulus was not considered due to the aforementioned problems in determining the contact stiffness.

Results of the composite hardness measurements are summarized in Fig. 4, which shows the depth dependence of the hardness computed in three ways. The filled circles are the hardnesses determined from standard nanoindentation procedures. These are in error not only because of the influences of pile-up on the contact area, but also because of the curve fitting

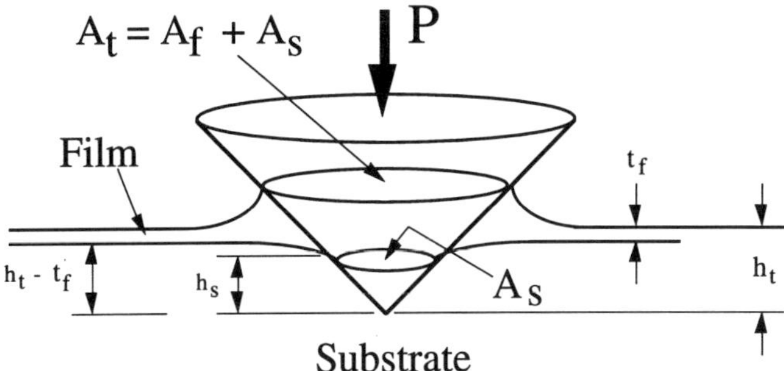

Fig. 5. Schematic illustration of the indentation of a soft film on hard substrate.

difficulties. The open circles are the hardnesses based on the actual area of contact. Comparing these two sets of data, it is clear that the curve fitting difficulties and failure to account for the pile-up result in the nanoindentation hardnesses being significantly greater than the actual hardnesses. The third set of data, shown as filled triangles on the plot, are the predictions of a simple model for the hardness of the composite film/substrate system. It is based on a rule of mixtures:

$$H_c = \left(\frac{A_f}{A_T}\right)H_f + \left(\frac{A_s}{A_T}\right)H_s \quad , \tag{3}$$

which states that the composite hardness, H_c, depends of the hardness of the film, H_f, and the hardness of the substrate, H_s, through the relative fractions of the projected indentation area in the film, A_f, and the substrate, A_s. A diagram illustrating the important parameters is shown in Fig. 5.

To implement this rule of mixtures requires that all of the parameters on the right hand side of Eqn. 3 be independently measurable. For the sake of calculation, we assume here that H_f is the hardness approached asymptotically at small depths and that H_s is 7.0 GPa, i.e., the independently measured value for the substrate. The small depth hardness of the 240 nm film was 1.0 GPa, somewhat higher than the 1700 nm film. To determine the area fractions, we use the SEM measurements as the total area, A_t, and partition this area into film and substrate portions using an approximate procedure which could be useful when A_f and A_s cannot be measured directly. The basic assumption, as illustrated in Fig. 5, is that at a given indentation load, P, the interface between the film and substrate sinks in to produce the same deflection geometry that would occur if there were no film on the substrate. Such an assumption should hold reasonably well when $H_f<<H_s$ and/or $h_t>>t_f$. With this assumption, the depth along which contact is made between the indenter and the substrate, h_s, can be computed from:

$$h_s = \alpha(h_t - t_f) \quad , \tag{4}$$

where the parameter α is the ratio of the contact depth to the total depth during indentation of the bare substrate. This parameter can be readily determined by standard nanoindentation measurements on the substrate. For the glass substrate used in this study, $\alpha=0.72$. Once h_s is established, A_s follows by evaluating the area function of the indenter at h_s, and A_f can be computed from $A_f = A_t - A_s$.

Using these procedures, the composite hardnesses predicted by the model are plotted in Fig. 5, where it is seen that the predicted hardnesses are in generally good agreement with the actual hardnesses. The model could prove useful in measuring the hardness and modulus of soft films on hard substrates when it is not possible to make hardness impressions small enough to give substrate independent properties.

ACKNOWLEDGMENTS

This research was sponsored by the Division of Materials Sciences, U.S. Department of Energy, under contract DE-AC05-96OR22464 with Lockheed Martin Energy Research Corp., and through the SHaRE Program under contract DE-AC05-76OR00033 between the U.S. Department of Energy and Oak Ridge Associated Universities. One of the authors (GMP) is grateful for sabbatical support provided by the Oak Ridge National Laboratory.

REFERENCES

1. W.C. Oliver and G.M. Pharr, J. Mater. Res. **7**, 1564 (1992).

2. G.M. Pharr and W.C. Oliver, MRS Bull. **XVII**, 28 (1992).

3. P.J. Burnett and D.S. Rickerby, Thin Solid Film **148**, 41-50 (1987).

4. P.J. Burnett and D.S. Rickerby, Thin Solid Films **148**, 51-65 (1987).

5. M.F. Doerner and W.D. Nix, J. Mater. Res **1**, 601-609 (1986).

6. M.F. Doerner, D.S. Gardner, and W.D. Nix, J. Mater. Res. **1**, 845-851 (1987).

7. M.F. Doerner and W.D. Nix, CRC Critical Reviews in Solid State and Materials Science **14**, 225-268 (1988).

8. B.D. Fabes and W.C. Oliver, Mat. Res. Soc. Symp. Proc. **188**, 127-132 (1990).

9. W.R. LaFontaine, B. Yost, and C.-Y. Li, J. Mater. Res. **5**, 776-783 (1990).

10. D. Stone, W.R. LaFontaine, P. Alexopoulos, T.W. Wu, and C.-Y. Li, J. Mater. Res **3**, 141-147 (1988).

11. D.S. Stone, Journal of Electronic Packing **112**, 41-46 (1990).

12. T.Y. Tsui, W.C. Oliver, and G.M. Pharr, J. Mater. Res. **11**, (1996).

13. A. Bolshakov, W.C. Oliver, and G.M. Pharr, J. Mater. Res. **11**, (1996).

MECHANICAL PROPERTIES OF CaF$_2$ SINGLE CRYSTAL SUBSTRATES DETERMINED FROM NANOINDENTATION TECHNIQUES

A. ARUGA*, R. B. INTURI**, J. A. BARNARD**, and R. C. BRADT**
*Department of Materials Science and Engineering, National Defense Academy, Yokosuka 239,
Japan
**Department of Metallurgical and Materials Engineering, The University of Alabama,
Tuscaloosa, AL 35487-0202

ABSTRACT

CaF$_2$ single crystals are interesting substrate materials for deposition of thin films. Its structure is cubic and it cleaves on {111} planes. CaF$_2$, whose hardness has been reported to be 4 on the Moh's scale, is plastic and soft. In this study, the mechanical properties such as hardness(H) and Young's modulus(E) of single crystal CaF$_2$ mineral were measured by using a nanoindenter with a Berkovich indenter normal to (100) and (111) planes. A normal indentation size effect (ISE) in accordance with the traditional power law and the proportional specimen resistance model (PSR) of Li and Bradt [1] was observed. The values of E and H on (100) plane are larger than those on (111) plane and these values on both planes decrease with increase in time during the hold segment. The effect of displacement rate on mechanical properties of (100) and (111) surfaces is also studied.

INTRODUCTION

The fluorite crystal structure occurs abundantly in nature. It is easy to get larger single crystals of fluorite with the (100) or (111) planes. Fluorite mineral is of industrial importance as a calcium source, of technical value as a fluorizing agent, and also of metallurgical significance as a flux. Most research studies of calcium fluoride CaF$_2$ have focused on their unique optical properties and related applications. Recently, CaF$_2$ has been developing to use as a substrate for electric conductive Langmuir-Blodgett films [2] and the growth of epitaxial films [3]. So it is getting necessary to measure mechanical properties like hardness and Young's modulus by use of nanoindentation technique.

The fluorite (CaF$_2$) structure, which is cubic, cleaves on the {111} planes. The primary and secondary slip systems are believed to be {100}⟨011⟩ and {110}⟨1$\bar{1}$0⟩ respectively [4]. On the Moh's scale, the hardness has been reported to be 4 for CaF$_2$. CaF$_2$ is plastic and soft materials. It is useful as a substrate for epitaxial growth, making high orientated thin films. First, we focus to determine mechanical properties of substrate which present an indentation size-load effect (ISE) by using a nano-indenter so that we can get accurate mechanical properties of thin films deposited on the (001) and the (111) planes of CaF$_2$.

The hardness of CaF$_2$ mineral was reported in some reference. O'Neill *et al.* [5] have published Knoop indentation hardness measurements, while Boyarskaya *et al.* [6] have carried out indentation measurements. These two sets of data are for the (001) and the (111) planes. Gil *et al.*

Mat. Res. Soc. Symp. Proc. Vol. 436 © 1997 Materials Research Society

[7] have also reported for various natural fluorite by use of a micro-Vickers indenter. They confirm that the anisotropy in hardness is essentially controlled by mechanisms of bulk plastic deformation on {001}<110> slip systems. Other measurements of hardness or microhardness for CaF_2 are related to other properties, dislocation mobility, interatomic bonding, and so on [8-11]. Young's modulus measurements of CaF_2 have been reported to the same extent about anisotropy of elasticity [12], the elasticity studied by using an excitation electrodynamic resonance method [13], and   determined by the ultrasonic pulse super-position technique [14]. However, there are no measurements of mechanical properties for CaF_2 single crystal substrates determined from nanoindentation technique, which can apply for thin films. Therefore, we carried out to measure microhardness and Young's modulus of CaF_2 single crystal at ultra low load by use of nanoindenter.

EXPERIMENTAL PROCEDURE

The fluorite single crystals obtained for this study were transparent, pale-violet natural crystals. These shapes are orthorhombic 25x22x18 mm in size and octahedral 15 mm on a side. So natural faces indicate {100} or {111} planes, respectively. Several specimens were cut from larger crystals guided natural (100) and (111) faces. These specimens were then fixed on aluminum blocks and mechanically ground with progressively #600 SiC abrasive papers, and followed by polishing with 0.3 μm Al_2O_3 and finally 0.05 μm κ-Al_2O_3 slurry to achieve a scratch free, mirrorlike surface for subsequent nanoindentation measurements.

The mechanical properties such as hardness and Young's modulus of single crystal CaF_2 minerals were measured by using a Nano Indenter™ II (Nano Instruments, Inc., Knoxville, TN) equipped with a Berkovich diamond indenter. Some specimens used had surfaces parallel to the (100) or the (111) crystal planes with orthorhombic 8x12x1 mm or hexagonal 6 mm on a side, respectively. Standard sample used was Corning 7940 fused silica glass. Prior to the experiments the indenter tip area calibration was carried out by using the data obtained on a standard fused silica sample provided by Nano Instrument, Inc. First, an usual constant displacement loading rate (10 nm/s) in the loading segment and a depth limit control were used. Different experiments were performed for each plane in the direction of some orientation with maximum depths of about 100, 200, 350, 700 and 1500 nm and then varied hold segment from 0 to 300 sec were carried out with 4 or 5 indentations for each depth. Different displacement rate of 1, 10, 25 and 100 nm/s also were applied. The hardness is derived as the maximum load divided by the indentation area calculated from the contact depth [15]. The modulus is determined from the slope (dP/dh) at the beginning of the unloading section. The Poisson's ratio used to calculate Young's moduli for CaF_2 is a default value (0.25).

RESULTS AND DISCUSSION

Figure 1 shows selected load-displacement curves for the largest loads applied in this study. The shape of the curve is to some extent similar to the typical of a metal — that is, unloading is accompanied by recovery of the small elastic component expected primarily to consist of recovery of surface flexure coupled with a small additional recovery of the indentation depth itself. And plastic deformation was observed during hold time segment described later. CaF_2 is soft and comparative highly elastic material , because the values of E are fifty or sixty times larger than those of H.

The load dependence of hardness data for a Berkovich indenter is shown in Figure 2 from which it is evident that the values of nano-hardness on the (001) plane are larger than those on the (111) plane and the (001) plane is little anisotropic in its nano-hardness and the (111) plane is apparently anisotropic, i.e. the value of H in the direction of [110] is less than that of [211]. Also the load dependence of hardness, or ISE, is clearly present. These results may be interpreted in terms of the PSR model of Li and Bradt [1].

Traditionally, the ISE has been described utilizing the power law relationship [16, 17]:

$$P = Ad^n \qquad (1)$$

where P is the applied indentation test load and d is the resulting indentation dimension. A and n in eqn.(1) are known simply as the power law coefficients, or parameters (listed in Table I). Figure 3 illustrates the selected logarithmic plots of indentation load-size relationship, mainly fitting with experimental data but not matching to them at the part of lower displacement. Because eqn. (1) lacks physical meaning, i.e. the n values provide no insight to the mechanism of the ISE, neither does the A value.

The PSR model of Li and Bradt divides measured microhardness into two parts: (i) the indentation load-dependent one, i.e. ISE ; and (ii) the indentation load-independent one. The relationship of applied test load (P) to the indentation size (d) is

$$P = a_1 d + a_2 d^2 \qquad (2)$$

In eqn. (2), the a_1 coefficient contributes to the former part of microhardness and the a_2 coefficient relates to the later, which is equal to (P_c / d_o^2). Here P_c and d_o are a characteristic load and indentation size, respectively. The load-independent microhardness is readily obtained from the slope of the linear plot of

$$(P/d) = a_1 + (P_c/d_o^2)d \qquad (3)$$

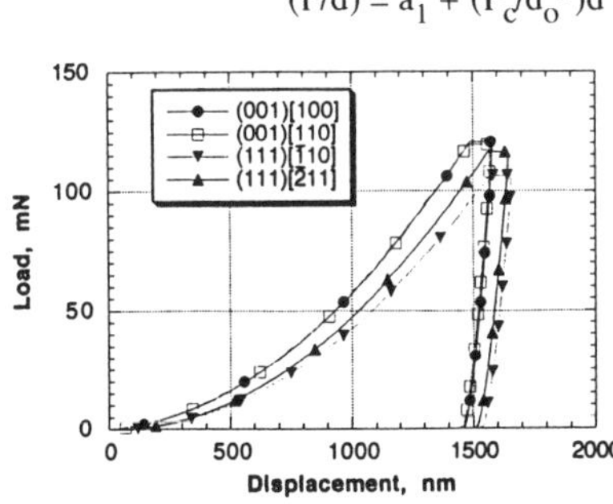

Figure 1. Load-displacement curves of Berkovich indentations into the (001) and the (111) faces of CaF_2 single crystal.

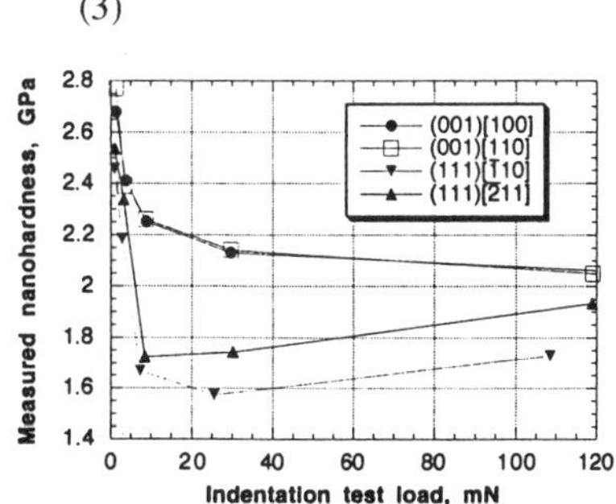

Figure 2. Load dependence of the Berkovich nano-hardness into the (001) and the (111) faces of single crystal CaF_2.

Figure 4 illustrates the plots of (P / d) versus d for two of the major crystallographic orientations on each of the (001) and the (111) planes of single crystal fluorite. These plots confirm the validity of the linear relationship, which is slightly different from those of Figure 3. Table I summarizes the parameters (A and n) for the power low and the PSR parameters (a_1 and a_2) with correlation factors (r^2), on both planes for the major crystallographic directions.

Typical load-displacement curves for the (001) plane of CaF_2 at five different applied loads are shown in Figure 5. It is evident that the extent of recovery appears to increase with decreasing maximum test load. Also plastic deformation was observed during hold time segment. Figure 6 shows that under conditions of creep and relaxation, i.e. maintaining the applied load for varying times, the behavior of calcium fluoride is the same as that of other crystalline solids [18] and also that the nature of anisotropy is unchanged. It is apparent that at least some plastic deformation occurs depend on hold time. There are a small amount of reverse plasticity upon unloading. Figure 7 illustrated the mechanical properties of CaF_2 depended upon holding time. The values of E and H on (100) plane are larger than those on (111) plane and these values on both planes decrease with increase in time during the hold time segment, even though the values are not so changed at longer hold time. Loading the surface to a load/deflection just below P and waiting for some times, produce the displacement discontinuity. The values of Young's modulus are remarkably higher than those of the effective modulus, normal to the {100} and {111} face of CaF_2 are 140 and 90 GPa, respectively.

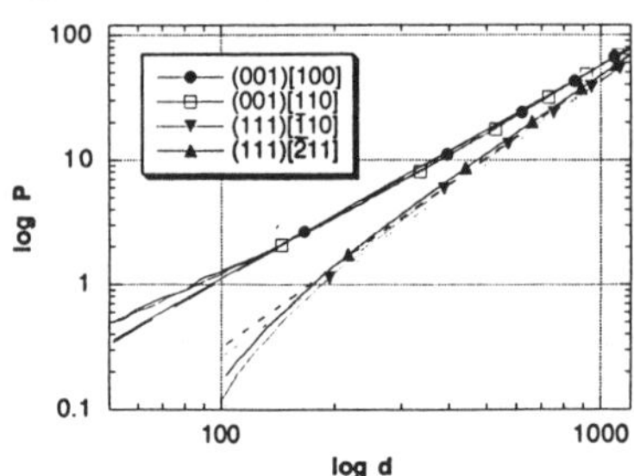

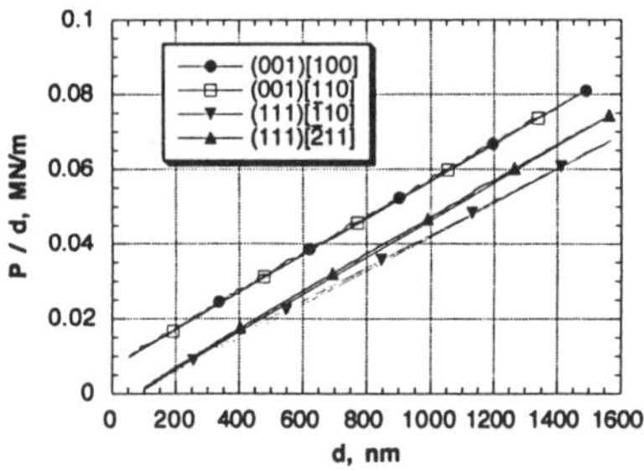

Figure 3. The single-crystal fluorite hardness results for the (001) and (111) planes presented as the logarithmic form of the power law.

Figure 4. The hardness of single-crystal fluorite on the (001) and (111) faces presented in the form of the PSR model.

Table I. Microhardness related parameters of CaF_2 for the (001) and the (111) planes

(hkl)[uvw]	A	n	r^2	a_1	a_2	r^2
	(mN m^{-n})			(mN m^{-1})	(mN m^{-2})	
(001)[100]	4.37x10^{-4}	1.70	0.9993	7.8625x10^{-3}	4.9368x10^{-5}	0.99989
(001)[120]	4.36x10^{-4}	1.70	0.9994	7.9503x10^{-3}	4.8888x10^{-5}	0.99987
(001)[110]	4.01x10^{-4}	1.72	0.9993	7.3347x10^{-3}	4.9597x10^{-5}	0.99996
(001)[210]	4.34x10^{-4}	1.70	0.9992	7.6096x10^{-3}	4.9298x10^{-5}	0.99997
(111)[110]	9.88x10^{-6}	2.21	0.9994	-2.4601x10^{-3}	4.4886x10^{-5}	0.99992
(111)[211]	1.36x10^{-5}	2.18	0.9994	-2.4515x10^{-3}	4.9533x10^{-5}	0.99990

Influence of the displacement rate against the value of H and E are illustrated in Figure 8. The values of hardness are not changed so much. On the contrary, the value of Young's modulus applied lowest displacement rate of 1 nm/s is extremely higher than those applied other faster displacement rate of 10, 25, and 100 nm/s. The latter are almost all same values, though to a little extent these decrease with increasing displacement rates. Load-displacement curve was curved at the beginning of unloading controlled at the lowest displacement rate. So an indenter is pushing into CaF_2 at the beginning of unloading segment at the lowest displacement rate, even if lower applied load of about 3.5 mN, because of plastic deformation of CaF_2.

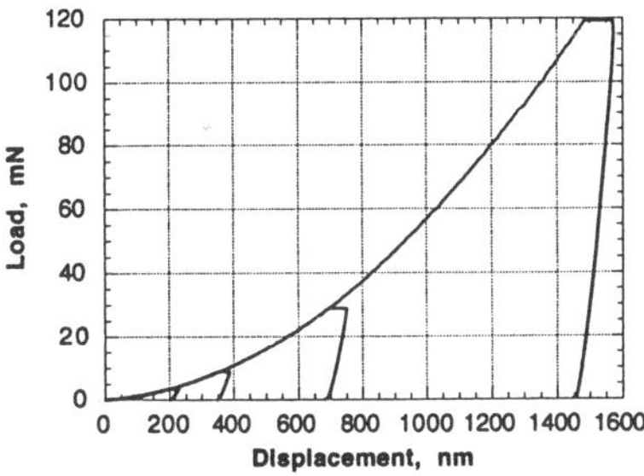

Figure 5. Load-displacement observations into the (001) face of CaF_2 single crystal at various loads.

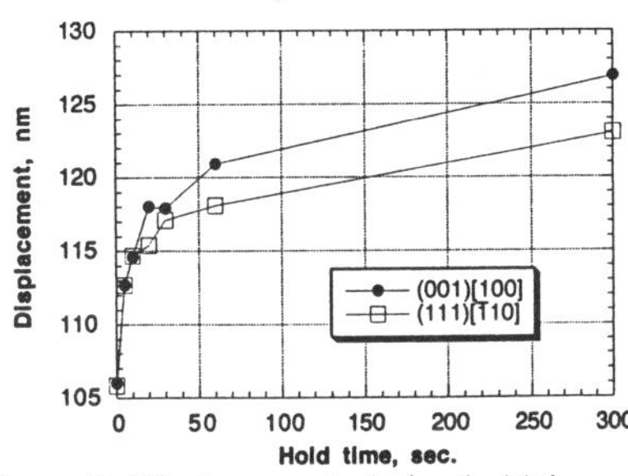

Figure 6. Displacements during hold time segment into the (001) and (111) faces of CaF_2 at lowest load applied (about 1.3mN).

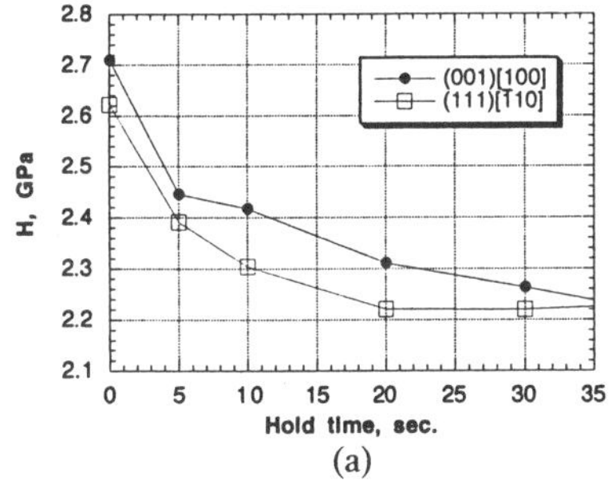

(a)

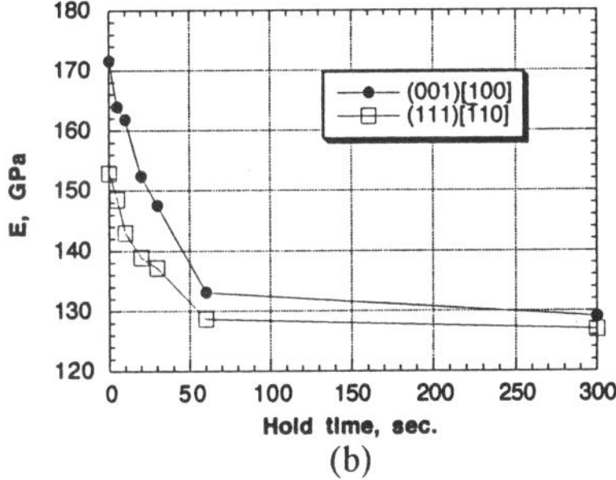

(b)

Figure 7. The changing of the measured (a) hardness and (b) Young's modulus by holding times.

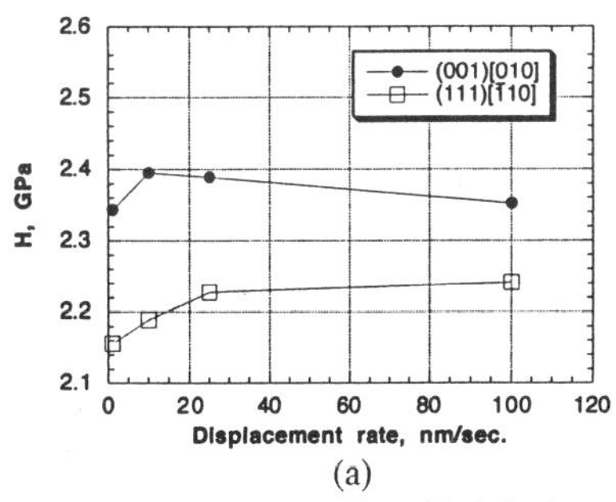

(a)

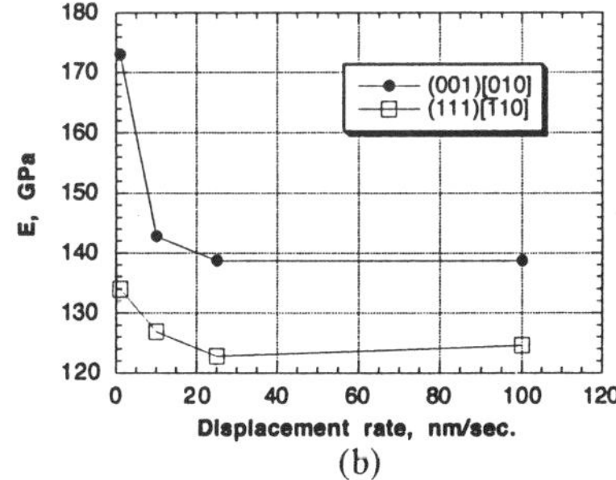

(b)

Figure 8. The influence of applied displacement rates upon the values of (a) hardness and (b) Young's modulus.

CONCLUSIONS

The mechanical properties of single crystal CaF_2 were determined from nanoindentation technique in order to measure the mechanical properties of thin films deposited on CaF_2 substrates later. Nano-indentation data exhibits ISE. A proportional specimen resistance (PSR) model was applied to explain the ISE. This approach is supported by experimental results of instrumented nano-hardness studies of the indentation penetration depth versus penetration test load. CaF_2 is soft and comparative highly elastic materials. The plastic deformation was observed during hold time segment, which needs for a time (about twenty or thirty seconds) to measure by use of nanoindenter. Also the nature of anisotropy is unchanged, i.e. the values of nano-hardness on the (001) plane are larger than those on the (111) plane and the (001) plane is little anisotropic in its nano-hardness and the (111) plane is apparently anisotropic. But the values of mechanical properties are higher than those measured by mirohrdness or other techniques.

REFERENCES

1. H. Li and R. C. Bradt, J. Non-Crist. Solids **146**, 197 (1992).
2. A. Barraud, A. Ruaudel, and M. Vandevyver, Fr. Patent No. 2 564 231 (15 November 1985).
3. H. Mizukami, K. Tsutsui, and S. Furukawa, Jpn. J. Appl. Phys. **30**, 3349 (1991).
4. A. G. Evans, C.Roy, and P. L. Pratt, Proc. Brit. Ceram. Soc. **6**, 173 (1966).
5. J. B. O'Neill, B.A. W. Redtern, and C. A. Brookes, J. Mater. Sci. **8**, 47 (1973).
6. Yu. S. Boyarskaya, M. I. Val'kovskaya, D. Z. Grabko, and N. I. Melent'ev, Fiz.-Khim. Yavleniya Shlifovanii <u>1976</u>, 155.
7. M. F. G. Gil, P. F. Alonso, and P. Francisca, Trab. Geol. **11**, 73 (1981).
8. Yu. S. Boyarskaya, D. Z. Grabko, M. P. Dintu, Cryst. Res. Technol **16**, 441 (1981).
9. K. K. Rao and D. B. Sirdeshmukh, Bull. Mater. Sci. **5**, 449 (1983).
10. W. Yang, R. G. Parr, and L. Uytterhoeven, Phys. Chem. Miner. **15** (2), 191-5 (1987).
11. K. K. Rao and D. B. Sirdeshmukh, Pramana **34**, 151 (1990).
12. I. I. Afanas'ev, Monokrist. Tekh. **3**, 146 (1970).
13. D. Vidal, C. R. Hebd. Seances Acad. Sci., Ser. B**279**, 345 (1974).
14. L. E. A. Jones, Phys. Earth Planet. Inter. **15**, 77 (1977).
15. M. F. Doerner and W. D. Nix, J. Mater. Res. **1**, 601 (1986).
16. E. Meyer, Phys. Z. **9**, 66 (1908).
17 P. M. Sargent and T. F. Page, Proc. Br. Ceram. Soc. **26**, 209 (1978).
18. A. G. Atkins, A. Silverio, and D. Tabor, J. Inst. Metals **94**, 369 (1966).

Part IV

Mechanical Property Methods and Modeling

IN-SITU TEM INVESTIGATION DURING
THERMAL CYCLING OF THIN COPPER FILMS

R.-M. Keller, W. Sigle, S.P. Baker, O. Kraft, and E. Arzt
Max-Planck-Institut für Metallforschung and Institut für Metallkunde, University of Stuttgart
Seestr. 71, D-70174 Stuttgart, Germany

ABSTRACT

In-situ transmission electron microscopy (TEM) was performed to study grain growth and dislocation motion during temperature cycles of Cu films with and without a cap layer. In addition, the substrate curvature method was employed to determine the corresponding stress-temperature curves from room temperature up to 600°C. The results of the *in-situ* TEM investigations provide insight into the microstructural evolution which occurs during the stress measurements. Grain growth occurred continuously throughout the first heating cycle in both cases. The evolution of dislocation structure observed in TEM supports an explanation of the stress evolution in both capped and uncapped films in terms of dislocation effects.

INTRODUCTION

Copper is a possible alternative to aluminum as a metallization for integrated circuits because of its better electromigration resistance and its higher electrical and thermal conductivity. However, the high elastic modulus and thermal expansion coefficient of Cu cause large thermal stresses and therefore potential reliability problems. The mechanical properties of Cu films with different barrier layers and cap layers have been investigated [1-3], but little *in-situ* information about the evolution of the microstructure and the dislocation motion during a thermal cycle is available. We have used a combination of substrate curvature measurements and *in-situ* TEM investigations to compare the stress evolution with microstructural features. We explain the deformation behavior of capped and uncapped copper films on the basis of dislocation motion.

EXPERIMENTAL DETAILS

The substrates used were 100 mm diameter Si wafers with a 50 nm thermal oxide layer. A 50 nm thick silicon nitride (Si_3N_4) layer was deposited onto both sides of each wafer by low pressure chemical vapor deposition (LPCVD) and was densified at 950°C for one hour. Next, a 0.6 μm thick copper film was deposited on one side by room temperature (RT) sputtering at 1020 Å/min in 2 mTorr Ar. Some films were additionally capped with a 50 nm thick Si_3N_4 layer. The cap layer was sputtered in order to avoid the high temperature anneal necessary for LPCVD Si_3N_4 and was deposited on both sides of the wafer. The sputter rate was 15 Å/min.

The substrate curvature technique was used to study the thermo-mechanical behavior of the Cu films by measuring the curvature changes during thermal cycles with a laser-based optical lever device. Stresses were calculated using the data analysis method described in reference [1]. Two stress-temperature cycles from RT to 600°C and back were conducted for films with and without the capping layer. All samples were cycled in a nitrogen atmosphere. The heating and cooling rate was 6 K/min. The arrangement of layers on the wafer allowed direct measurements of stresses in Cu alone. No curvatures are expected to be caused by the diffusion barrier or capping layer, since they are present on both sides of the wafer.

For the *in-situ* TEM investigations, a high voltage TEM (AEI/EM 7) operating at 1 MV was used. Plan view TEM samples were prepared by dimpling to create an electron-transparent composite window consisting of some residual silicon, diffusion barrier, Cu film and optional cap layer. The typical window was about 260 μm in diameter and about 700 nm thick. The high

Mat. Res. Soc. Symp. Proc. Vol. 436 © 1997 Materials Research Society

voltage of the TEM allowed for good transparency at relatively high film thicknesses, even with portions of the substrate left. The samples were heated and cooled twice between RT and 600°C in a special heating holder. This holder does not allow rate-controlled heating and cooling. In order to compare with the substrate curvature results, the temperature was increased in 10°C steps and dwell times were introduced after each step to get a similar total cycling time. One TEM temperature cycle took about 60-70 min. whereas the duration of the temperature cycle during a stress measurement was 100 min. Also, it must be realized that the stress state for the two methods is different because the "window" preparation removes the substrate constraint. The compressive stresses are lower because the film is free to buckle under compressive stresses while the tensile stresses are unchanged. Nonetheless, a simple thermoelastic calculation shows that the differential thermal expansion between the SiO_2, Si_3N_4 and Cu layers could result in strains high enough to cause yielding in Cu in compression even in the absence of the substrate.

Grain growth was investigated during the first cycle by taking a micrograph of the same area of the sample at 50°C intervals. The grain size distribution was determined by measuring the grain areas, A, of 50-100 individual grains from these pictures and converting the areas to equivalent diameters; $d = 2\sqrt{A / \pi}$. During the second cycle, dislocation motion was investigated by observing a single grain which contained dislocations from the beginning. Photographs were taken, as before, in 50°C steps.

RESULTS AND DISCUSSION

<u>Grain growth</u>

Grain growth occurred during the first heating in the capped and uncapped copper films. No further grain growth was observed during cooling or subsequent cycles. All grain size distributions are log-normal with some deviations for small and very big grains. Fig. 1 shows the median grain size with increasing temperature during the first heating cycle and the corresponding stresses. For both films, grain growth occurred continuously over a wide temperature range. A sharp drop in compressive stress was seen in both films. This stress drop occurs between 200 and 300°C for the uncapped and between 400 and 500°C for the capped film. For the uncapped film, grain size increases linearly with temperature. For the capped film, the grain size curve is S-shaped with maximum grain growth rate occurring at the maximum compressive stress. The grain sizes at the end of the first heating cycle agree well with *ex-situ* measurements reported previously [4].

The stress drop is commonly attributed to film densification resulting from grain growth. However, our results do not show a simple relationship between the grain growth and the stress drop. An estimate of the reduction in compressive stress in a film which is possible due to a reduction of grain boundary volume can be obtained as follows: The change in film strain due to the reduction of grain boundary volume is given by

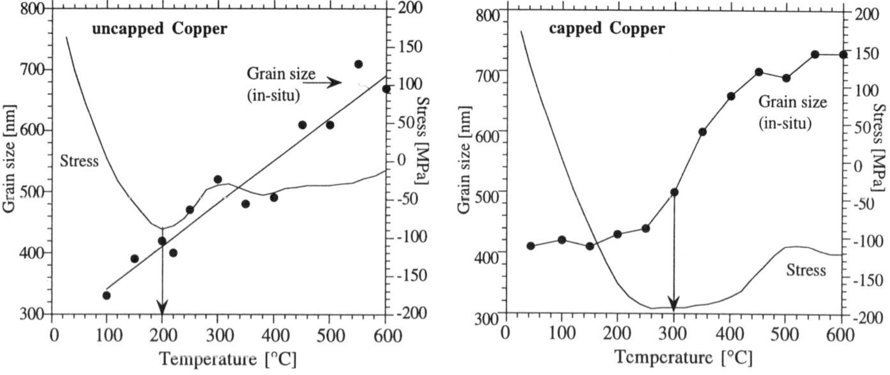

Fig. 1. Grain size evolution determined from *in-situ* TEM and the corresponding stresses calculated from substrate curvature for copper films with and without a cap layer during first heating.

$$\Delta\varepsilon = \left(\frac{1}{d_1} - \frac{1}{d_0}\right)\bigg/\left(\frac{1}{\delta} + \frac{1}{d_0}\right)$$

where d_0 and d_1 are the initial and the final grain sizes and δ the width of the grain boundary which is assumed to be material-free. The change in stress was estimated using the measured grain sizes at the beginning and end of the stress drop (200-300°C for uncapped and 260-500°C for capped films), one burgers vector for the grain boundary width and 179 and 185 GPa for the biaxial moduli of uncapped and capped films, respectively (difference due to texture). Even using this generous value for d, the estimated values are lower than the measured stress drops (see Table I) indicating that grain growth alone probably cannot account for the stress drop in either case.

Table I. Comparison of the estimated and measured stress changes in the area of the stress drop during the first heating cycle.

Metallization	d_0 [μm]	d_1 [μm]	$\Delta\varepsilon$	$\Delta\sigma_{estimated}$ [MPa]	$\Delta\sigma_{measured}$ [MPa]
no cap layer	0.42	0.52	1.17E-4	21	71
cap layer	0.44	0.69	2.11E-4	39	82

<u>Dislocation motion</u>

Figure 2a shows the initial dislocation structure in a grain oriented with [111] parallel to the film normal in the uncapped film. The majority of dislocations are trapped in a twin. Some threading dislocations are also visible inside the grain. With increasing temperature, more dislocations become visible in the twins. They pile up near the grain boundary. Above 300°C, the dislocation motion increases, while the dislocation density is reduced. Above 370°C, no further motion is visible (Fig. 2b). During cooling, the remaining dislocation structure is stable to 240°C (Fig. 2c). With further decreases in temperature, the dislocation density increases again. At RT the dislocation structure is nearly the same as at the beginning of the temperature cycle (Fig. 2d).

Figure 3 shows the sequence for the capped film. The initial dislocation structure is different (Fig. 3a). The dislocations are not trapped in twins but are strongly curved and pinned at the grain boundary and at an impurity particle. Some unpinned threading dislocations are visible, too. This dislocation structure is stable until 200°C. Above this temperature, the curved dislocations start to detach from the particle. The motion ceases at about 350°C (Fig. 3b) and above this, little motion of the threading dislocations is visible. Above 450°C, no further dislocation motion occurred (Fig. 3c). The dislocation structure is stable during cooling until 500°C, below which curved and pinned dislocations appear again at the same grain positions. Their number increases with decreasing temperature. In contrast to the curved dislocations, the threading dislocation structure is stable until 300°C. The dislocation structure at the end of the cycle is similar to that at the beginning (Fig. 3d). However the dislocations are less clearly affected by the particle and are not as strongly curved.

Fig. 4 shows the corresponding stress-temperature behavior. Both films show a plateau at very low compressive stress levels during heating which has been attributed [5] to a dislocation effect similar to the so-called Bauschinger effect [6, 7]. According to this explanation, dislocations are stopped at obstacles and create internal stresses in the direction opposite to the external load during cooling in the first stress-temperature cycle. The magnitude of the effect is bigger for the capped film because the cap layer prevents dislocations from escaping the film and is thus an effective obstacle. During heating in the second cycle, the critical stress required for dislocation motion is lower because the internal stresses now act in the same direction as the external load. The critical stress is lower for the capped film as expected. The very low and wide plateau in stress is caused by dislocations gliding away from obstacles. The plateau ends when all the dislocations are removed from their obstacles and the benefit of the internal stresses is consumed. Stress behavior at higher temperatures is determined by the relaxation mechanism acting in the film. Models of the stress hysteresis of uncapped copper films lead to reasonable agreement if power law creep is assumed to be the dominant deformation mechanism [8]. This allows the stress to relax to zero for the uncapped film. However, for the capped films the amount of thermal stress built up with increasing temperature is greater than the stress relaxation, resulting in an increase of the compressive stresses between 400 and 600°C. The dislocation effect acts on

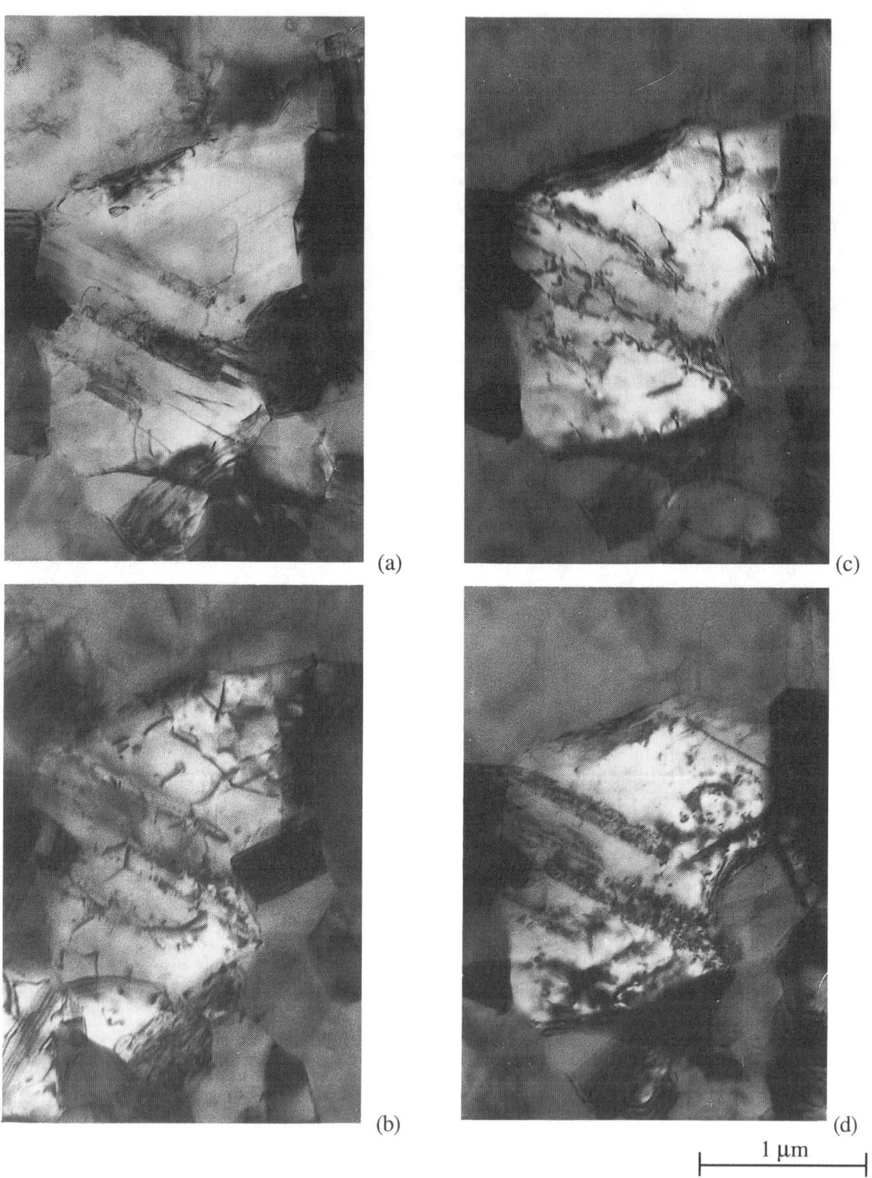

(a)

(b)

(c)

(d)

⊢——— 1 μm ———⊣

Fig. 2. Dislocation structure in the uncapped film during the second temperature cycle: (a) Initial dislocation structure with many dislocations trapped at twins. With increasing temperature the dislocations begin to disappear. (b) Stable structure at 400°C upon heating. No further dislocation motion occurred. (c) Stable structure at 250°C upon cooling With further decreasing temperature, the dislocations move again and their density increases. (d) Final structure at 100°C.

224

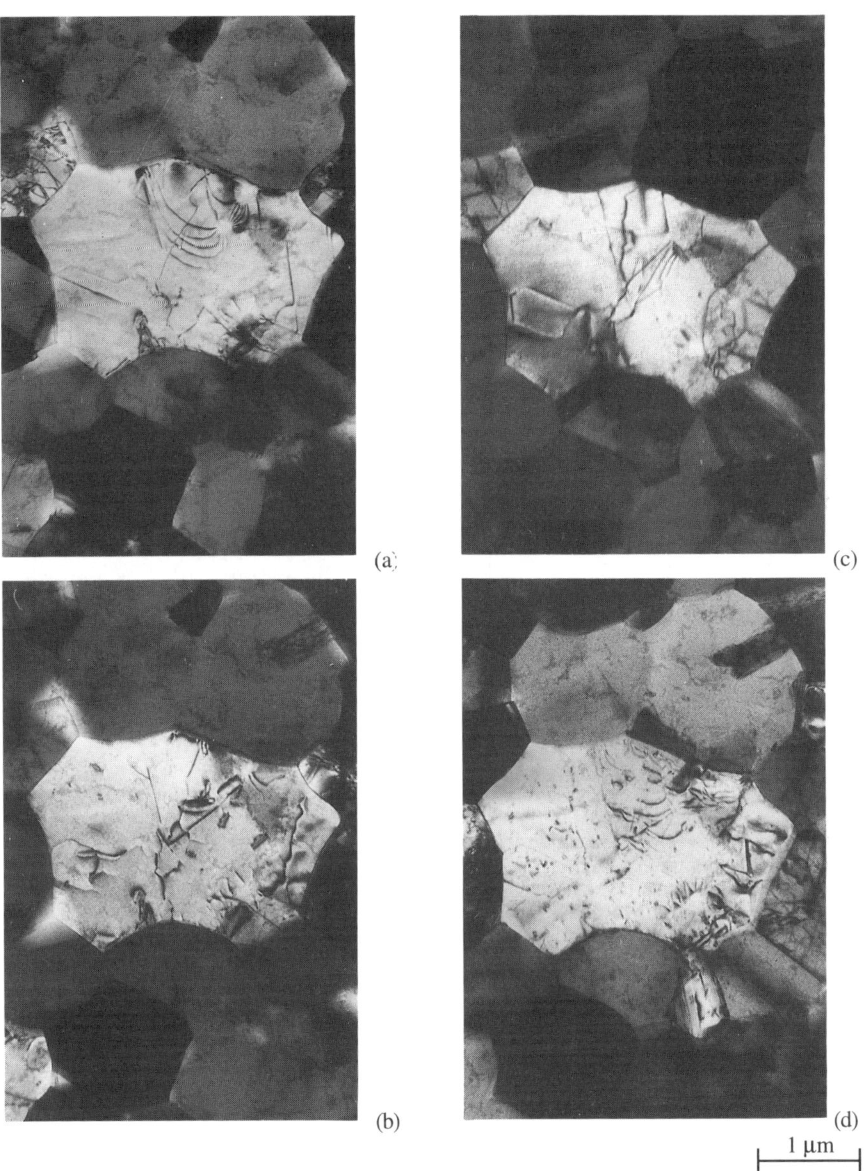

Fig. 3. Dislocation structure in the capped film during the second temperature cycle: (a) Initial state with strongly curved and pinned dislocations; above 200°C the dislocations start to detach and move away until 350°C (b) is reached. (c) Stable structure having a minimal number of dislocations at 600°C. During cooling the dislocation structure is stable until 500°C is reached. (d) Final dislocation structure. The dislocations are curved again, but are less pinned at particles

cooling as well as on heating but is less important. The reason is that thermally activated processes such as cross slip reduce the internal stresses which can be built up. This is more pronounced in uncapped films. Therefore, a corresponding tensile stress plateau, which is shorter and at a higher stress level, is evident only in the capped film. Below 200°C relaxation is no longer significant and dislocations glide and interact with obstacles again.

Our *in-situ* study supports this explanation of stress behavior influenced by a Bauschinger-like dislocation effect. Figures 2 and 3 illustrate our observations that all visible dislocation motion occurs between 200 and 400°C during heating and below 200°C during cooling. Dislocation motion is not observed between 400 and 600°C in the capped film because the compressive stresses in the TEM sample are too low. However previous x-ray results [5] indicate that dislocation density again increases in this region when the substrate is attached.

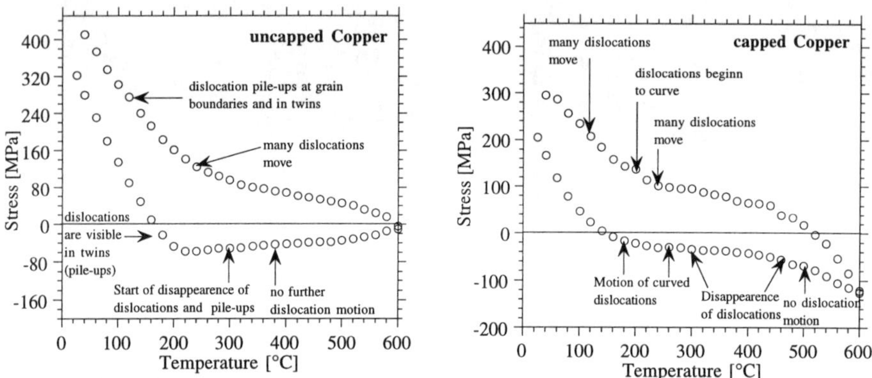

Fig. 4. Second stress-temperature cycles of the capped and uncapped films as determined by substrate curvature measurements, shown with observations from the *in-situ* TEM investigations.

CONCLUSIONS

It has been shown that in-situ TEM investigations and wafer curvature results can be correlated. Grain growth occurred continuously during the first heating cycle and no coincidence was found between the onset of grain growth and the characteristic compressive stress drop. The initial dislocation structure is different in the two films: for the uncapped films, most dislocations are associated with twins while for the capped films, pinning within the grains occurred. Most dislocation movement is visible between 200 and 400°C during heating and below 200°C during cooling. This is consistent with a dislocation effect controlling the stress evolution in both films.

ACKNOWLEDGEMENTS

This work was supported by the Deutsche Forschungsgemeinschaft under contract Ar 201/5-1. The authors wish to thank R.P. Vinci for sample production.

REFERENCES

[1]   Flinn, P.A., J. Mater. Res., **6** , 1498 (1991)
[2]   Vinci, R.P. and J.C. Bravman, MRS Symp. Proc. **309**, 269 (1993)
[3]   Vinci, R.P., Ph. D. thesis, Stanford University, 1994
[4]   Keller, R.-M., W.-M. Kuschke, A. Kretschmann, S. Bader, R.P. Vinci and E. Arzt., MRS Symp. Proc. **391**, (1995)
[5]   Keller, R.-M., S. Bader, R.P. Vinci and E. Arzt., MRS Symp.Proc. **356**, 453, (1994)
[6]   Sleeswyk, A.W. and G.J. Kemerink, Scripta Metallurgica, **19** , 471 (1985)
[7]   Scholtes, B. and O. Vöhringer, Zeitschrift für Metallkunde, **77** (9), 595 (1986)
[8]   Keller, R.-M., Ph.D. Thesis, University of Stuttgart 1996

MICROMECHANICAL TENSILE TESTING

S. GREEK, F. ERICSON, S. JOHANSSON, J.-Å. SCHWEITZ
Materials Science, Uppsala University, Box 534, S-751 21 Uppsala, Sweden,
Staffan.Greek@teknikum.uu.se

ABSTRACT

A method is described where tensile tests can be performed *in situ* on micromachined structures. The testing equipment consists of a testing unit mounted on a micromanipulator in a Scanning Electron Microscope (SEM). The fracture loads of micromachined beam structures made from thick and thin film polysilicon as well as from electrodeposited nickel and nickel-iron alloy were measured, and the fracture strengths then calculated via measurements of the test structures' initial cross-sectional areas. The statistical scatter of the polysilicon fracture strength values were evaluated by Weibull statistics. The mean fracture strength and the Weibull modulus, a measure of the scatter, were obtained

INTRODUCTION

The fracture strength of a microstructure can be derived from bending or tensile tests. Of the two methods, the tensile test is considered to have more reliable and more easily interpretable results. Linear tensile tests are more favourable from an error analysis viewpoint than bending tests, and the tensile stress distribution in the test structure during tension is more uniform than in bending tests. The tensile test is a popular method for macroscopic testing, but the mechanical properties of microstructures can not be fully and accurately determined by measuring on a macroscopic scale. The tensile test has, however, previously rarely been applied to micromachined structures because of their smallness [1,2]. A test on a film performed with the aid of micromechanical test structures will give a more correct prediction of the properties of an application made from the film. Furthermore, a test structure fabricated by the same process intended for a final structure will directly give information on how varied process parameters affect the structure. The problem of observing the micro structures can be overcome by using a Scanning Electron Microscope (SEM) and this paper describes a method where micromachined beams are pulled to fracture *in situ* using a micromanipulator developed at Uppsala University [3]. The study comprises tensile strength tests on two types of polysilicon films as well as on electrodeposited nickel and nickel-iron alloy.

EXPERIMENT

Scanning Electron micrographs of tensile test structures are seen in Figure 1. Test structures were realised in thick film [4] and thin film [5] polysilicon, as well as in nickel and nickel-iron alloy [6]. The thick film polysilicon was an epitaxially deposited film of 10 μm thickness, while

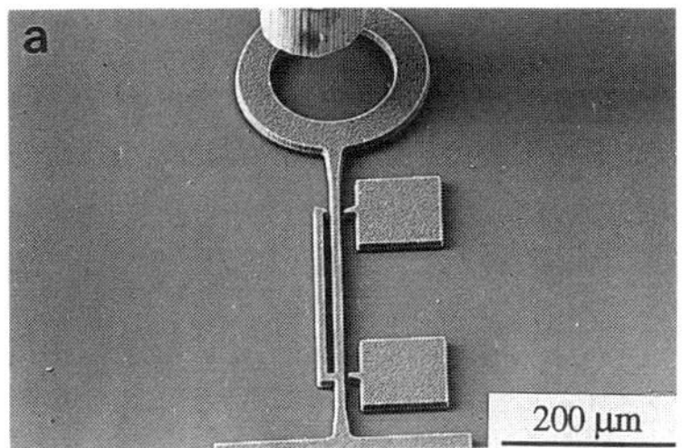
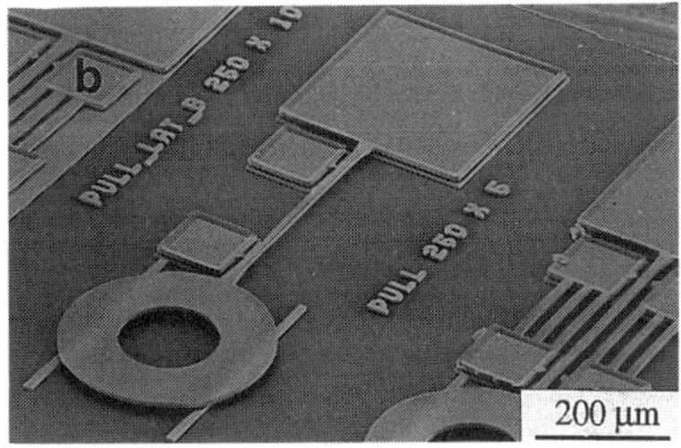

Figure 1. Tensile test structures realised in a) thick film polysilicon and b) electrodeposited nickel.

the thin film polysilicon was a 2 µm LPCVD film. The nickel and nickel-iron alloy structures were electrodeposited with a thickness of about 8 µm. The tensile test structure consist of a beam which is attached to a base plate at one end. The opposite end of the beam has a ring which is to be gripped by the testing apparatus. The ring and the beam are both released by the sacrificial layer etch while the base plate is fixed to the substrate. The beam lengths of the polysilicon test structures were 1000 µm, while the test structures made from electrodeposited nickel and nickel-iron alloy were of three different lengths: 250, 500, and 1000 µm. The nickel-iron alloy contained about 24% Fe.

The testing unit, described schematically in Figure 2, consists of an arm with a probe, a voltage driven piezoelectric actuator, a strain gauge force sensor and an optical encoder for the measurement of displacement. The probe is positioned in the ring of the test structure beam. The two parallel beams of the arm, connected with hinges to the arm and to the fixed support, restricts the arm's movement to a straight line, i.e. it moves without rotation. The design of the testing unit has been optimised for the typical tests performed, with a displacement resolution of 0.01 µm over a maximum testing stroke of 150 µm. The testing force resolution is 10 µN with a maximum of 1 N. The micromanipulator consists of three independently moveable xyz-tables mounted in a cradle. The cradle itself can be raised or lowered as well as tilted for better visualisation in the SEM. The tilting and translative motions are driven by electrical motors and all positioning is done from a remote control. The chip with the test structures was placed on the centre table and the testing unit on a side table. On the third table an assisting probe was placed. The micromanipulator can be placed in an SEM where the testing procedure can be monitored in high magnification. A computer based control and data acquisition programme was developed which slowly raises the voltage applied to the actuator, while sampling amplified signals from the force sensor and the optical encoder. The cross section of the beam was measured in the SEM, and the tensile strength calculated as the ratio of the maximum sampled force to the initial cross sectional area of the beam.

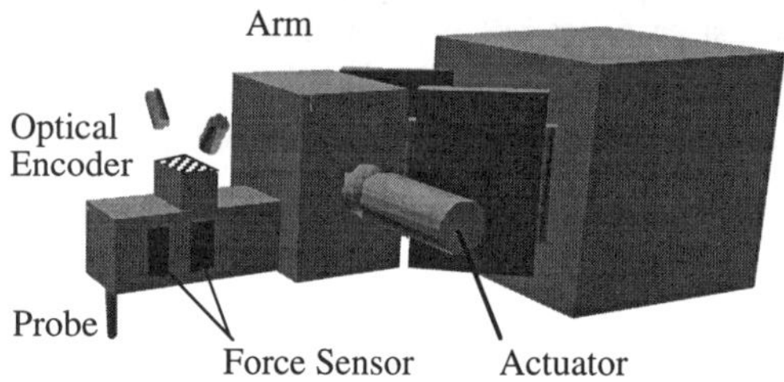

Figure 2. A schematic view of the tensile test unit

RESULTS

<u>Polysilicon</u>

Structures made of brittle materials, like silicon, fracture under tensile stress by a sudden and catastrophic growth of a crack or cracks through the structure. Cracks are initiated at defects within the structure, or at defects at the surface of the structure, due to a stress concentration at these points. In our particular test specimens, previous studies have shown the interior of the specimens to be free from voids or microcracks, while the vertical sides show a high surface roughness. Hence the cracking is assumed to initiate at surface defects. When a structure breaks at a given load it must contain a crack of a certain severity and orientation within a part of the structure where the stress is tensile. The orientation of the crack is important since a crack at an acute angle to the stress is not affected as much as a crack at right angles to the stress. Thus the fracture stress of a given structure depends not only on defect distribution but also on the geometry of the structure and the stress distribution within it.

The fracture stress is a stochastic parameter which needs a statistical treatment. When testing a large number of seemingly identical structures, it is found that the fracture strengths have a characteristic scatter around a certain mean value. To evaluate this characteristic behaviour Weibull [7] has described a probability function of fracture events as

$$P_f = 1 - \exp\left\{-\int_A \left(\frac{\sigma_a(x,y,z) - \sigma_u}{\sigma_0}\right)^m dA\right\} \tag{1}$$

where P_f is the fraction of the total number of structures that will fracture under the applied stress distribution $\sigma_a\,(x,y,z)$. σ_u is the lowest stress at which fracture will occur and σ_0 is a normalising factor. The exponent m is known as the Weibull modulus and is a materials parameter expressing the statistical scatter of fracture events: a high Weibull modulus indicates a low scatter. In equation (1) it has been assumed that the fracture cracks have originated from surface defects distributed over the area A of the detail under stress. If volume defects are considered to govern the crack initiation, the stress function must be integrated over volume instead. In brittle materials, σ_u is normally set to zero as is the case in the following analysis.

The fracture stress of a brittle material is a theoretical concept and in practice impossible to predict exactly. To measure an experimental value normally a large number of equally prepared samples have to be tested. If a "fracture stress" of the *material* is sought the samples have to be tested with a homogeneous stress distribution. For non-homogeneous stresses the value obtained will be a "fracture stress" of the *structure*. Essentially the Weibull analysis summarises the probability of fracture for all the differently stressed parts of the structure. To simplify the analysis, the stress distribution within a structure is expressed as

$$\sigma_a\,(x,y,z) = \sigma_a\, g \tag{2}$$

where σ_a is the maximum applied stress and g is a function depending on the structure geometry and the load distribution. Using (2), the probability function can be expressed as

$$P_f = 1 - \exp\left\{-\sigma_a^m \frac{1}{\sigma_0^m} \int_A g^m dA\right\} \tag{3}$$

$$= 1 - \exp\left\{-\sigma_a^m h\right\} \tag{4}$$

where the function

$$h = \frac{1}{\sigma_0^m} \int_A g^m dA \tag{5}$$

is introduced for simplicity. Since g depends on the geometry and load distribution, h also depends on these factors. The expected mean value of the fracture stress can be calculated from

$$\overline{\sigma_a} = \frac{1}{h^{1/m}}\, \Gamma(1 + \tfrac{1}{m}) \tag{6}$$

where $\Gamma(z)$ is the gamma function.

Returning to the tensile test structure, it is known that the stress is evenly distributed when the load is applied axially

$$\sigma_a(x,y,z) \equiv \sigma_a. \tag{7}$$

Thus

$$g = 1 \tag{8}$$

and

$$h = \frac{A}{\sigma_0^m} \tag{9}$$

where A is the area of the structure flanks, which are the surfaces where cracks are presumed to be initiated. The Weibull probability function for the tensile test structure becomes

$$P_f = 1 - \exp\left\{-A\left(\frac{\sigma_a}{\sigma_0}\right)^m\right\}. \tag{10}$$

The fracture probabilities resulting from the tensile tests are calculated as

$$P_f = \frac{i - \frac{1}{2}}{n} \tag{11}$$

where n is the total number of fracture tests and i is the index of the fracture stress result in an array where the values are written in an ascending order. The fracture probability as a function of applied stress for the experiments is presented in Figure 3. A curve of the form of equation (10) is fitted by the chi-square method to the measured data, and the Weibull modulus, m, as well as the normalising factor, σ_0, are derived from this fit. The expression for the expected mean fracture stress of the tensile test structure is

$$\overline{\sigma_a} = \frac{\sigma_0}{A^{1/m}}\, \Gamma(1 + \tfrac{1}{m}). \tag{12}$$

In Figure 3 it can be seen that some of the results from the thin film tests deviate from the fitted Weibull curve. These deviations are due to the relatively low number of performed tests, but do not affect the calculations of the Weibull modulus and mean fracture strength in any significant manner.

The beam that was actually tested in the tensile test had a length of 1000 µm and a width of 10 µm for all tests. The thick film beams had a thickness of 10 µm and the thin film beams had a thickness of 2 µm. Since the vertical flanks of both thick film and thin film structures were much rougher than the top and bottom surfaces, the fracture is assumed to have its origin from defects in the flanks. The area used in the exponent of the Weibull probability function (10) thus is 20×10^{-9} m^2 for the thick film and 4×10^{-9} m^2 for the thin film. From the curve fits of Figure 3, the Weibull modulus m and the normalising factor σ_0 were deduced. The results were used to calculate the expected mean fracture stress according to equation (12). The thick film polysilicon had a Weibull modulus of 7 and an expected mean fracture strength of 768 MPa. For the thin film polysilicon these values were 11 and 566 MPa, respectively.

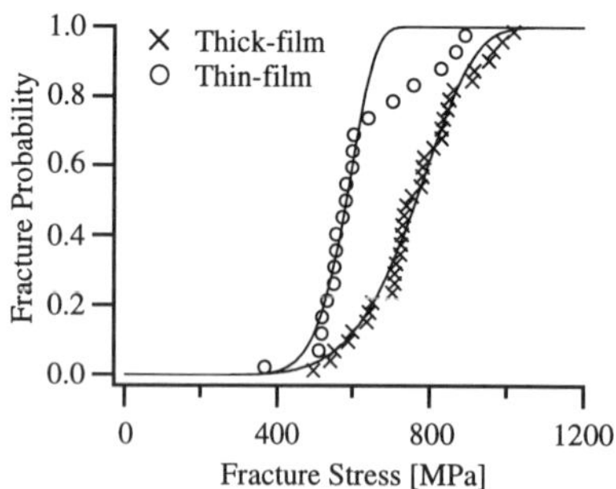

Figure 3. Results from tensile tests on beams of thick film and thin film polysilicon.

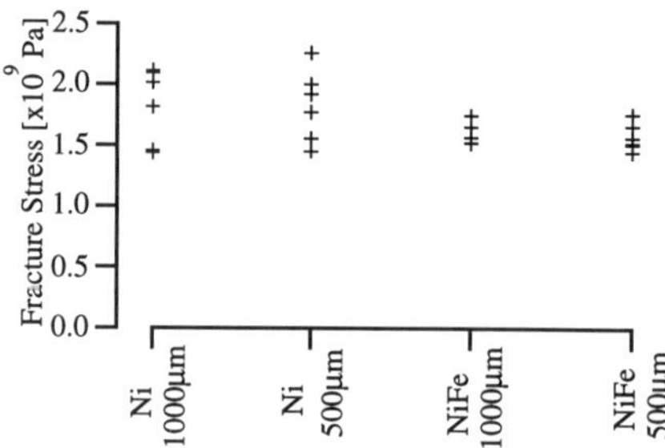

Figure 4. Results from tensile tests on beams of nickel and nickel-iron alloy with different beam lengths.

<u>Nickel and Nickel-Iron Alloy</u>

The fracture strength of electrodeposited nickel and nickel-iron alloy is seen in Figure 4. As expected, no significant variation in fracture strength between beams of different lengths was noted. The Weibull statistical model is not valid for ductile structures, instead, the fracture strengths follows the normal distribution. The fracture strength of nickel had a mean value of 1.83 GPa and a standard deviation of 0.29 GPa, while the fracture strength of nickel-iron alloy had a mean value of 1.59 GPa and a standard deviation of 0.10 GPa. Fracture strength was measured on 18 nickel samples and 12 nickel-iron alloy samples.

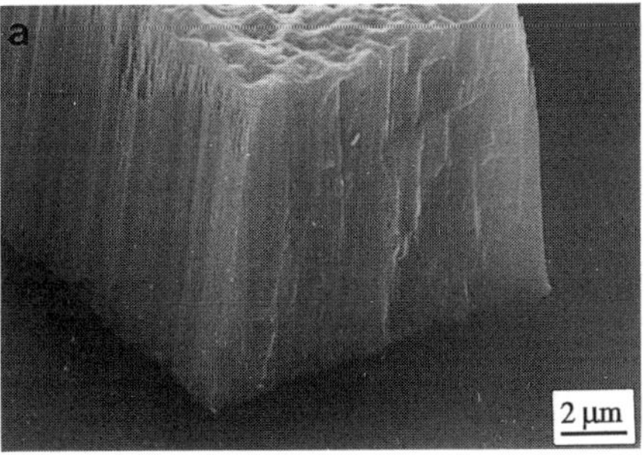

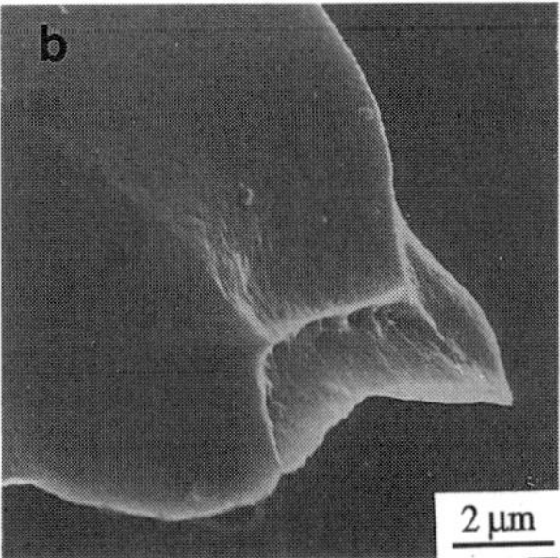

Figure 5. Electron micrographs of the fracture surfaces of structures made from a) thick film polysilicon and b) nickel-iron alloy.

Figure 5 shows the fracture surface of structures made from thick film polysilicon and nickel-iron alloy. The thick film polysilicon structure, which is a brittle material, shows a clean fracture surface without dimples or necking. The ductile nickel-iron alloy structure fractures after considerable necking and the fracture surface contains dimples.

DISCUSSION AND CONCLUSIONS

We can conclude that the method of fracture strength evaluation by *in situ* tensile testing of micromachined structures is easy and quick to perform. As a solution to the problem of handling small structures, the test structure was designed with a ring to match the cylindrical probe of the testing unit. The rounded surfaces interact to align the test structure in the direction of the motion of the tensile test, which facilitates easy handling. The testing unit is equipped with an optical encoder for the measurement of displacement with an accuracy of 1%. The tests indicate, however, that the elasticity of the whole testing system at the present is of the same order as the strain in the structure. Calibrations of the elasticity in the micromanipulator, test unit etc. have to be performed to calculate the strain in the microstructure. Once this calibration has been made, elastic and plastic parameters such as Young's modulus, yield strength etc., can be measured.

The mean tensile strength calculated for the polysilicon thick films was approximately one tenth of the tensile strength of bulk micromachined single-crystalline silicon (6 GPa) [8] measured by beam bending. This was recognised to be partly due to the higher surface roughness and the resulting density of defects in the dry etched polysilicon structures compared with single-crystalline silicon elements. The tested volume in the tensile test is large compared to the volume tested in a bending test, where maximum stress is located to a small layer close to the surface, where the bending moment is at a maximum. The probability of encountering a critical defect in a tensile test is therefore much greater than in a bending test, thus a higher maximum stress is expected to be found in a structure exposed to bending. Crack formation is highly dependent upon the local geometry, leading to a large scatter in the strength values. The

Weibull modulus is a parameter that reflects the scatter: a large Weibull modulus indicates that the critical defects are all alike. The Weibull modulus of the investigated films was measured to 7 and 11, respectively. As a comparison high-quality polished silicon wafers have a Weibull modulus of ~10 [9]. From this we can conclude that the micromachining processes result in a fairly well controlled distribution of defects. The Weibull modulus tells nothing about the severity of the defects, though, and therefore both the Weibull modulus and the mean fracture stress are needed to characterise the strength of a film.

The mean tensile strength of the electrodeposited nickel and nickel-iron alloy depend to a great extent on the electrodeposition bath. Factors that influence the tensile strength include film thickness, grain size, heat treatment etc. Typical values reported in literature are 0.34-1.46 GPa for nickel [9] and 1.40-1.80 GPa for nickel-iron alloy with 25-40% Fe [10]. The tensile strength of nickel measured in the present study is slightly higher. The values measured on nickel-iron alloy are in agreement with literature data.

In macro-scale mechanics, tensile tests are widespread and the theory and use of them common knowledge. To be able to transform this method and knowledge to micromechanical systems gives many advantages in terms of easy understanding of results and testing methods.

ACKNOWLEDGEMENTS

The authors would like to thank Fraunhofer-Institut für s of Berlin, Germany, Centro Nacional de Microelectrónica of Barcelona, Spain and CSEM of Neuchâtel, Switzerland for producing the test structures.

REFERENCES

[1]   M. Biebl and H. von Philipsborn, Digest of Techn. Papers, The 8th Int. Conf. Solid-State Sensors and Actuators, and Eurosensors IX, Stockholm, Sweden, June 25-29, **.2** 72 (1995).

[2]   P. Scafidi, M. Ignat, P. Mortini, M. Marty in <u>An Analysis by Scanning Acoustic Microscopy of the Mechanical Stability of PSG and Si3N4 Passivation Films,</u> (Mater. Res. Soc. Proc. **338**, Pittsburgh, PA, 1994) pp. 115-120.

[3]   S. Johansson, 1st IARP Workshop on Microrobotics and Systems, Karlsruhe, Germany, p. 72 (1993).

[4]   M. Kirsten, B. Wenk, F. Ericson, J.-Å. Schweitz, W. Riethmüller and P. Lange, Thin Solid Films, **259** 181 (1995).

[5]   M.A. Benitez, J. Esteve, M.S. Benrakkad, J.R. Morante, J. Samitier and J.-Å. Schweitz, Digest of Techn. Papers, The 8th Int. Conf. Solid-State Sensors and Actuators, and Eurosensors IX, Stockholm, Sweden, June 25-29, **2** 88 (1995).

[6]   J. Gobet, F. Cardot, J. Bergqvist, F. Rudolf, J. Micromech. Microeng. **3** 123 (1993).

[7]   W. Weibull, J. Appl. Mech., **18** 293 (1951).

[8]   F. Ericson and J.-Å. Schweitz, J. Appl. Phys., **68** 5840 (1990).

[9]   William H. Safranek, in <u>The Properties of Electrodeposited Metals an Alloys</u> (American Elsevier Publishing Co, New York, 1974) pp. 241-258.

[10] William H. Safranek, in <u>The Properties of Electrodeposited Metals an Alloys</u> (American Elsevier Publishing Co, New York, 1974) pp. 304-305.

TIME DEPENDENT DEFORMATION DURING INDENTATION TESTING

B. N. Lucas*, W. C. Oliver*, G. M. Pharr**, and J-L. Loubet***
*Nano Instruments, Inc., 1001 Larson Drive, Oak Ridge, TN 37830, nano@nanoinst.com
**Rice University, Department of Materials Science, Houston, TX, 77005
***Ecole Central de Lyon, LTDS, URA CNRS 855, BP 163, F-69131 Ecully Cedex, France

ABSTRACT

Constant loading rate/load indentation tests (1/P dP/dt) and constant rate of loading followed by constant load (CRL/Hold) indentation creep tests have been conducted on high purity electropolished indium. It is shown that for a material with a constant hardness as a function of depth, a constant (1/P dP/dt) load-time history results in a constant indentation strain rate (1/h dh/dt). The results of the two types of tests are discussed and compared to data in the literature for constant stress tensile tests. The results from the constant (1/P dP/dt) experiments appear to give the best correlation to steady-state uniaxial data.

INTRODUCTION

The response of a material to an applied stress will in general be a function of several parameters, e.g., the prior strain, the imposed strain rate, the microstructure and the temperature. At a given temperature and stress, given enough time, there is strong evidence that a steady state strain rate and microstructure can be reached in some materials. Any perturbations in these parameters will typically result in a transient period during which the microstructure will evolve to a new state representative of the new set of conditions. Given enough time, steady state conditions can again be reached. While these transient periods are very important, they are not very well understood.

The purpose of this study is to contribute to the understanding of how to use indentation testing to measure the time dependent mechanical properties of materials. The indentation strain rate is defined as the instantaneous descent rate of the indenter (dh/dt) divided by the displacement at that instant in time (h) [1]. Experimental observations indicate that the indentation strain rate may not be sufficient to completely predict the measured hardness. This is the expected result for any material for which the strength is not uniquely related to the strain rate as described above. The strain-time history that precedes the strain rate and resulting hardness measurements appears to be important as well.

In most types of indentation creep tests, the indentation strain rate and hardness continuously change during the experiment. For instance, in a constant load creep test the indenter is loaded at a specified rate and the load is then held constant for a period of time while the displacement is monitored. The indentation strain rate changes both during the loading and the subsequent hold segment under constant load. While the deformation under a pyramid shaped indenter has typically been viewed as being geometrically similar from a time-independent point of view, this type of test does not yield geometrical similarity from a rate-dependent point of view. That is to say, as the deformation proceeds under the indenter, the strain-rates experienced by geometrically similar portions of material change as the deformation proceeds. It is therefore desirable to perform an indentation experiment during which the indentation strain rate and therefore hardness remain constant.

DEVELOPMENT OF A CONSTANT INDENTATION STRAIN RATE EXPERIMENT

The technique for conducting constant indentation strain rate experiments can be developed from the equation for the hardness of a material,

$$H = \frac{P}{A} = \frac{P}{ch^2} \tag{1}$$

Mat. Res. Soc. Symp. Proc. Vol. 436 © 1997 Materials Research Society

where H is the hardness, P is the load, A is the projected contact area, h is the contact depth, and c is a constant that depends upon the geometry of the indenter (24.5 for the perfect Berkovich geometry). Equation (1) can be rewritten as

$$ch^2 H = P \qquad (2)$$

Differentiating with respect to time leads to

$$2ch\dot{h}H + ch^2\dot{H} = \dot{P} \qquad (3)$$

which can be simplified as follows:

$$2ch\dot{h}H = \dot{P} - ch^2\dot{H} \qquad (4)$$

$$\dot{h} = \frac{\dot{P}}{2chH} - \frac{ch^2\dot{H}}{2chH} \qquad (5)$$

$$\frac{\dot{h}}{h} = \frac{\dot{P}}{2ch^2 H} - \frac{ch^2\dot{H}}{2ch^2 H} \qquad (6)$$

substituting in from (2) yields

$$\frac{\dot{h}}{h} = \frac{\dot{P}}{2P} - \frac{\dot{H}}{2H} \qquad (7)$$

or

$$\frac{\dot{h}}{h} = \frac{1}{2}\left(\frac{\dot{P}}{P} - \frac{\dot{H}}{H}\right) \equiv \dot{\varepsilon}_i \qquad (8)$$

We assume a constitutive equation of the form:

$$\dot{\varepsilon}_i = A(H)^n \qquad (9)$$

where $\dot{\varepsilon}_i$ is the indentation strain rate and n is the stress exponent for creep. Equations (8) and (9) suggest an indentation experiment performed with a pyramid shaped indenter during which the loading rate is controlled so that the loading rate divided by the load is constant can result in a constant value of the indentation strain rate if a steady state value of the hardness can be reached and $\dot{H} = 0$. This type of test is very appealing as the loading rate is a directly controllable parameter in the Nano Indenter®II.

EXPERIMENTAL

Two types of experiments were conducted in an attempt to understand the behavior predicted by equations (8) and (9). The first type of experiment involved ramping the load at a constant loading rate to a maximum load of 10 mN in times of 1, 10, 30, and 100 seconds followed by a 10 minute hold segment at constant load (CRL/Hold). This experiment allowed the strain rate (1/h dh/dt) to be calculated during the loading segment while the (1/P dP/dt) term was changing and the (1/H dH/dt) term was expected to be changing. The load-time histories for these experiments are shown in figure 1. The second type of experiment involved loading at a constant (1/P dP/dt) to a maximum load of 10 mN at rates of 0.2, 0.1, 0.02, 0.01, and 0.005 (sec^{-1})

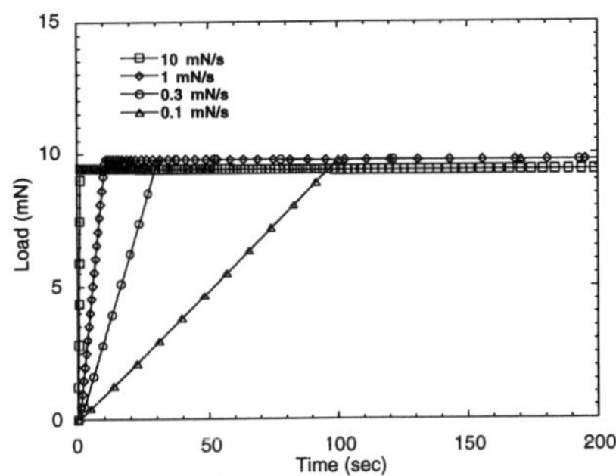

Figure 1. Load-time history for constant rate of loading indentation experiments.

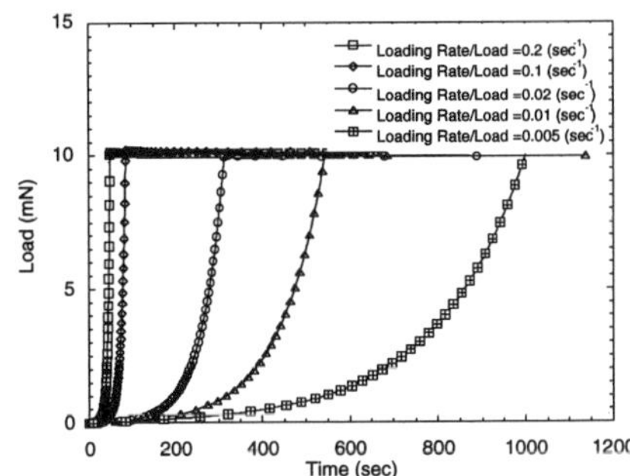

Figure 2. Load-time history for constant (1/P dP/dt) experiments.

followed by a 10 minute hold segment at constant load. This experiment was designed to test the hypothesis that for a material with a constant hardness as a function of depth, a constant (1/P dP/dt) would result in a constant indentation strain rate (1/h dh/dt). The load time histories for these types of experiments are shown in figure 2. Experiments were conducted on bulk indium which has a melting temperature of 156.8 °C. Data were acquired at depths of 2000 nm and greater to avoid any surface effects where the hardness of the material might be influenced by surface layers or environmental exposure. All of the experiments were conducted with a Nano Indenter®II mechanical properties microprobe.

RESULTS AND DISCUSSION

Figure 3 is a plot of the indentation strain rates versus displacement obtained using the CRL/Hold loading schemes shown in figure 1. The strain rates were calculated by taking the time derivative of the displacement during both the loading segment and during the subsequent hold segment and dividing the instantaneous rates by the displacement at that point in time. The indentation strain rate is observed to decrease continuously through the constant loading rate loading segment and, even more rapidly, during the constant load hold segment. Figure 4 is a plot of the calculated hardness versus displacement over the same displacement range. The hardness as plotted here is defined as the instantaneous load divided by the instantaneous contact area where the contact area was calculated using the appropriate area function for the indenter and the total depth of the indent. The total depth was used rather than an elastically corrected contact depth because elastic recovery of the indentation depth was small (<1% of the depth).

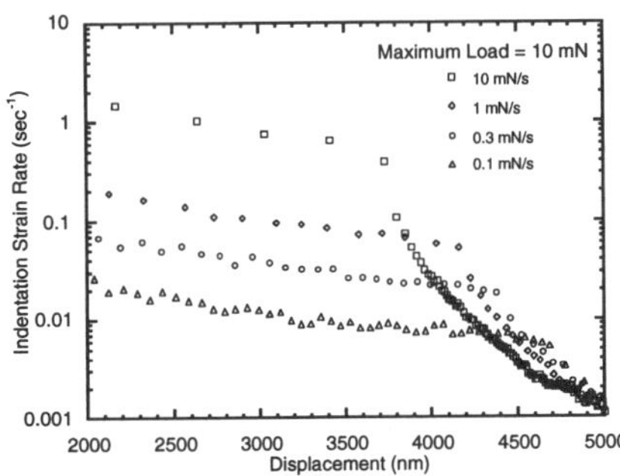

Figure 3. Indentation strain rates (1/h dh/dt) achieved using load-time histories in Figure (1).

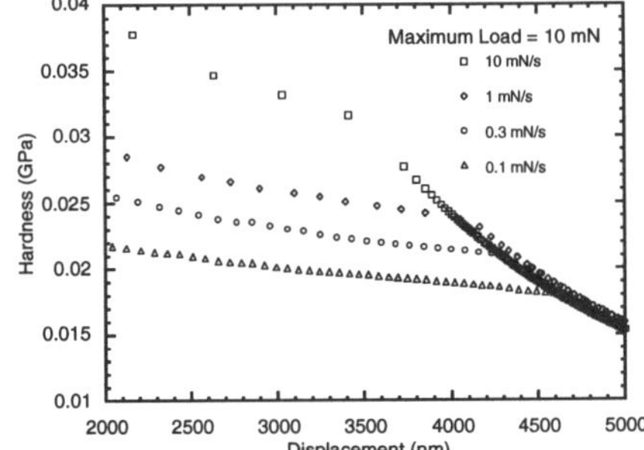

Figure 4. Hardness versus displacement for CRL experiments.

The hardness is observed to decrease as the experiment proceeds and the indentation strain rate decreases. The change in the hardness is gradual during the loading segment when the indentation strain rate is not changing rapidly and then becomes much more pronounced once the loading stops and the strain rate drops rapidly.

Figure 5 is a plot of the indentation strain rates vs. displacement obtained using the constant (1/P dP/dt) loading schemes shown in figure 2. The strain rates were again calculated by taking the time derivative of the displacement during both the loading segment and during the subsequent hold segment and dividing the instantaneous rates by the displacement at that point in time. The indentation strain rates are observed to be constant during the loading segment and equal to 0.5(1/P dP/dt) and are then again observed to decrease once the loading segment terminates. Figure 6 is a plot of the calculated hardness versus displacement for the constant (1/P dP/dt) experiments. The hardness is again defined as the instantaneous load divided by the instantaneous projected contact area calculated using the total depth of the indent. The calculated hardness is observed to remain constant during the constant (1/P dP/dt) loading segment and to decrease for each constant strain rate experiment. The hardness is then observed to decrease as the strain rate falls during the constant load hold segment.

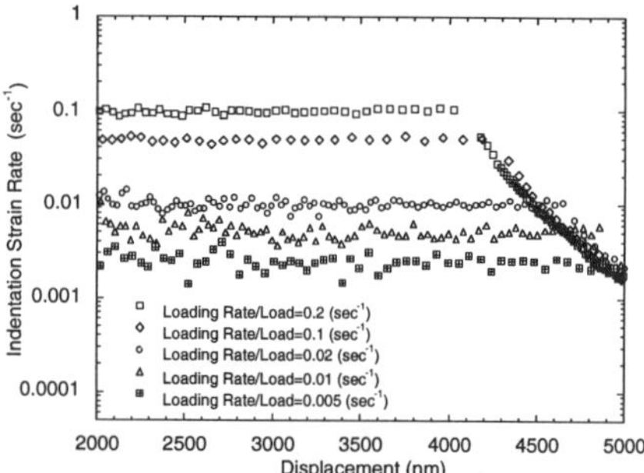
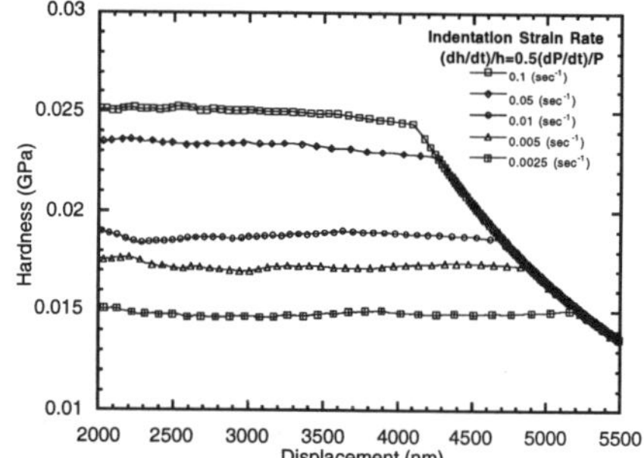

Figure 5. Indentation strain rates (1/h dh/dt) achieved using load-time histories in Figure (2).

Figure 6. Plot of hardness versus displacement showing constant hardness as a function of depth when using the constant (1/P dP/dt) experiment.

From these two types of data it is possible to tabulate strain rate-stress pairs analogous to uniaxial data. Figure 7 is a log-log plot of indentation strain rate vs. hardness for the data from the constant (1/P dP/dt) experiments. The data were generated by averaging the constant hardness/constant strain rate data between 2000 nm and the termination of the loading segment. Each data point represents the average of five experiments at each strain rate. A power law fit to the data yields a slope of 7.3 (R=0.995) which is the stress exponent for creep in the material. Over the range of strain rates investigated the power law relationship was observed to remain virtually constant. This stress-exponent for creep is very similar to the value of 7.6 obtained for indium by Weertman [2] for constant stress tensile data near room temperature.

Figure 8 is a log-log plot of indentation strain rate vs. hardness for the data from both the constant (1/P dP/dt) experiments as well as the constant rate of loading/hold experiments. Strain rate-stress pairs were calculated for both the loading segment and hold segment from the CRL/Hold experiments. Each set of data represents the average of five separate indentation experiments performed for each loading rate. Several interesting observations can be made about these data.

The data immediately following a constant rate of loading segment are observed to undergo a transient period immediately upon completion of the load ramp. These data are then observed to converge to a common curve over a period of time which is similar to the loading time. This transition region which appears to have a higher stress exponent is believed to be due to the transient microstructural evolution occurring immediately after the completion of the load ramp

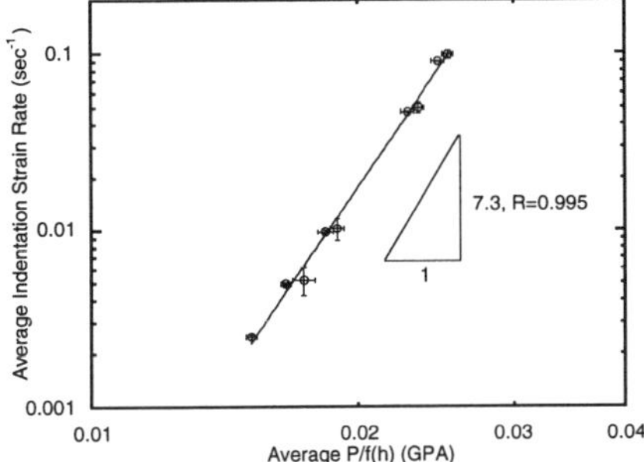

Figure 7. Average indentation strain rate versus average hardness for data obtained from constant (1/P dP/dt) experiments.

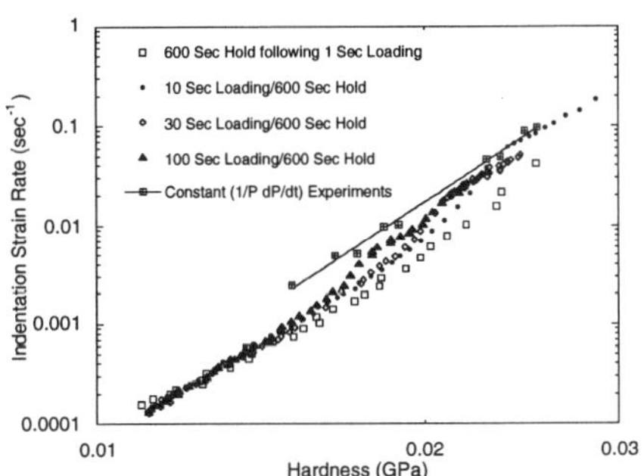

Figure 8. Indentation strain rates versus hardness for both constant (1/P dP/dt) experiments and CRL/Hold experiments.

and is not believed to be associated with power law breakdown. This is supported by the fact that even at higher strain rates the constant indentation strain rate-stress data obey a constant power law relationship.

As previously pointed out, the indentation strain rate-stress data during the hold segment are observed to converge to a single curve after an initial transient period. A power law fit to the linear portion of the data in this region yields an average slope of 6.6 which is also in good agreement with the constant stress uniaxial data of Weertman.

The similarity of the stress exponent obtained from the two types of experiments and the offset in the hardness leads to two conclusions. One, the dominant factor in both types of experiments seems to be the stress exponent for creep. While the two sets of data were acquired under very different conditions, the appropriate stress exponent seems to come out of both sets of data. Secondly, the data from the CRL/Hold experiments appears to have a stronger microstructure as indicated by the higher hardness determined at the same strain rate. This is not unexpected since the material has undergone a wide range of strain rates beginning at very high values and decreasing throughout the experiment. Therefore the material is evolving from a strong microstructure to a weaker microstructure as the experiment proceeds.

One final comparison that is worth making is that of comparing the constant (1/P dP/dt) indentation data to that obtained by Weertman using a constant tensile stress apparatus. Figure 9 is a log-log plot of both the indentation data and Weertman's tensile data at room temperature. While the offset between the two sets of data cannot be completely accounted for by either constraint factors relating the hardness to a flow stress for the material [3] or by predictions from time dependent finite element calculations [4], the utility of the indentation data is evident. The most often sought after parameter in these types of experiments is how the rate of deformation of the material changes in response to a change in the applied stress, i.e., the stress exponent for creep. From the comparison of the two sets of data it is obvious that the stress exponent obtained from the two types of experiments is very similar. Since the constant (1/P dP/dt) indentation experiments yield a constant hardness this type of experiment is believed to most closely approximate the results from the constant stress tensile experiments.

CONCLUSIONS

1. For a material with a constant hardness as a function of depth it is possible to obtain a constant indentation strain rate by controlling the loading rate in such a way that the instantaneous loading rate divided by the instantaneous load remains constant. The resulting indentation strain rate is then given by

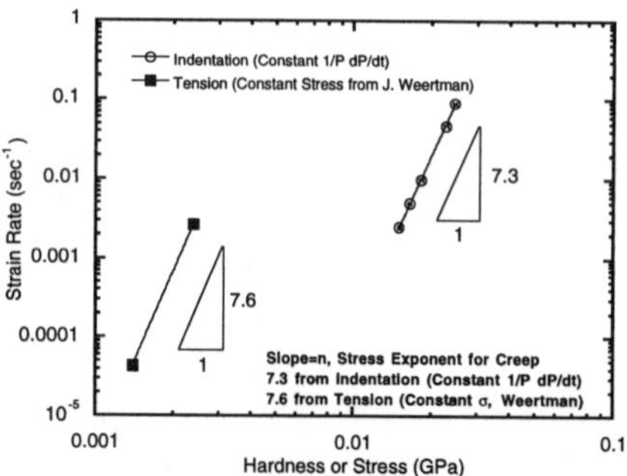

Figure 9. Plot of strain rate versus hardness or stress comparing constant (1/P dP/dt) data to constant stress tensile data from Weertman[2].

$$\frac{\dot{h}}{h} = \frac{1}{2}\left(\frac{\dot{P}}{P}\right) \equiv \dot{\varepsilon}_i \tag{10}$$

2. Comparison of the results from the constant (1/P dP/dt) experiments and the CRL/Hold experiments suggests that the apparent increase in the stress exponent for creep observed in the initial stages of the hold segment in the CRL/Hold experiments is due to the transient microstructural evolution in going from the loading segment to the hold segment and not with power law breakdown.

3. The constant (1/P dP/dt) indentation experiments on indium yield a stress exponent for creep of 7.3. This result is in excellent agreement with the stress exponent from constant stress tensile tests.

ACKNOWLEDGMENTS

Research sponsored by Nano Instruments, Inc. and the Division of Materials Sciences, U. S. Department of Energy, under contract DE-AC05-96OR22464 with Lockheed Martin Energy Research, Inc.

REFERENCES

1. M.J. Mayo and W.D. Nix, Acta. Met. **36**, 2183-2192 (1988).

2. J. Weertman, Trans. of AIME **218,** 207-218 (1960).

3. D.S. Tabor. *Hardness of Metals* (Clarendon Press, Oxford, 1951).

4. A.F. Bower, N.A. Fleck, A. Needleman, and N. Ogbonna, Proc. R. Soc. Lond. A **441,** 97-124 (1993).

STRESS DISTRIBUTION IN Si UNDER PATTERNED THIN FILM STRUCTURES

S.P. Wong*, L. Huang*[#], W.S. Guo*, W.Y. Cheung*, Shounan Zhao**
* Department of Electronic Engineering and Materials Technology Research Centre, The Chinese University of Hong Kong, Shatin, Hong Kong.
** Department of Applied Physics, South China University of Technology, Guangzhou, China.

ABSTRACT

We have employed the infrared photoelasticity (PE) method to study the stress distribution in Si substrates under patterned thin film structures such as thermal oxide layers partially covered by metal films and oxide layers with long trench openings. It is demonstrated that a lot of information on the two dimensional stress distribution in the substrate under patterned thin film structures can be obtained from PE experiments. The capability, limitation, and further development of the PE method for semiconductor applications are discussed.

INTRODUCTION

The photoelasticity (PE) method has long been a common and widely used technique in experimental mechanics to study stress distribution in isotropic materials. There are also a number of reports on the application of the PE method to study stress in semiconductors [1-5]. As the role of stress in microelectronics and optoelectronics technology has become more and more important, it is anticipated that this technique will attract increasing attention in the semiconductor technology community. In this work, we shall present some preliminary results of our attempt to apply the PE method in the study of stress distribution in Si substrates under patterned thin film structures.

THEORY AND MEASUREMENT METHODS

The principle and the experimental arrangements of the PE method have been described in detail in many standard texts (e.g. see [6] and references therein). Basically the photoelastic effect refers to the phenomenon that when an ordinarily optically isotropic material is under stress, there will be stress-induced birefringence. The PE method measures the changes in the polarization state of a polarized light after passing through the sample under stress. Hence, information on the stress state in the sample can be obtained. The arrangement of a simple photoelasticity measurement system used in this study is shown in Fig. 1.

The photoelastic effect in isotropic materials is characterized by the stress-optic law:

$$n_i - n_j = C\,(\sigma_i - \sigma_j), \qquad\qquad i, j = 1, 2, 3 \qquad\qquad (1)$$

where n_i's are the principal refractive indices, σ_i's are the principal stresses, and C is the relative stress-optic coefficient.

While the stress-optic law is valid for optically isotropic materials, it is not appropriate for crystals since the principal axes of the refractive index ellipsoid deviate from those of the stress ellipsoid in a crystal. However, it has been shown that for crystals with cubic symmetry, such as silicon, and under a two-dimensional state of stress, when the incident light is along certain directions, the stress-optic law can take a similar form as equation (1) but with the coefficient C now dependent on direction [1]. Shown in Fig. 2 are two appropriate choices of such incident light directions for (100) and (111) Si wafers. The corresponding values of C are respectively $1.87 \times 10^{-12} \mathrm{cm^2/dyne}$ and $1.59 \times 10^{-12} \mathrm{cm^2/dyne}$ for (100) and (111) Si wafers [1].

[#] Visiting from Dept. of Applied Physics, South China University of Technology, Guangzhou, China.

Mat. Res. Soc. Symp. Proc. Vol. 436 © 1997 Materials Research Society

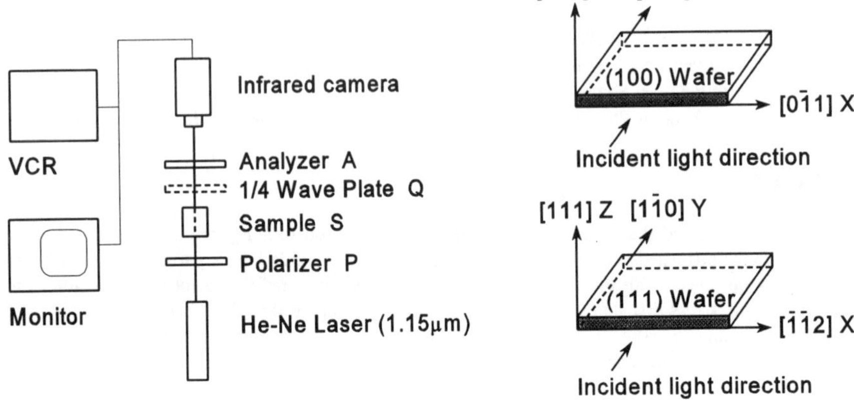

Fig. 1 A schematic showing of the infrared photoelasticity system.

Fig. 2 The sample orientations and incident light direction for (a) (100) and (b) (111) Si substrates for the photoelasticity experiments.

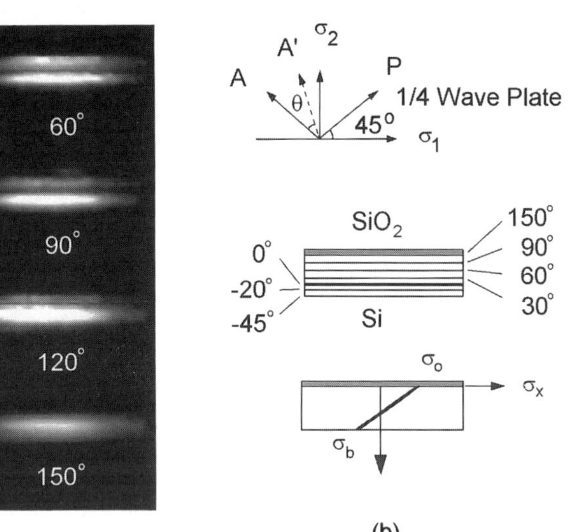

Fig. 3 (a) Stress photoelasticity patterns of an oxidized Si sample at various compensation angles. (b) A schematic showing of the arrangements of the polarizer P, quarter-wave plate Q, analyzer A and the compensation angle θ (top), the stress contours in the Si substrate (middle) and the stress distribution versus depth obtained from the patterns in (a) (bottom). The values of σ_0 and σ_b were determined to be 5×10^7 and -1.5×10^7 dyne/cm^2, respectively.

Depending on the detailed arrangements of the optical components in the PE system as shown in Fig. 1 where a monochromatic light source is used, different PE fringe patterns can be obtained. For example, when the quarter-wave plate Q is not included, and with the axes of the polarizer P and analyzer A crossed, the arrangement is known as a plane, dark-field polariscope in which the intensity I transmitted can be shown to be given by [6]

$$I = I_o \sin^2 2\beta \sin^2(\alpha/2), \tag{2a}$$

with the phase difference α determined by

$$\alpha = 2\pi C(\sigma_1 - \sigma_2)d/ \lambda \tag{2b}$$

where I_o is a constant, σ_1 and σ_2 are the secondary principal stresses in the plane perpendicular to the incident light, β is the angle between the direction of σ_1 and the axis of the analyzer, C is the stress-optic coefficient, d is the distance traveled by the light in the sample, and λ is the wavelength of the light in free space. Here it is assumed that σ_1 and σ_2 are constant throughout the distance traveled by the light in the sample.

The dark fringes corresponding to $\beta = 0°$ or $90°$ are known as isoclinics which provide information on the directions of σ_1 and σ_2. The dark fringes corresponding to $\alpha = 2n\pi$ are known as isochromatics of order n which provide information on the magnitude of $\sigma_1 - \sigma_2$. Note that by rotating P and A simultaneously, the isoclinics will move but the isochromatics will not. However, the isochromatics can be shifted by introducing optical retardation into the light path by means of various compensation methods. In this study, we have used the Sénarmont method [6] for its simplicity to achieve such compensation. In this method, a quarter-wave plate Q is inserted between the sample S and the analyzer A. The axes of P & A are aligned at an angle of $45°$ to the secondary principal stress axes and the axis of Q is aligned in parallel with the polarizer. Then the isochromatic fringes can be shifted to the point of interest by rotating A by some angle θ. If the order of the fringe is n before the shift, it is easy to show that the order of the shifted fringe is given by

$$N \equiv \alpha/2\pi = n + \theta/\pi. \tag{3}$$

From equations (3) and (2b), the magnitude of $\sigma_1 - \sigma_2$ at that point can be determined.

EXPERIMENTAL RESULTS AND DISCUSSIONS

The stress photoelasticity patterns of an oxidized (100) n-Si sample of 10-20 Ωcm and about 280μm thick by the Sénarmont method are shown in Fig. 3a for various compensation angles, β. The thickness of the thermal oxide layer is about 1μm. Before that, the principal stress directions have been determined to be parallel or perpendicular to the wafer surface for the whole substrate by the plane, dark field PE images. Hence the arrangement of the optical components as shown in Fig. 3b was used. The horizontal fringes are isochromatics. The order of the fringe corresponding to $\beta = 0°$ was determined to be zero by the observation that after subsequent thinning of the oxide layer, the position of this fringe remains unchanged. It is a well known fact from elasticity theory that if only the magnitude of the loading forces change, while the magnitude of the stress changes accordingly, the stress distribution pattern will remain unchanged. From the movement of the fringes under a slight press on the sample surface, the stress is determined to be tensile at the oxide/silicon interface and compressive at the bottom of the substrate. In this particular case, it is obvious that σ_1 is in the horizontal direction and σ_2 vanishes so that the difference $\sigma_1 - \sigma_2$ in fact gives σ_1. The stress contours and distribution versus depth are also shown in Fig. 3b. The magnitude of the stress at the interface, σ_o, and that at the bottom, σ_b, were determined to be 5×10^7 and -1.5×10^7 dyne/cm^2, respectively.

In another experiment, it is found that before annealing, the stress induced in the Si substrate by deposition of Al thin films is too small to be detected by the PE method. However, after annealing at 500°C for 15 minutes, similar PE fringe patterns as the oxidized sample have been observed. But the sign of the stress is opposite to that of the oxide case, i.e., the stress is compressive at the Al/Si interface and tensile at the bottom of the substrate. In Fig. 4a and 4b, the PE patterns showing zero stress fringes in Si substrates due to oxidation and due to Al film deposition are compared. It is noticed that while the position of the zero stress fringe for the oxidized sample is lower than that predicted by the bimetallic strip theory [7], which is expected to be located at 1/3 of the wafer thickness from the bottom, the opposite is true for the Al case. This is also schematically shown in Fig. 4c. It is noted in ref. 1 that for one-sided polished Si wafers, there will be stress induced due to the difference in surface finish at the two sides. The stress is tensile at the rough side and compressive at the other side. For the one-sided polished wafers we used, the tensile stress at the bottom was determined to be about 3×10^6 dyne/cm^2. The superposition of the stress due to this effect and that due to oxidation or Al deposition leads to the shifts in the zero fringe positions as observed in Fig. 4. As a matter of fact, the resolution in stress by the PE method can be as small as 10^6 dyne/cm^2.

In Fig. 4d, the compensation effect of the stress due to a 1μm thick oxide and that due to a 1.7μm thick Al film after annealing at 500°C for 15 minutes is clearly demonstrated by the PE pattern. While total extinction is observed at the side under the Al/SiO$_2$ double layer except at regions near the edge, the zero stress fringe is clearly seen at the side under the oxide layer. This PE pattern also provides some ideas on the extent of the range affected by the edge.

Shown in Fig. 5 are typical PE patterns showing the isoclinic fringes of an oxidized (111) p-Si sample of 10-15 Ωcm and about 400μm thick with a long strip oxide window of 200μm wide. The edge effect is again clearly seen. From the pattern for $\beta = 0°$, far from the oxide edges and in the dark region under the oxide window, the principal stress directions are in fact either parallel or perpendicular to the surface as expected. The failure of extinction in the regions under the edges shows that the stress states at these regions are not biaxial. These non-biaxial (NB) regions also show up in substrates with other oxide layer structures such as a long strip of oxide layer or a half plane of

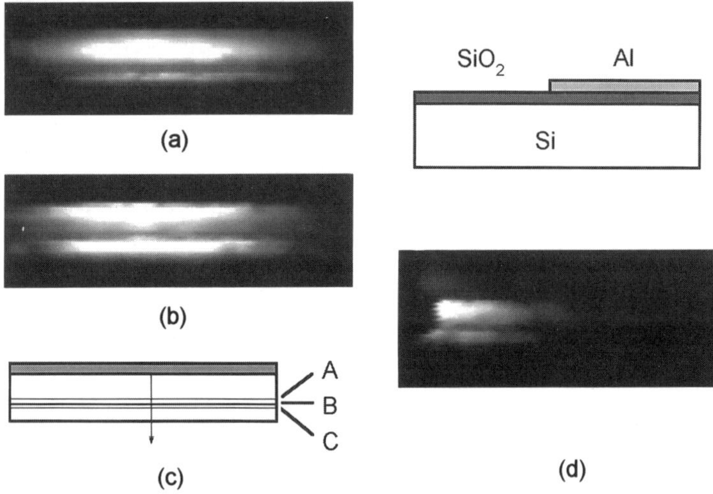

Fig. 4 Photoelasticity patterns showing zero stress fringes in Si substrates due to (a) oxide layer and (b) Al thin film grown on it. (c) A schematic showing of the relative positions of the zero stress fringes: A -- due to Al film, B -- according to bimetallic strip theory, C -- due to oxidation. (d) Photoelasticity pattern showing the compensation effect of substrate stress induced by oxide layers and Al films.

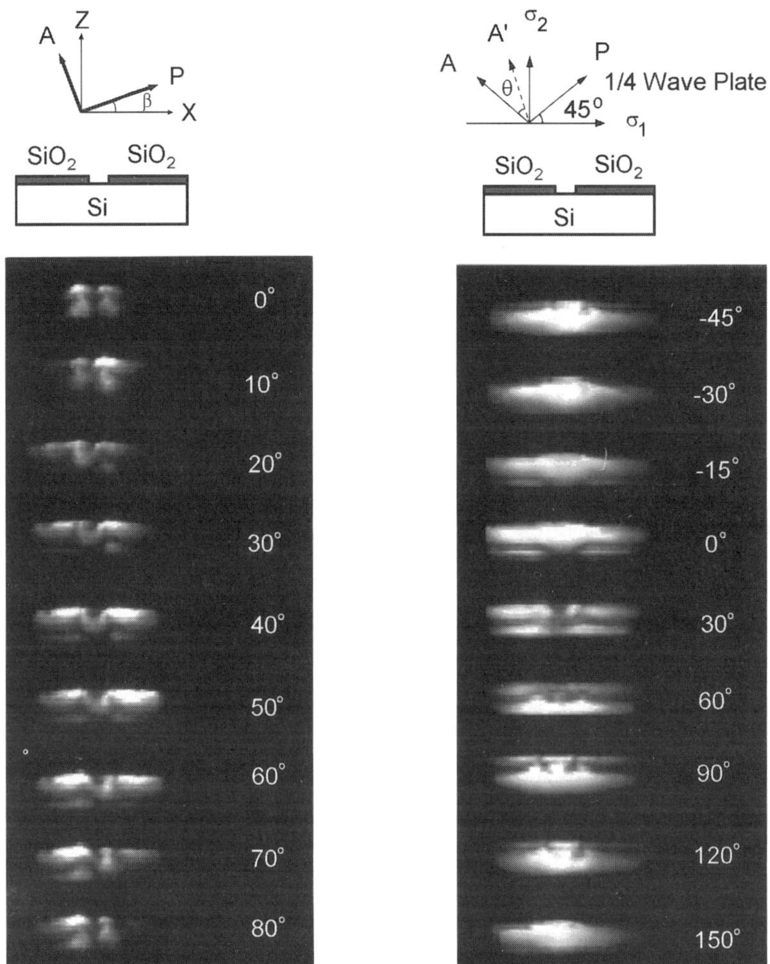

Fig. 5 Stress photoelasticity patterns showing the isoclinic fringes of an oxidized sample. The oxide layer has a long strip window 200μm wide as shown schematically. The arrangement of the polarizer P, analyzer A and the angle β are also shown schematically.

Fig. 6 Stress photoelasticity patterns of an oxidized Si sample for various compensation angles. The oxide layer has a long strip window 300μm wide as shown schematically. The arrangement of the polarizer P, quarter-wave plate Q, analyzer A and the compensation angle θ are also shown schematically.

oxide layer. The shapes of the NB regions under the oxide edges are quite complicated but have a common feature that they generally consist of a few leaves. However, the details of their shapes are observed to depend on the thickness of the oxide layer as well as the width of the oxide window. We need to refine our experimental set-up for higher quality PE images in order to study these edge effects in more details. In the patterns for other β angles in Fig. 5, the zero stress fringes clearly emerge at regions far from the oxide window. However, under the oxide edges, there are certain portions of the NB regions which seem to have no angle at which extinction will occur. This may indicate that the directions of principal stresses at these NB regions change very rapidly.

From the above discussion, the Sénarmont method is still applicable to study the stress distribution in the substrate except for the NB regions under the oxide edges. Typical PE patterns for another oxidized (100) n-Si sample of 10-20 Ωcm and about 280μm thick with a long strip window 300μm wide are given in Fig. 6 obtained with the arrangement of optical components as shown. The stress distribution far from the window is similar to that shown in Fig. 3 as expected. However, for the region under the oxide window, again we need to improve our experimental set-up for higher quality PE images to obtain more conclusive results.

A more elaborate optical system should be able to enhance the spatial resolution of the PE method to a few microns. If the PE method can be combined with techniques developed from near field scanning optical microscopy, we anticipate that it may be possible to push the limit to submicron resolution. The application of photometry methods and modern image processing techniques should also improve the lower limit of the minimum detectable stress as well as the spatial resolution. These are the two directions of improvements that are under way in our laboratory. Another topic of interest for further development is of course the study of edge effects on stress by the PE method.

CONCLUSIONS

In short, we have employed the photoelasticity method to study the stress distribution in Si substrates under a number of patterned thin film structures. Typical PE patterns are shown for various structures and the stress distribution can be determined quantitatively except for regions near and under the thin film edges in which the stress is found to be non-axial. With these demonstrated examples, we suggest that the PE method is a promising technique worth further development for the study of edge effects on stress in semiconductor structures.

ACKNOWLEDGMENTS

This work is partially supported by a Direct Grant for Research from the Faculty of Engineering of The Chinese University of Hong Kong. One of us (WYC) is supported by a Postdoctoral Fellowship from CUHK.

REFERENCES

[1] H. Liang, Y. Pan, S. Zhao, G. Qin, K.K. Chin, J. Appl. Phys. 71, 2863 (1992).
[2] Huang Lan, Liang Hancheng, Zhao Shounan, Chin. J. Infrared and Millimeter Waves 8, 203 (1989).
[3] M. Yamada, Appl. Phys. Lett. 47, 365 (1985).
[4] H. Kotake and Shin Takasu, J. Electrochem. Soc. 127, 179 (1980).
[5] R.O. DeNicola and R.N. Tauber, J. Appl. Phys. 42, 4262 (1971).
[6] H. Aben and C. Guillemet, Photoelasticity of Glass, Springer-Verlag, Berlin, Heidelberg, 1993.
[7] S.D. Brotherton, T.G. Read, D.R. Lamb, A.F.W. Willoughby, Solid State Electronics 16, 1367 (1973).

NONLINEAR ACOUSTIC RESPONSE IN THIN OXIDE LAYERS ON FUSED SILICA

C.R. OTTERMANN[1], S.U. FASSBENDER[2,3], W. ARNOLD[2], and K. BANGE[1]
[1]SCHOTT GLASWERKE, R&D, P.O. Box 24 80, D-55014 Mainz, Germany
[2]Fraunhofer Institute for NDT, University Bldg. 37, D-66123 Saarbruecken
[3]Q-NET Quality Management GmbH, Altenkesslerstr. 20, D-66115 Saarbruecken

ABSTRACT

Nonlinear mechanical properties of layered systems of Ta_2O_5 and TiO_2 films deposited on fused silica by reactive evaporation (RE), reactive ion plating (IP) and spin coating (SC) are investigated by means of an ultrasonic technique. The coatings with thickness of 100 nm possess differences in density and crystal structure, due to the different deposition conditions. The nonlinear acoustic response of the film/substrate systems depends on film material. Differences are observed in respect to film density as obtained by the alternate deposition methods. The origin of the differences in nonlinear acoustic response of the samples is discussed. The results are correlated to adhesion properties of the films determined by a scratch-test method.

INTRODUCTION

Coatings on glass surfaces are widely used to meet various objectives. Typical examples are coatings with desirable optical properties for decorative purpose, technical applications or with increases wear resistance. It is obvious that good adhesion is one of the most important key factors for coating on glasses. Inadequate adhesion causes detachment and consequently a catastrophic failure of the coated part. Therefore, the evaluation of adhesion between coating and substrate by means of a reliable measurement is of prime interest. Unfortunately, most of the commonly used semi-quantitative testing methods developed for adhesion determination are destructive, like topple, pull or scratch tests. Further, the application of these methods is normally time consumptive and requires additional investigations, such as electron microscopy, for a clear interpretation of the testing result [1].

From the practical point of view a testing method would be desired which is fast, nondestructive, at least semi-quantitative, and easy to use. Ultrasound (US), as a mechanical testing tool, is known to fulfill several of these demands, at least for samples with large dimensions and no need for high lateral resolution on a microscopic scale. A new US testing technique, developed and optimized in the last years for determining the adhesion properties of adhesive joins, shows promising results and potential for additional applications [2]. The aim of this work is an evaluation of this method in respect to adhesion determination of oxidic layers on glasses with film thickness much smaller than 1 μm. Single films of TiO_2 and Ta_2O_5 deposited on fused silica substrates by several standard techniques are chosen as a model for this investigations.

BASIC CONSIDERATIONS

A brief introduction is given here to the basic idea of the testing method. A more comprehensive description can be found in [2, 3]. The adhesion strength of joined materials is dominantly described by the behavior of their interfaces. The binding properties in the interface can

Mat. Res. Soc. Symp. Proc. Vol. 436 © 1997 Materials Research Society

be represented by a potential. The derivative of this potential energy is the restoring force, F(a), i.e. the force necessary to separate the interfacial surfaces of both materials by a certain distance, as sketched in Fig. 1. The maximum value, F_{max}, of the restoring force represents the binding force. An interface with distance, a_0, between the two surfaces is regarded to be well bonded. The real distance, a_i, can be larger than this equilibrium distance, due to additional boundary conditions such as residual film stresses or impurities in the interface region.

The resting position a_i is a non-equilibrium state, but, stabilized by the boundary conditions. It is shifted towards higher F(a) values and the amount of additional load needed to break the bonding of the two materials is reduced (see Fig. 1b). Therefore, this kind of interface is regarded as poorly bonded. It is obvious from Fig. 1b that the gradient of the restoring force is smaller for poor bonding than for the optimal case. Also the curvature, i.e. the nonlinear behavior, of the F(a) function is increased at the location a_i.

The interface can be tested by mechanical loading, due to its reaction to displacements from its resting position. An ultrasonic wave transferring the interface will force a periodical variation of the distance around the rest point a_i. Therefore, the propagation of the ultrasonic wave is modulated by the interface according to the shape of the restoring force around its resting position. The nonlinear distortion of the wave creates higher harmonics with amplitudes, A_n, and the total amount of transported energy is proportional to the slope of the F(a) curve. It can be shown [3] that the following summation of the amplitudes of the harmonics contain information on the interfacial properties:

$$F(a_i + s_0) = \Sigma_A = \sum_n (-1)^{n+1} A_n = -A_0 + A_1 - A_2 + A_3 - A_4 + \cdots \tag{1}$$

where s_0 is a displacement of the interface, due to the incident US wave with amplitude u_0.

The magnitudes of the amplitudes A_n of the higher harmonics depend on the nonlinearity of the curve of restoring force around the resting point and on the amplitude, u_0, of the incident US wave. For increasing u_0 the nonlinear part of the restoring curve becomes more dominant resulting in an increase of the amplitudes of the higher harmonics. A reduction in bond strength possesses a decrease in Σ_A and an increase in the higher harmonics for a given incident amplitude u_0.

The relation between Σ_A and the distance, a, can be expressed by the following much simplified parametrization [3]:

$$\Sigma_{A,fit} = C_1 \cdot u_0 \cdot \exp(-C_2 \cdot u_0) + C_0 \qquad \text{and} \qquad u_0 = f(a - a_0) \tag{2}$$

where C_1, and C_2 are a measure for the shape of the restoring force, and C_0 is an experimental offset. An increase in C_2 represents an increase in the curvature of the restoring force F(a), and, therefore, corresponds to a decrease in bond strength.

In real composites, not only the restoring force of the interface has to be taken into account, but also the restoring forces, i.e. the nonlinear properties, of both the joined materials. For poor bonding, usually the interfaces possess the weakest restoring forces and, therefore, the modulations of the ultrasonic wave traversing the sample are mainly given by the interfacial properties.

EXPERIMENTAL DETAILS

Fused silica was chosen as substrate material because its nonlinear properties causing modulations of ultrasonic waves are known to be rather small. Substrates were circular shaped and

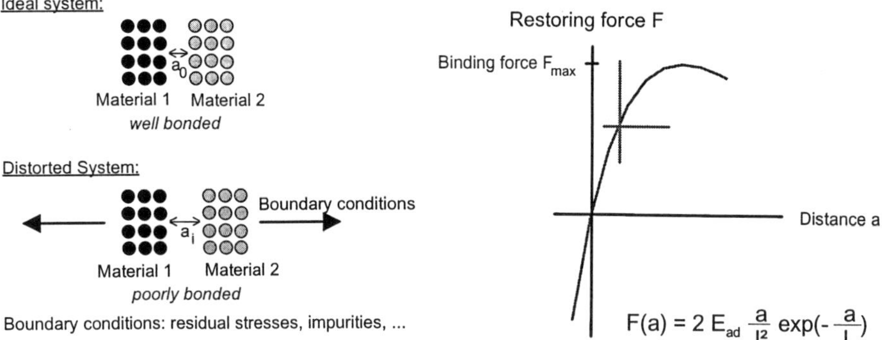

Fig. 1a: Schematic diagram of well and poorly bonded interfaces.

Fig. 1b: Schematic diagram of the restoring force, F(a), and the resting positions for well (a_0) and poorly bonded interfaces (a_i).

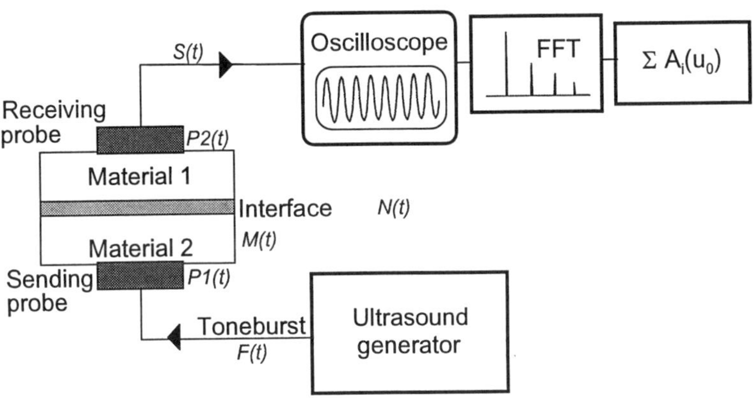

Fig. 2: Block diagram of the experimental arrangement for measuring the ultrasonic transmission. The transfer functions are K(t) of the apparatus and N(t) for the layer.

Table I.: Process parameters for TiO_2 and Ta_2O_5 films deposited by reactive electron-beam evaporation (RE) and reactive ion plating (IP).

coating process	deposition system	$T_{substrate}$ [°C]	$p_{tot}\,(O_2)$ [x 10^2 Pa]	dep. rate [nm/s]	I_{arc} [A]
RE	BAK 760	~ 250	1.6	0.3 - 0.47	-
IP	BAB 800	190	10 (Ar+O_2)	0.3	55

1 mm thick. Sample surfaces are polished to high quality and flatness. They were cleaned by using a standard four-step cleaning process [4] and treated by glow discharge directly before starting the coating procedure.

Approximately 100 nm thick TiO_2 films and Ta_2O_5 layers are obtained by reactive electron-beam evaporation (RE) and reactive ion plating (IP). The main deposition parameters for both film materials and deposition methods are summarized in Table I. Additionally, TiO_2 layers are also deposited by a spin coating process (SC) performed from a titanium alkoxide solution on substrates rotated with 1400 revolutions per minute. After hydrolyzation in moist air, the films were burned-in at 450°C for 1h in an electric heated oven. More information on the deposition conditions can be found in [1, 4-5]. Film thickness, d, and spectral refractive index, $n(\lambda)$, of the layers are controlled by optical means. All films are found to be amorphous by X-ray diffraction (XRD) investigations, with exception of SC TiO_2 layers possessing anatase structure. More information on film characterization is given elsewhere [1, 4, 6].

Adhesion properties of the thin oxide films is obtained conventionally by a scratch-test investigation based on a new testing method, the microtribolometry, developed by S. Baba [7]. A commercially available testing equipment (CSR-02 of RHESCA) was used for the investigations. Details on the method and data evaluation are given in [1].

Nondestructive testing of the film/substrate compound was performed by means of an ultra-sound technique. The experimental arrangement is shown in Fig. 2. Ultrasonic (US) waves with a frequency of 2 MHz are generated in the sending probe and transverse the sample under normal incidence. The US signal consists of a narrowband burst of about 30 oscillations. The transmitted ultrasonic wave is received by a broadband probe. The electrical signal is detected by an oscilloscope and Fourier transformed to determine the amplitudes, A_n, of the higher harmonics. The dc-part, A_0, cannot be detected by the present setup. The transfer function of both transducers was taken into account.

The dependence of Σ_A on the amplitude, u_0, is fitted by the parametrization given in Eq. (2). Information on the nonlinear properties of the film/substrate compound, i.e. the adhesion properties, is deduced from the fitting parameter C_2. This information is qualitatively, but allows a comparison of sets of samples of the same type with differences in bonding properties.

RESULTS

Fig. 3 shows the results on Σ_A as a function of the ultrasonic amplitude, u_0, for RE and IP tantalum oxide layers. The slope of Σ_A, and, according to Eq. (1), the shape of F(a), is different for both thin layers. The curvature of the Σ_A function obtained on the RE Ta_2O_5 film is larger than for the IP layer. The lines drawn in Fig. 3 represent best fits to the data by the predicted correlation given in Eq. (2). The fits describe the data well and possess correlation coefficients better than 0.997. Fig. 4 shows the respective results on the C_2 fitting parameter. C_2 is larger for RE Ta_2O_5 indicating a reduced adhesion strength of this film in respect to the layer produced by reactive ion plating.

In Fig. 5 C_2 fit results are compared for TiO_2 layers deposited by reactive evaporation and spin coating. Here, the difference in curvature is small and shows the tendency of better adhesion in the case of the RE film. A comparison of adhesion properties is depicted in Fig. 6 for TiO_2 and Ta_2O_5 films, produced by reactive evaporation. The C_2 fitting parameter is smaller in the case of TiO_2 implying a better adhesion behavior of this film.

Adhesion of the films is also investigated by scratch testing and is already published in parts [1, 8]. Results are summarized in Table II. Here the critical load, $L_c(film)$, is quoted at which detachment of the films occurs. For comparison this load is normalized to the critical load,

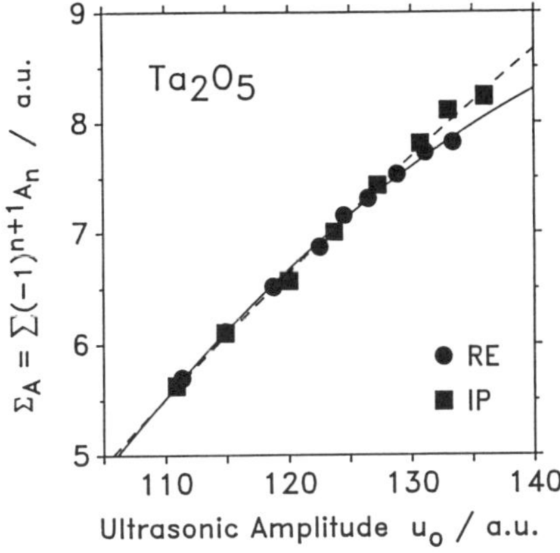

Fig. 3: Σ_A as a function of ultrasonic amplitude, u_0, for Ta$_2$O$_5$ films produced by reactive ion plating (IP) and reactive evaporation (RE). The lines are best fits according to Equ. (2).

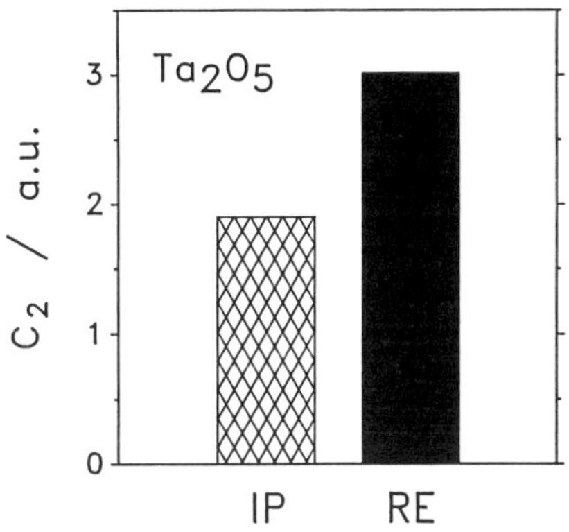

Fig. 4: Results for the C$_2$ fitting parameters of the best fits shown in Fig. 3.

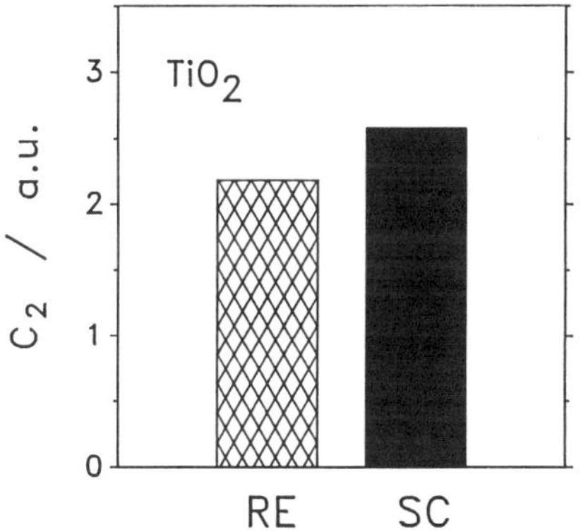

Fig. 5: C$_2$ best-fit parameters for the TiO$_2$ films deposited by reactive evaporation (RE) and spin coating (SC)

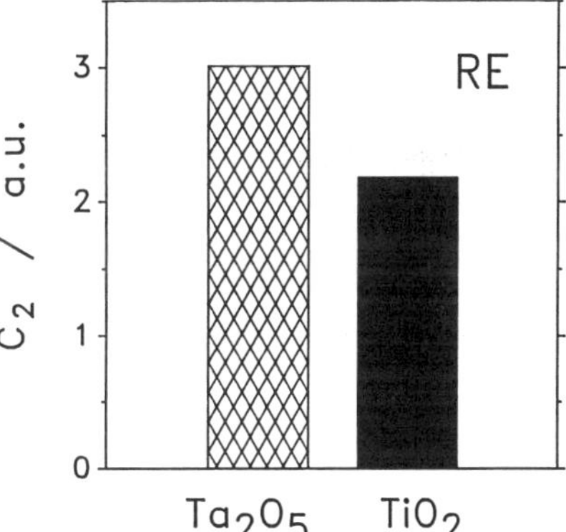

Fig. 6: Comparison of C$_2$ best-fit parameters for Ta$_2$O$_5$ and TiO$_2$ layers produced by reactive evaporation (RE).

Table II.: Scratch-test results obtained on similar films

material	TiO$_2$			Ta$_2$O$_5$	
dep. method	IP	RE	SC	IP[§]	RE
L$_c$(film)/L$_c$(substrate)	0.97	0.73	0.06	~ 0.9	0.22

[§]substrate material: AF 45 (a borosilicate glass)

L_c(substrate), needed to destroy the uncoated substrate surface by a cohesial failure. It is evident that the adhesion of TiO_2 films depends on deposition method and increases according to: SC < RE < IP. Also adhesion of RE TiO_2 films is better than for RE Ta_2O_5 layers.

DISCUSSION

The nonlinear ultrasonic signal obtained at thin oxidic single-films on fused silica show, in principle, that this method possesses sufficient sensitivity to detect differences in the acoustic response, even for rather thin layers. The investigations are fast and nondestructive. The rating of adhesion properties by the US method in respect to different film materials and different deposition methods are qualitative in agreement with findings obtained by the scratch test. However, both methods show remarkable differences on a quantitative scale. One of the reasons for this finding is may be the influence of the mechanical properties of the film material on the propagation of the US wave, which cannot easily be separated from the effect of the adhesion behavior.

The new ultrasonic testing method shows a potential for the nondestructive evaluation of adhesion properties of optical coatings. But, still several aspects have to be investigated for a routinely application of this technique. From the experimental point of view, the sensitivity has to be increased and the arrangement has to be optimized in respect to the demands of testing thin oxidic films. The theoretical understanding of the experimental situation has to be refined. Especially, this is valid for the interaction of the US wave with the mechanical properties of the interface region and the film material, both influencing the distortion of the wave. Here, information in required on the plastic/elastic film properties for a more comprehensive modeling of the propagation of the ultrasonic wave through the compound coating/substrate.

ACKNOWLEDGMENT

We would like to thank U. Jeschkowski and M. Plößer for preparing the oxide films, H. Koglin and U. Martens for technical assistance, and Y. Tomita and N. Tadokoro for performing the scratch-test investigations.

REFERENCES

[1] C.R. Ottermann et al., Mat. Res. Soc. Symp. Proc. Vol. **308**, 627 (1993); Vol. **356**, 839 (1995)

[2] S. Pangraz and W. Arnold, Review of Progress in QNDE **13B**, eds. D.O. Thompson and D.E. Chimenti, Plenum Press N.Y., 1995 (1994)

[3] S. Pangraz, PhD-Thesis, University of Saarbruecken, Technical Faculty and Fraunhofer IzfP-report 950117-TW (1995), in German

[4] K. Bange et al., BMFT-Abschlußbericht FKZ 13 N 5476/6 (1991)

[5] N. Arfsten, B. Lintner, M. Heming, O. Anderson and C.R. Ottermann, Mat. Res. Soc. Symp. Proc. Vol. **271**, 449 (1992)

[6] C.R. Ottermann, A. Temmink, and K. Bange, Proc. SPIE **1272**, 111 (1990)

[7] S. Baba, A. Kikuchi and K. Kinbara, J. Vac. Sci. Techn. **A4**, 3059 (1986); **A5**, 1860 (1987)

[8] C.R. Ottermann and K. Bange, in preparation

YOUNG'S MODULUS AND DENSITY OF THIN TiO$_2$ FILMS PRODUCED BY DIFFERENT METHODS

C.R. OTTERMANN[1], R. KUSCHNEREIT[2], O. ANDERSON[1], P. HESS[2], and K. BANGE[1]
[1]SCHOTT Glaswerke, R&D, P.O. Box 24 80, D-55014 Mainz, Germany
[2]Physikalisch-Chemisches Institut, Universität Heidelberg, D-69120 Heidelberg, Germany

ABSTRACT

The Young's modulus and density are analyzed for 280 and 500 nm thick TiO$_2$ layers deposited by reactive evaporation (RE) and ion plating (IP) by means of surface acoustic wave spectroscopy (SAWS) and grazing incidence X-ray reflection spectroscopy (GIXR). The layers are amorphous or polycrystalline, and have densities between 2.9 g/cm^3 and 3.8 g/cm^3, depending on the deposition conditions. The measured Young's moduli vary between 65 GPa for RE films and 147 GPa for IP layers. They are independent of film thickness, but correlate with the density. A change of the Young's modulus due to the phase transition from amorphous to anatase is described, which occurs at temperatures above 210°C.

INTRODUCTION

Thin films may possess chemical and physical properties which differ drastically from the properties of the corresponding bulk materials and vary in a broad range according to the deposition conditions [1]. A general description of the films is therefore difficult and time consuming, and complex analytical multi-method approaches have to be applied to determine the different physical quantities which are important for the various coating properties. To minimize the effort, knowledge of key quantities is desired, which will allow conclusions to be drawn about further materials properties.

TiO$_2$ is one of the most important materials for optical coatings where the increased interest is due to the wide field of applications of this high refractive-index material in interference systems. But, an understanding and optimization of the mechanical properties of these coatings is still one of the primary challenges in thin film technologies. Wear resistance and durability under mechanical, chemical, and thermal loading is strongly determined by the mechanical behavior of the films as one of the primary factors. In particular, the Young's modulus, E, and the Poisson's ratio, μ, which describe the elastic properties of the layer, are responsible for several quantities, for example hardness and changes in film stress due to temperature variations.

But, up to now, the data base on elastic properties of thin titania films is rather limited. Still nothing is known about the dependence of Young's moduli on deposition conditions or about further properties of the films. A Young's modulus of 90 GPa is reported for reactively evaporated TiO$_2$ films [2]. This differs markedly from the values of the corresponding properties in the bulk solid. For sintered polycrystalline plates of titania E is found to be of the order of 185 GPa [3], taking into account the density of the TiO$_2$ crystal state of anatase ($\rho = 3.84$ gcm^{-3}). The elastic constant varies between 120 and 480 GPa for the rutile state [4], depending on the crystal orientation.

In this work, we report on Young's modulus investigations on titania single layers deposited under various conditions. Relations are discussed between E and density, and a correlation is identified with other film properties such as refractive index and stress. Finally, the variation of

Mat. Res. Soc. Symp. Proc. Vol. 436 © 1997 Materials Research Society

Young's modulus is also investigated under annealing conditions in respect of the crystallization behavior of an originally amorphous TiO_2 film.

EXPERIMENTAL DETAILS

All coatings were deposited on fused silica substrates polished to high quality and flatness. Before coating all samples were cleaned by using a standard four-step cleaning process [5]. TiO_2 films with thickness of ≈280 nm and ≈500 nm were deposited by reactive e$^-$-beam vacuum evaporation (RE) and reactive ion plating (IP) with deposition rates of 0.15 nm/s and 0.3 nm/s. A detailed description of the deposition conditions is given in [1, 5]. The main process parameters are summarized in Table I.

The films were characterized by several analytical methods with respect to film thickness, optical properties, chemical composition, crystal structure, and film stress. A comprehensive description of the methods used, the experimental setups, and the data evaluation is given elsewhere [1, 5-7]. The results of these investigations are also summarized in Table. I.

Density depth profiles of the layers were obtained by means of grazing incidence X-ray reflectometry (GIXR) using a standard Philips diffractometer with PW 1830 generator and optically encoded PW 1820 goniometer. With this setup the intensity of the reflected X-rays was measured as a function of the angle of incidence in an angular range from 0° to 2°. In contrast to visible light, for X-rays the index of refraction in solid material is complex and smaller than 1 [8]. This means that X-rays are reflected away from the surface normal on entering the sample. Therefore, there exists a critical angle below which total reflection occurs. The mass density, ρ, can be determined absolutely from the critical angle of total reflectance of X-rays taking into account the stoichiometry of the film material [8]. The uncertainty of the density determination is less than 2 %. The X-rays pass through the films for angles of incidence larger than the critical angle, and the reflected intensities exhibiting an interference structure due to the two contributions being reflected from the front- and backside of the layer. The film thickness and interface roughnesses can be deduced from these interference patterns by employing the algorithm GIXA [9] for data analysis of X-ray reflectance spectra. Thickness values obtained by optical spectroscopy and GIXR are in agreement within the quoted uncertainty of both methods.

Surface acoustic wave spectroscopy (SAWS) was used to determine the elastic film proper-

Table I: Deposition conditions and results on morphology, stoichiometry, optical, and mechanical properties for the titanium oxide films on fused silica substrate investigated in this work.

process	reactive evaporation (RE)			reactive ion plating (IP)		
$T_{substrate}$ / °C	195	188	320	150	100	180
p_{tot} / (10^{-2} Pa)	1.4	1.4	1.4	10	8	8
I_{arc} / A	-	-	-	40	40	70
thickness / nm	514	291	279	499	271	286
morphology	amorphous	amorphous	anatase	amorphous	amorphous	amorphous
O/Ti ratio	2.1	2.0	2.0	2.0	2.03	2.0
H/Ti ratio	0.55	0.53	0.07	0.04	0.06	0.02
density / gcm^{-3}	3.0	3.0	3.3	3.8	3.8	3.8
refractive index	2.221	2.224	2.345	2.521	2.527	2.535
stress / MPa	216	226	220	-194	-251	-148
E / GPa	65	65-68	99-102	144	146-149	147

ties. A frequency-tripled Nd:YAG laser (wavelength 355 nm) with a pulse duration of 7 ns FWHM was employed to excite broadband coherent surface acoustic wave pulses. Fig. 1 presents a scheme of the experimental setup. The laser is focused onto the sample surface by a system of three cylindrical lenses to obtain focus dimensions of about 7 μm x 12 mm. Most of the laser beam energy is deposited within the film, due to the strong absorption of TiO_2 for $\lambda < 380$ nm [10]. This leads to the emission of nearly plane SAW pulses with a broad frequency distribution, due to the thermoelastic effect. A piezoelectric foil-transducer system is used to detect the acoustic pulse shapes at the sample surface at distances between 15 and 30 mm from the excitation line. The electrical signals obtained from the foil were recorded by a digital oscilloscope and Fourier transformed to amplitude spectra. The dispersion curve of the SAW pulse is deduced from the phase information of this pulse detected at different distances from the excitation source. More information on this technique are given elsewhere [11].

The dispersion of SAWs in layer/substrate systems is determined by the densities, ρ, and the elastic properties (Young's modulus, E, and Poisson's ratio, μ) of the substrate and film materi-

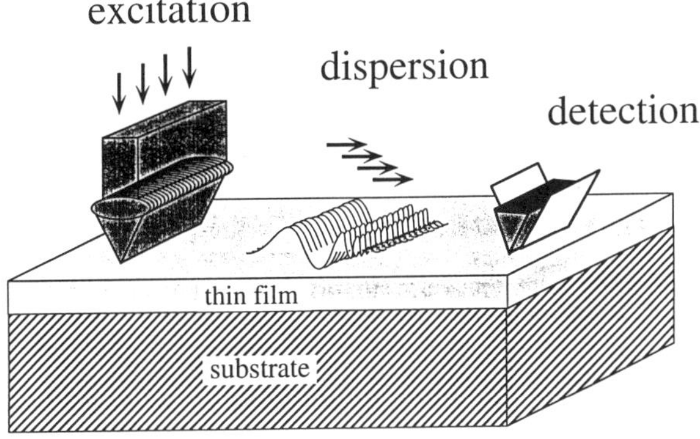

Fig. 1: Scheme of the experimental method with excitation laser (355 nm), cylindrical lenses, and foil-transducer system for the detection of the surface acoustic waves.

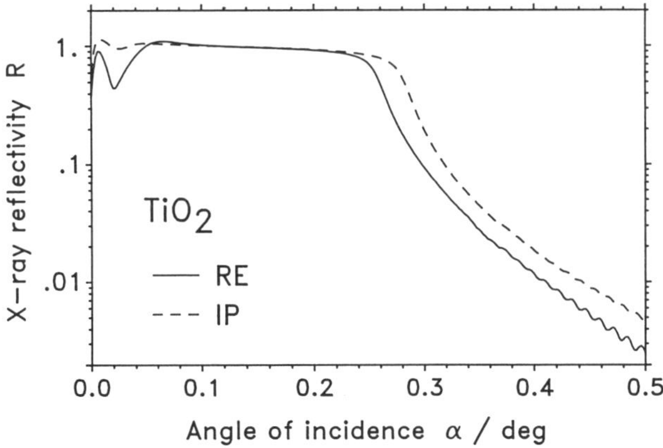

Fig. 2: X-ray reflectivity vs. angle of grazing incidence for TiO_2 films produced by reactive evaporation (RE) and reactive ion plating (IP).

als and the film thickness, d [12]. In principle, these four quantities can be deduced from a non-linear dispersion curve by a fitting procedure [11]. The accuracy of these fitting parameters is not sufficient for films thinner than 1 μm due to correlations and reduced nonlinearity. There-fore, d and ρ were fixed to the values obtained by GIXR, see Table I. Additionally, the variation of Poisson's ratio over a wide range was found to exert only a minor influence on the fit results for E. Therefore, μ was set to 0.2, a typical value for metal oxides.

RESULTS and DISCUSSION

All TiO_2 films are amorphous with the exception of RE layers deposited on substrates at temperatures above 300°C. These films exhibit polycrystal anatase structure. The titanium ox-ides are nearly stoichiometric, but the RE films contain hydrogen, due to uptake of H-containing species into their porous film structure [13]. Refractive properties for the different layers show the typical dependence on production conditions [1]. RE and IP titanium oxide films differ also in respect of their stress properties. The porous films produced by reactive evaporation possess tensile stress whereas IP layers show compressive stress, in agreement with results given in [1].

The film density is expected to vary also with production conditions, due to its strong corre-lation with the refractive index [1]. In Fig. 2 typical X-ray reflectivity spectra are shown for 280 nm thick RE and IP TiO_2 layers, obtained by GIXR. The critical angle of total reflectance is larger for the IP film than for the RE layer, indicating a difference in mass density. Results for ρ are summarized in Table I. Also for this work, the density and n_{550} possess the same correlation as published before [1]. The film thickness can be evaluated from the interference pattern. Find-ings of GIXR measurements are in agreement with the results obtained by optical investigations.

In Fig. 3a a typical SAW pulse-shape signal is shown, detected on a 286 nm thick IP TiO_2 film. The pulse consists of a superposition of coherent partial waves of the frequency spectrum up to 170 MHz, see Fig. 3b. The oscillations of the pulse shown in Fig. 3a are due to the inter-ference of these partial waves, which propagate with different phase velocities due to the dis-persion of SAWs on coated surfaces. The dispersion curve, as shown in Fig. 3c for the same IP film, can be deduced from a series of pulse shapes obtained at different distances. The solid line in Fig. 3c represents the result of a best fit optimizing the Young's modulus, E, as free fitting pa-rameter. The other model parameters (d, ρ; μ = 0.2) are fixed to the values given in Table I. The results on E for the different titania films are also shown in Table I. The accuracy and repro-ducibility of the E determination by the combination of SAWS and GIXR is of the order of $\Delta E = \pm 3$ GPa.

Young's moduli are shown in Fig. 4 as a function of film density for films with two different thicknesses. A nearly linear increase of E is obtained with increasing density. But, for a given density Young's modulus seems to be independent of film thickness in the thickness range of this investigation. The data base for these interpretations is still quite small and more extended in-vestigations are required to confirm the findings of this work.

The validity of the strong correlation between Young's modulus and film density would en-able an indirect determination of the elastic properties of TiO_2 films by optical means, as follows. The density of the layers can be estimated by an optical measurement of the refractive index, n, due to the strong correlation between n and ρ, as given in [1], and E is deduced from the density according to the dependence shown in Fig. 4. In addition, also hardness and thermal stress com-ponents could be estimated for TiO_2 films taking into account the prediction of the Young's modulus.

Upon annealing the amorphous titania films crystallize to the anatase state. The dynamics of the phase transition, i.e. the starting temperature, transformation speed and size of the formed

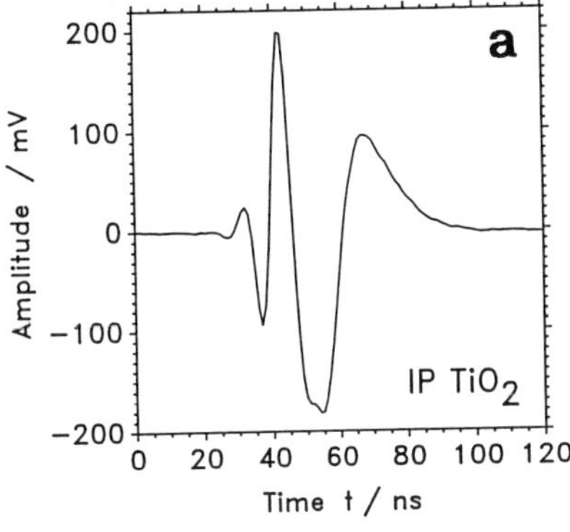

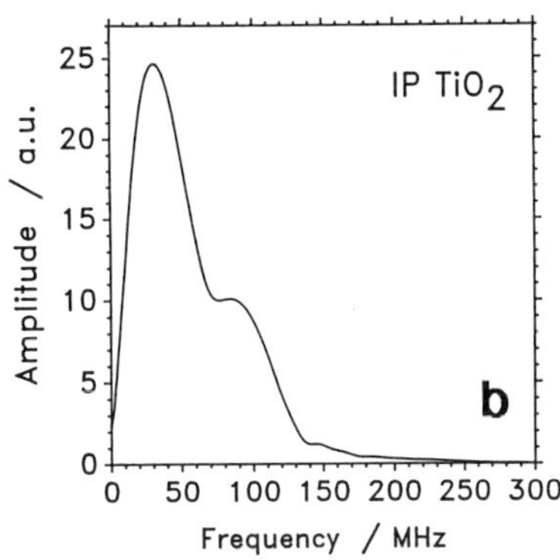

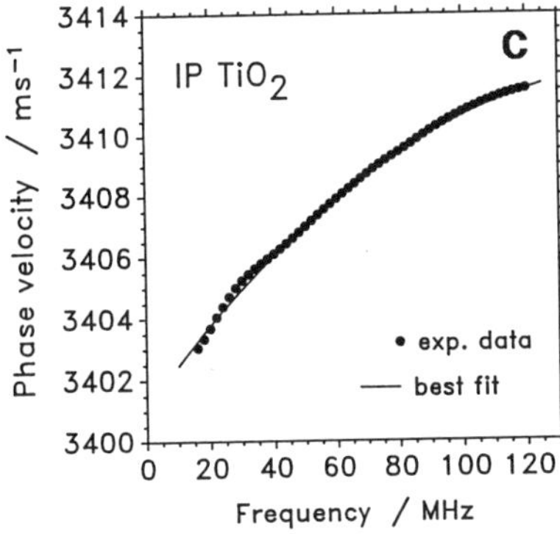

Fig. 3: (a) SAW pulse shape obtained on a 286 nm thick IP TiO_2 film, and

(b) corresponding amplitude spectrum.

(c) Experimental (●) dispersion curve and theoretical best fit (solid line) for the same film/substrate combination.

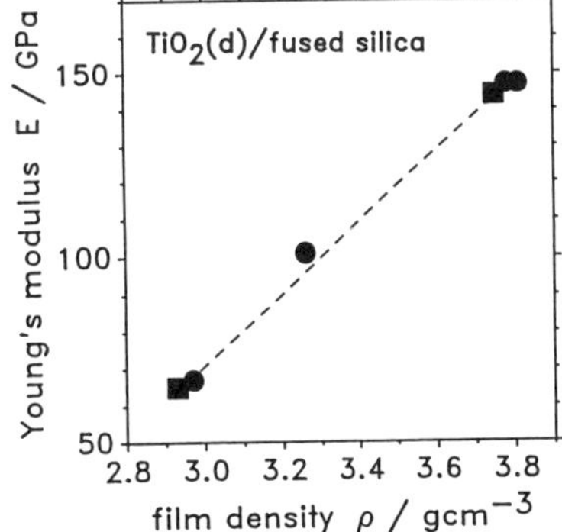

Fig. 4: Young's modulus, E, as a function of film density, ρ, for 280 nm (●) and 500 nm (■) thick TiO_2 layers.

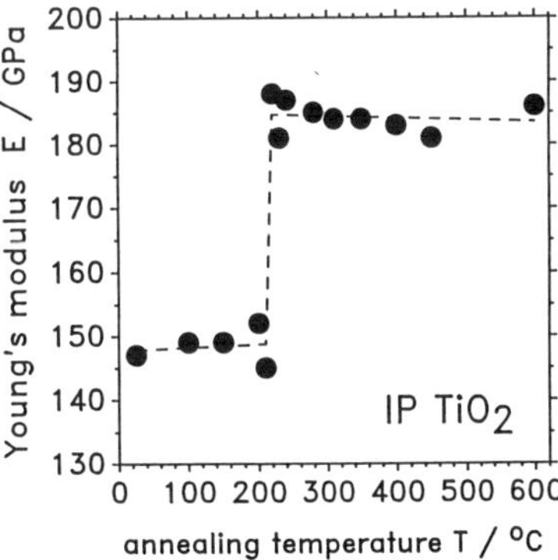

Fig. 5: Young's modulus, E, as a function of annealing temperature for a 286 nm thick titania film deposited by IP on a fused silica substrate heated to 180°C.

grains, depends strongly on the film density [1]. Fig. 5 depicts the Young's modulus of the 286 nm thick IP TiO_2 film as a function of the annealing temperature. For this investigation the samples were temperature treated at the different temperatures given in Fig. 5 for 20 min each time. The measurements of Young's modulus were performed after each treatment with samples cooled down to room temperature. A steep increase in E of about 40 GPa was obtained at an annealing temperature of 220 °C and correlates with the formation of crystal grains. No additional variations in E were obtained for larger temperatures after crystallization is completed.

SUMMARY

TiO_2 films with thicknesses of approximately 280 nm and 500 nm are deposited by reactive evaporation (RE) and reactive ion plating (IP). A multi-method approach is applied to characterize the film material. Specifically, the density, ρ, is obtained by Grazing Incidence X-ray Reflectivity measurements and the Young's modulus, E, by surface acoustic wave spectroscopy. E is of the order of 65 to 100 GPa for RE and 150 GPa for IP layers, correlating strongly with film density. Young's modulus does not depend on film thickness in the range investigated in this work. Annealing of amorphous titania films induces a phase transition to the crystal state of anatase. The Young's modulus increases strongly with crystallization.

ACKNOWLEDGMENT

We thank U. Jeschkowski and M. Plößer for the preparation of the TiO_2 films, and M. Laube and K.-H. Gießler for NRA and RBS investigations. This work is supported in part by BMBF under contract No. 13N6005, the European Union, and the Fonds der Chemischen Industrie.

REFERENCES

[1] C.R. Ottermann et al., Mater. Res. Soc. Symp. Proc. **308**, 69 (1993); ibid. **356**, 187 (1995)

[2] M.L. Scott, Natl. Bur. Stand. (U.S.) Spec. Publ. **688**, 329 (1985)

[3] W.P. Minnear and R.C. Bradt, J. Am. Ceram. Soc. **60**, 458 (1977)

[4] M.H. Grimsditch and A.K. Ramdas, Phys. Rev. **B14**, 1670 (1976)

[5] K. Bange et al., BMFT-Abschlußbericht FKZ 13 N 5476/6 (1991)

[6] C.R. Ottermann, A. Temmink, and K. Bange, Proc. SPIE **1272**, 111 (1990)

[7] C.R. Ottermann and K. Bange, in preparation

[8] F. Stanglmeier, B. Lengeler, W. Weber, H. Göbel, and M. Schuster, Acta Crystallogr. A **48**, 626 (1992)

[9] Philips Analytical X-ray. NL-Almelo

[10] L.-J. Meng and M.P. dos Santos, Thin Solid Films **226**, 22 (1993)

[11] H. Coufal, R. Grygier, P. Hess, and A. Neubrand, J. Acoust. Soc. Am. **92**, 2980 (1992)

[12] G.W. Farwell, <u>Acoustic Surface Waves</u>, edited by A.A. Oliner (Springer, Berlin, 1978), p. 42

[13] M.Laube, W. Wagner, F. Rauch, C. Ottermann, K. Bange, and H. Niederwald, Glass Sci. Technol. **67**, 87 (1994)

SIMULTANEOUS OVERALL AND LOCAL STRESS ANALYSIS OF THIN METAL FILMS BY LIGHT SCATTERING AND BEAM DEFLECTION MEASUREMENTS

C. KYLNER*,** , L. MATTSSON**
*Department of Physics II (Optics), Royal Institute of Technology, S-100 44 Stockholm, Sweden, carina@optics.kth.se
**Surface Evaluation Laboratory, Institute of Optical Research, S-100 44 Stockholm, Sweden.

ABSTRACT

We have developed an optical instrument for real-time overall stress as well as local stress relaxation measurements. The physical principle of the instrument is based on the fact that when a sample is exposed to heating the overall film stress bends the substrate and hillocks, the evidence of the local stress relaxation, form on the surface. Early hillock formation can be observed in real-time with integrated scattering technique, but the overall stress can be readily observed by laser beam deflection. Our instrument integrates laser beam deflection and light scattering techniques to achieve simultaneous overall stress and local stress relaxation measurements. A stress study is presented for an aluminium thin film on a silicon substrate.

INTRODUCTION

Integrated circuits and optical mirrors are generally described in terms of their electrical or optical properties. However, mechanical stresses often limit the lifetime of these devices [1-4]. Stresses are developed in integrated circuit structures during both processing and services. Short circuits caused by hillocks and open circuits created from metal cracking or voiding are phenomena related to mechanical stress. Similar problems with hillock and void formation are also well known in the optical device industry. For example, the surface of an aluminium coated mirror needs a protective layer of quartz to withstand harsh treatment and chemicals. To obtain good protective layers, the quartz needs to be deposited at elevated temperatures, well above 100 °C. However, slightly above 70 °C the performance of high quality aluminium coated mirrors begins to degrade due to light scattering from hillocks (Fig. 1) [4]. Thermally induced stress modifies the surface topography as mentioned above as well as the film structure (grain structure, dislocation structure, precipitates). The processes involved in these modifications depend on the magnitude of strain, the rate of strain, the temperature range and local inhomogeneities. Stress relaxation beyond the elastic limit is caused by several mechanisms of plasticity (dislocation glide and climb) and creep (grain boundary and lattice diffusion creep) [5]. These complex processes are difficult to assess independently and call for multiple and simultaneous recordings of several parameters. The objective of this study is to present an appropriate method for studying early hillock formation, the evidence of local stress relaxation, and its dependence on the overall stress in the film in real-time.

Mat. Res. Soc. Symp. Proc. Vol. 436 ® 1997 Materials Research Society

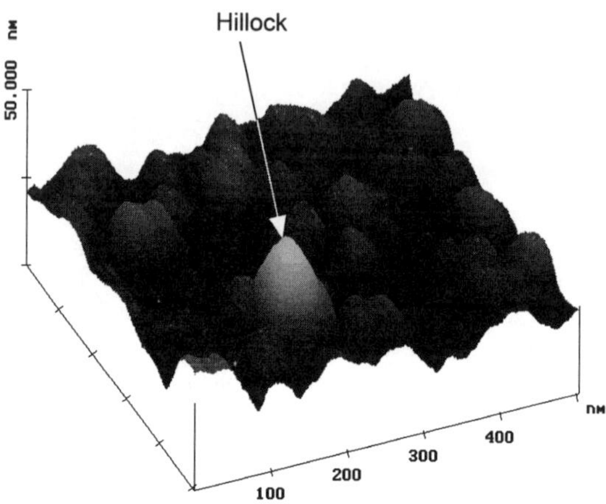

FIG. 1. Atomic force microscope image of early hillock formation in a thermally treated Al-film (130 °C) on a silicon substrate.

Laser beam deflection [6,7] and light scattering [4] are two experimental techniques which are widely used. To our knowledge, the two techniques have always been used separately. These two techniques are complimentary. Laser beam deflection gives information about the overall stress and integrated light scattering technique is very sensitive to surface roughening effects, like local stress relaxation. The combined overall stress and local stress relaxation measurement concept that we have adopted is based on the physical condition shown in Fig. 2. A metal film on a silicon wafer is illuminated by a collimated laser beam. At room temperature the laser beam reflects all light, see Fig. 2(a). In Fig. 2(b) the sample has been heated, which causes a large difference in thermal expansion between the metal film and silicon wafer. This strain yields an overall stress in the film, which bends the silicon wafer to a convex shape as viewed from the film coated side. For a sufficiently high temperature, about 70 °C for a room temperature evaporated aluminium film [4], the compressive stress in the film is high enough to create local relaxation phenomena, i.e. hillocks are formed at the grain boundaries. These hillocks will increase the light scattering by a considerable amount. The curved wafer deflects the reflected beam as well as scatters a fraction of the incident light. The deflection of the laser beam is easily monitored by a position sensitive detector. The deflection allows the curvature to be determined and the overall stress to be calculated. By simultaneously recording the scattered light and the deflection of the laser beam we obtain information on overall stress development and local relaxation causing hillock formation.

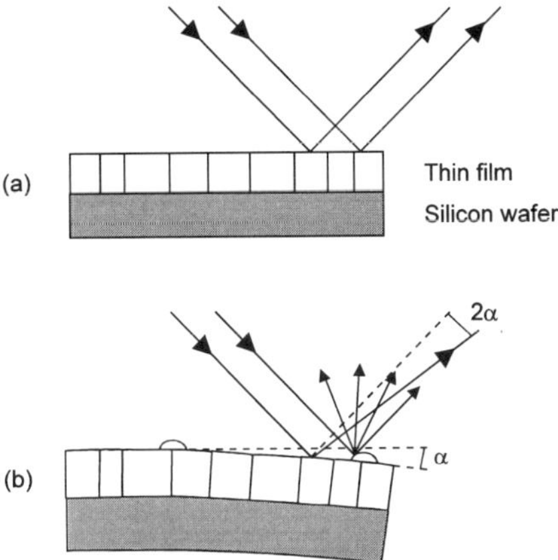

FIG. 2.   The physical principle of the instrument: (a) The laser beam is reflected specularly by a sample kept at room temperature. (b) On the heated sample the laser beam is deviated by 2α and a fraction of the light is scattered in different directions by hillocks formed on the surface.

EXPERIMENT

A brief description of the instrument

The Vacuum Compatible Partial Integrated Scattering and Deflection (VaCPISD) instrument is a stress measuring instrument for thin metal films. The instrument is shown in Fig. 3(a) and it is based on partial integrated light scattering from hillocks and voids and laser beam deflection caused by the film stress induced wafer bending. The design of the optical system (see Fig. 3(b)) makes the instrument relatively compact, easy to align and most important of all it combines the two different measurement techniques into one instrument using one and the same laser beam for both the deflection and the scattering analysis. The VaCPISD instrument details have been presented elsewhere [8].

By designing the deflection part in a way such that the beam hits the surface *twice* an improved performance is obtained. The instrument yields a precision in change in substrate curvature of approximately $3.5 \cdot 10^{-5}$ m^{-1} or an equivalent radius of curvature of 40 km. *Without* any knowledge of the initial substrate curvature, the change in curvature could be determined with an accuracy of about 4 %. The scattered light is collected by a Partial Integrated Scattering (PIS) system, with a dynamic range covering the scattering from super smooth surfaces to rough surfaces.

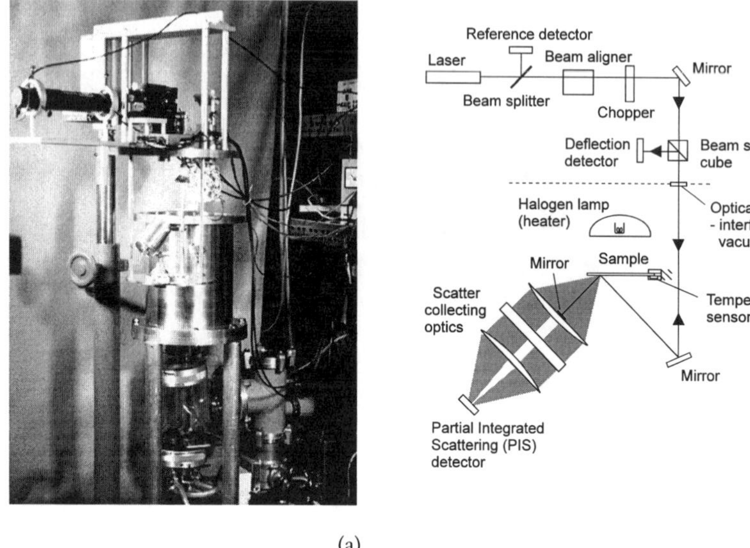

(a)   (b)

FIG. 3. (a) The Vacuum Compatible Partial Integrated Scattering and Deflection (VaCPISD) instrument. (b) The outline of the optical system and some parts of the thermal system.

<u>Sample preparation</u>

Aluminium films were evaporated by an e-beam gun onto one side of a 0.5 mm thick and natively oxidised (100)-oriented silicon wafer. The substrates were intentionally not heated. The films were deposited at a rate of 10 Å/s using a deposition pressure of $5 \cdot 10^{-5}$ Pa and the background pressure was $2 \cdot 10^{-6}$ Pa. During the evaporation the samples were rotated to achieve uniform films. The films were approximately 2500 Å thick. The grain size of the films was about 100 nm as determined by AFM. The thermally induced stress behaviour of these aluminium films on silicon substrates was investigated with the VaCPISD instrument. The thermal cycling experiments were performed in air.

RESULTS

Simultaneous recordings of temperature, overall stress and local stress relaxation have been performed on aluminium thin films on silicon wafers. Fig. 4 shows a measurement result from a thermal cycling between room temperature and 180 °C and back to room temperature. We have divided the thermal cycle into four regions (A-D). A schematic picture of the sample is given for each region. The initial condition before the heater was turned on corresponds to region A. An almost instantaneous bending occurred when the heater was turned on, confirming a compressive overall stress in the heated aluminium film. Region B only lasted approximately one minute and was characterised by the large change in curvature. When the first bending maximum was reached, the hillocks formed rapidly on the surface of the film.

This bending maximum corresponds to a minimum in the curve and is labelled "1ˢᵗ" in Fig. 4. Region C was characterised by the increased scattering caused by hillock formation. A slight initial reduction in bending was observed, before another maximum deflection was reached and a relatively slow continuous development of hillocks took place. This second bending maximum also corresponds to a minimum in the curve and is labelled "2ⁿᵈ" in Fig. 4. Note that in this particular measurement the deflection of the laser beam exceeded the range of the detector. Therefore the overall change in curvature tended to be saturated in region C. After the heater was turned off, the film developed a tensile stress and a very slight decrease in scattering occurred. In region D the light scattering was still high, but the change in curvature was opposite to that in region B. These first results tell us that the local stress relaxation by hillock formation does influence the overall stress in the initial stage to some extent. But for an extended heating period it does not affect the overall stress to a significant extent. The tensile stress developed after cooling can partly be explained by the formation of hillocks, causing loss of material in the thin film.

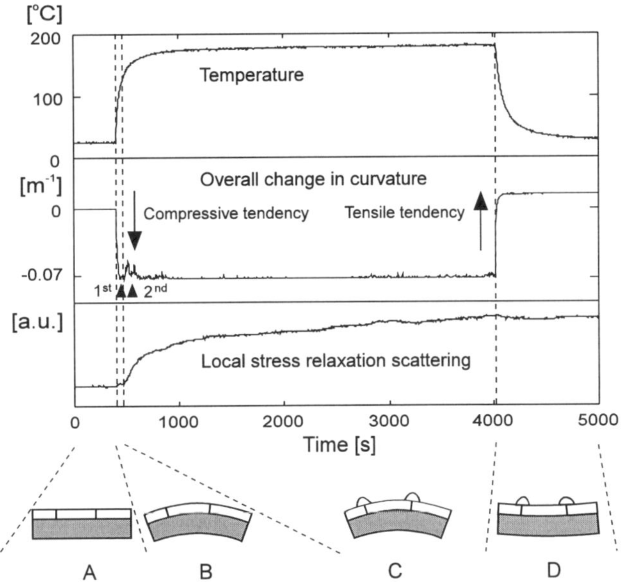

FIG. 4. Stress measurements as a function of temperature and time on an Al-film/Si-wafer with the VaCPISD instrument. The thermal cycle was divided into four regions (A-D).

CONCLUSIONS

Simultaneous overall and local stress analyses of thin metal films by scatter and beam deflection measurements were performed on aluminium films on silicon substrates. The results show the interplay between the high overall stress and the local stress relaxation. A new kind of real-time study can now be realised with the VaCPISD instrument, where the relation between overall stress and local stress relaxation in thin films can be thoroughly investigated both in the temperature and in the time domain.

ACKNOWLEDGEMENTS

The authors are very grateful to L. Kjellberg and S. Bolin for the electronic and mechanical workshop assistance and L. Krummenacher for thin film deposition and AFM measurements.

The research and development was financially supported by Swedish National Board for Technical Development and the Göran Gustafsson Association.

REFERENCES

1. W.R. Hoffman, Phys. Thin Film **3**, 211 (1966).
2. W.D. Nix, Metall. Trans. **20A**, 2217 (1989).
3. M.D. Thouless, Annu. Rev. Mater. Sci. **25**, 69 (1995).
4. L. Mattsson, Y.-H. Le Page, and F. Ericsson, Thin Solid Films **198**, 149 (1991).
5. D. Gerth, D. Katzer, and M. Krohn, Thin Solid Films **208**, 67 (1992).
6. A.K. Sinha, H.J. Levinstein, and T.E. Smith, J. Appl. Phys. **49** (4), 2423 (1978).
7. P.A. Flinn, D.S. Gardner, and W.D. Nix, IEEE Trans. Electron. Dev. **ED-34** (3), 689 (1987).
8. C. Kylner and L. Mattsson, (submitted to) Rev. Sci. Instrum. (1996).

Part V

Tribological Properties of Thin Films

SIMULATION OF MECHANICAL DEFORMATION AND TRIBOLOGY OF NANO-THIN AMORPHOUS HYDROGENATED CARBON (a:CH) FILMS USING MOLECULAR DYNAMICS

J.N. GLOSLI*, M.R. PHILPOTT**, and J. BELAK*

*University of California, Lawrence Livermore National Laboratory, Livermore CA 94550
**IBM Research Division, Almaden Research Center, 650 Harry Road, San Jose, CA 95120-6099

ABSTRACT

Molecular dynamics computer simulations are used to study the effect of substrate temperature on the microstructure of deposited amorphous hydrogenated carbon (a:CH) films. A transition from dense diamond-like films to porous graphite-like films is observed between substrate temperatures of 400K and 600K for a deposition energy of 20 eV. The dense a:CH film grown at 300K and 20 eV has a hardness (~50 GPa) about half that of a pure carbon (a:C) film grown under the same conditions.

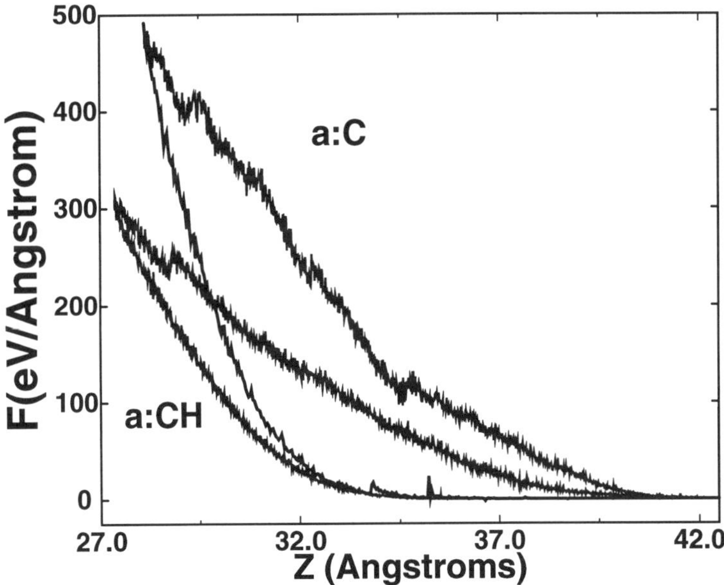

Figure 1. The indentation loading and unloading curves for simulated amorphous carbon films. The upper curve is for a:C and the lower curve is for a:CH. Both films were grown with 20eV deposition energy, 300K substrate temperature and 4nm surface cells. The estimated hardness of the a:C films is ~100GPa and that of the a:CH film is ~50GPa.

INTRODUCTION

Amorphous carbon films about 20nm thick are used throughout the magnetic disk industry as protective coatings on magnetic disks. Despite intense experimental and theoretical study [1,2], the microstructure of these amorphous films and the effect on mechanical properties such as hardness is not well understood. This is in part due to the variety of deposition methods used and to the difficulty in probing the state of the material at the nanometer length scale.

In this work we use a molecular dynamics computer model [3,4] with a reactive bond-order interatomic potential [5,6] to simulate the growth and resulting hardness of hydrogenated amorphous carbon films. This potential model represents the chemistry of the carbon/hydrogen system allowing the formation of both graphitic and diamond-like regions. The films are created by depositing a 50/50 mixture carbon and hydrogen atoms onto a diamond (100) surface (12 layers) at one atom per picosecond. The substrate temperature is controlled with a Nose-Hoover thermostat and with a Langevin thermostat. A time step of 0.5fs is used and the simulation cell is periodic in the plane of the surface. Cell sizes of 2nm and 4nm are considered. Indentation was performed at 35m/s using a tip cleaved from three (100) planes on the diamond lattice. The tip was blunted to create an effective radius of about 1nm. The tip atoms are held rigid during the indentation and interact with the surface atoms through a truncated Lennard-Jones potential. The indentation rate of 35m/s is comparable to the sliding speeds at the head-disk interface in magnetic recording disks.

RESULTS and DISCUSSION

Shown in Figure 1 is the loading and unloading curves (force on the indenter as a function of height) for simulations of both a:C and a:CH films. The results for the a:C films are presented in reference 4. Both films were grown with a 4nm surface cell, a substrate temperature of 300K, and a deposition energy of 20eV. The indentation computer experiment was performed at 35m/s as described above. The step-like kinks in the loading curves are due to the onset of plasticity. Unloading the tip prior to the first step does not give the hysteresis loop as shown. Our simulation cell is too small to observe the formation of cracks and plasticity occurs through changes in the chemical bonding network. The simulated pure carbon film is about twice as hard as the simulated hydrogenated carbon film. We estimate the hardness from the loading curves and the area of contact (H=L/A) to be ~100GPa for the pure carbon film and ~50GPa for the hydrogenated carbon film.

Substrate temperature is an important control variable determining the microstructure and hence properties of amorphous carbon films. To investigate the effect of substrate temperature we performed a series of deposition simulations ranging in temperature from 150K to 600K at 20eV using a 2nm surface cell. Snapshots from these simulations are shown in Figure 2. The microstructure of the low temperature (150K) film is qualitatively the same as the film grown at room temperature (300K). The bonding at 20eV is dominated by sp^3 hybridized carbon (diamond-like) with a small amount of sp^2 (graphitic) [3,4]. The presence of hydrogen tends to stabilize the diamond-like carbon and the graphitic carbon atoms tend to come in pairs. The microstructure at 400K is also qualitatively the same as at 300K. Between 400K and 600K we find a transition to a porous structure dominated by graphitic bonding. The snapshot at 600K displays large voided regions within the film. In a recent experimental study of the effect of substrate temperature on the

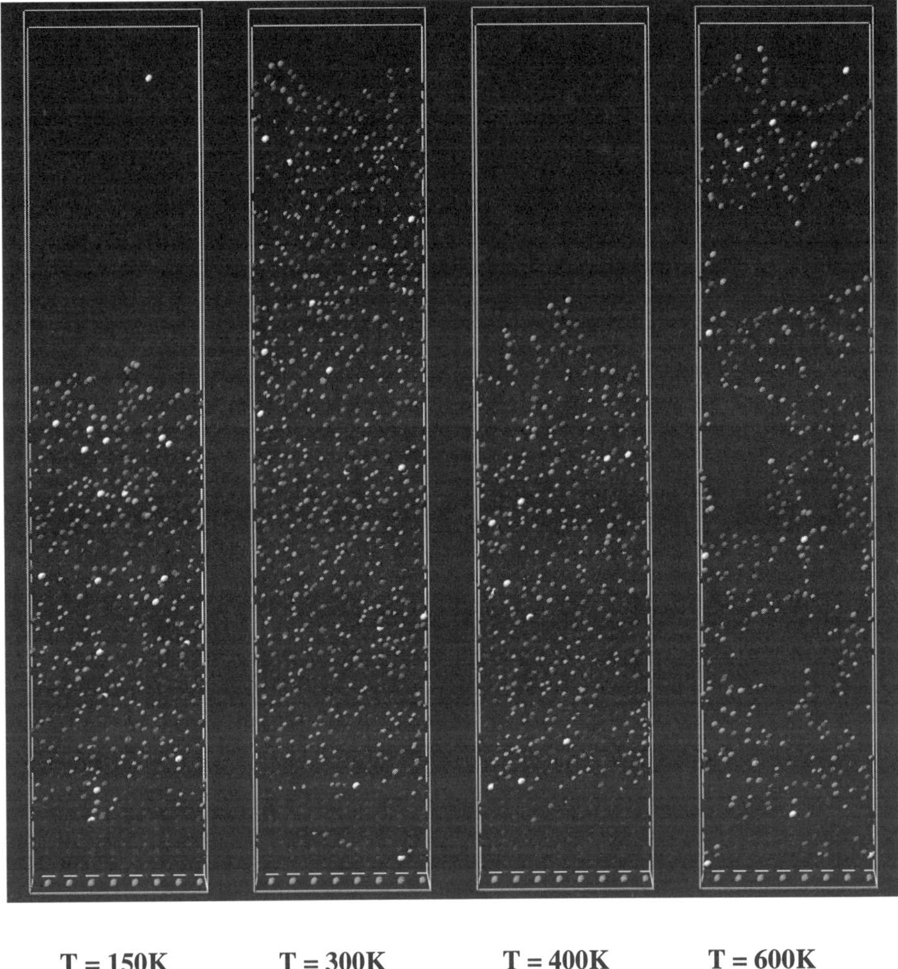

T = 150K   **T = 300K**   **T = 400K**   **T = 600K**

Figure 2. Snapshots from our computer simulations of the growth of hydrogenated amorphous carbon films. The four snapshots are for substrate temperatures of 150K, 300K, 400K, and 600K. All simulations were for 20eV deposition energy, a 2nm surface cell, 0.5fs time step and a 50/50 mixture of carbon and hydrogen atoms being deposited.

microstructure of nonhydrogenated carbon films, Bhargava et. al [7] found an observable difference between films grown at a substrate temperature of 400K and films grown at 800K. Their micro-Raman spectra suggest a significantly greater graphitic character for the high temperature films. The computer simulations presented here are consistent with that interpretation.

ACKNOWLEDGEMENTS

Work performed under the auspices of the U.S. Department of Energy by Lawrence Livermore National Laboratory under contract No. W-7405-ENG-48.

REFERENCES

(1) H-C. Tsai and D.B. Bogy, J. Vac. Sci. Tech., **A5**, 3287 (1987).

(2) A. Grill, Wear, 168, **143** (1993).

(3) J.N. Glosli, J. Belak, and M.R. Philpott, "Ultra-Thin Carbon Coatings for Head-Disk Interface Tribology," in **Thin Films: Stresses and Mechanical Properties V**, S.P. Baker, C.A. Ross, P.H. Townsend, C.A. Volkert, and P. Borgesen eds., MRS Symposium Proceedings 356, MRS, Pittsburgh, USA 1995.

(4) J.N. Glosli, M.R. Philpott, and J. Belak, "Molecular Dynamics Simulation of Mechanical Deformation of Ultra-Thin Amorphous Carbon Films," in **Mechanical Behavior of Diamond and Other Forms of Carbon**, M.D. Drory, M.S. Donley, D. Bogy, and J.E. Field eds., MRS Symposium Proceedings, MRS, Pittsburgh, USA, 1995.

(5) D.W. Brenner, Phys. Rev. **B42**, 9458 (1990).

(6) D.W. Brenner, J.A. Harrison, C.T. White, and R.J. Colton, Thin Solid Films **206**, 220 (1991).

(7) S. Bhargava, H.D. Bist, A.V. Narliker, S.B. Samanta, J. Narayan, and H.B. Tripathi, J. Appl. Phys **79**, 1917 (1996).

TRIBOLOGY STUDIES OF ORGANIC THIN FILMS BY SCANNING FORCE MICROSCOPY

G. BAR[*], S. RUBIN[**], A. N. PARIKH[**], B. I. SWANSON[**], T. A. ZAWODZINSKI[**]
*Freiburger Materialforschungszentrum, FMF, Albert-Ludwigs University, Stefan-Meier-Str. 21, 79104 Freiburg, Germany, gbar@fmf.uni-freiburg.de
**Los Alamos National Laboratory, Los Alamos, NM 87545, USA

ABSTRACT

Using the micro-contact printing method we prepared patterned self-assembled monolayers (SAMs) consisting of methyl-terminated alkanethiols of different chain lengths. The samples were characterized using lateral force microscopy (LFM) and the force modulation technique (FMT). In general, higher friction is observed over the short chain regions than over the long chain regions when a low or moderate load is applied to the scanning force microscopy (SFM) tip. For such cases the high friction (short chain) regions are also "softer" as measured by FMT. At high loads, a reversal of the image contrast is observed and the short chain regions show a lower friction than the long chain regions. This image contrast is reversible upon reduction of the applied load.

INTRODUCTION

The use of organic thin films as lubricants on solid surfaces is important in many modern technologies including magnetic storage and micromachines [1, 2]. Langmuir-Blodgett (LB) films and self-assembled monolayers (SAMs) are attractive candidates for lubricant layers and for model studies of lubrication because of their strong adsorption to the surface, i. e. they are expected not to migrate on the surface or to transfer from one solid surface to the other. The recent interest on the properties of LB films and SAMs has been also motivated by their potential applications in sensors [3], non-linear optical devices [4], lithography [5] and microelectronics [6].

Recently, considerable interest has been shown in characterization of LB films and SAMs (and in particular patterned SAMs) using scanning force microscopy (SFM). In particular, it has been demonstrated that lateral force microscopy (LFM), which measures the friction forces between the tip and the sample surface, can distinguish between chemically different surface regions. Such experiments providing material-contrast imaging are relevant to applications involving lubrication, adhesion and wetting. Knowing the interaction leading to different friction forces between the SFM tip and the sample surface is essential for the understanding of the image contrast relating the contrast to the tribological surface properties. Unfortunately, the interpretation of the experimental data is complicated because the mechanisms underlying the image contrast are not yet completely understood. As a consequence, conclusions based on results reported to date are far from being unequivocal. The LFM image contrast observed between different domains of phase-separated LB films is believed to be due to differences in the mechanical properties of the LB domains, i. e. elastic compliance [7]. The frictional properties of SAMs of methyl-terminated alkylsilanes of different chain lengths on mica substrates are seen to depend strongly on the chain length [8] such that higher friction was observed for short chain lengths. This effect was attributed to disorder in the layers formed by molecules with short chain length. On the other hand recent investigations of patterned SAMs consisting of regions of alkanethiols with chemically different terminal groups suggested that the friction contrast observed in LFM might be dominated by the chemical identity of the end groups of the surface and tip [9-10]. Thus, there is a need to study in more detail the effects of order, packing, crystallinity and chemical identity of the terminal end groups on the observed frictional contrast in LFM.

In this paper, we discuss the LFM image contrast between different regions of a patterned SAM consisting of alkanethiols having the same terminal end group, $-CH_3$, but different chain lengths. We investigate in particular the correlation between the observed frictional contrast and the measured local elastic properties of the different regions of the SAM using the force modulation technique (FMT). FMT is based on the principle that the vertical force between the

SFM tip and the sample surface is modulated by oscillating the probe and the relative elasticity of the sample is measured by recording the amplitude of the tip deflection versus position over the sample [11-12]. We show that the frictional contrast between the short and the long chain regions strongly depends on the applied load.

EXPERIMENT

Materials: The silicon wafers were obtain from Semiconductor Processing Co. The gold surfaces were prepared by sputter deposition of 10 nm of titanium, followed by sputter deposition of 100 nm of gold, on the native layer of the silicon oxide. The gold and the titanium purity was 99.999%. Octadecyl mercaptan (C18), hexadecyl mercaptan (C16), nonyl mercaptan (C9) and heptyl mercaptan (C7) were obtained from Aldrich and were used without further purification. The micro-contact printing process was applied as described elsewhere [13].

Instruments: A commercial Digital Instruments Nanoscope III was used for the SFM experiments. The experiments reported here were carried out in air under ambient conditions. The measurements were performed with commercial Si cantilevers with nominal force constants $C = 0.02 - 0.1$ N/m and commercial Si3N4 cantilevers with nominal force constants $C = 0.06 - 0.12$ N/m. Topographic ("height") images were obtained in the constant force mode and simultaneously recorded with the friction maps which were obtained in the lateral force mode (LFM). The elasticity measurements, performed using the force modulation technique (FMT), were obtained simultaneously with topographic images. The data are presented as gray coded images such that bright areas correspond to higher regions in topographic images, higher friction in LFM images and softer regions in FMT images.

RESULTS AND DISCUSSION

Figure 1 shows the SFM images of a patterned SAM sample on gold consisting of C18 (circular structure) and C7 arrays (surrounding the circles). The sample was prepared by first stamping the C18 component using a master consisting of circles (approx. 3 μm in diameter) and then exposing the stamped sample to C7 solution. The topography image (Figure 1a) shows clear image contrast between the more elevated C18 regions (bright gray) and the C7 surroundings. We measured ≈ 0.8 nm as height difference between the C18 and C7 surface regions which is in reasonable agreement with the expected value of 1 nm. One should note, however, that an exact measurement of the height difference is difficult due to the different penetration of the SFM tip into the C18 and C7 regions (see discussion of the local elastic properties below).

Figure 1b shows the LFM image (friction map) recorded simultaneously with the topography image shown in Figure 1a. In the LFM mode, very good contrast is observed between the circular C18 regions and the surrounding C7. The LFM map shows higher friction (bright gray) over the short chain C7 regions and lower friction over the long chain C18 regions (circles). The image presented was obtained applying a low load (≈ 17 nN) [14]. As shown below, the image contrast depends strongly on the force the tip exerts on the sample surface. Similar images were observed for patterned samples consisting of C18 and C9 arrays, C16 and C7 arrays and C16 and C9 arrays (data not shown). In all cases a higher frictional force was recorded over the short chain regions than over the long chain regions when a low or moderate load was applied (up to ≈ 190 nN).

Figure 1c shows the FMT image (elasticity map) of the patterned SAM sample. Imaging the patterned SAMs under ambient conditions produced contrast between the C18 regions (bright circles) and the C7 regions (dark surroundings). Bright regions correspond to softer sites which absorb more of the cantilever's energy, causing a reduced cantilever response and lower amplitudes. Figure 1c shows that the C7 surface regions are "stiffer" than the C18 circles. The same type of elasticity maps were observed for patterned samples consisting of C18 and C9 arrays, C16 and C7 arrays as well as for C16 and C9 arrays (data not shown). In all cases the short chain regions appeared "stiffer" than the long chain surface regions as measured by FMT. Studies of patterned SAMs consisting of surface regions with different chemical functionalities showed that the LFM image contrast generally seems to correlate with the surface free energy of

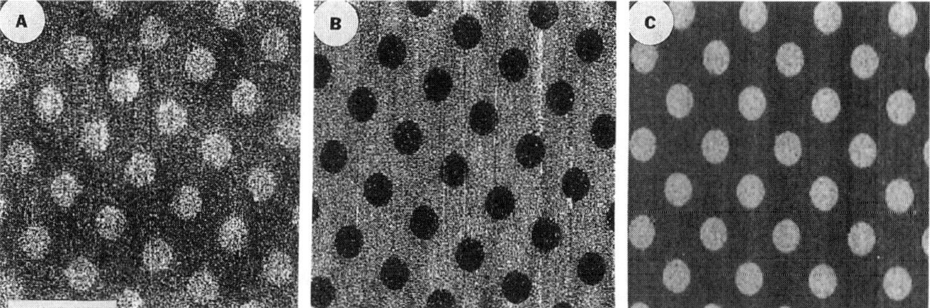

Figure 1: SFM images of a patterned C18/C7 alkanethiol SAM on gold. (a) Topographic image recorded with a low load of 17 nN. The horizontal bar corresponds to 10 µm and is representative for all images. The gray scale corresponds to a z-scale of 10 nm. (b) LFM image showing the friction map. Dark regions (C18) correspond to lower friction and bright regions (C7) to higher friction. (c) FMT image showing the local elasticity. Bright regions (C18) correspond to "softer" areas and dark regions (C7) to "stiffer" areas.

the arrays [9, 10, 15]. Higher friction has been observed over the hydrophilic -COOH, or -OH terminated surface regions than over the hydrophobic -CH3 terminated areas. However, these studies did not clarify to which extend other factors such as packing and disorder contribute to the observed image contrast. Contact angle measurements do not show pronounced differences between the short and long alkanethiols [16]. Thus one must conclude that the image contrast observed in the friction map (Figure 1b) originates from factors such as disorder and packing rather than from surface free energy/hydrophobicity. A recent study [8] of the chain length dependence of the frictionalproperties of alkylsilane SAMs on mica showed that friction forces strongly depend on the length of the alkyl chains, being higher for the short chains, in agreement with our study. The observed dependence of the friction forces on the chain length could be explained by a much higher alkyl chain disorder in the short chain SAMs. FTIR data suggest that SAMs of short chain alkanethiols (n <10) have liquid-like chains packed at lower densities than their longer chain counterparts [17]. Thus, the possibility of differences in surface coverage should also be taken into account.

Our data show that the friction maps correlate with the local elasticity maps: the long chain arrays exhibit lower friction and are more compliant than the short chain regions. A similar correlation was reported for patterned SAMs consisting of chemically different arrays, i.e. -COOH and -CH3 terminated alkanethiols, where the high friction -COOH regions appeared to be "stiffer" [10]. In the case of phase-separated LB films, a different correlation was observed: higher friction was observed for the more compliant fluorocarbon domains than for the "stiffer" hydrocarbon domains [12]. In that case, however, the fluorocarbon domains consisted of multilayers, the fluorocarbons sitting on top of the hydrocarbon monolayer. This seems to be important if the thickness of the organic film and the substrate plays a role. The liquid-like short chain alkanethiols are expected to be softer; however, one must take into account that the oscillating tip in FMT might "feel" the underlying substrate. In that case the short chain regions would appear "stiffer" as detected by FMT since the tip may penetrate more easily through the organic monolayer through the substrate. We note, however, the same type of image contrast was observed even with very small load and small oscillating amplitude.

We now turn to a study of patterned SAMs consisting of arrays of alkanethiols quite similar in chain length. This is assumed to minimize the effects arising from different film thickness and surface coverage. We prepared and investigated SAMs consisting of C18 circular regions and C16 regions surrounding the circles. Figure 2a shows the SFM topography image, which exhibits little image contrast. This is not surprising since the height difference between the C18 and C16 surface regions should be on the order of 0.2 nm. Figure 2b shows the friction map (LFM image) of the patterned SAM. The LFM image shows very good image contrast between the C18 and

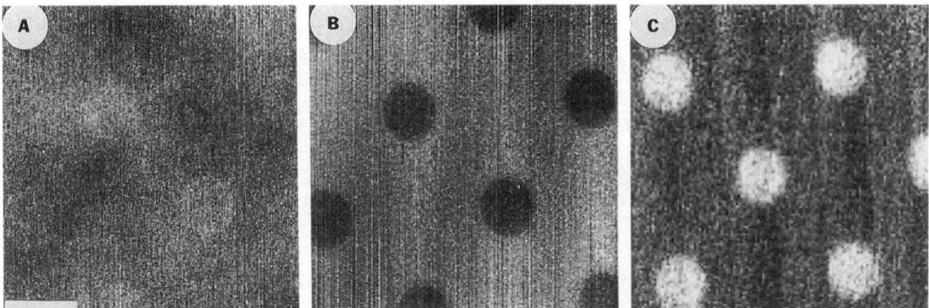

Figure 2: SFM images of a patterned C18/C16 alkanethiol SAM on gold. (a) Topographic image recorded with a low load of 10 nN. The horizontal bar corresponds to 5 μm and is representative for all images. The gray scale corresponds to a z-scale of 10 nm. (b) LFM image showing the friction map. Dark regions (C18) correspond to lower friction and bright regions (C16) to higher friction. (c) FMT image showing the local elasticity. Bright regions (C18) correspond to "softer" areas and dark regions (C16) to "stiffer" areas.

surrounding the C18 circles (dark gray), irrespective of which of the two components was adsorbed by stamping or from solution. This result is quite remarkable since it indicates that friction contrast can be observed between alkanethiol surface regions which differ by just two CH_2 groups: lower friction is always observed over the long chain surface regions. Figure 2c shows the FMT image (elasticity map) of the patterned SAM sample consisting of the C18 circular surface regions and their C16 surroundings. It shows that the C16 surface regions are "stiffer" than the C18 circles, again, independent of which component was stamped and which adsorbed from solution. Thus, as in the case of the patterned SAM consisting of quite different alkanethiol chain regions (e.g. C18 and C7), the short chain regions appeared "stiffer" than the long chain surface regions as measured by FMT.

It is instructive to explore the friction image dependence on the applied load. Figure 3 shows a series of LFM images for a patterned SAM sample consisting of C18 circular regions and C7 surroundings obtained with different loads. Figure 3a shows the friction map recorded applying a load of approx. $\approx$ 16 nN. As discussed above, lower friction is observed over the C18 surface regions (dark gray) than over the C7 surface regions at low load. Figure 3b shows the friction

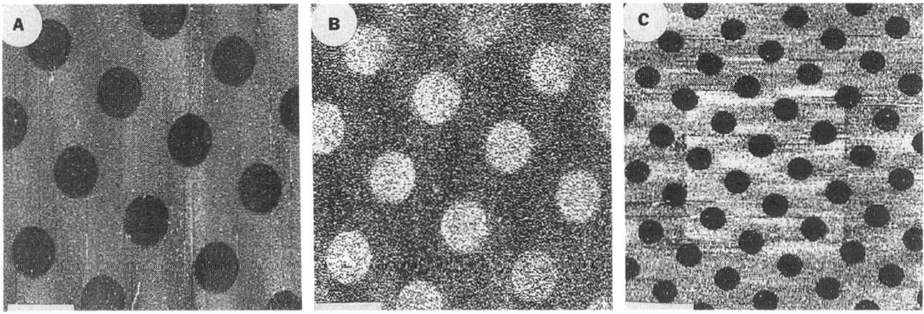

Figure 3: LFM images of a patterned C18/C7 alkanethiol SAM on gold. The horizontal bar corresponds to 5 μm in (a) and (b), and to 10 μm in (c). (a) LFM image recorded with a low load of 16 nN. The C18 regions show a lower friction (dark gray) than the C7 regions (bright gray) (b) LFM image recorded with a high load of 230 nN. The C18 regions show now a higher friction (bright gray) than the C7 regions (dark gray). (c) LFM image of a bigger area which has been previously scanned with a high force recorded with a low load of 11 nN. Now the C18 regions show again a lower friction (dark gray) than the C7 regions (bright gray).

map of the same surface region recorded applying a much higher load of $\approx$ 230 nN. Above a certain threshold, a reversal of the LFM image contrast is observed. At high loads lower friction is observed over the short chain C7 region (dark gray) and higher friction over the long chain C18 regions (bright gray). One might expect that at high loads, destruction of the adsorbed organic surface layers would occur. However, we observed that the frictional contrast obtained at different loads was reversible. Figure 3c shows the LFM image acquired with a low load of a surface region which has been previously scanned with a high load. One can clearly recognize the rectangular surface region which has been scanned with higher loads. But nevertheless, the original image contrast is restored and the long chain C18 surface regions show again a lower friction than the short chain C7 regions.

The dependence of friction forces on the tip load has been studied for alkylsilanes of different chain length self-assembled on mica [8]. Up to loads of 100 nN, elastic behavior was found and the frictional forces were always higher for the short chain molecules. Above a tip load of 100 nN, a substantial distortion of the alkylsilane chains was observed resulting in irreversible displacement of the molecules and damage of the organic film [8]. It is important to note that the situation is different for alkanethiols adsorbed on gold which undergo reversible displacement upon distortion at high loads. Atomic resolution imaging showed that the observed periodicity changed from a $(\sqrt{3}\times\sqrt{3})R30°$ structure at low loads (repeatedly observed for thiol layers adsorbed on Au(111)) to a (1x1) structure at high load (due to the underlying Au(111)) and back to $(\sqrt{3}\times\sqrt{3})R30°$ when the load was decreased again [18]. The mechanism underlying the observed behavior was unclear and three possibilities have been discussed. (i) desorption of the thiols at high loads and binding to the SFM tip, which are adsorbed to the gold again when the load is decreased; (ii) liquefying of the thiols under the tip pressure so that the tip penetrates the monolayers and images the gold; (iii) lateral displacement of the thiols, which are still adsorbed to the substrate [18]. Our findings exclude the first possibility since this mechanism would be expected to degrade the pattern quality, manifestly not the case here, as observed in Figure 3c. The same argument is true in a weaker sense for the second possibility since it is unlikely that the liquefied C18 and C7 molecules reassemble in a patterned ordered layer, especially at the phase boundaries. Thus some degradation of the pattern might be expected. However, the effect of liquefying the layer is likely to be felt quite locally, i.e. under the tip. We believe that the observation of an essentially unperturbed pattern is most consistent with the third possibility. The displacement is expected to be more difficult for the long chain molecules resulting in a higher friction which is experimentally observed (Figure 3b). This is also consistent with the differences observed between alkylsilanes on mica vis a vis thiols on gold: in the former case, the tip/monolayer interactions tend to be destructive at high force loadings while in the latter case, the monolayer more easily accommodates high pressure form the tip. This derives from the differences in structure between the two types of film: the lateral bonding in the silane-based layers implies that neighboring molecules are essentially connected to a perturbed molecule and thus moves together with it, while in the alkanethiol case the molecules are sufficiently mobile to readjust their positions under pressure.

CONCLUSIONS

We have shown that LFM and FMT provide excellent image contrast for patterned SAMs of alkanethiols of different chain lengths adsorbed on a gold substrate. The observed lower friction over the long chain surface regions correlates with a higher compliance detected for the long chain regions by FMT. This is even true for a patterned SAM consisting surface regions of quite similar chain length , i. e. C18/C16. To clarify the role of the substrate as well as the effects of disorder and coverage complementary measurements, including spectroscopic methods, are in progress. The observation of a reversible image contrast reversal in the friction map as the applied load is increased and decreased again strongly supports an underlying mechanism in which the thiols are adsorbed to the substrate but laterally displaced/bent.

ACKNOWLEDGMENTS

G. B. wishes to thank the Deutsche Forschungsgemeinschaft (DFG) and the Los Alamos National Laboratory (LANL) for the financial support. Work at LANL was supported by the

Department of Energy, Office of Basic Sciences and LANL Chemistry LDRD Funding. S. R. wishes to thank LANL for a Director's Funded Post-doctoral Fellowship.

REFERENCES

1. J. Seto, T. Nagai, C. Ishimoto and H. Watanabe, Thin Solid Films **160**, 453 (1985).
2. K. Deng, R. J. Collins, M. Mehregany and C. N. Sukenik, J. Electrochem. Soc. **142**, 1278 (1995).
3. I. Willner, R. Blonder and A. Dagan, J. Am. Chem. Soc. **116**, 9365 (1994).
4. A. Ulman, Introduction to Ultrathin Organic Films, (Academic Press, Inc., San Diego, (1991), pp. 339-362.
5. J. L. Wilbur, E. Kim, Y. Xia and G. M. Whitesides, Adv. Mat. **7**, 649 (1995).
6. T. J. Gardner, C. D. Friesbie and M. S. Wrighton, J. Am. chem. Soc. **117**, 6927 (1995).
7. R. M. Overney, E. Meyer, J. Frommer, H. J. Güntherodt, M. Fujihira, H. Takano and Y. Gotoh, Langmuir **10**, 1281 (1994).
8. X. Xiao, J. Hu, D. H. Charych and M. Salmeron, Langmuir **12**, 235 (1996).
9. C. D. Frisbie, L. F. Rozsnyai, A. Noy, M. S. Wrighton and C. M. Lieber, Science **263**, 2071 (1994).
10. J. L. Wilbur, H. A. Biebuyck, J. C. MacDonald and G. M. Whitesides, Langmuir **11**, 825 (1995).
11. M. Radmacher, R. W. Tillmann, M. Fritz and H. E. Gaub, Science **257**, 1900 (1992).
12. R. M. Overney, T. Bonner, E. Meyer, M. Rüetschi, R. Lüthi, L. Howald, J. Frommer, H.-J. Güntherodt, M. Fujihira and H. Takano, J. Vac. Sci. Technol. B **12**, 1973 (1994).
13. A. Kumar and G. M. Whitesides, Appl. Phys. Lett. **63**, 2002 (1993).
14. This is an estimate only assuming a force constant of 0.1 N/m for the used cantilever according to the value provided by the manufacturer.
15. A. Noy, C. D. Frisbie, L. F. Rosznyai, M. S. Wrighton and C. M. Lieber, J. Am. Chem. Soc. **117**, 7943 (1995).
16. C. D. Bain, Troughton E. B., Y.-T. Tao, J. Evall, Whitesides, G. M. and R. G. Nuzzo, J. Am. Chem. Soc. **111**, 321 (1989).
17. M. D. Porter, T. B. Bright, D. L. Allara and C. E. D. Chidsey, J. Am. Chem. Soc. **109**, 3559 (1987).
18. G.-y. Li and M. B. Salmeron, Langmuir **10**, 367 (1994).

HARDNESS AND DEFORMATION MECHANISMS OF HIGHLY ELASTIC CARBON NITRIDE THIN FILMS AS STUDIED BY NANOINDENTATION

S.V. HAINSWORTH, H. SJÖSTRÖM†, T.F. PAGE & J-E. SUNDGREN†

Materials Division, Department of Mechanical, Materials and Manufacturing Engineering, The University of Newcastle, Newcastle upon Tyne, UK
†Thin Films Division, Department of Physics, Linköping University, Linköping, Sweden.

Carbon nitride (CN_x) thin films ($0.18<x<0.43$), deposited by magnetron sputtering of C in a N_2 discharge, have been observed to be extremely resistant to plastic deformation during surface contact (i.e. exhibit a purely elastic response over large strains). Elastic recoveries as high as 90% have been measured by nanoindentation. This paper addresses the problems of estimating Young's modulus (E) and hardness (H) in such cases and shows how different strategies involving analysis of both loading and unloading curves and measuring the work of indentation each present their own problems. The results of some cyclic contact experiments are also presented and possible deformation mechanisms in the fullerene-like CN_x structures discussed.

INTRODUCTION

CN_x films are predicted to have high hardnesses and high elastic moduli if the β-C_3N_4 structure can be realised [e.g. 1]. In attempts to synthesise this structure, a series of films were grown using unbalanced magnetron sputtering of C in N_2 discharges on Si (100) substrates where the native oxide had been removed by thermal desorption. The substrate temperatures were maintained in the range 300°C-600°C and the substrate was kept at a floating potential of approximately -60V. The microstructure and chemical nature of the films have been characterised by HRTEM and XPS and the mechanical properties of the films assessed using low-load continuously recording indentation techniques (nanoindentation) (for details see [2,3]).

This paper now critically assesses the ways in which the mechanical properties of these CN_x films may be determined from nanoindentation data, coupled with AFM imaging, and accounts for these properties in terms of possible structurally-controlled deformation mechanisms.

METHODS FOR DETERMINING MECHANICAL PROPERTIES

Conventional methods for determining hardness and modulus

Hardness values can be calculated from nanoindentation load-displacement curves by several different methods. The most common procedure is to analyse the unloading portion of the nanoindentation load-displacement curve using the method proposed by Oliver & Pharr [4] which was based on an earlier approach by Doerner & Nix [5]. In this analysis, hardness is simply determined from the load divided by the contact area, i.e. $H = P_{max}/A_p$ where H is the hardness, P_{max} is the maximum applied load and A_p is the plastic area of contact. In turn, this contact area is determined from a knowledge of the tip area-to-depth end-shape function multiplied by the plastic contact depth δ_p which can be determined from $\delta_p = \delta_{max} - \delta_s$, where δ_{max} is the maximum indenter displacement at P_{max}, δ_s is the surface flexure calculated from $\delta_s = 0.72\,P_{max}/S$ where S is the contact stiffness calculated from the unloading curve [4]. An elastic contact modulus can also be calculated from the observed unloading stiffness and thus, if the Poisson's ratio of the sample is known, this can be used to calculate the Young's modulus of the sample itself. This method assumes that the unloading curve can simply be described by a power-law fit, which will be discussed in the following section.

<u>Using an approach based on the loading curve</u>

The continuously highly curved unloading curve for CN_x is not easily modelled to provide quantitative data using existing expressions since it cannot be approximated to either a linear fit over the top third [5] or a power-law fit with a single valued exponent [4]. In such cases, an alternative approach is to analyse the loading curve to obtain mechanical properties [6]. The relationship between load and displacement can be given by

$$P = E\left(0.930\sqrt{\frac{H}{E}} + 0.194\sqrt{\frac{E}{H}}\right)^{-2} \delta^2 = K_m \delta^2$$

where P is the load, E and H are Young's modulus and hardness respectively and δ is the displacement. The parameter K_m describes all the necessary materials constants [6,7] and can be matched to a value, K_{exp}, experimentally-determined from a graph of P versus δ^2. This value K_{exp} then allows either E or H to be calculated if the other is either known or is independently estimable.

<u>Using a work of indentation approach</u>

A further way in which hardness can be calculated is by using a *work of indentation* approach [e.g. 8-10] (which is often referred to as "dynamic hardness"). In this case, the hardness is calculated from the plastic work of indentation (i.e. the area enclosed by the nanoindentation loading and unloading curves) divided by the residual volume of the indentation itself. This residual volume can either be measured directly using 3-D topographic mapping techniques such as atomic force microscopy (AFM) or calculated from the load-displacement curve and the indenter geometry using

$$V = \frac{1}{3}(A_p \times \delta_{res})$$

where A_p is the plastic contact area and δ_{res} is the residual indentation depth measured as the final indenter displacement on unloading. However, estimating A_p from the load-displacement curve is subject to the same reservations concerning analysis of the unloading curve as were expressed earlier. Under such circumstances, one approach to estimating A_p is from a knowledge of the tip end-shape function and the maximum indentation depth (δ_{max}). However, since δ_{max} includes a very significant contribution from elastic flexure of the surface, this method is expected to give a sizeable *overestimate* of A_p and too low a value of H. Alternatively, A_p could be estimated from the residual displacement (δ_{res}) but this is likely to provide an *underestimate* especially if an unexpectedly large amount of elastic recovery of the indentation area has occurred (as seems likely with these materials). The previously discussed method of using δ_p from the Oliver & Pharr method is expected to work well for materials with well-behaved unloading curves - but not for CN_x. In such cases, AFM should be able to give good direct estimates of both A_p and the indentation volume but the tip geometry of the AFM cantilever will have to be deconvoluted from the data to ensure maximum accuracy, the z-piezo drive needs to be carefully calibrated and the finite image pixel size can create significant problems. (A fuller discussion of the problems associated with measuring volumes of indentations by AFM is outside the scope of the present paper but will form the basis of later publications).

EXPERIMENTAL

Nanoindentation experiments were performed using a Nano Indenter II^{TM} (Nano Instruments Inc, Knoxville, TN) to various peak loads in the range 1-100mN. The indentation cycles incorporated a hold segment to allow the correction of any thermal drift. The data was processed using proprietary software to produce load-displacement curves. AFM of the indentations was performed using a Park Scientific Instruments M5 scanning probe microscope using contact mode imaging, a proprietary 0.6μm sharp microlever and 256 x 256 pixel images.

RESULTS

Figure 1a shows typical load-displacement curves for indentations into a 300nm CN_x film on (100) silicon, a 300nm TiN film on (100) silicon and, for comparison, the silicon substrate itself. It is immediately clear that whilst δ_{max} is less for the TiN system, the CN_x system shows a very considerable amount of elastic recovery, a much smaller work of indentation and a residual indentation depth of only ~20nm. Figure 1b shows a similar load-displacement curve for a thicker 1μm CN_x film on (100) silicon. If the percentage elastic recovery (%R) is calculated using

$$\%R = \frac{\delta_{max} - \delta_{res}}{\delta_{max}} \times 100$$

then %R is 91% for this indentation and ~85% for the 300nm films.

While the low values of δ_{res} confirm that these films have very high hardnesses, the high values of %R also indicate a high proportion of elastically-recovered strain (~H/E) on unloading.

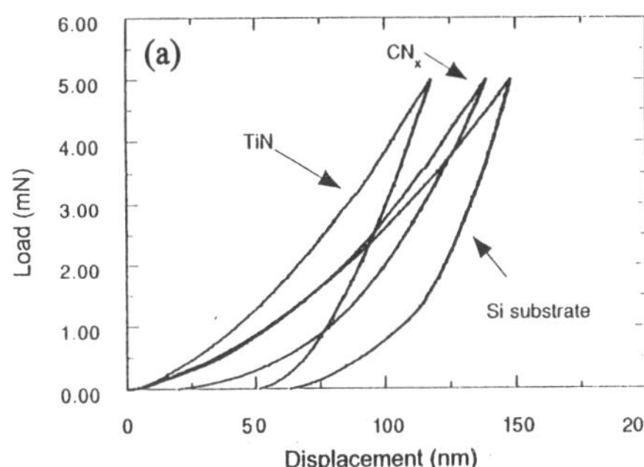

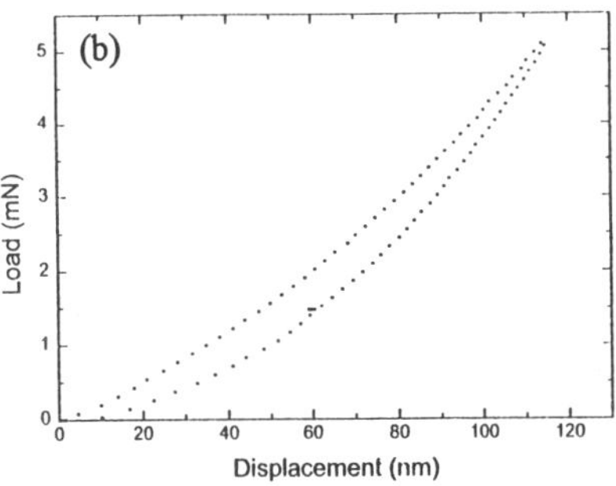

Figure 1: a) Load-displacement curves for a 300nm TiN coating on silicon, a 300nm CN_x coating on silicon and the (100) silicon substrate alone. Note the unloading curve for the CN_x system is highly curved and thus it is difficult to calculate parameters such as hardness and modulus for this system by conventional analysis of the unloading curves. b) shows a load-displacement curve for an indentation in a 1μm thick CN_x film on (100) silicon, note the extremely high %R (~91%) and, compared to both the silicon alone and the TiN coated samples, a smaller work of indentation.

For the 300nm CN$_x$ coating in Fig. 1a, if H is simply calculated from the plastic area associated with δ_{res} =25nm, then H = 326.5GPa, i.e. extremely high. By finding the best-fit value of K_{exp} from the P vs δ^2 plot (i.e. 2.68E+11Nm^{-2}) then a corresponding value of E = 334GPa is obtained which is not unreasonable and gives H/E ~ 1 (which implies that plastic deformation is occurring at, or near, the theoretical strength - something not unreasonable for these materials).

A different approach, described earlier, is to calculate the hardness from the work of indentation (readily measured from the P-δ plot) and the volume of the indentation as measured by AFM (see figure 2). This hardness value can then be used to calculate a value of E consistent with the measured value of K_{exp} for this system. This approach gives H = 24 GPa and E = 212 GPa. These values both appear to be too low - for example, they are less than corresponding values for sapphire - and also give a low value of H/E (indicating a lower fraction of elastically-recovered strain).

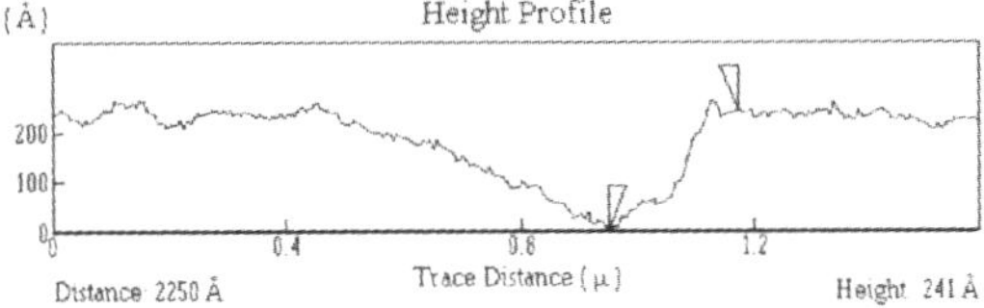

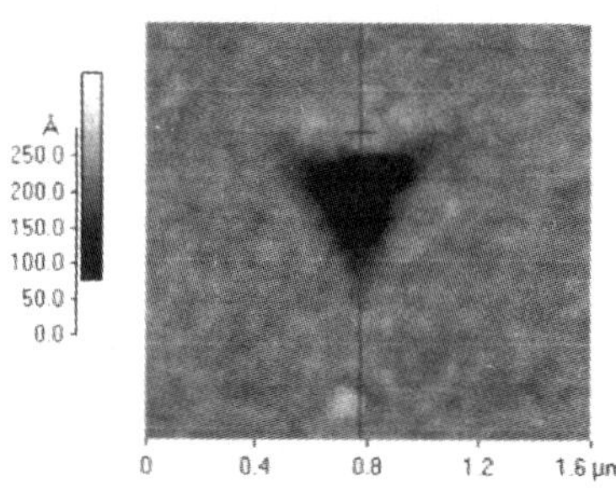

Figure 2: A contact mode AFM image of a 5mN indentation on a 300nm CNx film on (001) silicon. The profile through the indentation establishes that the depth recovery is far more than expected for an indentation of this width. Parameters such as the indentation area and volume can also be calculated from such images.

Alternatively, the hardness can also be calculated from the residual contact area measured by the AFM. This gives a value of hardness of 33 GPa for the indentation shown in figure 2 and a value of 196 GPa for the modulus corresponding to K_{exp}. While this hardness value is slightly higher than the previous estimate, both the modulus and H/E appear too low.

In both latter cases, it seems that the residual indentation depth is a better indicator of H than is the apparent residual contact area. Thus, further studies are in progress to explore the consistency of estimates of A_p between different calculative and measurement techniques. Our objective is to allow the most appropriate means of measuring H and E to also be identified. Our values of H = 326.5GPa and E = 334GPa are probably the best current estimates of property values since they are consistent, at least, with each other (through K_{exp}), with the rank position of the loading curve [6] and with the expected high value of H/E.

CYCLIC INDENTATIONS

Figure 3 shows an indentation where the loading and unloading segments have been cycled 4 times to 90% unload before complete unloading. Some mechanical hysteresis - and thus energy absorption on each cycle - is apparent. However, there is very little increase in either δ_{max} or δ_{res} at the extremes of successive cycles and this suggests that the elastic-plastic response of the films is very stable and that the plastic zone, once formed, supports the load on each cycle without growing significantly larger. With other materials, cycling in this manner often leads to a 'sawtooth' load-displacement fingerprint where the plastic enclave, δ_{max} and δ_{res} all become progressively larger [eg. 4]. Work is progressing at identifying the energy-dissipative mechanisms involved in CN_x and in exploring whether the hysteresis diminishes after some small number of cycles (thus, producing a totally stable deformed structure) as we have observed with silicon [11].

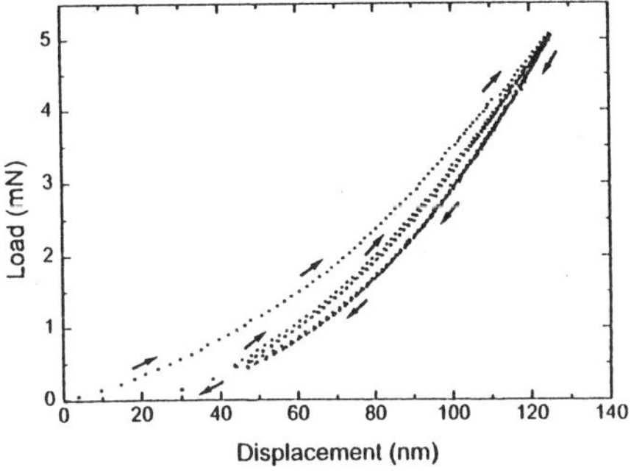

Figure 3: Load-displacement curve for a cyclic indentation on 1μm CN_x on a silicon (001) substrate

PROPOSED DEFORMATION MECHANISMS

Thus far we have shown that as deposited CN_x films show large amounts of elastic recovery (%R=85-91%) and only a small amount of plasticity when they are indented to depths less than the coating thickness. Previously, the structure of these films has been determined to be turbostratic-like [2,3] with the c-axis in the film plane. XPS studies showed that the N atoms in the films were bonded to both sp^2- and sp^3- coordinated carbon and thus the proposed structure of the films is buckled graphitic six rings (with substitutional N) that are crosslinked by sp^3 - coordinated carbon. The samples have a N:C ratio between 0.2-0.35 and thus a significant fraction of the C atoms have other C atoms as their nearest neighbours. However, the presence of substitutional N in the graphite rings of the basal plane should allow a degree of cross-linking to occur between these planes. Thus the films can be considered to have a fullerene-like microstructure. This cross-linking means that the expected mechanical responses of these films should be much stiffer than an equivalent structure with only van der Waals forces acting between the c-planes (e.g. graphite). Figure 4 shows the proposed mechanism by which the films respond to contact stresses and correlates well with molecular dynamics simulations of the response of fullerenes to an applied load [12]

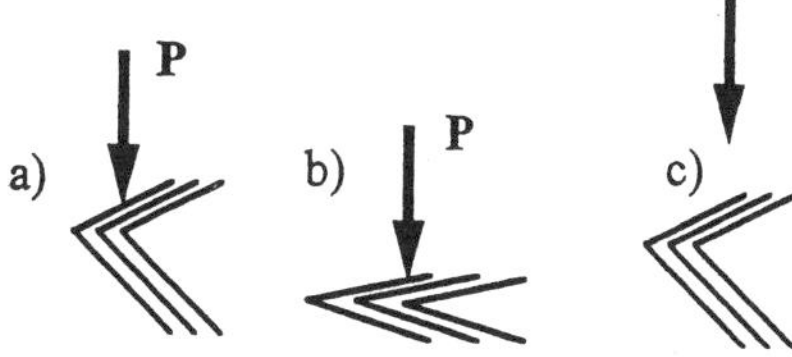

Figure 4: Indentation of cross-linked basal planes a) the load is applied to the basal planes which are compressed to the structure shown in b). c) shows the elastic recovery that occurs once the load is removed and the planes 'spring-back' to their original position.

Eventual significant plastic deformation of the films could be due to a number of mechanisms such as true bond rupture and block shear of the basal planes over one another. At this stage it is unclear exactly how the plasticity of these films arises and further extensive (TEM) studies are needed to explore whether dislocation motion or mechanisms, such as structural densification, are operating at these high contact stresses. This seems a fruitful area for further theoretical modelling.

CONCLUSIONS

The CN_x films studied show considerable elastic recoveries of between 85-91% with the higher %R being observed for $1\mu m$ thick films. The microstructures of the films are 'fullerene-like' with N crosslinking of the basal planes and N incorporation into the basal planes themselves. The proposed indentation mechanism is bond angle deformation of the structure which creates the considerable elastic recoveries observed. Analysing the mechanical properties of these films using nanoindentation is non-trivial due to the continuously highly curved unloading portion of the load-displacement curve. Three alternative methods have been proposed for calculating parameters such as hardness and Young's modulus from either the loading curve or AFM measurements. The difficulties currently associated with these approaches have been highlighted as has the need to ensure the self-consistency of any value of E and H so derived.

ACKNOWLEDGEMENTS

In Sweden, the NUTEK/NFR Materials Research consortium on Thin Film Growth is thanked for financial support. In the UK, the EPSRC are thanked for support of the Newcastle Nanoindenter, the Scanning Probe Microscope and SVH. Travel support from MRS is gratefully acknowledged.

REFERENCES

1.   A.Y. Liu and M.L Cohen, Science **245** 841 (1989)

2.   H. Sjöström, I..Ivanov, M. Johansson, L. Hultman, J.-E. Sundgren, S.V. Hainsworth, T.F. Page and L.R. Wallenberg, Thin Solid Films, **246** 103-109 (1994)

3.   H. Sjöström, L. Hultman, J.-E. Sundgren, S.V. Hainsworth, T.F. Page and G.S.A.M Theunissen, J.Vac. Sci. Technol., A **14** 1-7 (1996)

4.   W.C. Oliver and G. M Pharr, J. Mater. Res., **7** 1564-1583 (1992)

5.   M.F. Doerner and W.D. Nix, J. Mater. Res., **1** 601-609 (1986)

6.   S.V. Hainsworth, H.W. Chandler and T.F. Page, J. Mater. Res., 1996 (in press)

7.   S.V. Hainsworth and T.F. Page, Mat. Res. Soc. Symp. Proc. 1996, (this volume)

8.   N.A. Stillwell and D. Tabor, Proc. Phys. Soc. Lond. **78** 247-274 (1961)

9.   M. Sakai, Acta Metall. Mater., **47** 1751-1758 (1993)

10.   P.C. Twigg, M.R. McGurk, S.V. Hainsworth and T.F. Page, *in Plastic Deformation of Ceramics III* (Eds. R.C. Bradt, C.A. Brookes & J. Routbort), Plenum Publ. Corp., New York, 1995, p219-229.

11.   S.V. Hainsworth, A. J. Whitehead and T.F. Page, as Ref 10, p 173-186

12.   R. Smith, Vacuum, **46** 1195 (1995)

Effect of Film Thickness and Substrate Surface Treatment on Substrate Deformation : DLC on MgO

A. Strojny, E.T. Lilleodden, G. Wang. J.V. Sivertsen, W.W. Gerberich
Department of Chemical Engineering and Material Science, University of Minnesota,
Minneapolis, MN 55455.

ABSTRACT

Diamond-Like Carbon films with thicknesses controlled in the 10 to 100 nm range were sputter-deposited on single crystals of MgO. The substrates were prepared with varying roughness from 5 to 50 nm and varying surface dislocation densities from 100 to 2000 disl./μm^2. Mechanical properties of the film and film-substrate interactions were investigated with an atomic force microscope retrofitted with a diamond indenter.

The load bearing capacity of the substrate increased with increasing film thickness. As the indenter approached the film-substrate interface, load excursions occurred for the thinner films. No excursion was found for the thick film. Load excursions may be attributed to dislocations nucleating at the interface and/or film delamination.

INTRODUCTION

Diamond-like carbon coatings have received increased attention as protective coatings for tribological applications, especially in the magnetic recording industry. Several investigators [1,2] have shown that diamond-like carbon coatings on the order of 100 nm delay the deformation in the underlying substrate and therefore enhance the mechanical properties of the coated system. Coated systems and the nature of the pop-in [3] upon indentation is due to various deformation mechanisms and also dependent on the mechanical properties of the film .

EXPERIMENT

MgO treatments

The substrate material for the thin film deposition were single crystals of MgO cleaved on the 100 plane, as verified by Laue x-ray diffraction. Three substrate surface treatments were chosen which resulted in different surface roughnesses and dislocation densities. Surface roughness was measured with an atomic force microscope. In order to compare the dislocation densities of the substrates, indentations were performed on the MgO with a Triboscope^{TM}. The area and height of the resulting impression was measured and evaluated for geometrically necessary dislocations, ρ_G by [4]

$$\rho_G = \frac{3}{2}\frac{h}{b}\frac{1}{\beta^3 R^2} \tag{1}$$

where R is the measured radius of the indentation, h is the measured depth of the indentation, b is the Burgers vector of MgO (2.98Å) and β is a material constant(=2 for hard materials).

The substrate treatments were, 1) as cleaved 2) cleaved and mechanically polished, 3) cleaved, mechanically polished and chemically etched, and are hereby referred to as cleaved, polished and etched respectively. The average height of the cleavage steps was on the order of

Mat. Res. Soc. Symp. Proc. Vol. 436 © 1997 Materials Research Society

0.5 μm and spaced 1-2 μm apart as verified by optical microscopy. The polished substrates were also annealed at 1300°C for 5 hours. Samples were mechanically polished with various grades of diamond lapping film, finally polished using colloidal silica. Chemical polishing was accomplished using 85% phosphoric acid at 75°C for 3.5 minutes. Table 1 summarizes the measured surface roughnesses and dislocation densities.

Table 1: Surface roughnesses and dislocation densities obtained for various surface treatments on MgO and resulting

Surface treatment	Roughness (nm)	Dislocation Density (disl./μm^2)
polished	3-5	500
etched	5-10	1000
cleaved	30-50	2000

<u>Diamond-Like Carbon deposition</u>

A Facing Target Sputtering (FTS) System with two graphite targets was used to deposit the films at 0.2 mTorr of nitrogen and 0.3 mTorr of argon at a deposition rate of 30Å/min. The voltage applied to the target was 390 V and the AC discharge current was 0.2 A. Prior to film deposition, the specimens were sputter cleaned for 1 minute to remove any surface contamination. Three DLC-films were produced with thicknesses of 10, 30 and 100 nm. The films contained mainly amorphous carbon and approximately 10 at% nitrogen as characterized by Auger electron spectroscopy and Raman spectroscopy.

<u>Indentations</u>

In order to compare the various films and substrate surface treatments, indentations were performed with a Triboscope^{TM}, a load controlled instrument with load and depth resolution of 1 nN and 0.3 nm respectively. the instrument fits on a conventional atomic force microscope and enables the user to scan the surface before and after an indentation test and image the resulting indentation right after. A conventional STM pyramidal Boron-doped diamond tip with a tip radius of 100 nm, as imaged by field-emission gun SEM and an included angle of 72° was used for the indentation tests.

Indentations were performed on the substrates in order to investigate the effects of surface treatments on the load-displacement behavior. Load-displacement curves for polished MgO showed a consistent load excursion or "pop-in" at 750 μN and elastic behavior up to the pop-in (Fig. 1a). Pop-in loads for etched MgO (Fig. 1b) varied from 350 to 750 μN. The increase in depth at pop-in load was greater for the polished samples. Cleaved MgO did not exhibit consistent pop-in behavior.

The film thickness effect on substrate deformation was investigated by indenting films with different thicknesses deposited on the polished substrate. Samples with 10 and 30 nm films showed pop-in at loads of 123 and 220 μN and corresponding increases in depth of 0.019 μm and 0.045 μm respectively. The 100 nm film exhibited no pop-in at loads up to 2000 μN.

Indentations performed on 30 nm films deposited on substrates with all three surface treatments showed pop-in at the same load . The slopes of the load-displacement curves before pop-in were the same for all treatments; after pop-in the slope decreased for the cleaved substrate. The load-displacement curves prior to pop-in showed a hysteresis for all three films. This hysteresis was larger for the etched substrate than for the cleaved and polished substrates.

RESULTS AND DISCUSSION

Figures 1a and 1b show typical load displacement for polished and etched MgO. Polished MgO exhibits elastic behavior below pop-in.

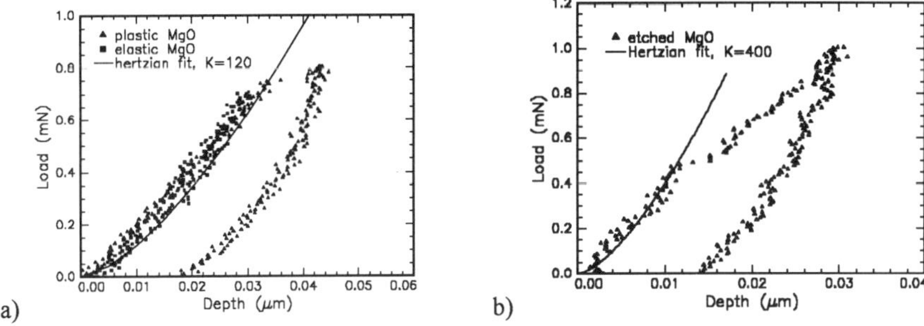

a)
b)

Figure 1: a)Load displacement curves for polished MgO. Below the pop-in the material is elastic and Hertzian. b) Load-displacement curve in etched MgO. Random pop-ins are due to surface roughness

In order to verify Hertzian contact theory for the loading portion below the pop-in, the elastic portion up to pop-in was curve fitted with the Hertzian equation

$$P = K\delta^{3/2} \qquad \text{where } K = c\,R\,E^* \qquad (2)$$

using a least square fitting routine. For polished MgO, K=120 using a reduced indentation modulus E^*=239GPa and c=4/3 for a spherical indenter. An effective tip radius of R=140nm was obtained. This radius is greater than the imaged radius because the indenter was not positioned perpendicular to the surface as seen from the imaged AFM area (Fig. 2a).

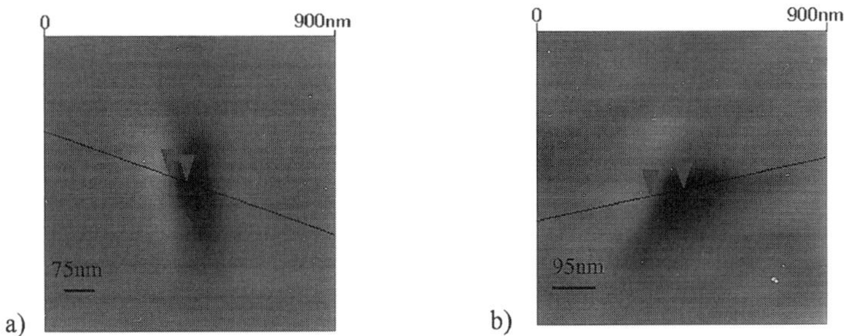

a)
b)

Figure 2: AFM-images. a)Indentation in polished MgO. b) Indentation in etched MgO

The maximum shear stress, τ_{MAX}, at pop-in can then be calculated from the maximum pressure, p_o, underneath the indenter by [6]

$$\tau_{max} = 0.3 p_0 \qquad \text{where} \qquad p_0 = \frac{3}{2}\frac{P}{\pi \delta R} \qquad (3)$$

The critical shear stress, $\tau_{crit.}$, for MgO ranges from 16 to 23.5 GPa depending on the orientation of the indenter to the slip systems. Using equation 3, the max. shear stress, τ_{max}, at pop-in for polished MgO is 22 GPa, which falls within the above range for $\tau_{crit.}$. The length of the pop-in, $\delta_{exec.}$, is related to the number of dislocation loops nucleated, which is calculated by[7]

$$N = \frac{\delta_{exec}}{b} = 70\text{Å}/2.98\text{Å} = 23 \qquad (4)$$

For etched MgO, the load displacement curve (Fig. 1b) showed a less pronounced pop-in with N=13 dislocations. The Hertzian curve fit below the pop-in yields K=400, resulting in an effective radius of 300 nm, clearly not possible based on the AFM image (Fig. 2b) and SEM-imaging of the tip. The maximum shear stress at pop-in was then calculated from the imaged area of the indent. The mean pressure was taken as the load at pop-in divided by the imaged area, resulting in a maximum shear stress of 8.2 GPa. This approach overestimates the shear stress, since imaging an impression with the tip that produced it, underestimates the area of the impression. The effect of chemical treatment changes the surface properties of a material[5] and in the case of MgO, absorption of species on the surface changes the plastic flow behavior. It is possible that this might also produce a residual stress which could change the contact area during indentation. The etched surface, because of its high surface roughness has more sites available for nucleation of dislocations which would decrease the shear stress needed to initiate plastic flow. From the AFM image (Fig. 2b) it can be seen that cracking occurred around the indentation. It is possible that absorbed species pin surface dislocations which may lead to cracking instead of plastic flow.[8]

Figure 3 shows a comparison of load-displacement curves for different film thicknesses.

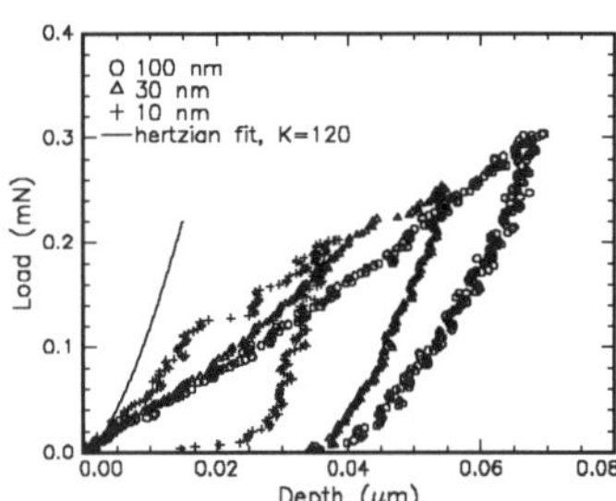

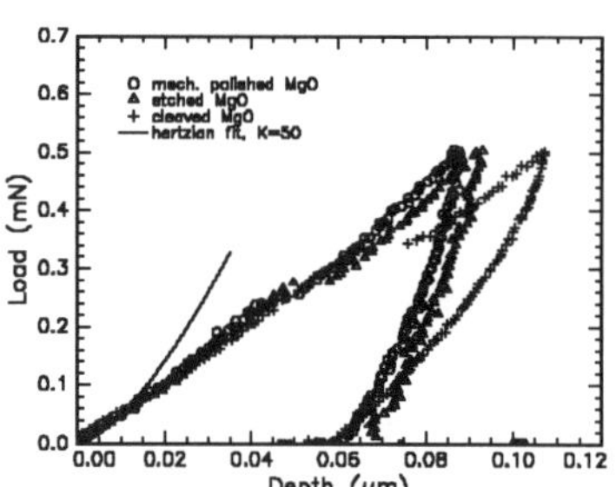

Figure 3: Load-displacement curves for three different film thicknesses

Figure 4: Load -displacement curves for 30 nm films on the three substrates. The curves deviate from Hertzian behavior

The behavior below pop-in deviated from Hertzian behavior, except in the initial portion of the loading curve. Indentation tests performed to a load less then the pop-in load showed hysteresis in the load-displacement curves suggesting that the pop-in was not the onset of plasticity, but could be due to compaction of the film. The maximum shear stress, τ_{max}, at pop-in for the 10 and 30 nm film was calculated using equation (4) and the imaged area rather than A=$\pi\delta$R and found to be 5.8 GPa and 6.0 GPa respectively. These values are about one-third of the critical shear strength for dislocation nucleation in MgO. For native oxide films, Gerberich et. al. [7] have shown that it is possible to nucleate dislocations at the oxide/substrate interface at stresses lower than the critical strength of the specimen, which is given by $\mu_s/2\pi$. This analysis predicts a critical load to nucleate dislocations through the following equations

$$\frac{P_{UYP}}{E^* R^2} = \frac{1}{6}\left(\frac{\pi\tau_c / E^*}{\tau_R / p_o}\right)^3 \tag{5}$$

$$\frac{\tau_R}{p_o} = \frac{1}{\sqrt{18}}\left\{-2(1+v)\left[1 - \frac{h_{fo}}{a}\tan^{-1}\left(\frac{a}{h_{fo}}\right)\right] + 3\left(1 + \frac{h_{fo}^2}{a^2}\right)^{-1}\right\} \tag{6}$$

where P_{UYP} is the load to nucleate dislocations, h_{fo} the film thickness and a the contact radius given by $(\delta_{exec}.R)^{1/2}$, using E^*=239 GPa, R=140 nm, τ_c= 16GPa, v=0.233 a nucleation load P_{UYP} of 432 and 260 µN were obtained for the 10 and 30 nm film respectively. For the 10 nm film, the difference between the observed and calculated value is 72%. This discrepancy could be due to load excursion caused by cracking of the film and/or delamination of the film prior to dislocation nucleation. For the 30 nm film the difference between the observed and calculated value is 15%, which is within the experimental error. Therefore the pop-in could be due to dislocation nucleation along the interface or in the substrate. Further experiments need to be performed in order to establish the driving force for the pop-in. The predicted nucleation load due to the presence of the 10 and 30 nm film is less than the nucleation load for the bare substrate, therefore the 10 and 30 nm film are ineffective as a protective coating. The predicted nucleation load for the 100 nm film is 1176 µN. Experiments to loads of 2000 µN showed no pop-in, therefore the 100 nm film increases the load bearing capacity of the substrate.

The maximum shear stresses, τ_{max} , at pop-in for the 30 nm films deposited on the three different substrates, calculated with equation 4 and the imaged area of the indent was determined to be ~5.8 GPa in all three cases. This observation suggests that there is a common mechanism governing the pop-in. Since the three films have very different substrate dislocation densities and roughnesses as seen in table 2, surface treatment seems to have little effect prior to the pop-in event, even though surface treatment had a large effect on the deformation of the bare substrate.

Table 2:Surface roughnesses of the film, substrate and measured shear stress at pop-in for 30 nm DLC-films

Surface treatment	τ_{max} (GPa)	DLC roughness (nm)	MgO roughness (nm)
polished	6.0	1.6	2.8
cleaved	5.5	1.1	5.7
etched	5.8	3.5	7.1

AFM-images of indents performed on 30 nm films on the three substrates, showed cracks emanating from the corners, which may be a possible cause for the pop-in. By equations 5 and 6, all films failed before the bare substrate would have failed. As a result of failure of the films, the load bearing capacity of the substrates is not altered by surface treatments.

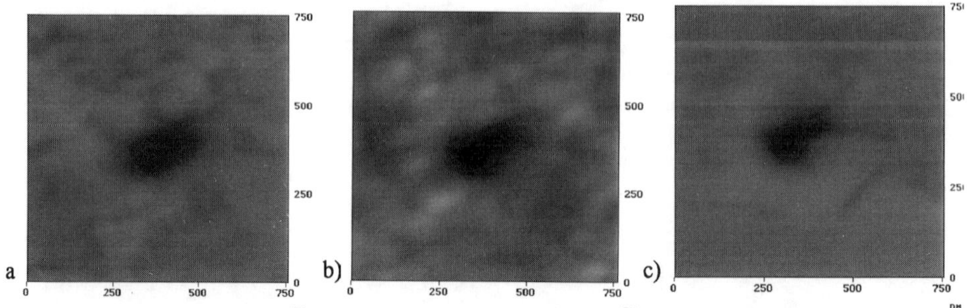

Figure 5: AFM-images of indents on 30 nm DLC-films on a) polished b) etched and c) cleaved substrates of MgO.

SUMMARY

1. Surface treatment changed the deformation behavior of the MgO-substrate.
2. An increase in film thickness increased the load bearing capacity of the substrate, however the 10 nm and 30 nm failed below the critical shear strength of the substrate.
3. Surface roughness has little effect on the failure mechanism of the films.

ACKNOWLEDGMENTS

The author would like to thank J. Hoehn for calculations on the indentation modulus of MgO, MTS Test systems, Inc., Mpls, St. Jude Medical, St. Paul and the Center for Interfacial Engineering at the Univ. of Minnesota under NSF/CDR-8721551 for financial support.

REFERENCES

1. S.V. Hainsworth, T. Bartlett T.F. Page, The Nanoindentation response of systems with thin hard carbon coatings, Thin Solid Films, 236 (1993) 214-218
2. J.C. Knight, A.J. Whitehead, T.F. Page, Nanoindentation experiments on some amorphous hydrogenated carbon (a-C:H) thin films on Silicon, J. of Mat. Sci, 27 (1992) 3939-3952.
3. T.F. Page, S.V. Hainsworth, Using Nanoindentation techniques for the Characterization of coated systems: a Critique, Surface and Coatings Technology, 61 (1993) 201-208
4. M. S. DeGuzman, G. Neubauer, P. Flinn, W.D. Nix, Role of Indentation Depth on the Measured Hardness of Materials, Mat. RES> Soc. Symp. Proc. 308(1993) 613-618
5. S.J. Bull, T.F Page, Chemomechanical effects in ion-implanted MgO, J. of Physics D, 22, no.7(1988) 941-947
6. K.L. Johnson, Contact Mechanics, Cambridge Univ. Press,, Cambridge (1985).
7. W.W Gerberich, J.C. Nelson, E.T Lilleodden, P. Anderson, J.T. Wyrobek, Indentation induced dislocation nucleation: The initial yield point, *to be published 1996*
8. M.O. Guillou, J.L. Henshall, R.M. Hooper, G.M. Carter, Indentation hardness and fracture in single crystal magnesia, zirconia and silicon carbide, Br. Ceramic. Proc. (UK), no. 49, 191-202.
9. A.B. Mann, J.B. Pethica, W.D. Nix, S. Tomiya, Nanoindentation of epitaxial films: A study of pop-in events, Mat. Res. Soc. Symp. Proc., 356 (1995) 271-276

HARD AMORPHOUS HYDROGENATED CARBON FILMS DEPOSITED FROM AN EXPANDING THERMAL PLASMA

J.W.A.M. GIELEN, M.C.M. VAN DE SANDEN, W.M.M. KESSELS and D.C. SCHRAM
Eindhoven University of Technology, Department of Physics,
PO Box 513, 5600 MB Eindhoven, The Netherlands - e-mail: m.c.m.v.d.sanden@phys.tue.nl

ABSTRACT

Diamondlike amorphous hydrogenated carbon is deposited from an expanding thermal argon/acetylene plasma. It is observed that the film quality improves with increasing deposition rate. To obtain the best material quality the admixed acetylene flow has to be of comparable magnitude as the argon ion flow from the plasma source (critical loading). A new method to determine the ion density in an argon/acetylene plasma, by probe measurements, is presented. They reveal that the deposition during critical loading is governed by radicals. It is suggested that acetylene is dissociated once and that the C_2H radical is formed dominantly.

INTRODUCTION

Amorphous hydrogenated carbon (a-C:H) is a non-crystalline form of carbon containing up to 40 % of hydrogen. It is generally produced via a Plasma Enhanced Chemical Vapour Deposition (PECVD) technique. The properties of the deposited a-C:H films can be adjusted by variation of the plasma deposition parameters and may vary from soft-polymerlike to hard-diamondlike. The C-C bonding types, i.e. the sp^2 and sp^3 bonds, determine the actual properties. The sp^3 bonds are responsible for the beneficial mechanical properties, which are for diamondlike material: high hardness (10-20 GPa) and high elastic modulus, low friction, chemical inertness and transparency to infrared light. The optical and electronic properties are mainly due to the sp^2 bonded carbon sites, which decrease the optical bandgap by enhanced cluster formation. Possible applications of a-C:H films are protective- and anti reflection coatings on glass plates in e.g. bar-code laser scanner devices or in flat panel displays [1].

The deposition of a-C:H is performed by several PECVD techniques [2]. All techniques have in common that a substrate bias of at least 50 V per deposited carbon atom is needed to obtain diamondlike material [1]. This ion bombardment results in the displacement of hydrogen and carbon atoms. The hydrogen recombines with atomic hydrogen from the plasma and desorbs. On the surface dangling bonds are created. These may either recombine with each other or bond to hydrocarbon radicals from the plasma. Both effects result in more cross-linking and denser diamondlike material [3]. An interesting question now is whether this ion energy is always necessary to obtain diamondlike properties or that plasma-surface chemistry also could play a role. In e.g. the synthesis of diamond, atomic hydrogen from the plasma passivates and etches hydrogen and sp^2 bonded carbon from the growing film and is responsible for the diamond nature of the deposited films. In this paper the quality of diamondlike a-C:H will be discussed when films are grown without additional substrate bias. The quality will be related to the plasma chemistry, i.e. the dissociation mechanism of acetylene molecules will be described.

a-C:H FILM DEPOSITION

Deposition of the a-C:H films is performed with an expanding thermal plasma set-up. In a cascaded arc plasma source a flowing argon plasma (flow = 100 standard cm^3/s (sccs); pressure

Mat. Res. Soc. Symp. Proc. Vol. 436 © 1997 Materials Research Society

≈ 0.5 bar; electron density $\approx 10^{22}$ m^{-3}; electron temperature ≈ 1 eV) is created. The plasma expands into a vacuum chamber, which is at low pressure (0.25 mbar). The electron density and temperature in pure argon decrease to $\approx 10^{19}$ m^{-3} and ≈ 0.2 eV, respectively. In the expansion zone acetylene is admixed. The films are deposited on crystalline silicon and glass substrates, mounted on a grounded holder, at low temperature (≤ 100 °C); thicknesses are about 2 µm. Due to the low electron temperature (≈ 0.2 eV) the sheath potential is low, i.e. on the order of 1 V, and thus deposition takes place without additional ion acceleration. Two process parameters are varied. The electrical current through the plasma source is changed from 22 to 88 A. This implies that the ionisation degree of the argon flow varies from 5 to 25 %. Also the admixed acetylene flow is varied from 2 to 20 scc/s. An extensive description of the set-up is given elsewhere [4].

RESULTS

The deposited films are characterised by its hardness and infrared refractive index. Via Nano-indentation the hardness is determined. The refractive index and thickness of the films are obtained via infrared absorption spectroscopy. In Figure 1a the hardness, which is a measure for the film quality, is given as a function of the infrared refractive index. It clearly correlates with the infrared refractive index. The film quality changes from soft-polymerlike (4 GPa), at low refractive index, to hard-diamondlike (13 GPa), at higher indices. The maximum observed hardness of about 13 GPa is in the range known for diamondlike material [1]. However, note that our diamondlike properties are obtained without applying a substrate bias. At the best film quality other properties like the hydrogen content, carbon-hydrogen bonding structure and mass density are also found to be in the diamondlike material range [5,6].

As is obvious from Figure 1a film quality improvement is equivalent to maximising the refractive index. In Figure 1b the refractive index is given as a function of the process parameters varied, i.e. the arc current and admixed acetylene flow. For each arc current applied a similar trend is observed in the refractive index with respect to the acetylene flow: with increasing acetylene flow the refractive index first increases until a maximum is reached, then a decrease occurs. From Langmuir probe measurements in a pure argon plasma the argon ion density as a function of the radial position is determined for various arc currents. These densities are transformed into ion flows by integrating over the beam area assuming a beam velocity of 1000 m/s. For arc currents of 22, 48 and 75 A argon ion flows are calculated of about 5, 12 and 25 scc/s. The maximum in the refractive index is observed at those acetylene flows which are of

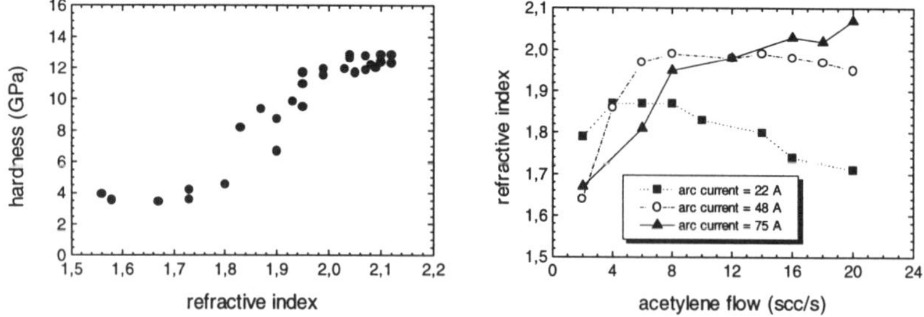

Figure 1 :   a.)   The film hardness vs. the refractive index.

b.)   The refractive index vs. the admixed acetylene flow at various arc currents.

the same magnitude as the argon ion flows appearing from the arc source. For the arc current of 75 A the maximum is not yet observed as the admixed acetylene flow is lower than 25 scc/s.

To understand the observed behaviour of the refractive index under plasma variation, the plasma chemistry has to be studied. Dissociation of the acetylene does not take place via electron induced processes as the electron temperature is too low (≈ 0.2 eV). Also dissociation by heating is excluded as it is slower than the here presented dissociation mechanism. The acetylene is dissociated in two steps. First a charge transfer reaction occurs between an argon ion and an acetylene molecule:

$$Ar^+ + C_2H_2 \rightarrow Ar + C_2H_2^+ \tag{1}$$

at a rate $k_{ce} \approx 10^{-16}$ m^{3}s^{-1} [7]. The formed acetylene ion rapidly recombines dissociatively with an electron:

$$\begin{aligned} C_2H_2^+ + e^- &\rightarrow C_2H + H \\ &\rightarrow CH + CH \\ &\rightarrow C_2 + H_2 \end{aligned} \tag{2}$$

at a rate $k_{dr} \approx 3 \cdot 10^{-13}$ m^{3}s^{-1} [8]. In this recombination process many different radicals are formed. Depending on the argon ion density the formed radicals can be ionised and dissociated further. Note that the ionisation degree decreases as a result of the proposed dissociation mechanism.

In literature no branching ratios are presently known for the various reaction paths in Eq. (2). From emission spectroscopy it is determined that the creation of C_2H is far dominant. Von Keudell [9] has shown that this radical plays an important role in the deposition of diamondlike a-C:H from an argon/acetylene plasma. The C_2H radical is chemisorbed on the growing surface in an sp^2 hybridised bond, with each carbon atom either bonded to a hydrogen or carbon atom or containing a dangling bond. Impact of energetic ions or hydrogen atoms transform the layer into its diamondlike form. In our plasma hydrogen atoms are created too during the dissociation of acetylene into the C_2H radical. These hydrogen atoms can either passivate dangling bonds on the surface or abstract hydrogen from the surface, depending on the substrate temperature. The influence of the formed hydrogen molecules on the plasma will be discussed later. No energetic ions are available for surface modification as there is no substrate biasing. Nevertheless, diamondlike a-C:H is obtained. This suggests that plasma-surface interactions may be as important as surface modification by energetic ion bombardment.

Qualitatively, the effect of arc current variation and acetylene flow admixture on the plasma composition is best explained from Eq. (1) and (2). When at a fixed arc current the acetylene flow is smaller than the argon ion flow from the source (underloading), enhanced dissociation of the radicals formed in Eq. (2) occurs. When the acetylene flow is equal to the argon ion flow from the source (critical loading), the acetylene can be dissociated only once, as for each dissociation step one argon ion is needed. Further increment of the acetylene flow (overloading) results in non-dissociated acetylene as there are not enough ions for full dissociation. Therefore, Figure 1b suggests that critical loading of the argon plasma with acetylene is necessary for the highest film quality and results in a one step dissociation. Both underloading and overloading proof to be unbeneficial for film quality.

As is observed from Eq. (1) and (2) the ion density decreases with increasing acetylene flow. To determine the ion density Langmuir probe experiments [10] have been performed. Measuring complete probe characteristics (probe current vs. probe voltage) in an argon/acetylene plasma by the method described by Flender at al. [11] is in our case not possible. On the probe an insulating film is grown at a high deposition rate (up to 75 nm/s), which changes the surface conditions continuously.

We have developed a new procedure to determine the argon ion density. We measure the ion saturation current (I_{ion}) which is related to the ion density (n_{ion}) by

$$I_{ion} \propto n_{ion} c_s .$$
(3)

In this equation c_s is the Bohm velocity which depends on the electron temperature and the mass of the collected ions. In our plasma the ion saturation current is dominated by argon ions as molecular hydrocarbon ions recombine fast into hydrocarbon radicals. By measuring the ion saturation current as a function of time we obtain time dependent information on the ion density. However, the measured current is influenced by variations in the probe conditions due to deposition of a-C:H.

In Figure 2 a typical example of the measured absolute ion saturation current as a function of time is given. At t_1 a pure argon plasma is ignited and a constant ion current is measured. At t_2 acetylene is admixed, which immediately decreases the current as ions are consumed by reaction Eq. (1). Although the plasma does not change, the probe current is observed to increase, which is caused by a change of the probe surface by deposition of a-C:H. At t_3 the acetylene is switched off and the ion saturation current increases. The plasma composition is the same as at t_1 but the measured ion saturation current is larger. This is due to the a-C:H film on the probe surface. It is observed that the ion current increases slightly with time. This might be due to changing probe conditions by either heating or modification by the argon plasma. Admixture of oxygen to the plasma (t_4) cleans the probe. When the current is constant, oxygen is switched off (t_5) and the ion saturation current in pure argon is again the same as at t_1. This demonstrates that the deposition of a-C:H does not change the probe irreversibly.

The argon ion density in an argon/acetylene environment is determined from the ion saturation current in argon/acetylene and pure argon under equal probe conditions: this is the case at t_3. In argon/acetylene the current is easily read from the measured current - time graph as the last current before switching off acetylene. As the current in a pure argon plasma is not immediately constant at t_3 a linear extrapolation to this point is performed. The ratio of the probe current for the argon/acetylene plasma and the pure argon plasma at t_3 is a measure for the argon ion density in an argon/acetylene environment. Multiplication of this ratio with the argon ion density in a pure argon plasma reveals the ion density in the argon/acetylene environment. The

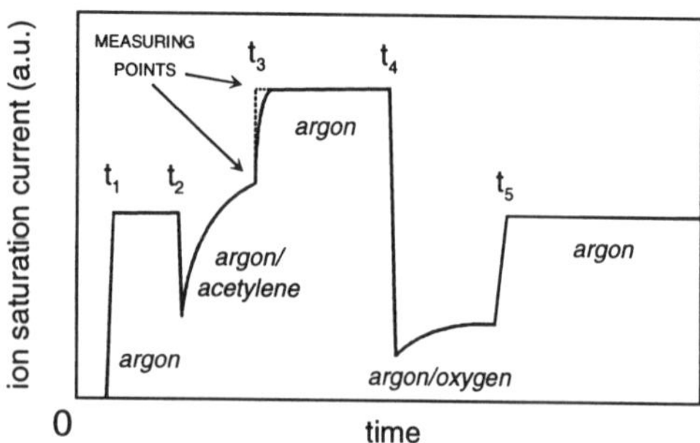

Figure 2 :   The measured ion saturation current vs. time for various plasma conditions (t_1: argon; t_2: argon/acetylene; t_3: argon; t_4: argon/oxygen; t_5: argon).

only assumption made is that the electron temperature is constant in all cases and that only argon ions are collected.

Via the described procedure the argon ion density is determined as a function of the admixed acetylene flow, for one specific arc current (48 A). The probe is positioned half way the expansion vessel at 30 cm from the arc exit and at $x = 25$ cm from the acetylene injection point. The result is shown in Figure 3. The argon ion density decreases from 2×10^{19} m^{-3} to 2×10^{18} m^{-3} when the admixed acetylene flow increases from 0 to 10 scc/s. An approximation for the description of the argon ion density with varying acetylene density is given by an exponential decay:

$$n_{ion}(n_{ac}) = n_{ion}(0)\exp\left[-\frac{k_{ce}x}{v}n_{ac}\right] \tag{4}$$

with n_{ion} the ion density and n_{ac} the acetylene density, which are calculated from their respective partial pressures; v is the beam velocity. The acetylene flow at which the ion density has decreased to 1/e times the original value, i.e. 7.4×10^{18} m^{-3}, is calculated from Eq (4) and is about 6 scc/s (a typical beam velocity of 1000 m/s is assumed). Comparing this result to Figure 3 the ion density is found to be about 7×10^{18} m^{-3} at an admixed acetylene flow of 6 scc/s.

The mean free path (v/nk_{ce}) for the acetylene molecules to react with argon ions for critical loading is about 20 cm ($v \approx 1000$ m$/$s, $n_{ion} = 6 \times 10^{19}$ m^{-3} at an acetylene flow of 12 scc/s). From Figure 3 it is determined that the ion density at 25 cm from the acetylene injection point has decreased more than 90 % for an acetylene flow of 12 scc/s. Enhanced dissociation of the formed radicals now is unlikely as no dissociative capacity is available anymore. This supports the suggestion that the acetylene needs to be dissociated once and that the formed radicals are responsible for good diamondlike material.

In the reasoning above it is assumed that only reactions with acetylene decrease the ion density. However, also molecular hydrogen, which is formed by abstraction of hydrogen from the growing film or via recombination of atomic hydrogen in the plasma with wall adsorbed hydrogen, can decrease the ion density. If this would be a dominant process, no full dissociation of acetylene occurs. But in this case the amount of hydrogen molecules formed is too low to explain the loss in ionisation observed in Figure 3. This observation supports the statement in [6] that full dissociation of acetylene occurs and that the formed C$_2$H radicals have a sticking probability of about 5 %.

More support for the suggestion that no surface modification by energetic ions is needed for the deposition of diamondlike a-C:H is obtained from a growth rate analysis. At increased

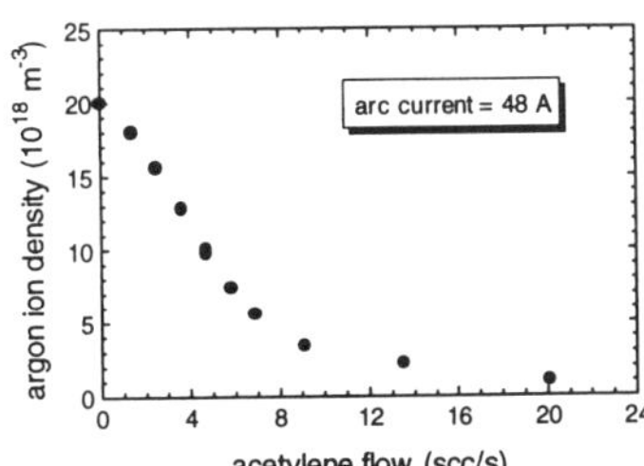

Figure 3 : The argon ion density vs. the admixed acetylene flow.

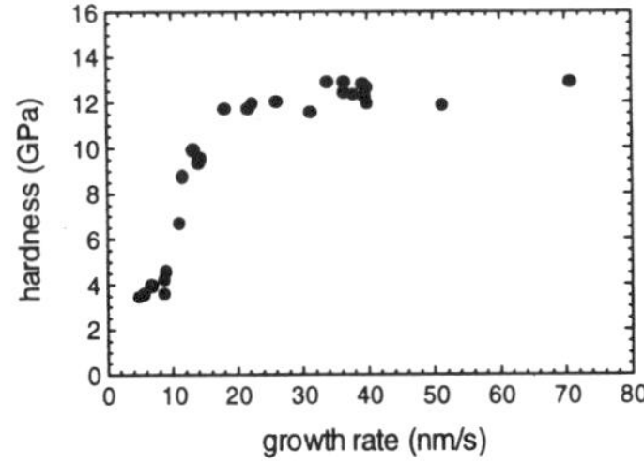

Figure 4 :   The hardness of the a-C:H films vs. the growth rate.

acetylene admixture the growth rate is observed to increase. The ion density then decreases. In Figure 4 the quality of the a-C:H material, given by its hardness, is plotted as a function of the growth rate. The quality is found to increase at increasing growth rate. Especially remarkable is that the highest rates are obtained at the highest acetylene flows and the lowest ion densities. Growth rate increment just means an increment of the C_2H radical flux towards the growing film surface, which only influences the film quality marginally. Currently, a maximum growth rate of 75 nm/s is obtained.

CONCLUSIONS

The use of an expanding thermal plasma allows the deposition of high quality diamondlike a-C:H. It is observed that increment of the deposition rate is accompanied by quality improvement. The best diamondlike a-C:H material quality is obtained when the argon plasma is critically loaded with acetylene. Then a one step dissociation of acetylene occurs. Under these conditions primarily C_2H is formed and is considered to determine the film quality. The plasma-surface chemistry is suggested to determine the film quality which implies that no substrate bias is needed for surface modification.

ACKNOWLEDGEMENT

The authors would like to thank E.H.A. Dekempeneer from VITO (Mol, Belgium) for performing the Nano-indentation measurements. D.A.J. Wouters is thanked for his part in the probe measurements. The Foundation of Fundamental Research on Matter (FOM) and the Royal Netherlands Academy for Arts and Science (KNAW) are thanked for their financial support.

REFERENCES

[1]   J. Robertson, Surf. Coat. Technol. **50**, p. 185 (1992).
[2]   Y. Catherine in <u>Diamond and Diamondlike Films and Coatings</u>, edited by R.E. Clausing et al. (NATO ASI Series B 266, Plenum Press, New York, 1991), p. 173.
[3]   A. von Keudell in <u>Film Synthesis and Growth Using Energetic Beams</u>, edited by H.A. Atwater et al. (Mat. Res. Soc. Proc. 388,Pittsburgh, PA, 1995), p. 355.
[4]   J.W.A.M. Gielen, M.C.M van de Sanden, P.R.M. Kleuskens and D.C. Schram, Plasma Sources Sci. Technol., accepted for publication.
[5]   J.W.A.M. Gielen, P.R.M. Kleuskens, M.C.M. van de Sanden, L.J. van IJzendoorn, D.C. Schram, E.H.A. Dekempeneer and J. Meneve, submitted for publication.
[6]   M.C.M. van de Sanden, R.J. Severens, J.W.A.M. Gielen, R.M.J. Paffen and D.C. Schram, Plasma Sources Sci. Technol., accepted for publication.
[7]   W.B. Maier, J. Chem. Phys. **42**, p. 1790 (1965).
[8]   P.M. Mul and J.W. McGowan, Astrophys. J. **237**, p. 749 (1980).
[9]   A. von Keudell, <u>Growth mechanisms during the plasma enhanced chemical vapour deposition of hydrocarbon films, investigated by in situ ellipsometry</u>, Thesis, Max - Planck Institut für Plasmaphyisk, Garching bei München, 1996.
[10]   N. Hershkowitz in <u>Plasma Diagnostics 1</u>, edited by O. Auciello and D.L. Flamm (Academic Press Inc., Boston, 1989)
[11]   U. Flender and K. Wiesemann, Plasma. Chem. Plasma. Proces. **15**, p. 123, (1995).

TRIBOLOGICAL PROPERTIES OF TETRAHEDRAL CARBON FILMS DEPOSITED BY FILTERED CATHODIC VACUUM ARC TECHNIQUE

X. Shi, B. K. Tay, D. I. Flynn and Z. Sun
School of Electrical and Electronic Engineering
Nanyang Technological University
Singapore 639798

ABSTRACT

Ta-C films have been deposited using FCVA technique. The hardness and Young's modulus of the films on both silicon and sapphire substrates are determined by an ultra low load depth sensing nanoindenter to examine their dependence on the carbon ion energy. An optimum ion energy around 80 to 90 eV has been found, which coincides with the energy at which the sp^3 content and film density reach maximum values. At this ion energy, the hardness, modulus and critical load of a 60 nm film on sapphire exhibit maximum values of 60 GPa, 580 GPa and 7 mN, respectively, whilst the frictional coefficient shows a minimum of 0.16.

I. INTRODUCTION

Filtered Cathodic Vacuum Arc (FCVA) is a promising technique for producing high quality hard thin films [1-7]. It employs electro-magnetic and mechanical filtering techniques to remove unwanted macroparticles and neutral atoms. Only ions within a defined energy range reach the substrate, thus producing films with good controllability and reproducibility [4-8,10-12]. Much research has been performed by various groups [1-7] using this FCVA technique to deposit Tetrahedral Amorphous Carbon (ta-C) films. The term ta-C is used to describe the nature of this amorphous carbon films with a large percentage of sp^3 bonding (up to 90%) [5,7,11], although recent research has showed that nano diamond crystallite may exist in ta-C films under certain deposition conditions [13]. The properties of ta-C films have been strongly correlated to the energy of the incident carbon ions during deposition [11-12]. The property of the film-substrate interface can also be influenced by the ion energy. Therefore by carefully controlling the ion energy achieved by dc or rf substrate bias, good interface qualities and desired film properties can be obtained. The high sp^3 content in the ta-C films results in unique properties such as extremely high hardness and Young's modulus, chemical inertness, high electrical resistivity and low coefficient of friction. As a result of these favorable properties the ta-C films could be useful for many electronic, optical and mechanical applications [8-10]. This film is particularly promising as a dense and thin protective layer with extremely high hardness and modulus for hard disk head and media. However, it is very difficult to derive the hardness and modulus values for such hard and very thin films [14].

In this paper, we report on the tribological properties of ta-C films as a function of impinging carbon ion energy. In particular, better tribological measurements are obtained for ta-C films with ultra thin thickness by using a continuous stiffness technique developed by Nano Instruments, Inc. The critical load and frictional coefficient are also reported on these films by the scratch method.

II. EXPERIMENTAL METHODS

Fig.1 shows the schematic diagram of the FCVA deposition system. The details of the system is described elsewhere [11]. Two sets of experiments were carried out. In the first set, ta-C films were deposited on silicon substrates at various negative dc bias, whilst in the second set, ta-C films were deposited on sapphire substrates at various rf bias. Application of the rf bias leads to the removal of the of the surface charge and thus creating an effective negative dc bias on the sapphire substrate. The original ion energy arriving at the substrate was carefully determined by a Faraday cup which was calibrated using a RF ion beam source with adjustable known energies. The relation between the substrate bias and ion energy is described elsewhere [11].

Mat. Res. Soc. Symp. Proc. Vol. 436 © 1997 Materials Research Society

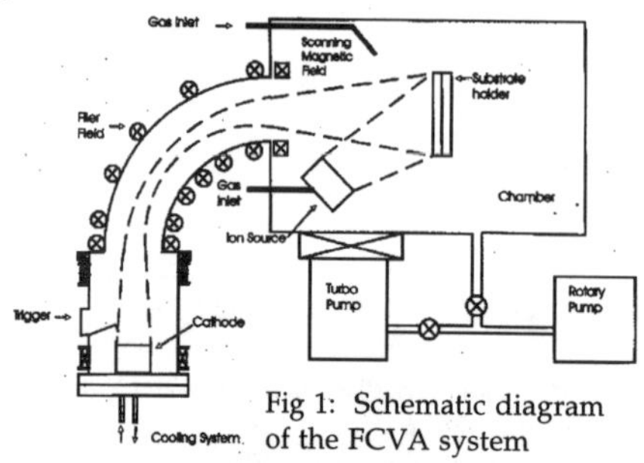

Fig 1: Schematic diagram of the FCVA system

In the experiments, the base pressure in the vacuum chamber was below 2.0×10^{-4} Pa but rose to about 1.5×10^{-3} Pa during deposition due to outgasing of the cathode. The arc current was fixed at 90 A, and the magnetic field at the centre of the toroid set at 40 mT. All depositions were carried out with the substrate at ambient temperature. The substrates used were <100> *n*-type silicon wafers of average thickness 0.28 mm and sapphire of average thickness 1.03 mm. Both the silicon and sapphire surfaces were precleaned using detergent and de-ionized water in an ultrasonic bath. Before deposition, an ion scrub was performed on the substrate surface with a rf ion source using argon as a working gas. In the case of the sapphire substrate, the films were produced using a two step process in which a thin layer, approximately 5 nm, was first deposited at a relatively high substrate bias of -300V followed by growth of the majority of the film at the desired voltage. The interfacial layer produced in this way serves to improve the film adhesion and reduce the stress at the interface. Two thickness of film were deposited, one about 60 nm, the other about 120 nm. The deposition rate at the substrate was typically 0.5 nm/s over an area of 25 cm^2. The peak ion energy in the plasma was 28 eV and the FWHM was 18 eV [11].

III. FILM CHARACTERIZATION

A. Nanoindentation Measurements

The hardness and Young's modulus of the films deposited on both silicon and sapphire substrates were investigated using nanoindentation techniques by Nanoindenter® II from Nano Instruments, Inc. [15]. Normally hardness and Young's modulus measurements of thin films are done by analyzing loading and unloading curves. A minimum indentation depth of around 100 nm is needed to obtain a good unloading curve from which the Young's modulus of the specimen can be calculated. The hardness of the specimen is then calculated based on the resulting modulus value. However, in the case of films with ultra thin thickness, the modulus obtained is inevitably influenced by the substrate. The problem is exasperated if the mechanical properties of the film and substrate are substantially different. To ameliorate the problem, we used the continuous stiffness measuring technique attached to the Nanoindenter® II developed by Nano Instruments. The continuous stiffness measurement provides a method of measuring continuously the film stiffness without the need for discrete unloading cycles. It therefore allows measurements to be made at small penetration depths to obtain hardness and Young's modulus continuously and simultaneously with the indentation displacement for ultra thin films. The indenter was operated in the constant-displacement-rate mode [15]. Sapphire substrates are used to further reduce the substrate effect and as a comparison, measurements were also made on films deposited on silicon substrates.

B. Scratch Tests

The adhesion strength and frictional coefficient of the films are another two parameters that can be evaluated using the scratch option available in Nanoindenter® II. The scratch is made by a diamond tip. A typical scratch profile consists of a surface profile scan across the full length of the scratch route with an extremely light load, the pre-scratch scan maintaining the light load but reverse the direction for 10% of the entire scratch length, and a scratch section with increasing load applied at a pre-selected constant force loading rate for the 80% of the entire scratch length with the maximum load attained just before the final length is about to be reached. A post-scratch profile scan returns the indenter to the initial start point to validate the measurement, and finally a second scan is performed over the entire length to reveal the surface profile after the scratch.

IV. EXPERIMENTAL RESULTS AND DISCUSSIONS

A. Nanoindentation Results

Fig 2 shows the hardness and Young's modulus of one of the ta-C films of 60 nm deposited on the silicon substrate as a function of the indentation depth. The carbon ion energy of this sample is 95 eV. Fig 3 shows the hardness and Young's modulus of one of the films of 60 nm deposited on the sapphire substrate. The carbon ion energy of this sample is 55 eV. For all films, as depicted in both figures, the hardness values increase initially from zero to a maximum at low indentation depth and gradually decrease to the values of the substrates at larger indentation depths. The initial rise from zero is due to the fact that the practical tip of the diamond indenter is of a finite radius instead of a sharp point. This effectively sets the limit on the minimum indentation depth (d_{min}) that is necessary to obtain reliable hardness data. When the indentation depth increases the effect of the substrate becomes more and more significant. This is, for example, observed in Fig. 2 where at a large indentation depth > 150 nm (compared to film thickness of 60 nm) the measured hardness (~15 GPa) and modulus (~170 GPa) are close to those of silicon. The same is also illustrated in Fig. 3 where at a large indentation depth (> 150 nm) the measured hardness (~36 GPa) and modulus (~470 GPa) are close to those of sapphire. The influence of the substrate effectively sets the limit on the maximum indentation depth (d_{max}) that can be made to obtain a more realistic hardness and modulus value of thin films. With the combined effects of the diamond tip and substrate, realistic hardness and modulus values are only possible by nanoindentation if d_{max} > d_{min} when a plateau will normally appear. When d_{max} < d_{min} as we have in our cases here, it is extremely difficult to obtain absolute hardness and Young's modulus values of ultra thin films. For a hard film on a softer substrate, the true hardness and Young's modulus should be higher than those measured. It is only sensible to compare results of different samples with the substrate and the film thickness clearly stated. It is obvious from Fig 2 and 3 that the effect of the substrate is exasperated if the difference of the hardness and modulus values between the film and substrate is larger.

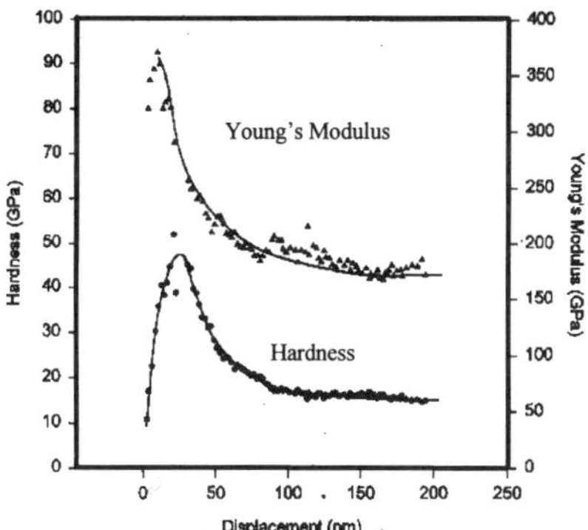

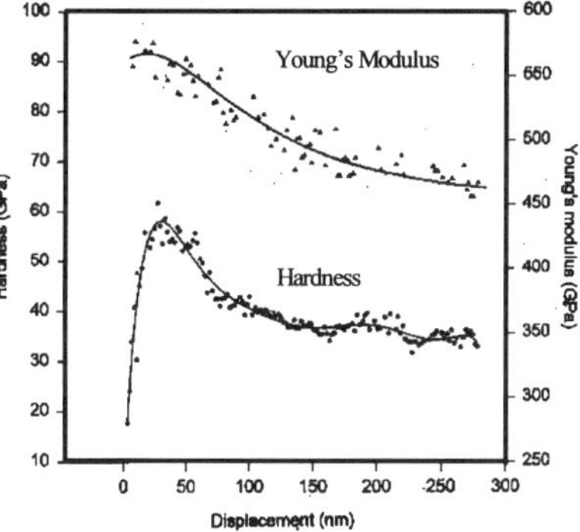

Fig 2: Graph of Hardness and Young's Modulus as a function of displacement (for ta-C film deposited on silicon substrates at ion energy of 95eV)

Fig 3: Graph of Hardness and Young's Modulus as a function of displacement (for ta-C film deposited on sapphire substrates at ion energy of 55eV)

When the maximum hardness and Young's modulus values are taken from the indentation experiments, the hardness increases from about 38 GPa at an ion energy of 15 eV to about 49 GPa at around 90 eV and then decreases to 37 GPa at 135 eV, as shown in Fig.4. The Young's modulus values ranging from 280 – 380 GPa were obtained for the various ion energies with the maximum value of about 380 GPa occurring at the ion energy about 90 eV. The modulus data shows good correlation with the hardness results. A better set of results is obtained for ta-C films on the sapphire substrate simply due to the much higher hardness and modulus values for sapphire compared to those of silicon. As shown in Fig. 5, the peak hardness of around 60 GPa occurs at around 80 eV and the modulus peak of 580 GPa at the same ion energy. It is seen that the influence of the silicon substrate on the hardness and modulus values of the film is significant. To reduce this substrate effect, a thicker film (~120 nm) was deposited at an ion energy of 90 eV and found that its hardness increased from 49 GPa to about 60 GPa and the modulus increased from 380 GPa to 520 GPa.

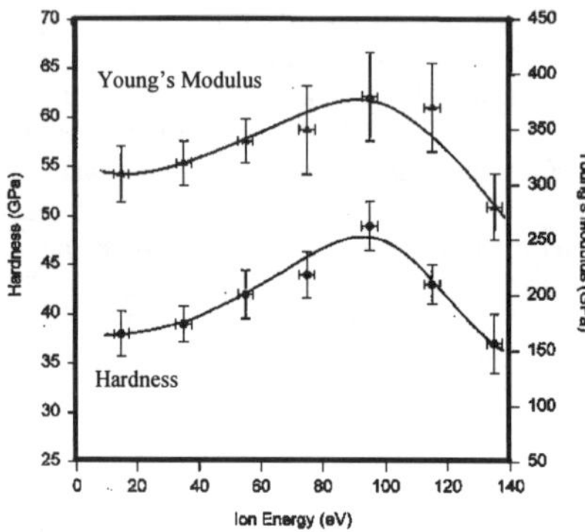

Fig 4: Graph of Hardness and Young's modulus as function of ion energy (for films deposited on silicon substrates)

Fig 5: Graph of Hardness and Young's modulus as a function of ion energy (for films deposited on sapphire substrates)

The ion energy corresponding to the peak position of hardness and Young's modulus are around 80 and 90 eV for rf biased sapphire substrate and dc biased silicon substrate, respectively. This agrees well with our sp³ and density data from the EELS measurement [11]. Our results also coincide with the indentation results (maximum 61 GPa at an optimum ion energy of 92 eV) for the amorphous hydrogenated-carbon films prepared by a gas plasma beam source [16]. For the sake of comparison, their film of 100 nm thick was measured having a hardness of about 33 GPa and modulus of about 240 GPa [17]. These are about half of the value for our films of 120 nm thickness on silicon. Compared to the results reported by Anders *et al* [18], their hardness and Young's modulus values for films of 250 nm thick are about 30% less than those of our ta-C films of thickness of 120 nm on silicon. The main reasons are due to two factors (a) the much more sensitive continuous stiffness measurement technique used in our present work and (b) the higher sp³ contents in our ta-C films, which can be seen from the respective film densities. Our ta-C film has a maximum density of 3.37×10^3 kg/m³ while their ta-C film has a density of 3.0×10^3 kg/m³ [11,17-18]. The hardness and modulus of a set of our ta-C samples of ~ 60 nm thickness on silicon substrate with dc bias has been investigated earlier by loading and unloading method [11]. A 16% increase in hardness and 11% increase in Young's modulus values is observed in our present work compared to that of our earlier. It is thus believed that factor (b) is more significant than factor (a).

Compared to hardness, which is a measure of the film's plastic deformation, the Young's modulus, which is the measure of the film's elastic property, is much more susceptible to the substrate influence. It is seen, from the experiment, Young's modulus obtained from the films on sapphire is about 1.6 times higher than that obtained from the film on silicon, whilst an average 1.3 times increase in hardness is observed. Another interesting indicator on the indentation experiment is the ratio of the measured Young's modulus to the measured hardness. It can be considered as a measure of the extent of reduction of the substrate influence on the modulus measurement. It is known that the ratio for diamond is about 11. For our ta-C films of 60 nm on sapphire it is 10 and slightly less than 8 for our films of 60 nm on silicon. It is about 8.8 in Ander's work (film thickness 250 nm) [18] and between 6 to 7 in Pharr's work (film thickness 100 to 320 nm) [17]. The low ratio associated with the samples even with high thickness are typical for the indentation experiment using loading and unloading curves.

B. Scratch Results

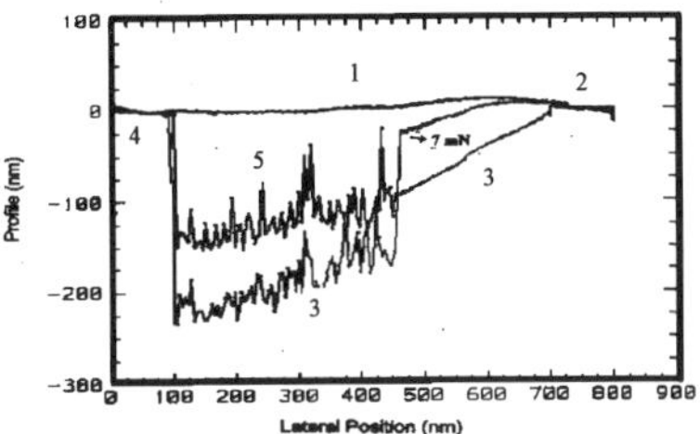

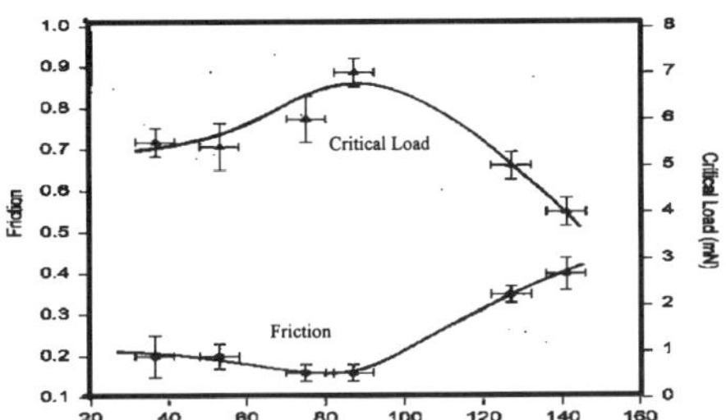

Fig 6: Scratch test measurement for ta-C film deposited on sapphire substrates at ion energy of 85 eV

Fig 7: Coefficient Friction and Critical Force as a function of Ion Energy for ta-C films deposited on sapphire substrates

Fig. 6 shows one of the scratch tests. It was made on the film deposited on sapphire substrate with 85 eV ion energy. Curve 1 is the pre-scratch scan to obtain the film profile. Curve 2 is the profile scan before the scratch. Curve 3 is the scratch with the force loading rate of 300 μN/sec at a scan rate of 10 μm/sec. Curve 4 is a profile scan after the scratch and curve 5 is the post-scratch profile scan along the whole length of the scratch. Curves 2 and 4 are mainly used to validate the experiment. They should overlap with the pre- and post-scratch profiles. It can be clearly seen that the film delaminates and is scratch off the substrate at around 7 mN which is the critical load of this film. By measuring the lateral force during the scratch section, the frictional coefficient can be obtained. Fig. 7 shows the frictional coefficient and critical load of the films deposited on sapphire at various ion energies. It is seen that the frictional coefficient decreases from 0.2 to 0.16 at an ion energy of around 80 eV and then increases to around 0.4 at 140 eV. This phenomenon is attributed to the change of sp^3 contents in the film deposited at different ion energies. A maximum sp^3 contents at around 80 eV [11] is associated with a minimum coefficient of friction. This result coincides with the findings from Coll [19]. However, it is seen that the critical load increases from 5.5 mN to a peak of 7.0 mN at an ion energy of around 80 eV and then reduces to around 4.0 mN at 140 eV. This maps the change of the hardness of the film deposited at various ion energies.

V. CONCLUSION

The tribological properties of ta-C films deposited by FCVA technique have been studied in terms of ion energy. An optimum ion energy around 80 to 90 eV has been found. This optimum energy is found to coincide with the ion energy at which the sp^3 fraction, plasmon energy and film density reach maximum values. The high concentration of the tetrahedral (sp^3) bonding gives the films their diamond-like properties. Hardness and Young's modulus values for ta-C films (60nm thickness) on silicon substrates vary from 37 – 49 GPa and 310 – 380 GPa respectively. A thicker ta-C film (~120 nm) deposited at an ion energy of 90 eV showed higher hardness and Young's modulus values of 60 GPa and 520 GPa, respectively. For ta-C films deposited on sapphire substrates, maximum hardness and Young's modulus values of 60 GPa and 580 GPa at around 80 eV were found, respectively. At the same ion energy, a maximum critical load of 7 mN and a minimum frictional coefficient of 0.16 were also obtained for the ta-C film on sapphire.

In our case of a hard and thin film on a soft substrate, the continuous stiffness is a much superior indentation technique to the conventional loading and unloading method, especially when $d_{max} < d_{min}$ (where d_{min} is the minimum indentation depth set by the sharpness of the diamond tip and d_{max} is the maximum indentation depth set by the substrate effect). There is an average between 10 – 20% improvement in the hardness and modulus measurement by using this technique. To minimize the substrate effect, a thicker film could be deposited. Unfortunately, it is not always possible to deposit thicker film without compromising other film properties. Neither is it always desirable to do so.

Choosing a substrate with high hardness and Young's modulus is thus very important. A 30% improvement in hardness values and a 60% improvement in modulus values is observed when films on sapphire substrates were measured compared to films of the same thickness on silicon substrates. For our samples, d_{max} is less than d_{min} which means the hardness and modulus values obtained are less than the actual values of the film. The real values are believed higher than those obtained experimentally, probably approaching those of natural diamond.

It is found that the hardness and Young's modulus values of our ta-C films are substantially higher than those reported elsewhere. This is due to two reasons; (a) better measuring technique, and (b) better film quality. It is believed that (b) is more significant. The ratio of Young's modulus to the hardness value for our films is close to 10, again substantially higher than results reported elsewhere. This means the substrate influence on the modulus measurement in our experiment is much reduced. This is attributed to both the continuous stiffness technique and by using sapphire substrate.

References

1. I. I. Aksenov, V. A. Belous, V. G. Padalka, V. M. Khoroshikh, Sov. J. Plasma Phys. **4**, 425 (1978).

2. I. I. Aksenov, S. I. Vakula, V. G. Padalka, V. E. Strelnitskii, and V. M. Khoroshikh, Sov. Phys. Tech. Phys. **25**, 1164 (1980).

3. P. J. Martin, S. W. Filipczuk, R. P. Netterfield, J. S. Field, D. F. Whitnall, D. R. McKenzie, J. Mater. Sci. Lett. **7**, 410 (1988).

4. P. J. Martin, R. P. Netterfield, T. J. Kinder, and L. Descotes, Surf. Coat. Technol. **49**, 239 (1991).

5. D. R. McKenzie, D. Muller and B. A. Pailthorpe, Phys. Rev. Lett. **67**, 773 (1991).

6. S. Falabella, D.B. Boercker, and D.M. Sanders, Thin Film Soilds, **236**, 82 (1993).

7. P. J. Fallon, V. S. Veerasamy, C. A. Davis, J. Robertson, G. A. J. Amaratunga, W. I. Milne, and J. Koskinen, Phys. Rev B, **48**, 4777 (1993).

8. D R McKenzie, D Muller, B A Pailthorpe, Z H Wang, E Kratvchinskaia, D Segal, P B Lukins, P D Swift, P J Martin, G Amaratunga, P H Gaskell and A Saeed, Diamond Relat. Mater., **1**, 51(1991).

9. V S Veerasamy, G A J Amaratunga, W I Wilne, P Hewitt, P J Fallon, D R McKenzie and C A Davis, Diamond Relat. Mater., **2**, 782 (1993).

10. V S Veerasamy, G A J Amaratunga, J S Park, W I Wilne, H S MacKenzie and D R McKenzie, J. Appl. Phys. Lett., **64** (17), 2297 (1994).

11. Shi Xu, B K Tay, H S Tan, Li Zhong, Y Q Tu, J. Appl. Phys., **79**, (1996) - to be published

12. Y Lifshitz, G D Lempert and S Rotter, Diamond And Related Materials, **3**, 542 (1994)

13. S.R.P. Silva, W.I. Milne, Shi Xu, B.K. Tay and H.S. Tan, J. Appl. Phys. - under review

14 G.M. Pharr and W.C. Oliver, MRS Bulletin **17**, 28 (1992)

15. W.C. Oliver and G.M. Pharr, J. Mater Res. **7**, 1564 (1992)

16. M. Weiler, S.Sattel, K. Jung, H. Ehrhardt, V.S. Veerasamy and J. Robertson, Appl. Phys. Lett. **64**, 2797 (1994)

17. G.M. Pharr, D.L. Callahan, S.D. McAdams and T.Y. Tsui, Appl. Phys. Lett. - to be published

18. S Anders, A Anders et al, Meeting of the Materials Research Society, LBL-36768, UC-404, San Francisco, April (1995)

19. B.F. Coll, M. Chhowalla, Surf. Coat. Technol. **79**, 76 (1996).

ADHESION STUDY OF DLC/Cr/DLC AND DLC/Ti/DLC SANDWICH STRUCTURES ON HARDENED STEEL

U. MÜLLER, R. HAUERT
Swiss Federal Laboratories for Materials Testing and Research (EMPA),
Überlandstrasse 129, CH-8600 Dübendorf (Switzerland)

ABSTRACT

Hard amorphous hydrogenated carbon, also referred to as diamond-like carbon films, are of technological interest as protection coatings due to their special properties such as high hardness, low friction, low wear, chemical inertness and electrical insulation. In this paper a study of the adhesion properties of multilayer structures composed of DLC/Metal/DLC/substrate is presented. As substrates a ball and roller bearing steel (DIN 1.3505, AISI 52100) was used. After Ar plasma cleaning of the substrate, DLC films were deposited from acetylene gas by rf plasma chemical vapor deposition at a self-bias of -600V. Metal films were deposited in the same chamber by DC magnetron sputtering.

Adhesion of the different layers on the underlying layer was investigated by optical analysis of the coating at the border of Rockwell C indentations made into the substrate. If necessary, additional thin layers are deposited at the particular interface to ensure good adhesion. Further information on the interface quality and hence the adhesion is gained by a second optical analysis of the Rockwell indentation after exposure to ambient air for two weeks.

The multilayer structure containing titanium as the metal layer showed no ablation of the coating around the indentation even after two weeks of exposure to ambient air. Chromium on the other hand already showed some ablation immediately after the indentation as well as still ongoing ablation after two weeks. These results are further corroborated by pin-on-disc tests.

INTRODUCTION

Diamond-like carbon (DLC) films are well known for their outstanding properties such as high hardness, extremely low friction coefficient, and good corrosion resistance [1]. These properties are due to the unique structure of a random network of graphitic and diamond types of carbon-carbon bonds where dangling bonds are terminated with hydrogen. Furthermore these coatings are wear resistant, chemically inert and electrically insulating. However, their friction coefficient is moisture sensitive and they are not stable at temperatures above 300°C. Therefore their main applications are in areas where low coefficients of friction and/or low wear and/or chemical inertness are needed but working temperatures are below 300°C. These applications for DLC coatings include sliding bearings, forming tools (for example deep-drawing dies for aluminum without lubrication), as well as magnetic data storage disks (more than 90% of all magnetic disks).

In this paper we show the possibility of producing extremely stable sandwich structures consisting of first a DLC layer as electrical insulation against the substrate, a metal layer as conduction layer and a second DLC layer for protection and insulation.

EXPERIMENT

Ball and roller bearing steel 100Cr6 (DIN 1.3505, AISI 52100) with a Rockwell C hardness of 60HRC is used as substrate. Diamond-like carbon films are produced by RF plasma assisted chemical vapor deposition (PACVD) from a hydrocarbon gas. This is the most widely spread method for DLC coatings because it allows the homogeneous coating of parts with different geometries. Using acetylene as hydrocarbon gas yields the densest and hardest films [1]. In this study the films were produced by rf plasma (13.56MHz) deposition from acetylene in an all stainless steel high vacuum system with a base pressure better than $2{\cdot}10^{-5}$Pa and the substrates placed on the rf powered electrode. The rf generator output was regulated to yield a constant sample self-bias of -600V. The samples were cleaned in an ultrasonic bath with an acetone/ethanol mixture, rinsed with pure ethanol, dried and immediately transferred into the vacuum system. Just prior to the deposition the substrates were plasma cleaned at an argon pressure of 3.4Pa for 30 minutes. The sputter rate during this plasma cleaning process was 5nm/min. When necessary an intermediate layer of amorphous hydrogenated silicon-carbon (Si-DLC) was deposited to ensure adhesion. This layer was grown from tetra-methylsilane (TMS, $Si(CH_3)_4$) for two minutes at a pressure setting of 2.0Pa, the Pirani pressure gauge being calibrated for air. The use of silicon compounds to increase adhesion has already been reported by others [2,3]. Then a 2.4µm thick diamond-like carbon film was deposited. The growth rate for the DLC coatings was 40nm/min. The first three steps were done by simply switching gases, keeping the total pressure as constant as possible and especially without interrupting the plasma. Using DC magnetron sputtering metal films of either chromium or titanium were deposited onto the DLC film. The DC magnetron sputtering was done using argon at a pressure of 0.85Pa resulting in deposition rates of 25nm/min for titanium and 30nm/min for chromium metal films. On top of this metal layer a second DLC layer was deposited using the same conditions as for the first one.

Adhesion of the coatings is examined according to VDI 3198 [4] performing an optical analysis of the film at the border of Rockwell C indentations made into the substrate. Pin-on-disc measurements were done on a CSEM tribometer in dry air and at a relative humidity around 36%, both at room temperature. The balls used were 100Cr6 balls, the same material as the substrate, with a diameter of 6mm.

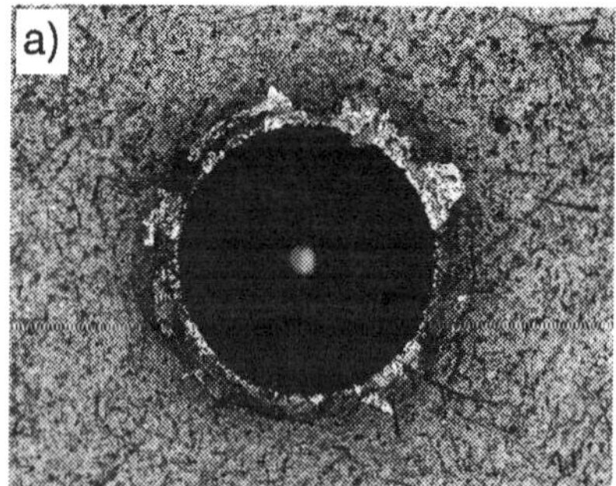
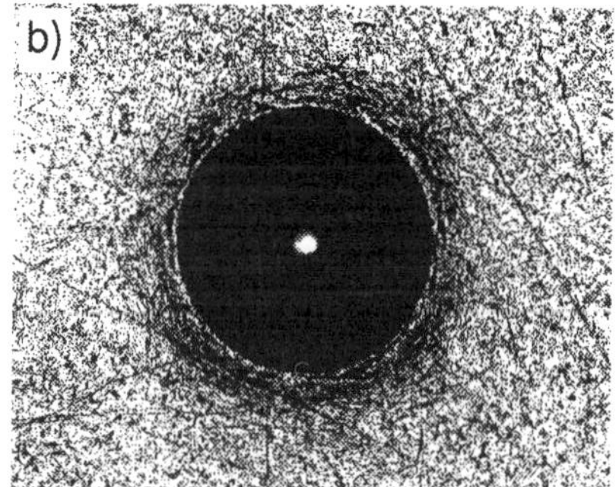

Figure 1:   Rockwell C indentations on a DLC coating deposited onto hardened bearing steel. The diameter of the indentations is approximately 500µm. a) DLC directly deposited onto the substrate (medium adhesion) and b) using an intermediate Si-DLC layer prior to DLC coating (good adhesion).

RESULTS

In a first step the adhesion of the DLC layer on the steel substrate had to be optimized. Figures 1a) and b) show a comparison between the adhesion of a DLC layer

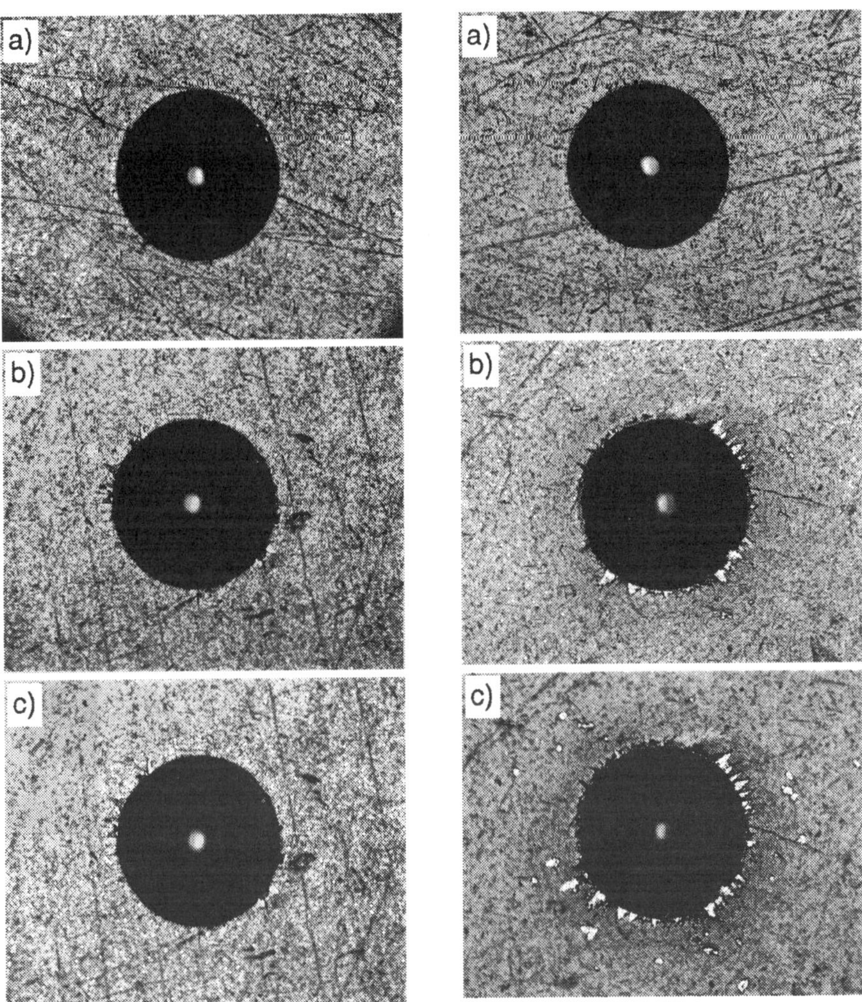

Figure 2: Rockwell C indentations on the titanium structures. The diameter of the indentations is approximately 500μm. a) Ti/DLC/Si-DLC/Steel, b) DLC/Ti/DLC/Si-DLC/Steel and c) the same as in b) but after 13 days of ambient air exposure.

Figure 3: Rockwell C indentations on the chromium structures. The diameter of the indentations is approximately 500μm. a) Cr/DLC/Si-DLC/Steel, b) DLC/Cr/DLC/Si-DLC/Steel and c) the same as in b) but after 14 days of ambient air exposure.

deposited directly onto the substrate (a) and the one using an intermediate layer of Si-DLC (b). In Figure 1a) bright areas are seen where the coating has flaked off exposing the substrate. The larger the areas where the film delaminates, the lower the adhesion quality of the coating. Figure 1b) on the other hand shows no areas where the substrate is exposed, thus proving the excellent adhesion of diamond-like carbon coatings on hardened steel when an appropriate intermediate layer is used. The optimum thickness of the intermediate layer was found to be approximately 80nm. In contrast to this finding, no interface layer was necessary to give good adhesion of DLC on stainless steel (DIN 1.4301, AISI 304) [5].

In the next step a metal film consisting either of titanium or chromium was sputtered on top of the first DLC layer and its adhesion was investigated. Figure 2a) and 3a) show the corresponding Rockwell C indentations made onto the titanium and chromium layer, respectively. In both cases perfect adhesion is obtained without the use of an intermediate layer. On top of these metal layers the third layer, again DLC, was deposited. In Figure 2b) and 3b) the corresponding Rockwell indentations are shown. Whereas in the case of titanium no ablations are observed a few small ablations are seen in the case of chromium. Looking at the same indentations about two weeks later yields information on the long term quality of the adhesion. A progressing delamination is due to a reduced chemical stability of the interface when it comes into contact with ambient air. It further results in a reduced failure tolerance of the system. In figures 2c) and 3c) the same indentations as in 2b) and 3b), respectively, but two weeks later are shown. No observable differences for the titanium sandwich structure but an ongoing ablation in the case of the chromium sandwich structure is seen. This is a further indication for a weaker interface in the case of chromium than of titanium. But in both cases best performance was obtained by continuously switching from metal deposition to DLC deposition without interrupting the plasma.

Pin-on-disc measurements in dry air were done on a sandwich structure consisting of 2.4µm DLC / 2.5µm Cr / 2.4µm DLC / 80nm Si-DLC / substrate. Figure 4 shows scanning electron microscope (SEM) images of the wear tracks from this multilayer system. At a load of 1N only little wear is seen after 43'000 rotations at a speed of 0.02m/s as shown in Figure 4a). The friction coefficient dropped from 0.15 in the beginning to

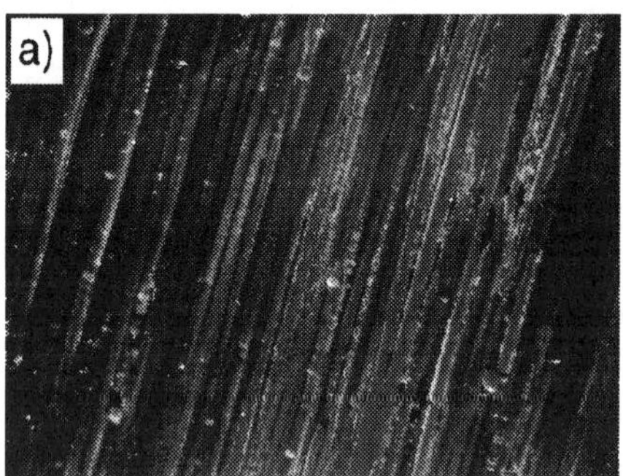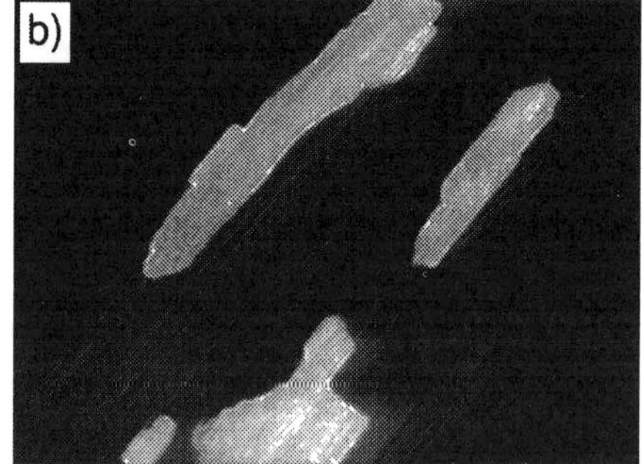

Figure 4: SEM images from the wear tracks of pin-on-disc measurements in dry air on the 2.4µm DLC / 2.5µm Cr / 2.4µm DLC / 80nm Si-DLC sandwich structure. a) After 43'000 rotations with a load of 1N at a speed of 0.02m/s. Only little wear is observable. The lateral size of the image is 230µm. b) After 35'000 rotations with a load of 5N at a speed of 0.02m/s. Ablations of the topmost DLC film can be seen. The bright areas consist of chromium. The lateral size of the image is 580µm.

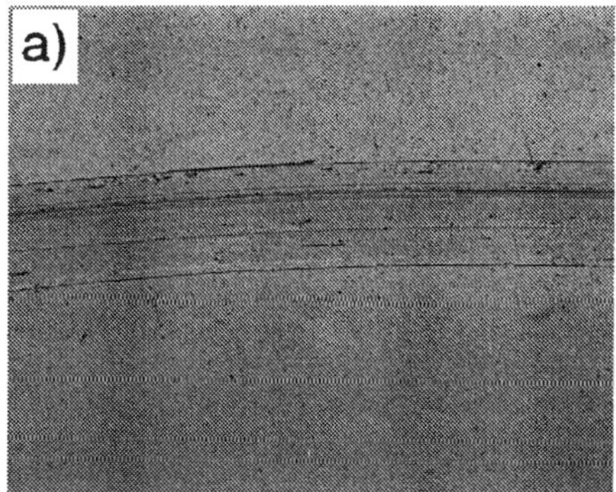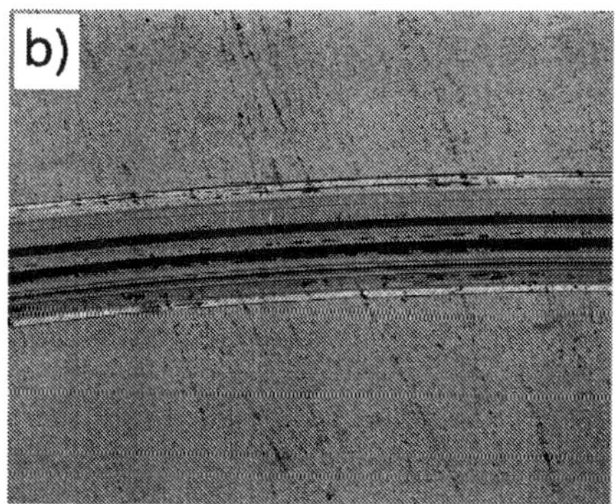

Figure 5: Micrographs of wear tracks of pin-on-disc measurements in humid air on the 100nm DLC / 100nm Ti / 2.0µm DLC / 80nm Si-DLC sandwich structure. a) After 19'000 rotations with a load of 10N at a speed of 0.26m/s. Only small wear scars but no ablations visible. b) After 89'000 rotations with a load of 10N at a speed of 0.97m/s. A few distinct and deep wear scars are visible but no film ablation occurs. The size of each image is 1.8x2.3mm.

below 0.05 after 30'000 rotations. This effect is often seen in the case of diamond-like carbon films [6]. After 35'000 rotations at a load of 5N and a speed of 0.02m/s areas where ablation of the topmost DLC layer occurs are seen in Figure 4b). The occurrence of these ablations is due to a weakness in the DLC / Cr topmost interface which was verified by EDX analysis of the bright areas seen in Figure 4b).

On a second sandwich structure consisting of 100nm DLC / 100nm Ti / 2.0µm DLC / 80nm Si-DLC / steel pin-on-disc measurements were done in air at a relative humidity of around 36%. In Figure 5a) the wear track after 19'000 rotations at a speed of 0.26m/s under a load of 10N is shown. No ablation and only little wear is seen. The friction coefficient remained quite constant at 0.1. After 89'000 rotations at a speed of 0.97m/s again under a load of 10N the wear track shown in Figure 5b) is obtained. A few distinct and deep wear scars are seen but despite the rough conditions used still no film delamination occurs.

CONCLUSIONS

We have shown that stable sandwich structures of DLC/metal/DLC can be deposited onto hardened steel. When the correct deposition conditions are chosen excellent adhesion of these structures on the substrate as well as in between the different layers is obtained. The system with titanium as intermediate metal layer shows a slightly better adhesion than the one with chromium. This results in a much better performance in the tribology experiments for the titanium containing sandwich structure.

ACKNOWLEDGMENTS

We thank A. Böll for performing the pin-on-disc measurements. Financial support by the Swiss Priority Program on Materials Research (PPM) is gratefully acknowledged.

REFERENCES

1.   J. Robertson, Prog. Solid State Chem. **21**(4/1991), 199-333

2.   L. Chandra, M. Allen, R. Butter, N. Rushton, A.H. Lettington, T.W. Clyne,
      Diamond Related Mater. **4**(05-6/1995), 852-856

3.   S. Miyake, R. Kaneko, Thin Solid Films **212**(1992), 256-261

4.   VDI 3198, in "VDI-Handbuch Betriebstechnik", Teil 3,
      Beuth Verlag GmbH, Berlin (Germany), 1992

5.   U. Müller, R. Hauert, B. Oral, M. Tobler, Surf. Coat. Tech. **71**(1995), 233-238

6.   K. Jia, Y.Q. Li, T.E. Fischer, B. Gallois, J. Mater. Res. **10**(06/1995), 1403-1410

TRIBOLOGICAL PROPERTIES OF NITROGEN IMPLANTED AND BORON IMPLANTED STEELS

K.T. Kern[1], K.C. Walter[2], S. Fayeulle[3], A.J. Griffin[2], Jr., H. Kung[2], Y. Lu[2], M. Nastasi[2], and J.R. Tesmer[2].

[1]Center for Materials Research, Norfolk State University, Norfolk, VA 23504, K_Kern@lanl.gov
[2]Materials Science and Technology Division, Los Alamos National Laboratory, Los Alamos, NM 87545
[3]Laboratore MMP URACNRS 447 Ecole Centrale De Lyon, Ecully, France.

ABSTRACT

Samples of steel with high chrome content were implanted separately with 75 keV nitrogen ions and with 75 keV boron ions. Implanted doses of each ion species were 2-,4- and 8×10^{17} /cm^2. Retained doses were measured using resonant non-Rutherford Backscattering Spectrometry. Tribological properties were determined using a pin-on-disk test with a 6-mm diameter ruby pin with a velocity of 0.94 m/min. Testing was done at 10% humidity with a load of 377g. Wear rate and coefficient of friction were determined from these tests. While reduction in the wear rate for nitrogen implanted materials was observed, greater reduction (more than an order of magnitude) was observed for boron implanted materials. In addition, reduction in the coefficient of friction for high-dose boron implanted materials was observed. Nano-indentation revealed a hardened layer near the surface of the material. Results from grazing incidence x-ray diffraction suggest the formation of Fe$_2$N and Fe$_3$N in the nitrogen implanted materials and Fe$_3$B in the boron implanted materials. Results from transmission electron microscopy will be presented.

INTRODUCTION

Ion implantation is one method for modifying the near-surface region of materials, in which the implanted species forms a layer that may be different from the substrate in chemical composition and in physical properties. Implantation may result in changes in the surface properties of a material, including hardness, wear, coefficient of friction and other properties Characterization of the implanted layer will allow for the use of this surface modification for control of surface properties, including tribological properties. Wear of mechanical systems is a $200 billion per year problem, and therefore control of tribological properties is important in the production of moving parts.

Previous studies of ion implanted steels have given evidence of the mechanisms for property changes due to ion implantation. In particular, nitrogen implanted 304 stainless steel has been extensively studied. The wear mechanism for 304 SS involves the conversion of austenitic (fcc) phase material by plastic deformation into a hard, brittle martensitic (bct) surface layer, which fractures under the load of the wearing body [1].

Mat. Res. Soc. Symp. Proc. Vol. 436 © 1997 Materials Research Society

Implantation of nitrogen into 304 SS stabilizes the austenitic phase and prevents transformation to the martensitic phase [2]. Evidence that boron implantation may modify the wear properties of 304 SS via the same mechanism is mixed [3, 4]. The modification of wear properties is sensitive to the phases existing in the material prior to implantation , as implantation of boron into steels with higher bcc content shows a greater hardening effect than in pure fcc materials [5]. In other steels, wear improvement of nitrogen implanted steels has been linked to the chromium content of the steel, implantation of N^+ into higher Cr content steels resulting in greater improvement of wear properties than low Cr content steels. It is expected that this observation is a result of chrome forming nitrides. No similar compositional effect has been reported for boron implantation.

White iron is a high chromium (about 30% Cr) steel with a small percentage of carbon. In this study, this high chrome steel was modified through nitrogen implantation and boron implantation. Surface characterization included wear testing, coefficient of friction determination, and hardness testing. The nature of changes in these surface properties will be correlated with changes in the microstructure of the material resulting from the implantation.

EXPERIMENT

Samples of white iron were prepared by mechanical polishing coupons (0.5 in x 0.5 in) of material to surface roughness of less than 20 nm. Samples were implanted with N_2 ions at 150 keV to incident fluxes of $2x10^{17}$ /cm^2, $4x10^{17}$ /cm^2, and $8x10^{17}$ /cm^2, and implanted with boron ions at 75 keV to incident fluxes of $2x10^{17}$ /cm^2, $4x10^{17}$ /cm^2, and $8x10^{17}$ /cm^2. Samples were cooled during implantation. Retained ion doses were determined using Rutherford Backscattering Spectrometry (RBS), including analysis at a nitrogen resonance (8.9 MeV) and at a boron resonance (6.6 MeV) [6]. The analysis software RUMP was used to determine implant species quantity and distribution.

Surface properties were determined by tribology and nano-indentation. Tribological properties were studied using a pin-on-disk tribotester, in which the pin was a 6-mm diameter ruby ball and the disk was the sample being tested. In these tests, the pin rests on the sample to be tested which is turned at a constant rate. The pin thus traces a circular path on the disk. In this study, the disk spun at 100 RPM and the pin traced a track 3-mm in diameter, for a relative velocity of 0.94 m/min. Weight was added to the pin for a total load of 4.5 Newtons and a maximum contact stress of 1.4 G Pa. All tribotests were conducted in a 10% humidity atmosphere. Tests were halted so that the resulting wear was within the implanted region, for most samples less than 6000 cycles. During tests, the coefficient of friction was calculated and recorded using an automated computer system. Following tests, the wear track was cleaned of wear debris and the wear track depth and cross-sectional area were measured using a profilometer. Wear rates were calculated as K = wear volume/(load x distance). Nano-indentation was used to measure the hardness of the implanted layers. Twelve indents were made at each of four depths, nominally 25, 100, 200, and 300 nm. The 12 hardness values at each depth were then averaged.

Grazing incidence x-ray diffraction (GXRD) was used to determine micro-structure of the implanted materials. The grazing incidence was used to probe only the near-surface. Resulting diffraction peaks were compared to a standard data base for identification of compounds. In addition, transmission electron microscopy (TEM) was used for micro-structural analysis.

RESULTS

RBS of non-implanted white iron indicated an atomic Cr to Fe ratio of 28 to 72. In addition, the non-implanted material contains a small percentage of carbon, which could not be accurately determined. RBS results for implanted materials are summarized in Table 1. Boron implanted materials retained the entire implant dose, while the nitrogen implanted material became sputter limited at the higher doses and thus did not retain the full doses. The implant distributions could be modeled using gaussian distributions, with the exception of the 8×10^{17} /cm^2 nitrogen implant, which was skewed toward the surface due to sputtering. In addition, the nitrogen implant distribution was shallower than the boron distribution. These effects can be seen in the 4×10^{17} /cm^2 dose results shown in figure 1, in which the decrease in the Fe-Cr edge is an indication of implanted atoms being substituted for the Fe and Cr atoms.

Implanted Species	Implant Dose (x 10^{17} ions/cm^2)	Retained Dose (x 10^{17} ions/cm^2)	Peak Depth (nm)	Peak FWHM (nm)
N	2.0	1.9	79	100
N	4.0	3.9	87	108
N	8.0	5.5	73	146
B	2.0	2.0	125	120
B	4.0	4.0	136	133
B	8.0	8.0	129	126

Table 1. Results of RBS analysis on implanted White Iron.

Wear rates for non-implanted and implanted white iron are shown in figure 2. Some reduction in wear rate is evident for samples implanted to 2×10^{17} N$^+$/cm^2 while samples implanted with 4- and 8×10^{17} N$^+$/cm^2 have wear rates that are less than half that of the non-implanted material. In contrast, all materials implanted with boron show significant reduction in wear rate, the rates for samples implanted to 2- and 4×10^{17} B$^+$/cm^2 being similar to the rates for the materials implanted to 4- and 8×10^{17} N$^+$/cm^2. The wear rate for white iron implanted to 8×10^{17} B$^+$/cm^2 , however, showed the largest improvement and was only 3% of the wear rate for the non-implanted material. This sample showed this wear resistance even when tested to 12,000 cycles.

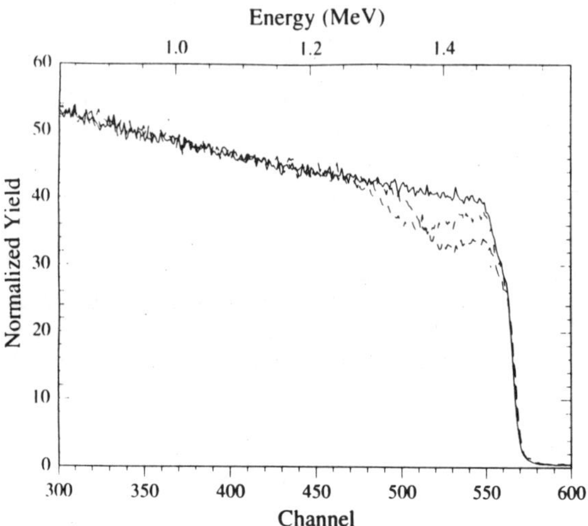

Fig. 1. RBS (2 MeV alpha) on white iron and white iron implanted to 4×10^{17} N^+/cm^2 and 4×10^{17} B^+/cm^2.

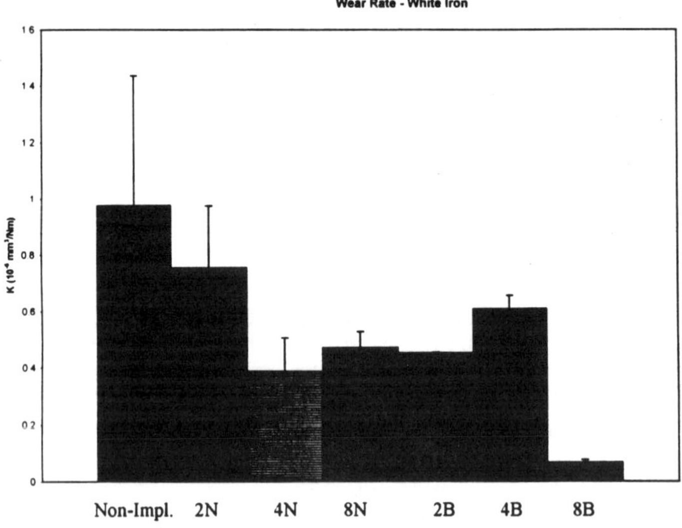

Fig. 2. Wear rates of non-implanted and implanted white iron.

The coefficient of friction for a ruby pin sliding on a white iron disk was recorded for each implant condition. Coefficient of friction for selected samples are shown in figure 3. No significant change in coefficient of friction was observed for nitrogen implanted white iron, although for the material implanted to $4x10^{17}$ N$^+$/cm^2 the coefficient of friction was 10% higher than the coefficient for the non-implanted material. This difference is not significant and is not interpreted as being a result of the implantation. For the boron implanted materials, the materials implanted to $2x10^{17}$ B$^+$/cm^2 and $4x10^{17}$ B$^+$/cm^2 showed a slightly higher coefficient of friction than the non-implanted material, but again this difference is not considered significant. For the $8x10^{17}$ B$^+$/cm^2 implanted material, however, the coefficient of friction was 20% of that for the non-implanted material. This low coefficient of friction, while corresponding to the low wear rate, requires further investigation as the RBS data indicated a thin layer of carbon on this sample which may be acting as a lubricant during the wear test. It is suspected that the carbon layer resulted from backstreaming during the ion implantation of boron.

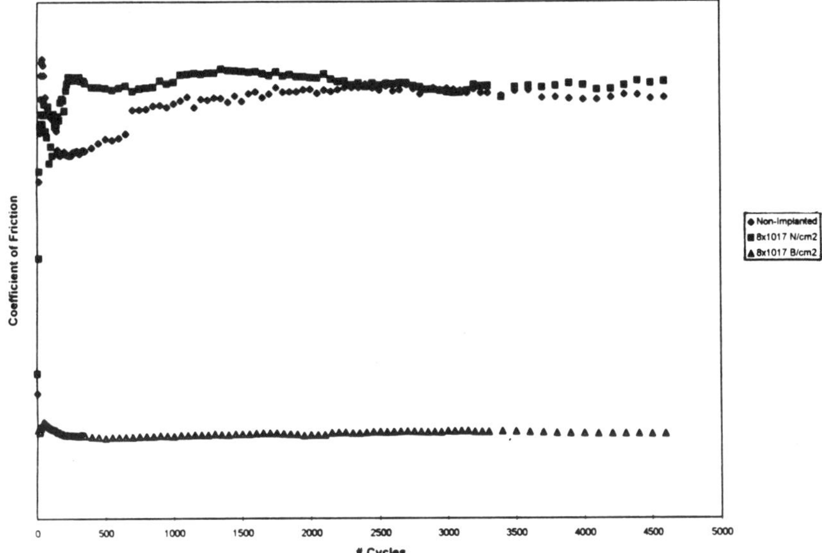

Figure 3. Coefficient of friction for non-implanted white iron and white iron implanted to $8x10^{17}$ N$^+$/cm^2 and $8x10^{17}$ B$^+$/cm^2.

Nano-indentation showed that the hardness of the implanted materials varied through the implanted region. Implanted materials, regardless of species implanted or dose, had a hardness at a depth of 23 nm that was 20% to 40% greater than the hardness of the non-implanted materials. Due to surface roughness, the hardness of white iron implanted to $8x10^{17}$ N$^+$/cm^2 could not be determined at this depth. At 102 nm, near the peak of the implanted species distribution, the hardness was not significantly different from that of the non-implanted material, while indents to depths greater than the depth of the implants showed that the implanted materials were slightly less hard than the non-implanted material.

GXRD and TEM were used to determine micro-structural information for the implanted materials. TEM and XRD of non-implanted materials showed a bcc structure. For the nitrogen implanted white iron, evidence of Fe_2N and Fe_3N was seen in these samples as has been reported in previous studies [7]. GXRD of boron implanted materials showed a peak corresponding to Fe_3B, which was particularly prominent in the $8x10^{17}$ B^+/cm^2 spectrum, although TEM results indicated an amorphous implanted region.

CONCLUSIONS

Implantation of nitrogen into white iron results in hardening of the surface layer and reduced wear for implanted doses greater than $4x10^{17}$ N^+/cm^2. Micro-structural analysis show the formation of Fe_2N and Fe_3N in the implanted materials. For boron implanted white iron, hardening of the surface layer was also observed, however improvements in wear resistance occurred for doses as low as $2x10^{17}$ B^+/cm^2. Additionally, boron implantation resulted in a low coefficient of friction for a dose of $8x10^{17}$ B^+/cm^2, although this result may be influenced by a layer of carbon on the surface of the material. The micro-structure of boron implanted white iron indicates Fe_3B is forming in the material, although TEM indicates a largely amorphous surface layer. Curiously, no evidence of hard compound formation with Cr has been found. The increased wear resistance of N and B implanted white iron is due to the confirmed presence of hard compounds (Fe_2N, Fe_3N, Fe_3B) in the implanted surface.

ACKNOWLEDGMENT

This work was supported by DOE grant DE-FG01-94EW11493.

REFERENCES

1. K.L. Hsu, T.M. Ahn, and D.A. Rigney in K.C. Ludema, W.A. Glaeser, and S.K. Rhee (eds.) *Wear of Materials*. ASME. New York, 1979.
2. R.N. Bolster and I.L. Singer, *Appl. Phys. Lett.*, **36** (1980) 208.
3. S. Raud, H. Garem, A. Naudori, J.P. Villain, and P. Moine, *Mater. Sci. Eng.*, **A115** (1989) 245.
4. S. Shrivastava, A. Jain, and C. Singh, *Acta Metall. Mater.*, **43** (1995) 59.
5. S. Shrivastava, A. Jain, R.D. Tarey, D.K. Avasthi, D. Kabiraj, L. Senapati, and G.K. Mehta, *Vacuum.*, **47** 3 (1996) 247.
6. Handbook of Modern Ion Beam Analysis, J.R. Tesmer & M. Nastasi, editors, Materials Research Society, Pittsburgh, PA, 1995, pgs. 497, 499.
7. S. Fayuelle, *Appl. Surface Sci.* **25** (1986) 288.

CHARACTERIZATION OF THE MECHANICAL AND TRIBOLOGICAL PROPERTIES OF SPUTTERED a:SiC THIN FILMS

T.W. Scharf, R.B. Inturi, and J.A. Barnard

Department of Metallurgical and Materials Engineering, The University of Alabama, Tuscaloosa, AL 35487-0202.

ABSTRACT

D.C. magnetron sputtering from a CVD β-SiC target has been utilized to deposit amorphous SiC thin films on various substrates (Corning 7059 glass, unoxidized Si (111), and sapphire). The approximately 1 μm thick films were grown under various Ar sputtering pressures and flow rates. *In situ* annealing during deposition in vacuum and *ex situ* post-deposition annealing in air, both at 500°C for two hours, were implemented to determine their effects on the properties of the films. The mechanical properties were assessed via nanoindentation. An accelerated sphere-on-flat(tape) wear tester was administered to measure the wear volume losses and resultant wear rates under 0.1 and 0.2N loads, a 0.024m/s tape speed, and a 1mm ruby sphere diameter. An atomic force microscope (AFM) established the wear scar volume losses as well as the surface arithmetic roughness (R_A) and root mean square roughness (RMS) of the films. The amorphous microstructure was verified by X-ray diffractometry. There was a decreasing trend in the plastic contact damage resistance, hardness, elastic modulus, and wear resistance of the films with increased amounts of Ar gas pressure; on the other hand, annealing of the lower Ar content films generated an increase in these properties compared to the as-deposited films. Atomic force microscopy revealed a more pronounced change in surface features and roughness for the *in situ* annealed films.

INTRODUCTION

Silicon carbide has traditionally attracted attention over the years based on its outstanding properties, such as chemical, corrosion, and oxidation resistance, thermal stability, high hardness, elevated strength at high temperatures, low coefficient of thermal expansion, and high stiffness. Silicon carbide films (amorphous and crystalline) are of technological interest in potential applications including high temperature electronic devices [1], sensors [2,3], solar cells [4], X-ray lithography masks [5], and tribologically hard protective coatings [6,7]. In the field of magnetic recording media, the last application has spurred recent interest in possible alternative protective overcoats on rigid magnetic disks in order to minimize contact damage between the head (read/write transducer) and the disk. Currently the overcoats contain some type of carbon; for example, amorphous C-H or diamond-like carbon is utilized in the industry today [8,9]. The necessity for thin film overcoats (typically 20-30nm thick) possessing high hardness, low friction coefficient, high wear resistance, high plastic contact damage resistance, low surface roughness, and adherence to the bulk material are paramount especially since the trend toward in-contact (<2nm head-disk interface spacing) recording is on the horizon. Ultra-high storage densities will also dictate even thinner protective overcoats, eventually 5-10nm thick. As the flying height decreases, the number of contacts during start/stop cycles between the head and disk increases. The need for thin film overcoats with excellent tribological and mechanical properties, as well as a smooth surface to minimize mechanical friction, is clear.

EXPERIMENTAL DETAILS

<u>Preparation of films</u>

Amorphous SiC films were deposited by dc magnetron sputtering (Key Vacuum Sputtering System) with the guns positioned in a sputter-up orientation. The commercially CVD fabricated SiC monolithic target material was fabricated by Morton Advanced Materials, Woburn, MA. The theoretically dense target disk (>99.9995% purity) was 2" in diameter and 1/4" thick. The target to substrate distance was 3.5". The substrates utilized were 2" diameter unoxidized single crystal Si (111) wafers $\approx$0.01" thick, 1/2" diameter single crystal sapphire (α-Al$_2$O$_3$) disks $\approx$0.02" thick, and 2" square Corning 7059 glass plates $\approx$0.016" thick.

The background pressure prior to deposition was on the order of 2×10^{-7} Torr (2.7×10^{-5} Pa). Ultra-high purity argon gas was fed to the chamber through ports in the sputtering gun (U.S. Thin Film Products Gun II). The Ar gas pressure ranged from 2 to 14 mTorr (0.27 to 1.87 Pa) and was controlled with a gate valve. The flow rate of the gas was regulated by an MKS mass flow controller such that the total flow rate was maintained at 15 or 20 sccm (standard cubic centimeter). Power density ranges of 4.9 W/cm^2 to 14.8 W/cm^2 were applied to the target resulting in 560 to 730 volts. A pre-sputtering step 5-10 minutes in duration was performed to eliminate any impurities, debris, etc. on the target. The power was ramped at a rate of $\approx$17W/minute to the final deposition power to avoid damage to the brittle target. The increasing erosion zone, i.e. wear, of the target at different stages of sputtering did not alter the mechanical properties of the films prepared under similar deposition conditions. The deposition times were varied in order to achieve $\approx$1μm thick films. The substrate fixed on a holder was heated to

311

500°C during deposition for *in situ* annealing. *Ex*-situ annealing at 500°C was implemented in a Flexus (model FLX 2320) furnace. Film thickness was measured by a Dektak IIa surface profilometer, in which a step between the substrate and film quantitatively established the height.

<u>Characterization of films</u>

The characterization of the structure of the films was carried by X-ray diffraction (Rigaku D/Max-2BX Diffractometer) with Cu Kα radiation using a thin film goniometer. Silicon and carbon contents were quantified using a Philips XL30 scanning electron microscope (SEM) coupled with a energy-dispersive X-ray spectroscopy (EDXS) package. A 0.5keV accelerating voltage was employed to minimize the substrate EDXS signal. The surface topography and roughness of the films were measured with an atomic force microscope (TMX 2000 Explorer, Topometrix, Co.) in the contact mode in air with a 2µm liquid scanner. The silicon nitride tip attached to the cantilever was scanned over a 1 x 1µm^2 range. In addition, the wear scar volume was quantified with a 100µm dry scanner in air over a 100 x 100µm^2 range.

The hardness and elastic modulus of the films were assessed with a mechanical properties microprobe (MPM) utilizing a Berkovich diamond indenter (Nano Indenter® II, Nano Instruments, Inc.). The apparatus can sense both load and displacement continuously while simultaneously the probe indents the film. A constant displacement loading rate in the load segment and a depth limit control were used. Three indenter penetration depths (40, 60, and 90 nm) were used on each film with 5 indentations for each depth for a total of 15 indents on every film. For these thin films deposited on a thick substrates, indentation depths < 20% of the film's thickness were generally sufficient to eliminate the Substrate Hardness Effect (SHE). The determination of the hardness and elastic modulus were calculated using the method of Oliver and Pharr [10].

The wear resistance was characterized by an accelerated sphere-on-flat wear tester developed in our lab [11]. The films were subjected to wear by pressing a 1mm diameter ruby sphere on the back of a moving metal particle (MP) magnetic tape which is rubbing against the film surface. The wear resistance was assessed by making a 2 x 5 array of circular wear scars on each film as a function of two loads (0.1 and 0.2 N) and five times (40, 60, 80, 100, 120 s) under a fixed tape speed of 0.024 m/s. The AFM quantified the average diameter and depth of the wear scars necessary to calculate the wear scar volume loss using the following equation:

$$V=\pi hd^2/8 \qquad\qquad\qquad (1)$$

where V is the wear scar volume, and h and d are the depth and diameter of the wear scar, respectively. In addition, the resultant specific wear rates (k) were derived from the normal test load (F_n), the tape speed (v), the test duration (t), and V, as follows:

$$k=V/F_n vt. \qquad\qquad\qquad (2)$$

RESULTS AND DISCUSSION

<u>X-ray Diffraction and EDXS</u>
A typical X-ray diffraction scan of an a:SiC film on a Si(111) substrate in the vitreous state is displayed in Figure 1. The characteristic tetrahedral short-range compositional order characteristic of amorphous thin films is evident from the scan. The expected positions of the common α and β-SiC peaks are noted. No crystalline SiC peaks were detected for any of the sputtering deposition conditions used. In addition, annealing up to 500°C for 2 hours caused no change in the XRD scans . Only the strong preferred orientation of the Si (111) substrate was evident at a 2θ of 29°. The transformation from the amorphous to crystalline phase in sputtered SiC films is strongly dependent upon the substrate temperature during deposition. Various crystallization temperatures for the amorphous to cubic β phase have been reported for SiC thin films. Previously the transformation temperature range was determined to be 850°-900°C for diode sputtering of a single crystal hexagonal SiC target under an Ar discharge [12]. However, other investigators rf sputtered from α-SiC and β-SiC targets in an Ar atmosphere and subsequently found 550°C and 700°C as the crystallization temperatures, respectively [13].

The discrepancy may be due to several factors including target composition, sputtering rates, pressures, etc. For example, a decrease in the deposition rate, in addition to the effect of substrate heating, induced the transformation to the crystalline phase for sputtered SiC thin films [14]. However, in the present study, a decrease in the deposition rate from 2.8Å/sec to 0.95Å/sec with *in situ* annealing at 500°C resulted in no observable crystalline transformation.

A preliminary EDX analysis of a common sputtering condition at room temperature was performed. The characteristic X-ray spectrum was analyzed and subsequently the Si/C ratio was quantified to be 46a/o C and 54a/o Si, i.e. near stoichiometric.

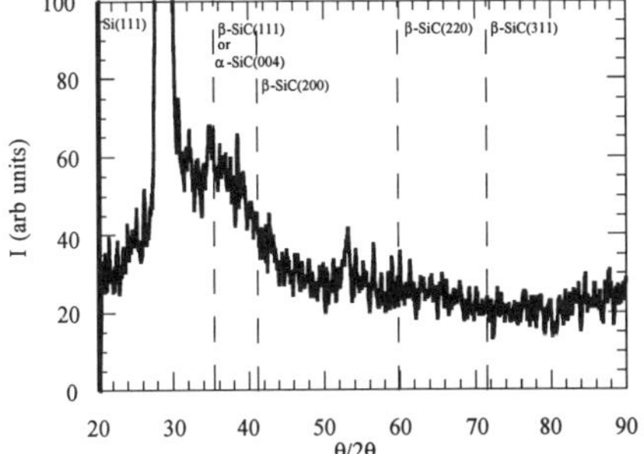

Fig. 1. XRD scan of amorphous SiC film.

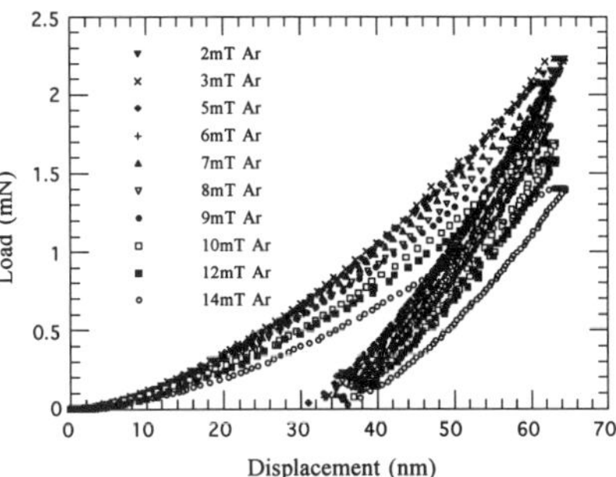

Fig. 2. Load-displacement curves for unannealed a:SiC films at 60nm penetration depths for varying Ar sputtering pressures.

Nanoindentation

1. Sputtering Gas Pressure Effects on Thin Film Mechanical Properties

Figure 2 displays typical loading vs. displacement curves for unannealed amorphous SiC films deposited on sapphire substrates with varying Ar sputtering pressures. The a:SiC film with a pressure of 3mTorr Ar was the hardest film, since this film required the highest load to reach a given indentation depth. Furthermore, as the Ar pressure was increased, the loads systematically decreased at a specified indentation depth, which translated to a decline in the hardness. Figure 3 illustrates this trend of decreasing hardness as well as elastic modulus and plastic contact damage resistance (H/E ratio) with increasing Ar pressure. After 6mTorr of Ar was introduced into the films, the values began to significantly decrease until a low of ≈15GPa in hardness, ≈215GPa in elastic modulus, and ≈0.069 in H/E ratio were obtained. The optimum films with high hardness (≈25GPa), elastic modulus (≈330GPa), and H/E ratio (≈0.8) values were those sputtered at low pressures (2,3, and 4 mTorr Ar). The explanation for the decline in mechanical properties may be related to a decrease in film density as the sputtering pressure increases [13].

The H/E ratio is an attractive assessment of the material's ability to resist plastic deformation in contact events. Less damage is obtained with high hardness and low elastic modulus materials, since the material can elastically deform to distribute the contact load over a larger area thus reducing the contact force [15]. Accordingly, the increase in the H/E ratio for the 2, 3, and 4mTorr Ar films translated to a slight increase in plastic contact damage resistance despite their high elastic modulus.

A similar characteristic load vs. displacement curve for a triode sputtered unannealed a:SiC film has been reported with an accompanying hardness of ≈28GPa and elastic modulus of ≈231Gpa [16]. In addition, another study determined a hardness the of ≈27GPa and elastic modulus of ≈250GPa for rf-sputtered SiC films [8]. The hardness values are comparable to the present study; however, the elastic modulus values are less stiff than those in

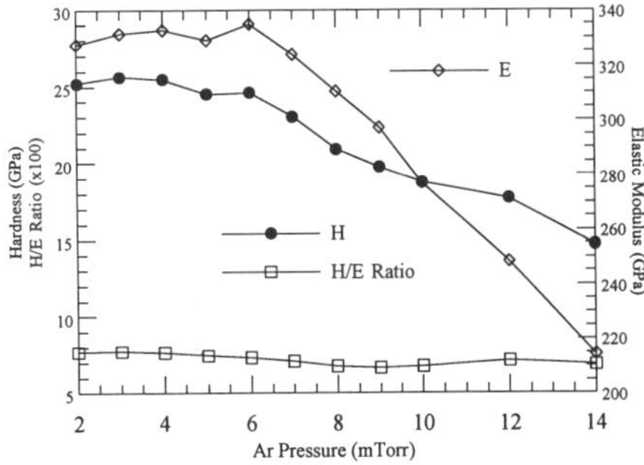

Fig. 3. Average H, E, and H/E ratio values as a function of Ar pressure for unannealed a:SiC films at an indentor displacement of 60nm.

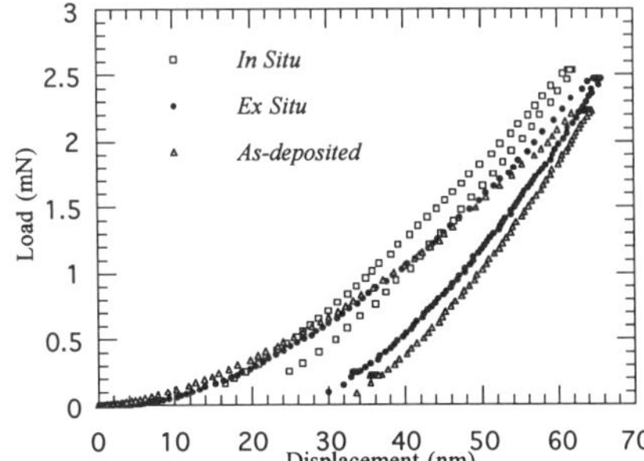

Fig. 4. Load-displacement curves for *in situ, ex situ* annealed, and as-deposited 2mTorr Ar pressure a:SiC films at 60nm depths.

this study by ≈100GPa and subsequently the plastic contact damage resistance is greater. There is a lack of experimental studies in the literature pertaining to sputtering pressure affects on the nanoindentation mechanical properties.

2. Heat Treatment Effects on the Mechanical Properties

Hardness, elastic modulus, and H/E ratio values were measured for films with 2mTorr Ar sputtering pressure in the as-deposited and annealed states. Figure 4 illustrates load vs. displacement curves for these conditions. An increase between the *ex situ* annealed and as-deposited films of up to 13%, 35%, and 21% in elastic modulus, hardness, and H/E ratio values, respectively were measured. The percent difference between the *in situ* annealed and as-deposited films was even more significant, i.e. a decrease of up to 11% and an increase of 51% and 75% in elastic modulus, hardness, and H/E ratio values, respectively. As a result of the decrease in elastic modulus for the *in* situ annealed films, the plastic contact damage resistance dramatically increased surpassing both unannealed and *ex situ* annealed films. Contrary to these results, a report on the annealing of the crystalline target material to 1000°C in N_2 and vacuum resulted in virtually no change in hardness values [17]. There was an increase in the as-deposited thin film hardness compared to the monolithic β-SiC target material, i.e. the film exhibited a nanohardness of ≈25GPa while the crystalline target possessed a Vickers microhardness of ≈21 GPa [17]; however, there was a decrease in Young's modulus from 466 GPa to ≈ 330GPa.

Nanoindentation studies have been performed on PECVD a:SiC:H films at substrate temperatures of 175°C and 600°C.[9] An accompanying increase from 18 GPA to 46 GPA and 145 GPA to 241 GPA in hardness and elastic modulus, respectively were measured for this increase in temperature and associated microstructural change. Another investigation of pulsed ArF laser deposited a:SiC and β-SiC films determined a microhardness of 42.6GPa for a 7μm thick a:SiC film at 700°C deposition temperature [6]. Contrary to the present results the microhardness the authors calculated was comparable for films deposited from 300°C to 950°C, i.e. no correlation between microhardness and microstructural change. However with the scattering of hardness values for each film, they surmised a weak trend existed of slightly increasing hardness with increasing temperature. In this investigation the *in-situ* annealed hardness results agree quite well with the latter investigation for films that were still in the amorphous state at 700°C.

Although there were no transitions from the amorphous to nanocrystalline state from XRD, a microstructural change to some extent has in all probability occurred. With increasing substrate temperatures thin films become more dense. Fourier Transform Infrared Spectroscopy (FTIR) analysis has been performed on unannealed a:SiC thin films and it was determined that the hardness and elastic modulus properties can be improved by an increase in Si-C bond density [16]. This increase will lead to an enhancement of the cross-linkage of Si and C atoms, resulting in a strengthened a:SiC material frame and ultimately an improved amorphous network stiffness. In addition, the increase in the mechanical properties during annealing may reside in the reduction of any free H atom impurities which form hydrocarbons (C-H bonds) on the films [18].

<u>Wear Resistance Properties</u>

Figure 5 depicts the wear volume loss and specific wear rates for films with varying amounts of Ar sputtering pressure. There is an increasing trend in these properties as the Ar pressure is systematically increased, similar to the behavior of the mechanical properties. The highest wear resistance and lowest specific wear rates were

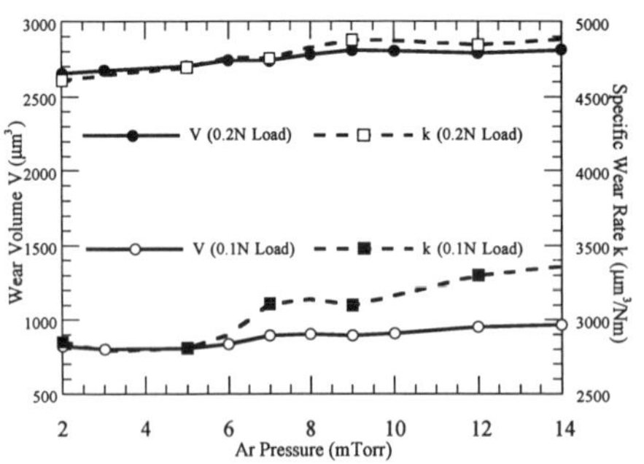

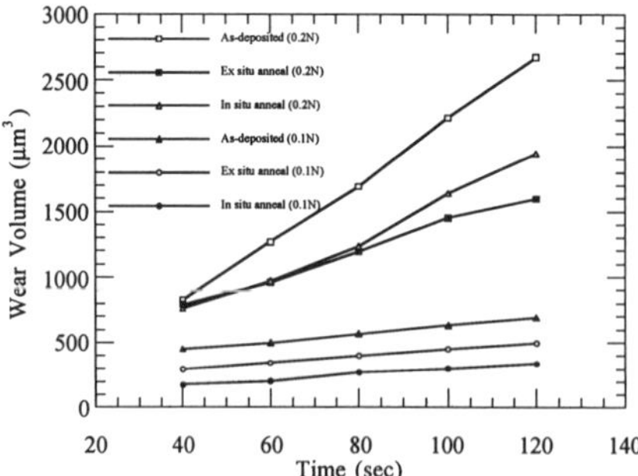

Fig. 5. Wear volume loss (V) and specific wear rate of as-deposited a:SiC films as a function of Ar sputtering pressure under 0.1 and 0.2N loads and 2 minutes.

Fig. 6. Wear volume loss (V) of as-deposited, *ex situ* and *in situ* annealed 3mTorr Ar pressure a:SiC films as a function of time under 0.1 and 0.2N loads.

observed for the low pressure ranges, i.e. 2-5mTorr Ar at both 0.1 and 0.2N loads. Figure 6 shows the effects of annealing on the wear properties for 3mTorr Ar pressure films. For both loads the annealed films exhibited higher wear resistance as a function of time. Figure 7 illustrates typical AFM wear scar images of *in situ* annealed and unannealed 3mTorrr a:SiC films. No transfer of the MP magnetic tape in the circular wear scar was evident. A similar study utilizing the same wear tester for protective overcoats of TiB$_2$ and TiB$_2$(N) films showed lower wear resistance under similar testing conditions compared to the a:SiC films [19].

<u>Surface Roughness Studies</u>

The previously stated premise that some microstructural change has occurred with annealing was investigated by atomic force microscopy. AFM images of the topography of the bare unoxidized Si(111) substrate and SiC film (unannealed and *in-situ* annealed 500°C) with similar sputtering parameters are illustrated in Figure 8. *Ex situ* annealed films were intentionally omitted due to surface contaminats which influenced the surface roughness parameters. A 1μm^2 area for each film was scanned three times in different positions on the film. The average roughness (R$_A$) and root mean square roughness (RMS) values were then calculated for each scan.

The ≈279μm thick substrate (Figure 8a) possessed a R$_A$=0.106nm and RMS=0.136nm. The unannealed 1μm thick SiC film (Figure 8b) exhibited a R$_A$=0.039nm and RMS=0.050nm. The sputtered SiC has apparently smoothed the Si substrate. However, the *in situ* annealed 1μm thick film (Figure 8c) with R$_A$=0.138nm and RMS=0.192nm exhibited a rougher surface accompanied by larger hill-type features. These protrusions may be nanocrystallites indicating the onset of crystallization too small to be detected by X-ray diffraction [6].

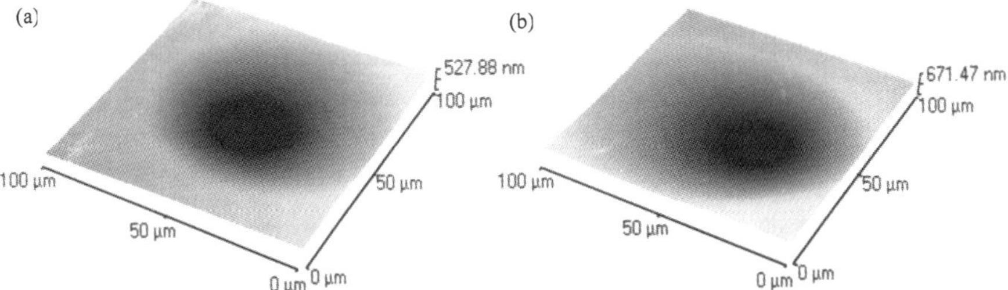

Fig. 7. AFM wear scar images of a) *in situ* annealed b) unannealed 3mTorr Ar pressure a:SiC films under a 0.2N load 2 minute test.

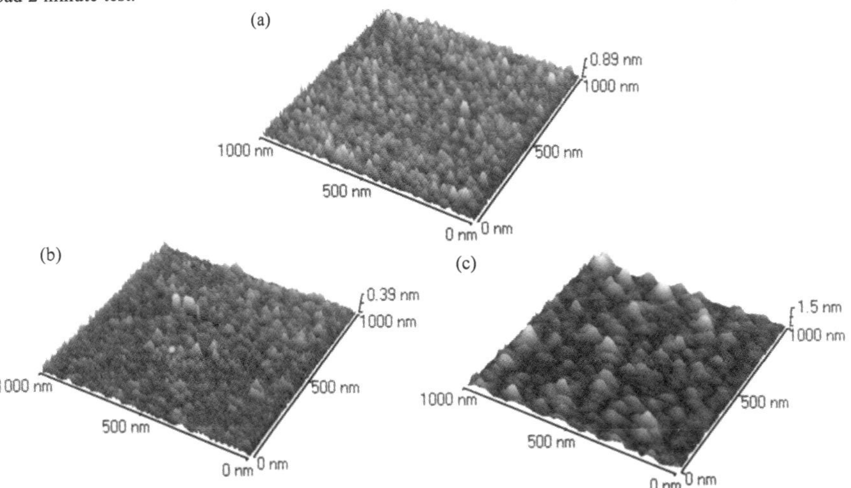

Fig. 8. AFM topography images of a) unoxidized Si(111) substrate b) unannealed and c) *in situ* annealed 500°C 3mTorr Ar pressure films.

CONCLUSION

Amorphous thin films of silicon carbide have been deposited on Si(111) and α-Al$_2$O$_3$ substrates by DC-magnetron sputtering. In retrospect, the effects of various argon sputtering pressures and thermal heat treatments of the films on the mechanical and tribological properties and subsequent surface characterization were assessed. The sputtering gas pressure effects revealed the 2,3, and 4mTorr films exhibited the most attractive properties, high hardness, plastic contact damage resistance, and high wear resistance and low wear rates. Similarly these identical pressure conditions resulted in even superior mechanical and tribological properties when the films were *in situ* annealed to 500°C.

ACKNOWLEDGMENT

This work is supported by the National Science Foundation Grant No. EHR-9108761 (EPSCoR-Alabama). Also, the use of the facilities at the Materials for Information Technology Center was appreciated.

REFERENCES

1. H.P. Philipp and E.A. Taft, in <u>Silicon Carbide, A High Temperature Semiconductor</u>, edited by J.R. O'Connor and J. Smiltens (Pergamon Press, New York, 1960), p. 371.

2. Y.H.C. Cha, S. Jou, S. Prakash, H.J. Doerr and R.F. Bunshah, Mat. Sci. & Eng. **A163**, 207 (1993).

3. Y.H.C. Cha, H.J. Doerr, and R.F. Bunshah, Surf. Coat. Technol. **62**, 697 (1993).

4. J. Bullot and M.P. Schmidt, Phys. Status Solidi, **B143**, 345 (1987).

5. M. Chaker, S. Boily, Y. Diawara, M.A. El Khakani, E. Gat, A. Jean, H. Lafontaine, H. Pepin, J. Voyer, J.C. Kiefer, A.M. Haghiri-Gosnet, F.R. Ladan, M.F. Ravet, Y. Chen, and F. Rousseaux, J. Vac. Sci. Technol. **B10** (6), 3191 (1992).

6. T. Zehnder, A. Blatter, J. Burger, C. Julia-Schmutz, and R. Christoph, in <u>Tribological Properties of Laser Deposited SiC Coatings</u>, edited by K. Barmak, M.A. Parker, J.A. Floro, R. Sinclair, and D.A. Smith (Mater. Res. Soc. Proc. **343**, Pittsburgh, PA, 1994), p.621.

7. J. Meneve, R. Jacobs, F. Lostak, L. Eersels, E. Dekempeneer, and J. Smeets, in <u>Micromechanical Behaviour of Amorphous Hydrogenated Silicon Carbide Films</u>, edited by P.H. Townsend, T.P. Weihs, J.E. Sinclair Jr., and P. Borgesen (Mat. Res. Soc. Symp. Proc. **308**, Pittsburgh, PA, 1993), p.671.

8. B.K. Gupta and B. Bhushan, IEEE Trans. on Magn. **31** (6), 3012 (1995).

9. M.J. Loboda and M.K. Ferber, J. Mater. Res. **8**, 2908 (1993).

10. W.C. Oliver and G.M. Pharr, J. Mater. Res. **7**, 1564 (1992).

11. H. Deng, V.R. Inturi, and J. A. Barnard, in <u>Mechanical Properties and Wear Resistance of High Moment Thin Film Head Materials</u>, edited by S.P. Baker, C.A. Ross, P.H. Townsend, C.A. Volkert, and P. Borgesen (Mater. Res. Soc. Proc. **356**, Pittsburgh, PA, 1995) pp. 773.

12. C.J. Mogab and W.D. Kingery, J. App. Phys. **39**, 3640 (1968).

13. K. Wasa, T. Nagai, and S. Hayakawa, Thin Solid Films **31**, 235 (1976).

14. Y. Hirohata, Y. Nemoto, T. Hino, and T. Yamashina, Thin Solid Films **214**, 150 (1992).

15. T.Y. Tsui, G.M. Pharr, W.C. Oliver, Y.W. Chung, E.C. Cutiongco, C.S. Bhatia, R.L. White, R.L. Rhoades, S.M. Gorbatkin, in <u>Nanoindentation and Nanoscratching of Hard Coating Materials for Magnetic Disks</u>, edited by S.P. Baker, C.A. Ross, P.H. Townsend, C.A. Volkert, and P. Borgesen (Mat. Res. Soc. Symp. Proc. **356**, Pittsburgh, PA, 1995), p.767.

16. M.A. El-Khakani, M. Chaker, A. Jean, S. Boily, J.C. Kieffer, M.E. O'Hern, M.F. Ravet, and F. Rousseaux, J. Mater. Res. **9**, 96 (1994).

17. H. Wang, R.N. Singh, and J.S. Goela, J. Amer. Cer. Soc. **78**, 2437 (1995).

18. S.M. Ojha, in <u>Physics of Thin Films 12</u>, edited by G. Hass, M.H. Francombe, and J.L. Vossen (Academic Press, Inc., New York, 1982), p.273.

19. H. Deng, J. Chen, R.B. Inturi, and J.B. Barnard, Surf. Coat. Technol. **76-77**, 609 (1995).

THERMAL PROPERTIES OF MAGNETRON SPUTTERED TiN AND TiAlN THIN FILMS ON HSS

E. Lugscheider, O. Knotek, C. Barimani, H. Zimmermann, M. Lake, Aachen University of Technology, Materials Science Institute, Aachen, Germany

ABSTRACT

Increasing demands on production processes in terms of performance, reliability and environmental compatibility shape the specification profile for modern wear resistant coating systems. The specification of PVD coated cutting tools by the parameters hardness, wear and frictional behavior, chemical structure or chemical resistance are already known and inspected by small scale tests, e.g. pin on disc or taber abraser test. Hitherto the thermophysical properties of thin film PVD coatings were not characterized completely although the investigation of the thermal diffusivity of coated substrates give information about the thermal barrier function of the deposited coating.

To determine the thermophysical properties of PVD coatings the Jenoptik Thermal Wave Inspection (TWIN) can be used. This nondestructive and contactless measuring system is based on the photothermal spectroscopy. In this paper fundamentals of the TWIN measuring technique are presented as well as first TWIN results of investigated MSIP TiN and TiAlN coated and oxidized high speed steels.

INTRODUCTION

In cutting processes more than 90% of the mechanical energy is transformed into thermal energy of the cutting edge. The chip formation area and the contact zone between the chip and the rake face of the tool are to be considered as the main heat sources. The temperature in last surnamed zone rises up to 1000 °C and distribution of temperature in the system is determined by the value of the thermal diffusivity of the coating. The lower it is, the more heat remains in the chip and in the area of chip formation, e.g. the shear plane. This affects the chip formation as well as the position and the width of the contact zone on the rake face of the tool. PVD coatings with a low thermal diffusivity can be used as a thermal barrier on the tool surface to increase the thermal loads. Therefore TiN and TiAlN amongst other coatings are deposited. TiN coatings are stable regarding their structure in a wide range of temperature during the cutting process in contrast to TiAlN coatings. Dependent on the process temperature aluminium extravasates from the TiAlN coating and forms an oxide film on the tool surface. This effect can be described by an oxidation model presented in [1]. Regarding the TiN and the TiAlN coating antagonistic behavior of thermal diffusivity dependent of the process temperature is expected.

The before-mentioned PVD coatings were investigated by using the Jenoptik Thermal Wave Inspection System (TWIN). This non destructive and contactless measuring system is based on a novel photothermal measuring technique. The TWIN system was applicated to determine the thermal conductivity of coated and oxidized HSS tips in a short measuring sequence.

EXPERIMENT

The Thermal Wave Analysis (TWA) can be used for in situ evaluations in the fields of PVD coating, semiconductor testing, microtechnology and electronics. This presented TWIN system was originally developed for testing semiconductors regarding to their grade of implantation, surface quality and for defect control. Recently this novel measurement system was adapted for

Mat. Res. Soc. Symp. Proc. Vol. 436 © 1997 Materials Research Society

the demands of the PVD coating technology for determination the thermophysical properties, in this case thermal diffusivity.

The TWIN system is designed as modular stand-alone measuring system instrument composed of a TWM measuring head with viewing system, stand, scanning stage with MCL-2 positioning system, ESG Electronic control unit, personal computer, TWA control and analysis software. The microscope unit can be used to detect surface defects, e.g. cracks or grind marks, or layer anomalies, e.g. droplets.

PRINCIPLE OF PHOTOTHERMAL MEASURING TECHNIQUES

The basis of all photothermal techniques is the excitation of a sample by an intensity modulated laser beam. Due to absorption of light by the sample surface so-called thermal waves are generated. The propagation of these highly damped thermal waves depends on the sample properties to be measured such as sample material or material combination, thermal conductivity, thermal diffusivity, transient thermal impedance etc. It is detected by an appropriate technique, e.g. optically by measuring the change in the reflection coefficient of the sample. Employing a thermal process as information carrier makes sample properties accessible that can not be detected by mere optical methods. The applied method of optical excitation and detection ensures non-contact and nondestructive measurements [2].

On the TWIN system the method of frequency double modulation is applied. Laser light of a single wavelength (λ=785 nm) is employed for both excitation of the sample and reading the sample information. For this purpose, the laser light is additively intensity modulated with two closely adjacent modulation frequencies (Ω_1 and Ω_2). The difference frequency between Ω_1 and Ω_2 counts 10 kHz for the whole frequency range. This excitation is effective on the sample surface with the arithmetic mean of the two frequencies. The triggered thermal wave response is contained in the reflected portion of the light in the form of mixed products of both frequencies ($2\Omega_1$, $2\Omega_2$, $\Omega_1+\Omega_2$, $\Omega_1-\Omega_2$) which are not existing in the excitation light. On the TWIN system, the low frequency difference $\Omega_1-\Omega_2$ is detected and interpreted. The operating principle of frequency double modulation is illustrated in Fig. 1 [2].

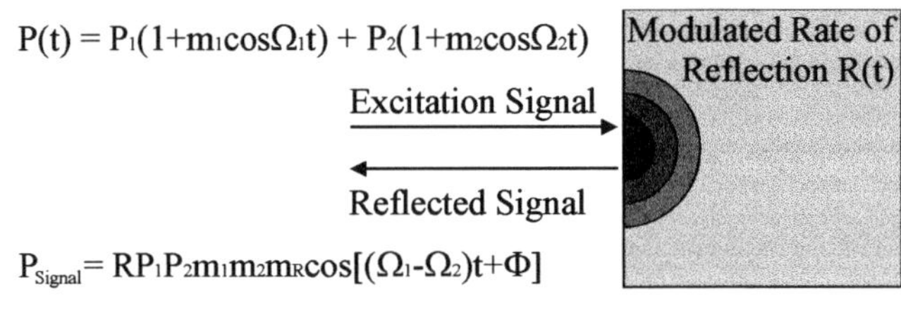

$$P(t) = P_1(1+m_1\cos\Omega_1 t) + P_2(1+m_2\cos\Omega_2 t)$$

Excitation Signal

Reflected Signal

$$P_{Signal} = RP_1P_2m_1m_2m_R\cos[(\Omega_1-\Omega_2)t+\Phi]$$

$$R(t) = R[1+m_RP_1m_1\cos(\Omega_1 t-\Phi) +m_RP_2m_2\cos(\Omega_2 t-\Phi)]$$

Fig.1 : Operating principle of frequency double modulation

A major advantage of this technique is the possibility of synchronous variation of modulation frequencies - the so-called frequency sweep - at constant detection frequency $\Omega_1-\Omega_2$. The frequency sweep (100 kHz - 1.2 MHz) extends the information obtainable about the sample, since the penetration depth of the thermal wave depends on the modulation frequency, thus permitting different depth ranges to be analyzed. In such a measuring process, the thermal penetration depth l_t of the highly damped thermal waves into the sample is of key significance.

Due to the frequency dependence, the penetration depth can be set on the measuring system and adapted to the lateral and vertical dimensions to be investigated. The wave property includes reflection, refraction and interference of thermal waves [2].

The constancy of detection frequency provides advantageous and comparatively simple electronic solutions by means of the lock-in technique. To obtain a measurement result that is independent of the reflection coefficient and the conditions of excitation, the conversion coefficient K was introduced as the basis for the interpretation. The conversion coefficient K represents that part of the reflected laser power that was converted per absorbed laser power density into the difference frequency by means of thermal reflection. In terms of measurement strategy it turned out to be useful to determine K as a complex quantity in form of equation 1:

$$K = \frac{Pe^{i\phi}w}{R(1-R)m_1m_2P_1P_2} \qquad (1)$$

with conversion coefficient   K   [m/W]

with	conversion coefficient	K	[m/W]
	effective laser beam waist on the sample	w	[µm]
	reflectivity	R	
	degree of modulation	m_1, m_2	
	laser power	P_1, P_2	[W]
	signal power	P_{Signal}	[W]
	phase shift	ϕ.	

The complex conversion coefficient K is optionally put out as real part Re {K} and imaginary part Im {K}, as well as amount |K| and phase Φ or if necessary as a circle plot [2].

In [3] is was already shown that by using the frequency sweep option of the TWIN system it is possible to carry out quantitative measurements of certain material parameters such as thermal diffusivity without a calibration process. Therefor the maximum peak of the frequency plot of Im {K} must be detected. In our investigation it was only possible to carry out a qualitative determination of the thermal diffusivity of the coatings by using the following dependency (2):

$$\kappa \sim \Omega_{max.}w^2 \qquad (2)$$

with	thermal diffusivity	κ	[cm^2/s]
	maximum frequency of Im {K}	$\Omega_{max.}$	[Hz]
	effective laser beam waist on the sample	w	[µm].

RESULTS

To testify the connection between the thermal diffusivity κ and maximum peak of the frequency of Im {K} $\Omega_{max.}$ three specimen with different thermophysical properties were investigated by the TWIN system. The reference materials were aluminium, copper and steel in bulk configuration. Their thermophysical properties and the detected maximum frequencies are presented in Table I [4].

Tab. I: Thermophysical Characteristics and TWA results of the Reference Samples

Material	Density [kg/m³]	Spec. Thermal Capacity [J/kgK]	Heat Conductivity [W/mK]	Heat Diffusivity [cm²/s]	Maximum Frequency [MHz]
Copper	8960	394	384	1,02	0,8
Aluminium	2700	945	238	0,94	0,6
Steel	7860	456	81	0,16	0,14

The modulation parameter for investigation bulk materials and PVD coatings, e.g. modulation frequencies and laser power, were ascertained in a various number of tests. Best results were achieved by using 12 mW as the laser power for excitation and a range from 100 kHz to 1.2 MHz in 100 kHz steps for modulation the intensitation of laser beam.

By TWIN investigation of the reference samples the real part Re {K} and the imaginary part Im{K} were detected. For determination the thermal diffusivity of the material only Im{K} was analyzed [5]. The investigated values are shown in the table above. The maximum peak of frequency of the steel sample is lower than the maximum peak of the copper and the aluminium material. It is obvious that the connection in form of the equation $\kappa \sim \Omega \cdot w^2$ is given. This fact admits the possibility to determine the thermal diffusivity in a qualitative manner on PVD coated substrates, e.g. cemented carbides, cermets, high speed steel, plastics or glass.

By determination the thermal diffusivity of PVD coated substrates the thermal barrier function of the film-substrate and their corresponded interface, can be predicted. A thermal barrier function of a PVD coating is characterized by its high thermal diffusivity value.

TWA EXAMINATION OF COATED AND OXIDIZED HIGH SPEED STEEL

For the test series commercial high speed steel (HS 6-5-2) from an industrial supplier was used. The HSS samples were coated with TiN and TiAlN by the MSIP PVD method. The coated samples were oxidized by an atmosphere annealing furnace on a temperature of 650 °C and 800°C for a duration of 1 hour.

By TWIN investigation of the described samples the real part Re {K} and the imaginary part Im{K} were detected but only Im{K} was analyzed. In Fig. 2 the imaginary part Im {K} depending on the frequency sweep is shown for the TiN coated/oxidized substrate material. The maximum peak of frequency of the coated sample (No. 1) and the coated and 650 °C oxidized sample (No. 2) is lower than the maximum peak of the coated sample (No 3) which was oxidized by 800 °C. The maximum peaks of the frequencies of the imaginary part of the coefficient K correlating with the thermal diffusivity. A lower frequency means a lower thermal diffusivity and a lower thermal conductivity. Between the non oxidized and the 650 °C oxidized sample there was no difference detectable. Reason for this effect is the high stability of the TiN coating under thermal load. The oxidization by 800 °C increases the thermal diffusivity of the tested sample.

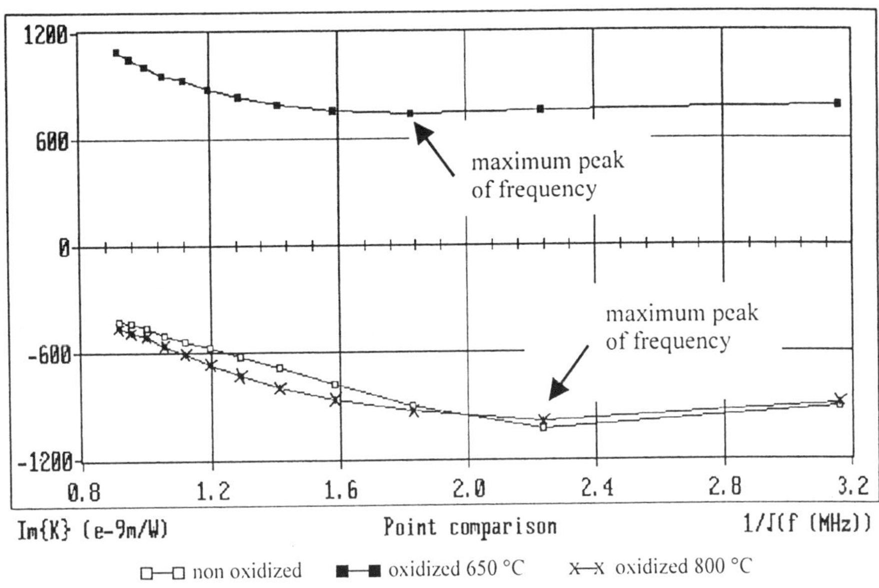

Fig. 2: TWA examination of TiN coated and oxidized HSS

In Fig.3 the imaginary part Im {K} depending on the frequency sweep is shown for the TiAlN coated and coated/oxidized substrate material.

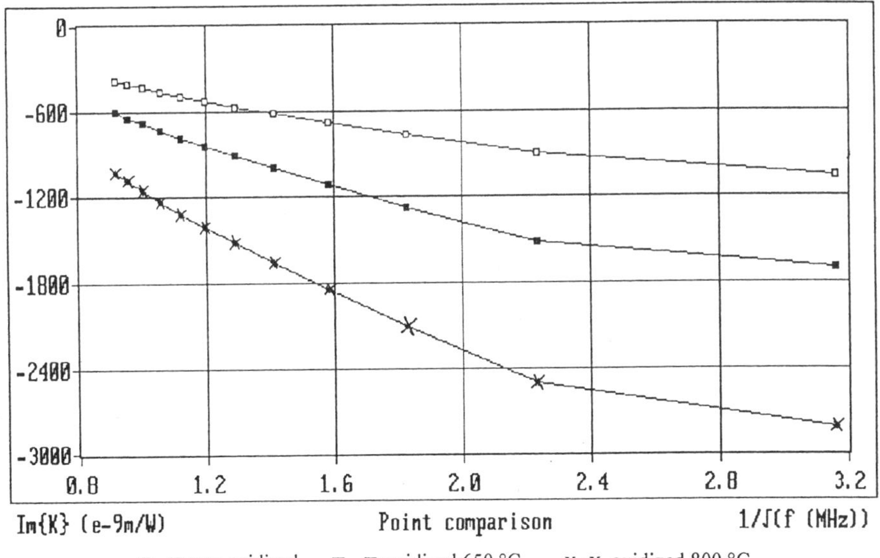

Fig. 3: TWA examination of TiAlN coated and oxidized HSS

Between the TiAlN coated and oxidized specimen there was no maximum peak of frequency detectable. Substantiated by this fact no thermal diffusivity can be determined. Reason for this effect is the bound of the lowest frequency by a value of 100 kHz. The frequency range of the presented TWIN system was originally developed for testing semiconductors regarding their grade of implantation or defect control. For this application the frequency range is sufficient. Recently this novel measurement system was adapted for the demands of the PVD coating technology for determination the thermophysical properties, in this case thermal diffusivity.

CONCLUSIONS

The thermal wave analysis is an non destructive and contactless process for testing materials. A detection of the generated thermal waves by modulated reflectivity measurement allows the characterization of the response on the laser excitation within a time of some milliseconds.

Due to the adaptation of the new TWIN system to investigate thin PVD coatings a lot of experiments were necessary to determine the exciting parameters, e.g. laser power or frequency range. Further the specimen preparation, e.g. grinding and polishing or the cleaning of the surface by ultrasonic is of a great interest because results were yield by these processes.

The presented results show that the thermal diffusivity of TiN coatings are stable under thermal load up to 650 °C. The thermal diffusivity of 800 °C oxidized samples increases regarding to the before-mentioned specimen.

The investigation of the thermal behavior of TiAlN coatings was insufficient because of the bounded frequency range.

This presented Thermal Wave Analysis system offers a great variety of opportunities regarding to the characterization of thin PVD coatings, e.g. determination of film thickness or investigation of the bonding. These problems will be item of further research and development activities.

REFERENCES

1. A. Barth, <u>Oxidationsverhalten metastabiler $(Ti_{1-x}Al_x)N$-Schichten</u> , doctoral thesis, Aachen University of Technology, 1994

2. N.N., <u>Instruction manual for the Jenoptik Thermal Wave Inspection system</u> (TWIN), Jenoptik GmbH, Jena, Germany, 1994

3. H.D. Geiler, Kontaktfrei photothermisch Messen, <u>Proceedings 2. Jenaer TWA-Workshop</u>, Jenoptik GmbH, Jena, Germany, 1995

4. K. Gieck, <u>Technische Formelsammlung</u>, Gieck-Verlag, Heilbronn, 1989

5. E. Lugscheider, M. Lake and H. Zimmermann, Bestimmung thermophysikalischer Kenngrößen von Arc-PVD-Verschleißschutzschichten auf Werkzeugen für die Trocken-zerspanung in <u>Proceedings zum 2. TWA-Workshop der Jenoptik Technologie GmbH</u>, Jena, Germany, 1995

COPPER GROWN DIAMOND FILMS

E. PEREIRA, QI HUA FAN, J. GRACIO*
Department of Physics, University of Aveiro, 3810 Aveiro, Portugal
*Department of Mechanical Engineering, University of Aveiro, 3810 Aveiro, Portugal

ABSTRACT

Diamond has been synthesised on copper using Microwave Plasma Chemical Vapour Deposition (MPCVD). The effect of substrate pre-treatment including different polishing and cleaning on diamond nucleation was investigated. It has been found that the residues from the polishing process played an important role in the diamond nucleation. Diamond films synthesised under optimised conditions presented a narrow $1332 cm^{-1}$ Raman peak. During the ramp down procedure, diamond films usually cracked due to the large thermal mismatch between copper and diamond. To avoid this problem, the effect of growth temperature and deposition time was investigated and a two-step-growth method for stress relief is proposed.

INTRODUCTION

Copper is a common and widely used material. It has a cubic crystal structure like diamond and its lattice parameter at room temperature is similar to that of diamond (diamond a=3.567Å, Cu a=3.608Å). Furthermore, copper almost does not dissolve carbon. These make copper a promising candidate for heteroepitaxial growth of diamond.

The main difficulty of synthesising diamond film on copper is that the internal stresses of the diamond film can be very large. They usually cause film cracking during post cooling procedure. These internal stresses are of two main types [1]. One is of intrinsic nature due to structural non-uniformity during the film growth. Another is the thermal stresses due to the difference of thermal expansion coefficients between diamond and copper. So, the control of the film growth conditions to reduce the thermal mismatch influence is essential.

It has been found that substrate pre-treatment as well as deposition conditions could greatly influence diamond nucleation and growth [2]-[4]. There have been a few reports on diamond growth on copper, but none of them mentioned the substrate pre-treatment effect in detail [5]-[8]. In order to understand the effect of substrate pre-treatment, we have undertaken a series of experiments with different pre-treated copper substrates. The results not only allow us to control the nucleation by pre-treatment, but also contribute for a better understanding of the nucleation mechanism on copper.

Diamond films were synthesised under optimised conditions and the main characteristics are also presented in this paper. To avoid the common problem of film cracking, low growth temperatures and different growth times have been tried and a two-step-growth method for stress relief is proposed.

EXPERIMENTAL DETAILS

The substrates were high purity polycrystalline copper foil (99.5%) about 1mm thick. They were cut into 1cm x 1cm in squares. Before pre-treatment, the substrates were polished with 2400 mesh sand paper. The pre-treatment included polishing on fine polishing cloth with diamond pastes, diamond powders, Al_2O_3 pastes, and different ultrasonic cleaning. The details

Mat. Res. Soc. Symp. Proc. Vol. 436 © 1997 Materials Research Society

of preparation are given in the corresponding sections for each case. Diamond deposition was conducted using an ASTeX PDS18, 2.45GHz MPCVD system with a cooled substrate stage. Each experiment included a ramp up of about 10min in hydrogen gas, a deposition in hydrogen-methane-oxygen gas system, and a ramp down in hydrogen gas of about 10~30min respectively. The deposition conditions were fixed as follows if not otherwise stated: microwave power, 2500W; gas pressure, 80Torr; H_2 flow rate, 478sccm; CH_4 flow rate, 30sccm; O_2 flow rate, 1sccm; deposition time, 30min. The substrate temperature is dependent on the deposition parameters, being ca 750~800°C in this case.

The deposits were analysed by a Renishaw 2000 Raman spectroscopy system with He-Ne laser and a Hitachi 4100 scanning electron microscope (SEM).

RESULTS AND DISCUSSION

Effect of Polishing Duration

The copper substrates were polished with 1μm diamond paste for 0.5-15min followed by an ultrasonic cleaning for 3min. It was found that the nucleation density increased with increasing the polishing time as shown in figure 1. However the increase is non linear indicating a saturation effect for times above 10min.

Effect of Particle Size and Polishing Materials

The copper substrates were polished with 0.25-15μm diamond paste. Polishing duration was 3min followed by a 3min ultrasonic cleaning in ethanol. Figure 2 shows that the nucleation density increases with decreasing the diamond particle size. The same tendency was found when the diamond paste was replaced by diamond powder. However, polishing with diamond powder results in a relatively higher nucleation density.

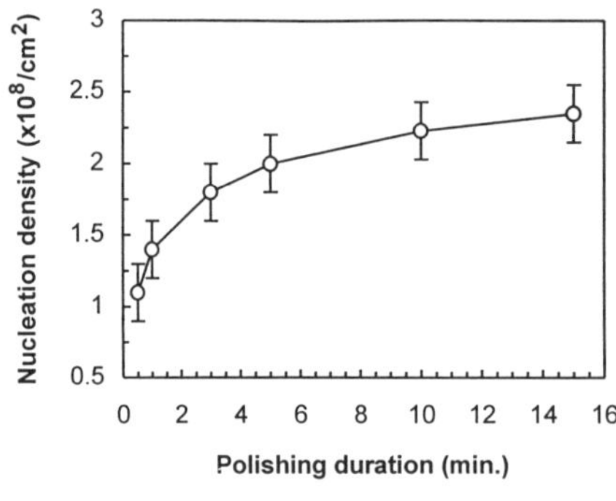

Fig.1 Nucleation density dependence on polishing duration

Fig.2 Nucleation density dependence on particle size in polishing materials

As a comparison, the copper substrates were polished with Al_2O_3 paste down to 0.25μm. The nucleation density on these samples was about 2 orders of magnitude lower than on those polished with 0.25μm diamond paste. In addition, the nucleation sites on samples polished with only sand paper were very few.

Two different polishing sequences were also performed. In the first sequence (sample 1) the substrate was polished with 7μm diamond paste, 0.25μm diamond paste and finally with

0.25μm Al₂O₃ paste. In the second sequence (sample 2) the substrate was polished with 3μm Al₂O₃ paste, 0.25μm Al₂O₃ paste and finally with 0.25μm diamond paste. Polishing duration for each sample was 3min for the first two steps and 0.5min for the final step. An ultrasonic cleaning for 3min was performed after each polishing step. Figure 3 shows that the nucleation density on sample 1 is much lower than on sample 2.

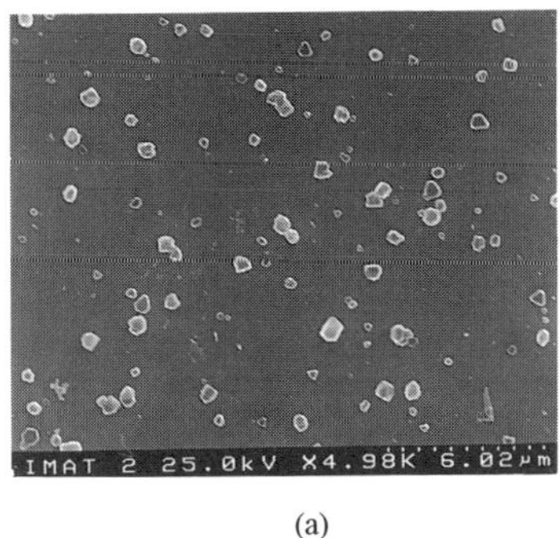

(a)

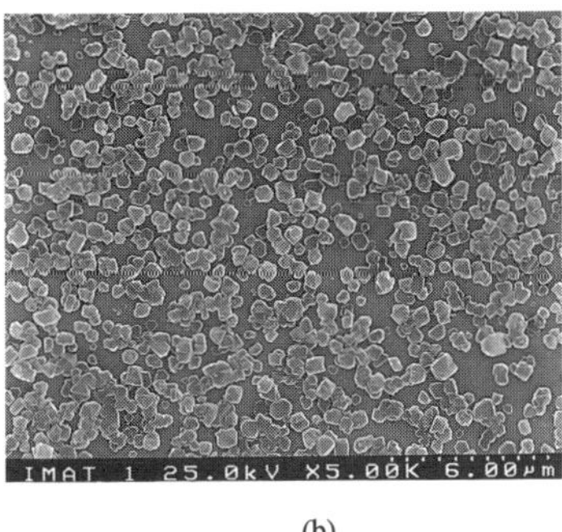

(b)

Fig.3 Diamond nucleation sites on Cu substrates polished in different sequence.
(a). Sample 1: sand paper + 7μm diamond paste + 0.25μm diamond paste + 0.25μm Al₂O₃ paste
(b). Sample 2: sand paper + 3μm Al₂O₃ paste + 0.25μm Al₂O₃ paste + 0.25μm diamond paste

Effect of Ultrasonic Cleaning Duration after Polishing

The copper substrates were polished with 7μm diamond paste for 3min and ultrasonically cleaned in ethanol for 3min. Then they were polished with 1μm diamond paste for 3min and ultrasonically cleaned for 3-90min. A small decrease of nucleation density with increasing ultrasonic cleaning time was noticed. For example, ultrasonic cleaning for 3min resulted in a nucleation density of about $1.8 \times 10^8/cm^2$; while for 90min the density decreased only to about $1.4 \times 10^8/cm^2$. On the sample ultrasonically cleaned for 3min no large remains from polishing materials were found. This time of ultrasonic cleaning seems therefore enough.

In view of these pre-treatment results, we see that the residues from polishing process play an important role in the diamond nucleation on copper. Figure 3 gives a strong support to this idea. For sample 1 shown in Fig.3(a), the final Al₂O₃ paste polishing removed a lot of diamond residues left in the copper surface from previous polishing steps. So the nucleation density decreased. On the contrary, a reverse polishing sequence, i.e. Al₂O₃ paste polishing followed by diamond paste polishing, leads to much higher nucleation density (Fig.3(b)). This experiment is also in agreement with those found on Si substrate [4].
Ultrasonic cleaning did not affect strongly the nucleation density. In fact, copper is a relatively soft material and the diamond residues could be "seeded" into the copper surface during polishing. So, ultrasonic cleaning does not remove these residues efficiently.

According to these experimental results, we may consider the optimised pre-treatment conditions as follows: polishing time, 3min; ultrasonic cleaning time, 3min; polishing material, diamond powder 0~1μm or 2~4μm, diamond paste 0.25μm or 1μm, depending on the required nucleation density.

<u>Film Growth Characteristics and Stress Relief</u>

Diamond films were successfully deposited on copper substrates pre-treated under the optimised conditions refered above. The final polishing was 0.25μm diamond paste. For a deposition time of 45min, the film thickness is about 1 μm. It was found that the adhesion of the diamond film to the copper substrate was poor. It could be easily removed from the copper substrate and became free-standing. Figure 4 shows SEM images of the diamond film. It can be seen that the diamond grain size of the film surface and back sides is similar, except that but the film back side seems denser. The Raman spectra of the film back and surface sides are shown in Fig.5. The sharp peak at about 1332cm^{-1} shows the existence of the diamond structure. The similarity of the spectra of the film surface side and back side demonstrates the uniform characteristics of the film and shows that almost no transition layer is formed between the copper substrate and diamond film.

Fig. 4 SEM images of diamond film grown on Cu. (a). film surface side, (b). film back side

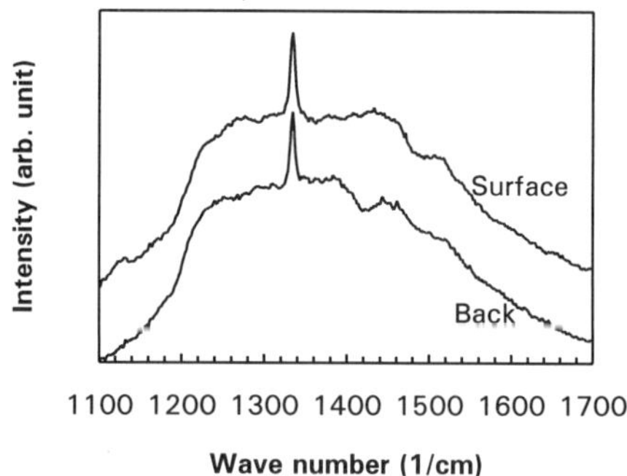

Fig.5 Raman spectra taken from the surface side and back side of the diamond film grown on Cu

We may notice that the Raman spectra shown in Fig.5 possess a high background, which is probably due to the grain boundaries and surface chemical bonds. Further investigation is currently under way to find out the origin of this background.

Our experiments showed that diamond films grown on copper usually cracked during the ramp down procedure, especially when the films were thin (<20μm). This is mainly because the mismatch of thermal expansion coefficients between diamond (α=0.8x10^{-6}/K) and copper (α=17.7x10^{-6}/K) is very large [7],[9]. To reduce this thermal mismatch, we tried to deposit diamond with a lower microwave power for decreasing the substrate temperature. When the microwave power was lower than 1500W, which corresponded to a substrate temperature of about 450°C, the diamond nucleation and growth became very slow and seemed not suitable for an efficient synthesis of diamond. Even with such a low microwave power, the diamond films still cracked. If we make an approximate calculation using the common formula [10]:

$$\sigma_{th}=E_d(\alpha_d-\alpha_{sub})\ (T_{sub}-T_r)/(1-v_d) \tag{1}$$

where E_d is Young's modulus of diamond (E_d=1210GPa), v_d is Poisson's ratio of diamond (v_d=0.1), α_d and α_{sub} are thermal expansion coefficients of diamond and substrate, T_{sub} is the substrate temperature during deposition, and T_r is room temperature, we may find that the value of thermal stresses σ_{th} is still of about 10GPa.

On the other hand, longer deposition was performed in order to obtain thicker films while the microwave power was kept at a low level (about 1800W). However, it was found that the diamond films grown on copper often cracked even during deposition. This can be explained as follows. As mentioned above, the polishing residues played a main role in diamond nucleation. These residues could be "seeded" into the copper surface and were difficult to move. With the increasing of film thickness, we found that diamond grain size increased obviously. So the difference of grain size between bottom and surface of the film increased. This led to the increasing of the intrinsic stresses in the film. As a result, the film cracked during growth.

To avoid this cracking problem, we propose a two-step-growth method. This method includes a short pre-growth followed by a quick ramp down for stress relief and a final growth. Continuous diamond film about 10μm thick and 10mm x 10mm in square, as shown in figure 6, was obtained by this method. The details will be reported elsewhere.

Fig.6 Complete diamond film grown on copper by a two-step method.

CONCLUSION

Diamond nucleation on copper is strongly affected by substrate pre-treatment. Smaller diamond particle size in polishing materials and /or longer polishing duration result in higher nucleation density. Ultrasonic cleaning does not significantly affect the nucleation density because copper is a relatively soft material and the residues can be "seeded" in the copper surface, even at room temperature. The fact that diamond polishing followed by Al$_2$O$_3$ paste polishing greatly reduces the nucleation density supports the idea that the residues left in the copper substrate surface play an important role in diamond nucleation.

The adhesion of the diamond film deposited on copper is poor because copper does not form carbide. Diamond films grown directly on copper substrate usually crack due to the large thermal mismatch. This problem may be overcome by releasing the stresses of diamond film. Free standing films can be obtained by careful control of the growth conditions.

ACKNOWLEDGEMENTS

The authors would like to thank Mr. A. Fernandes for his help in experiment and useful discussion. The financial support of NATO, research project NATO-SFS-PO-OPTOELECT is acknowledged.

REFERENCES

1. D.Schwarzbach, R.Haubner and B.Lux, Diamond and Related Materials, **3**, p.757 (1994)

2. K.Kobashi, K.Nishimura, Y.Kawate and T.Horiuchi, Physical Review B, **38** (6), p.4067 (1988).

3. W.Zhu, A.R.Badzian, R.Messier, SPIE, **1325**, Diamond Optics III, p.187 (1990).

4. E.J.Bienk and S.Eskildsen, Diamond and Related Materials, **2**, p.432 (1993).

5. Z.P.Lu, J.Heberlein and E.Pfender, Plasma Chemistry and Plasma Processing, **12**(1), p.35 (1992).

6. M.Kawarada, K.Kurihara and K.Sasaki, Diamond and Related Materials, **2**, p.1083 (1993).

7. M.L.Hartsell and L.S.Plano, J. Mater. Res., **9**(4), p.921 (1994).

8. S.S.Perry and G.A.Somorjai, J. Vac. Sci. & Tech. (A), **12**(4), p.1513 (1994).

9. M.N.R.Ashfold, P.W.May, C.A.Rego and N.M.Everitt, Chem. Soc. Rev., **23**, p.21 (1994).

10. R.W.Hoffman, in <u>Physics of Non-metallic Thin Films</u>, edited by C.H.S.Dupuy and A.Cachard (Plenum, New York, 1974), p.306

THE STUDY OF TRIBOLOGICAL PERFORMANCE AND SURFACE FILM CHARACTERIZATION OF BISMUTH DIOCTYLDITHIOCARBAMATE

Chen Ligong, Dong Junxiu, Chen Guoxu

Logistical Engineering College,Chongqing 630042, P.R.of China

ABSTRACT

In this study, bismuth dioctyldithiocarbamate has been synthesized, and its tribological behaviors, such as friction-reducing ability, antiwear property and extreme pressure performance have been respectively evaluated with a ring-on-block test rig and a fourball machine. In addition to correlate its tribological behaviors with the film formed on the metallic rubbing surface under boundary lubrication conditions, surface analyses have been conducted to characterize the surface film by means of Auger electron spectroscopy (AES), X-ray photoelectron spectroscopy(XPS) and energy dispersion of X-ray(EDX).

Test results show the additive compound can effectively improve the friction and wear of the rubbing couples. On the other hand, EDX confirmed the presence of carbon, oxygen, sulfur, nitrogen, bismuth and iron on the surface; AES revealed their depth distribution of atomic concentration percentages. Whereas XPS further disclosed that the composition of the surface film was composed of organic and inorganic species including iron sulfide and sulfate, metallic bismuth, bismuth oxide and sulfide etc. which are conducive to the reduction of friction and wear.

INTRODUCTION

Dithiocarbamates such as zinc dithiocarbamate have been widely used in lubricants for internal combustion engines, compressors, gas turbines, hydraulic systems,gears etc. due to their excellent anti-oxidation stability, antiwear and extreme pressure performances, higher decomposition temperature than dithiophosphates(such as ZDDP) and no phosphorus contained and thus without toxicity to the catalysts in exhaust gas recycling equipment of automobiles.on the other hand,bismuth is located at sixth period and VB group in periodic table of the chemical elements.Because of its location in transitional area between metal and non-metal elements in the table,bismuth and its compounds have many unique physical and chemical properties,and have been found some uses in nuclear and space technology, pharmaceutical and cosmetic industries etc..

Unfortunately,inadequate attention had been paid to bismuth compounds and their applications so far,especially as additives to lubricants.Concerning the fact that it follows lead in atomic number order and situates in the same VB group ,bismuth compounds are also assumed to find their places as agents for lubricants. In fact, this assumption was initially confirmed by the work of Rohr etc.(1,2). They studied the extreme pressure performances of several bismuth compounds and concluded they exceed those of corresponding toxic lead compounds,thus indicating the substantial application value of the compounds to lubricants.

In this paper,a bismuth compoud, bismuth dioctyl dithiocarbamate has been synthesized,its tribological performances were evaluated with a fourball machine and block-on-ring test rig.In addition,in order to correlate its tribological behaviors with the film formed on the metallic rubbing surface under boundary lubrication conditions,scanning electron microscope with energy dispersion of X-ray,Auger electron spectroscope and X-ray photoelectron spectroscope were utilized to characterize the surface film formed by the additive.

EXPERIMENTAL

Additive Synthesis

A quarter of a mole of bismuth oxide and 0.75 mole of dioctylamine were added in 200ml acetonitrile in a half liter,three-necked round bottom flask,The necks were sealed and fitted repectively with thermometer jacket, reflux condencer and droping funnel.After the contents were cooled to 10 ℃ ,0.85mole of carbon disulfide was droped in from the funnel,with constant stirring the contents.When the carbon disulfide was run out,keep agitating for 15 minutes.After maintaining at room temperature for half an hour,the reaction mixture was refluxed for some seven hours.The solvent was then distilled off under reduced pressure and the residue extracted five times with hexane.The combined hexane exteacts were again distilled off and the crude product was purified by solvent-solvent extraction.The final yellow solid product thus prapared was characterized by infra-red spectroscopy, inductively coupled plasma spectroscopy (ICP) for elemental analysis and C^{13} nuclear magnetic resonance spectroscopy(NMR).The results of elemental analysis are shown in Table 1:

Table 1. Contents of The Elements in The Additive Measured by ICP(%)

element	C	H	S	N	Bi
theoretical	52.90	8.82	16.59	3.63	18.06
measured	52.24	8.97	16.63	3.56	18.60

The measured data from the above analyses confirmed the following formula of the compound prapared:

$$\left[\begin{array}{c} C_8H_{17} \\ \phantom{C_8H_{17}} \end{array} \!\!N\!-\!C \!\! \begin{array}{c} S \\ S \end{array} \right]_3 Bi$$

The compound can be abbreviated as BiDTC for the sake of conciseness.

Friction & Wear Tests

The basestock for all the friction and wear tests in this study was ISO VG32 mineral paraffinic oil.

The firction coefficient of the compound was evaluated with a HQ-1 block-on-ring test rig.The block was of mild carbon steel with the size of 12.35(length) × 12.35(height) × 19.00 (width)mm.The ring was made of GCr15 bearing steel,similar to AISI SAE52100 steel with the hardness of HRC58-62 and the size of 49.24(diameter) × 12.70(width)mm.The relative sliding speed between the ring and the block was fixed to about 2.0m/s.The friction coefficients were respectively measured under the load of 600N and 900N with the additive at different concentrations.

The initial seizure load and welding load of BiDTC were evaluated with a MQ-800 four-ball machine.The diameter of the steel ball made of GCr15 was 12.7mm,The rotary speed of the upper ball was 1450rpm,and the friction time for each test was 10 seconds. Effect of additive concentration on the load-carrying capacity was provided,and antiwear performance of the additive was also investigated by measuring the wear scar diameters of lower balls after respectively tested in the sample oil with 2.0% BiDTC under the load of 294N and 588N for 1800,2700 and 3600 seconds.

Surface Analysis

In order to understand the surface film formed by the additive on the rubbing surface under boundary lubrication conditions,a PHI 610 Auger electron spectroscope (AES), scanning electron microscope with an energy dispersion of X-ray,and a PHI 5100 X-ray photoelectron spectroscope(XPS) were used to characterize the rubbing surface of the block after tested in the lubricant containing 5.0% BiDTC under the load of 900N when the lower ring sliding against it at the speed of 2.0m/s.Energy dispersion of X-ray was first used to disclose the elements on the surface.AES analysis was performed to conduct surface surveys,followed by an AES depth profile.This was done in the working vacuum below 1×10^{-9}torr,with an electron beam energy of 3KV and beam current of 0.1mA.The depth profile was accomplished by sputtering through the surface layer using a 25.0mA and $1 \times 1m^2$ Argon ion beam at an estimated rate of 250A/min,which was calibrated against Ta_2O_5.Surface surveys were also carried out by XPS,followed by argon ion etchig to achieve elemental depth profiles and the chemical states of the present elements.The XPS analysis was done at the take-off angle of 45 degrees in the working vaccum below 5×10^{-6}torr,with the X-ray of accelerating voltage of 15KV and the power of 400W to bombard the aluminium anode.The working potential of the ion gun for sputtering was 3.0KV and the ion beam strength 9.0mPa.Under this working condition,the sputtering rate was 15-20A/min as it was calibrated against SiO_2.

RESULTS & DISCUSSION

The results of friction coefficients of BiDTC were listed in Table 2.

Table 2. Friction Coefficients of BiDTC at Various Concentrations

concentration(wt.%)	0.0	0.5	1.0	2.0
friction coefficient	0.12	0.10	0.091	0.087
reduction, %	—	16.7	24.2	27.5

It can be seen from the table that the friction coefficient of BiDTC decreases with the increase of its concentration,but the reduction tendency slows down with the further addition of the same amount of BiDTC into the basestock.This may have relation with the surface film formed by the contact of the ring and block under boundary lubrication conditions.

In contrast to the influence of concentration on the friction coefficients,The four-ball test results show that the load-carrying capacity of the BiDTC increases with the increase of the amount of the additive in the basestock,as Fig.1 shows.The initial seizure load reaches the highest at the concentration of 3.0%,whereas the maximum welding load occurs at the concentration of 10.0%.

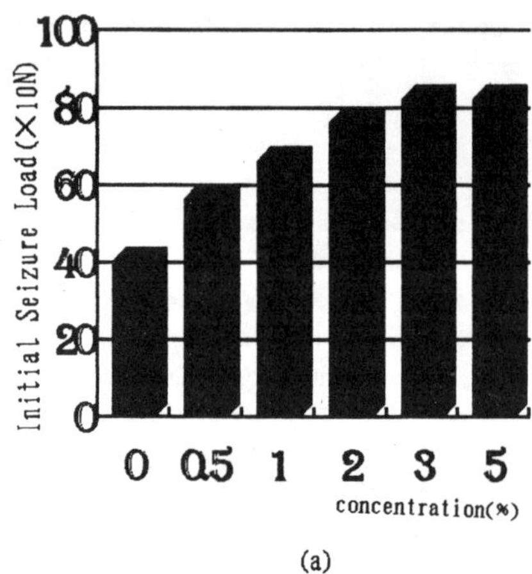

(a)

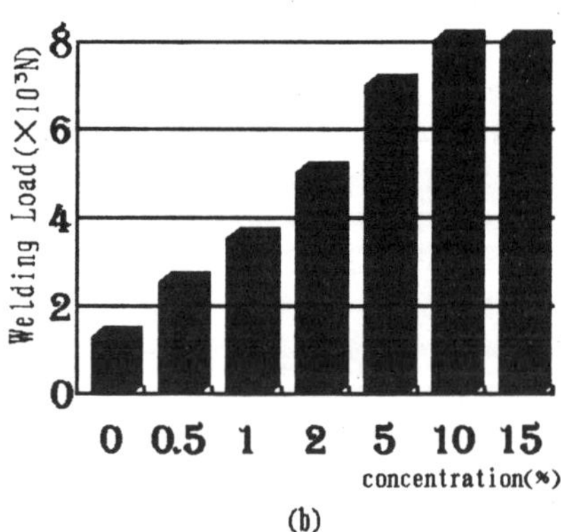

(b)

Fig.1 Initial Seizure Load and Welding Load of BiDTC and The Effect of Concentration

The wear scar diameters of the steel balls after tested for 1800,2700 and 3600 seconds under the load of 294N and 588N are listed in Table 3. As shown i n the table,　much improvement in the wear scar diameters of the steel ball has been obtained with the addition of the additive.

Table 3. The Change of Wear Scar Diameters(mm) with Load and Test Time

lubricant	load (N)	test time(sec.)		
		1800	2700	3600
VG32	294	0.61	0.68	0.74
	588	2.41※	※	※
VG32+2.0% BiDTC	294	0.33	0.37	0.40
	588	0.88	0.92	0.97

※——seizure occurs

Surface Film Characterization

EDX disclosed that the chief elements in the surface layer of the block after tested in the BiDTC-containing oil were iron,oxygen,carbon,sulfur,bismuth and nitrogen. Surface survey by AES shown in Fig.2 supported this discovery.

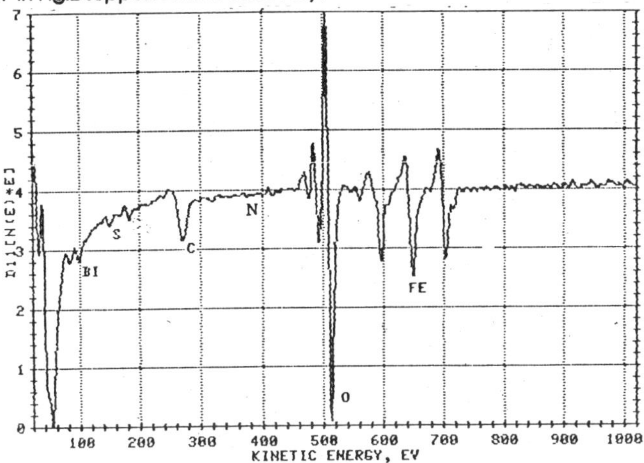

Fig.2 AES Survey to The Block Surface after Tested with BiDTC

Depth profile with argon ion sputtering gave the distribution condition of the elements shown in Fig.3 .

The atomic concentration of oxygen in the outmost layer is the highest among all the elements as the figure shows.It declines with sputtering which means less oxides were formed in the inner surface layer of the block.The atomic concentration of iron is the second in the outmost layer and increases with sputtering. After sputtering for about 9 minutes (to the depth of about 2250A as measured by the sputtering rate and the time), it exceeds oxygen.This suggests lower oxidizing state of iron in the inner layer.The atomic concentration of bismuth in the outmost layer follows iron and maintains about 8.0% with sputtering. Whereas the concentrations of the left elements,carbon,sulfur and nitrogen are very low.The phenomenon that the concentration of bismuth exceeds that of sulfur is especially unusual concerning the fact that the stoichiometric ratio of sulfur atoms to bismuth in the BiDTC molecules is 6 to 1, and tht fact that sulfur is an active element which is often found in antiwear and extreme pressure additive compounds to form chemical reaction film with the rubbing metal surface under boundary lubrication conditions.

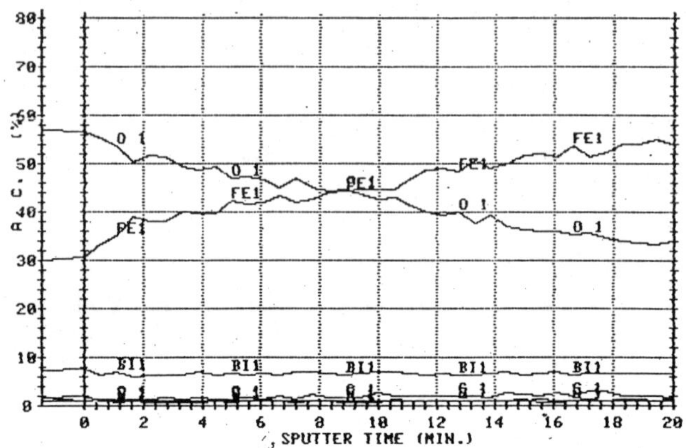

Fig.3. AES Profile of Oxygen, Iron, Bismuth, Sulfur, Carbon and Nitrogen

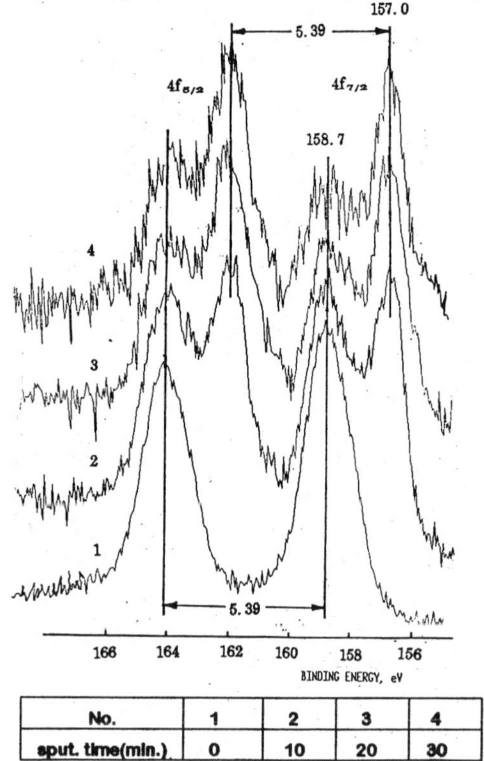

No.	1	2	3	4
sput. time(min.)	0	10	20	30

Fig.4 The XPS Spectra of Bismuth

334

Analytical results by XPS have not only confirmed the presence of the above elemental components, but also identified the chemical composition of the surface film on the block surface. Fig.4 is the XPS spectra of bismuth. The energy difference 5.39ev between the primary $4f_{7/2}$ peak and secondary $4f_{5/2}$ is coincident with the value reported in the literature(3,4).Before sputtering, one Bi $4f_{7/2}$ primary peak is present at 158.7ev which suggests the formation of bismuth oxide or sulfide.Because the difference between the binding energies of bismuth in its oxide and sulfide is only 0.1ev,it is very difficult to exactly identify which compound is formed.However,XPS spectra of oxygen and sulfur may provide more proof and the atomic concentration percentage of the elements in the surface layer suggests bismuth oxide is more likely. After sputtering for 10 minutes,another peak at 157.0ev occurs in the bismuth spectra.This is the characteristic peak of metallic bismuth(3),and it becomes stronger with further sputtering.At the same time,the peak at 158.7ev tends to weaker.This implies that bismuth oxide or sulfide exists in the outer film,whereas metallic bismuth is in the inner film.The generation of metallic bismuth from BiDTC is very conducive to the friction reduction and wear control because bismuth is a kind of metal with very low melting point of 271℃,and is inclined to form alloys of low melting points with other metals(1).

From the XPS spectra of oxygen in Fig.5,the oxide layer of the sample is chiefly identified as Fe_2O_3 .The data obtained after each sputtering is a complex O1s spectrum which can be at least fitted to three peaks at 529.8ev,531.2ev and 532.4ev,corresponding to the oxygen in Fe_2O_3,oxygen in hydroxyl groups(5) and the oxygen in $FeSO_4$(3).However, it is possible that other types of oxides have been generated on the surface as the difference among

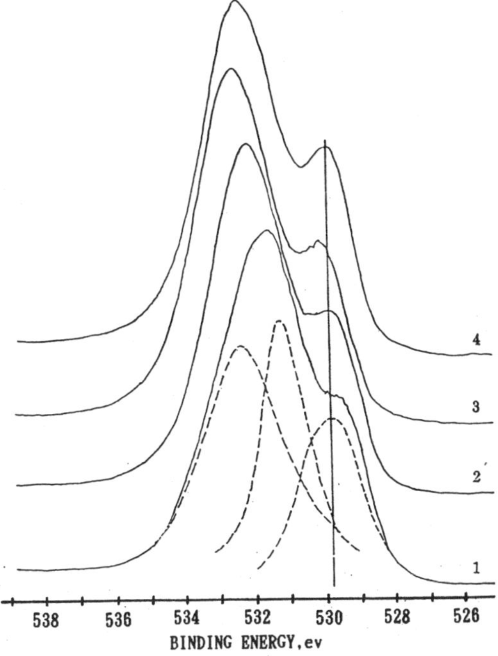

Fig.5 XPS Spectra of Oxygen

O1s binding energies in Fe_2O_3(529.8ev),Bi_2O_3(530.0ev),Fe_3O_4(530.0ev) and FeO(529.8ev) is very small.As a matter of fact,the XPS spectra of iron in Fig.6 confirm the presence of Fe_2O_3, which is evidenced by the characteristic $Fe_{2p3/2}$ peak 710.8ev and the $Fe_{2p1/2}$ peak at 13.6ev below that peak. The $Fe_{2p3/2}$ peak at 712.1ev represents the iron in $FeSO_4$, but it disappears in the later spectra after sputtering for 10 minutes. it must be added that the $Fe_{2p3/2}$ peak at 710.8ev shifted a little lower with sputtering, indicating other types of iron compounds, such as Fe_3O_4 (710.4ev), FeO(710.3ev) and FeS(710.2ev) are also possibly present.The atomic distribution condition of oxygen in Fig.4 can well be explained by this change of oxidation values of iron.

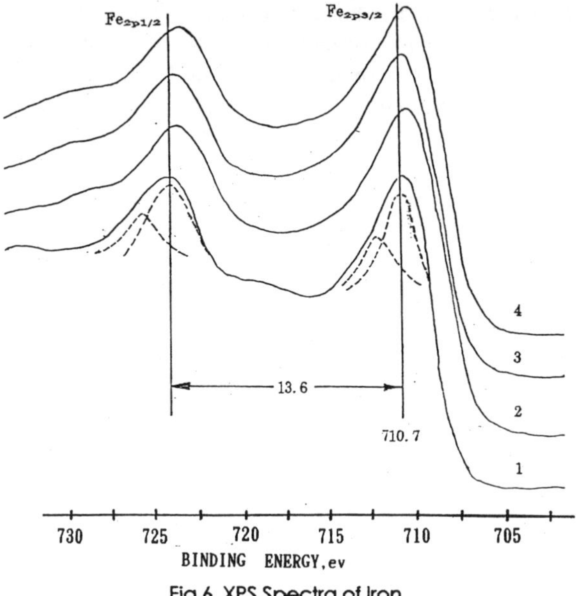

Fig.6 XPS Spectra of Iron

More evidence of the formation of $FeSO_4$ and FeS is given by the XPS spectra of sulfur in Fig.7. The S_{2p} peaks at 168.8ev and 161.1ev are apparently associated with the above two inorganic compounds which are commonly found in the products of sulfur-containing antiwear agents with steel surfaces(6,7). The S_{2p} peak at 168.8ev disappeared after 20 minutes' sputtering,impling sulfates only exist in the outer surface layer of about 500A depth below as calculated based on the sputtering rate and time.The N_{1s} peak is always at 399.0ev,representing the -CN group.

According to the above film characterization,the antiwear action mechanism of BiDTC can be postulated as follows: In boundary lubrication conditions, BiDTC molecules are decomposed under the action of mechanical shear, high temperature created by frictional heat, high pressure by the contact of asperities on the relatively moving surfaces and many other friction induced effects. The disintegrated debris of the additive further experience

336

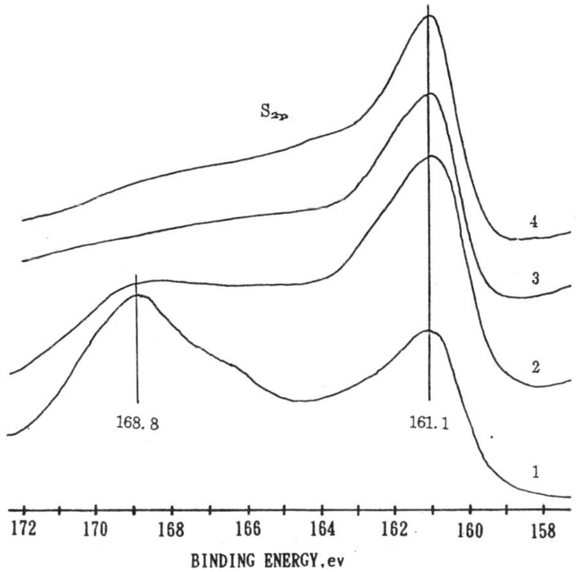

Fig.7 XPS Spectra of Sulfur

various chemical changes. Some organic radicals further combine or condence with each other to form polymeric materials. The decomposed active sulfur atoms may react with the freshly exposed metal surface to form FeS, and it can be further oxidized into $FeSO_4$ by atmosphere to deposit in the outer layer. The produced metallic bismuth is melted to liquid state by the local elevated temperature and high pressure at contacting asperities, and thus making the shear of the surfaces much easier within the liquid. On the other hand, the melted bismuth tend to flow to the concave sites to 'polish' the contact surfaces and the surface roughness is hence improved. In addition,oxidation and sulfuration of the produced bismuth and the iron surface under this condition are proceeded and the products thus produced are beneficial to the reduction of friction and wear.

CONCLUSIONS

According to the above results and discussion, some conclusions can be drawn as follows:

1.   A new oil-soluble additive for lubricants, bismuth dioctyl dithiocarbamate, has been successfully prepared.

2. The compound exhibits good friction-reducing, antiwear and extreme pressure abilities, and the effect of its concentration in the basestock on the tribological behaviors has been provided as well.

3. Surface film characterization discloses that various species had been formed by the additive under boundary lubrication conditions. In addition to oxides, sulfates and sulfides, metallic bismuth and bismuth oxide and sulfide played important roles in the reduction of friction and wear.

REFERENCES

(1) Rohr Otto, NLGI Spokesman, Vol.57, No.2, pp.6-13.

(2) Tuli D.K., Sarin R., Gupta A.K., Kumar A.H., Lub. Eng., Vol.51(1995), No.4, pp. 298-303.

(3) Wagner C.D. and Riggs W.M.,Handbook of X-Ray Photoelectron Spectroscopy, Perkin-Elmer Corporation,Eden Praire,1979. p.238.

(4) Briggs D. and Seah M.P., Practical Surface Analysis by Auger and X-Ray Photoelectron Spectroscopy, New York,John Wiley & Sons(1988).p.509.

(5) Baldwin B.A., Lub. Eng., 32(1975),3,pp.125-130.

(6) Godfrey D., Fundamentals of Tribology, edited by Suh N.P. and Saka N., The MIT Press(1980),p.955.

(7) Tamai Y., Fundamentals of Tribology, edited by Suh N.P. and Saka N., The MIT Press(1980), p.979.

IN SITU ELECTRON SPECTROSCOPIC IDENTIFICATION OF CARBON SPECIES DEPOSITED BY LASER ABLATION

E. C. SAMANO, GERARDO SOTO*, ARTURO GAMIETEA, LEONEL COTA
Laboratorio de Ensenada, IFUNAM, A. P. 2681, 22800 Ensenada, B. C., México.
* Also at Programa de Posgrado en Física de Materiales, Centro de Investigación
Científica y de Educación Superior de Ensenada, 22800 Ensenada, B. C., México.

ABSTRACT

Thin carbon films were grown on Si (111) substrates by ablating a graphite target
utilizing an excimer pulsed laser in a UHV Riber © LDM-32 system. Two kinds of films
were produced, a highly oriented pyrolytic graphite (HOPG) type and a diamond-like
carbon (DLC) type. A relationship of the films microstructure with laser power density
and substrate conditions was observed. The HOPG films were homogeneous but the
DLC films were heterogeneous, as shown by micrographs. The thin films are monitored
and analyzed *in situ* during the first stages of the deposition process. The monitoring
was done by RHEED and the characterization by several surface spectroscopic
techniques, AES, XPS and EELS. The formation of a SiC interface was observed for
both films due to the reaction of the first carbon species with the substrate surface.

INTRODUCTION

The study of carbon deposition by several techniques is one of the most active
research areas in thin films because they may result in hard non-crystalline carbon
films, named by the acronym DLC [1]. The ordinary technique to prepare DLC films has
been CVD, where hydrogen is used as a catalyst [1]. Recently, Pappas *et al.* [2], and
Voevodin *et al.* [3] have produced hydrogen-free carbon films by pulsed laser deposition
(PLD). They have reported that the atomic arrangements depend on the kinetic energy
of the ablated species. Their findings predict that depositions of DLC films with high
percentages of sp^3 bonds are possible at kinetic energies of the carbon species found in
the plume generated by PLD, a few hundred of eV.

The properties of DLC films are unique: good optical, chemical and mechanical
properties. Regarding the last property, they become ideal as hard coatings because of
their high hardness to wear resistance and shear stress. Strong adherence of a film to the
substrate is required to provide protection against wear and abrasion. Consequently, the
study of the interfacial growth is important in obtaining a high quality coating. In
particular, the C/Si system is important because it forms a very strong covalent bond
at the interface but there is a large difference between the lattice constants of the
materials involved. An investigation of the early growth stages of this system is needed
to find out if the very first layers are C/Si, with a lattice mismatch of about 34 %, or
SiC/Si, with a 20 % lattice mismatch.

Therefore, the object of the current work is the detailed study of both the kinetic
process which controls the chemical structure and the interface formed during the early
stages of carbon films by PLD. This investigation can be performed in a very precise
manner by spectroscopic studies in the same system where the deposition is performed.

Mat. Res. Soc. Symp. Proc. Vol. 436 © 1997 Materials Research Society

EXPERIMENT

The experiment is based on the photoevaporation of a pyrolytic graphite target placed in a laser ablation system, Riber © LDM-32. The system consists of three chambers: sample loading, deposition and analysis. Each chamber is independently pumped by an ion pump, and isolated by UHV gate valves. The target is ablated in the deposition chamber, base pressure of 10^{-10} Torr, by means of a pulsed KrF laser with a wavelength of 248 nm and a pulse length of 34 ns. The laser energy range is from 200 to 600 mJ, corresponding to a power density from 5.9×10^7 to 1.7×10^8 W/cm^2.

The photoevaporated carbon species are deposited on a clean Si (111) surface and monitored by RHEED, observing epitaxial growth in the first laser pulses. The HOPG and DLC films are grown under different experimental conditions. The HOPG films are achieved by depositing the carbon species at low laser power density, approximately 7×10^7 W/cm^2, and a few laser pulses, no more than 50 pulses, on a substrate kept at room temperature. The interface formed in carbon films is investigated *in situ* by interrupting the deposition process at selected time intervals, number of laser pulses, to transfer the resulting sample to the analysis chamber to be characterized by AES, XPS, and EELS. The DLC films are accomplished by depositing the carbon species at high power density, approximately 1.7×10^8 W/cm^2, and many laser pulses, from 900 to 1000 laser pulses, on a substrate kept at approximately 500°C. In order to produce DLC films in this application, defects on the surface of the single-crystal substrate have to be created, like scratching it with diamond powder.

RESULTS

Prior to any deposition, the clean substrate and target are analyzed. A survey scan and high resolution scans of the C(1s), Si(2p), and O(1s) regions are done for both specimens [4], as shown by curve (a) in Figs. 1 and 2 for the clean substrate. Similarly, the Auger transitions are investigated in the 40 to 350 eV energy range [5], as shown by the no pulse curve in Fig. 3 for the clean substrate. The characterization of the pyrolytic graphite target is exhibited by curve (a) in Fig. 4. This spectrum is used as a reference to determine the carbon phase of the overlayers deposited on the Si (111) surface under different conditions.

After analyzing the substrate and target, the PLD process is started. A time dependent-study of the different film growth stages is carried out to have a better understanding of the kinetic process which controls its composition and morphology. The time evolution of the deposition process is realized by XPS, as shown in Figs. 1 and 2, and AES, as shown in Fig.3, respectively. The experimental data obtained by XPS have been fitted using a recently developed computer code named XPSPEAK [6], available at Internet. The main features of the program are in peak fitting, which makes use of an algorithm based on the binary search method to iteratively find the best value of position, width, area and shape for each peak in every spectrum. The algorithm is found to be very stable, and provides information about the various species in the film for their identification and quantification. Plasmon losses in the target and films are studied by EELS using an electron beam with a 1 keV incident energy. This technique is particularly important in distinguishing very fine details in the sample, like chemical identification of several carbon phases and carbides [7]. Fig. 4 shows the main peaks due to plasmon energy losses for the two kinds of films grown in this work as compared to the graphite target. The films morphology is studied as well, Fig. 5 shows a sputtered area by Ar ions of a HOPG film exposing the different phases, and Fig. 6 shows the "cauliflower" type structure of a DLC film, as the one reported by Voevodin *et al.* [3].

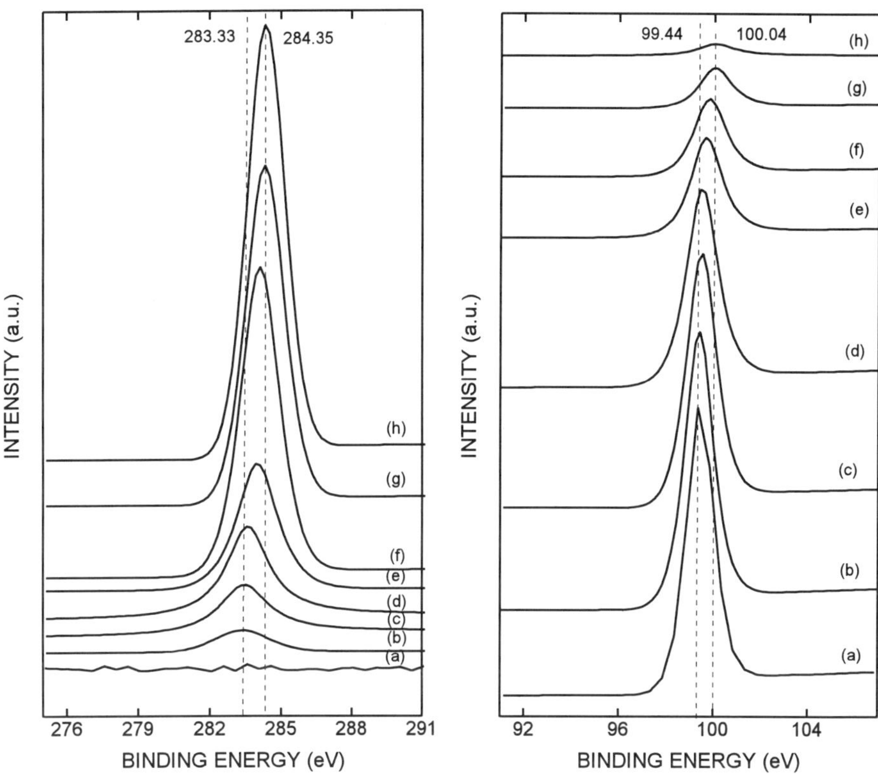

Fig. 1. XPS fitted spectra in the C (1s) region of the different stages of carbon film growth on Si (111) at room temperature. The curves (a), (b), (c), (d), (e), (f), (g), and (h) correspond to no laser deposition, 1, 2, 5, 10, 20, 30, and 45 laser pulses of deposition, respectively.

Fig. 2. XPS fitted spectra in the Si (2p) region of the different stages of carbon film growth on Si (111) at room temperature. The curves (a), (b), (c), (d), (e), (f), (g), and (h) correspond to no laser deposition, 1, 2, 5, 10, 20, 30, and 45 laser pulses of deposition, respectively.

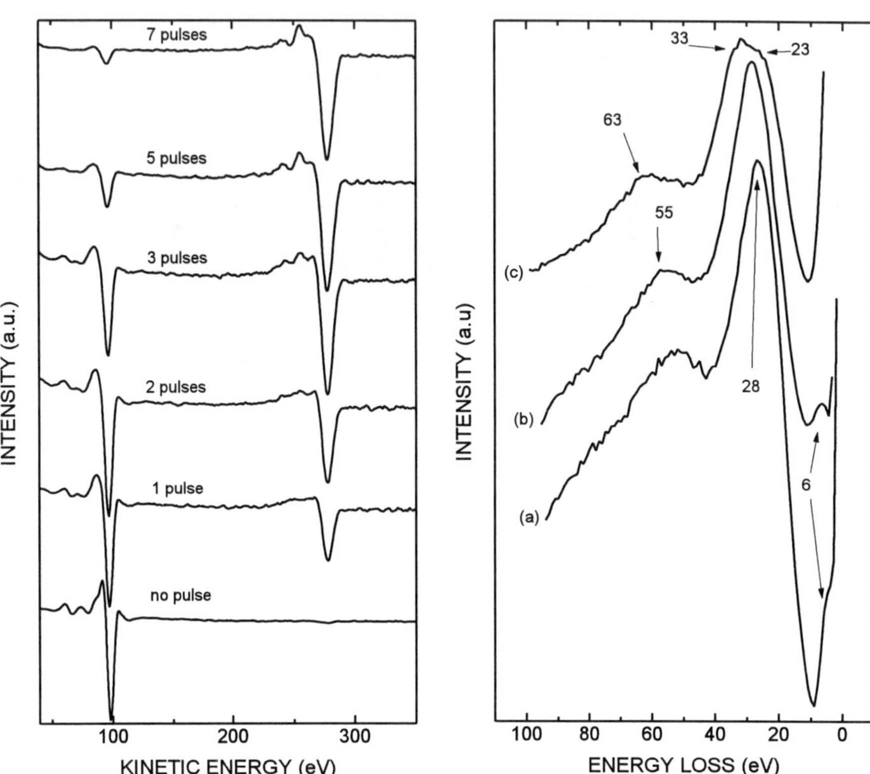

Fig. 3. AES spectra showing the Si *LVV* and C *KLL* Auger transitions of the different stages of carbon film growth on Si (111) at room temperature. The no pulse curve corresponds to the clean substrate.

Fig. 4. Comparison of EELS spectra of the (a) graphite target, (b) HOPG film after a deposition of 7 laser pulses and substrate at room temperature, (c) DLC film after approximately 900 laser pulses and at $T_s = 500$ °C.

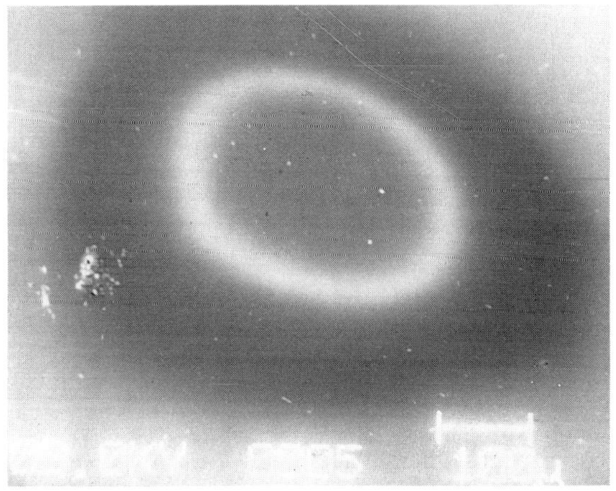

Fig. 5. A SAM micrograph showing a sputtered region of a HOPG film
deposited on a clean Si (111) surface with the coexisting phases.

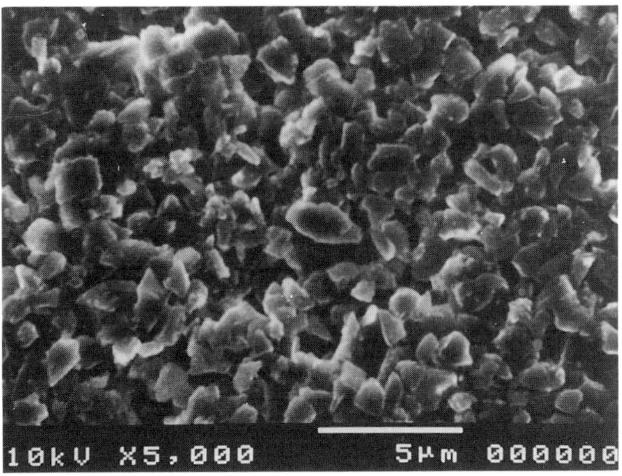

Fig. 6. A SEM micrograph showing the "cauliflower"-type structure of
a DLC film grown on a clean Si (111) surface at $T_s = 500\,^\circ$ C.

CONCLUSIONS

The growth evolution of hydrogen-free carbon films by PLD and *in situ* electron spectroscopies has been investigated. The XPS spectra in Fig. 1 show a binding energy change from 283.33 eV, for 1 pulse of deposition, to 284.35 eV, for 45 pulses of deposition, this corresponds to the chemical shift from SiC(1s) to C(1s) [4]. The XPS spectra in Fig. 2 also exhibit a change from 99.44 eV, for a clean substrate, to 100.04 eV, for 45 pulses of deposition, this corresponds to the chemical shift from Si(2p) to SiC(2p) [4]. Belton *et al.* [8] and Tench *et al.* [9] found similar values for these energy shifts. The existence of a SiC interface in carbon films is confirmed by observing the AES spectra in Fig. 3. The characteristic peak at 92 eV for Si LVV Auger transition of the clean substrate [5], no pulse curve, has almost vanished after 7 pulses of deposition. Similarly, the shape and intensity of the spectra for 1 and 2 pulses just resembles to the SiC LVV, flat shoulder at 85 eV, and SiC KLL transitions, sharp peak at 270 eV, reported in the literature [9]. A fully developed peak at 272 eV for C KLL of the HOPG type is identified at 7 pulses [5]. Finally, the EELS spectra in Fig. 4 typify the different kinds of carbon films. The curve (a) shows the typical strong peaks at 28 and 55 eV of graphite [7]. The curve (b) shows a well defined shoulder at 6 eV, in addition to the peaks in curve (a), a fingerprint of HOPG, [7]. The curve (c) lacks the shoulder at 6 eV, and it exhibits small humps at 23 and 33 eV, which are all characteristic of diamond [7]. Summarizing, this work shows how a very precise carbon deposition can be performed.

ACKNOWLEDGMENTS

The authors would like to thank CONACYT (México), thru research projects 4228-E and IF-0076, for financial support and I. Gradilla for technical assistance.

REFERENCES

1. J.C. Angus, and C.C. Hayman, Science **241**, 913 (1988).

2. D.L. Pappas, K.L. Saenger, J. Bruley, W. Krakow, J.J. Cuomo, T. Gu, and R.W. Collins, J. Appl. Phys. **71**, 5675 (1992).

3. A.A. Voevodin, S.J.P. Laube, S.D. Walck, J.S. Solomon, M.S. Donley, and J.S. Zabinski, J. Appl. Phys. **78**, 4123 (1995).

4. J.F. Moulder, W.F. Stickle, P.E. Sobol, and K.D. Bomben, *Handbook of X-ray Photoelectron Spectroscopy* (Perkin-Elmer, Eden Prairie, 1992), pp. 40, 44 and 56.

5. L.E. Davis, N.C. McDonald, P.W. Palmberg, G.E. Riach, and R.E. Weber, *Handbook of Auger Electron Spectroscopy* (Perkin-Elmer, Eden Prairie, 1978), pp. 25, 33 and 49.

6. R.W.M. Kwok, presented at 42nd AVS National Symposium, Minneapolis, MN, 1995 (unpublished).

7. D. N. Belton, and S.J. Harris, J. Vac. Sci. Technol. A **8**, 2353 (1990).

8. D.N. Belton, S.J. Harris, S.J. Schmieg, A.M. Weiner, and T.A. Perry, Appl. Phys. Lett. **54**, 416 (1989).

9. R.J. Tench, M. Balooch, A.L. Connor, L. Bernardez, B. Olson, M.J. Allen, W.J. Siekhaus in *Laser Ablation for Material Synthesis*, edited by D.C. Paine, and J.C. Bravman (Mater. Res. Soc. Proc. **191**, Pittsburgh, PA, 1990), pp. 61-66.

MECHANICAL PROPERTIES OF PURE CARBON AND CARBON-NITROGEN COATINGS ON THIN FILM HEAD SLIDERS

G. Wang, A. Strojny, J. M. Sivertsen, J. H. Judy[a] and W. W. Gerberich
Department of Chemical Engineering and Materials Science
[a]Department of Electrical Engineering
University of Minnesota, Minneapolis, MN 55455

ABSTRACT—The mechanical properties of pure carbon (C) and carbon-nitrogen (C:N) coatings on thin film head sliders were investigated by continuous drag testing (CDT) and nano-indentation. Comparisons were made in terms of wear protection, elastic modulus and hardness of these two types of carbon films. The C and C:N thin films with various thickness were deposited on thin film head sliders using a facing target sputtering (FTS) system. After 23,000 revolutions of CDT tests, all the testing head sliders which were uncoated and coated with 90 Å C or C:N exhibited some degree of wear damage as indicated in AFM micrographs where that of the uncoated head was the most severe and that of the C:N coated head was the least. Head sliders coated with 1000Å C and C:N were studied under the Triboscope™ nano-indenter, where load-displacement curves at different maximum loads were recorded. Elastic modulus and hardness were determined from those curves. The results show that elastic modulus and hardness of C:N are greater than that of C. Therefore, one may conclude that both C and C:N behave like a protective coating for the head slider where C:N is better than C, which could be well related to the larger elastic modulus and hardness of C:N.

INTRODUCTION

In competition with optical disks and semiconductor flash memory for computer storage devices, the rapidly growing demand for magnetic information storage has driven the magnetic recording industry to accelerate development of lower cost, smaller size, higher capacity and higher density recording systems. To achieve higher recording density, it is necessary that the head is flown closer to the magnetic media[1]. Since the head must be close to the magnetic layer under the overcoat, the height of the airbearing surface must be decreased and the overcoat must be as thin as possible. In this approach, the primary tribological challenge to overcome is to survive the increased number of contacts between the magnetic head and the media. If 10 Gbits/in^2 recording density is to be achieved, a durable overcoat thickness of the order of 100-50 Å has to be developed. As a consequence of these effects, the wear life of disk drives will be significantly decreased unless the durability of the head disk interface is enhanced substantially.

It has been demonstrated that a significant improvement in wear performance of the head-disk interface can be obtained by coating thin film head sliders with a carbon thin film[2,3,4,5]. The most widely used thin film head sliders now are made of Al_2O_3-TiC composites. The microstructure of the composite shows that Al_2O_3 and TiC areas are well separated from each other. It has been shown that Al_2O_3 areas more readily suffer wear during the wear test and a thin carbon overcoating can reduce wear. To serve as a protective coating, carbon films must adhere well to the head slider and have good

Mat. Res. Soc. Symp. Proc. Vol. 436 © 1997 Materials Research Society

mechanical properties. The purpose of this paper is to present the results of the comparisons of the wear performances and mechanical properties of pure C and C:N thin overcoatings on head sliders and the results are correlated.

EXPERIMENTAL

Thin film carbon coatings were deposited on thin film head sliders using a facing target sputtering (FTS) system. The background pressure of the FTS system was less than 5×10^{-7} Torr. The pure Argon, or Argon-Nitrogen gas mixture (40%-N_2) was used as the sputtering gas. The resulting carbon nitrogen film has a 22 at. % nitrogen which was determined by AES. No adhesion layer was deposited under the carbon coatings.

Continuous drag tests (CDT) were conducted with a computer-controlled friction-wear tester. The thin film head sliders were the 50% Al_2O_3-TiC type with a normal load of 4 gms and a z-height of 0.032". All heads used in these studies had similar head parameters (crown, twist and camber). Before each test, both the head slider and the disk surfaces were carefully observed in an SEM and were verified to be clean and free of contamination. The radius at which to perform the CDT tests was chosen to be 1" on the 3.5" disks. The linear velocity was fixed at 26.60 cm/sec. The average kinetic friction was measured every 1000 revolutions at a speed of 0.1 rpm for one complete revolution. After the test was terminated, the head slider surface was carefully studied by atomic force microscopy (AFM), and an image of the most severe wear damage was taken for each head.

Continuous nano-indentation tests were carried out with a Triboscope[TM]. The Triboscope is a load controlled instrument with a load resolution of 1.8 nN and a depth resolution of 0.2 nm. It fits on a conventional atomic force microscope and enables the user to scan the surface prior to indentation and image the resulting indentation right after. The indenter tip used was a conventional STM pyramidal Boron-doped diamond tip with a tip radius of 100 nm and an included half-angle of 72°. In each case, multiple indentations were made at different locations of the film surface at different loads. In a typical experiment the indenter was driven into the sample at a constant rate by a capacitive force/displacement transducer. The load and depth of penetration of the indenter into the sample were continuously recorded. Once a predetermined load was reached, the indenter was removed from the sample at the same rate.

RESULTS AND DISCUSSION

Fig. 1 shows a comparison of the friction versus the number of revolutions for the cases of an uncoated head slider and sliders coated with 90 Å of FTS sputtered pure carbon and carbon nitrogen. After a total of 23,000 revolutions in each CDT test, only that disk paired with an uncoated head exhibited any visible wear marks which indicate that the head-disk interface had failed at this point. For comparison purposes, we note that the coefficient of kinetic friction versus the number of revolutions increases rapidly from 0.12 initially to about 0.7 at 23,000 revolutions for the head slider which was uncoated. On the other hand, the kinetic friction coefficient of the head coated with 90 Å pure carbon increases from 0.1 initially to 0.5 at 15,000 revolutions and remained constant, whereas that for the head coated with 90 Å carbon nitrogen increases from 0.1 to a constant value of 0.45. Upon completing the CDT tests, the head surfaces were observed by AFM and the resulting micrographs are shown in Fig. 2. As shown in Fig. 2, all three testing heads exhibited some degree of wear damage where that of the uncoated head was the most severe and that of the carbon nitrogen coated head was the least. This is in good qualitative agreement with the CDT friction results.

In terms of the wear protection capability of carbon films, a well established common agreement is that the more sp^3 bonding the better the wear performance. Hydrogenated carbon (C:H) is better than pure carbon because of its larger sp^3/sp^2 bonding ratio. According to the studies of Torng et al.[6] and Yeh et al.[7], thin C:N films prepared by rf diode sputtering method are more diamond-like as compared to the pure C films. They found that the addition of nitrogen will stabilize the diamond sp^3 bonding and therefore can enhance it. Another important fact is that the atomic density of the film increased with increasing N_2 partial pressure during the deposition. Their result shows that the atomic density of the C:N films deposited at 2.5 mTorr and 10 mTorr N_2 partial pressures is 1.37 $\times$ 10^{23} atoms/cm^3 and 1.63 $\times$ 10^{23} atoms/cm^3, respectively, of which the later is comparable to that of diamond (1.77 $\times$ 10^{23} atoms/cm^3). In our study, we examined the carbon films (C and C:N) by electron energy loss spectroscopy (EELS). For this purpose, 300 Å C and C:N films were prepared by FTS on cover glass. The samples for EELS were prepared by dissolving the cover glass substrate. The carbon films on cover glass were checked under AES and XPS to be identical to that on head sliders. The resulting EELS spectra of carbon K-absorption edge are shown in Fig. 3. This high energy region of EELS spectrum consists of a peak at 285 eV corresponding to the $1s$-π^* transition[8,9,10] and followed by a broad main peak corresponding to the $1s$-σ transition. The intensity of the $1s$-π^* peak corresponds to the number of sp^2 coordinated atoms, which then is correlated to the number of sp^2 bonding sites. For the sake of comparison, the C K-absorption edge of graphite is also included in Fig. 3. As can be seen, C:N has the lowest $1s$-π^* peak intensity, which indicates that C:N has the lowest sp^2 bonding sites or in other words has the highest sp^3/sp^2 bonding ratio.

Three typical load-displacement curves, representing indentation test results on the uncoated head slider and sliders which were coated with 1000 Å pure C and C:N, respectively are shown in Fig. 4. It can be seen that the C:N film offered the most resistance to penetration. The load displacement curve of the pure substrate fell in between the C:N film and the C film. The modulus of elasticity and hardness of each case were calculated using the modified method by Doerner and Nix[11] by using the upper one-third of the unloading slope to calculate a reduced modulus. The method is schematically shown in Fig. 5 and the reduced modulus (E^*) and hardness (H) are

$$E^* = \frac{\sqrt{\pi}}{2}\frac{dP}{dh}\frac{1}{\sqrt{A}}$$

$$H = \frac{P_{max}}{A}$$

where $A = \pi(2h_c R - h_c^2)$, the area of indent; $h_c = h_{max} - \dfrac{\varepsilon P_{max}}{dP/dh}$, the contact of depth;

dP/dh is the slope of linear fit of the upper one-third of the unload curve; R is the radius

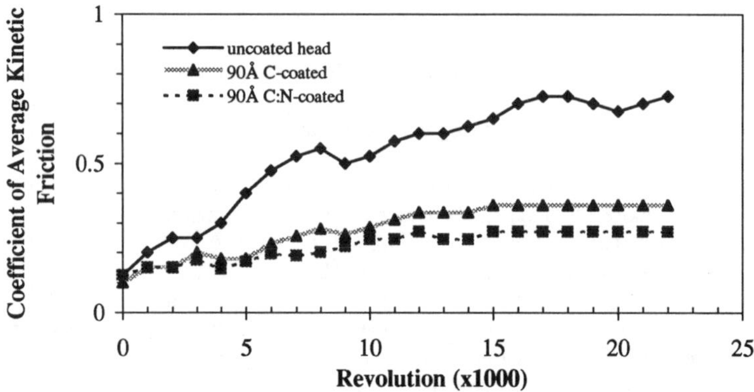

Fig. 1 The coefficient of friction versus number of revolutions with and without carbon coating on sliders.

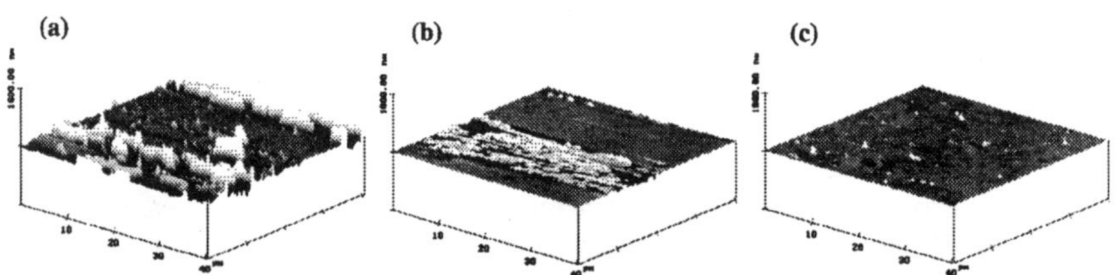

Fig. 2 The AFM images of the air bearing surfaces of (a) uncoated, (b) pure carbon and (c) carbon nitrogen coated head sliders after 23,000 revolutions of CDT test runs.

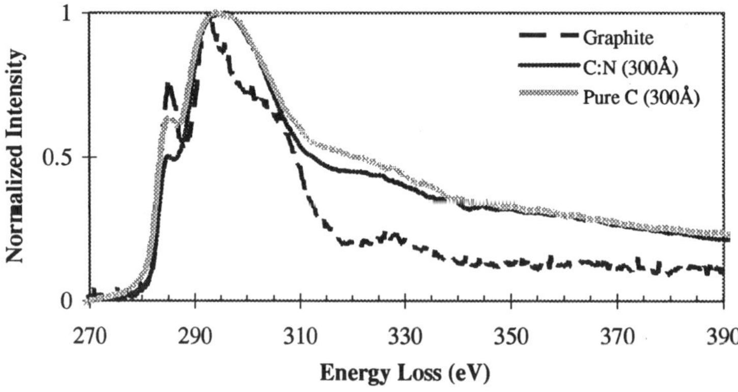

Fig. 3 The EELS C K-absorption edge spectra of carbon thin films and graphite.

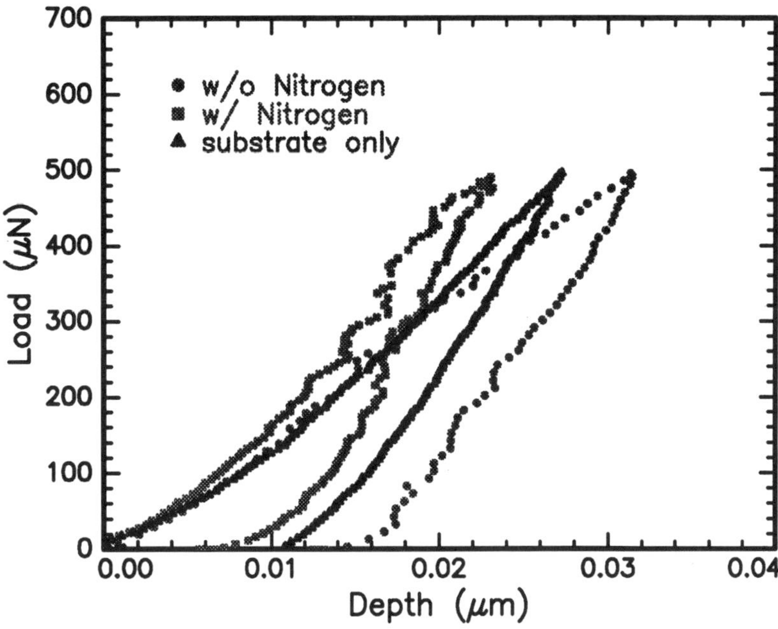

Fig. 4 Load -diaplcement curves that show the effect of nitrogen on the film. The nitrogenated film seems harder than the film without nitrogen. The waviness of the curves is due cracking of the film.

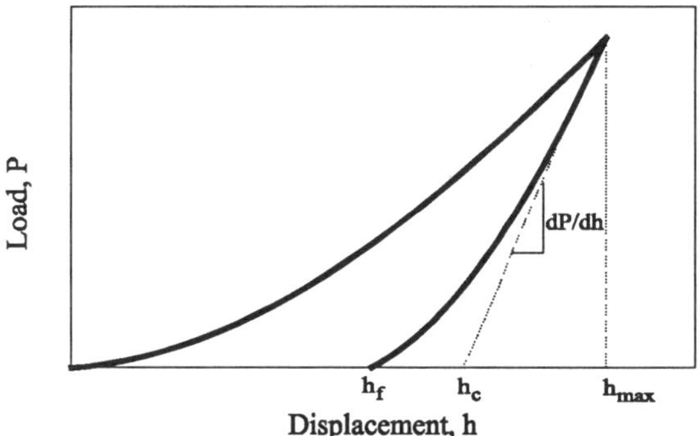

Fig. 5 A schematic representation of load-displacement curve and the Doener & Nix method.

of the indenter; h_{max} is the maximum penetration depth; P_{max} is the maximum load; ε is a parameter that depends on the indenter tip shape.

Table I summarizes the mechanical properties of both type of carbon films and the bare slider substrate. As can be seen, the elastic modulus of C:N (427 GPa) is much larger than that of C (272 GPa), while that of the substrate (330 GPa) is in between. It is worthwhile to point out that both type of carbon are harder than the slider substrate itself.

Table I Mechanical properties of pure C, C:N and head slider substrate.

	Modulus (GPa)	Hardness (GPa)	Contact Depth (nm)
Pure C	272	21.5	8
C:N	427	23.5	8
Head Slider (Al_2O_3-TiC)	330	18.5	9

This explains our CDT results well. We know that in the case of the uncoated head slider, the friction built up dramatically and the wear damage on the slider surface at the end of the test was quite severe. On the other hand, in the case of carbon coated head sliders, the friction remain relatively low during the tests and only slight wear is observable on the slider surfaces after the tests. Therefore, one may conclude that the improvement of wear performance of carbon coated head slider is partly due to the hardness increase of the overcoatings. Our EELS spectrum data shows that C:N has less sp^2 bonding than pure C, which indicates that C:N is more diamond-like. This is again in good agreement with the nano-indentation test results, i.e., C:N has a much larger modulus of elasticity and is slightly harder than pure C does. We believe that this is the reason why C:N is better than C in terms of protection on thin film head sliders.

CONCLUSION

Both FTS sputtered pure carbon (C) and carbon nitrogen (C:N) films can protect a head slider from wear failure to some extent during continuous drag test (CDT) with carbon nitrogen being better than pure carbon. This is demonstrated by both the low friction buildup and small wear damage of C:N coated head slider. EELS spectrum shows that the intensity of the $1s$-π^* peak of the C K-absorption edge of C:N film is the lowest, which indicates that C:N has the highest sp^3/sp^2 bonding ratio. Nano-indentation results show that both type of carbon films (prepared by FTS method) are harder than head slider substrate, with C:N being slightly harder than pure C. Also C:N has a much larger elastic modulus than C. This leads to the difference between both type of carbon thin films.

REFERENCES

[1] B. L. Wallace, Bell Syst. Tech. J., vol 30, pp. 1145-1173, Oct. 1951.

[2] Grill, C. T. Horng, B. S. Meyerson, V. V. Patel, and M. A. Russack, US Patent #5,159,508, October 27, 1992.

[3] D. B. Bogy, X. Yun, and B. J. Knapp, Digests of The Magnetic Recording Conference, September 1993.

[4] G. Wang, T-A. Yeh, J. M. Sivertsen, J. H. Judy and Ga-Lane Chen, Abstracts of The MMM-Intermag Conference, June 1994.

[5] S. S. Varanasi, J. L. Lauer, F. E. Talke, G. Wang and J. H. Judy, *Journal of Tribology*, 1996, in print.

[6] C. J. Torng, J. M. Sivertsen, J. H. Judy and C. Chang, J. Mater. Res. 5 (1990) 2490.

[7] T. Yeh, C-L. Lin, J. M. Sivertsen and J. H. Judy, IEEE Trans. Magn. MAG-27 (1991) 5163.

[8] H. Tsai, D. B. Bogy, J. Vac. Sci. Technol., A5, 3287, 1987.

[9] S. D. Berger, D. R. McKenzie, P. J. Martin, Phil. Mag. Lett., 57, 285, 1988.

[10] J. Fink, Advances in Electronics and Electron Physics, 75, 121, 1989.

[11] M. F. Doerner and W. D. Nix, J. Mater. Res. 1, 601 (1986).

Part VI

Properties of Polymer Films

Viscoelastic Properties of Healthy Human Artery Measured in Saline Solution by AFM-Based Indentation Technique

A. Lundkvist**, E. Lilleodden*, W. Siekhaus*, J. Kinney*, L. Pruitt** and M. Balooch*
* Lawrence Livermore National Laboratory, Livermore CA 94550, balooch1@llnl.gov
** University of California, Berkeley CA 94720

ABSTRACT

Using an Atomic Force Microscope with an attachment for indentation, we have measured local, *in vitro* mechanical properties of healthy femoral artery tissue held in saline solution. The elastic modulus (34.3 kPa) and viscoelastic response (τ_ε = 16.9 s and τ_σ=29.3 s) of the unstretched, intimal vessel wall have been determined using Sneddon theory and a three element model (standard linear solid) for viscoelastic materials. The procedures necessary to employ the indenting attachment to detect elastic moduli in the kPa range in liquid are described.

INTRODUCTION

Coronary artery disease or atherosclerosis is the leading cause of death in the United States [1]. Percutaneous Transluminal Coronary Angioplasty (PTCA) is the most common treatment for patients suffering from atherosclerosis. Although PTCA yields a greater than 90% immediate success rate, the 30-40% restenosis rate, or re-narrowing of the coronary arteries, hinders complete success [2]. A significant portion of this restenosis rate is due to geometric remodeling of the diseased artery wall incurred during balloon inflation (e.g. medial stretching [2], plaque dissection [3] and plaque compression [2,3]). If the plaque contains a significant amount of calcium deposits, preferential regions of high stress conducive to stretching or dissection may form during balloon inflation. Eighty percent of patients recommended for PTCA exhibit calcium deposits in the plaque [4]. Hence, there is particular clinical interest in understanding the local material properties and mechanical behavior of calcified atherosclerotic deposits and healthy arterial wall.

The present work describes the technique of using the AFM-based nanoindenter to determine the mechanical properties of soft tissues held in saline solution. The static elastic modulus and viscoelastic constants of the intimal surface of a healthy human femoral artery are reported. These results will be used as a base for further studies on atherosclerotic coronary tissues.

EXPERIMENTAL TECHNIQUE AND RESULTS

The Atomic Force Microscope (small stage Nanoscope III, Digital Instruments, Santa Barbara, CA) has been modified by replacing the conventional head assembly with a transducer-indenter assembly called a Micromechanical Testing Instrument (Hysitron Inc., Minneapolis, MN). Like its conventional counterpart, the instrument can image the topography of specimens by tracing the superficial contours of the sample with nano-Newton loads. In addition, the device is a force-generating and depth-sensing instrument (Nano-indenter) capable of generating load-displacement curves at specific locations in ambient and, with the corrections described here, liquid environments. The minimum load applied is less than 1 µN and the maximum displacement measured is 35 µm. Indenters of varying materials (diamond, tungsten carbide, aluminum); shapes (Berkovich, spherical, flat cylindrical punch) and a range of diameters (.17 to 2.0 mm) have been used for indentation of healthy human arteries.

For the experiments reported here, healthy femoral arteries supplied by the National Disease Research Interchange (NDRI) were scanned by intravascular ultrasound (Boston Scientific, Sunnyvale, CA) to ascertain that no plaque existed. The vessel was cut longitudinally, sectioned in 5mm x 5mm regions and placed unstretched in a .9% saline bath on the AFM stage.

The basic components of the Micromechanical Testing Instrument are shown in Figure 1. The force recorded and displayed by the instrument is defined as,

Mat. Res. Soc. Symp. Proc. Vol. 436 © 1997 Materials Research Society

$$F_{display} = A\varepsilon_0 \frac{V^2}{d_0^2} \tag{1}$$

where A is the plate area, V is the voltage applied between the drive plate and pickup electrode, ε_0 is the dielectric constant and d_0 is the spacing between the pickup electrode and drive plate when no weight (such as the indenter) is attached and no voltage is applied. The vender specifies d_0 as 120 μm. In many material science applications, equation (1) adequately represents the actual force applied to the material under investigation.

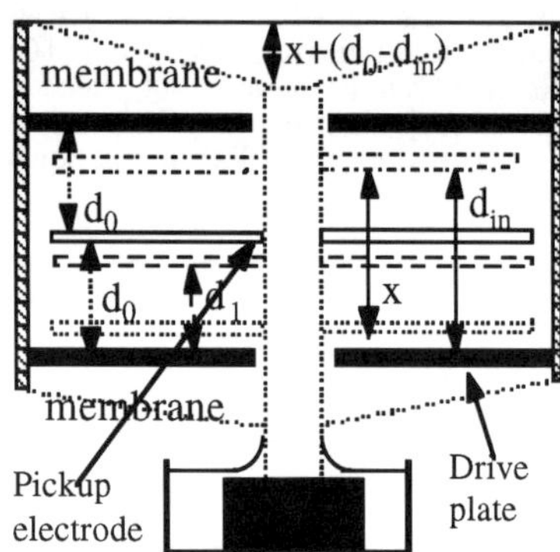

Figure 1. Schematic of Micromechanical Testing Instrument with parameters needed to determine the actual force applied to soft materials in saline solution. The combination of pickup electrode and drive plates is used to both generate a force, $F_{display}$, and to measure displacement, x.

For soft materials, however, the displacements could be large even for small applied voltages. In this case, the change in the plate spacing deviates substantially from that described by equation 1, and the force displayed does not equal the force actually applied to the sample. Additionally, the stiffness of the membrane, the weight of the indenter itself and the force exerted on the indenter's shaft as a result of surface tension should be included to accurately describe the applied force in liquid. The actual force applied to the sample is

$$F_a = F_{display}\left[\frac{d_0}{d_{in} - x}\right]^2 - K_s\left(x + d_0 - d_{in}\right) + F_{s.t.} + Mg \tag{2}$$

where x is the displacement of the pickup electrode from its initial position d_{in}, K_s is the spring constant of the membrane (ignoring possible nonlinearity), Mg is the weight of the indenter and $F_{s.t.}$ is the force due to surface tension. When a preload is applied to the sample, the force due to the initial displacement is

$$Preload = K_s\left(d_{in} - d_1^{liq}\right) \tag{3}$$

In the absence of an initial preload, the initial spacing, d_{in}, is replaced by

$$d_1^{liq} = d_1^{air} - \frac{F_{s.t.}}{K_s} \tag{4}$$

where d_1^{air} is the spacing between the pickup electrode and drive plate in air due to the weight of the attached indenter (Figure 1). K_s and d_1^{air} are determined by making a best fit to displayed force versus distance curves of several calibration runs done in air without a load applied to the sample. The results are shown in Figure 2a along with the deviation of experimental points from the optimum fit. The experiments were repeated in liquid with no load applied to the sample to obtain $F_{s.t.}$. The result of the curve fit and its deviation from data points in liquid is shown in Figure 2b.

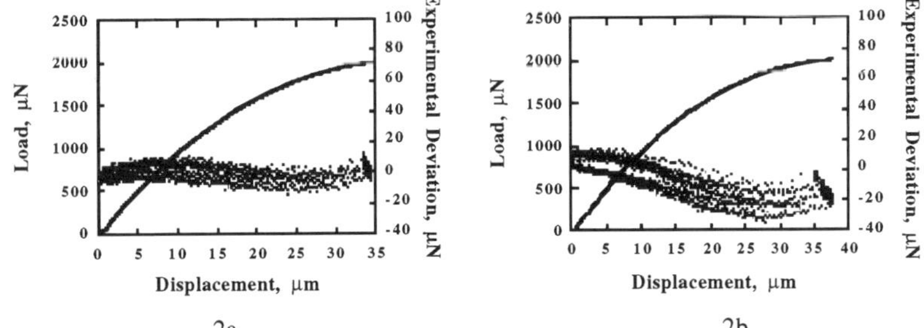

2a   2b

Figure 2. Displayed force vs. displacement with no load applied to the sample, and the deviation of experimental results from the optimum fit to force vs. displacement: a) in air b) in saline solution.

Since healthy human arteries are known to have high compliance, this study used a flat circular punch consisting of a thin aluminum disk 2mm in diameter, attached to a stem of .5 mm diameter (to minimize surface tension forces) held by the mechanical testing instrument. The "sine wave" appearance of the deviation of experimental results from the optimum fit to curves of force vs. displacement evident in both air and liquid calibration runs in Figure 2 can be eliminated when a nonlinear third-power term is added, as suggested by Timoshenko [5], to describe the force vs. displacement of the membrane. The load accuracy of the measurements is ±10 μN in air and ± 20 μN in saline solution, and by almost a factor of two better when a nonlinear spring is considered.

The data were analyzed using a simple three element model (standard linear solid [6])--a spring (μ_0) placed in parallel with a Maxwell spring (μ_1)-damper (η_1) system. The behavior of this model is characterized by three parameters:

$$\tau_\varepsilon = \frac{\eta_1}{\mu_1}\,, \quad \tau_\sigma = \frac{\eta_1}{\mu_0}\left(1+\frac{\mu_0}{\mu_1}\right) \quad \text{and} \quad E_R = \frac{\mu_0}{2\,r} \tag{5}$$

where τ_ε is the relaxation time for constant strain, τ_σ is the relaxation time for constant stress, r is the indenter radius and E_R is the relaxed elastic modulus of the system as time approaches infinity, based on contact stiffness and Sneddon indentation theory [7, 8]. The creep response [6], c(t), of this system to a constant force of magnitude unity, u(t), is described as

$$c(t) = \frac{1}{\mu_0}\left[1-\left(1-\frac{\tau_\varepsilon}{\tau_\sigma}\right)e^{-t/\tau_\sigma}\right]u(t) \tag{6}$$

The actual force and displacement vs. time for a healthy artery are shown in Figure 3. The instrument was programmed to generate a trapezoidal "display-force" vs. time. Applying the corrections described in equations (2) - (4) provided the time-dependent actual applied force shown in Figure 3. A second order polynomial was fit to the *loading portion* of the actual applied force vs. time curve and this fitted analytic function, F(t), was integrated using the Boltzmann superposition principle for creep response given an arbitrary load profile [6]. This results in an analytic displacement function, X(t), containing three undetermined model parameters μ_0, τ_ε and τ_0.

$$X(t) = \frac{1}{\mu_0}\left[F(t)\Big|_0^t - \left(1 - \frac{\tau_\varepsilon}{\tau_\sigma}\right) e^{-t/\tau_\sigma} \int_0^t e^{+\tau/\tau_\sigma}\, \frac{dF(\tau)}{d\tau}\, d\tau \right] \qquad (7)$$

This function was then fit to the measured displacement during the loading-only portion of the displacement vs. time data to determine the model parameters. These parameters were then used in the Boltzmann integral on the actual applied load curve F(t) to predict the displacement vs. time data, as plotted in figure 3.

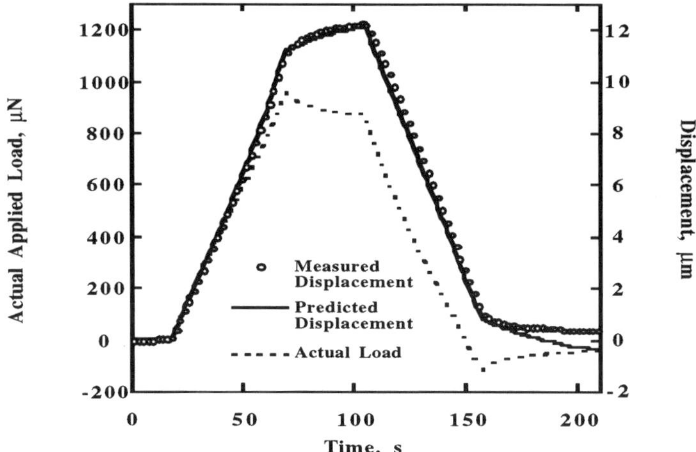

Figure 3. The actual force applied and measured displacement as a function of time for a healthy, femoral artery wall using a 2mm diameter, aluminum, flat punch. The Boltzmann superposition principle was used to predict the displacement vs. time based on the Kelvin model parameters and the time-dependent loading curve.

Table I compares the results of this study with previous indentation studies. Although the elastic modulus values reported by Gow et al. [9, 10] were obtained with indenters smaller in diameter than used here, information about the viscoelastic response is not provided. The technique used by Lee et al. [11] used a 7 mm diameter indenter, and hence the results represent an average over an area more than 10 times larger than the area used in this study. In addition to providing local measurements of the elastic modulus of healthy, femoral artery (intimal) wall, to the best of the authors' knowledge, this investigation represents the first *in vitro*, localized assessment of the viscoelastic response of healthy inner arterial wall.

	Aortic Intimal Surface [9,10]	Non-Fibrous Aortic Plaque Caps [11]	*Healthy Femoral Intimal Wall
Elastic Modulus (kPa)	90-152	41.2 ± 18.8	34.3
τ_ε (sec)	---	---	16.9
τ_σ (sec)	---	---	29.3

SUMMARY AND CONCLUSIONS

The add-on transducer to the standard AFM enhanced the capability of the instrument to measure local mechanical properties in addition to topographic imaging. To further increase its capability for use on soft tissues held in aqueous solution one also needs to consider the forces applied to the indenter, such as weight and surface tension, that are usually negligible for most materials used in engineering applications. By considering these forces, the viscoelastic properties of an unstretched healthy artery wall in saline solution have been measured. To the authors' knowledge, there is no previous 3 parameter-based model of the mechanical response of the intimal layer of the artery wall.

The instrumentation described here is being used with smaller indenters (.175 mm diameter) to measure the static mechanical and viscoelastic response of healthy and diseased arteries over a wide range of frequencies. These investigations will provide a sound engineering basis for future balloon angioplasty studies, and will help to improve the design of other clinical devices used to treat patients with coronary artery disease.

ACKNOWLEDGMENTS

This work was supported by the U.S. Department of Energy, through the LDRD program at Lawrence Livermore National Laboratory under contract No. W-7405-ENG-48.

We want to thank Dr. W. Gerberich, U. of Minnesota, Dept. of Mechanical Engineering and Material Science; Dr. P. Yock, Stanford University Medical Center for Research in Interventional Cardiology; Greg Hyde, Guidant Corp., Advanced Cardiovascular Systems; and Dr. Wayne Bonin, Hysitron Inc. for their contributions to the scientific and technical substance of this paper.

REFERENCES

1. Heart and Stroke Facts: 1994 Statistical Supplement, American Heart Association.

2. R. Virmani, A. Farb and A.P. Burke, American Heart Journal **127**: 163-179 (1994).

3. B.F. Waller, C.M. Orr, C.A. Pinkerton, J. Van Tassel, T. Peters and J.D. Slack, Journal of the American College of Cardiology **20** (3): 701-706 (1992).

4. L.A. Fitzpatrick, A. Severson, W.D. Edwards and R.T. Ingram, Journal of Clinical Investigation **94**: 1597-1604 (1994).

5. S. P. Timoshenko and J. N. Goodier. <u>Theory of Elasticity</u>, 3rd edition. (McGraw Hill Book Company, New York, 1970).

6. Y.C. Fung, <u>Biomechanics</u>, 2nd ed. (Springer-Verlag New York, Inc., New York, 1993), pp.41-46.

7. I.N. Sneddon, International Journal of Engineering Science **3**: 47 (1965).

8. G.M Pharr, W.C. Oliver and F.R. Brotzen, Journal of Materials Research **7** (3): 613-617 (1992).

9. B.S. Gow and R.N. Vaishnav, Journal of Applied Physiology **38** (2): 344-50 (1975).

10. B.S. Gow, W.D. Castle and M.J. Legg, Journal of Biomechanics **16** (6): 451-458 (1983).

11. R.T. Lee, G. Richardson, H.M. Loree, A.J. Grodzinsky, S.A. Gharib, F.J. Schoen and N. Pandian, Arteriosclerosis and Thrombosis 12 (1): 1-5 (1992).

MICRO-INDENTATION USING STRAIN RATE SENSITIVITY TO EXAMINE DEFORMATION OF POLYMERIC AND METALLIC SURFACE LAYERS

W.R. NEWSON, B.J. DIAK and S. SAIMOTO
Dept. of Materials and Metallurgical Engineering, Queen's University, Kingston, Ontario, Canada, K7L 3N6

ABSTRACT

A sensitive method to characterize the inelastic deformation of solids is to measure the thermodynamic response of the deforming volume by applying an inelastic strain rate change. Using the step ramp technique on a displacement controlled micro-indentor with Berkovich tip, the apparent activation volumes were determined for linear low density polyethylene, and eutectic Pb-Sn solder for depths less than 10 μm. The results show that the activation volume is a unique measure of a given microstructure.

INTRODUCTION

Hardness measurement in the form of indentation testing was one of the first accepted quality control tests for product strength assessment. The recent advent of inexpensive digitally controlled devices with unprecedented computing power has led to the development of automated micro-hardness machines and more recently to displacement controllable nano- and micro-indentation devices using different shaped indentors. However, quantifying materials behaviour with indentation, has developed more slowly primarily because pointed indentors invoke a large strain gradient within the plastic zone below the indentor [1]. Li and co-workers [2] have hence designed and quantitatively assessed impression creep testing with a flat indentor whereby materials relationships which correspond directly to bulk tensile and compressive tests can be derived. However, for very thin films in the micrometer range the flat indentor is not feasible and the pyramidal Berkovich indentor is the preferred choice. The recent workshop on indentation testing [4] elucidates the performance, materials characterization methods and applications of such devices.

Instrumented indentation testing to date has basically been automated hardness testing and the materials characterization is usually determined by creep or stress relaxation [5,6,7]. Our laboratory has promoted precision strain rate sensitivity tests to determine the thermodynamic response of various materials using tension tests [8] and has extended this method to nano-indentation [9]. The virtue of performing rapid rate changes in times less than 100 ms is that the fastest structural relaxation mechanism in polyethylene of 200 - 300 ms can be excluded from the thermodynamic response measurement. The present purpose is to illustrate that precision displacement rate changes can be performed using a Berkovich indentor and that the measured thermodynamic response can categorize the microstructures of both polymers and metals.

THEORY

Deformation in polymeric and metallic materials often exhibits rate dependent flow. These microstructures contain obstacles that can be thermally overcome to allow flow. Using the Eyring formulation of the rate theory, it is standard practice to describe in polymeric [10] or metallic [11] materials the inelastic strain rate using normal strain rate and stress as

$$\dot{\varepsilon}_{IN} = \dot{\varepsilon}_o \exp\left(-\frac{\Delta F - \sigma v}{kT}\right) \tag{1}$$

where ΔF is the free energy of activation to overcome the obstacles with thermal assistance; σ, the applied stress causing plastic flow; v, the apparent activation volume; $\dot{\varepsilon}_o$, the population density of strain centres and their frequency of vibration; T, the temperature and k, the Boltzmann constant. For indentation testing, the hardness (load divided by projected area of indent) is taken equivalent to 3σ [12] and the strain rate, $\dot{\varepsilon}$, is equivalent to $\dot{d}/d$ (the displacement rate divided by the depth of

Mat. Res. Soc. Symp. Proc. Vol. 436 © 1997 Materials Research Society

penetration) [13]. If the differential form of the equation is used, with the differentiation being performed at constant temperature, T, and structure, Σ, then

$$\frac{\partial \ln \dot{\varepsilon}_{IN}}{\partial \sigma}\bigg|_{T,\Sigma} = \frac{v}{kT} - \frac{1}{kT}\frac{\partial \Delta F}{\partial \sigma}\bigg|_{T,\Sigma} + \frac{\partial \ln \dot{\varepsilon}_o}{\partial \sigma}\bigg|_{T,\Sigma} \qquad (2)$$

This partial differentiation can only be experimentally performed if the elastic displacement transients of the load frame and specimen are eliminated from the rate change.

The Eyring formulation of deformation thermodynamics only applies to the inelastic strain rate. Unfortunately, it is the total strain rate, $\dot{\varepsilon}_T$, that is controlled during a test, where $\dot{\varepsilon}_T = \dot{\varepsilon}_E + \dot{\varepsilon}_{IN}$ and $\dot{\varepsilon}_E = \dot{\sigma}/E$. Since the equations of flow apply to the inelastic strain rate we desire an inelastic strain rate change like that in Fig. 1(a). From the theory we can also expect a drop in stress, $\Delta\sigma$, associated with the change in inelastic strain rate (Fig. 1(b)). Since the elastic strain is directly related to the stress there is also an elastic strain drop due to the stress drop (Fig. 1(c)). The total strain signal required to obtain the desired inelastic strain rate change is the sum of the elastic and inelastic strain profiles, or the step ramp of Fig. 1(d) [14]. In this way a controlled inelastic strain rate change can be applied and the right hand side of eqn. (2) determined.

To present the thermodynamic response data in the most concise form, we assume initially a simplified differential form (i.e. constant ΔF and $\dot{\varepsilon}_o$)

$$\frac{v}{kT} = \frac{\partial \ln \dot{\varepsilon}_{IN}}{\partial \sigma}\bigg|_{T,\Sigma} \qquad (3)$$

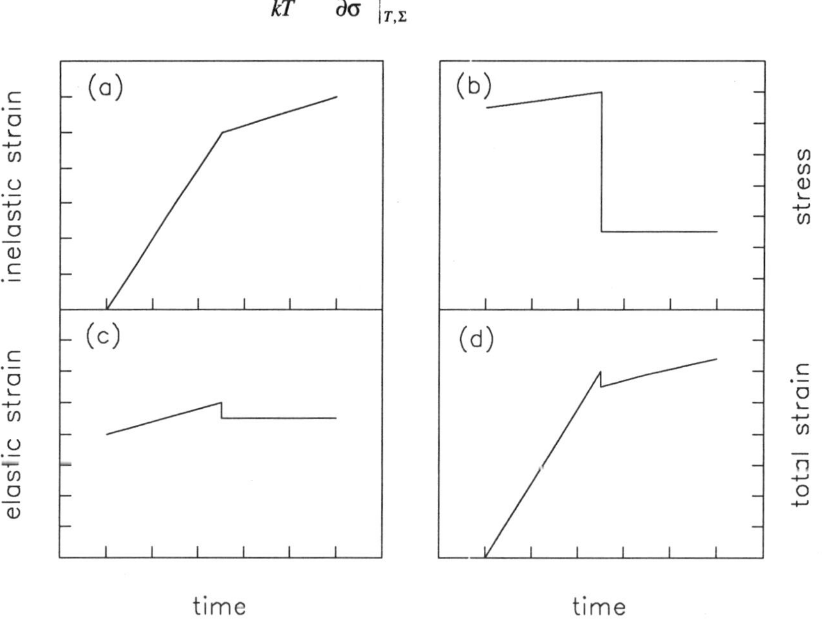

Fig. 1. Schematics of the step ramp testing method: (a) The desired inelastic strain rate change, (b) the theoretically predicted stress response to the rate change in (a), (c) the expected elastic strain response to the rate change, (d) the required total strain behaviour to achieve the response in (a).

such that the integral equation can be used in further analysis

$$\frac{\sigma v}{kT} = \ln \dot{\varepsilon}_{IN} - \ln \dot{\varepsilon}_o + \frac{\Delta F}{kT} \tag{4}$$

The integral relation, eqn. (4), has been previously applied to determine v at the yield point, σ_y, for polycarbonate [19,20] by testing many specimens at various temperatures and strain rates. The linear first order dependence of $\sigma_y \mid_T$ with $\ln\dot{\varepsilon}$ is assumed to justify that ΔF and $\dot{\varepsilon}_o$ are constant, and that Σ at yield is identical for the different specimens. As well, this procedure can not be extended to other points in the stress - strain curve since positions of identical structure for different specimens can not be identified. On the other hand, the derivative method has the potential to determine v at any point during the test at constant structure since the rate change occurs in less than 100 ms with compensation for machine and specimen compliance.

The following experiments attempt to show that this differential method can be applied to indentation testing to quantitatively measure the thermodynamic response and hence the evolving microstructure. $\sigma v/kT$ is determined from the product of v/kT and σ, or experimentally as $(\sigma/\Delta\sigma)\ln(\dot{\varepsilon}_1/\dot{\varepsilon}_2) = (L/\Delta L)\ln 10$, where ΔL is the instantaneous load drop measured after a 1/10 rate change at load, L. Thus a plot of $\sigma v/kT$ versus $\ln\dot{\varepsilon}$ characterizes the rate dependence of the microstructure during indentor penetration of the material. For a single obstacle system it is expected then that eqn. (4) can be represented as $(\sigma v/kT)/\ln\dot{\varepsilon}_{IN} = 1$, when v is determined from eqn. (3). Deviations from this theoretical prediction indicate a more complex system.

EXPERIMENTAL PROCEDURE

A stiff micro-indentor was developed with the displacement controlled to ±0.5 nm using a piezoelectric rod incorporating a capacitance sensor and a semiconductor strain gage 250 g load cell with a resolution near ±5 mg. Software developed by Upadhyaya [9] for nano-indentation has been applied to the micro-indentor utilized in this study. A Berkovich diamond was used as the indenter with a controlled displacement rate of 0.05 μm/s. The data was acquired every 0.5 ms and averaged over 10 to 100 ms depending upon the test material, and the displacement rate changes were imposed with a rate reduction of a factor of 10. The machine and contact compliance were determined using the steepest part of the unloading curve to evaluate the plastic indent depth. This procedure was verified using commercial purity 70-30 brass, by comparing the predicted area for a given depth to the residual area measured in the scanning electron microscope. Brass was chosen because its strain rate sensitivity is very near zero at room temperature. Two materials were used in this study: (1) eutectic Pb-Sn solder in an actual integrated circuit board, and (2) linear low density polyethylene (LLDPE) resin especially prepared by Dow Canada to vary the degree of crystallinity by varying the resin morphology and using the same plaque forming procedure [15]. The Pb-Sn solder was revealed by mechanically polishing off the Si chip attached to the copper base of the integrated circuit board and then electropolishing with a perchloric acid - alcohol solution at about -10°C. The polyethylene was mechanically polished using grit sizes down to 6000 [15] and all surfaces were relaxed at room temperature for at least one week prior to testing. Thermodynamic data is taken from at least two separate indents.

RESULTS AND DISCUSSION

A typical load versus depth plot with rate changes is shown in Fig. 2 for LLDPE. The insert shows the load - time response for one of the rate changes after almost ideal compensation for the sample and machine compliance by the step ramp. Rate changes at higher loads are over-compensated as unloading without relaxation is observed. Figure 3 summarizes the hardness versus depth behaviour for the eutectic solder and 74% crystallinity LLDPE. In the Pb-Sn solder the hardness initially increases with depth as the plastic volume increases, reaches a maximum, and gently decreases with depth. This is the typical behaviour observed for softer metals. Previous work [16] on aluminum had shown that at the maximum hardness range the strain rate sensitivity,

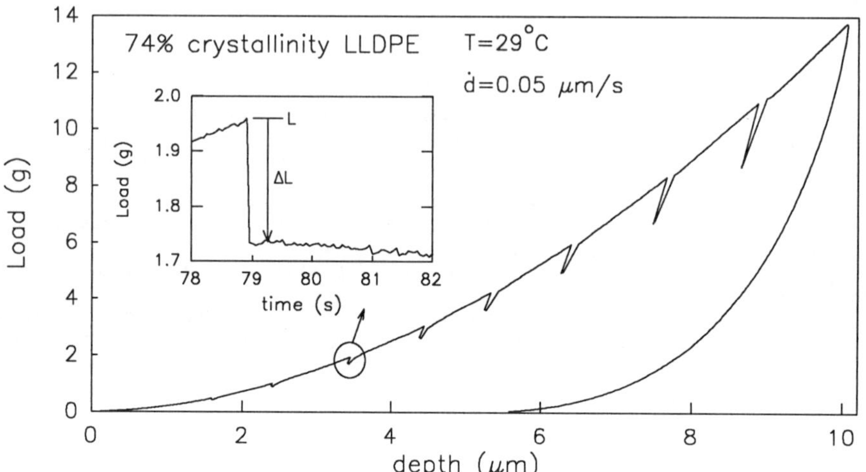

Fig. 2. Load versus depth behaviour for 74% crystallinity LLDPE showing 8 rate changes and the non-linear unloading behaviour. The insert is the actual load - time response to the step ramp as idealized in Fig. 1(b) showing the determination of L and ΔL.

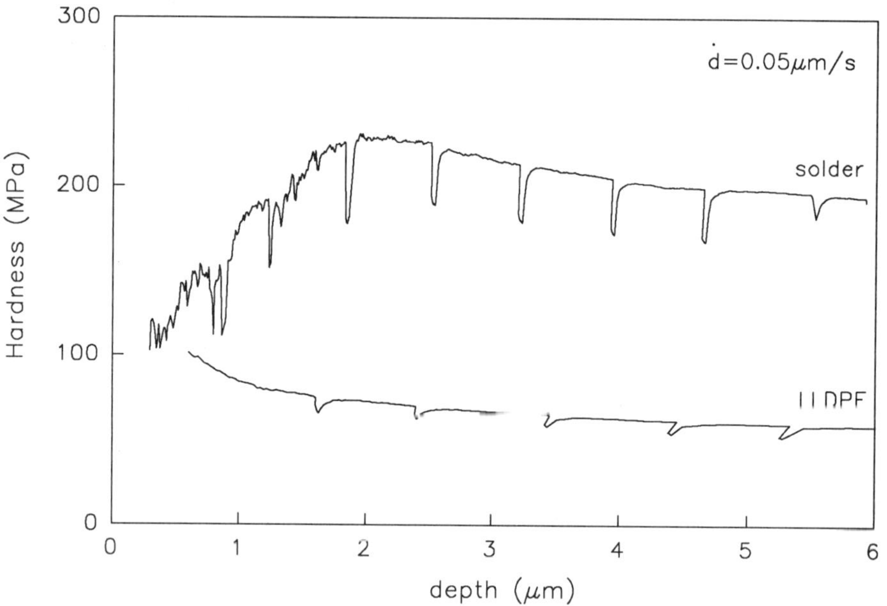

Fig. 3. Hardness versus depth behaviour for the Pb-Sn solder at 24°C and 75% crystallinity LLDPE at 29°C.

$d \ln \sigma / d \ln \dot{\varepsilon}$, approaches the value determined on bulk specimens by tensile testing. These differences must arise because the mechanisms of plastic flow in small confined volumes are unique to the details of dislocation loop formation and possible glide and climb modes. Although further study is required, the activation volume measurements described in the following section reveal how some of the parameters can be elucidated.

Figure 4 summarizes the $\sigma v/kT$ versus $\ln \dot{\varepsilon}$ behaviour for the LLDPE and the solder. Since the slope for the solder data approaches one according to eqn. (4), the assumption of constancy of ΔF and $\ln \dot{\varepsilon}_o$ with evolving microstructure seems applicable. Also, since the thermal energy is constant for a given temperature, $(\Delta F - \sigma v)/kT$ is constant if flow is occurring through thermal activation, which in turn means that $\sigma \propto 1/v$. Thus the Cottrell-Stokes relation holds [17] and is the expected result for a single obstacle system of monopole dislocation - dislocation interaction in all crystalline material [18]. Li *et al.* [10] have argued that such relations are applicable to amorphous material near or below the T_g. The 74% crystalline LLDPE tested at various temperatures shows two region behaviour. Despite the scatter due to the low loads involved and thus imprecision in determining ΔL, the $\sigma v/kT$ versus $\ln \dot{\varepsilon}$ appears to follow the theoretical line at fast strain rates (low depths) and curves slightly upward at the lowest strain rates (high depths). Moreover, all the data approach superposition suggesting a master curve and implying a similar deformation mechanism. The data points for 61°C indicate the start of a deviation detectable from the scatter of other data points. Figure 4 suggests that at high strain rates and ambient temperatures crystalline flow is predominant for the 74% crystalline LLDPE, but at lower strain rates ($<10^{-2}$/s) and higher temperatures, an upward deviation (increase in v) is detectable. Furthermore, the 55% crystalline case at 24 °C behaves similarly to the high temperature 74% case. This observation, together with the fact that T_g for completely amorphous LLDPE is much lower than ambient, strongly suggests that the inelastic flow mechanism can be attributed to those of crystalline and elastomeric components of the microstructure. These two modes of deformation have been reported earlier for tensile testing [21]. It has also been observed during constant strain rate tests under tension that v initially decreases with increase in stress, and at some stress, v starts to increase [15]. The inception of the v increase occurs at lower and lower stresses with increasing T, analogous to the present results. An increase

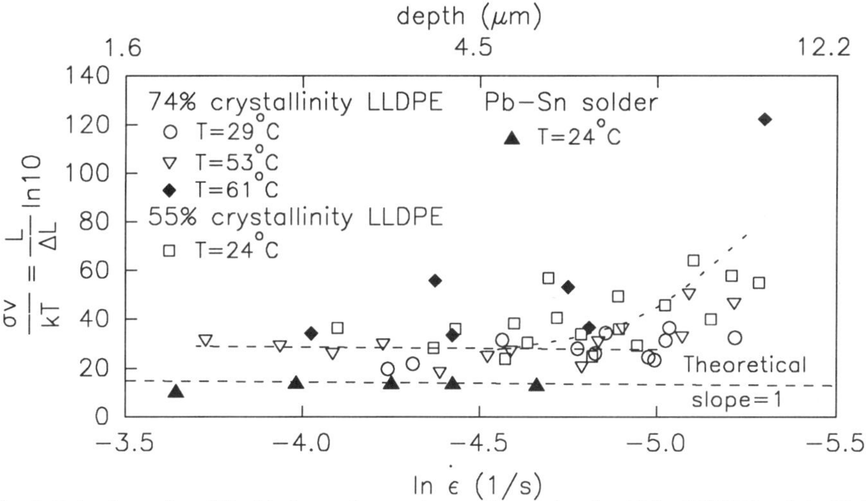

Fig. 4. Behaviour of $\sigma v/kT$ with decreasing strain rate (increasing depth) for LLDPE and the Pb-Sn solder. The theoretical slope of one from eqn. (4) is indicated for both the LLDPE and solder. The deviation for the LLDPE at $\ln \dot{\varepsilon} < -4.5$ is sketched in to indicate another region of behaviour.

in v with strain can be predicted for elastomeric structures from the freely jointed chain model. Thus the present technique can detect these mechanistic changes. The similarity in $\sigma v/kT$ values for solder and LLDPE indicates that the real microstructure characterizing parameter is v, which is large for polymers, but small for metals giving rise to the necessary differential in flow strength. The rate controlling mechanism in polymers at low strain rates and elevated temperatures requires much more detailed study. However, these preliminary results suggest that microstructures of various thin materials can be characterized using the v determined by micro-indentation testing with step ramp displacement rate changes.

CONCLUSIONS

Step-ramp displacement rate changes can be used to delineate the thermodynamic response of different materials (v) and ascertain if the microstructural obstacles are evolving with deformation and strain rate. This method using a Berkovich indentor permits microstructural evaluation of surface layers less than 10 μm thick by determining v as well as hardness.

ACKNOWLEDGEMENTS

The authors thank the Natural Science and Engineering Research Council of Canada and the Ontario Centre for Materials Research for financial support, and the J.P. Bickell Foundation and Natural Resources Canada for the equipment grants. We are grateful to K.R. Upadhyaya for modifications to the NanoRama software. We thank R.J. Collacott of Dow Chemical Canada Inc. for supplying the LLDPE plaques. One of us (W.R.N.) was the recipient of a W.W. King Graduate Fellowship during this study.

REFERENCES

1. L.E. Samuels, in <u>Microindentation Techniques in Materials Science and Engineering</u>, edited by P.J. Blau and B.R. Lawn (ASTM STP 889, Philadelphia, 1986), pp. 5-25.
2. S.N.G. Chu and J.C.M. Li, J. Mater. Sci. **12**, 2200-2208 (1977).
3. F. Yang and J.C.M. Li, Mater. Sci. Eng. **A201**, 40-49 (1995).
4. <u>Conference Proceedings: International Workshop on Instrumented Indentation</u>, edited by D. Smith (NIST STP 896, 1995).
5. C. Santa Cruz, F.J. Baltá Calleja, T. Asano and I.M. Ward, Phil. Mag. **A68**, 204-224 (1993).
6. O. Prakash and D.R.H. Jones, Acta Mater. **44**, 891-897 (1996).
7. K. Zeng, E. Söderlund, A.E. Giannakopoulos, and D.J. Rowcliffe, Acta Mater. **44**, 1127-1141 (1996).
8. S. Saimoto in <u>1st Pacific Rim Intl. Conference on Advanced Materials Processing</u>, edited by C. Shi, H. Li and A. Scott (TMS, 1992), pp. 87-95.
9. K.R. Upadhyaya, Ph.D. thesis, Queen's University, 1994.
10. J.C.M. Li, C.A. Pampillo and L.A. Davis, in <u>Deformation and Fracture of High Polymers</u>, ed. by H.H. Kausch, J.A. Hassell and R.I. Jaffee (Plenum Press, 1973), pp. 239-258.
11. U.F. Kocks, A.S. Argon, M.F. Ashby, <u>Thermodynamics and Kinetics of Slip</u>, Prog. Mater. Sci. **19** (1975).
12. D. Tabor, <u>The Hardness of Metals</u>, (Oxford Univ. Press, New York, 1951).
13. A.G. Atkins, A. Silverio and D. Tabor, J. Int. Metals, **94**, 369 (1966).
14. M. Carlone and S. Saimoto, Experimental Mechanics, accepted for publication.
15. W.R. Newson, M.Sc. thesis, Queen's University, in progress.
16. B.J. Diak, K.R. Upadhyaya and S. Saimoto, in <u>10th International Conference on the Strength of Materials</u>, edited by H. Oikawa *et al.* (Jap. Inst. Metals, 1994), pp. 211-214.
17. S. Saimoto and H. Sang, Acta Metall. **31**, 1873-1881 (1983).
18. F.R.N. Nabarro, Acta Metall. Mater. **38**, 161-164 (1990).
19. N.G. McCrum, C.P. Buckley and C.B. Bucknall, <u>Principles of Polymer Engineering</u> (Oxford Science, New York, 1991), pp. 172-175.
20. J.-C. Bauwens, in <u>Failures in Plastics</u>, edited by W. Brostow and R.D. Corneliussen (Hanser, New York, 1986), pp. 233-258.
21. S. Saimoto and D.R.M. Thomas, J. Mat. Sc., **21**, 3686-3690 (1986).

THIN FILM POLYMER STRESS MEASUREMENT USING PIEZORESISTIVE ANISOTROPICALLY ETCHED PRESSURE SENSORS

G. Bitko, R. Harries, J. Matkin, A.C. McNeil, D.J. Monk, M. Shah, and J. Wertz
Motorola, Semiconductor Products Sector, Sensor Products Division, M/D D138, 5005 E. McDowell Rd., Phoenix, AZ 85008

ABSTRACT

Silicon bulk micromachined piezoresistive pressure sensors are very sensitive to applied stresses: that is, applied pressure and/or packaging-related stresses. Device encapsulation has been observed to affect the electrical output of the pressure sensor significantly. The magnitude of the zero applied pressure output voltage (i.e., the offset voltage) that can be attributed to a thin film encapsulant is proportional to the magnitude of the room-temperature thermal stress of that film. Parylene C coatings have been used as encapsulants in this work. Finite element and analytical modeling techniques were used to evaluate the effect of material property variation on the offset of a pressure sensor. A simple, linear expression of offset as a function of a material property parametric group, that includes: parylene thickness, parylene biaxial modulus, parylene CTE, silicon thickness, and annealing temperature; has been established. Experimental analysis of parylene coated pressure sensors and parylene coated silicon and gallium arsenide wafers was performed to confirm the resulting model. Known variations in parylene material properties caused by processing (i.e., uncontrolled deposition, annealing, and high temperature storage) have been used as an experimental vehicle for this purpose. An empirical relationship between offset voltage on parylene coated devices and room-temperature thermal stress on parylene coated wafers that have been exposed to the same processing is a linear expression with a similar slope to the modeling results. Furthermore, stress measurements from parylene coated silicon wafers and parylene coated gallium arsenide wafers have been used to estimate the parylene biaxial modulus (approximately 5000 MPa) and the parylene CTE (approximately 50 ppm/°C) independently. These material properties were observed to shift following parylene annealing and high temperature storage exposure experiments in a manner that is consistent with the established model.

INTRODUCTION

Stress measurement in thin polymer films can be made by a variety of techniques. Radius of curvature measurements have been used for organic and inorganic materials that can be deposited on silicon wafers (e.g., [1]). Stress is calculated from these measurements by using Stoney's equation:

$$\sigma_f = \frac{E_s\, t_s^2}{6(1 - v_s)Rt_f},$$ (1)

where $\dfrac{E_s}{(1 - v_s)}$ is the biaxial elastic modulus of the substrate (1.805×10^{11} Pa for (100) silicon), t_s is the substrate thickness, t_f is the deposited films thickness, R is the radius of curvature, and σ_f is the calculated thin film stress. This measurement assumes that $t_f \ll t_s$. Radius of curvature can be observed by laser measurement of the wafer z-axis position (i.e., "flatness") when located on a referenced platen.

In traditional IC packaging, stress in polymer films only minimally affects the device electrical performance. However, silicon bulk micromachined piezoresistive pressure sensors (Fig. 1) are very sensitive to applied stress (i.e., applied pressure and/or packaging-related stresses) [2].

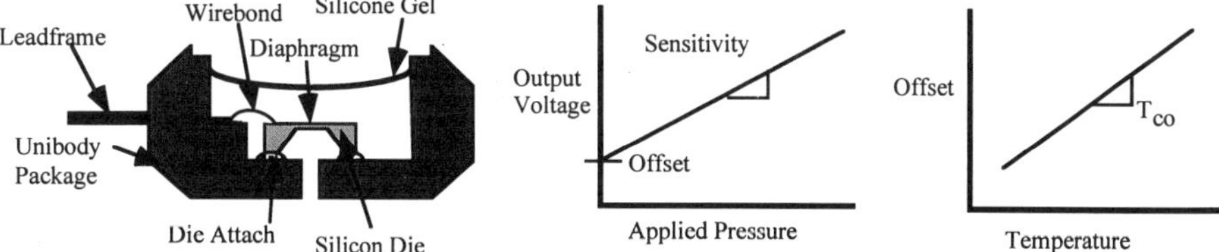

Figure 1.   A standard Motorola epoxy "unibody" pressure sensor package. The package contains a bulk micromachined silicon die with a piezoresistive transducer that is die bonded to the unibody package with an elastomeric silicone material. Gold wires are used to connect the die to the Au-plated, Ni-barrier layer, Cu leadframe. Device electrical output varies approximately linearly with pressure,

Mat. Res. Soc. Symp. Proc. Vol. 436 © 1997 Materials Research Society

where the slope is the sensitivity of the device and the intercept is the offset of the device. The slope of the offset as a function of temperature curve is the temperature coefficient of offset, T_{co}.

Parylene C (poly(monochloro-*para*-xylylene) coatings, which are room temperature, low pressure, vapor-deposited polymer thin films, have been used to protect the sensor from harsh chemical environment as an alternative to silicone gels [2-6]. Previous work has shown that the sensitivity of these devices is affected by the modulus and thickness of both the parylene thin film and the silicon diaphragm [2].

The current work concentrates on the modeling of the offset (Fig. 1) and the effect of material property changes on the stability of offset for a parylene-coated piezoresistive pressure sensor. Offset is proportional to the thermal stress in the passivation materials on the diaphragm at room temperature [2]. The temperature coefficient of offset, T_{co}, is proportional to the stress versus temperature curve, so offset is a single point on the T_{co} curve. However, offset can be affected by many of the fabrication and assembly processes; for example: transducer geometry and doping profile; resistor network and trimming; packaging, including, die attach and wirebonding; test fixturing; etc. These additional effects are outside the scope of this paper.

MODELING

Finite element analysis has been used to model the the effect of parylene on offset. The model is shown in Figure 2. It was assumed that there is 0.25 μm of thermal oxide on the diaphragm prior to coating. The silicon thickness for the diaphragm is assumed to be between 31.8 and 37.8 μm (Table 1). The die attach thickness was 0.010 inch of a silicone die attach material on an epoxy package. Parylene is assumed to coat both sides of the diaphragm with an equal thickness.

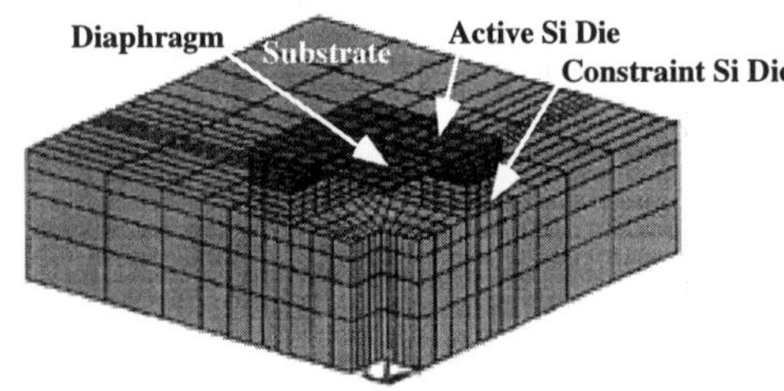

Figure 2. An example of the finite element model used to describe offset of a piezoresistive pressure sensor following coating of the diaphragm with a thin film (e.g., parylene). Quarter symmetry was used. This top view of the model shows the active silicon die with diaphragm as it is bonded to a constraint silicon die and the die is attached to a substrate. The substrate is epoxy and the die attach material is a silicone elastomeric adhesive. Thermal oxide was included as part of the diaphragm before parylene deposition. It was assumed that parylene coated the front and backside of the die equally.

Two finite element modeling experiments were performed to observe the effect of various material properties on the offset of a parylene coated piezoresistive bulk micromachined silicon pressure sensor. The variables of interest and the range of their values are given in Table 1.

Table 1. Offset Modeling Material Property Variables.

Material Property	Low	Nominal	High
t_p (μm)	5	10	15
E_p (10^5 psi)	3	4	5
α_p (ppm/°C)	29	39	49
t_{Si} (μm)	31.8	34.8	37.8
E_{Si} (10^6 psi)	17	18	19
α_{Si} (ppm/°C)	3.3	3.5	3.7
$t_{Die\ Attach}$ (mils)	10	15	20
$E_{Die\ Attach}$ (psi)		500	
$\alpha_{Die\ Attach}$ (ppm/°C)		300	
t_{Epoxy} (mils)		60	
E_{epoxy} (10^6 psi)		2	
α_{Epoxy} (ppm/°C)		21	
T_{anneal} (°C)	100	125	150

An initial finite element experiment determined that parylene thickness, t_p; parylene Young's modulus, E_p; parylene CTE, α_p; silicon diaphragm thickness, t_{Si}; and annealing temperature, T_{anneal} were the most important variables to describe offset and T_{co} for parylene-coated pressure sensors. It was assumed that the modulus and CTE of the die attach and the thickness, modulus, and CTE of the epoxy package were unimportant when predicting the offset shift caused by the addition of parylene to the pressure sensor. Therefore, these variables were held constant at the nominal values listed in Table 1. Output from the model was offset voltage at several temperatures between 150 °C to -40 °C assuming a 3 Vdc voltage drop accross the transducer [7]. The goal of this second FEA experiment was to determine a model for the effects of these material properties on the offset (and T_{co}) of a parylene-coated pressure sensor device.

An analytical modeling effort was also pursued to compare with wafer curvature experiments. The thermal stresses in the silicon substrate can be approximated by assuming that the silicon diaphragm is much stiffer than the parylene coating, and that the parylene is thinner than the silicon substrate. The silicon wafer thickness is 381 μm, and the parylene thicknesses are 5 to 15 μm. The parylene stress will generate both bending and tensile stresses in the silicon substrate. The bending stresses can be calculated by using the standard plate bending stress equation [8]. Two assumptions were made for this calculation: the force per unit length is the parylene stress/length multiplied by the film thickness, and the moment arm is approximated as half of the silicon thickness (again, assuming that the silicon is much thicker than the parylene). The tensile stresses in the Si are found by using a force equilibrium for the two layers, then solving for the silicon tensile stress. The bending and tensile stresses will have the same sign on the top surface of the silicon, and thus these two stresses add. Adding the bending stress and tensile stress equations yields the total stress at the top surface of the silicon:

$$\sigma_{Si} = 4 \frac{E_p}{(1 - \upsilon_p)} \; \frac{t_p \; \Delta T \; (\alpha_p - \alpha_{Si})}{t_{Si}} , \qquad (2)$$

where σ_{Si} is the stress in the silicon, υ_p is the parylene Poisson's ratio, ΔT is the change in temperature from the annealing temperature to room temperature, and α_{Si} is the silicon CTE. This equation can be used to estimate the stresses on an unrestrained Si layer, such as a wafer used in wafer curvature film stress measurement. A pressure sensor diaphragm has different boundary conditions (silicon diaphragm clamped at edges), but stresses will be proportional to this equation. Thus, eqn. 2 establishes a parametric group to calculate offset shift from parylene. Stress, and hence electrical output shifts, will be proportional to this parametric group (for the case of a thin parylene layer). Differentiating eqn. 2 with respect to temperature yields a similar parametric group that can be used to estimate the temperature coefficient of offset (T_{co}).

EXPERIMENTAL

Vapor-deposited parylene C coatings were used on standard Motorola pressure sensor packages and (100) silicon wafers. This experiment was performed to observe the effect of the parylene deposition temperature on the thermal stress stability of parylene coated silicon wafers and on the electrical offset stability of parylene coated devices during high temperature exposure (Table 2).

Table 2. Experiment for Analysis of Deposition Initial Temperature and Deposition Temperature Change during the Parylene Deposition Process on Offset Stability of Parylene Coated Pressure Sensors.

Deposition Designation	Initial Deposition Temperature	Deposition Temperature Change
ST.1.1	22 ± 3 °C	25 to 42 °C
ST.1.2	42 ± 3 °C	< 3 °C
ST.1.3	-10 - 0 °C	< 3 °C
ST.1.4	-10 - 0 °C	0 to 42 °C
ST.1.5	-10 - 0 °C	< 3 °C
ST.1.6	22 ± 3 °C	25 to 42 °C
ST.1.7	42 ± 3 °C	< 3 °C

Deposition was performed during seven separate parylene coating processes. The deposition temperature was controlled qualitatively by either having an external heater element on the outside of the deposition chamber, by not using the external heater, or by using dry ice applied to the outside of the deposition chamber. Temperature characterization of the deposition chamber was performed using thermocouples in several locations within the deposition chamber during a simulated deposition process. The deposition temperature was allowed to approach a steady state temperature that was a function of the chamber temperature control conditions. Additional details of the deposition process are described elsewhere (e.g., [9-11]). Annealing of parylene was performed at 125 °C for 90 minutes in N_2 [12]. In these annealing processes, nitrogen environments were used to minimize the effects of oxidative degradation of parylene [13, 14]. Offset stability was observed on the parylene coated devices during high temperature air storage at 105 °C.

Thickness measurements were made using a Nanometrics Nanospec AFT 210. Silicon diaphragm thickness was assumed to be the nominal production specification. Device electrical parameters were measured on calibrated production test systems before and after deposition, after annealing, and after each interval of high temperature storage. Wafer curvature measurements were performed on a FleXus F2300 system with hot plate temperature control on parylene C coated (100) silicon wafers. The slope of this stress versus temperature curve:

$$\text{Slope} = \frac{E_p}{(1 - v_p)}[\alpha_{Si} - \alpha_p(T)] , \tag{3}$$

where v_p is the Poisson ratio for parylene (assumed to be 0.4), provides only one equation for two unknowns: E_p and α_p. Gallium arsenide substrates were used in condition ST.1.7 (Table 2) to provide a second slope value (eqn. 3) so that E_p and α_p could be determined independently.

RESULTS AND DISCUSSION

Offset values from finite element analysis were calculated for an applicable range of the five most important material properties listed in Table 1: t_p, E_p, α_p, t_{Si}, and T_{anneal}. An implicit assumption that the silicon modulus and the silicon CTE do not change from device to device was made. The results condense into a linear equation when the parametric group shown in equation 2 is used to model offset:

$$\text{Offset (mV)} = -(1.1 \pm 0.13) - (0.74 \pm 0.024)\frac{E_p}{(1 - v_p)} \quad \frac{t_p \, \Delta T \, (\alpha_p - \alpha_{Si})}{t_{Si}} \quad R^2 = 0.975 . \tag{4}$$

The intercept is quite likely the result of non-parylene-related stresses in the model (e.g., silicon dioxide or silicon nitride stresses on the silicon diaphragm).

Two conclusions were necessary from our experimental work. The first was to demonstrate a similar function for offset voltage as a function of silicon stress that was caused by the addition of parylene to the silicon diaphragm (eqns. 2, 4). The second was to apply that model to identify which material properties were shifting during high temperature storage exposure of parylene-coated pressure sensors.

Parylene deposition rate is a strong function of temperature [15], so temperature profiling of the deposition chamber was performed (Table 2). Electrical data was recorded on all devices at seven discrete intervals: before parylene deposition, after parylene deposition, after parylene annealing, and after each of four high temperature storage intervals (96, 216, 368, and 758 hours at 105 °C). Annealing of parylene affected the electrical output of the device and the stress on wafers analogously [2]. Offset measurement shifted after each high temperature storage interval. This change in electrical parameters during HTS exhibited a trend when observed as a change in sensitivity versus a change in offset (Fig. 3).

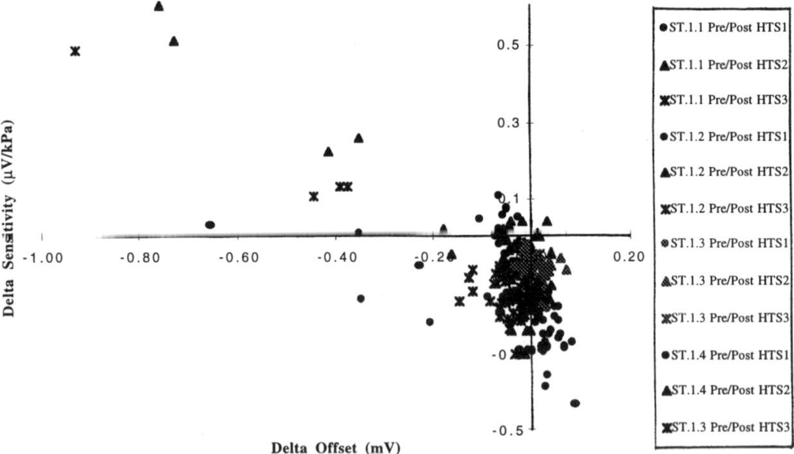

Figure 3. Summary of high temperature storage (105 °C in air) electrical parametrics for parylene coated devices from depositions ST.1.1-ST.1.4 (Table 2). In the legend, HTS1 represents 96 hours of exposure, HTS2 represents 216 hours of exposure, and HTS3 represents 368 hours of exposure. Most parts trend toward the origin, but some are obvious outliers.

Room temperature stress measurements on the parylene coated wafers exhibited a similar shift during HTS (Fig. 4). From similar deposition conditions, a change in stress on the parylene-coated wafers can be plotted against a change in offset voltage on the parylene-coated devices to confirm the relationship between the two measurements (Fig. 4). This further implies that the offset voltage is a function of the material properties listed in equation 2. The slope of this equation (Fig. 4b) is on the same order of magnitude as the slope in equation 4; thus, offset voltage in parylene-coated pressure sensors is caused, at least in part, by thermal stress generated in the silicon substrate. Moreover, a relationship among these material properties and the electrical offset has been established, so independent material property measurements can be used in the future to predict offset response for pressure sensor diaphragm coating.

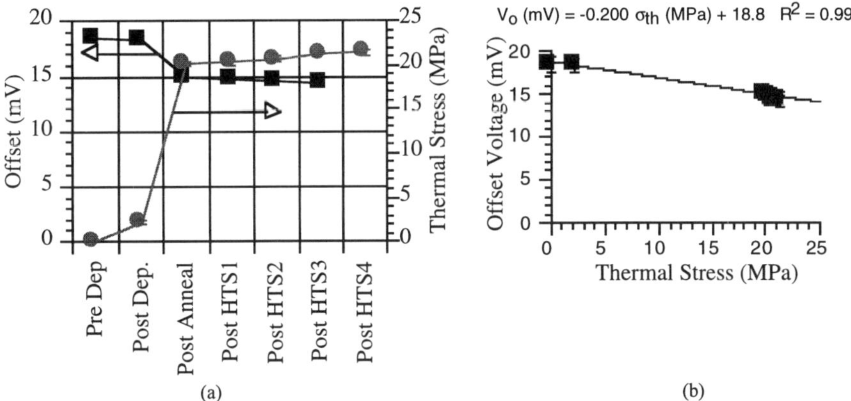

Figure 4.   (a) Offset voltage and thermal stress (i.e., room temperature parylene stress) plotted as a function of each condition during the experiment. (b) Offset voltage as a function of parylene thermal stress. Empirically, this demonstrates a relationship between the two measurable parameters.

Based on results from reference [2], Figure 5 can be used to show how variations in parylene modulus, CTE, and thickness are expected to affect the key device parameters.

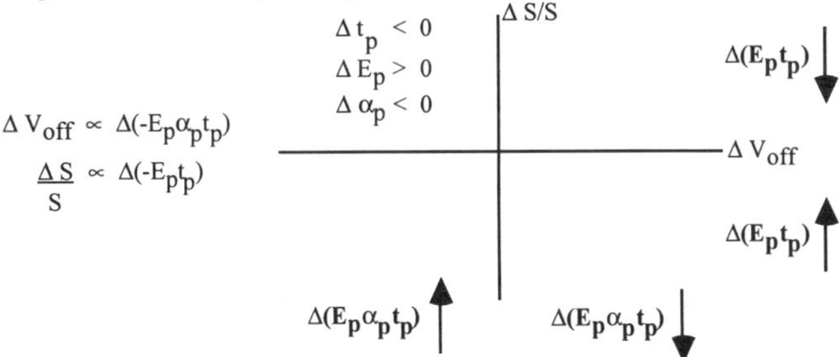

Figure 5.   The anticipated effect of changes in parylene material properties, on device parametrics. It is assumed that silicon material properties remain constant. Gauge limitations on our parylene film thickness measurement technique are such that thickness measurements are indeterminate for small variations.

Parylene coated gallium arsenide wafers supported these results (Fig. 5) when compared with parylene coated silicon wafers [16, 17]. From this comparison, it can be estimated that the parylene modulus is increasing during high temperature storage exposure for this deposition condition, and parylene CTE is decreasing (Table 3). Parylene thickness appears to be decreasing although our measurements are within the noise of the measurement technique. However, other researchers have observed parylene thickness to decrease during high temperature exposure [18].

Table 3. The Values of Slope and Intercept of the Stress vs. Temperature Curve (Assumed To Be Linear - All Correlation Coefficient Values Are Greater Than 0.998) during Cooling from 125 °C to 35 °C in 1 Hour.

Wafer Number/ Deposition Number	Elapsed High Temperature Storage Time (hrs)	Slope (MPa/°C)	Estimated $E_p/(1-\nu_p)$ (MPa)	Estimated α_p (ppm/°C)	Calculated Thermal Stress at Room Temp. (MPa)
Si 50/	0	-0.232	4760	57.7	22.7
ST.1.7	96	-0.243	5090	56.5	23.0
	216	-0.246	5480	53.4	23.1
	383	-0.257	7110	43.8	25.2
GaAs 2/	0	-0.248			24.6
ST.1.7	96	-0.258			24.9
	216	-0.263			25.1
	383	-0.270			26.5

The shifts observed in material properties account for the variation in offset that has been observed for these devices during the same exposure conditions. Because the modulus is increasing, one can postulate that the molecular weight of the polymer may be increasing. Additional work has shown that the stability of these parylene material properties is a function of location within the deposition chamber. High energy monomers that deposit onto substrates tend to contribute more to this phenomenon than lower energy monomers. For this application, mapping of the deposition chamber and/or baffling of the input monomer are necessary to provide stable parylene coated sensors.

CONCLUSIONS

Bulk micromachined piezoresistive silicon pressure sensor electrical output is affected by applied stress (e.g., applied pressure and/or packaging related stresses). For zero applied pressure, the output voltage (i.e., the offset voltage) is proportional to the room-temperature, thermal stress of a thin film that is placed on the silicon diaphragm. Modeling of this phenomenon was performed using parylene C as the thin film. A finite element analysis model and an analytical model have been used for this purpose. The finite element model was used in a designed experiment to simulate conditions where the most important factors: parylene thickness, parylene modulus, parylene CTE, silicon thickness, and annealing temperature; were varied from nominal values. The result was a simple, linear equation based on the parametric group that was determined from the analytical model of stress in the silicon substrate caused by the parylene thin film.

Experimental results support this relationship. Parylene thin films, when deposited in such a way that the energy of the input monomer was not controlled, caused a shift in the silicon stress as a function of high temperature storage exposure time. This type of parylene was deposited on pressure sensor packages, silicon wafers, and gallium arsenide wafers in the same process. *Ex situ* electrical offset measurement and wafer curvature measurements on both substrates exhibited shifts following high temperature storage intervals. A linear relationship between the offset and the room-temperature thermal stress has been established. The slope of the linear equation was on the same order of magnitude as the slope from the modeling results. This indicated that parylene stress contributes to the offset of these pressure sensor devices in a predictable fashion. Finally, this modeling effort is not limited to parylene coatings. Other passivation materials (e.g., PECVD silicon nitride, LPCVD silicon nitride, thermal oxide, etc.) can be evaluated with this modeling technique.

ACKNOWLEDGMENTS

The authors would like to thank Motorola Sensor Products Division for sponsoring this presentation. Fabrication was provided by the Motorola MEMS1 fab. Assembly of devices was performed by the Motorola MKL production facility and the SPD Prototype Lab. Test characterization of devices was conducted in the Motorola Phoenix production facility. High temperature storage of parylene coated devices was performed in the SPD/OSPD Reliability Lab and in the SPD Prototype Lab. Finally, FleXus wafer curvature measurements were performed in the ACT fabrication facility, and the authors are very grateful for the opportunity of using that equipment.

REFERENCES

[1] P. H. Townsend, T. P. Weihs, J. Sanchez, J. E., and P. Børgesen, "Thin Films: Stresses and Mechanical Properties IV," *Materials Research Society Symposium Proceedings*, vol. 308. Pittsburgh, PA: MRS, 1993.
[2] D. J. Monk and M. Shah, "Thin Film Polymer Stress Measurement Using Piezoresistive Anisotropically Etched Pressure Sensors," Mat. Res. Soc. Symp. Proc., San Francisco, CA, pp. 103-109, 1995.

[3] D. J. Monk, T. Maudie, D. Stanerson, J. Wertz, G. Bitko, J. Matkin, and S. Petrovic, "Media Compatible Packaging and Environmental Testing of Barrier Coating Encapsulated Silicon Pressure Sensors," 1996 Solid-State Sensors and Actuators Workshop, Hilton Head, SC, to be published, 1996.

[4] M. Noble, "Environmental Concerns for Integrated Circuit Sensors," *Measurement + Control*, vol. 19, pp. 210-213, 1986.

[5] K. E. Petersen, "Silicon as a Mechanical Material," *Proc. IEEE*, vol. 70, pp. 420-457, 1982.

[6] R. E. Sulouff, Jr., "Silicon Sensors for Automotive Applications," International Conference on Solid-State Sensors and Actuators: Transducers '91, San Francisco, CA, pp. 170-176, 1991.

[7] M. K. Shah, A. C. McNeil, M. D. Summers, and B. D. Meyer, "Application of Finite Element Analysis to Predict the Performance of Piezo-resistive Pressure Sensor," Sensors in Electronic Packaging, MED-Vol.3/EEP-Vol.14, 1995 ASME International Mechanical Engineering Congress and Exposition, San Francisco, pp. 79-85, 1995.

[8] W. C. Young and R. J. Roark, *Formulas for Stress and Strain*, 5th ed. New York: McGraw-Hill, 1982.

[9] W. F. Gorham, "A New, General Synthetic Method for the Preparation of Linear Poly-*p*-xylylenes," *J. Polym. Sci.*, vol. 4, pp. 3027-3039, 1966.

[10] B. J. Bachman, "Poly-P-Xylylene as a Dielectric Material," 1st International SAMPE Electronics Conference, Santa Clara, CA, pp. 431-440, 1987.

[11] W. D. Niegisch and W. F. Gorham, "Xylylene Polymers," in *Encyclopedia of Polymer Science and Technology*, vol. 15. New York: John Wiley & Sons, Inc., pp. 98-124, 1971.

[12] G. Bitko, D. J. Monk, H. S. Toh, and J. Wertz, "Annealing Thin Film Parylene Coatings for Media Compatible Pressure Sensors", Motorola Technical Developments, to be published, 1995.

[13] P. K. Wu, G.-R. Yang, J. F. Mcdonald, and T.-M. Lu, "Surface Reaction and Stability of Parylene N and F Thin Films at Elevated Temperatures," *J. Elect. Matl.*, vol. 24, pp. 53-58, 1995.

[14] D. J. Monk, H. S. Toh, and J. Wertz, "Oxidative Degradation of Parylene C (Poly (monochloro-*para*-xylylene)) Thin Films on Bulk Micromachined Piezoresistive Silicon Pressure Sensors," *Sensors and Materials*, to be published, 1996.

[15] R. Sabeti, E. M. Charlson, and E. J. Charlson, "Selective Deposition of Parylene," *Poly. Comm.*, vol. 30, pp. 166-169, 1989.

[16] W. A. Brantley, "Calculated Elastic Constants for Stress Problems Associated with Semiconductor Devices," *J. Appl. Phys.*, vol. 44, pp. 534-535, 1973.

[17] J. M. Hu and M. Pecht, "Temperature Dependence of the Mechanical Properties of GaAs Wafers," *J. Elect. Packaging*, vol. 113, pp. 331-336, 1991.

[18] S. Dabral, X. Zhang, B. Wang, G.-R. Yang, T.-M. Lu, and J. F. McDonald, "Metal-Parylene Interconnection Systems," Materials Research Society Symposium Proceedings: Low-Dielectric Constant Materials--Synthesis and Applications in Microelectronics, San Francisco, pp. 205-215, 1995.

CAPILLARY STRESS IN MICROPOROUS THIN FILMS

J. Samuel*, A. J. Hurd*, C.J. Brinker*†; L. J. Douglas Frink; F. van Swol
Ceramic Processing Science Department, Sandia National Laboratories, Albuquerque, NM 87185-0609*; Department 9225, Sandia National Laboratories, Albuquerque, NM 87185-1111†† UNM-NSF Center for Micro engineered Ceramics, University of New Mexico, Albuquerque, NM 87131†

ABSTRACT

Development of capillary stress in porous xerogels, although ubiquitous, has not been systematically studied. We have used the beam bending technique to measure stress isotherms of microporous thin films prepared by a sol-gel route. The thin films were prepared on deformable silicon substrates which were then placed in a vacuum system. The automated measurement was carried out by monitoring the deflection of a laser reflected off the substrate while changing the overlying relative pressure of various solvents. The magnitude of the macroscopic bending stress was found to reach a value of 180 MPa at a relative pressure of methanol, P/Po = 0.001. The observed stress is determined by the pore size distribution and is an order of magnitude smaller in mesoporous thin films. Density Functional Theory (DFT) indicates that for the microporous materials, the stress at saturation is compressive and drops as the relative pressure is reduced.

INTRODUCTION

When a porous material is brought into contact with a vapor, condensation will take place at a vapor pressure lower than the bulk saturation vapor pressure[1]. This capillary condensation induces stress in the material[2]. Capillary stress may cause cracking[3] and is one of the factors in dictating the final structure of a porous material that results from a drying gel[4]. The tensile stress induced in large pores can be understood in terms of bulk thermodynamics through the Kelvin and Laplace equations. For microporous materials, with pore diameters of a few solvent molecules, assumptions underlying the bulk thermodynamic description are expected to break down[5].

We have used microporous thin films prepared by a sol-gel route to study capillary stress in extremely small pores. These materials have the advantage of molecular sized pores and narrow pore size distributions as witnessed by the molecular sieving capabilities of similarly prepared membranes[6]. The stress in these films was measured under reduced vapor pressure of condensable vapors using a beam bending technique[7]. Experimental results show a large difference in stress for a microporous film under saturation versus vacuum, for example in methanol this difference reaches a value of 220 MPa. In order to understand the origin of capillary stress in small pores we have used a non-local density functional theory approach[8]. The results indicate that for pores on the order of a few adsorbate molecules in diameter, packing constraints on the adsorbate cause stress at saturation to be compressive, thus the stress in microporous materials is qualitatively different from that seen in large pores, in that it goes from compressive at saturation to zero under vacuum.

EXPERIMENTAL

Sol Preparation and Deposition: Sols were prepared from tetraethoxysilane (TEOS), water and HCl or NH_4OH in a two step process. In the first step, a stirred solution of TEOS was partially hydrolyzed under reflux for 90 minutes (TEOS: ethanol: water: HCl ratio of 1: 3.8: 1: 7×10^{-4}). In the second step of the preperation of the acid catalyzed sol, (A2), water, ethanol and acid were added to give a final TEOS: ethanol : water : HCl : Ratio of 1:19.6 : 5.1 : 0.056. For the B2 film, the second step entails adding water , NH_4OH and ethanol to give a final TEOS: ethanol: water: HCl: NH_4OH ratio of (1:21:3.7: 0.0007: 0.0009). The B2 sol was subsequently aged for 24 hours in a sealed container at 50°C. The films were deposited on 150μm thick <100> double polished silicon wafers by dip-coating in a dry atmosphere ([H20]~10ppm) at a speed of 1.7 mm/sec. The films were deposited on one side of the wafer by masking the opposite side with a layer of parafilm which was removed before heating the films to a temperature of 400 °C. The film thickness, measured by ellipsometry (Gartner L116C ellipsometer), was ~1000 Å. The films were

Mat. Res. Soc. Symp. Proc. Vol. 436 © 1997 Materials Research Society

deposited, under identical conditions, on a 390μm thick double polished <100> silicon substrate which was then used for IR experiments.

Stress Measurement: The stress measurement is carried out using a beam bending or cantilever technique[3], [7]. The thin porous film is deposited on a substrate (beam) of known thickness and modulus and the amount of stress exerted by the film is found from the change in curvature of the beam. The thin films were deposited on one side of 150μ or 75μ <100> single crystal double polished silicon wafers. A sample with dimensions of approximately 1 cm width and 5 cm length was clamped in a vertical position in the vacuum chamber. The sample was pumped down to a pressure of 1e-5 torr. Pressure was measured with a series of stabilized MKS transducers (0.1, 10 and 100 torr full scale). A 6mw HeNe laser is passed through a x40 beam expander and iris to reduce divergence and is then bounced off the uncoated side of the sample. The beam is bounced between 4 mirrors and is detected at a position sensitive detector (UDT model SL15). The path length of the beam can be changed to accommodate different extents of deflection. The solvent, after being degassed by freeze thaw cycles, is dosed into the chamber through a needle valve. The dosing is automatically controlled and the pressure and the height of the reflected laser spot are stored in the computer after a predetermined equilibration time. The stress measurement may be run in a kinetic mode by abruptly changing the partial pressure over the sample and following the stress at 1 second intervals, or an isotherm may be automatically collected for a series of pressures.

The lateral deflection of the wafer, δ, at the point that the laser hits the sample, is related to the change in height of the laser spot on the detector, h, by $\delta = L\, h\, /(4P)$ where P is the pathlength from the sample to detector, and L is the cantilever length. Under conditions of small deflection and film thickness relative to substrate thickness, δ is related to the stress in the film, s, by Stoney's equation: $s = E_w\, d_w^{\,2}\, /\, (3L^2\, (1-v_w)\, d_f)\, \delta$ where E_w is Young's modulus of the substrate, d_w is the substrate thickness, d_f is the film thickness and v_w is the substrate Poisson ratio.

IR measurements: Infrared adsorption spectra were collected on a Nicolet 800 FTIR equipped with a vacuum cell. Spectroscopic determination gave us sufficient sensitivity to measure the small amounts adsorbed on the thin film (up to 0.84 μgrams/cm^2 at saturation). The sample under vacuum was used as a background and the spectra from 975 cm-1 to 4000 cm-1 were collected under controlled vapor pressure. The spectra were corrected for the vapor phase contribution using an uncoated wafer at similar pressures.

RESULTS AND DISCUSSION

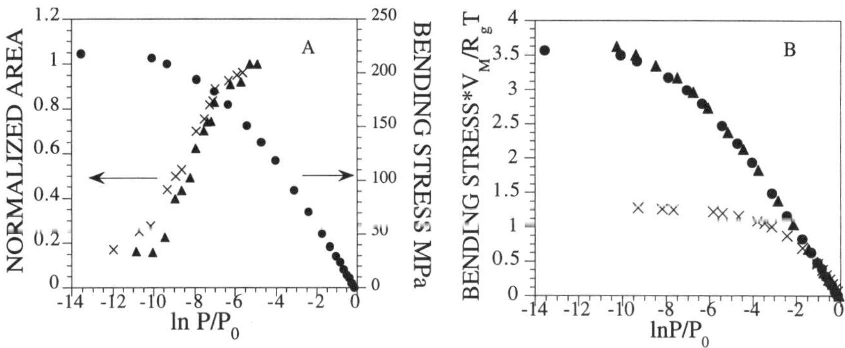

Figure 1: Adsorption and stress isotherms for an A2 film (pore diameter ~ 6 Å). **A:** Adsorption isotherm for methanol measured by FTIR (left axis) and stress isotherm (right axis); ▲ adsorption, x desorption, stress ● adsorption. The normalized amount adsorbed, plotted in the isotherm, is found from the peak area of 3600-2750 cm^{-1}, which encompasses a broad OH stretch band centered at 3330 cm-1 and CH$_3$ stretches at 2956.8 cm^{-1} and 2848.6 cm^{-1}. **B** stress isotherm for: ▲ acetonitrile, ● methanol, x water. The plots are scaled by V_m/R_gT, V_m is the molar volume , R_g is the molar gas constant and T the temperature.

Microporous films: In Figure 1A we compare the stress isotherm with the adsorption isotherm measured by FTIR spectroscopy. The most important feature is that the bulk of the change in stress takes place with no discernible change in the amount of methanol adsorbed. This means that the very appreciable stress induced in the film on altering P/P_0 is not simply a result of added adsorbate causing swelling but is a result of changes in the solvation force exerted by the adsorbate. For a series of adsorbed molecules, the initial part of the stress isotherm that is linear in $\ln(P/P_0)$ scales inversely with the molar volume of the adsorbate, V_m. This may be seen in Fig. 1B where we have plotted the stress isotherms of an A2 film for water , acetonitrile and methanol scaled by $V_m / R_g T$.

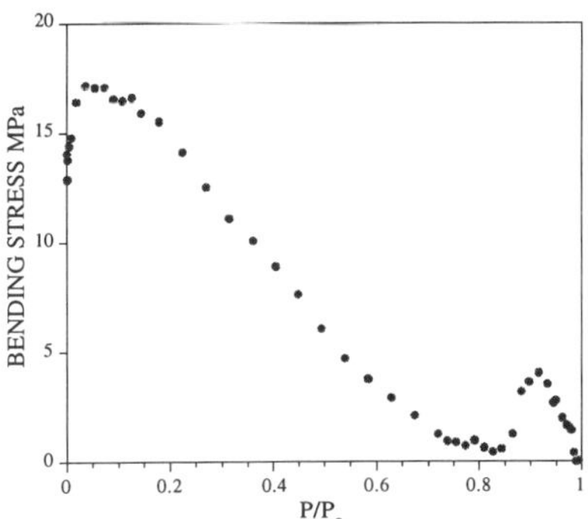

Figure 2 : Stress desorption isotherm measured for a mesoporous B2 film under methanol.

We determined the pore size of the microporous films by measuring the sieving behavior of the films towards a series of alcohols. the measurements were carried out by equilibrating the system at low P/P_0 ($\sim$10^{-7}) and observing the change in stress upon exposure (at P/P_0 = 0.8) to a series of alcohols of increasing kinetic diameters. For methanol vapor (σ = 5.6 Å), the change occurred instantaneously, while for 2-propanol (σ = 7.6 Å), no change in stress was observed. For ethanol vapor (σ = 6.2 Å), the stress changed with a half-life of 180 minutes, establishing the average pore diameter as approximately that of an ethanol molecule. Both the extremely sharp isotherm (Figure 1A) and size exclusion experiments indicate molecular sized pores. The ultramicroporous nature, and narrow pore size distribution of the silica films used is further supported by the fact that a microporous silica thin film, similar to those used in this experiment, when deposited on a porous support, has been used as a membrane to separate nitrogen and oxygen[6].

Mesoporous Films: Base catalyzed xerogels (B2 gels) are characterized by a larger average pore diameter than that seen in the acid catalyzed gels[9]. The B2 gels are mesoporous and are characterized by type IV isotherms with average pore diameters in the 30-40 Å range.
The B2 film develops considerably lower stress than the microporous film, and the isotherm is not monotonic, there is a peak in the stress at $P/P_0 \sim 0.9$ and an additional drop in stress at $P/P_0 \sim 0.05$.

DISCUSSION

Capillary condensation under reduced relative pressure induces tensile stress[2]. For large enough pores, the Kelvin and Laplace equations yield the following equation[10]

$$s_{\text{bend}} \propto P_c = \gamma/r_m = -\ln(P/P_0)\, R_g T/V_m \qquad (1)$$

essentially this equation is a chemical potential equation, with the capillary stress P_c following the chemical potential. Comparing equation 1 to the experimental results we see that the ratio of the slopes for various solvents, on the same film, should be equal to the inverse ratio of the molar volumes, as is indeed observed experimentally.

What we measure in the experiment is the bending stress s_{bend}, this is related to the solvation induced stress by equation 2[11]:

$$s_{\text{bend}} = C \upsilon \zeta P_c \qquad (2)$$

where υ is Poisson's ratio and $C_\upsilon = (1-2\upsilon)/(1-\upsilon)$[12]. C_υ stems from the biaxial nature of the stress developed due to attachment of the film to the substrate. For a typical value of $\upsilon = 0.2$ measured for a variety of silica gels, $C_\upsilon = 0.75$[12]. The parameter ζ for a <u>saturated</u> porous media is generally accepted to be 1- (K_n/K_S) [13] where K_n is the bulk modulus of the film and K_S is the bulk modulus of the silica skeleton. Based on the volume fraction porosity of the film (0.15 - 0.2, as measured by IR sorption and ellipsometry experiments) and literature data concerning the scaling of bulk modulus with porosity[14], we expect ζ to be in the range 0.4 -1. Comparing the predicted slope R_gT/V_m to the experimental slope in Figure 2b gives $C_\upsilon\zeta = 0.49$ and, using $C_\upsilon = 0.75$, $\zeta = 0.64$, well within the expected range

In the experiment, we measure the difference between the stress in the film at saturation and under vacuum. As the relative vapor pressure is reduced the force on the film becomes attractive but we know nothing about the absolute value of the stress at saturation. Attributing the results to capillary tension assumes that the stress is zero at saturation and becomes tensile as the pressure is reduced.

If the stress in the film is indeed caused by capillary tension induced by the adsorbate at reduced relative pressure this stress should relax under vacuum when desorption occurs, and a maximum in the absolute value of the stress should be seen. For microporous films this is not the case, the stress changes monotonicly with P/P_0. An additional question is raised by the magnitude of the stress change. The magnitude of the calculated tensile stress that the adsorbed fluid must exert in order to cause the measured bending stress is 380 MPa at $P/P_0 = 0.001$ (Equation 2). This is considerably beyond the predicted tensile strength of the bulk liquid[15],[10]. While the Kelvin approach assumes that the confined fluid has bulk properties at saturation, It is clear from surface force apparatus experiments that the characteristics of fluids confined in small dimensions differ greatly from bulk fluid characteristics[16]. Indeed, the density and solvation force of a fluid confined between to plates is seen to oscillate widely as a result of packing constraints.

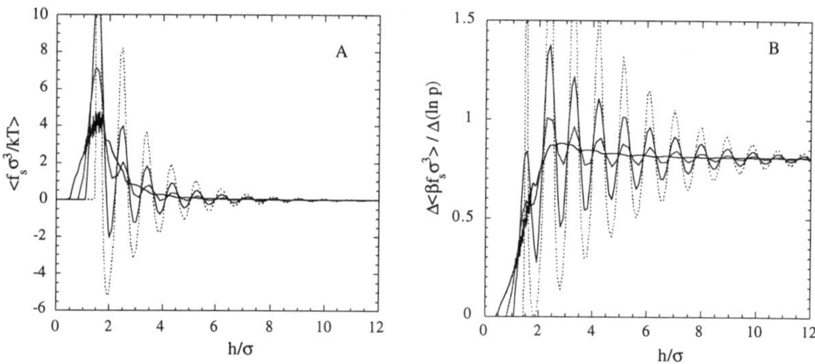

Figure 3: The solvation force (**A**) and the derivative of the solvation force with respect to lnP (B) for a saturated Lennard-Jones fluid in a slit pore obtained from DFT calculations. The dashed line is the result for monodisperse pores. The solid lines are the result for pores with a Gaussian distribution of pore sizes. In order of decreasing amplitude of oscillation these curves correspond to standard deviations of 0.2σ, 0.3σ, and 0.5σ. <h> is the slit width and T the temperature

In order to better understand the results, we have applied a non local density functional approach (DFT) [8]. The pores we consider are slits, and the fluid-fluid and fluid-wall interactions are characterized with 12-6 and 9-3 Lennard Jones potentials respectively. All potentials are cut and shifted at $r/\sigma = 10$ where pore sizes are characterized by the seperation between the walls, h/σ and

σ is the molecular diameter of the solvent. The ratio of the fluid-wall interactions (ε_{wf}) and fluid - fluid interactions (ε) are chosen to be $\varepsilon_{wf}/\varepsilon = 5.0$, where the fluid is wetting with a contact angle $\cos(\theta) = 1$[17]. In figure 3 we plot out the solvation force per unit area, f_s, (reduced by σ^3/kt) and the derivative of the solvation force with respect to ln P near saturation. f_s oscillates as a function of h/σ. This is a result of oscillations in the density caused by packing constraints at such small separations.

In real systems there is a certain amount of polydispersity. We imposed polydispersity on the DFT calculations by using a Gaussian distribution centered on h/σ with a standard deviation of γ. The oscillations in the solvation force are increasingly damped as the standard deviation is increased. Most importantly, polydispersity does not damp out solvation force to 0 and there remains a compressive force at saturation for γ=0.5 and <h/σ> < 4 (Figure 1A).

The derivative of the force with respect to lnP may be seen in figure 3B. The derivative, which may be compared to the initial slope of the isotherms near P/P$_0$ =1, oscillates for the monodisperse pore system, and reaches the Kelvin result for large pores. When polydispersity is imposed the oscillations, which are nearly symmetric around the Kelvin value, are damped. As a result the slope is predicted to be within 20% of the Kevin value for a moderately polydisperse system pore system in which h/σ > 1.

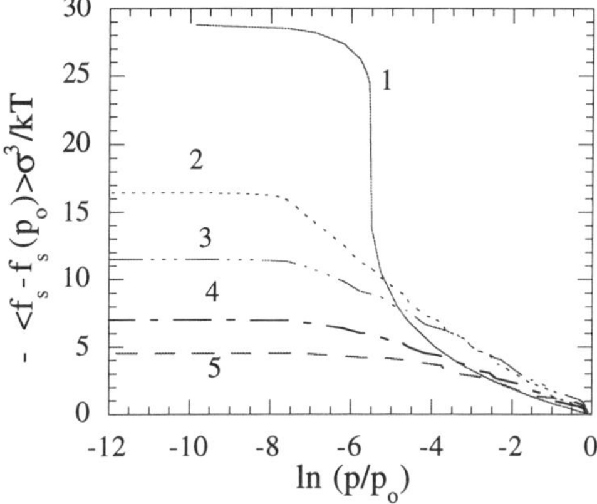

Figure 4: The solvation force as a function of relative pressure for an ensemble of pores characterized by Gaussian pore size distributions that differ primarily in their standard deviation . 1) γ =0, h =1.58; 2) γ =0.1, h =1.64; 3) γ =0.2, h =1.58; 4) γ =0.3, h =1.60; 5) γ =0.5, h =1.60;.

In Figure 4 we have plotted out an isotherm of the DFT result, i.e. the variation in solvation force as a function of pressure. These results may be compared to the experimental results in Figure 1B. In order to facilitate comparison, the DFT results have been shifted to 0 stress at saturation and -f$_s$ is plotted out versus ln P/P$_0$. The magnitude of the stress difference multiplied by σ^3 decreases as the standard deviation of h/σ decreases. This reduction is caused by a drop in the compressive force at saturation. The stress changes monotonicaly as lnP/P$_0$ is reduced, leveling off at low P/P$_0$. comparing this to the experimental results in 1B, we see that a standard deviation of γ =0.3 maps reasonably well onto the experimental results. Methanol and acetonitrile, which have similar molecular diameters, (5.5 and 5.6Å respectively) show similar plots in figure 1B. Water, which has a smaller diameter (σ = 4.3Å) shows a smaller stress difference both because of the larger value of <h/σ> and correspondingly, the larger degree of polydispersity. Note that the stress in the experimental plot is the bending stress and should be multiplied by a factor of~2 (from equation 2) to compare with the DFT results.

For sufficiently large pores (h/σ > 4) we would expect capillary tension at reduced P/P$_0$. The full isotherm for the mesoporous film (Figure 2), may be explained by a bimodal pore size distribution. A bimodal distribution of pores would be reasonable in the B2 films, which are formed by colloidal compaction of fractal clusters that takes place during film deposition. At high relative pressures a peak in stress, at P/P$_0$ = 0.9, is caused by capillary tension and subsequent emptying of the large pores. At lower pressures, a subset of smaller pores dominates the isotherm, causing an increase in the observed stress difference. The additional peak observed at low P/P$_0$ may be attributed to a component of tensile stress in the small pores, that relaxes when they empty.

CONCLUSION

The magnitude of stress induced by solvation forces in microporous materials can be quite considerable. For a microporous film the difference in bending stress between the state in vacuo and under a saturated atmosphere of methanol reaches a value of 220 MPa. The magnitude of the induced stress is influenced both by pore size and pore size distribution. The stress is clearly not caused by swelling because of added amount of solvent in the pores, as the the adsorption takes place in the range of P/P_0 from vacuum to $P/P_0=0.001$ while %80 of the change in stress takes place $1 > P/P_0 > 0.001$. The stress develops logarithmically as a function of P/P_0 and the stress induced by various solvents   scales with the bulk molar volume of the solvent, as would be predicted from the Kelvin and Laplace equations. Comparing the experimental results to density functional theory of micropores we conclude that (1) For the small pore materials the solvation forces are compressive at saturation dropping monotonicaly to zero in vacuo (2) That the pore sizes are indeed in the range of a few solvent molecules in diameter (3) That for the slope given by V_m to be observed there must be a pore size distribution and that the relatively high final stress observed is an indication of a relatively narrow distribution.

ACKNOWLEDGMENTS

The authors thank Rich Cairncross for helpful discussions and Hongbin Yan and Thomas M. Niemczyk for the IR measurements. This work was supported by the U.S. Department of Energy Basic Energy Sciences Program, the University of New Mexico/National Science Foundation Center for Micro-Engineered Ceramics and the National Science Foundation Division of Chemical and Transport Systems (CTS9101658). Sandia National Laboratories is a U.S. DOE facility operated under contract number DE-AC04-94AL 85000.

REFERENCES

1. S.J. Gregg, K.S. W. Sing, <u>Adsorption, Surface Area and Porosity</u> (Academic Press 1982).
2. G W. Scherer; J. Am. Ceram . Soc, **73**, 3 (1990)
3. G.W. Scherer, D.M.Smith ; J. Non Cryst Solids **189** (1995).
4. J.H.L.Voncken, C. Lijzenga, K.P.Kumar, K. Keizer, A.J. Burggraaf, B.C. Bonekamp; J. Mat. Sci. **27**, 472-478 (1992).
5. R. Evans in <u>Capillarity Today: Lecture Notes in Physics 386</u>, edited by G Petre and A Sanfeld (publishers: Springer-Verlag 1990) pp. 62-76.
6. R. Sehegal, J.C. Huling, C.J. Brinker; <u>Proceedings of the Third International Conference on Inorganic Membranes</u>, edited by Y.H. ma,1995 pp. 85-93.
7. E.M. Corcoran, Journal of Paint Technology **41** 635 (1969).
8. Y. Rosenfeld, J. Chem. Phys., **98**, 8120 (1993)
9. S.S. Prakash, C.J. Brinker, A.J. Hurd; J. Non Cryst Solids, **190**, 264 (1995).
10. C.G.V. Burgess, D.H. Everett; J. Colloid. Interface Sci. **33**, 611 (1970).
11. R.A.Cairncross, P.R Schunk, K.S. Chen, S.S. Prakash, J. Samuel, A.J. Hurd, C.J. Brinker IS&T Proceedings (1996).
12. C.J. Brinker, G.W Scherer, <u>SOL-GEL SCIENCE</u>, (Academic Press 1990).
13. S.K.Garg, A. Nur ; J.Geophysical Research, **78** 5911-5921(1973).
14. H. Hidach, T. Woignier, J. Phalippou, G. W. Scherer; J Non. Cryst Solids, **121,** 202, (1990).
15. O.Kadlec, M.M. Dubinin; J. Colloid Interface Sci. **31**, 479 (1969).
16. J Klein, E Kumacheva; Science, **169**, 816 (1995).
17. F. van Swol, J. R. Henderson, Phys. Rev. A, **40**, 2567 (1989).

MEASUREMENT OF LOW-K POLYMER/METAL INTERFACIAL TOUGHNESS USING 4-POINT BENDING METHOD

Qing Ma, Chuanbin Pan, Harry Fujimoto, Baylor Triplett, Peter Coon and Chien Chiang
Intel Corporation, Santa Clara, CA 95052

ABSTRACT

Four-point bending method offers significant advantages over more traditional techniques in measuring adhesion properties of thin polymer films. The former utilizes sandwich structure beams where the polymer film and the interface of interest are placed at the center of two elastic bulks. Such confined geometry closely resembles applications where polymer films are used as dielectric layers in IC interconnect technology. In this work, the bonding between a poly(arylene ether) (PAE) based polymer film and an Al film was studied. Ti layers of several different thickness were used as adhesion promoter. The sandwich samples were made by bonding bulk wafers using an epoxy. Fracture energies were seen to increase with the thickness of the Ti layer. Effects of heat treatment were also studied.

INTRODUCTION

Polymeric thin films are increasingly studied as low dielectric constant materials for IC interconnect technology. Besides process integration issues, reliability problems associated with polymers are also major challenges. In particular, adhesion between polymers and inorganic materials such as metals or ceramics are often relatively poor. Further, such interfaces tend to degrade over time or through thermal cycling because chemical species, such as water and small organic molecules, can segregate at the interfaces and weaken them by changing the nature of the interfacial atomic bonds. To address these problems, tools are needed to measure adhesion strength with respect to processing conditions, adhesive promoters or post-processing treatment.

One of the difficulties associated with testing polymeric interfacial strength is that the plastic energy dissipation in the polymer films is so dominant that measured results often depend more on the plastic properties of the film itself than on the interface properties. For example, peeling tests of polymer films on metals often gave "adhesion energies" of the order of 10^3 J/m^2 [1], three orders of magnitude larger than typical metal surface energies.

To make measurements that are more relevant to the real interconnect structure, where films are confined and can not flow freely, a 4-point bending method recently developed for measuring multilayer thin film adhesion [2] is particularly advantageous. Briefly, a silicon wafer with thin film layers of interest is bonded with another silicon wafer face to face, placing the film stack at the center of the sandwich structure. Such sample geometry places the thin films under the confinement of the elastic substrates and therefore significantly reduces plastic flow. The 4-point bending configuration is illustrated in Fig. 1. As the bending moment increases, a pre-crack initiates from the top surface (usually facilitated by a machined notch) and propagates vertically to the interface. If the interface is sufficiently weak, the crack deflects into the interface and propagates along it. When the crack tip is sufficiently far away from the vertical pre-crack (a>2h), the strain energy release rate becomes independent of the crack length,

Mat. Res. Soc. Symp. Proc. Vol. 436 © 1997 Materials Research Society

characteristic of steady-state crack growth. Applying beam theory, the strain energy release rate is related to measurable quantities [3]:

$$\mathcal{G} = \frac{21(1 - v^2)M^2}{4Eb^2h^3}$$

(1)

where the bending moment $M = Pl/2$, with P being the load and l the spacing between the inner and outer loading lines, b is the beam width, h is the half thickness, and E and v are the elastic modulus and Poisson's ratio of the bulk substrate, respectively.

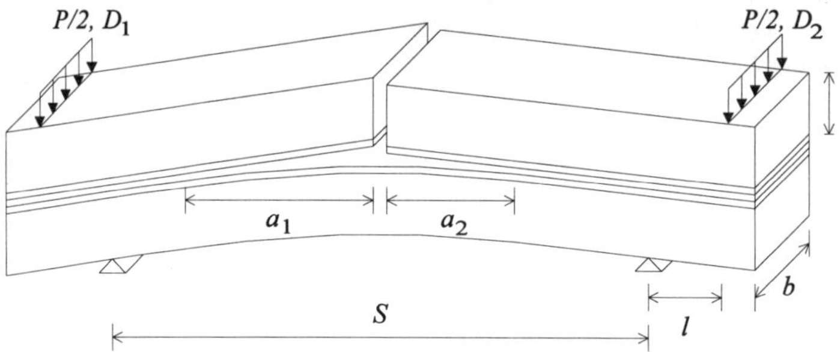

Fig. 1. Schematic illustration of the 4-point bending sandwich structure. The strain energy release rate is independent of crack length, a_1 and a_2, as long as the crack tip is not close to the vertical pre-crack or the inner loading lines.

In this work, this technique was applied to interfaces between aluminum and a poly(arylene ether) (PAE) based polymer, which has shown some desirable qualities such as good thermal stability and gap filling characteristics. Besides establishing the 4-point bending technique as a tool for studying polymeric interface adhesion, the emphasis was the effects of Ti on interface adhesion strength. This is because Ti is known to promote adhesion in many interface systems and it is already widely used as an interconnect shunt layer to improve electromigration resistance. Several Ti thickness and post deposition treatment conditions were studies, because previous experiences of Ti adhesion to oxides indicated that these parameters can change the interface adhesion significantly.

EXPERIMENTAL

<u>Sample Preparation</u>

PAE films were spun on wafers with a thermal oxide film, and cured at 400°C for 1 hour in N_2. Empirical calibration was used so that the cured film thickness was about 1 micron. Metal films were then deposited on PAE films by sputtering. Three splits were made to test the effect of Ti as an adhesive layer: 900 nm Al, 20 nm Ti/900 nm Al and 100 nm Ti/900 nm Al.

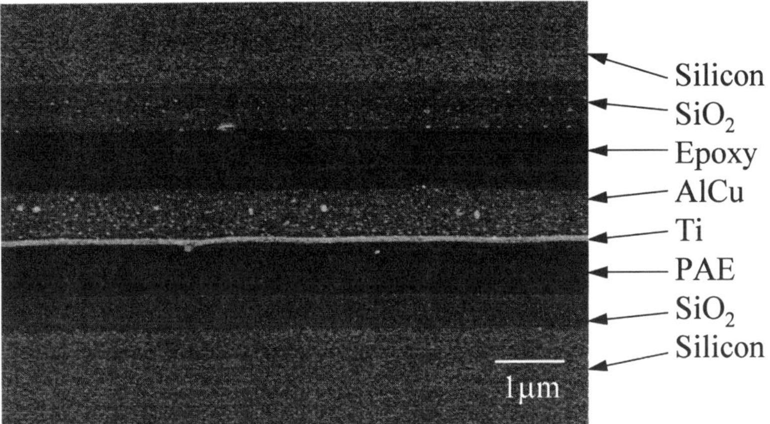

Fig. 2. Cross-section SEM of the film stack and bonding structure.

Two different heat treatment conditions were used, annealing at 300°C for one hour and without annealing. Finally, bending beams were fabricated by cutting the wafers to about 35x7 mm rectangular pieces, and bonding them face to face with Si pieces of matching size using an epoxy. The epoxy was cured at 150°C for about 1 hour. Figure 2 is a cross-section SEM micrograph showing the film stack and bonding structure.

<u>Four-Point Bending Test</u>

A computer controlled bending fixture was used for the tests. The load was measured by a load cell and the displacement was incremented using a motorized actuator. A typical load-displacement curve, obtained from an Al/PAE sample, is plotted in Fig. 3 as an example. Following the curve, at the beginning, the system was adjusting itself and then gradually became linearly elastic. At about 22 N, a pre-crack initiated from the top machined notch and propagated to the interface. Because it required relatively large loads to propagate the interface crack away from the pre-crack, a relatively high displacement rate was used to drive the crack. When the crack tip was sufficiently far away from the pre-crack, the displacement was intentionally fixed for a few minutes to lower the load. Then a displacement rate of 0.05 micron/s was used for the measurement of the plateau load.

For samples without a Ti layer and those with the 20 nm Ti layer, the plateau measurements were made with the displacement rate of 0.05 mm/s as described above. Because the polymer film was viscoelastic and therefore time dependent, it was important to consistently use the same incrementing rate for all samples. However, for the samples with the 100 nm Ti

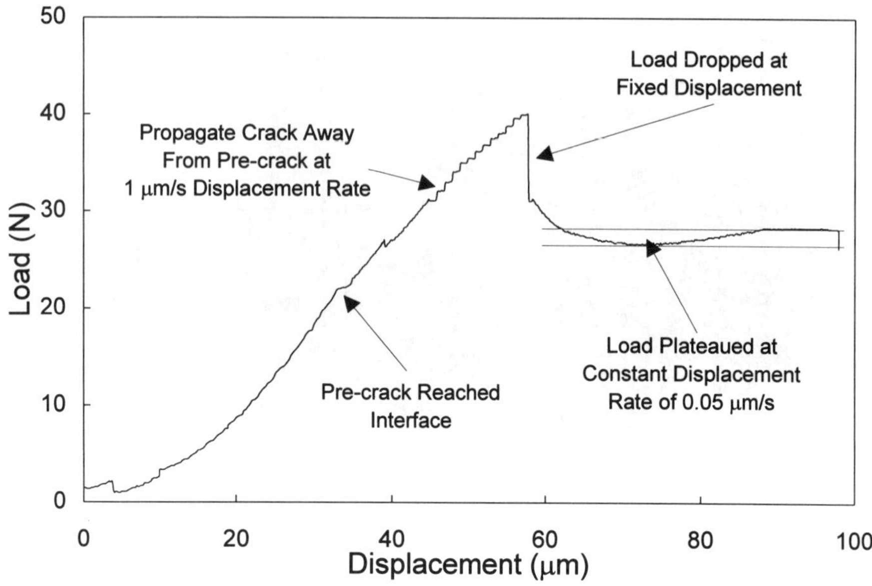

Figure 3. A load-displacement curve from an Al/PAE sample.

layer, the load required to propagate a crack exceeded the maximum load the actuator could supply. As a result, a manual micrometer head was used to control the displacement with approximately the same rate.

RESULTS AND DISCUSSIONS

The results from samples with all three different metal stacks and both with and without post deposition annealing are summarized in Fig. 4. All samples were debonded at the PAE/metal interfaces as verified using Auger spectroscopy. A general trend is evident from the figure: the interface strength improves with Ti layer thickness. Comparing results from samples with and without annealing, the fracture energies were about the same for the samples without Ti layer and with 100 nm Ti layer, annealing did seem to have any appreciable effect. However, remarkable differences were seen for the samples with 20 nm Ti layer. Without annealing, the fracture energy was significantly larger than the Al only samples, but smaller than the samples with 100 nm Ti layer. But after annealing, the interface was seen to be weakened to the strength of the Al only samples.

Besides its ability to form strong covalent bonds with a variaty of materials, the adhesion enhancing capability of Ti is also believed to be due to its ability to react with chemical species that are detrimental to adhesion and to absorb some of the reaction products. For example, when Al is deposited on CVD SiO2 that contains water, the adhesion is poor, because water can segregate at the interface and form weak hydrogen bonds. But if Ti is use as an adhesion layer, segregated water will react with Ti to form Ti oxide and releasing hydrogen which is absorbed by the remaining Ti layer [4]. Such mechanism keeps the interface dry and mechanically strong. In the PAE/metal samples studied in this work, similar process may be responsible to the observed behavior. Besides water, small organic molecules (from residual solvent or decomposition of polymer chain) can segregate at the interface. These species tend to weaken the interface by replacing covalent bonds with Van der Waals bonds. However, Ti can react with them and absorb the reaction products thus retaining covalent bonds. Auger analyses showed that both Ti oxide and Ti carbide were formed near the interface. Clearly, the reaction and absorption capacity of the Ti depends on the layer thickness. The 100 nm Ti layers were apparently thick enough so that the interfaces were kept clean. But for the 20 nm Ti samples, the Ti layer was probably partially saturated before the annealing, so that the interface was weakened compare to the 100 nm Ti samples, but still stronger than the samples without Ti. Annealing may have induced more segregation and consequently fully saturated the Ti layer, resulting weaker interfaces dominated by Van der Waals bonds similar to those in Al only samples.

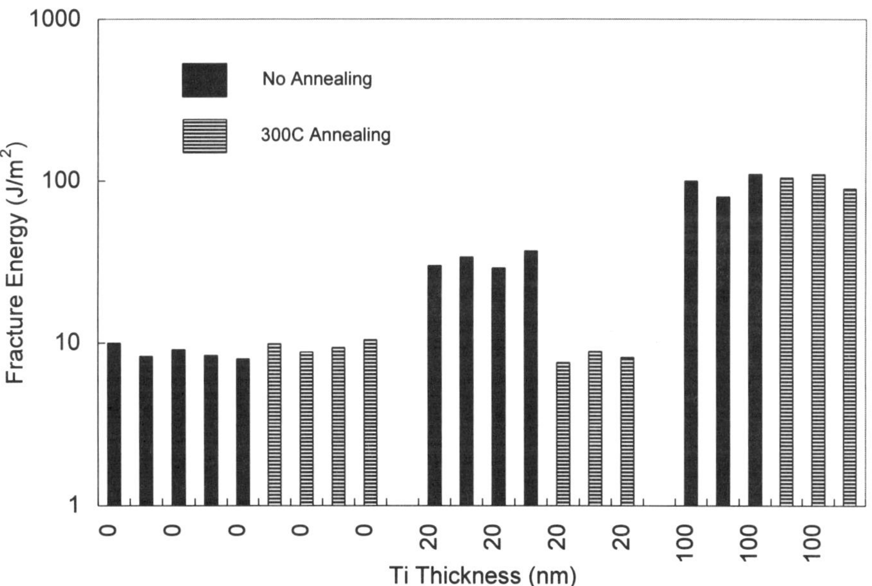

Fig. 4. The interface fracture energies of the PAE/metal interfaces.

SUMMARY

The 4-point bending technique was used to study PAE/metal interfaces. This technique is rigorously based on fracture mechanics so that the fracture energy can be obtained without ambiguities. Due to its confined geometry, plastic deformation was limited. For the relative weak interfaces of PAE/Al, the fracture energy was of the order of 10 J/m^2, about two orders of magnitude smaller than a typical peeling test result. More importantly, the geometry is similar to that in a real interconnect structure. Therefore, the measured data were directly relevant to the real reliability problems. Experiments performed on a matrix of PAE/metal interfaces showed interesting effects of Ti on interface adhesion strength. It was seen that a thin Ti layer (20 nm) could be saturated by low energy molecules in the PAE film and lose its adhesive ability, but thicker Ti (100 nm) films might have sufficient capability of absorbing such chemical species to retain strong adhesion. More analytical analyses are needed to provide a better understanding.

ACKNOWLEDGMENTS

The authors wish to thank Rick Booth of Intel for automation of the 4-point bending fixture, Ray Vrtis of Schumacher and Bill Burgoyne of Air Products for providing PAE films.

REFERENCES

1. R.J. Farris and J.L. Goldfarb, "An Experimental Partitioning of the Mechanical Energy Expended During Peel Test," pp. 265-281, in Adhesion Measurements of Films and Coatings Edited by K.L. Mittal, VSP, Utrecht, The Netherlands, 1995.

2. 2. Q. Ma, H. Fujimoto, P. Flinn, V. Jain, F. Adibi-Rizi, F. Moghadam and R.H. Dauskardt, in Materials Reliability in Microelectronics V, edited by A.S. Oates, W.F. Filter, R. Rosenberg, A.L. Greer and K. Gadepally (Mater. Res. Soc. Symp. Proc., **391**, Pittsburgh, PA, 1995) pp. 91-96.

3. P.G. Charalambides, J. Lund, A.G. Evans and R.M. McMeeking, "A Test Specimen for Determining the Fracture Resistance of Bimaterial Interfaces," *J. Appl. Mech.*, **111**, 77-82, 1989.

4. M. Yoshimaru, T. Yoshie, M. Kageyama, K. Shimokawa, Y. Fukuda, H. Onoda, and M. Ino, "Deoxidization of Water Desorbed From APCVD TEOS-O3 SiO2 by Ti Cap Layer," in IEEE International Reliability Physics Proceedings, 1995, pp. 359-364.

DETECTION OF SIMILAR ELASTIC PROPERTIES USING A MAGNETIC FORCE CONTROLLED AFM

Shin-ichi Yamamoto, Hirofumi Yamada[*], Suzanne P. Jarvis, Makoto Motomatsu, and Hiroshi Tokumoto[*]
Joint Research Center for Atom Technology (JRCAT), Angstrom Technology Partnership (ATP), 1-1-4 Higashi, Tsukuba, Ibaraki 305, Japan
[*]JRCAT, National Institute for Advanced Interdisciplinary Research (NAIR),
 1-1-4 Higashi, Tsukuba, Ibaraki 305, Japan

ABSTRACT

We have investigated regional variations of elastic properties using a magnetic force controlled AFM. A piece of small magnet was fixed at the end of the backside of the AFM cantilever so as to apply forces directly to the tip through the external magnetic field of an electromagnet. By modulating the applied forces to the tip and measuring the resulting amplitude of oscillation, a sensitive measurement of the local contact stiffness can be made. We have applied this technique to phase-separated films of polystyrene/polymethylmethacrylate (PS-PMMA) which have almost identical Young's moduli.

INTRODUCTION

The atomic force microscopy (AFM) [1] has been applied to the nanometer-scale investigation of various materials including hard ceramics and soft polymer samples. Several techniques using force microscopy to investigate surface mechanical properties [2,3] have been developed such as the force modulation technique [4,5,6,7] and indentation[8]. The measured properties include hardness [9], adhesion [10], friction [11] and the energy dissipation [12]. At the same time, elastic deformation is becoming important, especially, for the investigation of soft materials like polymer specimens [13]. An AFM tip in contact mode deforms both the tip and sample, and the resulting image may thus not correspond to actual surface structure and also cause the loss of true atomic resolution [14]. Recently, attempts were made to extract the substrate's elastic properties from the sample deformation during AFM scans by vibrating the sample or the cantilever at high frequencies [4,5,6,7]. These results suggest that force modulation imaging can be used in a wide range of applications including indentifying and mapping differences in stiffness or elasticity, and evaluating materials homogeneity.

Although the indirect force modulation technique in which the sample position is modulated has been successfully used, there are three problems. First the cantilever stiffness have to be chosen according to the materials stiffness (see eq.(2) given later). Secondly indentation amplitude can not be measured directly. Thirdly the vertical displacement of the sample causes the lateral motion of the cantilever since the indirect force technique requires the large amplitude of displacement for obtaining appreciable signals. On the contrary, the direct force modulation technique has several advantages overcoming these problems: the cantilever stiffness has nothing to do with the materials stiffness, the depth of surface deformation can be measured directly, and lateral forces effect can be minimized.

In this paper we demonstrate that direct force modulation is useful for detecting the difference between phase-separated surface films (polystyrene/polymethylmethacrylate) which have almost identical Young's moduli.

EXPERIMENTS

Apparatus of AFM using an electromagnet

The experimental set-up for the force modulation measurement is shown in Fig. 1. Cantilever deflection is measured using an optical lever method [16]. Prior to the measurement, the deflection of the cantilever was calibrated by lifting up the sample by 10 nm and measuring the signal change of the photodetector. Samples are mounted on a piezoelectric tube scanner (20×20 μm^2), inside of which a coil for an electromagnet is fixed. The coil is suspended by an aluminum holder so as to touch neither the piezotube nor the backside of the sample which is serious to avoid mechanical vibrations of the coil. The coil consists of 1030 turns with a core diameter of 3 mm and an inductance of 5.4 mH. All experiments were carried out in an ambient air atmosphere.

<u>Magnetized cantilevers and their procedure process</u>

The tip is a pyramid with an apex radius of less than 20 nm microfabricated on the cantilevers. For the present experiment, two different cantilevers with spring constants of 0.022 and 0.68 N/m (Olympus Opt. Inc.) were used. First a large SmCo magnet was crushed, and a piece of SmCo magnet (less than 10 μm in diameter) was glued at the end of a microfabricated cantilever with epoxy resin using a three-way micromanipulator under an optical microscope. The scanning electron microscope (SEM) picture of the cantilever is shown in Fig. 2.

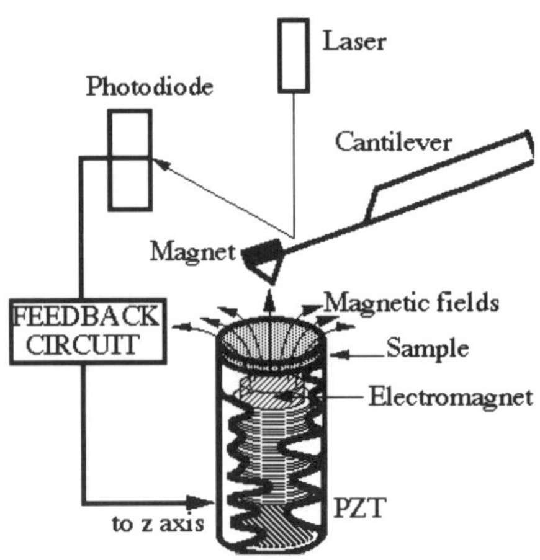

_10 μm

Fig.1   Schematics of the cantilever and coil assembly

Fig. 2   Scanning electron micrograph of an AFM cantilever with a piece of SmCo magnet.

<u>Direct force modulation technique</u>

With the direct force modulation technique, the sample assembly is scanned while a modulating force is applied directly to the tip using a magnetic field. The modulation frequency (5kHz) is higher than the bandwidth of the feedback loop of 0.5kHz. The response of the cantilever to this oscillation is detected with a lock-in amplifier and is used to obtain images related to local elastic properties of sample surface.

When the tip is brought into contact with a sample, the vibration amplitude is reduced. The oscillation amplitude is larger on the soft area than on the hard area with direct force modulation while the response becomes smaller on the soft area in the case of indirect force modulation as shown in Fig. 3. Then the contrast of the stiffness images between the direct and indirect force modulation method is reversed. A topographic image is also measured simultaneously with the force modulation data.

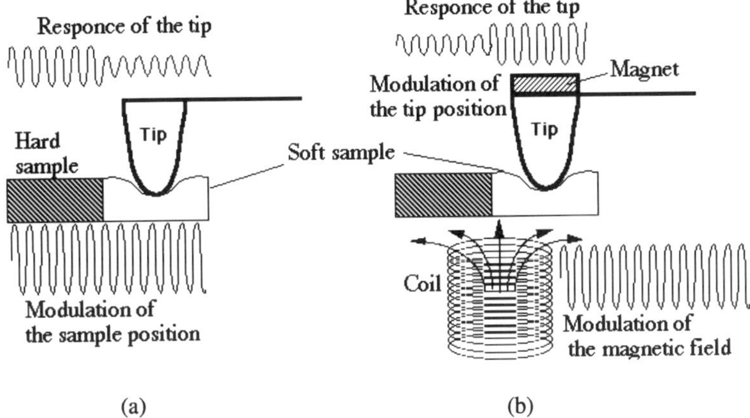

Fig. 3   Schematic drawings for the elastic measurements by two techniques : the indirect (a) and the direct (b) force modulations.

<u>Sample preparation</u>

To compare the sensitivity between the direct force modulation and indirect force modulation, we used polystyrene (PS) sample which was prepared by dropping a toluene solution of PS onto mica substrate rotating at a speed of 2000 rpm for 60 seconds. Then it was annealed at 170 °C for 3 hours. The molecular weight of the PS is 19600. The average film thickness was 100 nm.

Monodisperse polystyrene/polymethylmethacrylate (PS/PMMA) sample was prepared from diblock copolymer (Polymer Laboratories, UK) whose total molecular weight is 58,500. The films were prepared by spin-coating toluene solution onto a mica surface. The thickness of the film after annealing at 170 °C for 24 hours under vacuum was 200-300 nm.

RESULTS AND DISCUSSION

First we measured the stiffness of PS on mica to compare the sensitivity between the direct force modulation and indirect force modulation by using a cantilever of 0.68 N/m. Figures 4(a) and (b) show typical AFM topographic image and simultaneously recorded stiffness image of PS on mica with the indirect force modulation, respectively. Figure 4(c) is an image corresponding to Fig. 4(b) measured by the direct force modulation. In the topographic image, there appear holes with a diameter of 1-3 μm and the depth of about 40 nm, which is consistent with the results reported by Nie et al [7]. Therefore we found that bright area in the image (b) corresponds to the mica and the dark area to the PS films. As expected, the amplitude of the cantilever's motion is small when the tip is on the PS in the case of direct force modulation.

From Fig. 4(b) we found that the ratio of the difference between the amplitude of the cantilever on the PS and on the mica to the amplitude of the cantilever on mica was 0.84 %. In the case of Fig. 4(c), the ratio of the difference between the amplitude of the cantilever on the PS and on the mica to the amplitude of the cantilever on the PS was found to be 20.3 %. Note that the contrast ratio of stiffness for these two cases shown in Figs. 4(b) and (c) is about 24.2. This means that the sensitivity is increased by as much as a factor of 20 over the conventional sample oscillation method with the cantilever of 0.68 N/m, indicating that the small difference in the elasticity for the phase-separated PS/PMMA films can be detected using the direct force modulation technique.

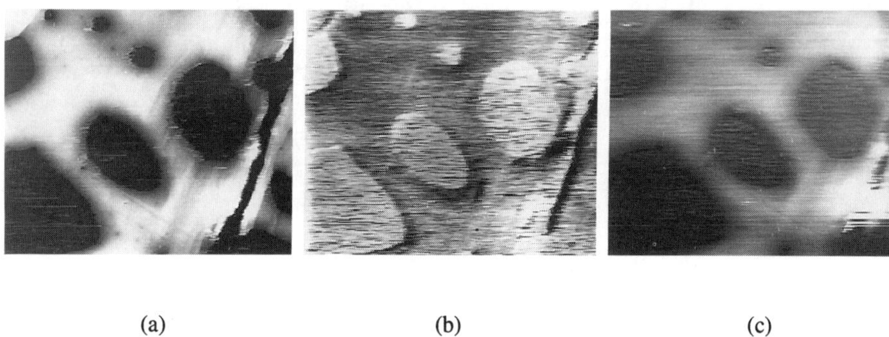

(a)   (b)   (c)

Fig. 4　Topographic,(a), and stiffness, (b) and (c), images (4×5 $\mu m^{2)}$ of PS on mica obtained simultaneously using a cantilever of 0.68 N/m spring constant at a repulsive force of 4.93 nN. The stiffness were measured by the indirect (b) and direct (c) force modulation techniques.

The contrast of stiffness image corresponds to the difference of the cantilever amplitudes on each phases. Here we defined the difference $\Delta x = x_{ps} - x_m$, where x_{ps} and x_m are the cantilever amplitudes on the PS and the mica, respectively. In the case of indirect force modulation, Δx can be written as follows,

$$\Delta x = x_{ps} - x_m = [S_{ps}/(k+S_{ps}) - S_m/(k+S_m)]X_0. \qquad (1)$$

Here X_0, k, S_{ps}, and S_m denote the driving oscillation amplitude, the spring constant of the cantilever, the tip-PS stiffness, and the tip-mica stiffness, respectively. Δx has its maximum value, when the spring constant k is given by

$$k = (S_{ps}S_m)^{1/2}. \qquad (2)$$

Equation (2) suggests that the spring constant k is determined by the sample stiffnesses.
　　　In the case of direct force modulation, Δx is given by

$$\Delta x = x_{ps} - x_m = F/(k+S_{ps}) - F/(k+S_m). \qquad (3)$$

Here F denotes the small modulation of force which is applied by the magnetic force. When the spring constant k approach zero, Δx becomes a maximum value,

$$\Delta x = x_{ps} - x_m = F/S_{ps} - F/S_m. \qquad (4)$$

This means that the absolute value of Δx could be increased by decreasing the spring constant k. At the same time, when the soft cantilever is used, the high sensitivity could be realized for any samples.

388

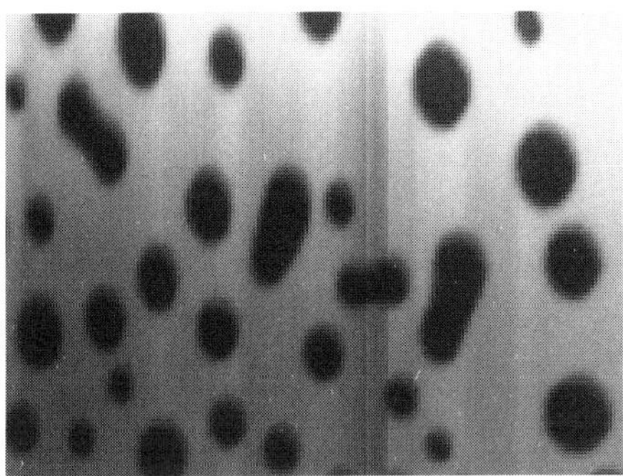
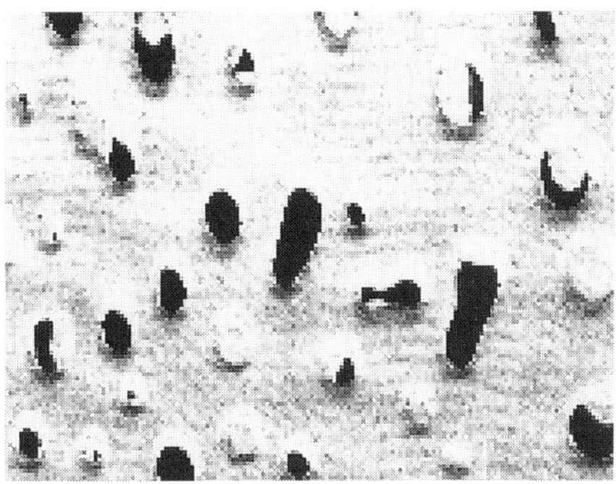

(a)   (b)

Fig. 5　Topography(a) and stiffness (b) images (10×14 μm^2) of PS/PMMA diblock copolymer obtained simultaneously with direct force modulation of a cantilever with 0.022 N/m. The topography in (a) shows holes of 1-3 μm in diameter and 30 nm in depth. In (b), the part with the bright contrast is the PS film and the remaining part is PMMA. It shows that all holes do not contain PMMA.

We observed phase-separated structures of PS/PMMA diblock copolymer to demonstrate the sensitivity of direct force modulation. The spring constant which we used was 0.022 N/m (near zero) in order to obtain the highest sensitivity. Figure 5 (a) is an AFM image of the PS/PMMA diblock copolymers surface, showing depressions with a diameter of 1-3 μm and the depth of 30 nm, while a simultaneously recorded stiffness image is shown in (b). At this moment, we can not assign whether the higher amplitude of the cantilever's motion corresponds to PS or PMMA since there are several uncertainties: (1) The PS and PMMA films have almost identical bulk Young's Moduli, 3.4-3.6 GPa for PS and 3.7 GPa for PMMA[17]. The indirect force modulation could not detect the difference between PS and PMMA films[15]. (2) There would be some difference in Young's moduli between the bulk and nanometer scale area. To check these two factors, we prepared samples consisting of either PS or PMMA in the same manner. The topographic image of these surfaces exhibited flat surfaces without any hillocks or depressions. The direct force modulation measurements found that the ratio of amplitude of the cantilever between the PS and PMMA films to the amplitude of the cantilever on the PS is 48.4±7.5 %. This value is much higher than 10 % which is expected from the corresponding bulk values. Although the reason of this difference is not clear at this moment, we can say that Young's modulus of PS film on nanometer scale is also smaller than PMMA film.

From the result shown in Fig. 5(b), the ratio of the difference between the amplitude of the cantilever on PS and on PMMA films to the amplitude of the cantilever on the PS is found to be 55.3±8.9 % which is very close to the above value of 48.4 %. Then we can conclude that the bright region in Fig. 5(b) corresponds to the PS film and the dark one to the PMMA film.

It should be noticed that all the holes in Fig. 5(b) do not necessary exhibit the same contrast: some holes or a part of holes exhibit the same contrast as that for the PS film. This means that those parts are partially filled by the PS molecules. However we have no conclusive reasons for these at this moment. To solve this problem, more extensive work is highly required.

CONCLUSIONS

We succeeded in taking a stiffness image of the distribution of phase-separated surface films with almost identical Young's moduli of PS and PMMA using the direct force modulation. The direct force modulation ensures that a higher contrast stiffness image can be realized by using a soft cantilever than the indirect force modulation. Especially when we take the stiffness image of unknown soft materials, the direct force modulation is useful.

ACKNOWLEDGMENT

This work was partly supported by the New Energy and Industrial Development Organization (NEDO), and by the National Research Laboratory of Metrology.

REFERENCE

1. G. Binnig, C. F. Quate, Ch. Gerber : Phys. Rev. Lett. **56** 930 (1986).
2. J. B. Pethica and W. C. Oliver, Mat. Res. Soc. Symp. Proc. **130,** 13 (1989)
3. T.P. Weihs and J. B. Pethica,, Mat. Res. Soc. Symp. Proc. **239,** 325-330 (1992)
4. P. Maivald, H. J. Butt, S. A. C Gould, C. B. Prater, B. D.rake, J. A. Grurley, V. B. Elings, and P. K. Hansma, Nanotechnology. **2,**103, (1991)
5. M. Radmacher, R. W. Tillmann, M. Fritz, and H. E. Gaub, Science. **257,** 1900 (1992)
6. H.-Y. Nie, M. Motomatsu, W. Mizutani, H. Tokumoto, J. Vac. Sci. Technol. **B13,** 1163(1995)
7. H.-Y. Nie, M. Motomatsu, W. Mizutani, H. Tokumoto, to appear in Thin Solid Films
8. B. Bhushan, and V. N. Koinkar, Appl. Phys. Lett. **64.** 1653 (1994)
9. S.M. Hues, C.F. Draper, and R.J. Colton, to be published in J. Vac. Sci. Technol.
10. N. A. Burnham and R. J. Colton, and H. M. Pollock, Nanotechnology. **4,** 64, (1993)
11. E. Mayer, H. Heinzelmann, P. Grutter, T. Jung, H. Hidber, and Guntherrodt, Thin Solid films. **181,** 52 (1989)
12. E. Boschung, M. Heuberger, and G. Dietler, Appl. Phys. Lett. **64,** 3566 (1994)
13. B. Nysten and R. Legras, J. Appl. Phys. **78,** 5953 (1995)
14. T. P. Weihs, Z. Nawaz, S. P. Jarvis, and J. B. Pethica, Appl. Phys. Lett. **59.** 3536 (1991)
15. M. Motomatsu, H-Y. Nie, W. Mizutani, and H. Tokumoto, Jpn. J. Appl. Phys. **33,** 3775 (1994)
16. G. Meyer and N. Amer, Appl. Phys. Lett. **53,** 1045 (1988)
17. L. E. Nielsen, <u>Mechanical Properties of Polymers</u> (Reinhold, New York, 1967)

Part VII

Stress Effects in Thin Films and Interconnects

THE MECHANICS OF A FREE-STANDING STRAINED FILM / COMPLIANT SUBSTRATE SYSTEM

L. B. FREUND
Division of Engineering, Brown University, Providence, RI 02912

ABSTRACT

The deformation of film/substrate systems due to a through-the-thickness distribution of mismatch strain is discussed for cases when the layers are thin, compliant and free-standing. The connection between substrate curvature and mismatch strain is reviewed within the framework of small strains and small deflections. These results are applied to determine the second-order correction to the Stoney formula and to note the connection between mismatch strain distribution and curvature history. Then, for small strains but large deflections, it is observed that a symmetric system with equi-biaxial mismatch strain can undergo a bifurcation in deformation mode. It is also shown that the bifurcation response is sensitive to system imperfections. Finally, the critical condition for introduction of an interface misfit dislocation in an epitaxial system is established, thereby extending the Matthews-Blakeslee criterion to the case of free-standing compliant systems.

INTRODUCTION

Attention is focussed on a class of film-on-substrate material configurations which is characterized by several features. The total thickness of the system is denoted by h, and rectangular coordinates are introduced with the plane $z = 0$ coincident with the bottom of the substrate; see Figure 1. Some results included here do not depend on the particular plan view shape of the system, but a circle of radius R is chosen when a specific shape is required.

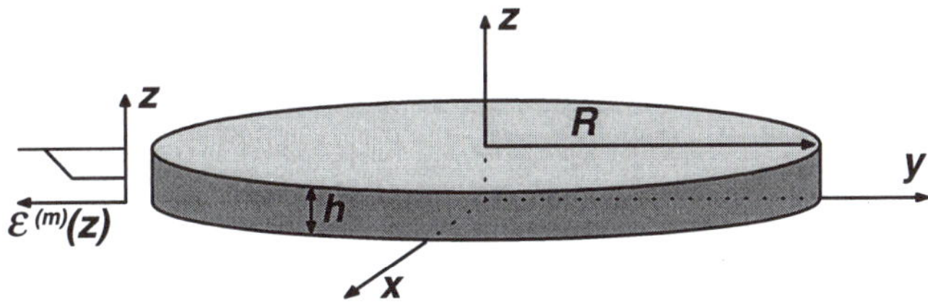

Figure 1. Schematic of a thin free-standing circular film/substrate system, indicating a general variation of elastic mismatch strain across the thickness of the layer. It is assumed that $h/R \ll 1$ for all cases considered.

One feature of the material system is the presence of an incompatible mismatch strain in the film which can vary with z in an arbitrary way and which may arise from a variation in thermal expansion coefficient, from the constraint of epitaxy, from a stress-free strain associated with a chemical reaction or phase change, or from processes of microstructural evolution as the film is grown. Suppose that laminae of infinitesimal thickness are removed from the layer at $z = 0$ and at some other level z. Then, the mismatch strain at z is defined

Mat. Res. Soc. Symp. Proc. Vol. 436 © 1997 Materials Research Society

to be that strain $\epsilon_{xx}^{(m)}(z)$, $\epsilon_{yy}^{(m)}(z)$, $\epsilon_{xy}^{(m)}(z)$ which must be imposed on the lamina at z to make it compatible with the reference lamina at $z = 0$.

A second feature of the class is that the total thickness of the film/substrate systems is much smaller than the lateral dimensions. Thus, $h/R \ll 1$ for the particular case of a circular configuration depicted in Figure 1. In addition, the substrate is compliant in the sense that h_s, the thickness of the substrate, may be comparable to h_f, the thickness of the film. This feature eliminates the special status of either the film or substrate within the system, which is reflected in the symmetry of analytical results with respect to the parameters h_s and h_f. Lastly, the present discussion is restricted to film/substrate systems which are free-standing, that is, unconstrained by external supports.

If both the strains and the deflections are very small, if the material response is elastic and if edge effects are ignored, then the curvature and stress distribution within the system can be expressed in terms of arbitrary mismatch strain variation $\epsilon_{ij}^{(m)}(z)$ and arbitrary modulus variation by means of elementary formulas.[1,2] This result is independent of the lateral dimensions or shape of the substrate. With the exact relationship in hand, it is possible to examine some special cases of interest in interpreting substrate curvature measurements of practical significance.

For thin, compliant structures of the kind under consideration the deflections may become moderately large while the strains remain small and the response of the material is elastic. This geometric effect is manifested through a coupling between the in-plane extensional strains and the out-of-plane deflection of the system. Within this regime of behavior, the overall size of the system (the value of R for the circular configuration depicted in Figure 1) comes into play and a perfectly symmetric system can undergo a bifurcation in deformation mode from a symmetric spherical shape to an asymmetric shape.[3-6] The conditions for this stable bifurcation to occur in a perfect system with equal biaxial mismatch strain are examined. Then, it is demonstrated that the system response is quite sensitive to small system imperfections by considering the response for slightly unequal biaxial mismatch strains.

Finally, the case of epitaxial single-crystal free-standing film/substrate systems is considered. A condition for the formation of an interface misfit dislocation as a strain relief mechanism in such a system is suggested. This condition, which reduces to the Matthews-Blakeslee condition[7] for the case of a thin film on a relatively very thick substrate, is illustrated for a cubic material system with a (100) interface.

CURVATURE AND STRAIN FOR SMALL DEFLECTIONS

The total elastic strain in the system in this case is the sum of the mismatch strain plus a compatible strain due to curvature. Following the reasoning outlined by Freund,[1] the form of the strain matrix for total thickness h is

$$\epsilon_{ij}(z) = -z\kappa_{ij}(h) + \epsilon_{ij}^{(r)}(h) + \epsilon_{ij}^{(m)}(z), \quad i,j = x, y \tag{1}$$

where κ_{ij} is the matrix of curvatures of the reference plane and $\epsilon_{ij}^{(r)}$ is the in-plane strain of the reference plane. These quantities are determined by means of Hooke's law, assuming the material response is linearly elastic, and by requiring the system to be in equilibrium.

For isotropic elastic response with the elastic modulus and Poisson ratio varying through the thickness as $E(z)$ and $\nu(z)$, respectively, the result of imposing the condition that the net force and net bending moment on any cross section must vanish is

$$\kappa_{ij}(h) = \frac{H^{(1)} J_{ij}^{(0)} - H^{(0)} J_{ij}^{(1)}}{H^{(1)2} - H^{(0)} H^{(2)}} \,, \quad \epsilon_{ij}^{(r)}(h) = \frac{H^{(1)} J_{ij}^{(1)} - H^{(2)} J_{ij}^{(0)}}{H^{(1)2} - H^{(0)} H^{(2)}} \tag{2}$$

where

$$H^{(k)}(h) = \int_0^h z^k \overline{E}(z)\, dz \,, \quad J_{ij}^{(k)}(h) = \int_0^h z^k \overline{E}(z) \epsilon_{ij}^{(m)}(z)\, dz \tag{3}$$

and $\overline{E}(z) = E(z)/(1 - \nu(z)^2)$. This provides an exact description of the curvature for any mismatch strain distribution and, through Hooke's law, an exact description of the stress distribution across the thickness of the system. A number of special cases are considered next which are relevant to the interpretation of measurements made by means of the wafer curvature technique.[8]

The most famous curvature-mismatch strain result is Stoney's formula[9] for the curvature of a substrate due to a spatially uniform equi-biaxial mismatch strain in a bonded film which has thickness much less than the thickness of the substrate. With the exact curvature result (2) in hand for any relative thickness of the film, the assumption of equi-biaxial mismatch strain, that is, $\epsilon_{xx}^{(m)} = \epsilon_{yy}^{(m)} = \epsilon^{(m)}$ and $\epsilon_{xy}^{(m)} = 0$, the deformed shape of the substrate is spherical with the curvature κ given by

$$\kappa = \frac{6\epsilon^{(m)}}{h_s} \frac{h_f}{h_s} \frac{M_f}{M_s} + \frac{6\epsilon^{(m)}}{h_s} \left(\frac{h_f}{h_s}\right)^2 \frac{M_f(M_s - 4M_f)}{M_s^2} + O\left(\frac{h_f}{h_s}\right)^3 \tag{4}$$

in powers of the ratio h_f/h_s, where $M = E/(1 - \nu)$ is the biaxial elastic modulus. The first term on the right side of the equal sign in (4) is Stoney's expression for curvature, which is based on the assumption of a very thin film. Thus, the second term on the right side of (4) provides a second order correction. It is interesting to note that, for example, when the moduli of the film and substrate are essentially the same ($M_f = M_s$) then the *ratio* of the second term to the first term is $-3h_f/h_s$. It follows that if the film thickness is 10% of the substrate thickness, then the error in neglecting the correction term to the Stoney formula included in (4) is 30%.

As a second special case, consider an equi-biaxial mismatch strain $\epsilon^{(m)}(z)$ which varies through the thickness of the film and elastic properties which are uniform throughout, that is, $E(z)$ and $\nu(z)$ are both constant. In this case, the general expression $(2)_1$ for curvature reduces to $\kappa_{xx} = \kappa_{yy} = \kappa$, $\kappa_{xy} = 0$ with

$$\kappa(h) = -\frac{6}{h^2} \int_0^h \epsilon^{(m)}(z)\, dz + \frac{12}{h^3} \int_0^h z\epsilon^{(m)}(z)\, dz \tag{5}$$

Clearly, knowledge of the curvature of the substrate at the final thickness h sheds no light on the variation of mismatch strain through the thickness of the film. On the other hand, suppose that the curvature *history* is known for all thicknesses up to the final thickness of the system. Then, multiplication of (5) by h^3 and repeated differentiation of the result leads to the inverse of (5) in the form of an expression for the mismatch strain *distribution* in terms of the curvature *history*, that is,

$$\epsilon^{(m)}(h) = \tfrac{2}{3}h\kappa(h) + \tfrac{1}{6}h^2\kappa'(h) + \tfrac{1}{3}\int_0^h \kappa(z)\,dz \tag{6}$$

where the prime denotes differentiation with respect to the argument. Thus, continuous measurement of curvature can provide information on the distribution of mismatch strain; this basic idea is being exploited to infer interface stress values from continuous curvature methods.[10] This result is based on the tacit assumption that no strain relaxation occurs once material is deposited during film growth. Other special cases are discussed by Freund.[11]

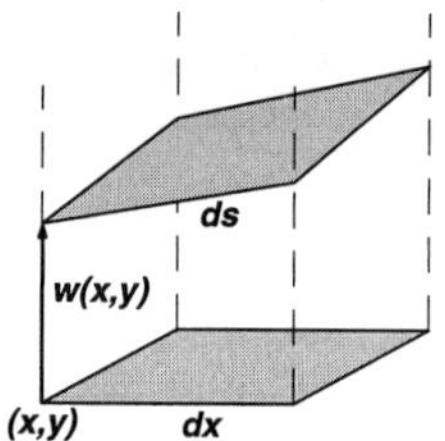

Figure 2. Sketch indicating the source of extensional strain due to transverse deflection of the reference plane. This extensional strain is approximately $(ds - dx)/dx \approx \tfrac{1}{2}(\partial w/\partial x)^2$.

STABLE BIFURCATION IN DEFORMATION MODE FOR LARGE DEFLECTIONS

The deformations considered in the preceding section fall within the range of response for which both the strain and the rotation at each material point are small. In this case, the stress at each point depends linearly on the local strain, and the strain depends linearly on the displacement gradients. It is well known that, for thin compliant systems such as plates or shells, the mechanical response commonly falls outside the range for which the displacement gradients are small, even though the strains may remain sufficiently small to justify continued use of Hooke's law to represent material response. In the context of the present discussion, the influence which is perhaps most significant is that transverse deflection of a lamina within the system, say the reference plane, can generate in-plane strains of that lamina which are comparable in magnitude to the direct in-plane strains. With reference to the sketch in Figure 2, an infinitesimal line element lying in the x direction in the initially flat element of the reference plane undergoes an extensional strain of magnitude $\tfrac{1}{2}(\partial w/\partial x)^2$ due to transverse deflection $w(x,y)$ alone. Most importantly, this strain can be as large in magnitude as the direct extensional strain, and therefore should be taken into account. From the sketch, it is clear that the transverse deflection induces a similar extensional strain in elements initially lying in the y direction. Likewise, the angle between the line elements, which are initially at right angles, changes due to transverse deflection alone, thus inducing an in-plane shear strain of the layer due to transverse deflection alone. If the components of displacement of a point in the reference plane are $u_x(x,y)$, $u_y(x,y)$ and $w(x,y)$ then the nonlinear strain-displacement relations which incorporate this additional geometric effect to account for "large" deflections and which provide the strain at any point x, y, z are

$$\epsilon_{xx} = \frac{\partial u_x}{\partial x} + \frac{1}{2}\left(\frac{\partial w}{\partial x}\right)^2 - z\frac{\partial^2 w}{\partial x^2}\,, \quad \epsilon_{yy} = \frac{\partial u_y}{\partial y} + \frac{1}{2}\left(\frac{\partial w}{\partial y}\right)^2 - z\frac{\partial^2 w}{\partial y^2} \tag{7}$$

$$\epsilon_{xy} = \frac{1}{2}\left(\frac{\partial u_x}{\partial y} + \frac{\partial u_y}{\partial x} + \frac{\partial w}{\partial x}\frac{\partial w}{\partial y}\right) - z\frac{\partial^2 w}{\partial x \partial y} \tag{8}$$

For example, the first term of ϵ_{xx} is the direct extensional strain of the reference layer, the second term accounts for the additional extensional strain due to transverse deflection of the reference plane, and the third term is the additional extensional strain due to bending.

It is not possible to state precisely and unambiguously the circumstances under which the additional nonlinear term should be included in the strain-displacement relations. In the literature on deflections of circular elastic plates under various loading conditions, it is commonly stated that the deflections are "finite" once the midpoint deflection relative to the perimeter deflection is comparable to the total thickness. Among other things, this statement tacitly presumes that the deflected shape is spherical, or at least roughly so. In the present study, the curvature will be computed on the basis of both the linear and the nonlinear strain-displacement relations so that a direct comparison can be made.

The equilibrium curvature of a film/substrate system with mismatch strain is now calculated by adopting the Ritz type approximate procedure introduced for the same problem class by Masters and Salamon[4]. The basic idea is to adopt a parametric family of shapes for the reference plane, along with a strain distribution which is consistent with this family of shapes and which incorporates the mismatch strain, and then to determine values of the parameters which represent stationary points of the total potential energy. The transverse deflection of the reference plane is assumed to be

$$w(x,y) = \tfrac{1}{2}\left(\kappa_x x^2 + \kappa_y y^2\right) \tag{9}$$

where κ_x and κ_y are the (assumed constant) curvatures in the x and y directions. The deformed shape of the reference plane is locally an ellipsoidal surface. A strain distribution consistent with this deflection is then

$$\epsilon_{xx} = a_1 + a_2\frac{x^2}{R^2} + a_3\frac{y^2}{R^2} - z\kappa_x + \epsilon_{xx}^{(m)}(z)$$

$$\epsilon_{yy} = b_1 + b_2\frac{y^2}{R^2} + b_3\frac{x^2}{R^2} - z\kappa_y + \epsilon_{yy}^{(m)}(z) \tag{10}$$

$$\epsilon_{xy} = a_3\frac{xy}{R^2} + b_3\frac{xy}{R^2} + \tfrac{1}{2}xy\kappa_x\kappa_y$$

where $a_1, a_2, \ldots, b_3$ are six additional dimensionless parameters characterizing the deformation.

With the explicit spatial dependence of the strains in (10), the strain energy density, say U, can be integrated over the volume of the film/substrate system, yielding the total potential energy of the system, say P, as a function of the eight characterizing parameters, that is,

$$P(\kappa_x, \kappa_y, a_1, \ldots, b_3) = \int_{\text{Volume}} U(\epsilon_{xx}, \epsilon_{yy}, \epsilon_{xy})\, dV \tag{11}$$

Then, equilibrium configurations are determined by finding those values of the characterizing parameters for which

$$\frac{\partial P}{\partial \kappa_x} = 0\,, \quad \frac{\partial P}{\partial \kappa_y} = 0\,, \quad \ldots\,, \quad \frac{\partial P}{\partial b_3} = 0 \tag{12}$$

The steps outlined above can be carried out for anisotropic material response, with the moduli varying in some prescribed way with z, and for an arbitrary variation in mismatch strain with z. As has been demonstrated by Masters and Salamon[4] and others, this can be accomplished by application of numerical methods. The goal here is to extract fairly simple analytical results of a general nature which are representative of the phenomena of interest, so some of the complexity in the model must be abandoned. Attention is restricted to systems for which the elastic response of both the film and substrate materials is isotropic, and the moduli of the film and substrate are the same. Furthermore, the mismatch strain is taken to be uniform throughout the film, but no restrictions are placed on the relative thickness h_f/h_s.

Suppose for the time being that the mismatch strain is an equi-biaxial strain, that is, $\epsilon_{xx}^{(m)} = \epsilon_{yy}^{(m)} = \epsilon^{(m)}$, $\epsilon_{xy}^{(m)} = 0$. For definiteness, and without loss of generality, $\epsilon^{(m)}$ is presumed to be positive in the discussion to follow. For the present case, the equilibrium conditions (12) can be obtained in closed form. In particular, for a given mismatch strain and geometric parameters, it is possible to express κ_x and κ_y in terms of $\epsilon^{(m)}$. If $\epsilon^{(m)}$ is then eliminated, a relationship between κ_x and κ_y is obtained which represents the locus of equilibrium states for the system in the plane of κ_x versus κ_y,

$$(\kappa_x - \kappa_y)\left[\kappa_x\kappa_y R^4(1+\nu) - 16(h_s + h_f)^2\right] = 0 \tag{13}$$

where ν is the Poisson ratio. This result is exact, within the class of deformations (9). An important consequence of admitting the possibility of finite deflections is evident in this expression. It is clear that a spherical deformed shape of the system, with $\kappa_x = \kappa_y$, is still an equilibrium shape (although the magnitude of the spherical curvature differs significantly from that predicted on the basis of the small deflection theory for a given mismatch strain, as will be seen subsequently). The new feature is the possibility of a second *asymmetric* equilibrium shape represented by the vanishing of the term in square brackets in (13).

The locus of possible equilibrium curvatures is plotted in Figure 3 for $\nu = 1/4$. The straight line bisecting the quadrant represents the spherical shape with $\kappa_x = \kappa_y$. The curved branch of the locus is obtained by setting the second factor in (13) equal to zero, and it represents asymmetric deformations, that is, $\kappa_x \neq \kappa_y$. The intersection point of these two branches is a bifurcation point. For values of spherical curvature on the branch with $\kappa_x = \kappa_y$ which are less than the curvature at the bifurcation point, it is found that the equilibrium value of potential energy is a local minimum under variations in curvature. Thus, that part of the symmetric branch represents *stable* equilibrium configurations. On the other hand, for values of spherical curvature which are larger than the curvature at bifurcation, the stationary value of potential energy is found to be a local maximum, so that part of the symmetric branch represents unstable equilibrium configurations. All equilibrium configurations on the asymmetric branch in Figure 3 (except that corresponding to the bifurcation point itself) are found to be stable configurations.

If the mismatch strain $\epsilon^{(m)}$ is viewed as a loading parameter, then the following scenario can be imagined. Initially, the mismatch strain is zero and the film/substrate system is flat. As the mismatch strain takes on positive values, the system deforms into a spherical shape with the curvature increasing with mismatch strain as the equilibrium point in Figure 3 moves from the origin toward the bifurcation point along the symmetric branch. If the strain is increased beyond the value corresponding to the bifurcation point (a value

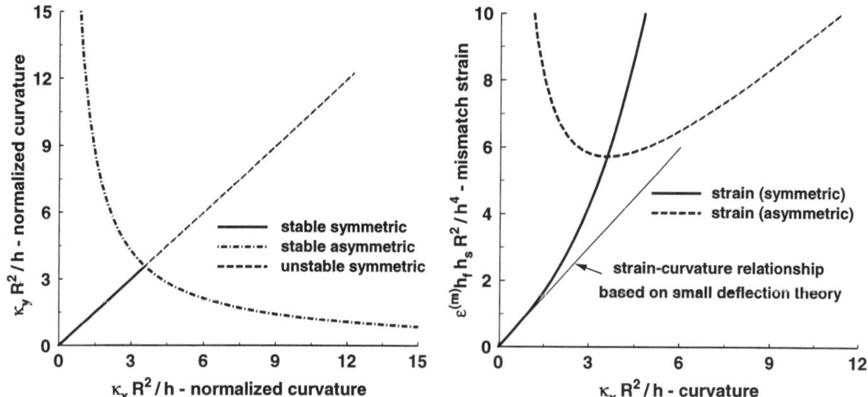

Figure 3. The curves on the left represent the locus of equilibrium configurations with constant curvature for a film on a circular substrate of radius R for Poisson ratio of 1/4. The intersection of the branches defines a state of stable bifurcation of equilibrium configurations. On the right, the corresponding level of uniform equi-biaxial mismatch strain $\epsilon^{(m)}$ at a given equilibrium curvature κ_x is given. The straight line indicates the relationship between film mismatch strain and substrate curvature implied by the small deflection theory. In both cases $h = h_s + h_f$.

which will be be established subsequently) then the deformation becomes asymmetric with the curvature increasing in one direction and decreasing in the orthogonal direction. The principal directions of the asymmetric deformation are completely arbitrary, of course. The bifurcation is stable, is the sense that an increasing strain is required to move the equilibrium configuration away from the bifurcation point along the asymmetric branch in Figure 3. As the strain magnitude increases further, the equilibrium shapes become more and more asymmetric, approaching a cylindrical limiting shape.

The locus of equilibrium shapes depicted in Figure 3 provides no information on the actual variation of mismatch strain $\epsilon^{(m)}$ along the equilibrium paths, so this is illustrated on the right in Figure 3. Again, the mismatch strain is normalized in such a way that the graph is universal, for the choice of $\nu = 1/4$. A particularly significant aspect of the relationship between strain and curvature is the portion between the origin of Figure 3 and the bifurcation point. It is obvious that this dependence is nonlinear, and this is a direct consequence of the inclusion of the nonlinear terms in the strain-displacement relationship. If only the linear terms had been used, the relationship would have followed the line labelled 'small deflection theory' in Figure 3.

The results in Figure 3 are valid for the full range of geometrical parameters for systems which meet the general characteristics of compliant free-standing layers. To give an impression of the magnitudes of parameters, consider the state represented by the bifurcation point in Figure 3. If the geometry of the system is characterized by h_s/R, the ratio of substrate thickness to radius, and by h_f/h_s, the ratio of film thickness to substrate thickness, then the spherical curvature $\kappa = \kappa_x = \kappa_y$ at bifurcation is given by

$$\kappa_{\text{bif}} R = \frac{4}{\sqrt{1+\nu}} \frac{h_s}{R} \left(1 + \frac{h_f}{h_s}\right) \tag{14}$$

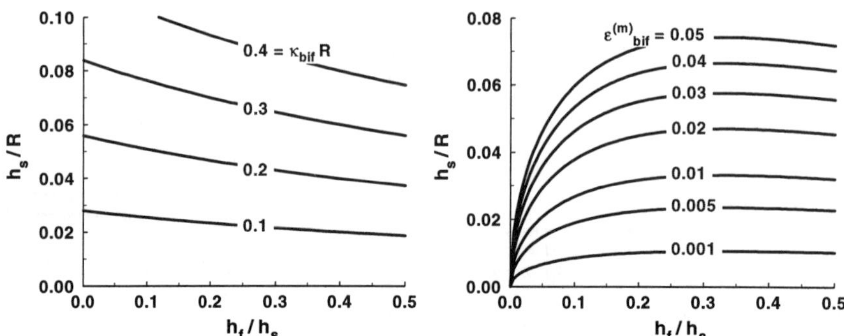

Figure 4. Level curves of substrate curvature (left) and mismatch strain in the film (right) at the bifurcation point in the plane of the geometrical parameters h_f/h_s and h_s/R. Each curve in the plot on the right can be viewed as a critical thickness condition for bifurcation at the level of mismatch strain corresponding to that curve.

and the critical magnitude of mismatch strain in the film which leads to bifurcation is given by

$$\epsilon_{\text{bif}}^{(m)} = \frac{4}{3(1+\nu)^{3/2}} \frac{h_s^2}{R^2} \frac{h_s}{h_f} \left(1 + \frac{h_f}{h_s}\right)^4 \tag{15}$$

Level curves of critical values for a range of values of the parameters h_s/R and h_f/h_s are shown in Figure 4.

SENSITIVITY OF RESPONSE TO SYSTEM IMPERFECTIONS

The bifurcation in deformation mode of this system arises from the competition between bending strain effects and extensional strain effects. The thin layer is relatively compliant in bending but stiff in extension, and the energy costs of bending deformation are relatively low compared to extension. For low levels of mismatch, the deformation can be accommodated by bending which is uncoupled to the extensional deformation in the system; in this case, the deflection is symmetric or spherical if the system is symmetric. However, if the magnitude of the deflection becomes large in some sense, these deformation modes become coupled. The system can then 'choose' between bending and extensional deformation in searching out minimum free energy equilibrium configurations. For sufficiently large mismatch strain, the system prefers the lower energy bending deformation of an asymmetrical deformation mode to the higher energy symmetrical or spherical mode. This trade-off accounts for the appearance of mode bifurcation at a certain level of mismatch strain. The wide difference in the stiffnesses for bending and extension is also a hint that the overall response may be sensitive to system imperfections. In other words, while the bifurcation in deformation mode discussed in the preceding section is a characteristic of a perfectly symmetric system, any slight imperfection in the system will result in significant departures from this bifurcation behavior.

To illustrate the influence of a small imperfection in the system, consider the case of a biaxial mismatch strain with $\epsilon_{xy}^{(m)} = 0$ and a slight anisotropy in the coordinate directions, that is, $\epsilon_{xx}^{(m)} \neq \epsilon_{yy}^{(m)}$. The imperfection will be represented by the deviation of the mismatch strain from its mean value through the parameter

$$\eta = \frac{\epsilon_{xx}^{(m)} - \epsilon_{yy}^{(m)}}{\epsilon_{xx}^{(m)} + \epsilon_{yy}^{(m)}} \tag{16}$$

The total strain distribution is again represented in the form (10) with the anisotropic mismatch strain incorporated. The locus of equilibrium curvatures for this imperfect system, analogous to (13) for the perfect system, is found to be

$$(\kappa_x - \kappa_y)\left[\kappa_x\kappa_y R^4(1 + \nu) - 16h^2\right] + \eta\left(\kappa_x + \kappa_y\right)\left[\kappa_x\kappa_y R^4(1 - \nu) + 16h^2\right] = 0 \tag{17}$$

For $\eta = 0$ this expression is identical to (13).

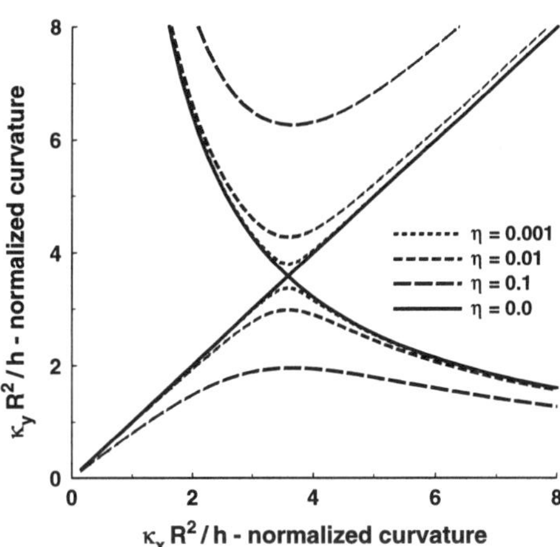

Figure 5. Locus of equilibrium curvatures according to (17) for several values of the mismatch strain imperfection η defined in (16) with $\nu = 1/4$.

For several values of $\eta > 0$, the locus of equilibrium curvatures is illustrated in Figure 5; results for $\eta < 0$ are identical except that κ_x and κ_y must be interchanged. It is evident from Figure 5 that even a slight deviation of 0.1% in mismatch strain from its mean value results in a perceptible separation of the branches of the equilibrium curves. A deviation of only 1% results in a large separation of the branches. Of the two branches in Figure 5, it is the lower one that is more likely to be followed by the system. While this result is merely an illustration of imperfection sensitivity of the system, it does suggest that bifurcation behavior would be very difficult to observe in real material systems and that actual response would nearly always be a gradual transition from symmetric to asymmetric deformation as the effective mismatch strain is increased.

MISFIT DISLOCATION FORMATION IN EPITAXIAL SYSTEM

As a final topic relevant to free-standing systems, the issue of misfit dislocation formation at the interface of a film grown epitaxially on a compliant substrate with a slightly differing lattice parameter is considered. The system is assumed to be free-standing, with both the surface of the film and the surface of the substrate away from the shared interface

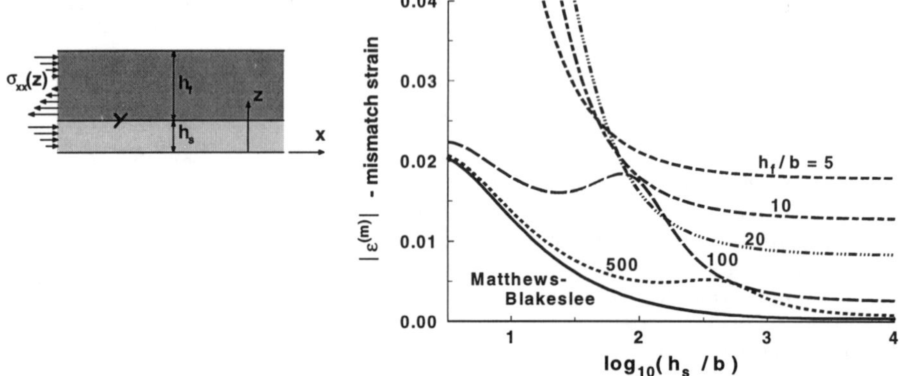

Figure 6. A schematic diagram of a segment of a free-standing film/substrate system with a mismatch strain is shown on the left. The stress distribution is indicative of the nonuniform equilibrium distribution prior to formation of an interface misfit dislocation. The plots on the right show the mismatch strain magnitude necessary for spontaneous formation of a misfit dislocation versus the substrate thickness for several fixed values of the film thickness. The thicknesses are normalized by b, the length of the Burgers vector. The solid curve is the Matthews-Blakeslee critical thickness condition for a thin substrate on a thick film.

being free of traction. The mismatch strain of the film is taken to be an equi-biaxial mismatch strain with $\epsilon_{xx}^{(m)} = \epsilon_{yy}^{(m)} = \epsilon^{(m)}$. Both the strains and the deflections are assumed to be small for this calculation. The calculation is carried out for formation of a straight dislocation with Burgers vector $\{b_x, b_y, b_z\}$ lying in the interface, with its line parallel to the y axis, as shown in Figure 6.

First, the stress acting on any plane $x = constant$ must be computed for the state of deformation prior to formation of the misfit dislocation. This is the stress which does work as the dislocation is formed. The strain is given by (1), and the stress distribution is the biaxial elastic modulus times the strain component $\epsilon_{xx}(z)$ obtained from (1),

$$\sigma_{xx}(z) = 2\mu\epsilon^{(m)}\frac{1+\nu}{1-\nu}\left[\frac{h_f}{h^3}\left(-6zh_s + 2h_s^2 - h_f h_s - h_f^2\right) + H(z - h_s)\right] \qquad (18)$$

where $H(z - h_s) = 0$ or 1, depending on whether $z < h_s$ or $z > h_s$, respectively. The work done by the elastic field due to the mismatch strain in forming the interface dislocation is

$$b_x \int_0^{h_s} \sigma_{xx}(z)\, dz = b_x \int_{h_s}^{h} \sigma_{xx}(z)\, dz \qquad (19)$$

As can be confirmed by direct calculation, the work done is the same no matter whether the dislocation is viewed as being formed from within the substrate or the film.

The critical thickness condition adopted here is that a misfit dislocation can form spontaneously if the work done by the background elastic field in its formation, as given by (19), is at least equal to the self-energy of the formed dislocation, including the effects of the free surfaces. Thus, this self-energy, which is the work required to create the dislocation in the bi-layer in the absence of any other loading, must be determined. Apparently, no simple mathematical solution of general applicability is available for a dislocation in a plate with free surfaces. Thus, following Freund and Nix,[12] an estimate of this self energy is made for general Burgers vector in a way which permits an assessment of the quality of the approximation for the special case of a screw dislocation. It is well-known that a dislocation parallel to a flat free surface in an otherwise unbounded elastic material is subject to an energetic force which is inversely proportional to the distance of the dislocation line from the free surface and which tends to pull the dislocation toward the surface; call this the *half-space force*. The estimate here is based on the assumption that the total force on a dislocation in the bi-layer is the sum of the half-space forces acting on the dislocation due to the two free surfaces separately. The self-energy of the dislocation at the interface is then the total work which must be done in overcoming this net force as the dislocation is moved from the surface of either the substrate or the film to the interface, which is

$$\frac{\mu[b_x^2 + b_y^2 + (1 - \nu)b_z^2]}{4\pi(1 - \nu)} \ln \frac{h_s h_f}{r_o(h_s + h_f)} \tag{20}$$

where r_o is the atomic-scale core cut-off radius.

The criterion which discriminates between conditions under which misfit dislocations can or cannot form spontaneously is obtained by equating the work expressions in (19) and (20). For given thicknesses and Burgers vector, the strain defined by this criterion is

$$\epsilon^{(m)} = \frac{[b_x^2 + b_y^2 + (1 - \nu)b_z^2]}{8\pi(1 + \nu)b_x} \frac{(h_s + h_f)^3}{h_s h_f(h_s^2 - h_s h_f + h_f^2)} \ln \frac{h_s h_f}{r_o(h_s + h_f)} \tag{21}$$

A sign convention implicit in this expression is that the product $\epsilon^{(m)}b_x$ must always be positive for strain relief by dislocation formation. A noteworthy feature of (21) is that it is symmetric in h_s and h_f; there is no fundamental distinction between the film and the substrate so their roles should be interchangeable. A second feature is that, if the condition is examined in the limit of $h_f/h_s \to \infty$, then the familiar Matthews-Blakeslee (MB) condition[7] is recovered, namely,

$$\epsilon^{(m)} = \frac{[b_x^2 + b_y^2 + (1 - \nu)b_z^2]}{8\pi(1 + \nu)b_x} \frac{1}{h_s} \ln \frac{h_s}{r_o} \tag{22}$$

This result is shown as the solid curve in Figure 6. Taking the limit as $h_s/h_f \to \infty$ yields the same result but with h_s replaced by h_f.

As an illustration of the condition (21), consider the SiGe/Si(100) system, or any other cubic system, which relaxes by formation of so-called 60° dislocations. If $\epsilon^{(m)} < 0$ then a possible choice of Burgers vector is $b_x = -b/2$, $b_y = b/2$, $b_z = b/\sqrt{2}$. For this case, graphs of the magnitude of mismatch strain at which dislocations can begin to form in the bi-layer versus substrate thickness are shown in Figure 6 for several values of film thickness. The MB critical thickness condition on h_s is also shown for the case of $h_f/h_s \to \infty$, $h_s \neq 0$. It is evident from the result that if h_s is less than the MB critical thickness of the substrate as determined from (22) then dislocations should not form no matter how thick the film becomes. Furthermore, if the thickness of the substrate exceeds the MB critical thickness, it is still possible to grow a strained film of substantial thickness on that substrate before the bi-layer critical thickness condition (21) is violated. For example, if the mismatch strain is 0.01 then the MB critical thickness is approximately $(h_s)_{MB} \sim 30\,b$. But if the substrate has a thickness of $h_s = 50\,b$, say, then the compliance of this substrate implies that a film can be grown up to a thickness of about $300\,b$ before the bi-layer critical thickness condition (21) is exceeded.

A complete analysis of the self-energy of a dislocation in a plate is required in order to assess the quality of the approximation made in estimating the self-energy as in (20), and such an analysis based on the finite element method has been carried out. Furthermore, exact results are available for the special case of a screw dislocation. This work provides support for the use of (20) as a good approximation. Finally, it is noted that the present analysis is based on the so-called energy approach to critical thickness. This approach is characterized by a comparison of energy states without concern for the physical mechanism by which one state is actually carried into the other. A mechanism by which an interface misfit dislocation of the kind considered here can be formed is the glide of a threading dislocation through either layer, which leaves behind an ever increasing length of interface misfit dislocation. It is by now well established that the alternate approach based on balance of forces acting on such a threading dislocation would lead to precisely the same critical thickness condition as found in (21).[13,14]

ACKNOWLEDGEMENTS

The research support of the Office of Naval Research, Contract N00014-95-1-0239, and the National Science Foundation, Materials Research Group Grant DMR-9223683 is gratefully acknowledged. I also benefitted from enlightening discussions with S. Suresh and M. Finot of MIT.

REFERENCES

1. L. B. Freund, *J. Crystal Growth* **132**, 341 (1993).
2. F. Kroupa, Z. Knesl and J. Valach, *Acta Tech. CSAV* **38**, 29 (1993).
3. W. H. Wittrick, *Quart. J. Mech. Appl. Math.* **6**, 15 (1953).
4. C. B. Masters and N. J. Salamon, *Int. J. Engrg. Sci.* **31**,915 (1993).
5. M. W. Hyer, *J. Compos. Mater.* **16**, 318 (1982).
6. M. Finot and S. Suresh, *J. Mech. Phys. Solids* **44**, 683 (1996).
7. J. W. Matthews and A. E. Blakeslee, *J. Cryst. Growth* **27**, 118 (1974).
8. P. A. Flinn, in *Thin Films: Stresses and Mechanical Properties*, edited by J. C. Bravman, W. D. Nix, D. M. Barnett and D. A. Smith (Mater. Res. Soc. Proc. **130**, Pittsburgh, PA, 1989), pp. 41-51.
9. G. G. Stoney, *Proc. Roy. Soc. (London)* **A82**, 172 (1909).
10. F. Spaepen, *J. Mech. Phys. Solids* **44**, 675 (1996).
11. L. B. Freund, *J. Mech. Phys. Solids* **44**, 723 (1996).
12. L. B. Freund and W. D. Nix, submitted for publication, (1996).
13. W. D. Nix, *Metall. Trans.* **20A**, 2217 (1989).
14. L. B. Freund, *J. Appl. Mech.* **54**, 553 (1987).

This article also appears in Volume 420.

AN ANALYSIS OF VOID NUCLEATION IN PASSIVATED INTERCONNECT LINES DUE TO VACANCY CONDENSATION AND INTERFACE CONTAMINATION

R. J. GLEIXNER AND W. D. NIX
Department of Materials Science and Engineering, Stanford University, Stanford, CA 94305-2205

ABSTRACT

Nucleation of voids due to vacancy condensation in passivated aluminum lines is analyzed within the context of classical nucleation theory. A discussion of sources of hydrostatic tensile stress in such lines provides a reasonable upper limit of 2 GPa. The void nucleation rate is then calculated at various sites within the line. Results suggest that nucleation rates are far too low to account for observed rates of voiding. Void nucleation at a flaw at the line/passivation interface is then considered as an alternative nucleation mechanism. Such flaws may be created by contaminants introduced during fabrication of the line. In this case, nucleation is feasible at greatly reduced stresses, well within the observed values. Furthermore, a simple model of void growth indicates that a fast atomic transport path, such as a grain boundary, must intersect the void for an appreciable growth rate. These results suggest that void nucleation in aluminum interconnect lines occurs at flaws at the sidewall of the line and that stress-induced and electromigration-induced voiding can be controlled by eliminating interfacial contamination.

INTRODUCTION

The failure of aluminum interconnect lines due to thermal and electromigration stress voiding has become a major issue in the design of reliable integrated circuits.[1,2] Voids are observed to nucleate within a line, then grow until the line is severed and the interconnect fails. Although the driving force for void growth is well understood, the phenomenon of void nucleation has received little attention. The lack of well-established models of void nucleation remains a limitation in modeling the failure of interconnect lines. In the case of thermal stress voiding, the line is subjected to large hydrostatic tensile stresses upon cooling from high passivation deposition temperatures. Here voids are often assumed to have nucleated during this cooling process, and the growth process is of concern.[3,4] Under electromigration conditions, the stress in a line increases slowly with time. In this case, voids are not observed for a substantial fraction of the time to failure. This is frequently modeled as an incubation period for void nucleation, ending when the maximum stress in the line reaches a critical value and a void forms.[5,6] In both of the these cases it is assumed that the critical stress required for void nucleation as well as the location of the void are known. A better understanding of the nucleation process would allow improvements to be made in these models.

This paper examines the feasibility of void nucleation within passivated aluminum lines on silicon substrates. Since the nucleation rate is highly dependent on the hydrostatic stress in the line, we first discuss the sources of this stress. We then calculate the nucleation rate at various sites within the line using classical thermodynamic and kinetic arguments. These analyses show that nucleation is not expected under typical service conditions within a reasonable time period. As a possible alternative site, we examine the case of void formation at a flaw on the metal/passivation interface and the growth rate of such a void.

STRESSES IN ENCAPSULATED METAL LINES

The driving force for void nucleation and growth in passivated lines is proportional to the hydrostatic tensile stress. In order to estimate the nucleation rate of various mechanisms, the magnitude of this stress must be known. Since this analysis is concerned with finding the maximum nucleation rate at various sites, an upper bound on the hydrostatic stress is desired.

When considering the stress in passivated lines, two major sources must be considered. The first is the thermal stress, resulting from the difference in thermal expansion between the passivation and metal upon cooling from high deposition temperatures. The second is the electromigration stress, which develops as metal atoms are swept from one end of the line to the

Mat. Res. Soc. Symp. Proc. Vol. 436 © 1997 Materials Research Society

other by the electron wind. Here the rigid passivation constrains the line from expanding or contracting in regions where the atomic flux diverges, and stresses thereby develop.

Table I contains the results of both x-ray diffraction experiments and F.E.M. calculations of thermal stresses in aluminum metallization lines.[7] This data corresponds to Al-0.5% Cu metallization lines, 1μm thick, 1.5μm wide, and 240μm long, deposited on silicon. The lines were passivated with 0.5μm of silicon nitride at 380 °C, and then cooled to room temperature (25 °C). These results indicate that the hydrostatic stress component, σ_h, is close to 500 MPa.

Technique	σ_{xx}	σ_{yy}	σ_{zz}	σ_h
X-Ray Measurements	538	453	362	451
F.E.M. Calculations	559	489	376	475

TABLE I. Thermal stresses (MPa) in passivated Al interconnects as found in [7]. The subscripts xx, yy, and zz refer to normal stresses, while the subscript h refers to the hydrostatic component.

Electromigration stresses, though similar in magnitude to thermal stresses, have not been measured using the above methods. Since they result from local accumulation or depletion of atoms, they average to zero along the line, and thus can not be measured by conventional x-ray diffraction techniques. In polycrystalline lines, where the fast path for atomic motion is along grain boundaries, atomic flow is effectively blocked at cross sections which contain no grain boundaries. As a result, large tensile stresses develop at the "upwind" end of a grain boundary segment, while compressive stresses develop at the "downwind" end. Analytical models have been developed to calculate the maximum hydrostatic stress, σ_h^e, that can develop in a continuous grain boundary segment in a line due to electromigration. The results give[8]

$$\sigma_h^e = \frac{1}{2}\frac{\lambda}{\Omega}eZ^*\rho j, \tag{1}$$

where λ is the length of the grain boundary segment, Ω is the atomic volume and ρ the resistivity of the metal, eZ^* is the effective charge of the metal ions, and j is the current density in the line. For the case of an aluminum line subjected to a current density of 1 MA/cm^2, a maximum hydrostatic stress of $\sigma_h^e = 160$ MPa is expected to develop in a 20μm grain boundary segment.

While the above results provide values of the average hydrostatic stress in the line, stress concentrators may further increase the stress in local regions. For the case of a passivated line, stress amplification is expected at the corners of the cross section. In addition to calculating the average stresses mentioned above, Sauter used finite element simulations to calculate the stress distribution over this cross section.[9] For the case where plastic flow is permitted to occur, the maximum stress concentration was found to be approximately two. Furthermore, shear stresses in the lines are expected to drive dislocations through the aluminum. If dislocation motion into the passivation is hindered, the resulting dislocation pileup will introduce further stress concentrations into the material. Based on the stresses in Table I, we find the maximum additional hydrostatic stress due to a dislocation pileup to be less than 500 MPa.

Assuming a worst-case scenario and summing the contributions from the above components, we find that 2 GPa is a reasonable upper limit for the hydrostatic stress in an interconnect line.

NUCLEATION THEORY

Detailed discussions of void nucleation by vacancy condensation are widely available, and will not be discussed in detail.[10,11] Here the term embryo describes a vacancy cluster which may or may not be energetically stable, while the term nucleus is reserved for clusters which are stable. For the case of a passivated Al interconnect under a hydrostatic stress σ, the free energy change upon creation of an embryo of volume V_e is given by

$$\Delta F = -\sigma V_e + \gamma_{Al}A_{Al} + (\gamma_{Al_2O_3} - \gamma_{Al-Al_2O_3})A_i - \gamma_{gb}A_{gb} \tag{2}$$

where the γ's refer to interfacial free energies and the A's refer to the areas of interfaces created or destroyed upon formation of the embryo. The quantity $(\gamma_{Al_2O_3} - \gamma_{Al-Al_2O_3})A_i$ applies in the case of sidewall nucleation. Here A_i is the area of the sidewall interface exposed by the existence of the embryo. The use of interfacial energies corresponding to Al_2O_3 rather than the SiN_x passivation can be explained as follows: before the passivation is deposited, a thin coating of Al_2O_3 is expected to form on the Al surface due to oxygen exposure. Electromigration testing of unpassivated lines has shown the Al_2O_3 film to remain intact as the void passes beneath.[12] This observation implies that void nucleation occurs at the Al-Al_2O_3 interface and not the Al_2O_3-passivation interface, and thus the energies corresponding to the former materials are appropriate.

The barrier to void nucleation, ΔF^*, is given by the maximum value of ΔF, and can be expressed in general by

$$\Delta F^* = \Delta F\big|_{\partial(\Delta F)=0}.$$ (3)

An embryo having this size is stable with respect to shrinkage and is referred to as a critical embryo. We define the quanitity n^* as the number of vacancies in a critical embryo.

The Becker-Döring theory of nucleation provides a link between the thermodynamics and kinetics of void nucleation by determining the rate at which critical embryos become nuclei.[13] Here the number of critical embryos per unit volume is given by

$$Z^* = \frac{N}{n^*}\left(\frac{\Delta F^*}{3\pi kT}\right)^{1/2} e^{-\Delta F^*/kT}$$ (4)

where N is the number of possible nucleation sites per unit volume (here the atomic density) and kT has the usual meaning. Considering a nucleation event to occur when a vacancy sticks to a critical embryo, the nucleation rate per unit volume, I, is given by

$$I = Rn_s^*Z^*,$$ (5)

where R is the sticking rate of vacancies on each atomic site and n_s^* is the number of vacancies in the matrix at the surface of a critical embryo. As an approximation, R can be given by the exchange frequency of the diffusive process, $R = v\exp(-U_D/kT)$. Here v is the frequency of vibration of the atoms and U_D is the activation energy of the jump process. n_s^* can be determined by the geometry of the embryo and the density of vacancies, such that

$$n_s^* = A_{Al}^* n_v\delta_s$$ (6)

where n_v is the number of vacancies per unit volume and δ_s is the thickness of the surface layer. The value of n_v is highly dependent on temperature, approximately given by

$$n_v = \frac{1}{\Omega}e^{-U_v/kT}e^{\sigma\Omega/kT}$$ (7)

where U_V is the formation energy of a vacancy in the absence of hydrostatic stress and Ω is the vacancy volume, assumed equal to the atomic volume.

If more than one vacancy path to the critical embryo exists, then the nucleation rate should then be summed over the paths. In the case at hand, we consider an intersecting grain boundary as well as the lattice as a path for vacancy incorporation. Combining equations (4) - (7), the nucleation rate is then given by

$$I =\left(\delta_{gb}l_{gb}^* e^{-U_{D,gb}/kT} + A_{Al}^* e^{-U_{D,l}/kT}\right)\frac{v\delta_s}{n^*\Omega^2}\left(\frac{\Delta F^*}{3\pi kT}\right)^{1/2} e^{-U_v/kT}e^{\sigma\Omega/kT}e^{-\Delta F^*/kT},$$ (8)

where the quantities δ_{gb} and l_{gb}^* refer to the width and length of the grain boundary intersecting the critical embryo, respectively.

NUMERICAL VALUES

Table II contains values of the above parameters which are independent of embryo geometry. Of particular note is the surface energy $\gamma_{Al\text{-}Al_2O_3}$. Since published values could not be found, this quantity was estimated using the force balance relation

$$\gamma_{Al-Al_2O_3} = \gamma_{Al_2O_3} + \gamma_{Al} \cos\theta. \qquad (9)$$

Here θ is the contact angle between a void and the passivation sidewall. By examining T.E.M. micrographs of actual voids, θ was found to be approximately 85°.[21] Inserting values from Table II gives $\gamma_{Al\text{-}Al_2O_3} \approx 2.28$ J/m².

Parameter	Value
T	200 °C
σ	2 GPa
v	10^{13} (1/sec)
N	10^{21} (1/m³)
δ_{gb}, δ_s	3×10^{-10} m
Ω	16.60 Å³
$U_{D,gb}$[14]	0.87 eV
$U_{D,l}$[14]	1.47 eV
U_v[15]	0.71 eV
γ_{Al}[16]	1.08 J/m²
$\gamma_{Al_2O_3}$[17]	2.19 J/m²
$\gamma_{Al\text{-}Al_2O_3}$	2.28 J/m²
γ_{gb}[16]	0.354 J/m²

Table II. Numerical values used in the analyses.

CALCULATION OF NUCLEATION RATES

The results of energy barrier (ΔF^*), critical number of vacancies (n^*), and nucleation rate (I) calculations are given in Table III.[18] For all cases, the assumptions of an equilibrium embryo surface and isotropic surface energy are made. As a result, the embryo surface is spherical in shape, characterized by a radius of curvature r. Furthermore, all calculations assume a hydrostatic stress of 2 GPa in the lines, and a temperature of 200° C.

The nucleation of a void per month in a 100µm x 1µm x 1µm line segment would require a nucleation rate of 10^{10} m⁻³ s⁻¹. For the cases given in Table III, the nucleation rates range from 10^{-340} m⁻³ s⁻¹ in the case of homogeneous nucleation to 10^{-57} m⁻³ s⁻¹ in the case of a sidewall notch. Even under the extreme conditions given here, we would not expect vacancy condensation to act as an observable source of voids.

	Homogeneous	Line Sidewall	Grain Boundary	70° Sidewall Notch
Nucleation Site				
Energy Barrier (ΔF^*) (eV)	32	15	12	6.5
Critical number of vacancies (n^*)	318	141	105	62
Nucleation Rate (I) (m⁻³ s⁻¹)	10^{-340}	10^{-145}	10^{-108}	10^{-57}

TABLE III. Calculated energy barriers, critical numbers of vacancies, and nucleation rates at various sites in passivated interconnect lines. These calculations assume a hydrostatic stress of 2 GPa and a temperature of 200°C.

FLAW ASSISTED NUCLEATION

As the above analyses show, void nucleation by vacancy condensation is not expected in passivated interconnect lines. Since experiments indicate otherwise, a nucleation mechanism other than that described by the classical theory is suggested. Such an alternative site is now considered.

Recently, the possibility of contaminants at the metal/passivation interface acting as void nucleation sites has been proposed.[19] Experimental evidence reveals that additional cleaning processes can dramatically suppress voiding in passivated lines.[20] The possibility of contaminants acting as nucleation sites provides an explanation of this observation. During patterning, the aluminum surfaces are subjected to dry etch processing. Such processing is known to leave behind resist residues, the source of the contamination.[21]

To analyze the effect of contaminants on void nucleation, consider the behavior of a flaw at the sidewall interface when subjected to a hydrostatic stress σ. As a simple case, the flaw can be viewed as a patch on which there is no adhesion between the line and passivation. Here the patch is assumed to be disc-shaped, having a radius a (figure 1 gives a side view of the situation envisioned). Under an applied hydrostatic stress σ, vacancy equilibrium between the free aluminum surface and the stressed solid requires the surface to adopt a radius of curvature r, where

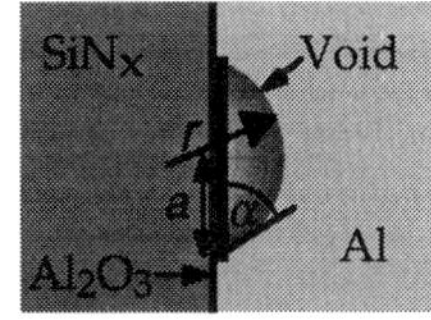

FIGURE 1. Schematic of an embryo growing on a contaminant patch.

$$r = \frac{2\gamma_{Al}}{\sigma}. \tag{10}$$

As the stress increases in the line, the radius of curvature of the void surface decreases and the surface bows out. In the early stages of growth (small α) the embryo is constrained to grow on the patch. If it were to leave the patch, imbalanced surface tensions would act to pull it back. This is true up to a critical value of the angle α. When α reaches the equilibrium contact angle θ, given by (1), surface energies pull the embryo off the patch at the interface and a void is formed. From then on, void growth limited only by the rate at which vacancies diffuse to the void.

According to this model, there exists a critical hydrostatic tensile stress sufficient to bow the void surface to the angle θ and nucleate a void. Noting that $r = a \sin \alpha$, the critical radius of curvature $r^* = a \sin \theta$. Substituting this into (10) and rearranging, the critical stress is given by

$$\sigma^* = \frac{2\gamma_{Al}}{r^*} = \frac{2\gamma_{Al} \sin \theta}{a}. \tag{11}$$

This equation shows the strong inverse relationship between σ^* and the flaw size. While a patch with $a = 10$ nm requires a hydrostatic stress of 200 MPa to develop into a void, a patch 100 nm in diameter requires only a 20 MPa stress.

Since it is reasonable to assume that contaminants are distributed uniformly over the surface of the interconnect, why are voids observed predominantly at the intersection of the grain boundary with the sidewall? This restriction on void location arises due to the large difference between the grain boundary and lattice diffusivities. For void growth to occur, atoms must be removed from the void surface and placed into the bulk of the line. A grain boundary acts as an extremely fast path for material removal relative to the lattice.

To estimate the time required for these processes we have developed a model which calculates the growth rate of a void on the sidewall of the line which is intersected by a single grain boundary cutting across the width of the line. Material is assumed to be removed solely via the grain boundary, which relaxes the tractions along the boundary. By assuming an initial thermal stress of 500 MPa in the line and a temperature of 200°C, we find that a time on the order of minutes is required for the void to grow to an observable size (~0.1μm). This agrees with experimental observations of thermal-stress induced voids.[22] However, at 200°C, the product of the lattice diffusivity and the line width ($D_l w$) is approximately 1000 times smaller than the product of the grain boundary diffusivity and effective grain width ($D_{gb}\delta_{gb}$).[14] Since these products provide a measure of the rate at which a path can transport atoms away from the void, we would expect a void growing by lattice diffusion alone to grow 1000 times slower. Therefore, the void would take several days to reach an observable size, and thus would not be seen during the duration of the experiment. As a result, we find that voids develop at the intersection of a grain boundary and the sidewall not because of a lack of contaminants, but instead due to slow matter transport.

CONCLUSIONS

Using classic thermodynamics and kinetics, void nucleation due to vacancy condensation is not expected to occur in passivated aluminum interconnect lines. At each of the sites considered, the nucleation rate is far too low to act as a practical source of voids.

Since voids in fact readily form under such conditions, a mechanism other than classical nucleation is suggested. Contaminants at the metal/passivation interface may provide such a mechanism. Analysis of void formation at a circular flaw at the interface shows that void formation at such a site would require a considerably smaller stress than the other mechanisms examined. In this case, thermal stresses alone are sufficient to cause void nucleation under typical processing conditions. A further limitation is placed on void growth due to the kinetics of the growth process. For a void to grow to an observable size in the observed time periods, it must be intersected by a fast diffusion path, such as a grain boundary.

ACKNOWLEDGMENTS

The authors would like to thank P.A. Flinn of Stanford University and Intel Corporation for suggesting the concept of contaminant assisted nucleation. We would also like to acknowledge the support of the Office of Naval Research through a Graduate Fellowship (RJG) and Grant No. N00014-92-J-4094 (WDN).

REFERENCES

[1] *Materials Reliability in Microelectronics III*, edited by K. P. Rodbell, W. F. Filter, H. J. Frost, and P. S. Ho (Materials Research Society, Pittsburgh, PA 1993), vol. 309.

[2] *Stress-Induced Phenomena in Metallization: 2nd Intl. Workshop*, edited by P. S. Ho, C.-Y. Li, and P. Totta (American Institute of Physics, New York, 1994).

[3] M. A. Korhonen, C. A. Paskiet, and C.-Y. Li, J. Appl. Phys. **69**, 12 (1991).

[4] A. F. Bower and L. B. Freund, in *Stress-Induced Phenomena in Metallization: 2nd Intl. Workshop* , edited by P. S. Ho, C.-Y. Li, and P. A. Totta (American Institute of Physics, New York, 1994), pp. 137-152.

[5] J. T. Trattles, A. G. O'Neill, and B. C. Mecrow, J. Appl. Phys. **75**, 7799 (1994).

[6] E. Arzt, O. Kraft, and U. E. Möckl, in *Thin Films: Stresses and Mechanical Properties IV*, edited by P. H. Townsend, T. P. Weihs, J. E. Sanchez, Jr., and P. Børgesen (Materials Research Society, Pittsburgh, PA 1994), vol. 388, pp. 397-408.

[7] B. Greenbaum, A. I. Sauter, P. A. Flinn, and W. D. Nix, Appl. Phys. Lett. **58**, 1845 (1991).

[8] W. D. Nix and E. Arzt, Met. Trans. A **23A**, 2007 (1992).

[9] A. I. Sauter, Ph.D. Dissertation, Stanford University, 1991, chap. 4.

[10] R. Raj and M. F. Ashby, Acta Metall. Mater. **23**, 653 (1975).

[11] J. P. Hirth and W. D. Nix, Acta Metall. Mater. **33**, 359 (1985).

[12] E. Arzt, O. Kraft, W. D. Nix, and J. E. Sanchez, Jr., J. Appl. Phys. **76,** 1563 (1994).

[13] J. W. Christian, *The Theory of Transformations in Metals and Alloys: Part I* (Pergamon, Oxford, 1975), chap. 10.

[14] H. J. Frost and M. Ashby, *Deformation-Mechanism Maps* (Pergamon, Oxford, 1982), p. 21.

[15] P. Shewmon, *Diffusion in Solids* (TMS, Warrendale, PA, 1989), p. 78.

[16] L. E. Murr, *Interfacial Phenomena in Metals and Alloys* (Addison-Wesley, Reading, MA, 1975), p. 124.

[17] P. Nikolopoulos, Journal of Materials Science **20**, 3993 (1995).

[18] R. J. Gleixner and W. D. Nix, submitted to J. Appl. Phys.

[19] P. A. Flinn, MRS Bulletin **XX** (11), 70 (1995).

[20] H. Abe, S. Tanabe, Y. Kondo, and M. Ikubo, in *Japan Society of Applied Physics, 39th Spring Meeting*, (1992), p. 658.

[21] S. P. Murarka and M. C. Peckerar, *Electronic Materials: Science and Technology* (Academic Press, San Diego, 1989), pp. 522-523.

[22] T. Marieb, Ph.D. Dissertation, Stanford University, 1994.

FINITE ELEMENT MODELING OF GRAIN ASPECT RATIO AND STRAIN ENERGY DENSITY IN A TEXTURED COPPER THIN FILM

R.P. VINCI, J.C. BRAVMAN
Department of Materials Science and Engineering, Stanford University, Stanford, CA 94305-2205

ABSTRACT

We have modeled the effects of grain aspect ratio on strain energy density in (100)-oriented grains in a (111)-textured Cu film on a Si substrate. Minimization of surface energy, interface energy, and strain energy density (SED) drives preferential growth of grains of certain crystallographic orientations in thin films. Under conditions in which the SED driving force exceeds the surface- and interface-energy driving forces, Cu films develop abnormally large (100) oriented grains during annealing. In the elastic regime the SED differences between the (100) grains and the film average arise from elastic anisotropy. Previous analyses indicate that several factors (e.g. elimination of grain boundaries during grain growth) may alter the magnitude of the SED driving force. We demonstrate, using finite element modeling of a single columnar (100) grain in a (111) film, that changes in grain aspect ratio can significantly affect the SED driving force. A minimum SED driving force is found for (100) Cu grains with diameters on the order of the film thickness. In the absence of other stagnation mechanisms, such behavior could cause small grains to grow abnormally and then stagnate while large grains continue to grow. This would lead to a bimodal grain size distribution in the (100) grains preferred by the SED minimization.

INTRODUCTION

Grain growth in thin films is often nonuniform, with preferential growth of certain crystallographic orientations due to minimization of surface energy, interface energy and strain energy density (SED)[1]. In Cu thin films on Si substrates the (111) texture is favored by surface and interface energy minimization, whereas the (100) orientation is favored by the SED driving force[2,3]. The SED difference between the (100)-oriented grains and the remainder of the film develops as a result of elastic and plastic anisotropy. For a given thermal strain, imposed by the difference in thermal expansion between the substrate and the film, different biaxial stresses will develop in grains with different orientations. Under conditions in which the SED driving force exceeds the surface- and interface-energy driving forces, Cu films develop abnormally large (100) oriented grains during annealing[2,4].

In the analysis of preferential grain growth it has been common to assume that SED differences can be derived by measuring the behavior of well textured or single crystal blanket films with the orientations of interest. Grains of a certain orientation within a single film of imperfect texture are then assumed to behave the same as the corresponding blanket films. A simple mechanical analysis, however, shows that interactions between adjacent grains of different orientation may modify the stress state in each grain and, hence, the SED[5]. Furthermore, it is expected that the extent of these interactions will depend on the aspect ratio of the grains of interest and the degree to which they approach the plane stress condition that exists, on average, in a textured film. Thus, as grain growth occurs, the SED is expected to change as the aspect ratio of the grains increases. We present a finite element analysis of the elastic behavior of a single columnar grain of (100) orientation within a (111)-textured film that demonstrates the extent to which the SED is altered by grain aspect ratio.

Mat. Res. Soc. Symp. Proc. Vol. 436 © 1997 Materials Research Society

MECHANICAL INTERACTIONS

Elastic strain energy is the total work performed during elastic deformation of a body. The SED for a grain with a texture (hkl) and a biaxial modulus M_{hkl} under an equal biaxial strain ε is given by $U_0 = M_{hkl}\varepsilon^2$. For a general triaxial stress distribution, the SED, in tensor notation, is given by $U_0 = \frac{1}{2}\sigma_{ij}\varepsilon_{ij}$

In a blanket film, on average, a plane stress condition obtains. Upon heating or cooling, due to the difference in thermal expansion between the substrate and the film, a uniform biaxial thermal strain is imposed (assuming isotropic thermal expansion). In Cu on a Si substrate the SED is largest in (111)-oriented grains and smallest in (100)-oriented grains under these conditions. The SED difference is therefore usually calculated as $\overline{U} = M_{111}\varepsilon^2 - M_{100}\varepsilon^2$.

The strain that develops in the direction normal to the film surface is due to the Poisson contraction or expansion resulting from the in-plane strain. Because there is no force in the normal direction under plane stress conditions, the normal strain is not included in the SED expression. In an elastically anisotropic polycrystalline material, grains of different crystallographic orientation may have large differences in Poissonian expansion normal to the film plane. For columnar grains, if perfect grain boundary sliding is allowed then the plane stress condition may apply to the individual grains despite the differences in out-of-plane strain. In the absence of complete grain boundary slip, the boundary between adjacent grains should provide some constraint in the direction normal to the film, thereby causing local normal and shear stresses not present, on average, in a blanket film. The general triaxial solution for the SED must be used in this case. This effect decreases with distance from the grain boundary, so a grain that is sufficiently large essentially still exists under plane stress conditions.

In an elastically anisotropic polycrystalline material, application of a uniform thermal strain results in in-plane stresses in the individual grains proportional to the appropriate biaxial moduli. If perfect adhesion exists at the interface between the film and the substrate then these conditions always apply at the base of the grains. However, elsewhere the grain boundary is free to move in response to the in-plane strains. Away from the interface, the film will relax at the expense of the grain, altering the in-plane strains and stresses. The change in in-plane strain will also lead to changes in the normal Poisson strain, thereby affecting the normal stresses close to the grain boundaries. Like the normal stresses, the in-plane relaxation decreases with distance from grain boundary and will not affect the average stresses in the film.

FINITE ELEMENT ANALYSIS

A finite element model of a single (100) grain in a (111)-textured film was constructed to evaluate the effect of grain aspect ratio on SED. The grain was modeled as a right circular cylinder of radius **a**. The grain was surrounded by a right circular cylinder of outer radius **b** and inner radius **a**, representing a textured film. All dimensions were normalized by the film thickness and **b** » **a**. Normalized grain widths from 0.25 to 16 were simulated. A commercial finite element code[6] was used for a two-dimensional analysis of 8-node axisymmetric quadrilateral elements. A zero radial displacement constraint was applied at the center axis of the grain (r = 0) and at the outer periphery of the film (r = **b**). Zero radial and normal displacement constraints were also applied along the bottom edge of the grain and film to simulate the presence of a substrate.

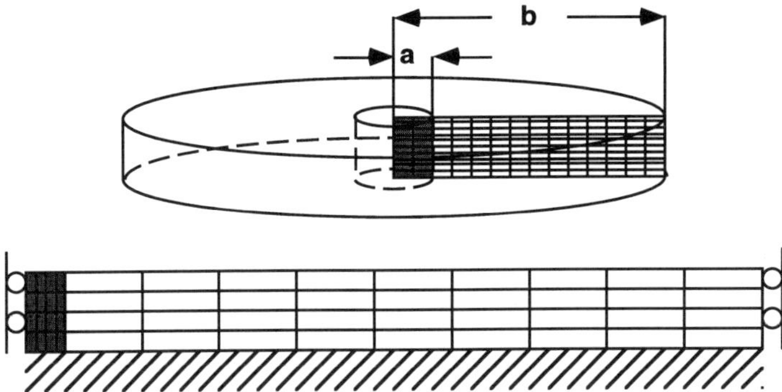

Figure 1. A single (100) grain is modeled as a cylinder of radius **a** surrounded by a film cylinder of radius **b** (top). An axisymmetric finite element mesh is used for the analysis (bottom). The mesh is free to expand in the vertical direction. Constraints are applied along the substrate interface and at the center axis and periphery of the mesh.

In a fiber textured film the grain orientation normal to the film is fixed, but the in-plane orientation is random. Thus, it is possible to calculate average in-plane moduli when the normal orientation is known, even for an orientation that is anisotropic in-plane. The resulting moduli are then suitable for evaluating the average behavior of grains of a certain orientation (e.g. as measured by x-ray techniques). For the (111)-fiber texture, the in-plane moduli are isotropic and calculation was direct[7]. An integrated average was used for the (100) fiber texture because it is anisotropic in-plane. The average elastic constants are listed in Table I.

Von Mises yield for the (111) orientation was 400 MPa. The (100) orientation was set to yield at 300 MPa to allow for clear differences in the yield temperature and stress level between the orientations. While elastic, perfectly plastic behavior and yield stresses of 400 and 300 MPa are not accurate for real copper films, these simplified parameters were chosen to avoid obscuring nuances in the elastic regime. The stress free temperature was set to 100 °C so that cooling to room temperature is purely elastic.

The coefficient of thermal expansion (CTE) for each cylinder ($\Delta\alpha_{grain}=\Delta\alpha_{film}=13.3 \times 10^{-6}/°C$) was equal to the difference between the CTE of the grain or film and the CTE of the substrate. Modeled in this way, the portion of thermal strain common to both the film and the substrate, which does not lead to film stresses, was not included. The model ignored factors such as grain boundary sliding, interface sliding, and grain boundary grooving that could also affect the stress state.

RESULTS AND DISCUSSION

Simulated heating from 20 °C to 420 °C spanned both the elastic and plastic regimes. The volume average elastic and plastic behavior of the (111) film matches that of a blanket film, as

Table I. Average Young's Modulus, Poisson's Ratio, and Shear Modulus for {111} and {100} fiber textured films. The Z axis is normal to the films while X and Y lie in the plane.

	E_z GPa	$E_{x,y}$ GPa	$\nu_{xz,yz}$	ν_{xy}	$G_{xz,yz}$ GPa	G_{xy} GPa
(111)	191	129	0.18	0.50	41	58
(100)	67	87	0.56	0.23	76	49

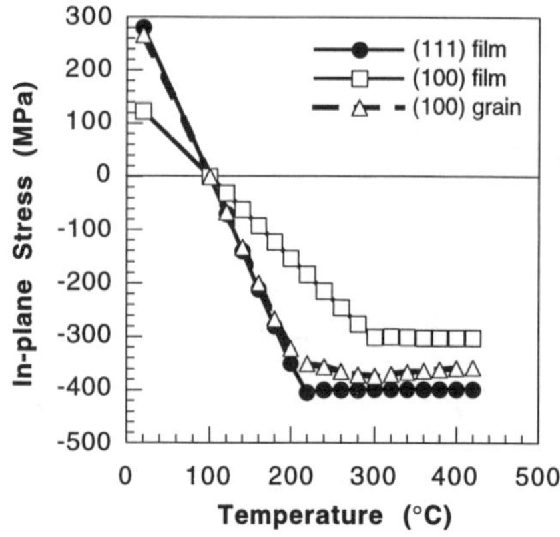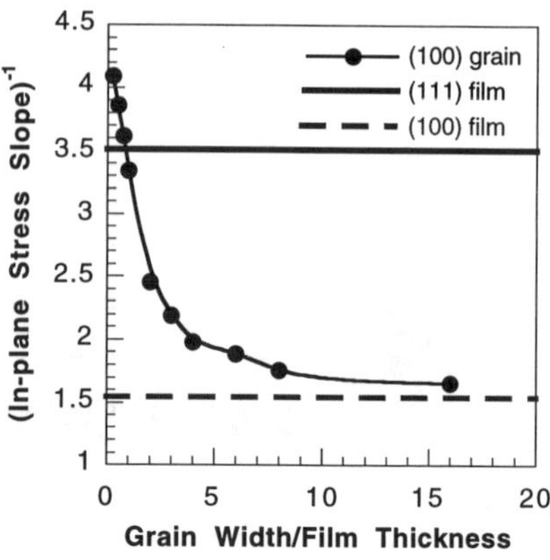

Figure 2. (a) The in-plane elastic behavior of a 1:1 aspect ratio (100) grain resembles that of a (111) continuous film rather than a (100) film. (b) The in-plane elastic behavior of a (100) grain becomes more like a continuous (100) film with increasing grain width.

expected when $\mathbf{b} \gg \mathbf{a}$. In agreement with previous results, the elastic behavior of the (100) grain deviates significantly from that of a (100) film over a wide range of grain aspect ratios. As shown in Figure 2a, the in-plane stress in a 1:1 grain is similar to that in the surrounding film. The change in slope with grain width is plotted in Figure 2b, demonstrating that the (100) slope can exceed that of the (111) film for narrow grains, whereas it approaches that of a (100) film for wide grains, as expected.

Once yield occurs in the (111) film the difference in SED declines, so the maximum SED will be found in the elastic regime just prior to the onset of yield. The normal and in-plane stresses and strains at 220 °C are plotted in Figures 3a and 3b. The stresses are large and compressive for the narrowest grains, and decrease monotonically with increasing grain width. The rate of change of the normal stresses is slightly larger than that of the in-plane stresses. The strains increase rapidly at the smallest widths, then slowly decrease. The normal strain undergoes the largest change at small grain widths as the effect of the normal constraint at the grain boundary falls away. The effect on the SED difference between the (100) grain and the surrounding film is shown in Figure 4, where the difference in SED is normalized to that determined from continuous film calculations. The mechanical constraints imposed by the surrounding film increase the strain energy density of the (100) grain relative to that of the film, decreasing the difference and, therefore, the driving force for preferential growth of the (100) orientation. A minimum in SED difference, just over one third of the film prediction, is seen for grains similar in diameter to the film thickness, designating a critical size for abnormal grain growth.

If the surface- and interface-energy driving forces are smaller than the SED driving force for all aspect ratios then the energy balance will, of course, favor the preferential growth of the (100) orientation regardless of grain size. If the surface- and interface-energy driving forces exceed the minimum SED driving force, however, preferential (100) growth will only be favored in the smallest and largest grain size regimes. For films with an initial grain size below the critical size, initial growth of the (100) orientation will be followed by stagnation. Grains that are larger than those at the SED minimum will be allowed to grow with limits imposed only by other possible stagnation mechanisms such as impingement on other (100) grains, grain boundary grooving, and stress changes due to the change in volume associated with elimination of grain boundaries. Thus

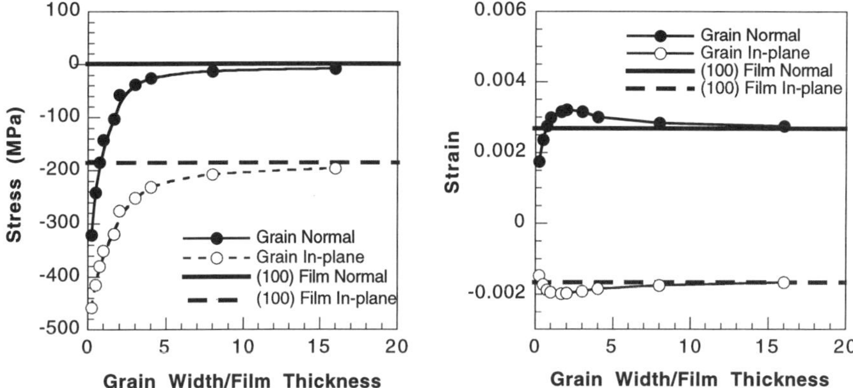

Figure 3. (a) Compressive stresses in a (100) grain decrease monotonically with increasing grain width. Normal and in-plane stresses in a (100) continuous film are plotted for reference. (b) Strains in a (100) grain pass through a maximum with increasing grain size. Normal and in-plane strains in a (100) continuous film are plotted for reference.

if a film with a lognormal grain size distribution has a small population of (100) grains larger than the critical size a bimodal grain size distribution could develop within the (100) population. If this is the case, complete conversion of the microstructure to a (100) orientation is unlikely.

Several significant limitations of the finite element model limit its direct applicability to abnormal grain growth prediction. First, the columnar model is not applicable to many small-grained films, including some Cu films. In noncolumnar grains a significant normal constraint would be present in all but the surface grains. Second, the model only treats the case in which (100) grains are sparsely dispersed throughout the film. If the (100) grains are close together then the "surrounding film" behavior will no longer match that of a continuous (111) film. Third, plasticity in single grains is still not fully understood, and the effect of the triaxial stress state on yield should be assessed. If some set of conditions can cause early stress relief in the (100) grains then the SED difference would increase until (111) yield occurs. Finally, the exact nature of the mechanical behavior of the grain/film interface and the grain/substrate interface is not known and is therefore not modeled. Nonetheless, to a first approximation the mechanical interactions between grain and film in an anisotropic film can cause a reduction in the SED driving force of significant magnitude in microstructures of interest.

CONCLUSIONS

A finite element model was constructed to provide insight into the effect of mechanical interactions between misoriented grains and a textured film. The model indicates that the behavior of a misoriented grain does not simply duplicate that of a blanket film of the same orientation. The slope of the in-plane stress in a Cu (100) grain in the elastic regime can exceed that of the surrounding (111) textured film despite the larger (111) biaxial modulus. This is due to the normal stresses at the grain boundary and grain boundary induced by elastic anisotropy. The strain energy density driving force for preferential growth of Cu (100) grains is smaller than that predicted by elasticity and blanket film behavior. A minimum in driving force is predicted for grains with diameters on the order of the film thickness. In the absence of other stagnation mechanisms, such a

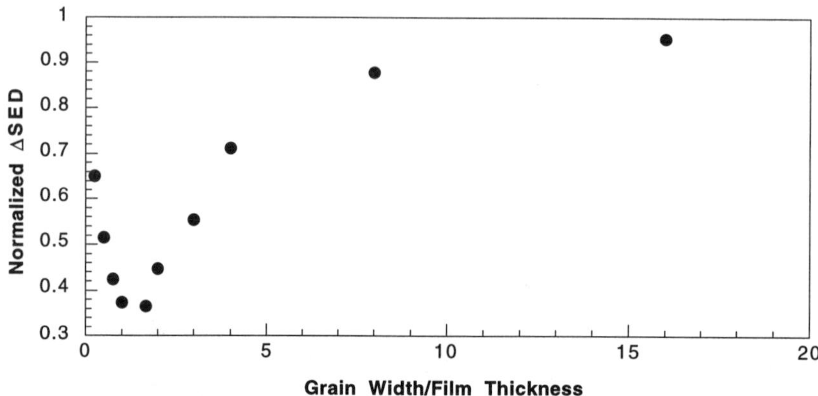

Figure 4. The strain energy density difference between the (111) film and (100) grain is normalized to the value expected from the continuous film assumption. Wide grains approximate the behavior of the continuous film, whereas the strain energy driving force for (100) growth is significantly reduced for smaller grain sizes.

behavior could limit preferential growth of small grains while large grains continue to grow. This would lead to a bimodal grain size distribution in the (100) grains preferred by strain energy density minimization.

ACKNOWLEDGMENTS

The authors would like to acknowledge the support of the Semiconductor Research Corporation (SRC).

REFERENCES

1. C. V. Thompson, in *Annual Review of Materials Science*; Vol. 20, edited by R. A. Huggins, J. A. Giordmaine, and J. B. Wachtman, Jr. (Annual Reviews, Inc., Palo Alto, 1990), p. 245.
2. E. M. Zielinski, R. P. Vinci, and J. C. Bravman, Appl. Phys. Lett. **67**, 1078-1080 (1995).
3. E. M. Zielinski, PhD Thesis, Stanford University, 1995.
4. E. M. Zielinski, R. P. Vinci, and J. C. Bravman, in *Materials Reliability in Microelectronics V*; Vol. 391, edited by A. S. Oates, W. F. Filter, R. Rosenberg, A. L. Greer, and K. Gadepally (Materials Research Society, San Francisco, 1995), p. 103.
5. R. P. Vinci, T. P. Weihs, E. M. Zielinski, T. W. Barbee, and J. C. Bravman, in *Materials Reliability in Microelectronics V*; Vol. 391, edited by A. S. Oates, W. F. Filter, R. Rosenberg, A. L. Greer, and K. Gadepally (Materials Research Society, San Francisco, 1995), p. 97.
6. MARC and MENTAT (MARC Analysis Research Corporation, Palo Alto, CA,).
7. U. A. Brantley, J. Appl. Phys. **44**, 534-535 (1973).

This article also appears in Volume 420.

MEASUREMENTS OF STRESS EVOLUTION DURING THIN FILM DEPOSITION

E. CHASON AND J.A. FLORO
Sandia National Laboratories, Albuquerque, NM 87185-1415

ABSTRACT

We have developed a technique for measuring thin film stress during growth by monitoring the wafer curvature. By measuring the deflection of multiple parallel laser beams with a CCD detector, the sensitivity to vibration is reduced and a radius of curvature limit of 4 km has been obtained in situ. This technique also enables us to obtain a 2-dimensional profile of the surface curvature from the simultaneous reflection of a rectangular array of beams. Results from the growth of SiGe alloy films are presented to demonstrate the unique information that can be obtained during growth.

INTRODUCTION

Understanding and controlling stress in thin films is critical for tailoring their optical, electronic and mechanical properties. Precise control of layer strain is required for the production of compound semiconductor heterostructure devices, and stress induced process can lead to the failure of interconnects and delamination of films. In general, most studies of thin film stress are performed after the films are grown. In this work, we discuss a new approach for measuring stress evolution *in situ* during the growth of thin films. We present results from experiments during epitaxial growth of SiGe alloy layers on Si(001) substrates with particular emphasis on the new information about the growth process that can be obtained from real time stress measurements.

WAFER CURVATURE MEASUREMENTS USING LASER BEAMS

A thin film under stress on a substrate will lead to bending of the substrate [1]. The radius of curvature of the substrate that results will be due to the balancing of the externbal bending moment applied and the bending moment of the curved substrate. The resulting curvature can be detected by the deflection of a beam of light incident upon the sample. If the substrate is flat, then the angle of reflection will be the same anywhere on the surface. However if the substrate is curved, then the reflection angle will change as the beam is move across the surface.

Various experimental approaches have been devised to measure the curvature of the surface. The scanning mirror technique [2,3,4,5] uses a rotating mirror and lens to scan the laser beam across the sample without changing the angle of incidence. A position sensitive detector measures the deflection of the beam as it is scanned. Alternatively, a beam splitter has been used [6,7,8] to produce two parallel beams whose deflections are measured independently with position sensitive detectors.

We have developed a variation of this technique with some features that simplify its use as an in situ diagnostic during growth. The experimental setup is shown in figure 1. An etalon that has been coated with highly reflective layers on both sides is placed at an angle to the laser beam. The non-normal incidence leads to multiple internal reflections inside the etalon so that a linear array of parallel beams is formed. The high degree of parallelism of the etalon faces ensures that the exiting beams are all parallel. These multiple parallel beams are then reflected simultaneously from the sample surface and measured with a CCD camera. The objective lens is used to focus

Mat. Res. Soc. Symp. Proc. Vol. 436 © 1997 Materials Research Society

the beams directly onto the CCD array so that no camera lens is necessary. Typically, five beams can be imaged on the CCD array simultaneously.

The use of multiple beams has several beneficial features for *in situ* measurement. First of all, the optics are very simple. The spatial filter and objective lens can be aligned before mounting on the deposition system and generally require only minimal subsequent alignment. Because a CCD camera is used, the reflected spots are fully imaged and can be viewed on a television monitor for focusing of the objective lens. In addition, because the CCD array has a relatively large active area, if the spots move due to sample motion (for instance during heating) this does not generally require realignment of the camera.

Simultaneous measurement of the multiple spots and no beam scanning make the system inherently less sensitive to sample vibration than the scanning mirror technique. Since all the laser spots move together, noise due to position changes or tilt of the sample do not appear as changes in the curvature. A quantitative representation of this is shown in figure 2. The sample in this run was unconstrained (held by gravity) and would sometimes exhibit a rocking motion, possibly driven by vibrations in the vacuum pumps used on the MBE system. In figure 2a, the variation in time of one of the spot centroids is shown; the excursions correspond to an RMS noise of 4.3 pixels. However the variation in time of the difference between the centroid of this spot and the spot adjacent to it (separated by approximately 120 pixels) is much smaller. Shown in figure 2b, the spacing between the centroids has an RMS deviation of only 0.09 pixels. So even though the positions of the spots may not be stable, the difference between the spots is much less sensitive to vibration.

2-DIMENSIONAL CURVATURE PROFILES

From the relative deflection of adjacent beams, we are able to determine the profile of the curvature across the sample. This is an advantage over the measurements using a beam splitter where only a single beam spacing is obtained. By using a pair of reflective optics oriented orthogonally to the laser beam, we are able to produce a two-dimensional grid of spots on the sample to obtain a two-dimensional curvature profile simultaneously. The results of this technique are shown in figure 3. The sample configuration is shown in figure 3a; the sample is clamped at one end in a cantilever arangement. The spots are incident on the sample at the postions shown in the figure. The reflected spot positions are shown in figure 3b for the as-prepared sample (o) and the sample after growth of 72 angstroms of $Si_{65}Ge_{35}$ (+). The centroids of the two beam profiles have been made to coincide to remove the effect of tilting of the sample. The difference between the spot positions of the flat sample and the curved sample can be related to the surface normal of the sample at the point of impact of the beam. From the surface normals, we can reconstruct a map of how the surface has deformed. In figure 3c, we show an image of the surface after the SiGe alloy growth. Note that the curvature is significantly larger along the unconstrained x-axis than along the y-axis where the clamp prevented the wafer from curving. Also note the difference in scales of the z-axis relative to the in-plane x- and y-axis. The maximum vertical deflection of the surface is less than 0.3 microns.

IN SITU MEASUREMENTS OF STRESS EVOLUTION

We have used the in situ wafer curvature technique for measuring the evolution of stress during growth of Si_xGe_{1-x} alloys on Si(001) substrates. An example of the stress measurement is shown in figure 4 where the evolution of the product of film stress (σ) and thickness (h) during

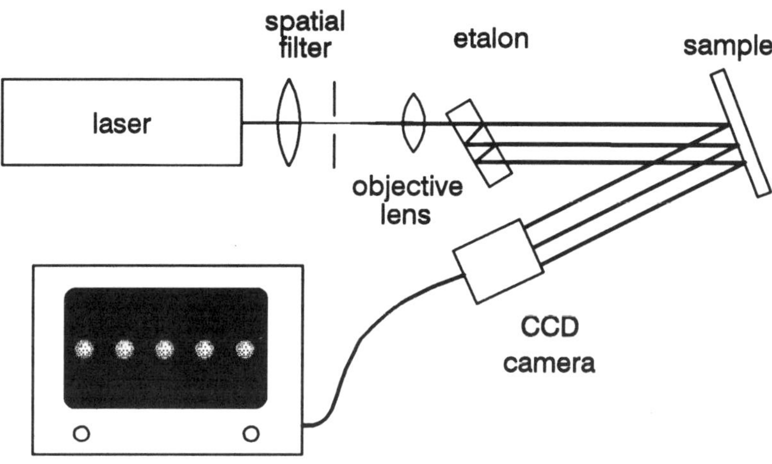

Figure 1. Schematic of system for measurement of wafer curvature using multiple parallel beams.

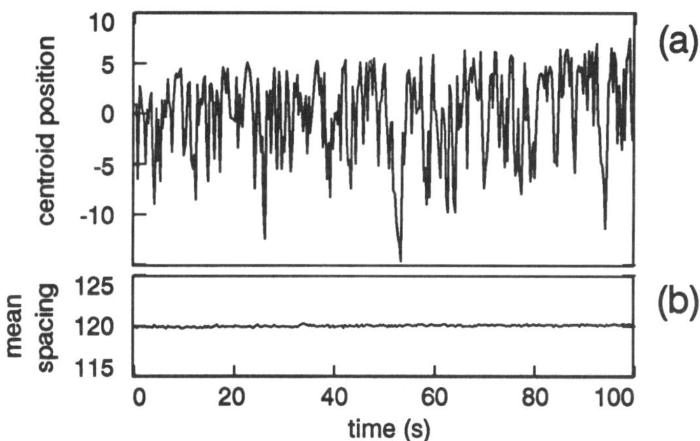

Figure 2. (a) Measurement of the positional variation in the centroid of one of the beams reflected from the sample. The RMS noise corresponds to 4.3 pixels. (b) Measurement of the variation in the difference between the centroids of adjacent beams reflected from the sample. The RMS noise (0.09 pixels) is reduced by a factor of 40 from the noise in the centroid positions.

growth is demonstrated. The three distinct regions of behavior observed during growth are discussed below.

Stress Offset

In the early stages of growth (figure 4a, 0 - 10 Å), the stress increases much more slowly than expected for a strained epitaxial film. We attribute this behavior to segregation of the Ge to the surface during the early stages of growth.   After a Ge rich surface layer is formed, the film grows at the nominal composition determined by the growth fluxes. The Ge-rich layer remains on the surface during subsequent growth, presumably by exchanging places with adatoms arriving from the deposition flux. This interpretation is supported by earlier work [9,10] and is also consistent with our measurements of the offset dependence on alloy composition. For decreasing Ge fraction in the film, the offset before the linear elastic regime begin increases since it takes longer to form a surface layer. We also find that growing pure Si after the growth of a SiGe layer results in increasing compressive stress as the surface Ge is re-incorporated into the Si layer. Without the presence of Ge on the surface, the growth of Si would not lead to additional strain since the films are fully pseudomorphic.

Two other possible explanations for this behavior, interface stress and surface morphology, have also been considered. Although interfacial stress can contribute to wafer curvature, it is probably not the dominant source of the offest in figure 4a since the offset is much longer than the time it takes to deposit one monolayer. Alternatively, if the surface is being covered with small islands, it is possible for the islands to be partially strain relaxed without being dislocated. However, simultaneous RHEED (reflection high energy electron diffraction) meaurements of the surface morphology during growth indicate that the surface does not develop sufficient roughness to enable significant relaxation by this mechanism.

Linear Elastic Regime

After the initial offset, σh increases linearly with the film thickness. For thicknesses below the limit for introduction of dislocations, the rate of change of σh is equal to $M(\varepsilon)\varepsilon\, dh/dt$ where ε is the film strain, $M(\varepsilon)$ is the biaxial modulus of the strained alloy film and dh/dt is the growth rate. We have measured $d(\sigma h)/dt$ for different alloy compositions to determine the dependence of $M(\varepsilon)$ on strain. The growth rates and compositions were calibrated using Rutherford backscattering spectrometry (RBS), double crystal X-ray diffraction and X-ray reflectivity. We find that the biaxial modulus can be explained by a simple rule of mixtures over the range of 15 - 60% Ge concentration where the endpoints are the bulk unstrained elastic constants for Si and Ge [11].

Strain Relaxation

As epitaxial films become thicker during growth, they reach a critical thickness beyond which dislocations can form to decrease the coherency strain energy. The onset of strain relaxation is seen in figure 4b as a decrease in the slope of σh vs. thickness at approximately 400 Å. We have measured the onset of strain relaxation in $Si_{71}Ge_{29}$ alloys grown at temperatures of 450, 550 and 650 °C [12] .Since the formation of dislocations is thermally activated, the metastable region before the onset of relaxation is larger at lower temperatures. Although strain relaxation kinetics in SiGe alloys have been extensively studied, the laser curvature technique

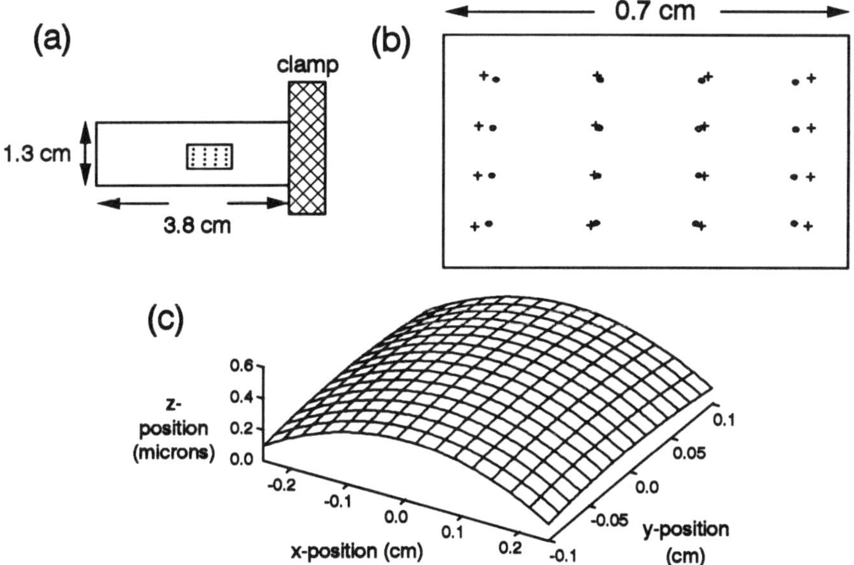

Figure 3. Reflection of a 2-dimensional array of parallel beams from the sample provides a 2-dimensional profile of the surface curvature. (a) Schematic of the sample configuration with one end clamped showing where the laser beams intersect the sample. (b) Position of the centroids of the reflected beams for the as-prepared sample (o) and after the growth of 72 Å of a $Si_{65}Ge_{35}$ strained film (+). (c) Profile of the curved sample surface after alloy growth reconstructed from the measured change in the surface normal.

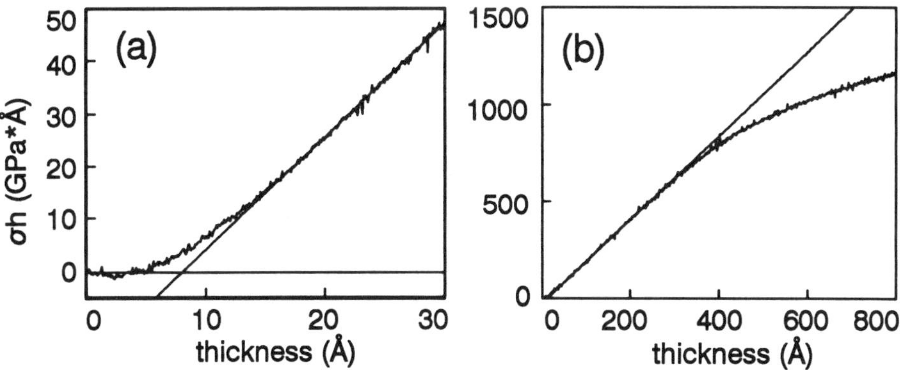

Figure 4. Evolution of σh during the growth of a $Si_{71}Ge_{29}$ alloy shows three distinct regions: initial stress offset, linear elastic and strain relaxation. (a) Early stage growth (0-10 Å) shows an initial period of essentially stress-free growth before the linear increase due to coherency strain. (b) Strain relaxation occurs above the critical thickness for dislocation formation.

enables the onset of relaxation and the subsequent kinetics to be measured much more easily than the ex situ techniques that have been previously used.

ACKNOWLEDGEMENTS

We thank Mike Sinclair, Carl Seager and R.C. Cammarata for useful discussions. This work was performed at Sandia National Laboratories and supported by the U.S. Dept. of Energy under contract DE-AC04-94AL85000.

REFERENCES

1. M.F. Doerner, W.D. Nix, CRC Crit. Rev. in Solid State and Mater. Sci. **14**, 225 (1988).

2. P. Flinn, Gardner and Nix. IEEE Trans. Elec. Dev. **ED-34**, 689 (1987).

3. C.A. Volkert, J. Appl. Phys. **70**, 3521 (1991).

4. J.A. Ruud, A. Witvrouw and F.A. Spaepen, J. Appl. Phys **74**, 2517 (1993).

5. A.L. Shull, H.G. Zolla and F.A. Spaepen, Mat. Res. Soc. Symp. Proc. **356**, 345 (1995).

6. R. Martinez, A. Augustyniak and J.A. Golovcenko, Phys. Rev. Lett. **64**, 1035 (1990).

7. A. Schell-Sorokin and R. Tromp,Phys. Rev. Lett. **64**, 1039 (1990).

8. Geisz et al., J. Appl. Phys. **75**, 1530 (1994)

9. K. Fujita, S. Fukutsu, H. Yaguchi, Y. Shiraki and R. Ito, Appl. Phys. Lett. **59**, 2103 (1991).

10. D.J. Godbey and M.G. Ancona, Appl. Phys. Lett. **61**, 2217 (1992).

11. J.A. Floro and E. Chason, unpublished.

12. J.A. Floro and E. Chason, Mat. Res. Soc. Symp. Proc. 1996 (in press).

This article also appears in Volume 420.

RELATIONSHIP BETWEEN THE VOID AND HILLOCK FORMATION AND THE GRAIN GROWTH IN THIN ALUMINUM FILMS

O.V. KONONENKO and V.N. MATVEEV
Institute of Microelectronics Technology & High Purity Materials, Russian Academy of Sciences, Chernogolovka 142432, Moscow District, Russia

ABSTRACT

Void and hillock formation during annealing was studied depending on the deposition conditions. Aluminum films were deposited onto oxidized silicon substrates by the self-ion assisted technique. The bias 0 or 6 kV was applied to the substrate during deposition. The films were then annealed in vacuum for 1 hour in the temperature range from 150° to 550°C. The structure of the films was investigated by transmission electron microscopy. The void and hillock formation was studied with optical and scanning electron microscopes.

It was found that recrystallization and void and hillock formation in the films depend on the bias during deposition. Normal grain growth occurred in the films deposited without bias. Abnormal grain growth was observed in the 6 kV-films. It was also found that the mechanism of stress relaxation during thermal cycling depends on the self-ion bombardment. In the films prepared without bias, stress relaxation proceeds by diffusion creep. In the films deposited at the 6 kV bias, stress relaxation proceeds by plastic deformation.

INTRODUCTION

The self-ion assisted technique is very attractive for deposition of thin metal films because it allows control over film properties such as structure, resistivity and electromigration resistance [1]. Recently we have shown [2,3] that the structure and properties of pure aluminum films can be varied in a wide range using bombardment by self-ions. It has also been demonstrated that the immunity to electromigration can be improved by increasing the bias during film deposition [3,5]. Mechanisms for electromigration and stress migration damage formation were found to be different for films fabricated with or without bias [5].

In this work we investigate the influence of the self-ion bombardment on evolution of the film structure and stress relaxation during an annealing cycle in the temperature range from 150 to 550°C.

EXPERIMENT

High purity aluminum was deposited using the self-ion assisted technique [1] at a rate of about 2 nm/s onto unheated substrates. Oxidized silicon wafers of the (100) orientation were used as substrates. The ion-to-atom ratio was about 6%. The ions were accelerated by potentials 0 and 6 kV applied to the accelerating electrode located near the substrate. The thickness of the films was about 350 nm.

After deposition, the films were annealed in a vacuum of about 10^{-6} torr for 1 hour in the temperature range from 150° to 550°C. The structure of the films was investigated in a HB501 transmission electron microscope. Samples for TEM examination were prepared by etching a crater on the back side of the substrate with a 5:3:3 mixture of HNO_3, HF and

Mat. Res. Soc. Symp. Proc. Vol. 436 © 1997 Materials Research Society

CH$_3$COOH. The void and hillock formation was studied with optical and scanning electron microscopes.

RESULTS

Structure

TEM investigations revealed substantial differences in the structure of the films deposited with the bias and without it. Figure 1 shows an average grain size as a function of the annealing temperature.

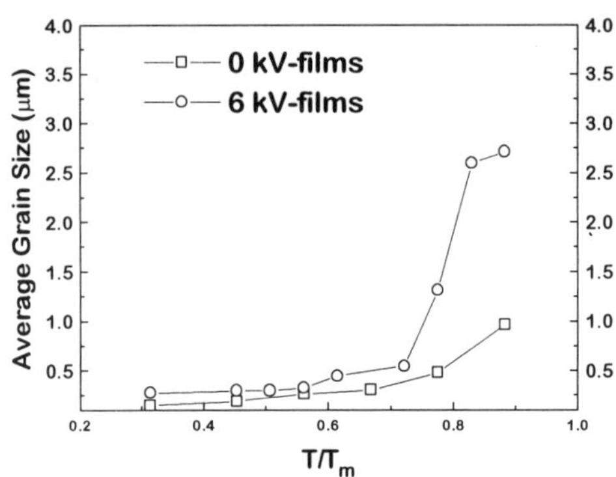

Fig.1 Average grain size in the Al films deposited at different biases as a function of the annealing temperature.

The average grain size in as-deposited films prepared without bias is about half of the film thickness. After annealing, the average grain size increases with annealing temperature and reaches the value equal to 1μm at 550°C. Figure 2a shows grain size distribution as a function of annealing temperature in the films deposited without bias. The distribution is log-normal and shifts to larger grain sizes as the annealing temperature increases. In other words, normal grain growth occurs in these films.

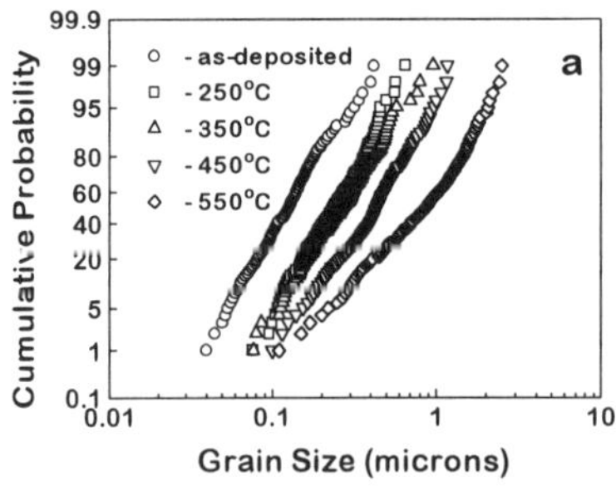

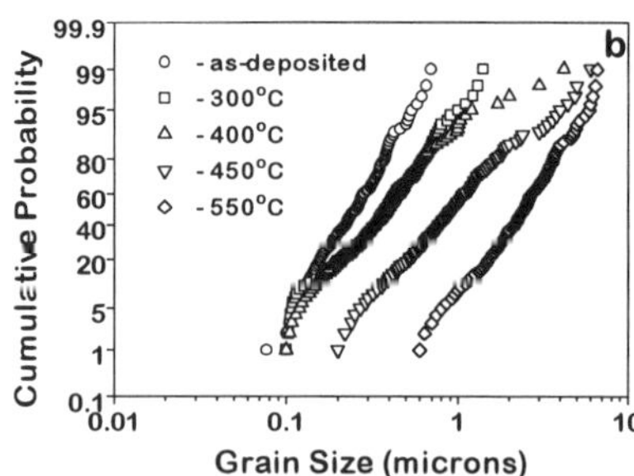

Fig.2 Grain size distribution as a function of annealing temperature in (a) the film deposited without bias and (b) the film deposited at 6 kV bias.

TEM studies of the 0 kV-films showed contrast from a high density of small features about 10-40 nm in diameter (Fig.3). They are formed at 350°C and higher annealing temperatures. Similar features were observed by Arzt and co-authors in oxygen implanted aluminum films [6].

Fig.3 Small features in the film deposited without bias after annealing for 1 hour at 350°C.

The behavior of films deposited at 6 kV bias differs from that of the 0 kV-films. The average grain size in as-deposited films is 0.8 of the film thickness. As the annealing temperature is increased up to 300-350°C, the average grain size increases slightly to values about equal to the film thickness. Further increase of the annealing temperature leads to an abrupt increase of the average grain size. It is eight times as large as the film thickness. Figure 2b shows the grain size distribution in the 6 kV-films at different annealing temperatures. It is seen that normal grain growth proceeds at temperatures up to 350°C. A subpopulation of large grains appears in the film at the annealing temperature 400°C. The number of such grains increases with annealing temperature. The distribution of grain size after 550°C annealing is log-normal again. Thus, abnormal grain growth takes place in the 6 kV-films.

Extensive slip was found in a number of grains with diameters greater than the film thickness (Fig.4). No such dislocations were found in the films deposited without bias.

Fig.4 TEM image of an annealed 6 kV-film. Extensive slip has taken place, primarily in the larger grains.

Hillocks

Two types of hillocks were found in the films deposited without bias (Fig.5). The first type is large hillocks about 5-10 µm in diameter and 1-2 µm high. The second type is small rounded hillocks. Their diameter and height are about 0.2-1 µm.

On the surface of the films deposited at 6 kV bias, only the second type of hillocks were observed. Many of them displayed a crystalline appearance (Fig.6).

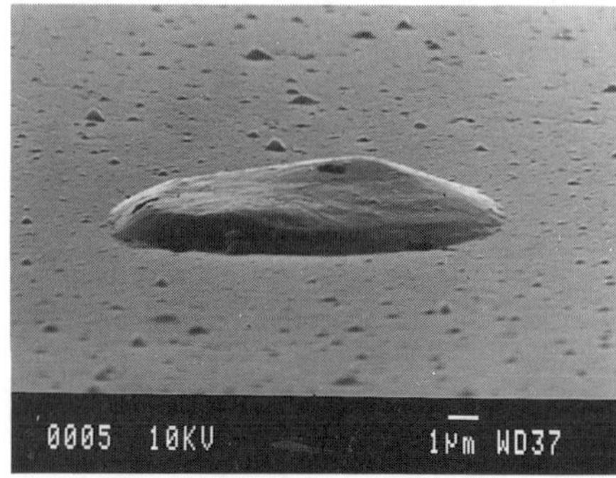

Fig.5 SEM image of the surface of an annealed Al film deposited without bias.

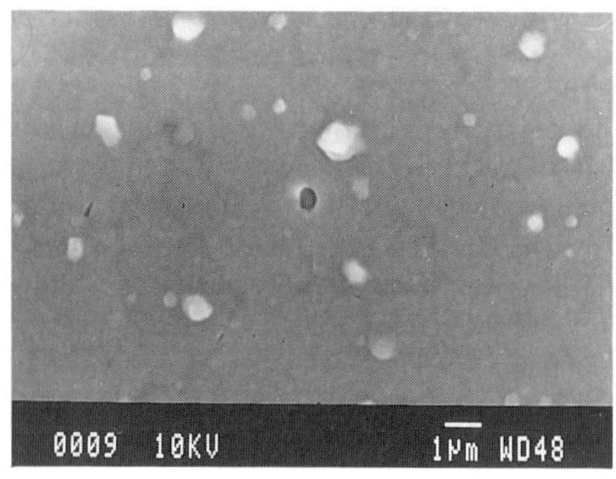

Fig.6 SEM image of the surface of an annealed Al film deposited at 6 kV bias.

The density of the second type of hillocks in 0 kV and 6 kV-films was about the same. It increased with annealing temperature. The density of large hillocks in the 0 kV-films was smaller by a factor of 10^2-10^3 than that of small hillocks. Their number also increased with annealing temperature.

<u>Voids</u>

Voids of branched-out shape were observed on the surface of the films deposited without bias (Fig.7). The density of voids was about $10^5/cm^2$ and increased with the annealing temperature.

A few submicron voids, looking like grain collapse, were found on the surface of the 6 kV-films.

DISCUSSION

Our observations show that films deposited under different conditions, demonstrate different mechanisms of stress relaxation during annealing. In the films deposited without bias the stress relaxation mainly proceeds by diffusion. In the 6 kV-films, relaxation by plastic deformation takes place.

From our viewpoint, these distinctions can be explained on the basis of the self-ion bombardment effect. It is known from literature[7,8] that self-cleaning of residual gases (as a rule

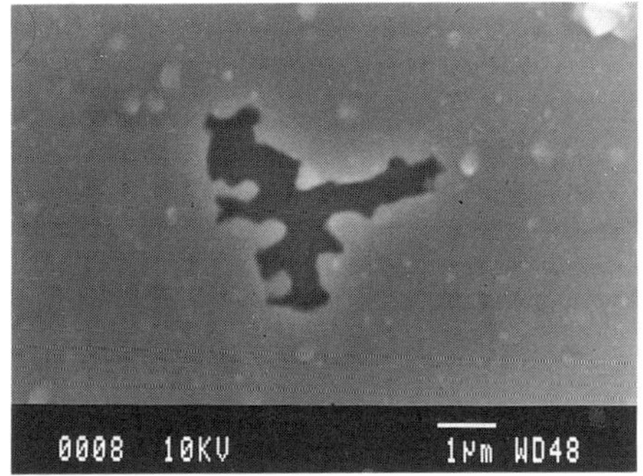

Fig.7 SEM image of the branched-out void in an annealed Al film deposited without bias.

it is oxygen [9]) adsorbed by the surface of a growing film occurs owing to sputtering by energetic self-ions. Films with pure grain boundaries are formed as a result of this process. Stress relaxation in these films occurs by plastic deformation. In the films deposited without bias, oxygen accumulates in grain boundaries and forms oxide precipitates at higher temperatures. These precipitates suppress plastic deformation, therefore the stress relaxation proceeds by diffusion in these films.

Our measurements showed that the resistivity of the films deposited at 6 kV bias was virtually equal to the resistivity of bulk aluminum (2.65 $\mu\Omega\cdot$cm). The resistivity of the films deposited without bias was about 4 $\mu\Omega\cdot$cm. These data confirm the supposition that grain boundaries in the 6 kV-films have higher purity than those in the films deposited without bias.

CONCLUSION

Aluminum films were deposited by the self-ion assisted technique at different biases. Grain growth was studied in the films at different annealing temperatures. Normal grain growth was observed in the films deposited without bias. Abnormal grain growth occurred in the 6 kV-films at annealing temperatures higher than 400°C.

It was also found that stress relaxation in the films prepared without bias proceeds by diffusion creep. In the films deposited at 6 kV bias, stress relaxation proceeds by plastic deformation.

ACKNOWLEDGMENTS

The research described in this publication was made possible in part by Grant No NJW300 from the International Science Foundation and the Russian Government.

REFERENCES

1. O.V. Kononenko, V.N. Matveev, A.Yu. Kasumov, N.A. Kislov, and I.I. Khodos, Vacuum **46,** 685 (1995).
2. L.K. Fionova, O.V. Kononenko, and V.N. Matveev, *Scripta* Metallurgica et Materialia **27,** 329 (1992).

3. L.K. Fionova, O.V. Kononenko, and V.N. Matveev, Thin Solid Films **227** (1993) 54.
4. O.V.Kononenko, E.D.Ivanov, V.N.Matveev, and I.I.Khodos, *Scripta* Metallurgica et Materialia **33**, 1981 (1995).
5. O.V. Kononenko, and V.N. Matveev in <u>Thin films: Stresses and Mechanical Properties V</u>, edited by P.S. Baker, P. Borgesen, P.H. Townsend, C.A. Ross, C.A. Volkert (Mat. Res. Soc. Proc. **356**, Pittsburgh, PA, 1995)
6. E. Arzt, O. Kraft, J. Sanchez, S. Bader, and W.D. Nix, Mater.Res.Soc.Symp.Proc. **239**, 677 (1991).
7. T.C. Nason, L. You, G.-R. Yang, and T.-M. Lu, J. Appl. Phys., **69**, 773 (1991).
8. A.S. Yapsir and T.-M. Lu, Appl. Phys. Lett., **52**, 1962 (1988).
9. G.-R. Yang, P. Bai, T.-M. Lu, and W.M. Lau, J. Appl. Phys., **66**, 4519 (1989).
10. A. Barna, P.B. Barna, G. Radnoczi, F.M. Reicha, and L.Toth, Phys. stat. sol. (a), **55**, 427 (1979).

This article also appears in Volume 420.

ANALYSIS OF STRESSES AND STRAINS IN PASSIVATED METAL LINES

I. Eppler *, H. Schroeder **, U. Burges **, W. Schilling **
*Daimler Benz AG, Frankfurt am Main, Germany
**Institut für Festkörperforschung, Forschungszentrum Jülich GmbH, 52425 Jülich, Germany

ABSTRACT

Passivated metal lines, commonly used in integrated circuits, show thermally induced stresses due to the difference of the thermal expansion coefficients of the lines and their surroundings. These stresses cause voidage and plastic flow of the lines. Aim of the analysis was to derive equations connecting experimentally measured strains or stresses by the X-ray diffraction and wafer curvature techniques with the magnitude of voidage and plastic shear deformation of the lines.

Using the concepts of linear elasticity the volume averaged stresses of an array of parallel interconnects embedded in a passivation layer on a flat substrate are analysed. Equations are derived connecting the volume averaged stresses in the metal and in the passivation with the "eigen-strains" of the metal which characterize the true (stress free) thermal strains and plastic deformation strains of the metal. The coefficients entering these equations are determined from (elastic) finite element method (FEM) calculations performed for various geometries and aspect ratios of the metal lines. Choosing the proper values of the coefficients allows the eigen-strains to be determined from the experimental data.

By comparison of the evaluated eigen-strains with the purely elastic eigen-strains $\Delta\alpha\Delta T$ the extent of voidage and/or plastic shear deformation of passivated metal lines caused by thermally induced stresses can be determined model independently.

1. INTRODUCTION

With continuing miniaturization of microelectronic circuits the stresses in metallization systems become of increasing technological concern. E.g. large tensile stresses are created in Al-interconnects during cooling down from elevated processing temperatures due to the thermal mismatch between Al and the underlying Si-substrate and the surrounding ceramic passivation.

There are two general techniques to measure stresses/strains in thin films on substrates: the wafer curvature and the X-ray diffraction techniques. A general goal of these stress and strain measurements is to find out the extent of voidage and precipitation and plastic shear deformation. But only in the case of an infinitely rigid passivation one can determine the amount of plastic deformation by comparison of the measured elastic metal strains with the strains $\Delta\alpha\Delta T$, induced by the difference of the thermal expansion coefficients of the metal lines (M) and their surroundings (S) ($\Delta\alpha = \alpha^M - \alpha^S$; $\Delta T = T - T_0$; T_0: temperature, where the metal is stress free, normally close to the deposition temperature of the passivation). In real situations the elastic behaviour of the metal lines depends not only on their elastic constants and their aspect ratio but also on the elastic constants, the thickness and the geometry of the passivation due to the interaction of metallization and passivation. Therefore quantitative estimates of the plastic deformation from measured metal stresses and stress-changes of embedded metal lines are rather difficult.

Because of the coupling of the metal lines and the passivation a formalism is developed based upon the principle of eigen-strains [1] to derive general equations connecting metal strains and stresses with the extent of the plastic behaviour of the lines and using FEM-calculations to allow for different sample geometries. Due to space limitations in this paper only the most important features of this methode can be outlined. For more details the reader is referred to refs. 2, 3.

Mat. Res. Soc. Symp. Proc. Vol. 436 ©1997 Materials Research Society

2. INTRODUCTION OF EIGENSTRAINS

For our analysis we consider an assembly of many parallel metal lines deposited with re-peat distance d on a flat silicon substrate (Fig. 1). The lines have a rectangular cross section with width w^M, thickness t^M and a length which is very large compared to w^M and t^M. For the passivation we consider, as limiting cases, a planarized (Fig. 1a) and a conformal one (Fig. 1b). The thickness of the passivation over the metal is denoted by t^P.

In the following we consider only volume averaged stresses and strains for the metal and the passivation because these are the only values which can be determined by the present experimental techniques. As a coordinate system we use the directions along the line (lower index 1), perpendicular to the line but parallel to the substrate surface (lower index 2) and per-pendicular to the substrate (lower index 3) which are the principle axes of the volume averaged stresses and strains in the metal lines [2].

It is further sufficient - to a very good approximation - to consider the stresses and strains in the metal and the passivation (upper index M and P) only. If the substrate is very flat and also very thick compared to the film-thickness the stresses in the substrate along the 1 and 2 di-rection are negligible small. Because of the free surface the volume integral of the stresses in the 3 direction in the substrate plus the metal lines and the passivation has to be exactly zero. FEM-calculations show that this volume integral in the substrate is nearly zero so that the stresses in the 3-direction of the metal lines and the passivation are equal but of opposite sign.

Table I summarizes the definitions of the different combinations of stress- and strain-components used in the following analytical treatment and their relations for an elastically isotropic material.

2.1 Definition of eigen-strains

To recall and to visualize the definition of the eigen-strains, ε^E_i, we consider in Fig. 2a a si-tuation in which the metal after cooling down is thought to be debonded from its surrounding passivaton and from the substrate. If in the bonded situation the metal was under tensile stress the line would shrink and the passivation would retreat. As shown in Fig. 2a a gap opens then between the - now stress free - metal line and a corresponding channel in the passivation.

In contrast to the metal the passivation has not to be stress free in this situation. It may still contain intrinsic (deposition) stresses and stresses from thermal mismatch to the substrate. However, those elastic passivation strains induced by the force balance of metal and passiva-

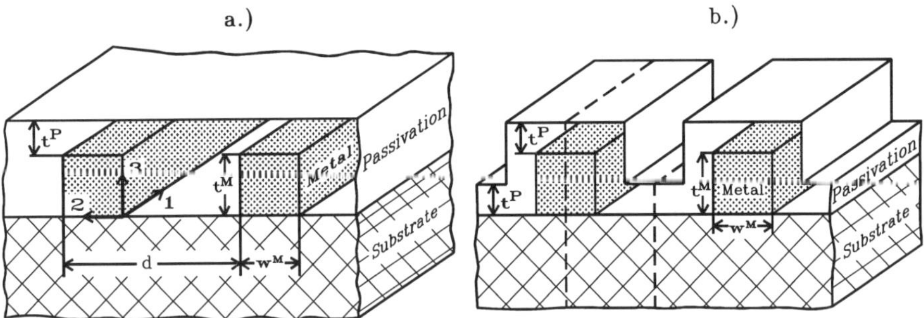

Figure 1: *Schematic picture of parallel metallic interconnects on a substrate covered by (a.) a planarized and (b.) a conformal passivation layer. The sample dimensions and the coordinate system used for the definition of the principle stresses and strains are indicated in Fig. 1a. In Fig 1b. the dashed lines indicate planes of symmetry used in the FEM calculations.*

$\varepsilon_V = \varepsilon_1 + \varepsilon_2 + \varepsilon_3$	$\sigma_V = 1/3^*(\sigma_1 + \sigma_2 + \sigma_3)$	$= K\,\varepsilon_V$
$\varepsilon_\perp = 0.5(\varepsilon_2 + \varepsilon_3)$	$\sigma_\perp = 0.5(\sigma_2 + \sigma_3)$	$= 2\mu/(1\text{-}2\mu)^*(\nu\,\varepsilon_1 + \varepsilon_\perp)$
$\varepsilon_\Delta = 0.5(\varepsilon_2 - \varepsilon_3)$	$\sigma_\Delta = 0.5(\sigma_2 - \sigma_3)$	$= 2\mu\,\varepsilon_\Delta$
$\varepsilon_S = \varepsilon_1 - \varepsilon_\perp$	$\sigma_S = \sigma_1 - \sigma_\perp$	$= 2\mu\,\varepsilon_S$

Upper Index:

M = Metal
P = Passivation
FS = Free film surface
F = Film (Metal plus Passivation)
S = Substrate

$\mu = E/2(1 + \nu) =$ shear modulus
$K = E/3(1 - 2\nu) =$ bulk modulus
$\nu =$ Poisson number

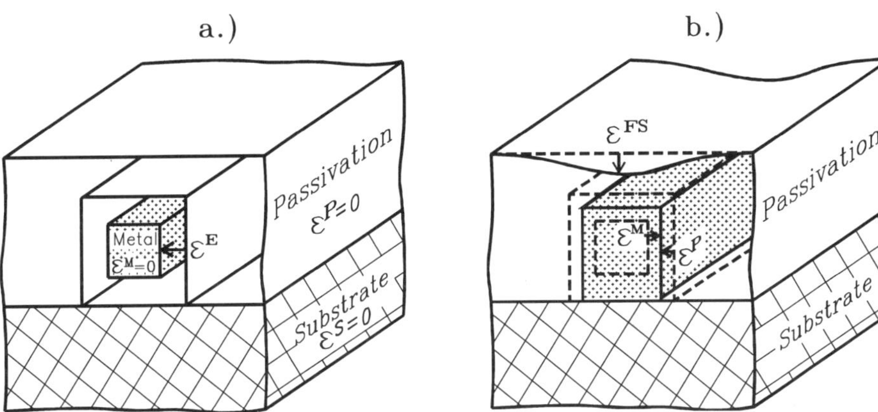

Figure 2: *Definition of the eigen-strains of an interconnect which has come under tensile stress during cooling down. a) Metal debonded from the passivation. The elastic strains ε^M in the metal and ε^P in the passivation are zero (aside of residual intrinsic strains not counted here). The eigenstrains ε^E measure the dimensions of the gap relative to the dimensions of the inclusion. b) Situation with the interconnect bonded to the passivation. During cooling down the metal and the passivation have come under tensile stress leading to elastic strains ε^M and $\varepsilon^P > 0$. In addition the whole film consisting of metal plus passivation, due to ist free surface, has undergone a shape change in the 3-direction leading to a strain component ε_3^{FS} with respect to the situation in Fig. 2a.*

tion stresses at the interface in the bonded situation (Fig. 2b) are relieved. In the following weconsider only these contributions to the total elastic strains/stresses of the passivation and call them ε^P_i and σ^P_i. By this definition ε^P_i and σ^P_i are zero in the debonded situation.

As shown in Fig. 2 and 3 the eigen-strains ε^E_i are characterized by the magnitudes of the gap in different directions in the debonded situation compared to the stress free dimensions of the metal lines. E.g. the volumetric eigen-strain ε^E_V (see definition in Table I) characterizes the

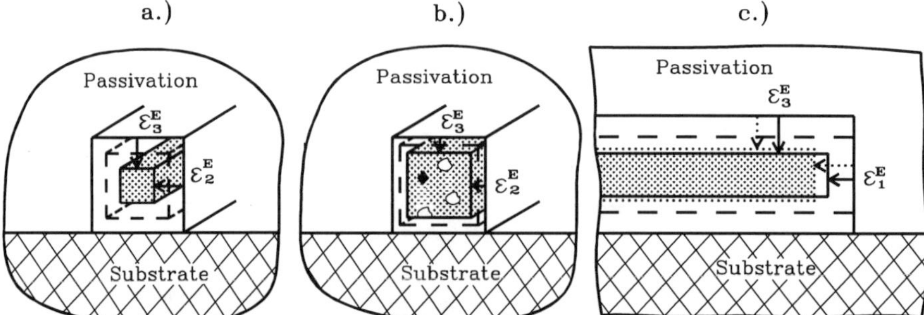

Figure 3: *Visualization of the eigen-strains of an metallic interconnect which was under tensile stress after cooling down from the passivation deposition temperature. Note: The length of the arrows shown in Fig. 2 and 3 do not represent the true displacements between metal and passivation but the displacements per unit length in this direction. a) Pure thermal contraction ($\alpha^S = \alpha^P$ assumed): $\varepsilon^E_1 = \varepsilon^E_2 = \varepsilon^E_3 = \Delta\alpha(T-T_0)$ => $\varepsilon^E_v = 3\Delta\alpha(T-T_0)$ and $\varepsilon^E_s = 0$. b) Metal in which voidage and precipitation had occurred during cooling down: => $\varepsilon^E_1 = \varepsilon^E_2 = \varepsilon^E_3 = \Delta\alpha(T-T_0) + (\Delta V/V)_{void} + (\Delta V/V)_{ppts}$ and $\varepsilon^E_s = 0$. c) Interconnect which has plastically yielded during cooling down in the 1-direction: $\varepsilon^E_v = 3\Delta\alpha(T-T_0)$ and $\varepsilon^E_s > 0$. For comparison the dotted metal line represents the situation of pure thermal contraction and no plastic elongation in 1-direction.*

relative misfit volume between the metal and the surrounding. By common definition the ε^E_i are measured from the inner surface of the passivation channel as starting point.

2.2 Examples of the relationship between plasticity and eigen-strains

Figs. 3a, b and c consider specific examples: At temperature T_0 the metal is stress free and fits well into the channel, therefore $\varepsilon^E_i = 0$. At a lower temperature T, without voidage or plastic shear (see Fig. 3a), the eigen-strains are then equal to the thermal strains: $\varepsilon^E_i = \Delta\alpha_i(T-T_0)$; $\Delta\alpha_i$ is the relative dimensional change per unit temperature in direction i between the metal line and the hollowed passivation channel. E.g. for the planar passivation holds: $\Delta\alpha_1 = (\alpha^M - \alpha^S) \approx \Delta\alpha_2$ and $\Delta\alpha_3 \approx (\alpha^M - \alpha^P)$. For simplicity we assume : $\alpha^S = \alpha^P$. If during cooling down voidage and /or precipitation has occurred in the metal, ε^E_v would be smaller in absolute magnitude than the volume change $3\Delta\alpha(T-T_0)$ given by the thermal strains (see Fig. 3b). An additional (eigen-volume conserving) plastic shear of the metal line would show up as non-zero deviatoric (shear) components ε^E_s and/or ε^E_Δ of the eigen-strains (see definition in Table I). E.g. for an uniform plastic elongation of the line in the 1-direction ε^E_s would be positive (see Fig. 3c). If on the other hand the line would plastically yield in 2-direction and contract in 3-direction ε^E_Δ would be positive. Thus, the eigen-strains calculated from experimental data directly give the extent of plastic shear deformation and the extent of voidage and/or precipitation.

3. GENERAL EQUATIONS

A general equation connecting the eigen-strains (E) to the metal (M)-, passivation (P)-, and "free surface" (SF)-strains can be obtained by comparing figs. 2a and 2b: As for two bodies of same size and shape the volume integrals over all strains (elastic, plastic, thermal) have to be equal we get for the volume averaged strains in each direction i (i=1,2,3):

$$V^M \varepsilon_i^E + V^M \varepsilon_i^M + V^P \varepsilon_i^P + (V^M + V^P)\varepsilon_i^{FS} = 0 \qquad (1a)$$

or (dividing by V^M)
$$\varepsilon_i^E + \varepsilon_i^M + \tilde{\varepsilon}_i^P + \tilde{\varepsilon}_i^{FS} = 0 \qquad (1b)$$

V^M and V^P are the total volumes of metal and passivation, respectively; ε_i^M and ε_i^P are the corresponding elastic strain components; ε_i^E are the eigen-strains, while ε_i^{FS} accounts for a possible shape change of the outer surface of the film (see fig. 2b). Of course, eq. 1 also holds for all kind of combinations of the strain components, e.g. those listed in table I.
For further evaluation of eq. 1 we make use of several boundary conditions. (Note: They are only listed here; for details see refs. 2, 3)

i) In the 1-direction the film is rigidly constrained by the substrate. For an elastically behaving passivation then holds: $\qquad \varepsilon_1^P = 0; \qquad \varepsilon_1^{FS} = 0; \qquad \varepsilon_1^M = -\varepsilon_1^E ; \qquad (2)$

ii) Due to the force balance at the interface metal/passivation no stresses and strains can be induced in the passivation if $\quad \sigma_2^M = \sigma_3^M = 0$; this implies eigen-strains of the form $\varepsilon_2^E = \varepsilon_3^E = -v^M\varepsilon_1^E$; or in other words: $\varepsilon_{2,3}^P$ and $\varepsilon_{2,3}^{FS}$ are independent on σ_1^M (or $\sigma_1^E = C_{1j}^M \varepsilon_j^E$).

iii) The film surface is stress free. Hence, $\quad V^M \sigma_3^M + V^P \sigma_3^P = 0; \qquad (3)$

iv) Planar passivation (see fig. 1a) implies additionally: $\quad \varepsilon_2^{FS} = 0.$

For all the involved strains we make a linear "Ansatz":

$$\text{a) } \varepsilon_i^M = M_{ij}\,\varepsilon_j^E ; \qquad \text{b) } \varepsilon_i^P = P_{ij}\,\varepsilon_j^E ; \qquad \text{c) } \varepsilon_i^{FS} = S_{ij}\,\varepsilon_j^E ; \qquad (4a,b,c)$$

From eq. (1) we obtain: $\qquad \delta_{ij} + M_{ij} + P_{ij} + S_{ij} = 0; \qquad (5)$

Using eq. (5) and all the listed boundary conditions, the 27 coefficients of the second rank tensors M_{ij}, P_{ij} and S_{ij} are reduced to 4 and 6 independent coefficients for planar and conformal passivation, respectively [3].

It should be noted that for the evaluation of the eigenstrains from X-ray data (ε_i^M) one needs only eq. 4a. In this case the tensor M_{ij} has only 4 independent coefficients because of boundary conditions i) and ii). These boundary conditions are reflected best in a special transformation of eq. 4a:

$$-\begin{pmatrix} \varepsilon_1^M \\ \sigma_\perp^M \\ \sigma_\Delta^M \end{pmatrix} = \begin{pmatrix} 1 & 0 & 0 \\ 0 & M_{\perp\perp} & M_{\perp\Delta}/(1-2v^M) \\ 0 & M_{\Delta\perp} & M_{\Delta\Delta}/(1-2v^M) \end{pmatrix} \bullet \begin{pmatrix} \varepsilon_1^E \\ \sigma_\perp^E \\ \sigma_\Delta^E \end{pmatrix} \quad \text{and} \quad \sigma_i^E = C_{ij}^M \varepsilon_j^E; \ (i=1,2,3) \qquad (6)$$

The stress components $\sigma_\perp$ and σ_Δ are defined in table I.

4. FINITE ELEMENT METHODE (FEM) CALCULATIONS

In order to determine the coefficients entering eq. (4) FEM calculations were carried out using the commercial code "Permas" [4]. Plane strain conditions were used for the 1-direction and for the symmetry planes perpendicular to the 2-direction as shown in fig. 1b simulating an infinite array of lines on a substrate. We also assured that the boundary conditions listed above were well approximated by the FEM data. In the FEM calculations the ε_i^M, ε_i^P and ε_i^{FS} were determined starting from different sets of eigenstrains, e. g. $\varepsilon_i^E = (1, 1, 1)$ and $\varepsilon_i^E = (0, -1, 1)$. Applying eqs. (4a, b, c) numerical values of the coefficents M_{ij}, P_{ij}, S_{ij} could then be derived.

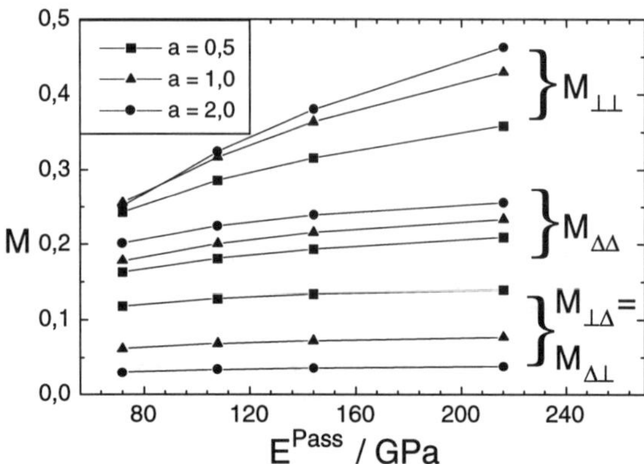

Figure 4: *Coefficients for an infinite array of parallel metal lines with distance $d = 2w^M$, and thickness $t^P = 0.6t^M$ as a function of the passivations Youngs Modulus and the aspect ratio t^M/w^M of the metal lines.*

The geometries chosen were those sketched in figs. 1a and b with different values of aspect ratio $a = t^M/w^M$, passivation thickness t^P, repetition distance d of neighbouring lines and elastic constants of the passivation which are rather uncertain.

Figure 4 shows as an example the coefficients defined by eq. (6) and calculated for the planar passivation situation using $E^M = 72$ GPa and $v^M = 0{,}343$ for the aluminum lines. Our FEM-calculations showed that the coefficients $M_{\perp\Delta}$ and $M_{\Delta\perp}$ are exactly equal for all situations explored. For the planar passivation the M_{ij} coefficients are also found to depend only very little on t^P, down to $t^P/t^M = 0{,}3$.

A compilation of all the coefficients M_{ij} for the conversion of the ε_i^M into ε_i^E and of analogous coefficients for the calculation of ε_1^E and ε_2^E from wafer curvature stresses will be published elsewhere [3]. An example for the application of the described evaluation method to experimental stress data from X-ray diffraction is given in [5].

REFERENCES

[1]   H.Reißner, Zeitschrift für Angewandte Mathematik und Mechanik, Band 11 (1931) 1 S.1

[2]   U. Burges, I. Eppler, W. Schilling, H. Schroeder and H. Trinkaus, Third International Workshop in Stress- Induced Phenomena In Metallization, AIP Proceedings

[3]   I. Eppler et al., to be published

[4]   E. Schrem, PERMAS Handbook for Linear Static Analysis, Intec Publication UM 404, REV B, Stuttgart 1988

[5]   D. Beckers, H. Schroeder, I. Eppler, W. Schilling, this volume

This **article** also appears in Volume 420.

PROCESSING-INDUCED STRESSES AND CURVATURE IN PATTERNED LINES ON SILICON WAFERS

Y.-L. SHEN*, S. SURESH* AND I. A. BLECH**
*Department of Materials Science and Engineering and **Materials Processing Center, Massachusetts Institute of Technology, Cambridge, MA 02139, U.S.A.

ABSTRACT

The evolution of stresses due to the patterning and thermal loading of thin lines on Si wafers, and the consequent changes in the overall curvature of the wafer are studied theoretically and experimentally. The analysis involves finite element simulations within the context of generalized plane strain models. The analysis is capable of predicting the wafer curvature in directions parallel and perpendicular to the lines. These predictions compare reasonably well with experimental measurements of curvature made on model systems. The thickness, width and spacing of the patterned lines have been varied systematically, and the associated changes in the evolution of stresses and curvature have been determined. The non-uniform stress field within the fine lines is also analyzed.

INTRODUCTION

In an interconnect structure composed of a series of parallel thin lines on a thick substrate, the evolution of stresses as a result of deposition, patterning and subsequent thermal loading induces different values of wafer curvature along the directions parallel and perpendicular to the lines. In recent years the finite element method has found increasing application in the modeling of stresses and deformation in interconnect structures [1-6]. These studies dealt with deformation induced by the thermal expansion/contraction mismatch of the interconnect systems, and they did not consider the role of line patterning in causing stress changes. Furthermore, these prior simulations invoked idealized two-dimensional formulations (such as the plane strain condition, where no deformation is allowed in the line direction). A more serious drawback of such idealizations is that the commonly used curvature measurements performed on wafers cannot be used directly to evaluate the accuracy of the predictions.

The objective of this work was to carry out a combined numerical and experimental investigation which circumvents the aforementioned drawbacks. The model system used in the experiments is a (111) Si wafer with the thermally grown SiO_2 film (or lines etched from the SiO_2 blanket film). This system is ideal for the fundamental study of processing-induced deformation because the thermo-mechanical response of each layer can be accurately described by linear elasticity. The numerical model is based on the generalized plane strain formulation which accounts for the deformation in the line direction, so that curvatures in directions both perpendicular and parallel to the lines can be directly predicted. In addition to simulating the deformation induced by thermal mismatch between dissimilar materials, the present model is capable of predicting the curvatures and stresses due to the etching of a film into parallel lines.

NUMERICAL MODEL

A computational model employing the finite element method was used. Figure 1(a) shows schematically the SiO_2 lines on the Si wafer and the nomenclature of the geometry and coordinate axes; the symbols t (line or film thickness), w (line width), p (pitch) and h

(substrate thickness) describe the structure throughout this paper. Due to the periodicity and symmetry of the arrangement, only a unit segment is needed for calculations. Figure 1(b) shows a representative top portion of such a segment. The full substrate thickness is accounted for in the model. The regions $z > 0$ and $z < 0$ correspond to the SiO_2 film and Si substrate, respectively. When simulating the etching process the material elements within $w/2 < x < p/2$ and $0 < z < t$ are removed while the static equilibrium condition is enforced. When simulating pure thermal loading of the patterned line structure, this part of SiO_2 is excluded from the model. A generalized plane strain formulation, which is an extension of the plane strain framework (with the xz-plane being the plane of deformation) is used in the calculations. This is done by superimposing a longitudinal strain, ϵ_{yy}, on the plane strain state without any change in the stresses in the xz-plane. To properly simulate the actual response of the patterned parallel lines on the substrate, the strain field ϵ_{yy} is constrained to induce a constant rotation about the x axis. A full description of the model can be found in [7].

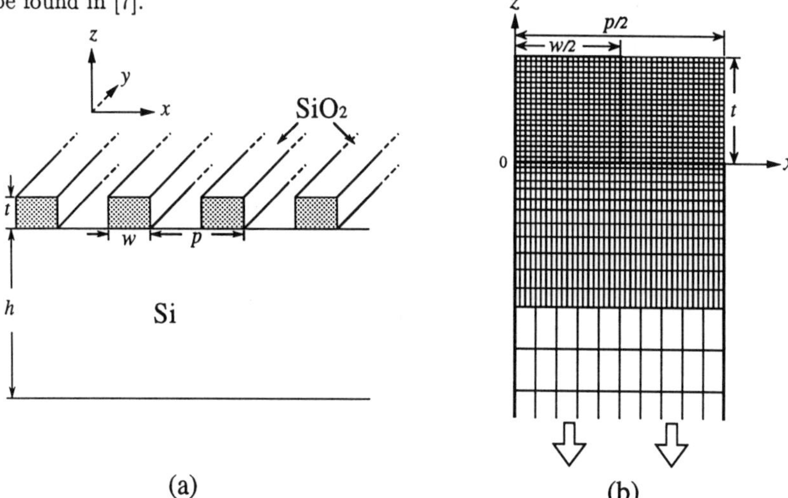

Fig. 1 (a) A schematic of the patterned SiO_2 lines on Si substrate. (b) A representative top portion of the unit segment and finite element discretization used for the calculations.

RESULTS

Since the present numerical model is capable of yielding more realistic field quantities than the strict plane strain formulation, an examination of the non-uniform field of various stress components can be made. An example is shown in Fig. 2, where the predicted contours of constant stresses, (a) σ_{xx} and (b) σ_{yy}, in the lines and the region of the substrate near the lines are plotted, for one of the specimens (with $t = 0.710$ μm, $w = 1.323$ μm, $p = 3.024$ μm and $h = 525$ μm) used in experiments. The stress values are normalized by the equibiaxial stress in the blanket film, σ_{film}. It should be noted that the normalized stresses obtained from the calculations for etching and for thermal loading are identical because linear elasticity produces a thermo-mechanical response which is independent of the loading path [7]. For etching, the contours represent the appropriate stress components normalized by the equibiaxial stress in the film ($\sigma_{xx} = \sigma_{yy} = \sigma_{\text{film}}$) before etching. For thermal loading, the contour values represent the ratio of the stress of the line–substrate structure to that of the film in the film–substrate structure. Figure 2(a) reveals that the stresses σ_{xx} in lines are greatly relieved, except in an extremely narrow area adjacent to the substrate. This is due to the existence of the vertical side walls of the lines; on the wall surface, the traction-free boundary condition has to be satisfied. Along the direction parallel to the lines, the extent of stress relaxation in the lines is much smaller, as shown in Fig. 2(b). Here the values of σ_{yy} are below 90% of σ_{film} within the majority of the line area, but still remain above 80% throughout. Very close to the corner adjacent to the substrate, $\sigma_{yy}/\sigma_{\text{film}} > 1$. It is also noticed that the substrate regions near the lines can be stressed to a significant extent.

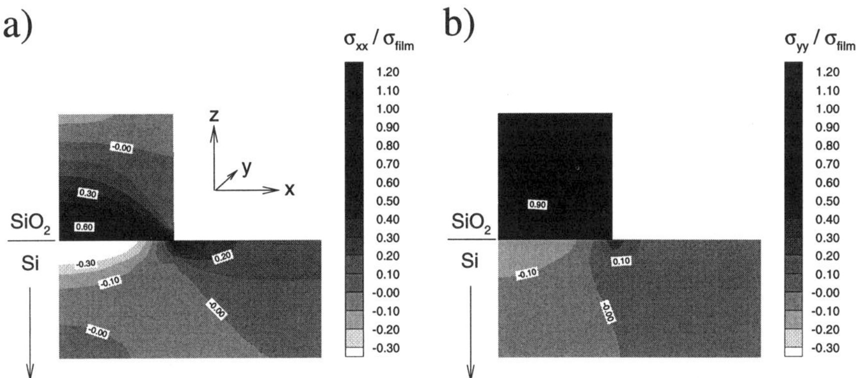

Fig. 2 Predicted contours of the constant stresses, (a) σ_{xx} and (b) σ_{yy}, within the lines and the top portion of the substrate for a specimen. All stress values are normalized by the film stress, σ_{film}, in the case of a blanket SiO$_2$ film on a Si substrate.

A parametric analysis of the evolution of stresses was performed. Figure 3 shows the predicted normalized average line stresses as a function of t/w, for various values of p/w. The normalized stresses are obtained by dividing the volume-averaged normal stresses in lines, $< \sigma_{xx,\text{line}} >$ and $< \sigma_{yy,\text{line}} >$, by the corresponding biaxial stress in the blanket film, σ_{film}. It is seen that $< \sigma_{xx,\text{line}} > = < \sigma_{yy,\text{line}} > \rightarrow \sigma_{\text{film}}$ as $t/w \rightarrow 0$. This behavior arises because, for $t \ll w$, the lines become segments of a very thin film firmly attached to the substrate with the effects of the line edges negligible. As t/w increases, the value of $< \sigma_{xx,\text{line}} > /\sigma_{\text{film}}$ drops precipitously due to the strong influence of the vertical side walls of lines. It is expected that $< \sigma_{xx,\text{line}} > \rightarrow 0$ as $t/w \rightarrow \infty$. The average normal stress parallel to the lines, $< \sigma_{yy,\text{line}} >$, also shows a decrease with increasing t/w. The extent of such a decrease, however, is much less pronounced. In the limiting case of $t/w \rightarrow \infty$, $< \sigma_{yy,\text{line}} >$ can be estimated by considering

$$\sigma_{\text{film}} = \frac{E_{\text{SiO}_2}}{1 - \nu_{\text{SiO}_2}} \cdot \epsilon_{\text{film}} , \tag{1}$$

and

$$\left[< \sigma_{yy,\text{line}} > \right]_{\frac{t}{w} \to \infty} = E_{\text{SiO}_2} \cdot \epsilon_{yy,\text{line}} \, , \tag{2}$$

where ϵ_{film} and $\epsilon_{yy,\text{line}}$ are the strain in the film and the strain in the lines in the direction parallel to the lines, respectively, upon etching or thermal loading; E and ν stand for Young's modulus and Poisson's ratio, respectively. Combining Eqs. (1) and (2) and taking into account that $\epsilon_{\text{film}} = \epsilon_{yy,\text{line}}$, one obtains

$$\left[\frac{< \sigma_{yy,\text{line}} >}{\sigma_{\text{film}}} \right]_{\frac{t}{w} \to \infty} = 1 - \nu_{\text{SiO}_2} = 0.84 \, . \tag{3}$$

This is in agreement with the trend seen in Fig. 3. An important feature observed in Fig. 3 is that the pitch-to-width ratio p/w does not have a strong influence on the average normal stresses in the patterned lines along both directions.

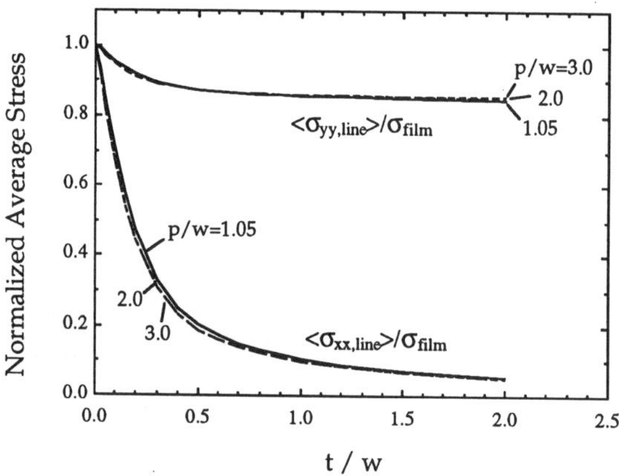

Fig. 3 Predicted average stresses in lines, $< \sigma_{xx,\text{line}} >$ and $< \sigma_{yy,\text{line}} >$, normalized by the corresponding film stress, σ_{film}, as a function of t/w for various ratios of p/w.

The significance of the evolution of multiple curvatures can be illustrated by plotting the variation of $\kappa_{x,\text{line}}/\kappa_{y,\text{line}}$ (the ratio of the curvature values in the x and y directions) as a function of t/w for different values of p/w, as shown in Fig. 4. The curves are essentially independent of the pitch-to-width ratio. Results from direct curvature measurements using the laser scanning technique on specimens of various t/w for etching and thermal loading are also indicated in the figure. It can be seen that the experimental results agree with the numerical predictions well. It is noticed that a reversal of curvature exists at $t/w \approx 0.4$ beyond which negative values of $\kappa_{x,\text{line}}/\kappa_{y,\text{line}}$ are observed. This is due to the Poisson effect arising from the anisotropic strain coupling in the patterned structure [7]. In the limiting case of $t/w \to \infty$, $\sigma_{xx,\text{line}} \to 0$. Then, by considering the problem of plate bending under the moment M_x [8],

$$M_x = \sigma_{xx,\text{line}} \cdot t(h/2) = \frac{E_{\text{Si}} h^3}{12(1 - \nu_{\text{Si}}^2)} \cdot (\kappa_{x,\text{line}} + \nu_{\text{Si}} \kappa_{y,\text{line}}) \, , \tag{4}$$

one obtains

$$\frac{\kappa_{x,\text{line}}}{\kappa_{y,\text{line}}} \to -\nu_{\text{Si}} \quad \text{as} \quad \frac{t}{w} \to \infty \; . \tag{5}$$

The line representing $-\nu_{\text{Si}} = -0.262$ is also drawn in Fig. 4, and the predicted curves of $\kappa_{x,\text{line}}/\kappa_{y,\text{line}}$ are seen to approach this line as t/w increases.

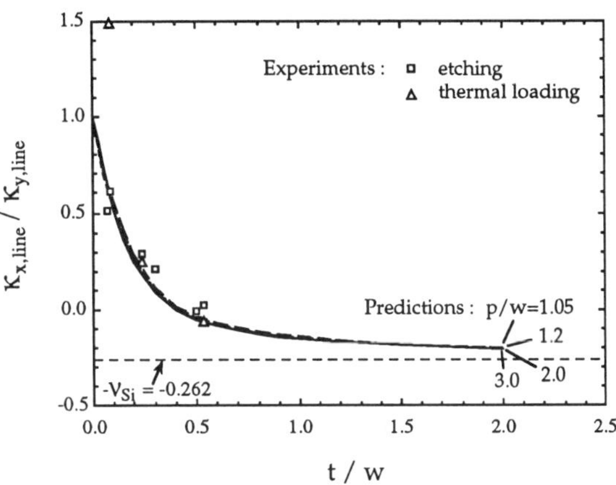

Fig. 4 Predicted $\kappa_{x,\text{line}}/\kappa_{y,\text{line}}$ as a function of t/w, for the cases of p/w=1.05, 1.2, 2.0 and 3.0. The experimental results are also included.

Figures 5(a) and 5(b) show the direct comparisons between the predicted and measured curvature changes in the x and y directions, respectively, as functions of temperature during thermal loading for three specimens with different geometries. Due to the relatively small range of total curvature change, a large scatter in the experimental data is seen in all cases. It is clear, however, that the salient trends observed experimentally are captured by the numerical predictions.

CONCLUSIONS

A finite element model was developed for simulating the evolution of stresses and curvature due to the patterning of parallel thin lines of an elastic material on a thick substrate. The model, based upon a generalized plane strain formulation, is capable of predicting the wafer curvature in directions parallel and perpendicular to the lines. Systematic experimental measurements of curvatures were performed on the model system of thermally grown SiO_2 films on the Si wafers. The results for both the etching of the film into lines and the thermal loading of the etched structure compare reasonably well with numerical predictions. Extensions of the present analyses to patterned metallic lines on Si wafers are currently being persued by the authors.

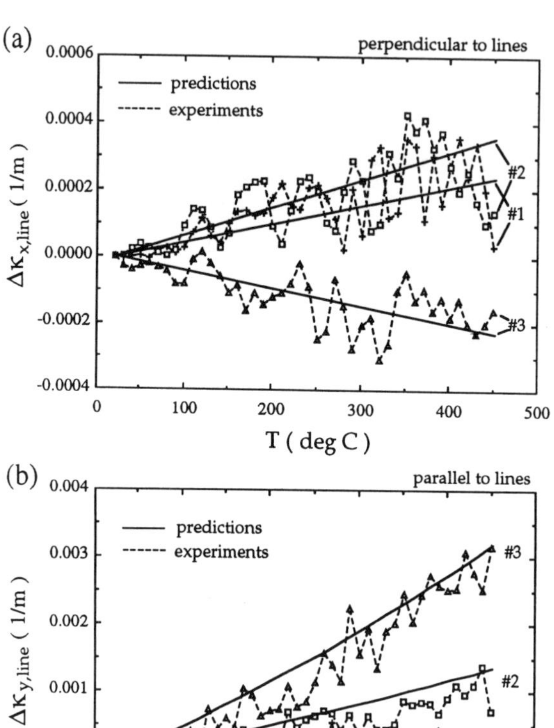

Fig. 5 Comparisons of the experimental measurements and numerical predictions of the curvature changes in (a) the x direction and (b) the y direction, for three different patterned structures (#1: $t = 0.102$ μm, $w = 1.243$ μm, $p = 2.878$ μm and $h = 525$ μm; #2: $t = 0.302$ μm, $w = 1.274$ μm, $p = 2.924$ μm and $h = 515$ μm; #3: $t = 0.710$ μm, $w = 1.323$ μm, $p = 3.024$ μm and $h = 525$ μm). The specimen numbers are indicated.

REFERENCES

1. R. E. Jones and M. L. Basehore, *Appl. Phys. Lett.* **50**, p. 725 (1987).
2. B. Greenebaum, A. I. Sauter, P. A. Flinn and W. D. Nix, *Appl. Phys. Lett.* **58**, p. 1845 (1991).
3. A. I. Sauter and W. D. Nix, *IEEE Trans. Comp. Hybrids Manufact. Technol.* **15**, p. 594 (1992).
4. Chidambarrao, K. P. Rodbell, M. D. Thouless and P. W. DeHaven, in *Materials Reliability in Microelectronics IV*, edited by P. Borgesen, J. C. Coburn, J. E. Sanchez, Jr., K. P. Rodbell and W. F. Filter (Mater. Res. Soc. Symp. Proc. 338, Pittsburgh, PA, 1994), p. 261.
5. A. S. Mack and P. Flinn, in *Thin Films: Stresses and Mechanical Properties V*, edited by S. P. Baker, C. A. Ross, P. H. Townsend, C. A. Volkert and P. Borgesen (Mater. Res. Soc. Symp. Proc. 356, Pittsburgh, PA, 1995), p. 465.
6. P. R. Besser, T. M. Marieb, J. Lee, P. A. Flinn and J. C. Bravman, *J. Mater. Res.* **11**, p. 184 (1996).
7. Y.-L. Shen, S. Suresh and I. A. Blech, submitted to *J. Appl. Phys.*, 1996.
8. S. Timoshenko, *Strength of Materials*, 3rd edition, Robert E. Krieger Publishing Company, Huntington, New York, 1976, p. 88.

ISOTHERMAL STRESS RELAXATION IN Al, AlCu AND AlVPd FILMS

J.P. Lokker, J.F. Jongste, G.C.A.M. Janssen, S. Radelaar

DIMES, Delft University of Technology, P.O. Box 5046, 2600 GA Delft, The Netherlands
Lokker@Dimes.tudelft.nl

Abstract
Mechanical stress and its relaxation in aluminum metallization in integrated circuits (IC) are a major concern for the reliability of the material. It is known that adding Cu improves the reliability but complicates plasma etching and increases corrosion sensitivity. The mechanical behavior of AlVPd, AlCu and Al blanket films is investigated by wafer curvature measurements. During thermal cycling between 50°C and 400°C the highest tensile stress is found in AlVPd. In a subsequent experiment, the cooling was interrupted at several temperatures to investigate the stress behavior during an eight hour isothermal treatment. Isothermal stress relaxation has been observed in the three types of films and is discussed.

Introduction

The two intrinsic reliability concerns for aluminium line metallization in integrated circuits (IC) technology are electromigration and stress migration. Extensive research all over the world is performed on these subjects separately and in combination. The aim of this effort is to obtain a physical correlation between the resistance against electromigration and stress migration.

In aluminum IC-metallization mechanical stress is induced during production as a result of the difference in thermal expansion coefficients of the metallization and the underlying silicon substrate. Several mechanisms may be operative in the material to relax the stress and could lead to degradation of the metallization. For instance open and short failures are phenomena related to mechanical stress relaxation. Since dimensions in ICs are becoming smaller and smaller, stress induced failure plays a more important role in the degradation [1]. Therefore, it is necessary to understand the behavior of stresses during production when, especially lower level metallization experiences many thermal cycles. Furthermore, study of stress as a function of time after production is required for a better understanding of the relaxation behavior.

The stress state in the material is determined by the thermal history [2], the temperature and the micro structure of the material. The latter can be influenced by addition of alloying elements. Addition of copper is known to improve the reliability but hampers the etching process and increases the corrosion sensitivity. Addition of vanadium and palladium results in a promising alternative concerning the mechanical properties.

In this work isothermal stress relaxation in Al, AlCu and AlVPd films has been studied by *in situ* wafer curvature measurements. The experimental results for AlVPd are compared to those obtained for Al and AlCu films. The distribution of the alloying elements is characterised using Rutherford Back-scattering Spectroscopy . In literature, dislocation glide is often proposed as one of the atomic processes governing the stress relaxation [1,2,3]. The validity of the dislocation glide model to describe the observed stress relaxation is discussed.

Experimental

Aluminum alloy films are deposited (DC sputtering at 4 kW and 3 µbar Ar) on 100-mm-diam. two-sided polished silicon (100) wafers of 0.525 mm thickness, using a magnetron sputter apparatus (Balzers 450). The wafers are thermally oxidized prior to deposition (8 hours, 1000°C) to provide a 200 nm thick SiO_2 barrier layer on both sides of the substrate. The sputter apparatus has a base pressure better than 10^{-7} mbar which results in an average resistivity $\rho = 3.03$ µΩcm for the pure aluminum films. Targets of Al (99.9995), AlCu(1 at%) and AlV(0.1 at%)Pd(0.1 at%) are used to produce different batches Al (alloy) films. The thickness of the films is determined by weight measurements and was 500 ± 20 nm. The thickness variation ($\Delta t/t$) over a single wafer is 4%, determined by four point probe resistance measurements.

Mat. Res. Soc. Symp. Proc. Vol. 436 ® 1997 Materials Research Society

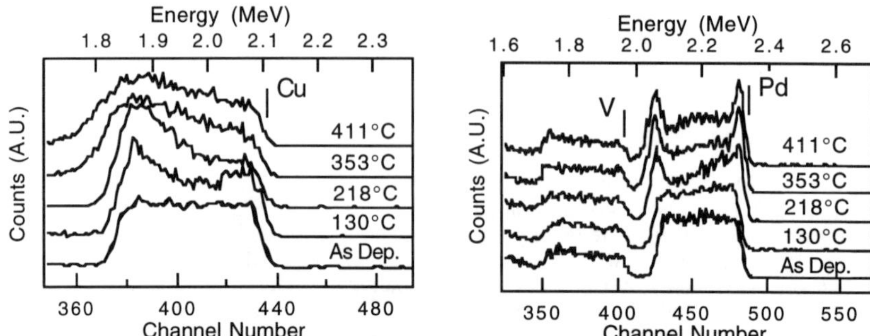

Figure 1: RBS data of AlCu (left) and AlVPd (right) after several thermal treatments. The angle of incidence was normal to the sample plane. The detection angle was 170°. The graph shows enlarged parts taken from the whole spectrum. For clarity, a arbitrary offset was given to the data. The vertical bars indicate the surface channels for Cu, V and Pd.

The processed wafers are heated to 400°C at a heating rate of 5°C/min and cooled down again to 50°C, in a 150-mm-diameter tube furnace. The temperature is measured using a Pt-PtRh(13%) thermocouple. A Balzers BPVA9510 turbo pump and a Membrane vacuum pump are used in series to obtain a base pressure of 10^{-5} mbar. The measurements are performed in nitrogen (1.7 mbar).

During successively applied thermal cycles the change in average stress in the films is determined, *in situ*, as a function of time and temperature, with an accuracy of 2 MPa. In order to do so, two laser beams are directed on the back side of the wafer. The reflected laser beams are projected on a screen at a distance of 6.67 m. Curvature of the wafer will result in a displacement of these projected spots. The spot displacement is continuously monitored and is converted in an average stress in the film using Stoney's equation [4]. In a subsequent experiment, the cooling was interrupted at several temperatures to investigate the stress behavior during an eight hour isothermal treatment.

To determine the depth profile of the alloying elements concentration, Rutherford Back-scattering Spectroscopy is used. Samples are annealed at five different temperatures for 30 minutes. Afterwards, RBS spectra using a 2.7 MeV He$^+$ beam are obtained. The results for AlCu(1 at%) and AlV(0.1 at%)Pd(0.1 at%) are shown in Figure 1. It can be observed that in as-deposited AlCu films the Cu concentration is homogeneous through the film thickness. However, after the different thermal treatments an enrichment is found at the SiO_2/Al interface. The results obtained for the AlVPd films show that the V concentration is homogeneous through the film thickness, even after the different thermal treatments, while for Pd an enrichment is found at both the surface and the SiO_2/Al film interface.

Results

Figure 2 shows the average stress for all Al (alloy) films investigated as a function of temperature during heating and cooling. These results are obtained during the second thermal cycle. It must be noted that these curves reproduce during further consecutive thermal cycles. During cooling of the film, the stress in the AlVPd is higher than in the AlCu and Al films in the temperature range between 370°C and 50°C. Although the stress in AlCu is close to that observed in Al in the temperature range between 400°C and 200°C it is found that cooling below 200°C results in a fast increase in the average stresses in the AlCu film, eventually ending at a tensile stress as high as in the AlVPd film.

Isothermal stress relaxation has been observed for all three Al (alloy) films. The results obtained from the AlVPd films are shown in Figure 3. A gradual change of the stress is observed during

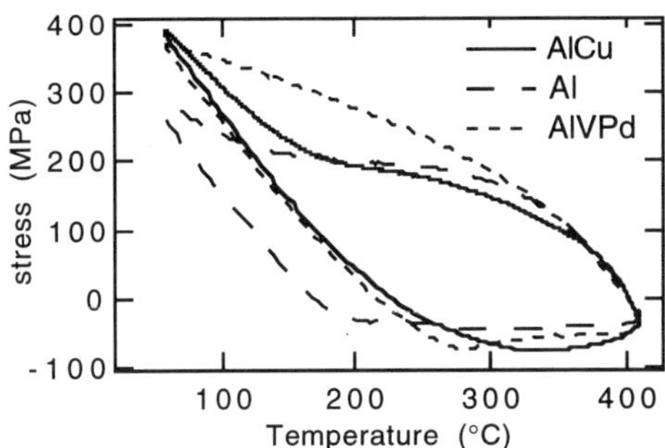

Figure 2: Stress in Al, AlCu and AlVPd films as a function of temperature.

each annealing in the whole temperature range between 59°C and 353°C. The absolute change in tensile stress first increases with increasing temperature up to a temperature of 216°C. During isothermal anneal at the higher temperatures the absolute change in tensile stress is lower than at the lower temperatures. Furthermore, it is most clearly observed that below 216°C the stress decrease does not level out after eight hours.

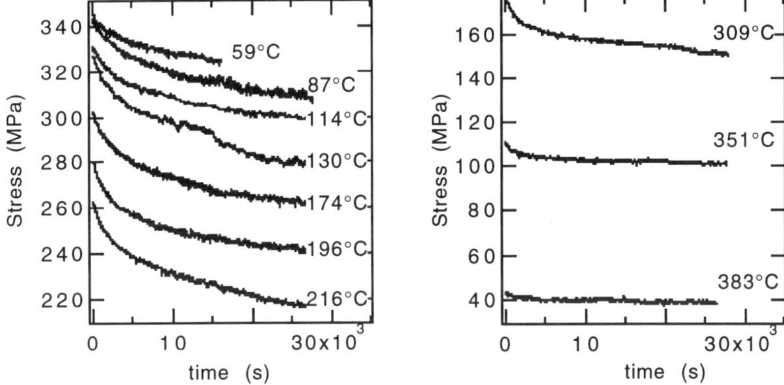

Figure 3: Isothermal stress relaxation in AlVPd films.

For comparison the same experiments have been performed on the AlCu and Al films. The results are shown in Figure 4 and 5. It is observed that for the AlCu films at 86°C and 91°C the stress decreases almost linearly with time. Then the relaxation at 115°C and 130°C shows a more pronounced concave shape. The absolute change in tensile stress after eight hours is larger at these temperatures compared with the lower temperatures. During annealing at higher temperatures (175°C to 386°C) the stress during the first 2000 seconds decreases rather fast after which the relaxation rate is very low. At 218°C, the stress does not seem to relax at all after 5000 seconds. The absolute decrease in tensile stress is lower than that obtained at lower temperatures.

In the Al films a somewhat different behavior is observed. At lower temperatures (57°C to 173°C) it can be seen that the relaxation rate after eight hours tends to decrease with increasing temperature. At the temperatures 200°C and 218°C a fast relaxation is observed during approximately 2000 seconds, after which the stress decreases very slowly in time. It is surprising that at higher temperatures (310°C and 352°C) again a larger stress decrease is observed.

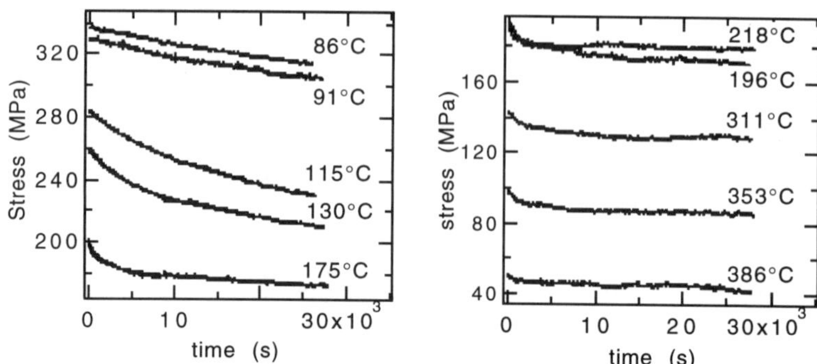

Figure 4: Isothermal stress relaxation in AlCu films.

In Figure 6 the initial relaxation rate in the various Al alloys is plotted as a function of the reciprocal temperature. The initial relaxation rate at lower temperatures is highest in the AlVPd film. The AlCu film shows the lowest relaxation rate at these temperatures. Furthermore, it is observed that the initial relaxation rate in the AlCu films has the strongest temperature dependence. In the AlVPd films the relaxation rate also decreases with decreasing temperature. However, a weaker temperature dependence of the initial relaxation rate is observed.

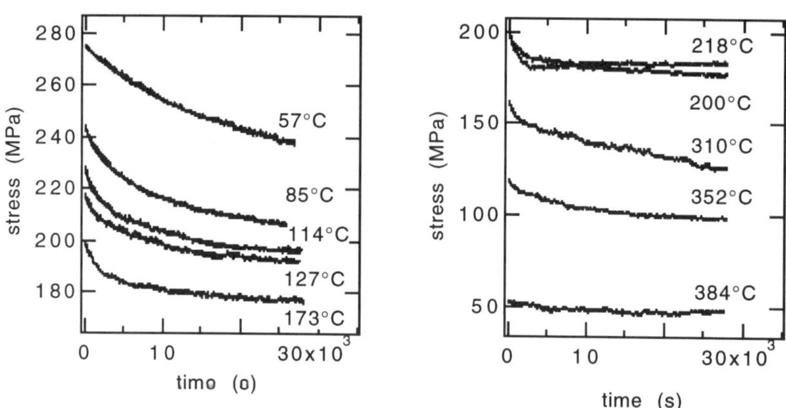

Figure 5: Isothermal stress relaxation in Al films.

Discussion

It has become clear from the experimental data that the various materials cannot sustain the stresses that are thermally induced in the film. First it is noticed that during cooling the tensile stress in the Al and the AlCu film remains lower than the stress in the AlVPd film in a considerably large temperature range (100°C to 300°C). Only during the last part in the cool down of the AlCu film the tensile stress increases to the final tensile stress obtained in the AlVPd film. In Figure 3, 4 and 5 it is shown that the three films also show a considerably large relaxation of the tensile stresses during isothermal anneals. Relaxation of tensile and compressive stress has been reported before for Al [2] AlCu [5] and AlSiCu [6]. It is often proposed that the decrease in the tensile stress is governed by

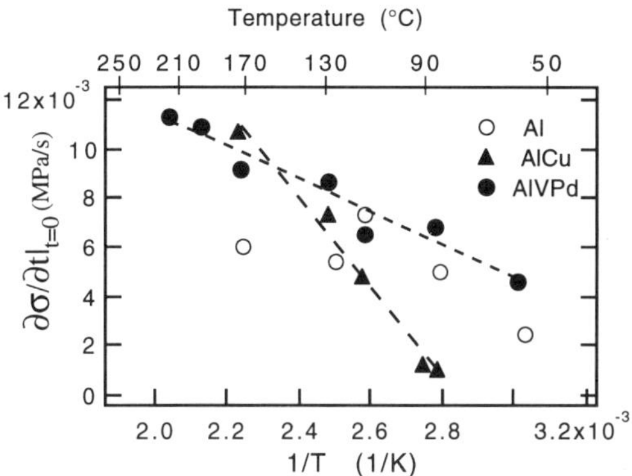

Figure 6: Initial tensile stress relaxation rate for Al, AlCu and AlVPd films. The dashed lines have to be considered as 'guides to the eye' and are not based on any model yet.

gliding of dislocations through the grains of the material. The dislocation glide model is based on the assumption that gliding is hindered by obstacles (being other dislocations, substitutional atoms in solid solution or precipitates). A minimum athermal stress is required to overcome these obstacles. The driving force for gliding is the thermally induced stress and the process is supposed to be thermally activated [1,3]. According to this model the stress relaxation exhibits a logarithmic time dependency.

Although perfectly good fits were obtained for Al and AlCu up to 175°C and for AlVPd up to 310°C for the individual measurements, it was not possible to deduce an activation energy or an athermal flow stress from the fit parameters. The same observations were reported by R. Venkatraman et al. for AlCu [5]. However, an activation energy of 2.0 eV was reported for Al [2] and 1.7 eV for AlSiCu [6]. In our case the model did not fit the data well for Al and AlCu above 175°C.

If in a given temperature range an activated process causes stress relaxation (e.g. dislocation glide, climb or diffusional flow) then the relaxation rate is expected to increase with increasing temperature. However, the relaxation rate also depends on the film stress present at that time which in turn depends on the dislocation density. Since the tensile stress decreases with increasing temperature we have two counteracting influences on the relaxation behavior. Furthermore, the RBS data shown in Figure 1, indicate that the distribution of the alloying elements changes when the material has experienced a thermal treatment. In fact it is known for bulk AlCu systems that several AlCu phases can exist in the temperature range between 50°C and 400°C [7]. If we consider the alloying elements to act in some way as barriers for dislocation glide, a change in size and distribution would result in a changing influence on the resistance against gliding of the dislocations. This of course complicates the quantitative analysis of the data.

However, a qualitative discussion may very well be possible. In Figure 2 it was observed that cooling the AlCu films below 180°C results in a fast increase in the tensile stress. It is observed in Figure 6 that the initial stress relaxation rate decreases with decreasing temperature at lower temperatures (86°C to 175°C). This may imply that the AlCu films are strengthened in this temperature range. It is proposed in literature [8] that strengthening is due to precipitation hardening.

In the AlVPd films the initial relaxation rate also decreases with decreasing temperature, but the temperature dependence is weaker. This implies that, if in AlVPd the stress relaxation is also governed by a thermally activated process, the activation energy for the relaxation mechanism in AlVPd is lower than for the relaxation mechanism in AlCu, i.e. the resistance against initial tensile stress relaxation in the AlCu film is higher than in the AlVPd film between 86°C and 175°C. According to Figure 2, cooling of the Al films below 200°C increases the resistance against stress relaxation. Since the initial stress relaxation rate decreases with decreasing temperature for Al, this is confirmed by the results obtained on isothermal stress relaxation.

It is striking as seen in Figure 6 that for both AlVPd and AlCu, the initial relaxation rate depends linearly on the reciprocal temperature. This is a good start for a more detailed quantitative analysis. Although the data in Figure 6 imply a high initial relaxation rate for AlVPd below 200°C, the tensile stress observed in the AlVPd films during cooling from 400°C till 50°C was always larger than in the AlCu and Al films. Therefore, it is very likely that the AlVPd films have a higher resistance against stress relaxation at the higher temperatures (above 200°C). This could be the result of the relative stable vanadium distribution which strengthen the interior of the grains in the Al matrix.

Conclusions

1. Isothermal stress relaxation has been observed in Al, AlCu and AlVPd blanket films.
2. Although good fits were obtained with the dislocation glide model at lower temperatures for the three types of films investigated, no clear trend was found in the fit parameters.
3. RBS data showed a stable depth profile of the V distribution. Segregation of Pd was found at the surface and at the interface of the film and the SiO_2 barrier layer. Segregation of Cu was found at the interface.
4. Most probably, the resistance against stress relaxation is highest for the AlVPd films above 200°C. At lower temperatures the highest resistance against initial stress relaxation was obtained in the AlCu films.

References

1. W.D. Nix, Metallurgical Transactions A, **20A**, 2217 (1989)
2. C.A. Volkert, C.F. Alofs, J.R. Liefting, J. Mater. Res., **9**, 1147 (1994)
3. H.J. Frost, in *Materials Reliability in Microelectronics II* eds. C.V. Thompson and J.R. Lloyd, Mater. Res. Symp. Proc. **264**, 3 (1992)
4. Hoffman R.W., in *Physics of Thin Films*, edited by G. Haas and R.E. Thun, (Academic, New York, 1966), **3**, 222
5. R. Venkatraman, P.R. Besser, J.C. Bravman, J. Mater. Res., **9**, 328 (1994)
6. A. Witvrouw, J. Proost, B. Deweerdt, Ph Roussel, K. Maex, Mater. Res. Soc. Symp.Proc., **356**, 441 (1995)
7. P. Haasen, in *Physical Metallurgy* (Cambridge University Press, London 1978) p. 201
8. P.A. Flinn, D.S. Gardner, W.D. Nix, IEEE Trans. on El. Dev., **34**, 689 (1987)

Acknowledgements

The authors are grateful to W.M. Arnold Bik for the RBS measurements. This work is part of the research program of the Stichting voor Fundamenteel Onderzoek der Materie (FOM), which is supported financially by the Nederlandse Organisatie voor Wetenschappelijk Onderzoek (NWO). We gratefully acknowledge financial support by the ESPRIT IV project ADEQUAT+.

THE EFFECT OF THE PASSIVATION MATERIAL ON THE STRESS AND STRESS RELAXATION BEHAVIOR OF NARROW AL-SI-CU LINES

A. WITVROUW[1], P. FLINN[2] AND K. MAEX[1]

[1] IMEC, Kapeldreef 75, 3001 Leuven, Belgium

[2] Department of Materials Science and Engineering, Stanford University, Stanford, CA, U.S.A.

ABSTRACT

800 nm thick and 1 µm wide Al-1wt.%Si-0.5wt.% Cu parallel lines with 1 µm spacing were passivated with PECVD oxide, oxynitride or nitride. Substrate curvature measurements as a function of temperature and XRD-measurements at room temperature were used to characterize macroscopic samples of these parallel Al-Si-Cu-lines. By using both techniques the average in-plane stresses for the Al-Si-Cu lines as well as for the covering passivation material can be determined as a function of temperature. The highest and lowest stresses in the Al-Si-Cu are observed for lines with nitride and oxide passivations, respectively. Also the number of voids in the lines after a storage test at 250 °C is clearly highest for a nitride passivation and lowest for an oxide passivation.

The stress in the passivation itself and its temperature dependence is found to be very different from the stress in a blanket passivation film.

INTRODUCTION

Al-alloys are used in the microelectronics industry as interconnect material because of their low resistivity and good etch ability. Al-Si-Cu is a very popular Al-alloy as the addition of Si prevents spiking [1] and the addition of Cu substantially improves the electromigration behavior of the Al [2]. However, over recent years it has become clear that Al-alloys can be susceptible to stress induced void formation [3]. The driving force for stress induced void formation is the high tensile stress in the passivated Al-lines, which is caused by the difference in thermal expansion between the Al and the Si-substrate on one hand and the passivation material on the other hand. FEM [4] predicts that this stress rises with increasing passivation stiffness.

In a previous work [5] the stress in nitride covered Al-Si-Cu lines was already studied. In this work we compare these results with results for oxide and oxynitride covered lines. Also the relation between stress relaxation and stress induced voiding and the dependence of the latter on the passivation material is examined.

EXPERIMENTAL PROCEDURE

800 nm thick Al-Si-Cu layers are sputter deposited from an Al-1 wt.% Si-0.5 wt.% Cu target in a CLC9000/Galaxy100 cluster tool on 5" thermally oxidized Si (100)-wafers. The layers

Mat. Res. Soc. Symp. Proc. Vol. 436 © 1997 Materials Research Society

were deposited with a three step process (to minimize the amount of etch residues and optimize the step coverage) at a chuck temperature of 300 °C. 1μm parallel lines and spaces were defined by reactive ion etching. The different passivation layers were all deposited at 380 °C by Plasma Enhanced Chemical Vapour Deposition. The passivation thickness was determined by reflectivity measurements on a Leitz MPVSP and is 1.26 ± 0.05 μm for the oxide, 1.63 ± 0.01 μm for the oxynitride and 1.19 ± 0.09 μm for the nitride. The sputtered Al-Si-Cu film thickness is calibrated on a regular basis by profilometer measurements. The thickness of the substrate is measured at 25 points on a MX203 stress meter.

Stress was determined from *in-situ* substrate curvature measurements during annealing of 24×6 mm^2 samples in vacuum by laser scanning. The middle of the samples were over a length of 12.5 respectively 6 mm completely covered with 1 μm lines and spaces parallel respectively perpendicular to the longest side of the sample. The curvature measurements were performed on this middle part only. The sample rests in a quartz holder on three points, two quartz pins on one side and the end of a chromel-alumel thermocouple on the other side, and is heated by a 150 W halogen lamp with gold-coated mirror. Base pressures down to the high 10^{-8} Torr are possible with a turbopump TSU 050 Pfeiffer with mechanical backing pump [5]. Thermal cycling was performed up to 425 °C with a 25 °C/min heating and cooling rate and a 10 min stay at the highest temperature. After completion of the temperature cycles for the stress measurements, the measured layers are etched off and the curvature of the substrate is measured at room temperature and during thermal cycling. These substrate curvature measurements are subtracted from the data. Al-Si-Cu layers are etched off in H_2O_2:H_2SO_4 (1:4) and nitride, oxynitride and oxide layers in HF 40%.

STRESS IN PASSIVATED LINES

To measure the stress in lines, Stoney's equation [6] cannot be used. Instead the following equations can be deduced from the equilibrium of forces and momenta for a macroscopic sample of parallel lines covered by a passivation:

$$\sigma_x f d_f + \sigma_{x,pass} d_{pass} = \frac{1}{6} * \left(\frac{A}{R_x} + \frac{B}{R_y} \right) * d_s^2$$

$$\sigma_y f d_f + \sigma_{y,pass} d_{pass} = \frac{1}{6} * \left(\frac{B}{R_x} + \frac{A}{R_y} \right) * d_s^2$$

(1)

with σ_x and σ_y the volume averaged stresses in the length and the width direction of the lines respectively, R_x and R_y the radii of curvature in the length and the width direction of the lines respectively, f the fraction of area covered by the lines, A and B the plane stress elastic constants of the Si [7], d_s the thickness of the substrate, d_f the thickness of the film, d_{pass} the thickness of the passivation and $\sigma_{x,pass}$ and $\sigma_{y,pass}$ the volume averaged stresses in the passivation. The curvature changes are thus due to stress changes in the lines <u>and</u> in the passivation. Assuming that the passivation stresses are equal to the biaxial stress in a blanket film of passivation did not lead to satisfactory results [5]. To be able to separate the contributions of lines and passivation, X-ray measurements were carried out at room temperature on as-deposited samples [8]. The results are shown in Table 1. Clearly the stresses in the lines increase with increasing passivation stiffness. Implementing these stresses at 20 °C and assuming zero stress at 380 °C during heating for the Al-

passivation	σ_x (MPa)	σ_y (MPa)	σ_z (MPa)	
oxide	645	476	437	*Table 1*: triaxial stress in the lines
oxynitride	735	551	466	for different passivations
nitride	783	643	564	

Si-Cu lines, implies that the passivation stress and stress changes as a function of T in the passivated sample are different from those in the blanket film. This is summarized in Table 2 for all passivations.

passivation	oxide	oxynitride	nitride
σ_{film} (MPa)	-202+0.21*T(°C)	-116+0.01*T(°C)	-397+0.08*T(°C)
$\sigma_{x,pass}$ (MPa)	-162+0.18*T(°C)	-45+0.01*T(°C)	-253+0.22*T(°C)
$\sigma_{y,pass}$ (MPa)	-8-0.05*T(°C)	15-0.14*T(°C)	-72-0.06*T(°C)

Table 2: Stresses in the length (x) and width (y) direction of the different passivations covering Al-Si-Cu lines compared to the stress in a blanket film of that passivation during a second thermal cycle. The values for the oxide and nitride are valid for heating and cooling as these films are stabilized after the first thermal cycle. The thicker oxynitride however still has stress relaxation at 425 °C and the stresses during cooling are slightly less compressive than those listed here.

$d\sigma_{x,pass}/dT$ is always positive and $d\sigma_{y,pass}/dT$ is always negative, but neither $\sigma_{x,pass}$ or $\sigma_{y,pass}$ has the same behavior as the stress in the film. This is not so surprising as a passivation

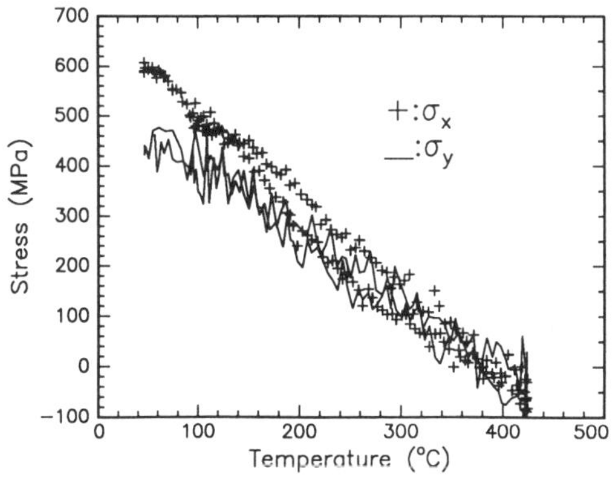

Figure 1:Stresses in the length (x) and width (y) direction of 1 μm oxide passivated Al-Si-Cu lines after correction for the oxide stress as explained in the text.

layer deposited on top of Al-Si-Cu lines has different degrees of freedom compared to a layer on a flat wafer. Also, it has been shown [9] that the stress in the passivation is actually a function of the stress in the lines as both have a common interface.

Figure 1 shows $\sigma_{\text{Al-Si-Cu},x}$ and $\sigma_{\text{Al-Si-Cu},y}$ during a second thermal cycle for oxide passivated lines after a correction is made for the oxide stress as determined above.

The stress during cycling in the lines covered with the other passivations is very similar. Only the room temperature stresses differ as shown in Table 1. Almost linear stress behavior is seen with very little hysteresis during cycling. The plot for nitride covered lines was shown in [5].

STRESS RELAXATION IN LINES

Stress relaxation experiments in passivated lines at temperatures lower than 100 °C for 10h only resulted in very small curvature changes [see also 5]. Also, the curvature in the width direction actually increased instead of decreased. No clear evidence of voiding was seen, neither in the stress measurements or in the optical microscope. However, a stress relaxation experiment at 233 °C for 100h in oxide passivated lines during cool down of a fourth thermal cycle clearly resulted in stress induced voiding. The measured values for $\sigma_x + \sigma_{x,\text{pass}}*(d_{\text{pass}}/fd_f)$ and

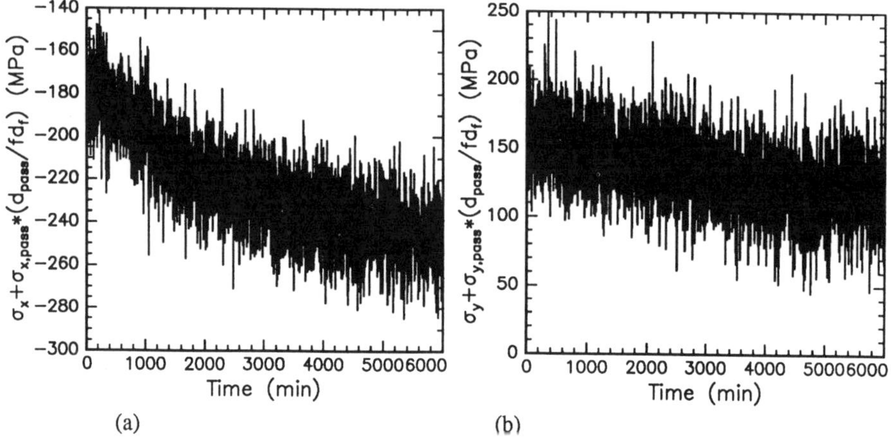

Figure 2: *Stress relaxation in the (a) length and (b) width direction of oxide passivated lines at 233 °C.*

$\sigma_y + \sigma_{y,\text{pass}}*(d_{\text{pass}}/fd_f)$ decreased with 75 and 45 MPa respectively during the isothermal stay. If the passivation were completely rigid during this isothermal stay these stress changes would be due to the metal stresses only, resulting in a 54 MPa change in the metal hydrostatic stress if it is assumed that $\sigma_z = 0.9*\sigma_y$. Using the elastic constants of Al with a <111> texture [8] an 0.07% volume change can be calculated in that case. However, it has been shown that both the stresses in and the shape (including the cavity for the metal) of the passivation change when the metal stresses relax [9,10]. Using the equations for a conformal passivation found in ref [9] a hydrostatic stress change of 51 MPa and a 0.2 % volume change for the metal lines can be calculated. The latter value seems in agreement with the amount of voiding seen in the optical microscope with Nomarski interference.

STRESS INDUCED VOIDING DURING A STORAGE TEST

As optical microscopy inspection of these parallel lines can easily reveal voiding, this method was used to qualitatively investigate the effect of the passivation material on the stress induced voiding. Samples with the different passivations were annealed together in a N_2-filled box furnace at 250 °C for 100h. Figures 3a-3c are the optical microscopy pictures of the lines covered with the different passivations after the storage test. Clearly the nitride covered lines have more voids than the oxynitride covered lines and the latter have more voids than the oxide covered lines. This was also confirmed quantitatively by counting the number of voids in a 20μmx20μm area for each passivation type. The average number of voids seen in the optical microscope for 5 squares is 3 ± 2 for the oxide covered lines, 15 ± 3 for the oxynitride covered lines and 41 ± 10 for the nitride covered lines. The number of voids are higher for lines with higher initial stress. Although only the number of voids visible in the optical microscope can be counted, it does not seem that the linear relation between void volume and initial stress, as determined by simple models [11], holds for these lines. Possibly other parameters, such as the interfacial diffusion coefficient, play a role as well.

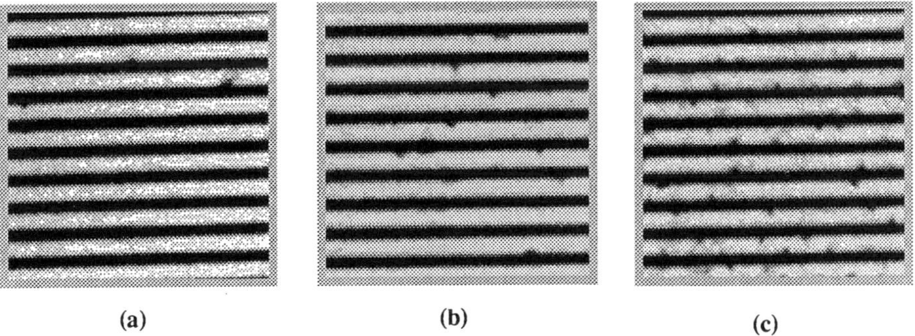

(a)   (b)   (c)

Figure 3: optical microscope pictures of 1 μm lines and spaces passivated with (a) oxide, (b) oxynitride and (c) nitride after a storage test for 100h at 250 °C.

CONCLUSION

Stress and stress induced voiding was studied for Al-Si-Cu lines passivated with oxide, oxynitride and nitride. The highest and lowest stresses in the Al-Si-Cu are observed for lines with nitride and oxide passivations, respectively. Also the number of voids in the lines after a storage test at 250 °C is clearly highest for a nitride passivation and lowest for an oxide passivation. Although hydrostatic stress relaxation clearly leads to voiding and the highest number of voids are seen in the lines with the highest initial stress, a simple linear dependence of the void volume with the initial hydrostatic stress does not seem to hold. Possibly other factors play a role as well.

The stress in the passivation itself and its temperature dependence is found to be very different from the stress in a blanket passivation film.

ACKNOWLEDGEMENT

The authors thank J. Steenbergen for help with the stress measurements and the void counting, F. Loosen for the image processing, the IMEC-pilot line for processing the wafers and the Jülich group for interesting discussions. This work was performed under a contract with Alcatel Bell, Antwerp, Belgium and with the support of the Flemish Institute of Scientific and Technological Research.

REFERENCES

[1]   J. F. Smith and I. Wagner, <u>Advanced Aluminum Metallization, Part I: Aluminum Thin Film Properties</u>, (1991 MRC Thin Film School), 2.

[2]   F.M. d'Heurle and P.S. Ho in <u>Thin Films: Interdiffusion and Reactions</u>, ed. by J. Poate, K. Tu and J. Mayer (N.Y.), 293 (1978).

[3]   H. Okabayashi, Mat. Sci. Eng. <u>R11</u> (5), 191-241 (1993).

[4]   A.I. Sauter, W.D. Nix, IEEE Trans. Comp. Hybr. Man. Tech., <u>15</u> (4), 594-600 (1992).

[5]   A. Witvrouw, J. Proost, B. Deweerdt, Ph. Roussel and K. Maex, Mat. Res. Soc. Symp. Proc. <u>356</u>, 441-446 (1995).

[6]   W.D. Nix, Metall. Trans A <u>20A</u>, 3317 (1989).

[7]   A. Witvrouw and F. Spaepen, JAP <u>73</u> (11), 7344 (1993).

[8]   P.A. Flinn and Chieng Chiang, JAP <u>67</u> (6), 2927 (1990).

[9]   U. Burges, I. Eppler, W. Schilling, H. Schroeder and H. Trinkaus, 'Analysis of Stresses in Passivated Metal Lines', presented at the 3rd Conf. on Stress Phenomena in Metallizations, Stanford, June 1995.

[10]   P.R. Besser, T.N. Marieb, J. Lee, P.A. Flinn and J.C. Bravman, J. Mater. Res. <u>11</u>, 184 (1996).

[11]   A.I. Sauter and W.D. Nix, J. Mater. res. <u>7</u> (5), 1133 (1992).

THERMAL- AND ELECTROMIGRATION- INDUCED STRESSES IN PASSIVATED AL- AND ALSICU-INTERCONNECTS

D. Beckers, H. Schroeder, I. Eppler*, W. Schilling

Institut für Festkörperforschung, Forschungszentrum Jülich, D-52425 Jülich, Germany;
*Daimler Benz AG, D-60528 Frankfurt

ABSTRACT

Al and Al- alloys are commonly used as interconnect materials in integrated electronic devices. Stress induced voiding and degradation of metal lines by electromigration are closely related to the stresses in the lines.

We have studied the strain and stress evolution during thermal cycling, isothermal relaxation and due to electromigration in passivated Al and AlSi(1%)Cu(0.5%) lines by X-Ray diffraction with variation of experimental parameters such as the aspect ratio and the electrical current density. Furthermore the extent of voiding and plastic shear deformation has been determined from the experimental metal strains with the help of finite element calculations.

Main results are: 1) During thermal cycling the voiding is less than $2 \cdot 10^{-3}$. The extent of plastic shear deformation increases with increasing line width and with decreasing flowstress. 2) During isothermal relaxation void growth occurs but no significant change in the plastic shear deformation. 3) An electric current in the lines causes no measurable additional change of the volume averaged stresses up to line failure.

INTRODUCTION

Mechanical stresses in microelectronic devices are of special interest because of degradation effects in microelectronic circuits such as stress induced voiding or electromigration. E.g. the thermal mismatch between the metal and the underlying Si-substrate and the surrounding passivation causes large tensile stresses in the Al-interconnects during cooling down from elevated passivation temperatures. These high tensile stresses can relax by the formation of voids. The main aim of the strain measurements is to find out the extent of stress relaxation in the metal by void growth or by dislocation plasticity.

The plasticity induced change in volume or shape of the metal line embedded in a passivation layer cannot be measured directly [2]. Therefore we applied a formalism described elsewhere [1, 2, 3] using eigen-strains to evaluate the extent of voiding and plastic shear deformation from the experimentally determined elastic metal strains. We present thermal cycling and isothermal relaxation data for different aspect ratios without and with applied electrical currents.

EXPERIMENT

Samples

The samples were an assembly of many parallel Al or Al(1wt.%)Si(0.5wt.%)Cu-lines on oxidized Si-wafers. Some samples had an additional 0.12μm thick Ti/TiN-underlayer. The metal lines are passivated with SiN_X. The detailed sample geometries are given in Table 1. The films were strongly $< 111 >$-textured and the grain sizes were about equal to the film thicknesses. At the same thickness the grain sizes of the samples with the Ti/TiN-layer are about 40% smaller than the grain sizes of the samples without Ti/TiN-layers. Before

Mat. Res. Soc. Symp. Proc. Vol. 436 © 1997 Materials Research Society

Sample No.	AlSiCu	Ti/TiN	t$[\mu m]$	w$[\mu m]$	a	d $[\mu m]$	t$^P[\mu m]$
1	x		1.2	2.0	0.6	4.0	0,5
2	x		1.6	2.0	0.8	4.0	0,65
3	x		1.7	1.8	0.95	4.0	0,5
4	x		1.2	0.7	1.6	2.0	0,4
5	x	x	0.8	0.9	0.9	2.0	0,5
6		x	0.85	1.0	0.85	2.0	0,4

Table 1: *Measured samples and their geometry; t, w, a = t/w are the thickness, the width, the aspect ratio of the metal-lines and d the repetition distance of the metal lines; t^P is the median thickness of the conformal SiN_X-passivation.*

passivation the samples were annealed 20 minutes at $450°C$ in H_2-N_2 and cooled down to room temperature, in order to stabilize the microstructure of the lines. For simplicity in the following all lines are called Al-lines.

To describe stresses and strains we use a coordinate system which is defined by the following directions. 1-direction: parallel to the lines, 2-direction: perpendicular to the lines in the film surface plane and 3-direction parallel to the surface normal.

<u>Strain Measurement and Stress Determination by X-Ray Diffraction</u>

X-Ray strain measurements were carried out using a so-called Ψ-type diffractometer geometry with K_α radiation from a Cu rotating anode tube. The scattered intensity was detected by a linear position sensitive detector. In these experiments we used the {333}- and the {511}- reflections of Al, which were observed at the highest scattering angles of about $2\theta \approx 156$ to 164 degree. The strain in the passivated lines was determined by the so-called "$\sin^2\Psi$-technique" [4], where Ψ is the angle between the surface normal and the normal of the diffracting planes. Because of the strong $< 111 >$-texture of the films the reflections could be studied only at few inclinations Ψ. The unstrained lattice parameters were received from continuous Al films of the same wafer by X-Ray measurements assuming an equi-biaxial stress.

The stresses were evaluated with consideration of the strong $< 111 >$-texture of the film [5, 6]. Because of the small anisotropy factor $s_{11}^M - s_{12}^M - s_{44}^M/2$ in Al both evaluations of stresses from the measured strains - using Reuss average (assuming constant stress in the lines) or Voigt average (assuming constant strain in the lines)- differ by only 5 MPa at the most. The presented strains we determined from measurements at selected values of Ψ in the 1, 3 - and the 2, 3 - plane using a least square fit and the Voigt average.

EVALUATION OF THE VOIDING AND THE PLASTIC SHEAR DEFORMATION

The determination of the extent of voiding and plastic shear deformation from the experimental results is described in detail in [1, 2, 3].

The volumetric eigen-strain ϵ_V^E characterizes the relative misfit volume between the metal and the surrounding. For elastic thermal contraction: $\epsilon_V^E = 3\Delta\alpha(T - T_0)$. In the nonelastic case the difference $\Delta\epsilon_V^E = \epsilon_V^E - 3\Delta\alpha(T - T_0)$ describes the volume change by voiding or precipitation. $\epsilon_S^E = \epsilon_1^E - 1/2 \cdot (\epsilon_2^E + \epsilon_3^E)$ describes the plastic yielding along the 1- from the 2- and 3-direction and $\epsilon_\Delta^E = 1/2 \cdot (\epsilon_2^E - \epsilon_3^E)$ the plastic yielding from the 3- into the 2-direction.

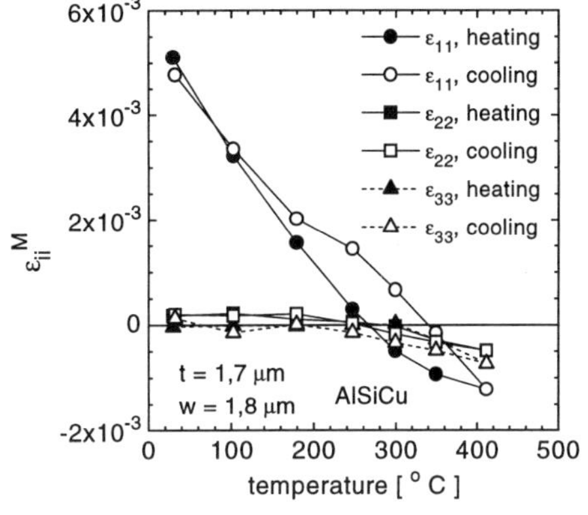

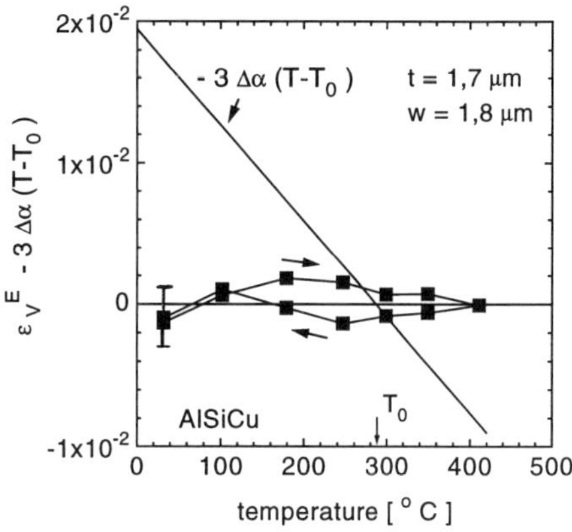

Fig. 1: *Temperature dependence of the metal strains ϵ_{11}, ϵ_{22} and ϵ_{33} of sample 3 between RT and 410°C during thermal cycling.*

Fig. 2: *Volumetric plasticity of sample 3 during thermal cycling. Within an error bar of $2 \cdot 10^{-3}$ only small amount of voiding or precipitation occur compared to the thermal volume change.*

RESULTS AND DISCUSSION

Thermal Cycles

As an example Fig.1 shows the strain evolution during thermal cycling of sample 3. There is high tensile strain in 1-direction at room temperature because of the rigid substrate under the line. With increasing temperature the strain in 1-direction decreases almost linearly up to the transition temperature between tensile and compressive strains. The strain has its minimum at the highest temperature of 410°C. During cooling down the strain in 1-direction increases and reaches at RT almost the starting value. The strain in 1- direction shows a hysteresis above 150°C. In contrast the strains in 2- and 3-direction are almost zero in the tensile range and become only slightly compressive in the temperature range above 300°C.

The evaluated volumetric plasticity of this sample is shown in Fig. 2. T_0 is the mean transition temperature between tensile and compressive stresses in the different directions. As shown in Fig. 2 the volumetric plasticity is nearly zero during the thermal cycling within the experimental errors (compared to $-3\Delta\alpha(T - T_0)$). The fact that the volumetric eigen-strain ϵ_V^E is not different from the elastically calculated value $3\Delta\alpha\Delta(T - T_0)$ implies little change in the Al-volume by voiding or precipitation during cooling within an error of $1 \cdot 10^{-3}$ at RT. This error consists first of the errors of the measurement and second of the uncertainties of the FEM calculations and of the knowledge of the elastic modulus of the passivation ($E^P \approx 100$ GPa).

The plastic shear deformation of the same sample is shown in Figs. 3 and 4. ϵ_S^E describes the plastic elongation of the line along the 1-direction and ϵ_Δ^E describes the plastic yielding between the 3- and the 2-direction. As shown in Fig. 3 plastic yielding along 1-direction indeed occurs during cooling of the sample below T_0. The yielding increases almost linearly with increasing distance from the transition temperature T_0. The plastic elongation at RT

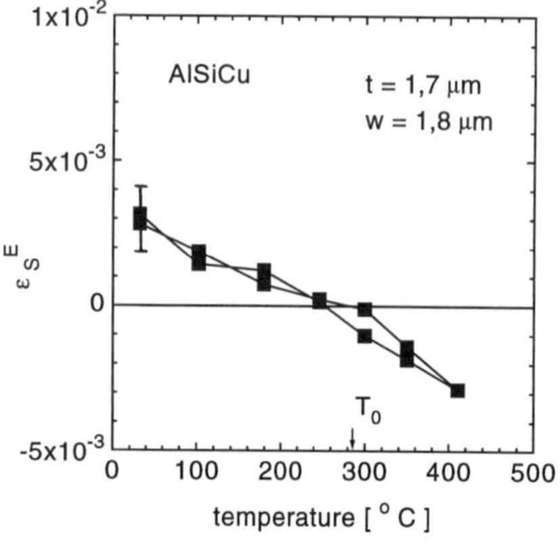

Fig. 3: *Plastic yielding from 1- into the 2- and 3-direction of sample 3 during thermal cycling.*

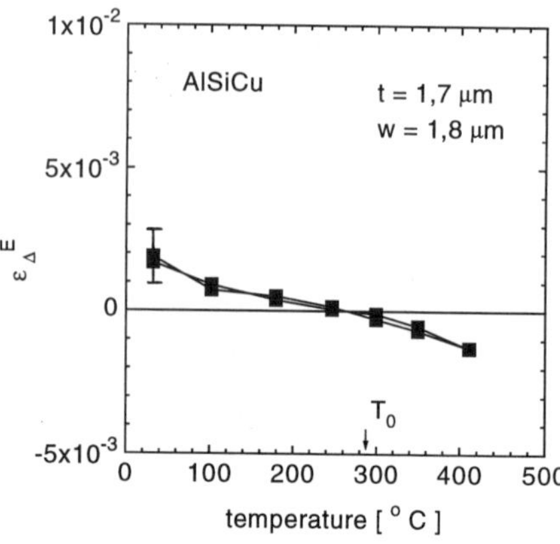

Fig. 4: *Plastic yielding from 2- into 3-direction of sample 3 during thermal cycling.*

is almost the same as the plastic compression at 410°C, reflecting the much lower flow stress at higher temperatures. The plastic yielding from the 3- into the 2-direction also increases with increasing distance from the transition temperature. In the whole temperature range applies: $\epsilon_S^E \approx 2 \cdot \epsilon_\Delta^E$.

Figs. 5 - 8 show the shear plasticity depending on different sample parameters. In Figs. 5 and 6 the plastic shear deformation of two samples with different line width and the same line thickness is shown. In the sample with aspect ratio a < 1 plastic yielding occurs from 3- into the 2-direction. In contrast to this in the sample with a > 1 the plastic yielding is in the opposite direction, because in this sample the stress in 3-direction is greater than the stress in the 2-direction. Beside this the shear plasticity ϵ_S^E is higher for the sample with the smaller aspect ratio a. This could be explained by the higher elastic shear strain in this sample during cooling down which leads to more plastic yielding. Additionally the flow stress - which increases with decreasing line dimensions - in the sample with the line width $w = 0,7 \mu m$ is probably higher leading to less plasticity.

In Fig. 7 the line thickness is varied at almost constant width. The plastic elongation in the 1-direction increases somewhat with decreasing aspect ratio. This could also be due to the increasing elastic shear strain with decreasing aspect ratio which is obviously not compensated by the also increasing flow stress.

The comparison of an Al-sample with an AlSiCu-sample with almost the same line geometry shows a higher plasticity in the Al-sample (Fig. 8). This can be explained by the smaller flow stress in pure Al.

Summarizing the results of the thermal cycles of all samples [7] only very small volume changes due to voiding and precipitation occur, while the plastic yielding depends strongly on the sample parameters.

<u>Isothermal Relaxation and Electrical Current Stressing</u>

Fig. 9 shows the results of an in-situ electromigration experiment with sample 6. The aim is to clarify the interrelation between current- and stress-induced degradation by voiding

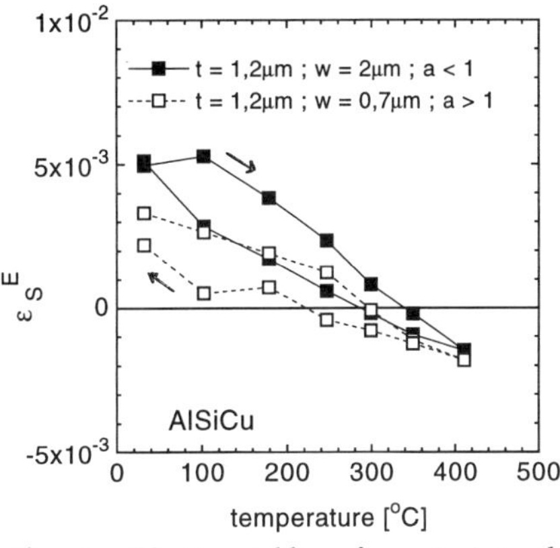
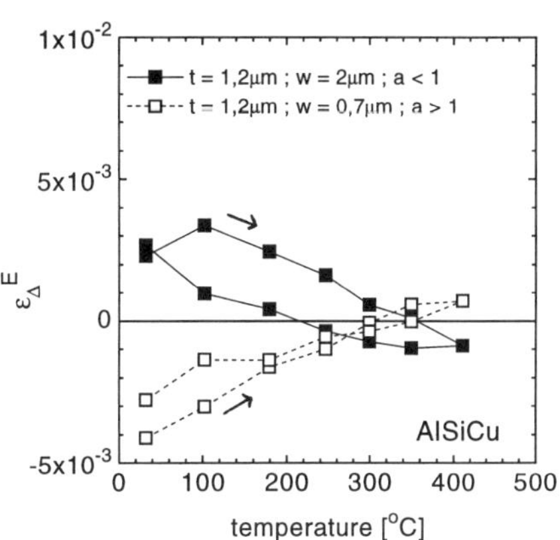

Fig. 5: *Plastic yielding from 1- into the 2- and 3-direction of sample 1 and sample 4 during thermal cycling. The yielding increases with decreasing aspect ratio.*

Fig. 6: *Plastic yielding from 2- into 3-direction of sample 1 and sample 4 during thermal cycling.*

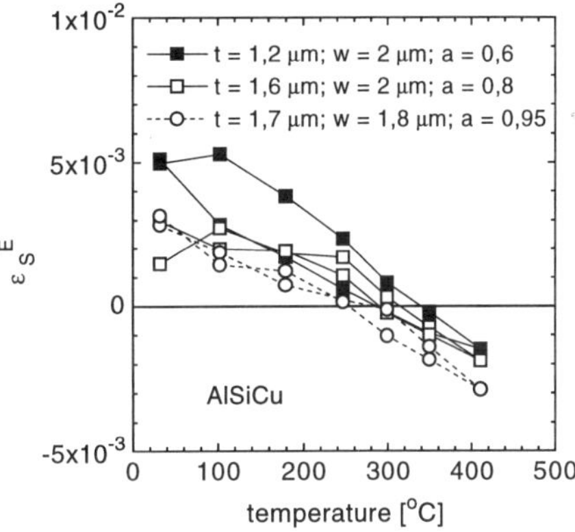
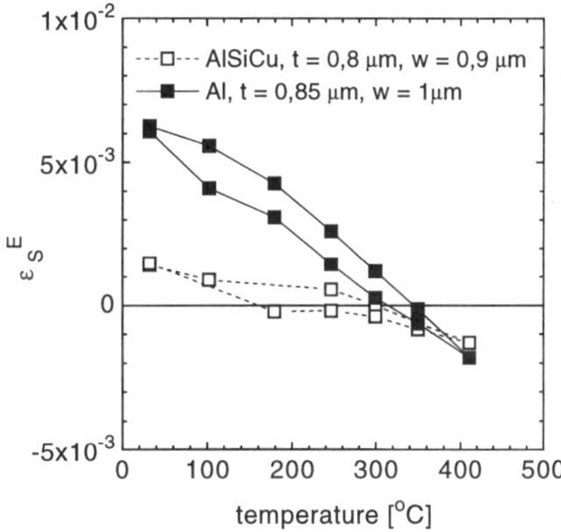

Fig. 7: *Plastic yielding from 1- into the 2- and 3-direction of sample 1, 2 and 3 during thermal cycling. The yielding increases with decreasing aspect ratio.*

Fig. 8: *Plastic yielding from 1- into the 2- and 3-direction of sample 5 and sample 6 during thermal cycling. Because of the smaller flow stress in pure Al the yielding in the Al-sample is much smaller than the yielding in the AlSiCu-sample.*

or crack formation in passivated lines. Material flux divergencies cause local depletion and accumulation of the metal. This leads to rising or decreasing stresses at these points and to the formation of stress gradients. If the mass depletion causes the formation of voids this results in a lower average volumetric stress.

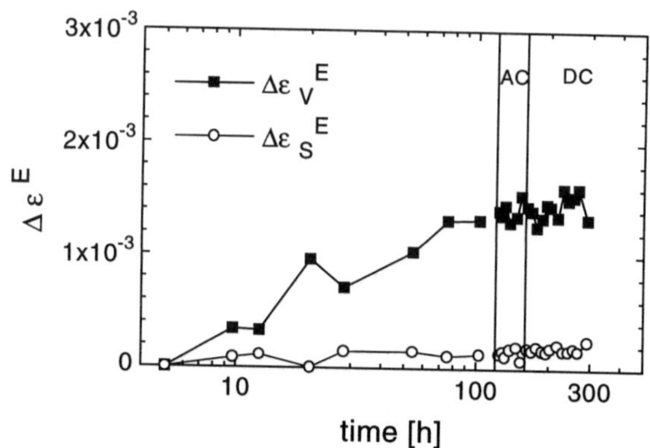

Fig. 9: *Relative change of the volumetric eigen-strain and the shear plasticity ϵ_S^E during isothermal relaxation at 180°C. After 120 h an electrical alternating current of 0,8 MA/cm² has been turned on which later has been changed into direct current.*

In the experiment shown in Fig. 9 the sample was heated up to 420°C, cooled down to 180°C and left at constant temperature. After an isothermal relaxation without electrical current of about 120 h an alternating current of about 0.8 MA/cm² has been turned on to separate temperature effects from heating the sample during the electrical current stressing. Afterwards a direct current of the same magnitude has been turned on.

During the isothermal relaxation voiding occurs but no change in plastic shear. The same result was obtained for all samples and all temperatures for which isothermal relaxations have been measured. The error in the measured change of the volumetric eigen-strain is about $4 \cdot 10^{-4}$.

During the 150 h isothermal relaxation with electric DC-current no additional change of the void volume due to electromigration can be detected within the experimental errors. The experiment has been repeated at higher temperatures and higher current densities up to failure of samples with the same results. Also in this case there is only an insignificant change of the volume averaged stresses. From this observation we have to conclude that electromigration damage is not accompanied by an enhanced overall nucleation and growth of voids.

References

[1] U. Burges, I. Eppler, W. Schilling, H. Schroeder, H.Trinkaus, Third International Workshop in Stress- Induced Phenomena In Metallization, AIP Conference Proceedings

[2] I. Eppler, H. Schroeder, U. Burges, W. Schilling, this volume

[3] I. Eppler, U. Burges, H. Schroeder, W. Schilling, to be published

[4] C. Noyan and J.B. Cohen, Residual Stress, Measurements and Interpretation, Springer 1987

[5] B.M. Clemens and J.A. Bain, MRS Bulletin, **17**, No. 7 (1992) 46

[6] R.Pollak, H. Huck, H. Haselier, P. Ehrhart and W. Schilling, Proceedings of the 3rd International Workshop in Stress-Induced Phenomena in Metallizations, June 1995

[7] D. Beckers et al., to be published

This article also appears in Volume 420.

STRESS IN SPUTTERED Co$_{90}$Fe$_{10}$/Ag GMR MULTILAYERS

J.D. JARRATT, V.R. INTURI, J.L. WESTON, AND J.A. BARNARD
Department of Metallurgical and Materials Engineering and The Center for Materials for Information Technology, The University of Alabama, Tuscaloosa, Alabama 35487-0202.

ABSTRACT

Stress, giant magnetoresistance (GMR), structure, and magnetic properties of sputtered (Co$_{90}$Fe$_{10}$ X Å/Ag Y Å)x20 multilayer films have been investigated at room temperature where X ranges from 7.5 to 25 Å and Y from 10 to 60 Å. These films exhibit distinct GMR behaviors dependent on individual layer thicknesses, including layered granular-type GMR in CoFe 7.5 Å samples and 'discontinuous' GMR (DGMR) in CoFe 15 and 25 Å samples with Ag thicknesses over 30 Å. No antiferromagnetic coupling was observed. CoFe 10 Å samples act as a transition between GMR behaviors. Compressive stress decreases with increasing Ag thickness in the CoFe 7.5 Å samples. In the CoFe 15 and 25 Å samples the stress fluctuates similarly depending on Ag thickness. The difference in stress and MR behavior between the CoFe 7.5 Å and the 15 and 25 Å samples is thought to be due to incomplete CoFe layering in the CoFe 7.5 Å samples. In the CoFe 15 Å DGMR samples, high temperature annealing resulted in tensile stresses large enough to cause film detachment. X-ray diffraction reveals a strong (111) growth texture as well as satellite peaks from coherent layering. This (111) texture is also evidenced by patterns with hexagonal symmetry formed by the detached films.

INTRODUCTION

GMR, a phenomenon found in magnetic/nonmagnetic systems was first observed in Fe/Cr multilayers grown by molecular beam epitaxy (MBE) where antiparallel alignment of Fe layers (antiferromagnetic (AFM) coupling) in zero field resulted in a high resistivity state. Application of a magnetic field resulted in parallel alignment (ferromagnetic (FM) coupling) of the Fe layers and a lower resistivity state. Even though these first films were MBE-grown, many sputtered systems also exhibit this AFM coupling which was found to oscillate with nonmagnetic spacer layer thickness [3]. The Co/Ag multilayer system is unusual in that no sputtered films have exhibited the oscillating AFM coupling and associated GMR. The effect has only been observed in electron-beam evaporated multilayers (both single crystal (100) and polycrystalline) [4]. GMR in the Co/Ag multilayer system was first observed in sputtered films with Ag thicknesses well beyond what is now considered conducive to AFM coupling, although that was the mechanism first assumed [5]. GMR has also been observed in granular systems (GGMR) [6-8] which consist of random immiscible ferromagnetic particles in a nonmagnetic matrix, including the Co-Ag system [6]. Recently, sputtered Co and CoFe alloy/Ag multilayers with ultrathin ferromagnetic layers have exhibited granular-type GMR [9-12].

Previously, we have surveyed the GMR in Co$_{90}$Fe$_{10}$/Ag sputtered multilayers [13-16] in which we have observed granular-type GMR (referred to as layered-GGMR) in ultrathin CoFe layered samples and 'discontinuous' GMR (DGMR) [17] in thicker CoFe and Ag layered samples. This report discusses the stress and structure in these films as a function of layer thickness (which also dictates GMR behavior) and annealing conditions.

EXPERIMENTAL METHODS

The films were DC magnetron sputter deposited at 1.2 and 0.31 W/cm^2 (100 W) for the Co$_{90}$Fe$_{10}$ (4" diameter) and Ag (8" diameter) targets, respectively in 2 mTorr of ultra-high purity argon in a computer-controlled Vac-Tec Model 250 side sputtering system with a base pressure of 3 x 10^{-7} Torr . The geometry of the films is [Ta 120 Å/(CoFe X Å/Ag Y Å)x20/Ta 50 Å/ ~1 µm SiO$_2$/Si<111>] where X ranges nominally from 7.5 to 25 Å with six different thicknesses, and Y is varied nominally from 10 to 60 Å. A permanent magnet was positioned behind each substrate providing a 90 Oe parallel field resulting in in-plane magnetic uniaxial anisotropy.

461

The stress measurements were performed on a Flexus Thin Film Stress Measurement System Model FLX 2300 which employs a laser to measure the substrate curvature before and after film deposition. The film stress, σ_f, is calculated using the equation originally developed by Stoney [18] and later modified by Hoffman [19]:

$$\sigma_f = (E_s t_s^2) / (6(1-v)t_f R),$$

where E_s is the elastic modulus of the substrate, t_s is the substrate thickness, v is Poisson's ratio for the substrate, t_f is the deposited film thickness, and R is calculated from the change in the substrate radius of curvature due to the film. A negative stress value indicates compressive stress and a positive stress value indicates tension.

Magnetic properties were measured on a Digital Measurement Systems Vibrating Sample Magnetometer (VSM) Model 880. X-ray diffraction was performed on a Rigaku D/Max-2BX XRD System with thin film attachment using Cu Kα (λ=1.5418 Å) radiation. The optical photographs were taken using an Olympus microscope. The SEM micrographs were taken using a Philips XL30 electron microscope. The reported MR measurements were made in the plane of the films using a 4-point in-line probe assembly with the current perpendicular to the applied magnetic field. The MR% is defined as: MR%=((ρ(H)-ρ_{max}) /ρ_{max})x100, where ρ(H) is the sample resistivity at a given field value and ρ_{max} is the resistivity at the maximum applied field. The 20 minute anneals were performed in a pre-evacuated quartz tube with an overpressure of flowing purified Ar. Helmholtz coils were positioned outside the tube providing a 50 Oe field in-plane in the films' easy axis direction.

RESULTS AND DISCUSSION

Figure 1 shows the as-deposited MR% and stress vs. Ag thickness for three series of multilayers with (a) 7.5 Å CoFe and (b) 15 and 25 Å CoFe. In the CoFe 7.5 Å series the compressive stress decreases with increasing Ag layer thickness. The CoFe 15 and 25 Å films both show a tensile stress maximum in the Ag 20 Å samples and also a similar fluctuation of stress with Ag thickness. The maximum as-deposited MR% in the CoFe 7.5 Å series occurs in the Ag 30 Å sample. The as-deposited MR in the CoFe 15 and 25 Å films is a maximum in the thicker Ag layered samples. Even though there is a similar trend toward maximum MR occurring in thicker Ag layered samples as the CoFe thickness increases, there is a distinct difference in the type of GMR observed between the CoFe 7.5 Å series and the CoFe 15 and 25 Å series.

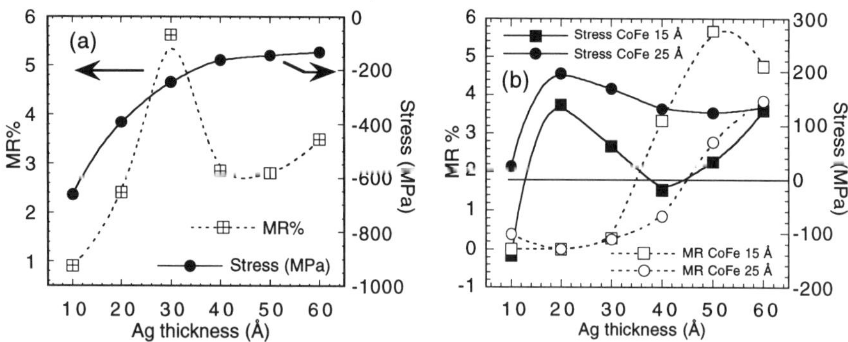

Fig. 1. As-deposited MR% and stress vs. Ag thickness for a) CoFe 7.5 Å and b) CoFe 15 and 25 Å samples [Ta 120 Å/(CoFe Y Å/Ag X Å)x20/Ta 50 Å/1 μm SiO$_2$/Si<111>].

All the MR profiles in this CoFe 7.5 Å series are reminiscent of those of single layer GGMR films [6-8]. That is, these layered-GGMR films have very broad MR profiles (unsaturated in the

12.5 kOe available field) with little hysteresis. This behavior corresponds to multilayers with ultrathin (incomplete) ferromagnetic layers. Figure 2a shows the MR profiles for as-deposited and annealed (CoFe 7.5 Å/Ag 20 Å) samples and below them is the schematic of the film morphology. Note the field values for the MR profiles. The structure of these layered-GGMR films is that of a stratified granular film. Annealing greatly increases the MR as the degree of phase separation increases over that of the metastable as-deposited state. Figure 2b shows MR profiles for as-deposited and annealed (CoFe 10 Å/ Ag 45 Å) samples. These samples display what is called 'transition' behavior where *complete* CoFe layering occurs at a minimum thickness of ~10 Å, but annealing results in the immediate agglomeration or clustering of the CoFe layers into stratified 'particles' that leads to a film structure (and MR behavior) similar to that of layered-GGMR films. The MR profiles of Fig. 2b show the dramatic broadening with annealing and the layering schematics below illustrate the film agglomeration. The third type of GMR observed in these films occurs in both the CoFe 15 and 25 Å samples. It is known as DGMR [20] and is characterized by sharp MR profiles that are very sensitive to annealing temperature. Initial annealing results in MR magnitude increase and dramatic profile sharpening whereas higher temperature annealing results in a loss of both magnitude and field sensitivity. Figure 2c shows the evolution of MR profiles with annealing temperature for (CoFe 15 Å/ Ag 50 Å) samples and the schematic below shows the layer evolution. This profile sharpening with annealing is due to ferromagnetic platelet formation from continuous layer breakup by grain boundary diffusion of Ag. The platelets in adjacent layers are negatively exchange coupled in a weak fashion that is easily overcome by small fields resulting in the sharp MR profiles. Figure 2c shows the sharpened profile and decreased hysteresis in the sample annealed at 350 °C. Annealing at 400 °C results in profile broadening and decreased MR magnitude. No GMR attributed to antiferromagnetic coupling of the CoFe layers was observed.

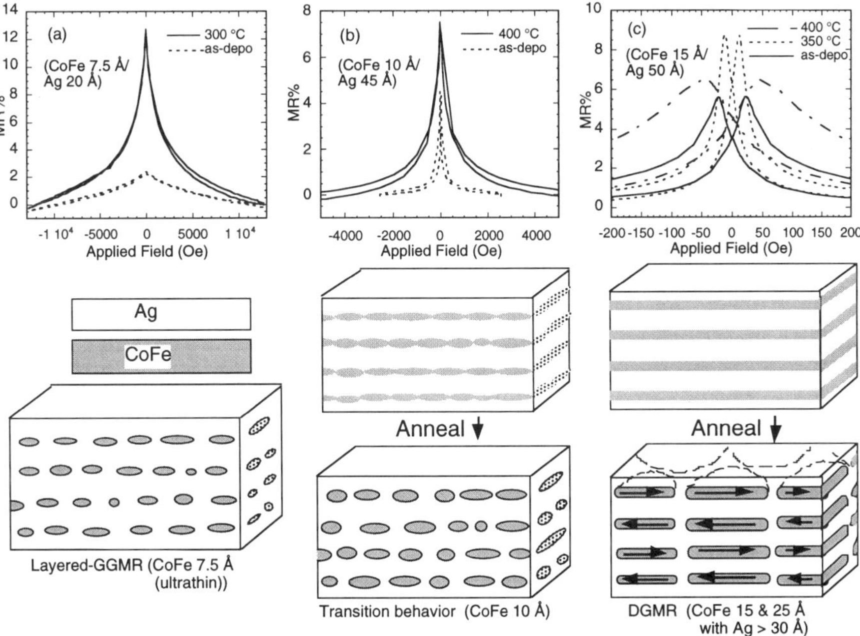

Fig. 2. Schematic of morphological evolution in multilayers as a function of layer thickness and annealing and the resulting MR profiles.

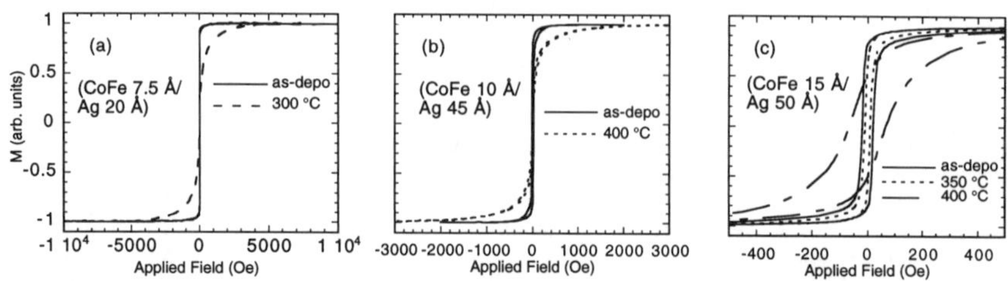

Fig. 3. *M*-H loops for as-deposited and annealed samples.

Figure 3 shows magnetization (*M*-H) loops for as-deposited and annealed a) (CoFe 7.5 Å/Ag 20) layered-GGMR , b) (CoFe 10 Å/Ag 45) transition, and c) (CoFe 15 Å/Ag 50 Å) DGMR samples all in the in-plane easy axis direction. Anisotropy was reduced with annealing in all samples. Notice the similarity between the annealed loops for the CoFe 7.5 and 10 Å samples where the saturation field increases and a significant loss of squareness was observed (increased superparamagnetic component). The coercivity remained low (<10 Oe) in these as-deposited and annealed *M*-H loops. In the DGMR *M*-H loops (Fig 3c) a decrease in coercivity after annealing at 350 °C is observed that corresponds to the decrease in hysteresis in the sharp MR profile of Fig. 2c. 'Over' annealing at 400 °C resulted in a substantial increase in coercivity and a loss of uniaxial anisotropy.

Table 1 lists the MR magnitudes and stress values with annealing in films that show each of the three MR behaviors. The transition and DGMR samples both show a tensile stress increase with annealing, but the layered-GGMR sample (CoFe 7.5 Å/ Ag 20) experiences increasing

Table 1. MR% and stress values for selected as-deposited and annealed samples

Layered-GGMR	(CoFe 7.5 Å/ Ag 20 Å)	As-depo: MR% 2.4% stress: -392.9 MPa	300 °C: MR%: 12.7% stress: -651.8 MPa	
Transition	(CoFe 10 Å/ Ag 45 Å)	As-depo: MR%: 4.5% stress: -81 MPa	400 °C: MR%: 7.5% stress: 114 MPa	
DGMR	(CoFe 15 Å/ Ag 50 Å)	As-depo: MR%: 5.3% stress: 33 MPa	350 °C: MR%: 8.8% stress: 217.8 MPa	400 °C:MR% 6.5% stress: 365.8 MPa

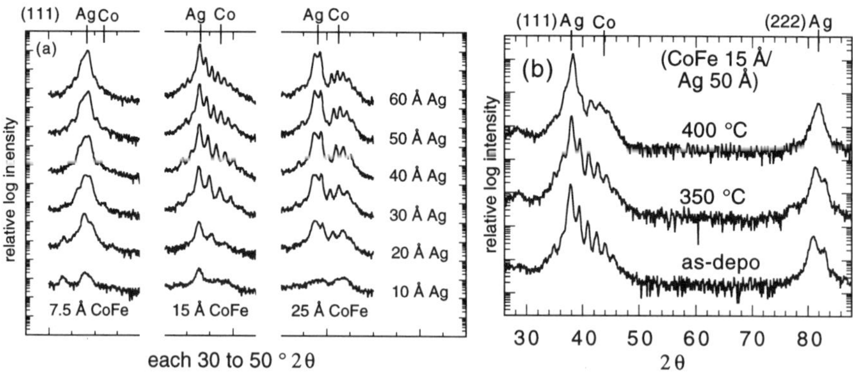

Fig. 4. HXRD scans of a) (111)-texture and satellite peaks for (CoFe 7.5, 15, and 25 Å/Ag *Y* Å) samples, and b) as-deposited and annealed (CoFe 15 Å/ Ag 50 Å) DGMR samples.

compression with annealing. This different trend may be explained by considering the layering coherency via high-angle x-ray diffraction (HXRD).

Figure 4 shows HXRD scans for a) the (111)-textured (CoFe 7.5, 15, and 25 Å/Ag 10 to 60 Å) as-deposited samples and b) as-deposited and annealed DGMR samples (CoFe 15 Å/Ag 50 Å). The satellite peak intensities from artificial layering coherency at the interfaces increases as the layer thicknesses increase. A distinct system of satellite peaks around the Co (111) peak is not observed until the CoFe 25 Å samples. Annealing the (CoFe 15 Å/Ag 50 Å) samples results in decreased satellite peak intensity as the bulk peaks begin to dominate.

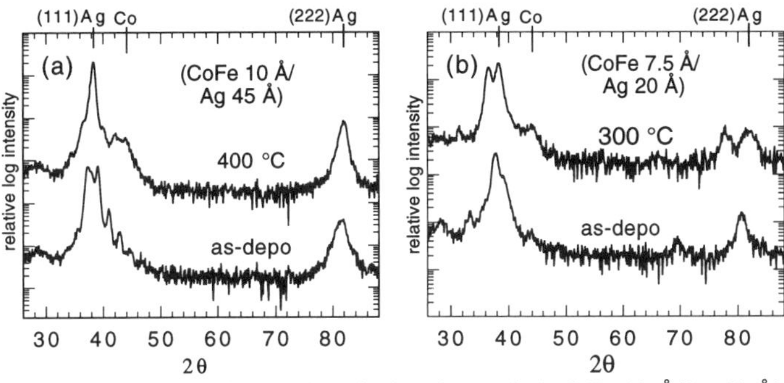

Figure 5 shows the HXRD scans for as-deposited and annealed CoFe 10 Å transition samples and the CoFe 7.5 Å layered-GGMR samples. The satellite peak intensities decrease with annealing in the CoFe 10 Å samples, similar

Fig. 5. HXRD scans of a) as-deposited and annealed (CoFe 10 Å/Ag 45 Å) transition samples and b) (CoFe 7.5 Å/Ag 20 Å) layered-GGMR samples.

to the (CoFe 15 Å/Ag 50 Å) DGMR samples. In the (CoFe 7.5 Å/Ag 20 Å) samples, however, there are more pronounced satellite peaks around the Ag (111) and (222) peaks after annealing, indicating an increase in artificial layering coherency. This increased layering coherency between the Ag and CoFe layers may inhibit Ag expansion with annealing since the thermal expansion coefficient of Ag is higher than Co. This may result in overall increased film compression since free expansion of Ag is disallowed by the constraining coherency at the interfaces with the CoFe layers. This source of film compression must outweigh trapped gas evolution and defect elimination that is usually associated with annealing and the subsequent increase in film tension. Low-angle XRD (LXRD) scans reveal layer degradation with annealing in all samples (not shown). Since LXRD results are more of a function of layer flatness and periodicity, it is possible that the increased presence of bonds at the CoFe and Ag interfaces (increased coherency) with annealing may occur even as the 'quality' of the layering (i.e. flatness and periodicity) degrade.

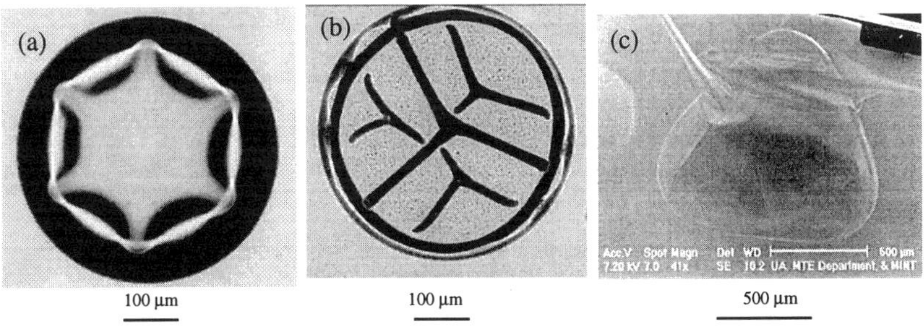

Fig. 6. a) and b). Optical photographs and c) SEM micrograph of [Ta 120 Å/(CoFe 15 Å/Ag 50 Å)x20/Ta 50 Å/1 µm SiO$_2$/Si<111>] sample annealed at 400 °C for 20 min.

Annealing the (CoFe 15 Å/Ag 50 Å) DGMR sample at 400 °C resulted in such an increase in tension that film detachment from the substrate occurred. The patterns of the detached regions have 3 and 6-fold symmetry which would agree with a (111) growth texture in the films. An additional and possibly more cogent argument would be that the thermal expansion of the Si<111> substrate caused the symmetrical film patterns. Figure 6 shows examples of the observed patterns. When the sample was placed in the SEM the patterns of Fig. 6a) and b) were not as prevalent due to the fact that the optical photographs show stretched regions of film but no actual holes in the film. The vacuum condition of the SEM resulted in bubble formation under these stretched regions. Most bubbles were circular although Fig. 6c shows a triangular bubble which may be due to the (111) film texture and/or substrate effects. Figure 7 shows an SEM micrograph of a burst bubble of stretched film. Notice the circular patterns on the substrate from two separate bubbles that coalesced.

200 μm

Fig. 7. SEM micrograph of detached film (CoFe 15 Å/ Ag 50 Å) sample annealed at 400 °C.

CONCLUSION

We have investigated the stress and GMR in CoFe/Ag multilayers. In samples with ultrathin (incomplete) CoFe layers (7.5 Å), annealing increased the granular-type GMR magnitude and the compressive stress. In thicker CoFe layered samples where the layering is complete, DGMR was observed and film tension increased with annealing. The as-deposited stress in (CoFe 15 and 25 Å/ Ag X Å) multilayers fluctuated with Ag thickness whereas in the CoFe 7.5 Å samples compressive stress monotonically decreased with increasing Ag thickness. Film detachment occurred in a recognizably 6-fold pattern in DGMR films that had large tensile stresses after annealing at high temperatures.

ACKNOWLEDGMENT

This work is supported by NSF-DMR-9301648 and NSF-PYI DMR-9157402.

REFERENCES

[1]   M.N. Baibich et al., Phys. Rev. Lett. **61**, 2472 (1988).
[2]   G. Binasch, P. Grünberg, F. Saurenbach, W. Zinn, Phys. Rev. B. **39**, 4828 (1989).
[3]   S.S.P. Parkin, N. More, and K.P. Roche, Phys. Rev. Lett. **61**, 2304 (1990). S.S.P. Parkin, Phys. Rev. Lett. **67**, 3598 (1991).
[4]   S. Araki, K. Yasui, and Y. Narumiya, J. Phys. Soc. Japan **60**, 2827 (1991).
[5]   D.H. Mosca et al., J. Magn. Magn. Mater. **93**, 480 (1991).
[6]   J.A. Barnard et al., J. Magn. Magn. Mater. **114**, L230 (1992).
[7]   J. C. Xiao, J.S. Jiang, and C.L. Chien, Phys. Rev. Lett.**68**, 3749 (1992).
[8]   A. E. Berkowitz, et al., Phys. Rev. Lett. **68**, 3745 (1992).
[9]   R. Loloee, P.A. Schroeder, W. P. Pratt Jr., J. Bass, and A. Fert, Physica B. **204**, 27 (1995).
[10] E.A.M. van Alphen and W.J.M. de Jonge, Phys. Rev. B. **51**, 8182 (1995).
[11] O. Redon, et al., J. Magn. Magn. Mater. **149**, 398 (1995).
[12] D.V. Dimitrov, A.S. Murthy, and G.C. Hadjipanayis, Mater. Sci. Engin. **A204**, 25 (1995).
[13] J.D. Jarratt and J.A. Barnard, IEEE Trans. Mag. **31**, 3952 (1995).
[14] J.D. Jarratt and J.A. Barnard, J. Appl. Phys. (in press).
[15] J.D. Jarratt and J.A. Barnard, Mater. Res. Soc. Symp. Proc. 1995, (in press).
[16] J.D. Jarratt and J.A. Barnard, IEEE Trans. Mag. (to be published).
[17] T.L. Hylton, K. R. Coffey, M.A. Parker, and J.K. Howard, Science **261**, 1021 (1993).
[18] G.G. Stoney, Proc. Roy. Soc. **82**, 172 (1909).
[19] R.W. Hoffman, <u>Physics of Thin Films</u>, Vol. 3, Academic Press, New York, 211 (1966).

REVERSIBLE FORCE-RESISTIVITY BEHAVIOR OF THIN FILMS OF THE TTF-TCNQ FAMILY

W.VOLLMANN, H.-U. SONNTAG
Department of Physics, Technical University of Chemnitz, 09107 Chemnitz, Germany,
vollmann@physik.tu-chemnitz.de

ABSTRACT

The electrical properties of vacuum sublimed thin films of TTF-TCNQ and its derivatives mainly are determined by electron barriers at grain boundaries. The electrical conductivity is thermal activated and exhibits a significant dependence on a force acting perpenticularly to the film plane. The sample resistance R decreases continiously with increasing force F. TCNQ thin films on steel show a similar R-F relation. The effect has been observed already at forces of 1 N, but also up to about 60 kN. An explanation of these phenomena is given by a grain boundary limited hopping mechanism with pressure dependent potential barrier width and height. Morphology investigations by SEM support the model.

INTRODUCTION

A decrease of the electrical resistivity with an increasing quasihydrostatic pressure in single crystals of the organic 1d conductor Tetrathiafulvalene-Tetracyanoquinodimethan (TTF-TCNQ) has already been observed by Chu, Harper et.al. [1]. Since then various pressure dependend investigations on the TTF-TCNQ family have been reported [2..5]. Explanations of it differ somewhat, and the discussions are not finished. It is noteworthy that in single crystals a pressure effect has been found to be significant first beyond pressures above 1 kbar.

Thin films, which are found to be polycrystalline, exhibit a very pressure sensitive electrical resistivity within the film plane [6] as well as perpenticular to the film plane [7], but already at smaller pressures. At TTF-TCNQ sandwich samples between metal plates an increase of the electrical current I could been observed by application of a force F between 20 N and 10 kN acting perpenticularly to the film plane. With the assumption of a homogeneous force distribution this corresponds to pressure changes between 100 mbar and 5 kbar. Similar results were obtained, when TTF was substituted by a TTF derivative [7].

In this work we report the dependence of the current through TCNQ-steel sandwich samples on the force in a range from 1 N up to 60 kN at temperatures between 250 K and 345 K and compare it with I-F relations of TTF-TCNQ and HMTTF-TCNQ. Because of the amazing similarity we try to explain the results by a film specific model, which is based on the polycrystalline morphology of all samples.

EXPERIMENTAL DETAILS

(TCNQ) possesses a bulk conductivity of $\sigma \leq 10^{-12}\Omega^{-1}cm^{-1}$ [6], but the values of its charge transfer (CT)-complexes with an organic donor can reach values up to $10^{2}..10^{3}\ \Omega^{-1}cm^{-1}$ for TTF-TCNQ and HMTTF-TCNQ single crystals and up to $2..70\ \Omega^{-1}cm^{-1}$ for thin films [8].

The TCNQ thin films have been prepared by vacuum evaporation at about 10^{-6} mbar and subsequent deposition on polished steel plates.

Mat. Res. Soc. Symp. Proc. Vol. 436 © 1997 Materials Research Society

For deposition of the CT-complexes the acceptor TCNQ and the donor (TTF or HMTTF) have been evaporated and deposited on steel substrates one after the other. Optimal evaporation temperatures were 330 K for TTF, 395 K for HMTTF and 390 K for TCNQ [9]. The substrate was kept at room temperature, only for TTF deposition a cooling to 250 K was necessary. Beginning at the interface the respective TCNQ complex is generated. This could been established for thin films by diffraction patterns and UV-vis spectrosgraphs [10]. Thicker films (> 500 nm) contain remaining neutral TCNQ. Therefore the practicable film thickness ranges from 300 nm up to 500 nm and only for pure TCNQ thin films up to 10 μm.

For pressure dependent electrical measurements two steel substrates coated with TCNQ or a TTF(derivative)-TCNQ complexe are pressed together face-to-face resulting in a sandwich sample with film thicknesses between 600 nm and 1 μm (20 μm for TCNQ, respectively). The sample area mostly is 60 mm², only for the high force measurements 100 mm². A perpenticularly acting force F in the ranges of 1N...15N, 100N...10kN and 1kN...60kN and a voltage U of 1V..10V were applied in an arrangement like demonstrated in figure 1. High forces were provided by an electromechanical presser and measured by a strain gauge sensor, whereas small forces were produced and measured by a simple lever-arm equipment with a shiftable weigth.

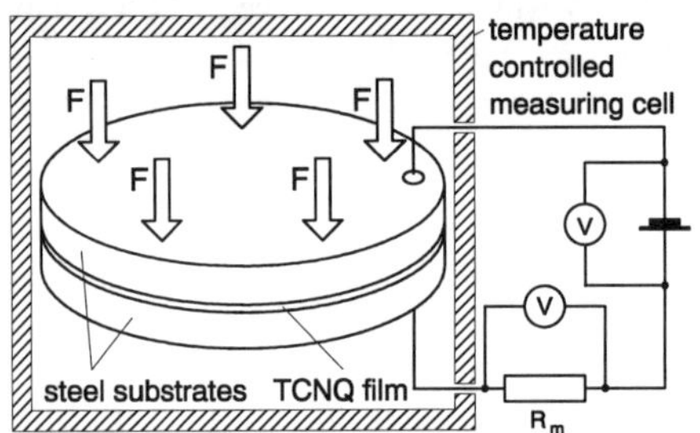

Fig.1: Schematic arrangement for force dependent measurements

RESULTS

Even though the sample resistance is the primary physical property, which is changed by the acting force, we prefer to demonstrate our results by current-force dependences. The reason is a better possibility of analytical description of that curves.

All of the current-force dependences measured at CT-complex (figures 2 and 3) as well as at TCNQ (figures 4 and 5) thin film samples show a continuous increase of the current I with increasing forces F, which can descripted by the relation
$$I = a + b\,F^{1/2}.$$
The sensitivity $b = dI/dF^{1/2}$ rises with the voltage, as shown in figure 2 for HMTTF-TCNQ and in figure 3 for TTF-TCNQ. For other derivatives of TTF-TCNQ the results are similar.

The I-U characteristics of TTF-TCNQ and HMTTF-TCNQ thin films are non-linear at high voltages always without an force [11]. It could been found a strong correlation between the crystallite length and the onset voltage U_0 of the non-linear I-U relation [8]. HMTTF-TCNQ thin films consist of long needle-shaped and TTF-TCNQ of bar-shaped crystallites in the μm range oriented mainly within the film plane (figure 6), but at thicker films sometimes out of it. The needle (bar) axis is the high conductive b-axis direction and the onset voltage U_0 decreases with increasing crystallite length.

In TCNQ thin films the crystallites grow on as stacks nearly perpendicularly to the film plane (figure 7). Here the I-U characteristics mostly are linear up to high voltages and high pressures, independent on the film thickness [12]. Thatswhy in figures 4 and 5 the results are given for only one voltage. The insert of the figures show the plotting of ln I against ln F. The slopes n of these plots correlate to the exponent of F and are always found to be nearly n = 1/2.

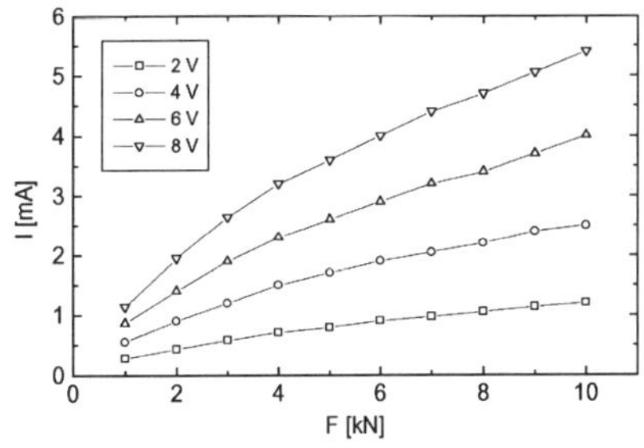

Fig.2: Current-force-relation of HMTTF-TCNQ film thickness: 500 nm

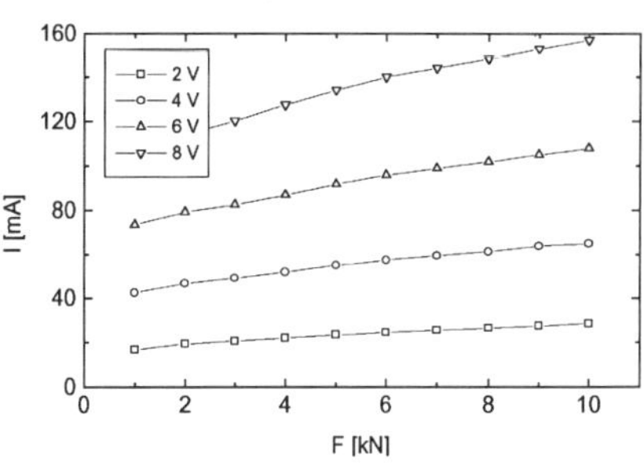

Fig.3: Current-force relation of TTF-TCNQ film thickness: 550 nm

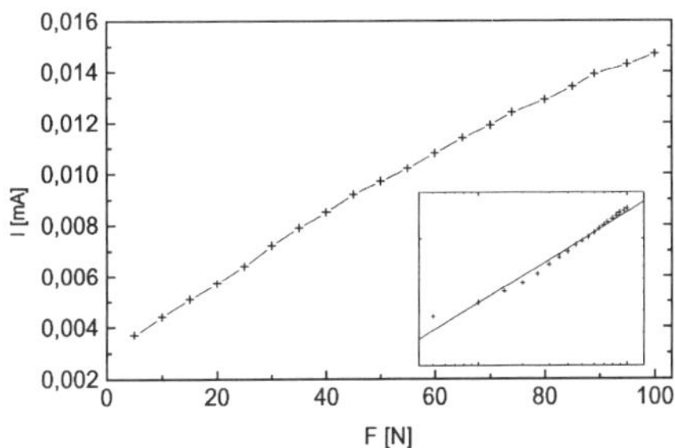

Fig.4: Current-force relation of TCNQ at low forces, film thickness: 2 μm, (insert plot: ln$I$ against ln$F$ with slope n=0.50±0.02)

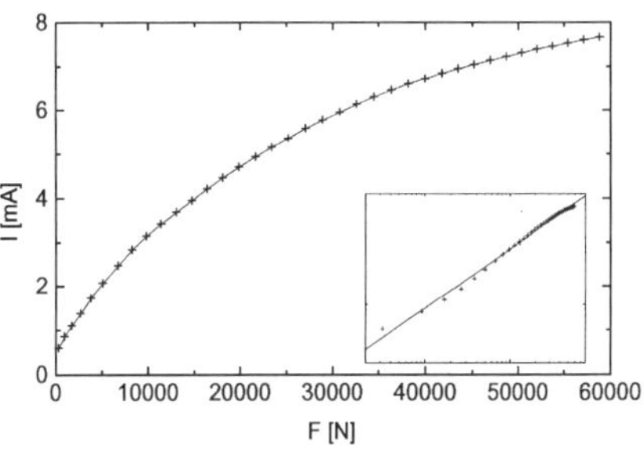

Fig.5: Current-force relation of TCNQ at high forces, film thickness: 6μm, (insert plot: ln$I$ against ln$F$ with slope n=0.53±0.01)

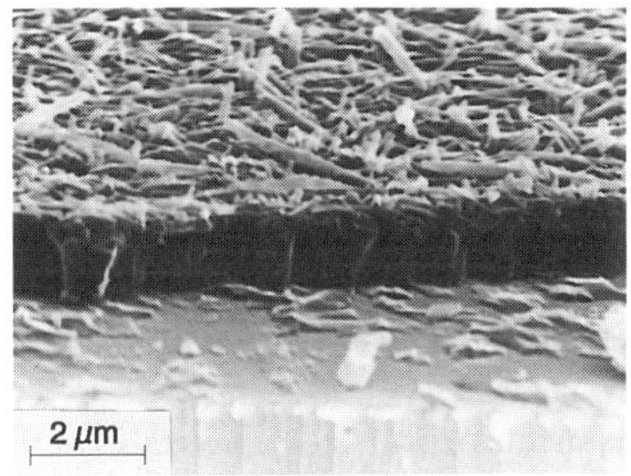

Fig.6: SEM picture of a TTF-TCNQ thin film, cross-section and top view

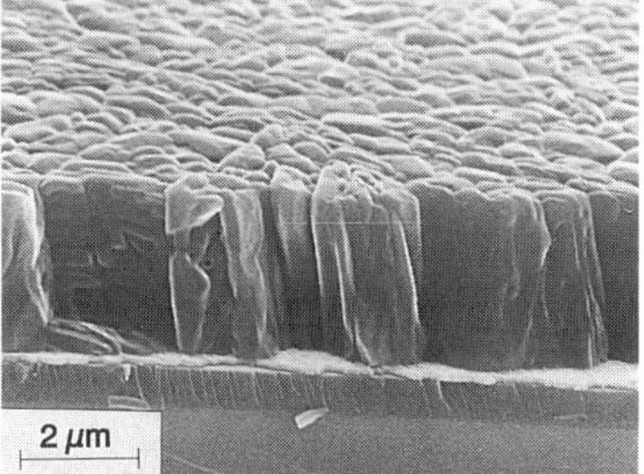

Fig.7: Cross-section SEM electronmicrograph of a TCNQ thin film,

Temperature dependent measurements at TCNQ thin films between 250 K and 345 K with forces from 1 kN up to 6 kN show that the known I-F relation is maintained in the investigated range. But the activation energies taken from an arrhenius plot of ln$I$ against $10^3/T$ (figure 8) decrease from 0,36 eV at 1 kN down to 0,24 eV at 6 kN (figure 9).

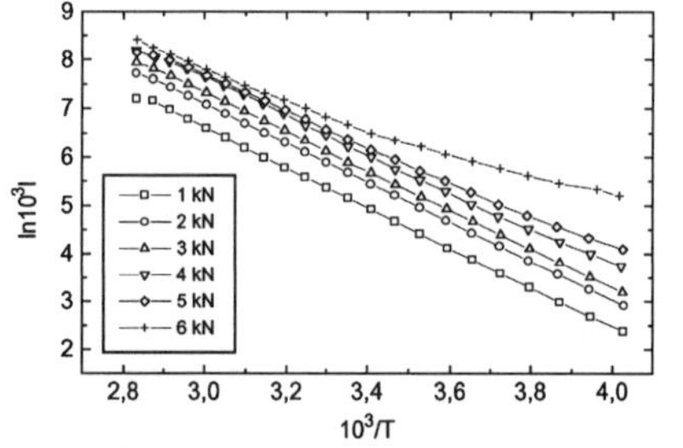

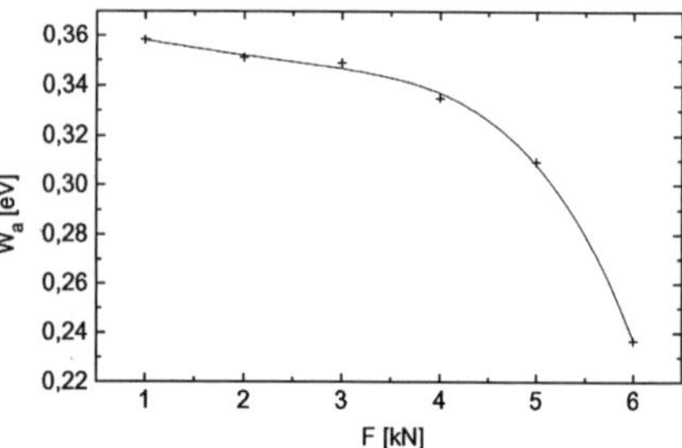

Fig. 8: Temperature dependence of the current of TCNQ at various forces (10 μm)

Fig.9: Force dependence of the activation energy (resulting from figure 8)

THEORY

The thermal activated and field enhanced conduction in polycrystalline thin films of TTF-TCNQ as well as in some other one-dimensional conducting TCNQ complexes (in opposite to the single-crystal behaviour of the same materials!) is mainly based on the grain boundaries, which produce barriers for electrons in the TCNQ stack (and for holes in the TTF stack, respectively), and these are controlling the electrical current [9, 12]. The similarity of the *I-F* relations of TCNQ and TTF suggest the supposition that the TCNQ stacks are more responsible for the pressure dependence than the TTF stacks. It would be in agreement with the observation of Bouveret and Megtert [5], who found by X-ray structural studies that TTF molecules do not exhibit any significant intramolecular deformations, whereas the TCNQ molecules undergo a large out-of-plane intramolecular distortion under pressure. On that reason for a better clearness we only consider the conduction through TCNQ stacks, that means an electron transport.

Starting from the thin film morphology (see for instance figure 6) the current path can be considered as a series connexion of high conductive crystallites of the mean length L and of the non-conducting grain boundary regions of a mean extension s shown in figure 10.

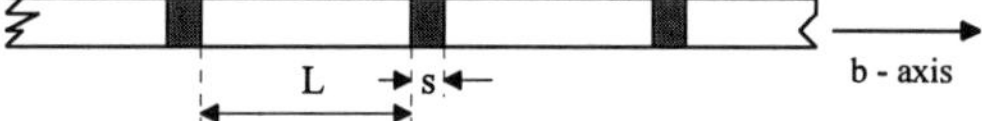

Fig. 10: Linear approximation of the current path along an array of crystallites

The overlap of the potential gradients of neighbouring crystallites with opposite signs (figure 11, curve $W(F=0)$) results in a dependence of the barrier heigth W_{A0} on the width of the disturbed range s. Additional barrier lowering by image forces is neglected in the scheme.

When the charge carrier transport within the crystallites takes place preferably along the molecul stack axis, which corresponds in TTF-TCNQ and HMTTF-TCNQ to the direction of the 1d metallic conduction, and the charge carriers overcome these barriers by thermal activated constant-distance hopping [12], the calculation results in current(I)-voltage(U) characteristics of

$$I = I_0 \exp\left(-\frac{W_{A0}}{kT}\right) \sinh \frac{eU}{2NkT} \tag{1}$$

(I_0 - current parameter, depending on various sample properties like geometry, substrate surface roughness, deposition temperature etc., N - number of crystallites in the current path between the electrodes, T - Temperature, k - Boltzmann constant) [9].

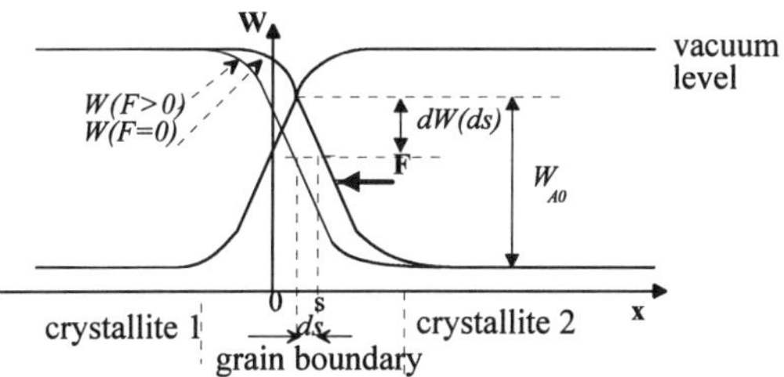

Fig.11: Scheme of the potential distribution at a grain boundary and the effect of an force F on it

An external force or a compressive load presses the crystallites together, enhances the potential overlap at the grain boundaries, and consequently, it causes the decrease of the barrier height $dW(ds)$, as can be seen in figure 11. This results in an increasing current:

$$I = I_0 \exp\left[-\frac{W_{A0} - dW(ds)}{kT}\right] \sinh\frac{eU}{2NkT} \qquad (2)$$

With the assumptions that

· $dW(ds)/kT$ is small and $\exp\dfrac{W(ds)}{kT} \approx 1 + \dfrac{dW(ds)}{kT}$, $\qquad (3)$

· the reduction of the barrier width ds is proportional to the force F, and
· the potential distribution between $x = 0$ and $x = s$ can be described by

$$W(x) = W_1 + gx^n \qquad \text{with } n \cong 0.5 \qquad (4)$$

the observed $I \sim F^n$ - dependences like in figures 2, 3, 4 and 5 can be explained. (W_1, g are parameters). The exponent n mostly is found to be about 0.5. It softly depends on the slope steepness of the potential and this depends on the nature of the grain boundary, which differs for various TCNQ complexes and for different deposition controlled crystallite orientation and sizes. The lowering of the activation energy dW with increasing forces could been measured like pointed out in figure 9 for TCNQ.

CONCLUSIONS

The observation that high conductive TTF(derivative)-TCNQ complex and TCNQ thin film sandwich samples, inspite of their difference in the electrical resistivity by many orders, exhibit similar current-force relations of the form $I = a + b\,F^{1/2}$ suggest us to describe this phenomena by a more general, but not material specific barrier limited hopping model. The most essentual assumption for justifying an application of this model is a series connexion of nearly one-dimensional conducting long crystallites (chains), which are separated by small "insulating" barriers.

A possible explanation of the similar behaviour could be that the charge carrier density (electrons) in TCNQ thin films is enhanced by a high level injection from the metal electrodes, in opposite to its bulk material. The transport of these injected electrons takes place preferably along the TCNQ stacks, which are oriented in the direction of that crystallite axis with the longest dimension. On the other hand, this is oriented nearly perpenticular to the film plane (figure 7) and therewith in the direction of the electrical field at a sandwich arrangement. The most interesting barriers in TCNQ thin films probably get out of the steel-TCNQ interfaces and somewhat less of crystallite defects. Such oriented TCNQ crystallites fill nearly the whole sample area and therewith their contribution to the conducting area is large.

TTF-TCNQ and HMTTF-TCNQ posses a much higher conductivity even in the case of thin films, but only a few of their long high-conductive needles or bars are oriented out of the film plane. The really well-conducting cross-section is only a fraction of the whole sample area. The charge transfer between TTF and TCNQ causes a high concentration of free charge carriers. These carriers move along a zigzag path from one conductive crystallite to the next. Their transport is limited by electron barriers at grain boundaries. Bulk properties, which are responsible for the pressure dependent resistivity in single crystals, are overshadowed from them and were not discussed.

REFERENCES

1. C.W. Chu, J.M.E. Harper and T.H. Geballe, Phys. Rev. Lett. **31,** 1491-1494 (1973)

2. E.M. Conwell, Solid State Commun. **33** (1), 17-19 (1980)

3. M.Weger, <u>Recent developements in condensed matter physics</u>, Vol.1, edited by J.T.Devereese (Plenum Press, New York, 1981) pp. 311-326

4. S. Klotz, J.S. Schilling, M. Weger and K. Bechgaard, Phys. Rev. **38**, 5878-5886 (1988)

5. Y. Bouveret and S. Megtert, J. Phys. (Paris) **50** (13), 1649-1671 (1989)

6. T.H. Chen and T.H. Schechtman, Thin Solid Films **30**, 173-193 (1975)

7. W. Vollmann,G. Adam and L. Thronicke, Wissenschaftliche Tagungen der Technischen Universität Karl-Marx-Stadt **6**, 267-271 (1989)

8. W. Vollmann and G. Adam, Wissenschaftliche Zeitschrift der Technischen Universität Karl-Marx-Stadt **32**, 85-93 (1990); Proceedings of the congress SENSOR 91 in Nürnberg, Vol.II, pp. 207-218 (1991); International Journal of Electronics **73**, 893-895 (1992)

9. W. Vollmann, C. Hamann, L.Libera and C. Reinhardt, Wissenschaftliche Zeitschrift der Technischen Universität Karl-Marx-Stadt **30**, 107-114 (1988)

10. W. Vollmann, studia biophysica **132,** 119-126 (1989)

11. W. Vollmann, L. Libera and G. Schukat, presented at the 1989 meeting of the Chemical Society of DDR, Karl-Marx-Stadt, 1989 (unpublished)

12. C. Reinhardt, W. Vollmann, C. Hamann, L. Libera and S. Trompler, Kristall und Technik **15**, 243-251 (1980)

13. W. Vollmann and H.-U. Sonntag, presented at the congress SENSOR 93 in Budapest, 1993 (unpublished); presented at the 1994 DPG spring-meeting in Münster, 1994 (unpublished)

Part VIII

Epitaxy and Strain-Relief Mechanisms, Measurements

STRAIN, STRUCTURE AND ELECTRONIC STATES IN MBE GROWN (NB, TI)O$_2$ MIXED RUTILE

S.A. CHAMBERS[1], Y. GAO[1], S. THEVUTHASAN[2], S.WEN[2], K.L. MERKLE[2], N. SHIVAPARAN[3], AND R.J. SMITH[3]

1 - Environmental Molecular Sciences Laboratory, Pacific Northwest National Laboratory, Richland, Washington. 2 - Materials Science Division, Argonne National Laboratory, Argonne, Illinois. 3 - Department of Physics, Montana State University, Bozeman, Montana.

ABSTRACT

We have grown and characterized epitaxial $Nb_xTi_{1-x}O_2$ on TiO_2(110) and (100) for the purpose of investigating the role of chemically-inequivalent metal atoms on the thermal and photocatalytic properties of TiO_2. Our goal is to introduce, in a highly controlled fashion, a Group VA transition metal into the lattice of a Group IVA transition metal oxide without altering the crystallographic structure. So doing would alter the electronic structure in interesting and potentially useful ways by the addition of one valence electron per substituted metal atom. However, strain builds in the film as more Nb is added at a rate which depends on the crystallographic orientation of the growth direction. Films grown along (110) can accommodate Nb mole fractions as high as ~0.3 without forming misfit dislocations, whereas those grown along (100) are limited to ~10 at. % Nb. Nb-O bond lengths in $Nb_xTi_{1-x}O_2$ are the same as Ti-O bond lengths in pure TiO_2 prior to the onset of dislocation formation. The extra 4d valence electron per Nb atom forms a nonbonding band which is degenerate with bonding states in the valence band region.

INTRODUCTION

TiO_2 is a very effective photocatalyst for the destruction of chlorinated organics [1]. Electron-hole pair production by UV light absorption coupled with low recombination velocities allows the electron and the hole to promote reduction and oxidation, respectively, of species sorbed on the surface of the particle. There are reports in the literature that the addition of a few atomic percent of Nb_2O_5 to TiO_2 powder significantly increases the extent of photocatalytic destruction of dichlorobenzene when using 5.5 eV UV light as an excitation source [2]. However, the enhancement mechanism is unknown. It is difficult to extract mechanistic information from studies of powders because each particle exposes several crystal faces, and neither the surface structure nor composition are unique or well characterized. In order to perform definitive studies on such systems, it is helpful to prepare well-defined, single-crystal surfaces which enable detailed investigations of the geometric and electronic structure to be carried out. Then, the observed surface thermal- and photochemistry of the material can be correlated with the properties of the surface. Molecular beam epitaxy is a desirable way to make such surfaces in that one can maintain control over the film composition, structure and morphology by varying the relative fluxes, growth rates, and diffusion rates across the surface. In this paper, we describe such an investigation for mixed $(Nb,Ti)O_2$ rutile films. We describe the growth and detailed characterization of the geometric and electronic structure of these materials.

EXPERIMENTAL

All MBE growth and *in-situ* measurements were carried out in a system described elsewhere [3]. *Ex-situ* high-resolution transmission electron microscopy (TEM) measurements were carried out using a JEOL 4000 EX2 microscope. Polished TiO_2 rutile single crystals oriented to within ± 1° of (110) were used as substrates. The substrates were ultrasonically cleaned in acetone and methanol prior to insertion into the MBE chamber. Once under ultrahigh vacuum (UHV), substrates were initially rid of carbon by electron beam heating at 600°C for 10 minutes while being exposed to activated oxygen from an electron cyclotron resonance (ECR) plasma source at an oxygen partial pressure of about 4×10^{-5} torr and a plasma power of 200W. However, after this treatment, the TiO_2 surfaces were found to contain trace amounts of Ba and Mg impurities as measured by x-ray photoelectron spectroscopy (XPS). Thus, a number of cycles of argon-ion sputtering and annealing in activated oxygen at 600°C were carried out until no impurities were detected. The sputtering and annealing treatment resulted in stoichiometric surfaces as shown by sharp Ti 2p peaks with no indication of Ti^{3+} by XPS. Ti^{3+} reveals the presence of oxygen vacancies, which leaves behind electrons that partially reduce nearest-neighbor Ti^{4+} cations in the lattice. In addition, the surface was well ordered after this treatment, as judged by the quality of 1x1 reflection high-energy electron diffraction (RHEED) and low-energy electron diffraction (LEED) patterns.

$Nb_xTi_{1-x}O_2$ films were grown by coevaporating Ti and Nb in an overpressure of activated oxygen from the ECR plasma source at a substrate temperature of 600°C [4-6]. Several films were grown on a single substrate with increasing Nb mole fraction in each successive film. Typical growth rates were 0.3-0.4 monolayer (ML) per second, and the oxygen partial pressure in the growth chamber during growth was typically $3\text{-}5 \times 10^{-5}$ torr with the ECR plasma source running at 200 watts. These settings constitute oxygen-rich growth conditions, and result in atomically flat and stoichiometric $Nb_xTi_{1-x}O_2$ surfaces. Structure was monitored during growth by RHEED. The relative fluxes measured from two water-cooled quartz crystal oscillators revealed Nb concentrations ranging from a few to ~40 at.%. After growth, the $Nb_xTi_{1-x}O_2$ films were transferred into the analytical chamber and characterized *in situ* by LEED, XPS, and scanned-angle x-ray photoelectron diffraction (XPD). One specimen was removed from the MBE chamber and prepared for high-resolution TEM lattice imaging.

RESULTS

Film Structure

$Nb_xTi_{1-x}O_2$ epitaxial films grow as single-domain single crystals for all compositions investigated, although misfit dislocations nucleate at higher values of x. We show in figures 1 and 2 representative LEED patterns for growth on $TiO_2(110)$ and (100) substrates, respectively. Patterns marked as x=0 are for pure TiO_2 buffer layers grown by homoepitaxy. A (3x1) reconstruction, brought about by microfacetting, is visible in the pattern for the (100)-oriented substrate [7-11] This pattern arises from annealing in ultrahigh vacuum at temperatures in excess of ~600°C. Surface reduction occurs through oxygen loss in this temperature range, and is accompanied by the formation of stable (110) microfacets. The surface reverts to a (1x1) structure upon annealing in the oxygen plasma, and after growth of TiO_2 buffer layers and mixed $(Nb/Ti)O_2$ films under oxygen rich conditions [6]. Sharp, bright patterns with very low background reveal that surfaces with excellent long-range order and smooth morphologies grow up to x = ~0.3 and ~0.1 for growth along (110) and (100), respectively. Above these mole fractions, high background develops, indicating the formation of point defects, and transmission patterns emerge in the RHEED (not shown), revealing significant roughening of the surfaces. It is at these Nb concentrations that the strain energies in the films exceed the energies required to nucleate misfit dislocations. This transition can be clearly seen in high-resolution TEM lattice images. One such image is shown in figure 3, which shows the interface of $Nb_{0.28}Ti_{0.72}O_2$ and $Nb_{0.43}Ti_{0.57}O_2$, both grown on a (110)-oriented substrate The primary beam was oriented along

$[\bar{1}11]$, and the bright spots at the corners of each parallelogram are rows of Nb and Ti cations. Oxygen rows are not visible because of the weak scattering strength of oxygen anions. Dislocations begin to appear after the growth of several monolayers of $Nb_{0.43}Ti_{0.57}O_2$. Although point defects are not visible in TEM lattice images due to a lack of long-range order, the appearance of high background in the RHEED and LEED patterns at mole fractions in excess of ~0.3 on (110)-oriented substrates reveals that such defects are generated along with the misfit dislocations.

Interestingly, there is no measurable change in metal-oxygen bond lengths surrounding Nb cations when Nb substitutes for Ti in the lattice. We have performed scanned-angle XPD at

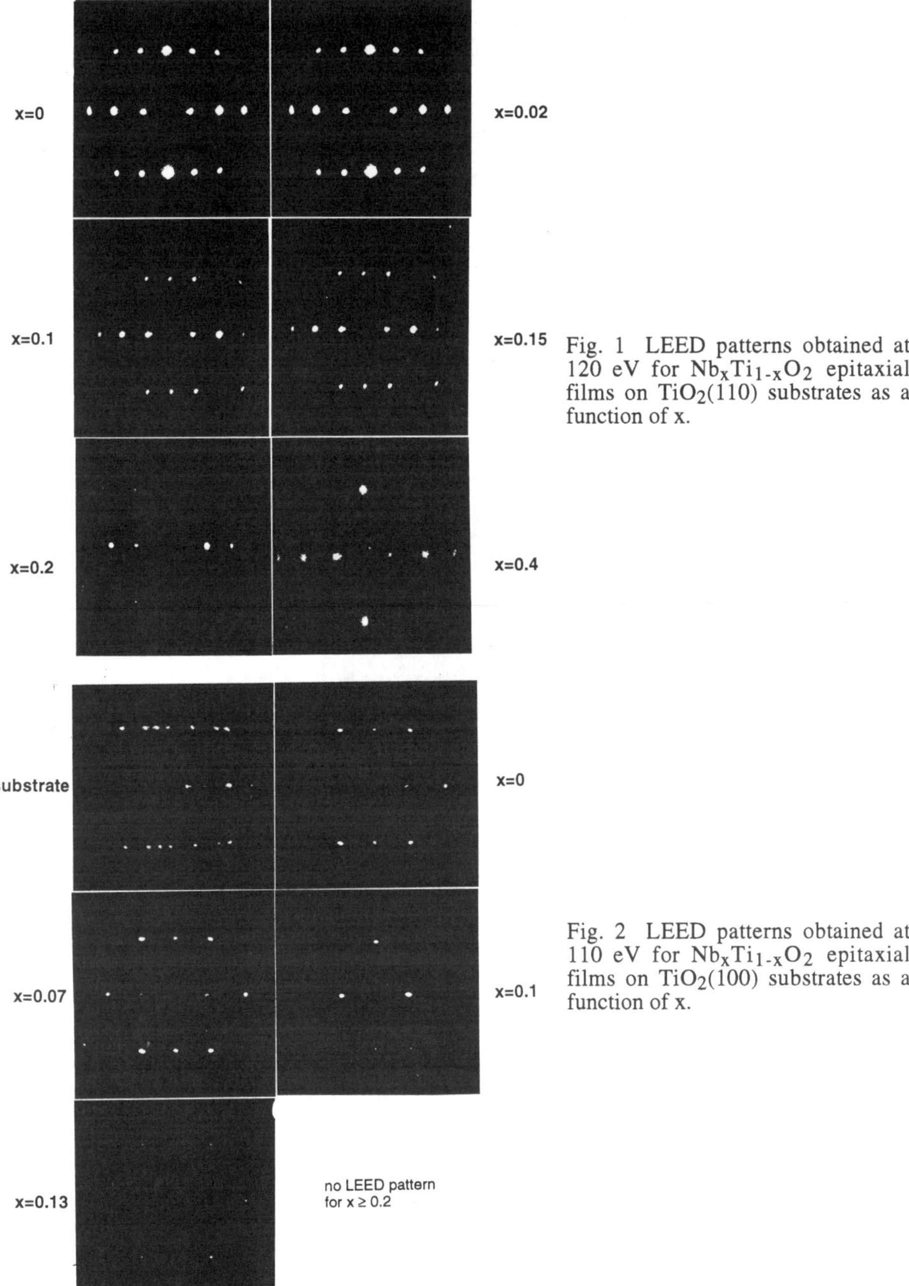

x=0

x=0.02

x=0.1

x=0.15

Fig. 1 LEED patterns obtained at 120 eV for $Nb_xTi_{1-x}O_2$ epitaxial films on $TiO_2(110)$ substrates as a function of x.

x=0.2

x=0.4

substrate

x=0

Fig. 2 LEED patterns obtained at 110 eV for $Nb_xTi_{1-x}O_2$ epitaxial films on $TiO_2(100)$ substrates as a function of x.

x=0.07

x=0.1

x=0.13

no LEED pattern
for $x \geq 0.2$

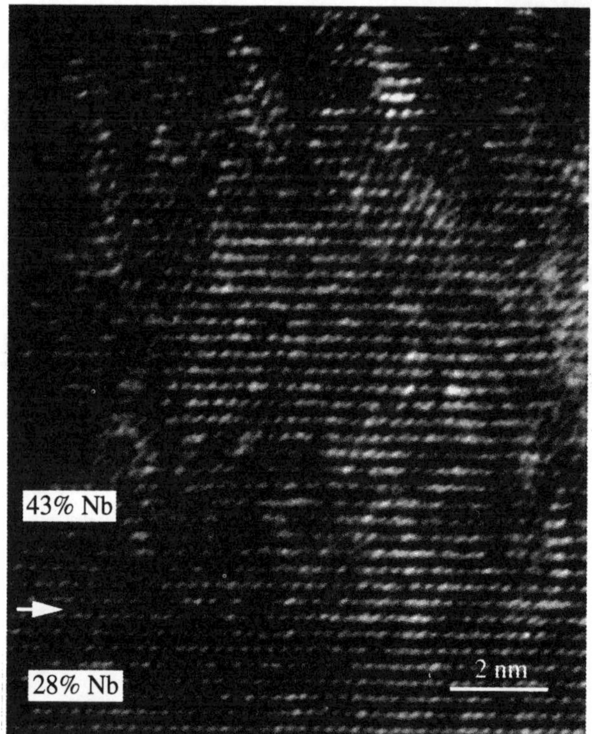

Fig. 3 High-resolution TEM lattice image viewed along [$\bar{1}$11] of the interface of epitaxial films of $Nb_{0.28}Ti_{0.72}O_2$ and $Nb_{0.43}Ti_{0.57}O_2$ grown on $TiO_2(110)$.

high angular resolution and ion channeling experiments on (110) and (100) oriented films, respectively. Analysis of these data reveal that the Nb-O bond lengths in $Nb_xTi_{1-x}O_2$ are within 0.1Å of Ti-O bond lengths in TiO_2 for x less than the critical value for misfit dislocation generation. The axial Nb-O bond length in NbO_2 rutile is 12% larger than the Ti-O axial bond length in TiO_2. The equatorial bond lengths are within 1% in the two oxides [12, 13]. Thus, the Nb-O bond length may elongate in order to relieve the interatomic repulsion resulting from Nb substituting in the smaller TiO_2 lattice. In order to do so, however, the adjacent O-Ti equatorial bond length would have to compress, which would result in considerable interatomic repulsion in that bond. This situation is denoted in the crystal diagram seen on the right side of figure 4. Also shown in figure 4 are angular distributions of Ti 2p and Nb $3p_{3/2}$ photoemission obtained at an angular acceptance of ±1° for several polar angles relative to the surface for a 5 at. % Nb-doped (110)-oriented film. The diffraction modulation is a sensitive function of the local structural environment of the emitting atom [14]. In the dilute limit (< ~10%), the modulation of Ti 2p photoemission will be identical to that observed in pure TiO_2. This result comes about because the diffraction modulation originates in scattering events within a few near neighbor shells of the emitter. In all likelihood, these shells will not contain any Nb for a given Ti emitter if x = 0.05. The Nb $3p_{3/2}$ diffraction modulation is essentially the same as that seen in the Ti 2p scan. This result reveals that there are negligible differences in the local structural environments surrounding the two kinds of atoms. This conclusion is corroborated by comparison of Nb $3p_{3/2}$ scans with those calculated with multiple scattering theory in which we assume: (i) no change in axial and equatorial bond length compared those found in TiO_2, and (ii) a 0.1Å expansion in the axial bond length surrounding substitutional Nb. These two calculated scans (not shown) are virtually the same, and agree reasonably well with experiment, suggesting that at least to within 0.1Å, there is no change in bond length upon substituting Nb for Ti [15]. This result is further

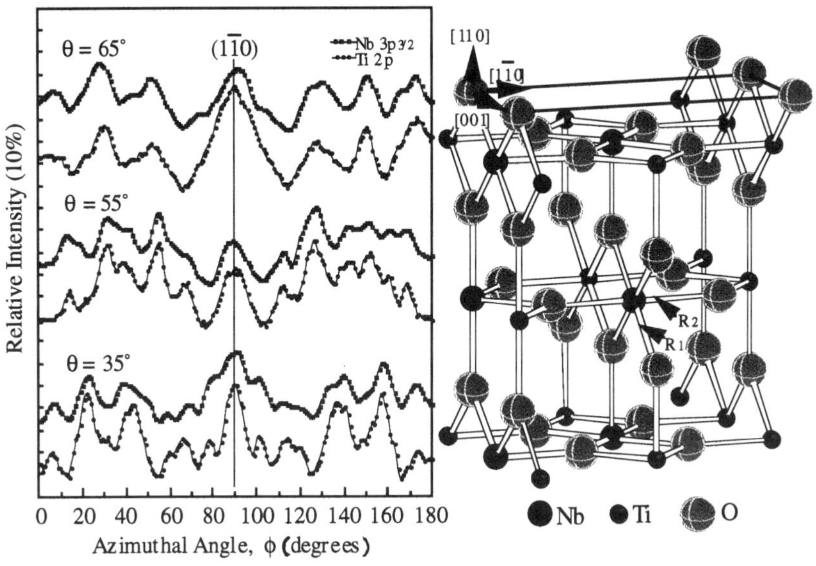

Fig. 4 Ti 2p and Nb $3p_{3/2}$ angular distributions for epitaxial $Nb_{0.05}Ti_{0.95}O_2$ on TiO_2(110).

corroborated by ion channeling rocking curves for (100)-oriented $Nb_{0.10}Ti_{0.90}O_2$, which are shown in figure 5. These scans show very narrow (~1.3°) angular widths for both Ti and Nb backscattering when the incident He beam is oriented along (100). Although we have not yet performed a detailed quantitative analysis of these results and the associated minimum yield data, it appears that to within ~0.1Å, there is no displacement of substitutional Nb from bulk cation lattice positions [16].

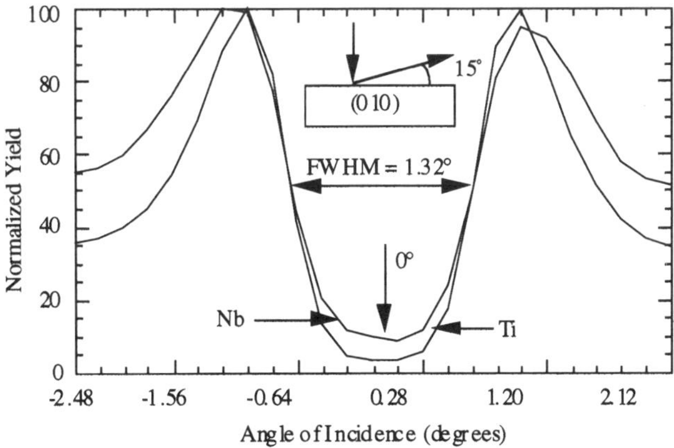

Fig. 5 Ion channeling rocking curves for epitaxial $Nb_{0.10}Ti_{0.90}O_2$ on TiO_2(100).

Surface Enhancement of Nb

Although Nb substitutes for Ti at lattice sites and does not form secondary phases, there is compelling evidence that Nb segregates to lattice sites in the near-surface region, rather than distributing equally throughout the film. XPS Ti 2p and Nb $3p_{3/2}$ core-level intensities obtained at normal emission ($\theta=90°$) and at shallow-angle emission ($\theta=35°$) suggest near-surface segregation of Nb. For instance, XPS derived values of x for a (110)-oriented film determined to be ~20 Nb at. % by the QCO are 0.25 and 0.35 for $\theta=90°$ and 35°, respectively. To further quantify this result, we have used ion channeling results to determine the extent of near-surface Nb enhancement. This analysis is summarized in figure 6 for a nominally 10 Nb at. % film grown on TiO_2(100). We have extracted the number of Nb cations per unit area as a function of depth from the channeling and backscattering spectra. The unit of area consists of a slab of twelve unit cells, as shown in fig. 6. There are twenty four cations lattice sites per slab. The analysis reveals that seven of these cation sites are occupied by Nb in the top slab, compared to three in the second slab and only one in the third (and deeper) slab. Thus, the surface-layer composition is $Nb_{0.3}Ti_{0.7}O_2$, whereas the bulk composition is $Nb_{0.04}Ti_{0.96}O_2$ [16]. This result has interesting implications for the effect of substitutional Nb on the surface chemistry of this material.

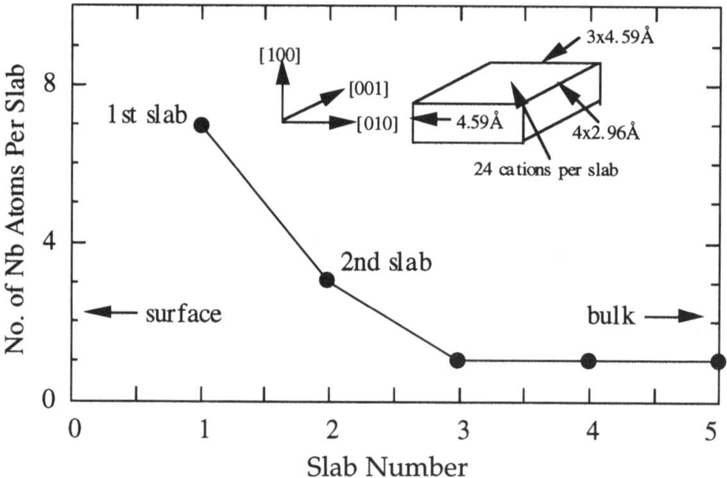

Fig. 6 Composition as a function of depth for epitaxial $Nb_{0.10}Ti_{0.90}O_2$ on TiO_2(100) based on ion scattering experiments.

Strain Accumulation and Relief

The amount of strain which accumulates in the film as a function of Nb mole fraction (x) is readily calculated using elastic theory. The calculation is particularly simple for growth along (110). In this case, the strain is biaxial, and one of the strain axes is [110], the growth direction. Strain in this direction is readily taken up by the free surface. Thus, the strain is actually uniaxial. We can estimate the strain energy along [$\bar{1}$10] in the plane of the surface by recognizing that each O-Nb-O bond pair surrounding a substitutional Nb cations is stressed by an amount (dy) equal to 2(2.23-1.98Å) = 0.52Å in the [11bar0] direction. This stress can be thought of as being distributed over some amount of length along [$\bar{1}$10] which decreases with increasing x. In the limit of a 1 at. % Nb film, this distance (y) is equal to $50R_1 + 50R_2$, where R_1 and R_2 are the equatorial and axial bond lengths in rutile, respectively. For some general Nb mole fraction x, y is given by $(50R_1 + 50R_2)/100x$. The elastic energy is then readily calculated using

Hooke's law by integrating the stress-strain product in the [$\bar{1}$10] direction. The result is $E_{el} = (1/2)Y_y\varepsilon_y^2$, where Y_y is Young's modulus and $\varepsilon_y = (dy/y)$, as given above. In doing this calculation, we have used Young's modulus for polycrystalline TiO_2 since the crystallographic dependence of Y_y is not known for TiO_2. The elastic energy is then estimated to be $E_{el} = 9.9x^2$ (GPa), and is equal to 0.89 GPa at x=0.30, the point at which dislocations form during growth along the (110) direction.

The factor-of-three difference in the amount of Nb that can be accommodated in (110)- and (100)-oriented films before relaxation occurs is a result of anisotropic strain. As mentioned above, the axial bond length, R_2, increases by 12% in going from TiO_2 to NbO_2, whereas the equatorial bond length, R_1, increases by only 1% [12, 13]. Thus, the amount of in-plane stress in each film depends on the orientations of R_1 and R_2 with respect to the growth plane. In $(Nb,Ti)O_2(110)$, alternating rows of cations have R_2 lying entirely in the growth plane; R_2 is oriented along the growth direction in the other rows. In contrast, *every* cation row has a component of R_2 in the growth plane in $(Nb,Ti)O_2(100)$. Thus, only half the cations are subjected to in-plane stress when the growth plane is (110) whereas all cations are experience in-plane stress when growing along (100). These geometries are shown in figure 7. As a result of this anisotropic lattice strain, we expect that more Nb can be accommodated during growth along (110) than along (100). In principle, the in-plane stress would be negligible up to x=0.5 if all the incident Nb accumulated in the unstrained rows on the (110) growth front. However, we expect a statistical distribution of Nb in both strained and unstrained rows on (110), so that film strain would build and relaxation would presumably occur at mole fractions less than 0.5 along (110), which is consistent with our observations [5].

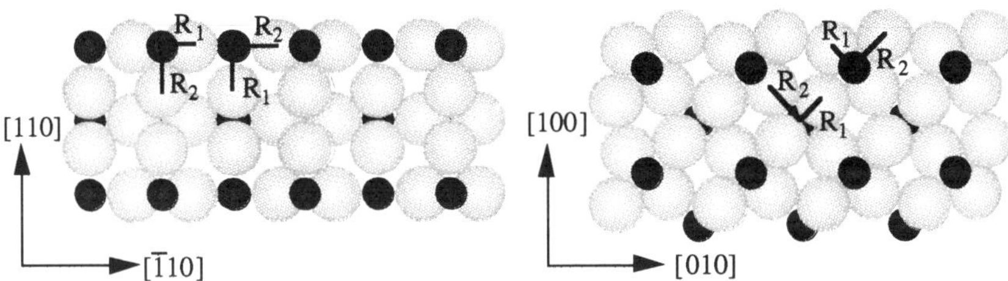

Fig. 7 Side views of $TiO_2(110)$ (left) and (100) (right) lattices showing the orientations of the axial (R_2) and equatorial (R1) bond lengths relative to the growth planes.

Electronic Structure of $Nb_xTi_{1-x}O_2$

Having established the geometric structure of $(Nb,Ti)O_2$ mixed rutiles, it is of interest to determine the electronic structure. We have done so with a combination of XPS and UPS valence band measurements. When present at the parts-per-thousand level, substitutional Nb creates a shallow donor level in the forbidden gap, making the material an n-type semiconductor. It is expected that this impurity state will evolve into a Nb-derived band which will overlap the conduction band as the amount of Nb increases, making the material metallic. However, the material was observed to remain transparent and insulating as more Nb was added. Valence band measurements reveal that the new occupied state density introduced by the addition of one electron per substitutional Nb does not develop near the conduction band, but rather in the valence band region. We show in figure 8 valence band measurements at normal emission excited with MgKα x-rays (1253.6 eV) and He I ultraviolet radiation (21.2 eV) as a function of x for growth along (110). The new occupied state density starts ~4-5 eV below the conduction band minimum (CBM) and extends to ~10 eV below the CBM. There is some new state density seen ~1 eV below the CBM in the He I excited spectra for x=0.10, but this feature was due to oxygen vacancy defects in the film created by growing under slightly oxygen poor conditions.

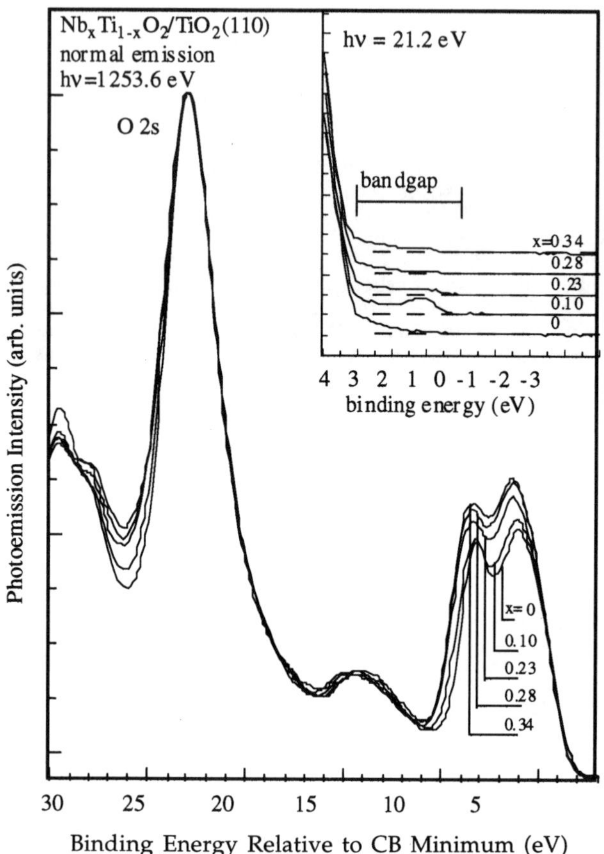

Fig. 8 Valence band spectra for epitaxial Nb$_x$Ti$_{1-x}$O$_2$/TiO$_2$(110) as a function of x.

This experiment was done to see if the Nb$_{0.1}$Ti$_{0.9}$O$_2$ films could be made conductive by creating a low concentration of oxygen vacancies. A control experiment in which growth was carried out under oxygen-rich conditions showed no new state density anywhere in the gap at x–0.10 relative to that for pure TiO$_2$.

The substantial movement of the Nb-derived impurity level from an energy close to the conduction band minimum at the parts-per-thousand level to well within the valence band for Nb concentrations of at least several at. % raises interesting questions about the nature of this state. To appreciate these questions more fully, it is useful to consider the electron counting rule for compound semiconductors and insulators [17, 18]. This rule states that the number of electrons donated by each kind of atom in a particular bond is such that the bond will have a total of two electrons. In TiO$_2$, each Ti ion donates four valence electrons to each of six ionic bonds, or 0.67 electron per bond. Similarly, each O anion donates four valence electrons to each of three ionic bonds, or 1.33 electron per bond. Thus, there are 2.00 electrons per bond. On the (110) surface, the coordination of each bridging oxygen and each "exposed" Ti is reduced by one relative to the bulk. This reduced coordination produces a dangling bond at each exposed cation and bridging oxygen site. There is nominally 0.67 electron per dangling cation bond, but the principle of autocompensation requires that this charge be transferred to the bridging oxygen dangling bond state. Since bridging oxygens and exposed Ti cations are present in a 1:1 ratio on the surface, all bridging oxygen dangling bonds are completely filled and all exposed cation dangling bonds are completely empty.

When Nb substitutes for Ti to form $Nb_xTi_{1-x}O_2$, the $3d^24s^2$ electrons which would normally be donated by Ti to the O 2p-derived valence band are replaced by four of the five electrons from the $4d^45s^1$ Nb free-atom electron configuration. To first approximation, the valence band character is not expected to change with Nb substitution due to its largely O-2p character. The central question is - what happens to the extra Nb valence electron? Based on the electron counting rule, this electron cannot go into bonding states in the valence band, since each bond in the valence band already possesses a full complement of two electrons. Therefore, based on the electron counting rule, these extra electrons are expected to form a Nb-derived nonbonding band. Similarly, on the surface, the extra valence electron per Nb ion cannot transfer to bridging oxygen dangling bonds, because the latter are completely filled. Therefore, these electrons remain localized in Nb dangling bond states. The data presented in figure 8, together with the fact that these materials are insulators, make it clear that the new Nb-derived states are degenerate with the valence band. However, compliance with the electron counting rule precludes these states from being bonding in nature. Therefore, we conclude that these electrons are in localized, Nb-derived nonbonding states.

DISCUSSION

The electronic structure model presented above has interesting implications for small-molecule chemisorption on $Nb_xTi_{1-x}O_2(110)$. CO is amphoteric toward metals in a Lewis acid/base sense, and can adsorb by either electron donation from the CO-σ orbital to the metal, or via back donation of metal d electrons to unoccupied CO-π^* molecular orbitals. In interacting with oxides, CO may act as a Lewis acid, provided the cation has some available d electrons to donate to the π^* molecular orbital. However, CO does not adsorb on $TiO_2(110)$ at temperatures as low as 110K, presumably due to the lack of available d electrons at Ti cation sites. However, CO adsorption at 110K may occur on $Nb_xTi_{1-x}O_2$ because of the availability of the nonbonding Nb electrons, which may increase the Lewis basicity of Nb cation sites. In contrast, Lewis bases such as H_2O should not bind any more strongly to Nb cation sites on $Nb_xTi_{1-x}O_2(110)$ than to Ti cation sites on pure $TiO_2(110)$. Preliminary experiments indicate that these two predictions are accurate, although more extensive experiments are planned for the near future.

The additional state density in the valence band region brought about by the formation of a Nb-derived, nonbonding band which is degenerate with the valence band is not sufficient to account for the factor-of-two increase in extent of photochemical destruction of dichlorobenzene (DCB) [2]. The effect was seen for only a few atomic percent addition of Nb_2O_5 to TiO_2, and actually decreases with the addition of more Nb_2O_5. Higher state density 4 to 10 eV below the conduction band minimum is observed by electron energy loss spectroscopy to give rise to increased dipole oscillator strength for transition energies of 4 to 8 eV, but this increase is not as great as that observed in the extent of DCB destruction [2]. Thus, we tentatively conclude that in addition to adding oscillator strength in the ultraviolet, which would increase the rate of electron-hole pair excitation, substitutional Nb in the lattice also decreases the recombination velocity. The combined effect of higher electron-hole pair production rate and longer lifetimes may explain the 100% increase in extent of DCB destruction. However, more work is needed to further elucidate this mechanism.

Finally, substitutional Nb in TiO_2 has been found to reduce the extent of interfacial reaction with thin films of Pt metal to almost zero compared to the case of pure TiO_2. These reactions give rise to the strong-metal support interaction (SMSI) effect, which poisons metal catalysts which are supported by oxide particles [19]. In this effect, interfacial reaction between the metal and the oxide gives rise to suboxide formation and subsequent diffusion over the surface of the metal. The presence of a thin suboxide film on the catalyst surface poisons the latter. Based on the relative extents of reaction of Pt with bulk reduced and fully stoichiometric TiO_2, we have proposed that oxygen vacancy migration from the bulk to the surface (or, equivalently, oxygen migration from the surface to the bulk) through interdiffusion is a major driver in the formation of suboxides at the Pt/TiO_2 interface [20]. The tendency of Nb-doping to further impede suboxide formation may come about because of lower oxygen vacancy diffusion coefficients in $Nb_xTi_{1-x}O_2$ compared to those in pure TiO_2. The fact that the vacuum annealing process which makes TiO_2 sufficiently semiconducting to do STM leaves $Nb_xTi_{1-x}O_2/TiO_2(100)$ in an insulating state supports this hypothesis. Indeed, we have found that it is exceedingly difficult to bulk reduce $Nb_xTi_{1-x}O_2/TiO_2$ specimens by annealing in vacuum. The inability of oxygen vacancies to diffuse to the surface would make the interface more inert with respect to TiO_x suboxide formation, which is what we observe.

CONCLUSIONS

We have explored the synthesis of model, single-crystal phases of $Nb_xTi_{1-x}O_2$ by MBE for the purpose of definitive surface thermal- and photochemical research on mixed-metal oxides. We have found that it is possible to selectively modify the electronic structure of the oxide without changing its geometric structure by adding Nb up to the point that misfit dislocations form. This concentration is as high as 30 at. % in the (110) orientation. The addition of Nb to the lattice at concentrations below the onset of dislocation formation produces strained but virtually defect free rutile with the same lattice parameters as TiO_2 (to within ~0.1Å), and a new Nb-derived nonbonding band which is degenerate with the valence band. The modified electronic structure changes the extent and nature of surface chemistry with small molecules and metal overlayers, as well as the photochemical activity of the surface with respect to the destruction of chlorinated organics.

ACKNOWLEDGMENTS

The authors gratefully acknowledge partial support from the US Department of Energy, Office of Basic Energy Sciences, Materials Science Division under Contracts DE-AC06-76RLO 1830 (PNNL) and W-31-109-ENG-38 (ANL). The ion channeling analysis at Montana State University was supported in part by NSF Grant DMR 9409205.

REFERENCES

[1]   A. Mills, R. H. Davies and D. Worsley, Chem. Soc. Rev., 417 (1993).

[2]   H. Cui, K. Dwight, S. Soled and A. Wold, J. Solid State Chem. **115**, 187 (1995).

[3]   S. A. Chambers, T. T. Tran and T. A. Hileman, J. Mat. Res. **9**, 2944 (1994).

[4]   Y. Gao, Y. Liang and S. A. Chambers, Surf. Sci. **348**, 17 (1996).

[5]   Y. Gao and S. A. Chambers, J. Mater. Res. **11**, 1025 (1996).

[6]   S. A. Chambers, Y. Gao, S. Thevuthasan, Y. Liang, N. R. Shivaparan and R. J. Smith, J. Vac. Sci. Technol. to appear, (1996).

[7]   P. Zschack, J. B. Cohen and Y. W. Chung, Surface Sci. **262**, 395 (1992).

[8]   P. W. Murray, F. M. Leibsle, H. J. Fisher, C. F. J. Flipse, C. A. Muryn and G. Thornton, Phys. Rev. B **46**, 12877 (1992).

[9]   P. W. Murray, F. M. Leibsle, C. A. Muryn, H. J. Fisher, C. F. J. Flipse and G. Thornton, Phys. Rev. Lett. **72**, 689 (1994).

[10] P. W. Murray, F. M. Leibsle, C. A. Muryn, H. J. Fisher, C. F. J. Flipse and G. Thornton, Surf. Sci. **321**, 217 (1994).

[11] P. J. Hardman, N. S. Prakash, C. A. Muryn, G. N. Raikar, A. G. Thomas, A. F. Prime, G. Thornton and R. J. Blake, Phys. Rev. B **47**, 16056 (1993).

[12] B. G. Hyde and S. Andersson, *Inorganic Crystal Structures*, (John Wiley & Sons, New York, 1989).

[13] R. W. G. Wyckoff, *Crystal Structures*, (John Wiley & Sons, New York, 1963), 1.

[14] S. A. Chambers, Adv. in Phys. **40**, 357 (1991).

[15] S. A. Chambers, Y. Gao, Y. J. Kim, M. A. Henderson, S. Thevuthasan, S. Wen and K. L. Merkle, Surf. Sci. , to appear (1996).

[16] S. Thevuthasan, N. Shivaparan, R. J. Smith, Y. Gao and S. A. Chambers, unpublished.

[17] J. P. LaFemina, Crit. Rev. Surf. Chem. **3**, 297 (1994).

[18] M. D. Pashley, Phys. Rev. B **40**, 10481 (1989).

[19] S. J. Tauster, S. C. Fung and R. L. Garten, J. Amer. Chem. Soc. **100**, 170 (1978).

[20] Y. Gao, Y. Liang and S. A. Chambers, Surf. Sci., to appear, (1996).

ANISOTROPIC BEHAVIOUR OF SURFACE ROUGHENING IN LATTICE MISMATCHED HETEROEPITAXIAL THIN FILMS

Cengiz S. Ozkan and William D. Nix
Materials Science and Engineering Department, Stanford University, Stanford, CA 94305

Huajian Gao
Mechanical Engineering Department, Stanford University, Stanford, CA 94305

ABSTRACT

Heteroepitaxial $Si_{1-x}Ge_x$ thin films deposited on silicon substrates exhibit surface roughening via surface diffusion under the effect of a compressive stress which is caused by a lattice mismatch. In these films, surface roughening can take place in the form of ridges which can be aligned along <100> or <110> directions, depending on the film thickness. In this paper, we investigate this anisotropic dependence of surface roughening and present an analysis of it. We have studied the surface roughening behaviour of 18% Ge and 22% Ge thin films subjected to controlled annealing experiments. Transmission electron microscopy and atomic force microscopy have been used to study the morphology and microstructure of the surface ridges and the dislocations that form during annealing.

INTRODUCTION

Strained layer semiconductor structures provide possibilites for novel electronic devices. When semiconductor layers are deposited epitaxially on a single crystal substrate with the same structure but a slightly different lattice parameter, the semiconductor layer grows pseudomorphically with a misfit strain that can be accomodated elastically below a critical thickness. When the critical thickness is exceeded, the elastic strain energy builds up to a point where it becomes energetically favorable to form misfit dislocations at the interface. In the case of $Si_{1-x}Ge_x$ films deposited on silicon substrates, the lattice misfit is negative which results in a compressive stress in the film. These films exhibit surface roughening via surface diffusion.

In this paper, we present observations of the anisotropic behaviour of surface roughening via controlled annealing experiments in both subcritical and supercritical heteroepitaxial $Si_{1-x}Ge_x$ thin films deposited on silicon substrates. The films were initially flat prior to annealing. Surface roughening takes the form of ridges with a characteristic amplitude and wavelength depending on the film thickness and composition. In the case of subcritical films, surface roughening results in the formation of ridges along <100> directions. Previous observations of surface roughening [1] in supercritical films have shown that the ridges form along <110> directions. Here we show that in the early stages of roughening, the ridges are aligned along <100> directions even for supercritical films. Only in the later stages of roughening do the ridges align along <110> directions. In the following, we first present an analysis which shows that ridge formation should always occur preferentially along <100> directions in the absence of any dislocation formation, for any film thickness. During surface roughening in supercritical films, the film surface evolves into a cycloid or cusp-like shape in which the valley regions act as dislocation generation sites in the latter stages, whereupon the formation of a misfit dislocation network causes the ridge formation to rotate from <100> to <110> directions.

EXPERIMENTAL WORK

Heteroepitaxial films containing 22% Ge and 50 nm in thickness were deposited on 100 mm (100) type bare silicon substrates in a single wafer chamber ASM Epsilon-1 CVD system. This CVD system is equipped with a rotating susceptor (35 rpm) which improves thickness uniformity across the substrate. An RCA clean was done on the substrates; this was followed by a high temperature bake in the reactor for thermal desorption of the passivating oxide and a surface etch using HCl at 1185 °C. A 500 nm thick Si buffer layer was deposited at 800 °C prior to $Si_{1-x}Ge_x$ deposition to ensure a high quality nucleation layer. $Si_{1-x}Ge_x$ deposition was carried out at 550-600 °C by thermal decomposition of silane and germane at 15 Torr total pressure, using hydrogen as a carrier gas. Film

Mat. Res. Soc. Symp. Proc. Vol. 436 © 1997 Materials Research Society

deposition was followed by annealing in the reactor for the required temperature and time in a hydrogen evironment. Atomic force microscopy was used to study the surface morphology of the films using a Park Scientific CP system. Plan view transmission electron microscopy was done using a Philips EM 430 operating at 300 KeV to study the morphology and microstructure of surface ridges or islands and to investigate any defects formed. Thin foils for XTEM were fabricated using a standard procedure of stack construction, slicing, thinning, disc cutting, dimpling and ion milling [2].

ANISOTROPIC ELASTIC ANALYSIS

Following a perturbation analysis on a undulating surface under general anisotropic conditions using the formalism of Stroh [3], the energy change associated with surface roughening can be given by [4]

$$\frac{\partial U}{\partial A} = -\pi A Y \sigma_0^2 = -\pi A Y M^2 \varepsilon_0^2 = -(\pi Y M) W_0 A \tag{1}$$

where $(\partial U/\partial A)$ is the energy release rate with respect to amplitude A of surface undulation, M and ε_0 are the biaxial modulus and the strain in the film respectively, W_0 is the strain energy density and Y is a component of the surface admittance tensor in the direction of undulation, which was described by Hirth and Loth [5] (1982). The surface admittance tensor is a very useful quantity in describing crack tip energy release rates and dislocation energies (Gao, [4]). Hirth and Lothe have provided an explicit expression for Y for cubic crystals, in directions of two fold symmetry. The result is given by,

$$Y = \frac{1}{\left(\sqrt{C_{11}'C_{22}'} + C_{12}'\right)} \left[\frac{C_{22}'\left(\sqrt{C_{11}'C_{22}'} + C_{12}' + 2C_{66}'\right)}{C_{66}'\left(\sqrt{C_{11}'C_{22}'} - C_{12}'\right)} \right] \tag{2}$$

where C_{11}', C_{22}' and C_{66}' are stiffness terms for a specific crystallographic orientation. For a $Si_{0.8}Ge_{0.2}$ alloy and <100> direction,

$$C_{11}' = C_{22}' = C_{11} = 152 \text{ GPa}$$
$$C_{12}' = C_{12} = 57 \text{ GPa} \tag{3}$$
$$C_{66}' = C_{44} = 75 \text{ GPa}$$

Substituting these values, one can obtain

$$Y_{<100>} = 0.01324 \text{ GPa}^{-1} \tag{4}$$

On the other hand, for the <110> direction,

$$C_{11}' = C_{11} - \left(\frac{C_{11} - C_{12}}{2} - C_{44}\right) = 179.5 \text{ GPa}$$
$$C_{22}' = C_{11} = 152 \text{ GPa}$$
$$C_{12}' = C_{12} = 57 \text{ GPa} \tag{5}$$
$$C_{66}' = C_{44} = 75 \text{ GPa}$$

and,

$$Y_{<110>} = 0.01188 \text{ GPa}^{-1} \tag{6}$$

Finally,

$$\frac{\partial U}{\partial A} = \begin{cases} -6.92 W_0 A & <100> \\ -6.20 W_0 A & <110> \end{cases} \tag{7}$$

Therefore, <100> roughening is favored.

We now would like to present a more comprehensive analysis of the anisotropic behaviour of surface roughening. Using Hooke's law, the biaxial stress state in a thin film can be expressed as,

$$\sigma_0 = \left(C_{11} + C_{12} - \frac{C_{12}^2}{C_{11}} \right) \varepsilon_0 \quad , \tag{8}$$

where the term in brackets is the biaxial modulus of the film. The strain energy density W_0 at any point in the film can be expressed as

$$W_0 = \left(\tfrac{1}{2}\right) \sigma_i \varepsilon_i = \sigma_0 \varepsilon_0 \quad . \tag{9}$$

We first assume that the film undergoes strain relaxation due to roughening along <100> type directions at the film surface (figure 1). Furthermore, we will assume that strain relaxation occurs only along X_2 by an amount $\Delta\varepsilon$,

$$\Delta\varepsilon_2 = -\Delta\varepsilon \quad . \tag{10}$$

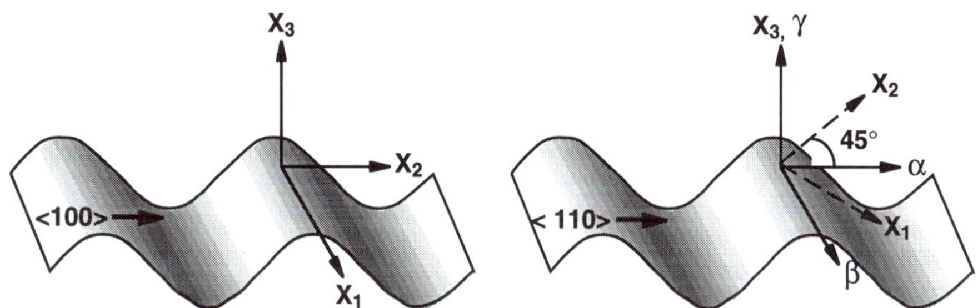

Figure 1.   Figure 2.

The corresponding stress changes associated with this strain relaxation are

$$\Delta\sigma_1 = -\left(C_{12} - \frac{C_{12}^2}{C_{11}} \right)\Delta\varepsilon \quad , \qquad \Delta\sigma_2 = -\left(C_{11} - \frac{C_{12}^2}{C_{11}} \right)\Delta\varepsilon \tag{11,12}$$

The change in the strain energy density upon relaxation is given by

$$\Delta W_{(100)} = W - W_0 \quad , \tag{13}$$

where W is the strain energy density after relaxation and can be expressed as

$$W = \frac{1}{2}\left(\sigma_1\varepsilon_1 + \sigma_2\varepsilon_2\right) = \frac{1}{2}\left\{ (\sigma_0 + \Delta\sigma_1)\varepsilon_0 + (\sigma_0 + \Delta\sigma_2)(\varepsilon_0 + \Delta\varepsilon_2) \right\} \quad , \tag{14}$$

Substituting equations (10)-(12) into equation (14), one can obtain,

489

$$\Delta W_{\langle 100\rangle} = -\varepsilon_0 \left\{ C_{11} + C_{12} - \frac{2C_{12}^2}{C_{11}} \right\} \Delta\varepsilon + \frac{1}{2}\left(C_{11} - \frac{C_{12}^2}{C_{11}} \right) \Delta\varepsilon \qquad , \qquad (15)$$

Now, we need to look at strain relaxation due to <110> roughening. Referring to figure 2, we define a new coordinate system (α, β, γ), which is oriented at 45° to the original coordinate system in the (001) plane. Furthermore, we assume that, strain relaxation in this direction can be characterized by $\Delta\varepsilon_{\alpha\alpha} = -\Delta\varepsilon$, $\Delta\varepsilon_{\beta\beta} = 0$ and $\Delta\varepsilon_{\gamma\gamma} = \Delta\varepsilon_3 \neq 0$. Transfoming these strains to the cube coordinate system and following a similar analysis, one can obtain

$$\Delta W_{\langle 110\rangle} = -\varepsilon_0 \left(C_{11} + C_{12} - \frac{C_{12}^2}{C_{11}} \right) \Delta\varepsilon + \frac{C_{44}}{2}(\Delta\varepsilon)^2 + \left(C_{11} + C_{12} - \frac{2C_{12}^2}{C_{11}} \right)\left(\frac{\Delta\varepsilon}{2} \right)^2 ,$$
$$\ldots\ldots\ldots(16)$$

Finally, we look at the difference between the two modes of surface roughening:

$$\delta\Delta W = \Delta W_{\langle 110\rangle} - \Delta W_{\langle 100\rangle} \qquad . \qquad (17)$$

With strain relaxation, $\Delta W_{\langle 110\rangle}$ or $\Delta W_{\langle 100\rangle}$ are negative values. Substituting equations (15) and (16) into equation (17), we have

$$\delta\Delta W = \frac{(\Delta\varepsilon)^2}{2}\left\{ C_{44} - \left(\frac{C_{11}-C_{12}}{2} \right) \right\} \qquad , \qquad (18)$$

from which we can conclude the following:

$$\delta\Delta W = \begin{cases} = 0, \ \Delta W_{\langle 110\rangle} = \Delta W_{\langle 100\rangle} : \ \text{Material is isotropic} \\ > 0, \ \Delta W_{\langle 110\rangle} > \Delta W_{\langle 100\rangle} : \ \langle 100\rangle \ \text{roughening is favored} \\ < 0, \ \Delta W_{\langle 110\rangle} < \Delta W_{\langle 100\rangle} : \ \langle 110\rangle \ \text{roughening is favored} \end{cases} \qquad (19)$$

In the case of heteroepitaxial $Si_{1-x}Ge_x$ films deposited on Si substrates, C_{44} is always greater than $(C_{11}-C_{12})/2$; hence, <100> roughening is always favored. This result is in qualitative agreement with the first analysis given above.

EXPERIMENTAL RESULTS AND DISCUSSION

Figure 3 shows AFM images from a series of annealing experiments on $Si_{1-x}Ge_x$ films with 22% Ge and 50 nm in thickness. All the images are representative of an area 5x5 μm^2 in size, and the features are magnified several times in the height direction. After 10 seconds at 750 °C in figure 1-a, ridges have formed along <100> directions. At this stage, the average wavelength and amplitude for the features are 12 nm and 220 nm respectively. After 10 seconds at 800 °C in figure 1-b, the <100> ridges are slightly disregistered at several locations along <110> directions. These features have formed upon dislocation half loops running along <110> directions and cutting through them at the surface while surface roughening continues.The average wavelength and amplitude for these features are 25 nm and 350 nm respectively. Deep grooves and island groups start forming along <110> directions with after 1 minute at 800 °C, as shown in figure 1-c. As expected, the features continue to grow and the average wavelength and amplitude values are 51 nm and 470 nm respectively. Finally after 5 minutes at 800 °C in figure 1-d, further roughening takes place along <110> directions, where the average wavelength and amplitude are 95 nm and 62 nm respectively.

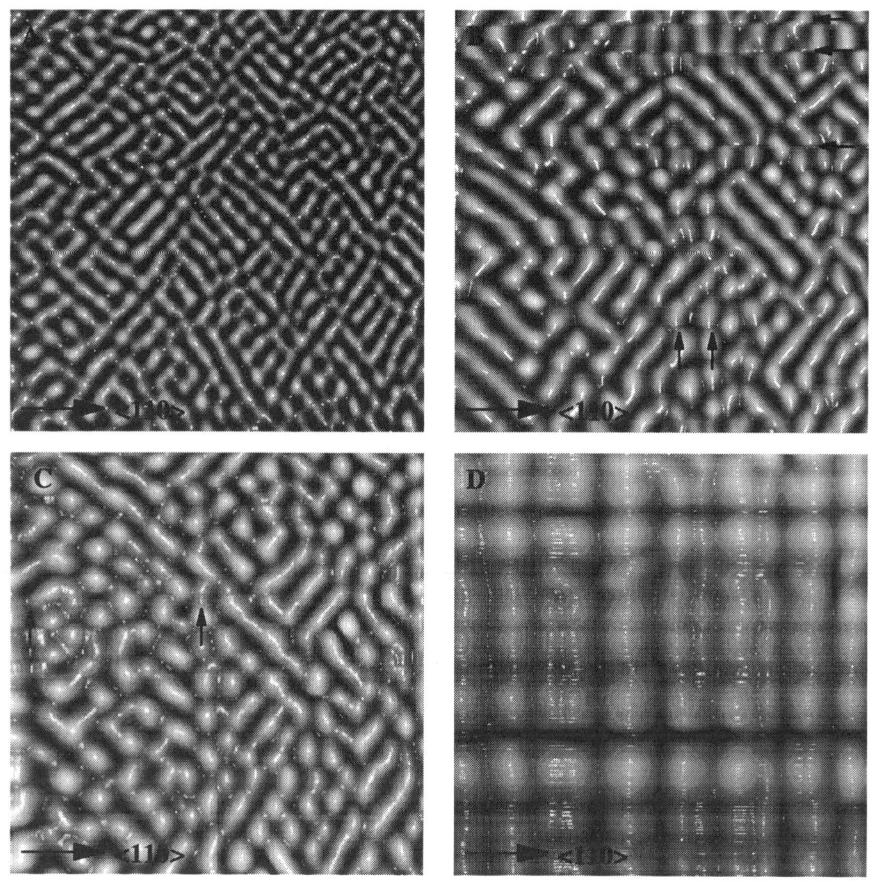

Figure 3.

Figure 4 shows a series of plan view bright field transmission elecron micrographs for the same samples. In all the micrographs, the g vector is along <220>. It is important to note that, after 10 seconds at 750 °C, no misfit dislocations have formed yet. This is an important result; no matter what the film thickness is, strain relaxation occurs by surface roughening first, in the absence of dislocation formation. Figures 2-b and 2-c show that, the misfit dislocation network formed is oriented at 45° to the surface ridges. Finally in figure 2-d, the misfit dislocation network and the surface ridges are both aligned along <110> directions. At longer anneal times or higher anneal temperatures, the dislocation density increases, as expected.

CONCLUSIONS

We have investigated the anisotropic behaviour of surface roughening in heteroepitaxial $Si_{1-x}Ge_x$ thin films. Controlled annealing experiments conducted on samples containing 22% Ge and 500 Å in thickness have shown that surface roughening begins with the formation of ridges along <100> directions, in the absence of any dislocation formation. In the later stages of roughening,

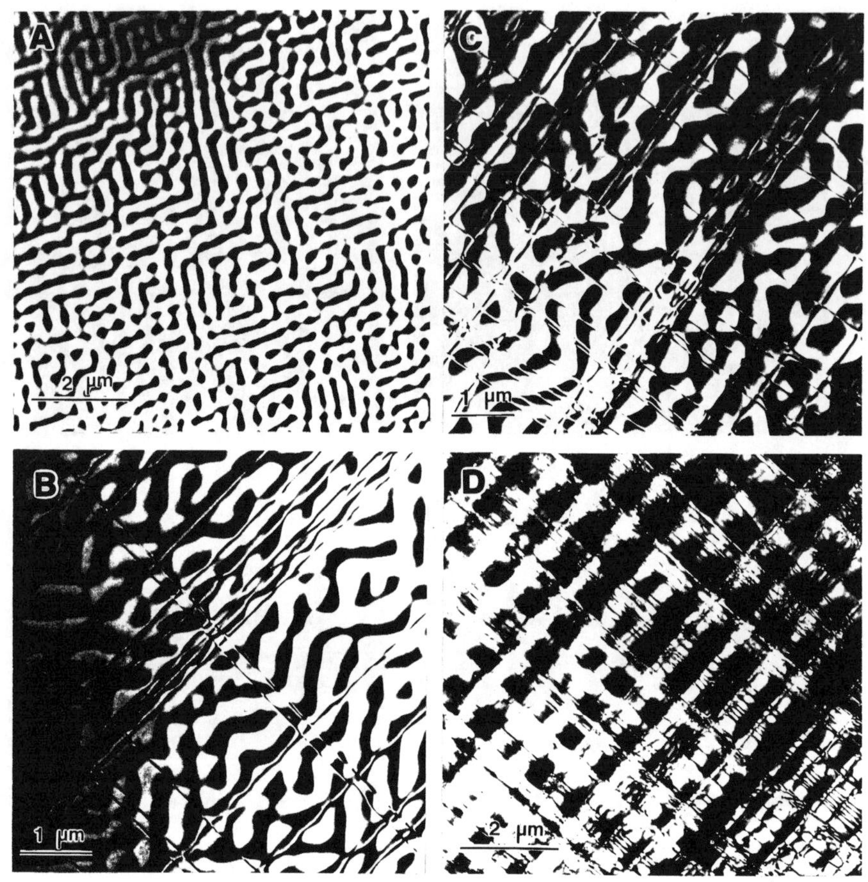

Figure 4.

misfit dislocations form first at the film/substrate interface and then the ridges at the surface rotate from <100> to <110> directions and become more perfectly aligned with the misfit dislocation network. Furthermore, the anisotropic elastic analysis has shown that surface roughening is favored along <100> type directions. Future studies will include an analysis of the effect of dislocations in rotating the ridge formation to <110> directions in the later stages of roughening.

ACKNOWLEDGEMENT

The support of the U.S. Office of Naval Rersearch under the grant ONR-N00014-92-J-4094 is greatfully acknowledged.

REFERENCES

1. C.S. Ozkan, W. Nix and H.Gao, MRS Proc., Symp. D, (1995), in print.
2. J.C. Bravman and R. Sinclair, J. Elec. mic. Tech., **1**, 53 (1984).
3. A.N. Stroh, Phil. Mag., **7**, 625 (1958).
4. H. Gao, Modern Theory of Anisotropic Elasticity and Applications, SIAM, 139 (1991).
5. J.P. Hirth and J. Loth, Theory of Dislocations, 2nd Ed., 120 (1982).

CALCULATION OF EQUILIBRIUM ISLAND
MORPHOLOGIES FOR STRAINED EPITAXIAL SYSTEMS

R. V. KUKTA and L. B. FREUND
Division of Engineering, Brown University, Providence, RI 02912

ABSTRACT

Strained islands grown coherently on a relatively thick substrate with similar elastic properties are considered within the framework of continuum mechanics. The condition of uniform surface chemical potential is imposed to calculate two-dimensional equilibrium island shapes. The stress distribution in the equilibrium islands is shown to be highly non-homogeneous. The effects of introducing a single misfit dislocation at the island-substrate interface are considered. It is found that there is critical island volume above which a dislocation decreases the total free energy of the system. The dislocation alters the stress distribution in the island, causing the island to relax via mass transport to an equilibrium shape with a lower height-to-width aspect ratio and a smaller surface chemical potential than the island prior to the introduction of the dislocation.

INTRODUCTION

The physical system discussed here is a periodic array of two-dimensional islands epitaxially grown on a relatively large substrate with similar elastic properties but with a slightly different lattice parameter. The mismatch strain is defined as $\epsilon_m = (d_S - d_I)/d_I$ where d_S and d_I are the lattice parameters of the substrate and island respectively. The islands and the substrate are modeled as isotropic elastic solids having a shear modulus G and a Poisson ratio ν. The free energy of the system is assumed to consist of the elastic strain energy and the free surface energy. Consequently, the chemical potential at a location s on the surface is[1]

$$\mu(s) = [U(s) - \kappa(s)\gamma]\,\Omega \tag{1}$$

where $U(s)$ is the local elastic strain energy density, $\kappa(s)$ is the local curvature of the surface, γ is the surface energy (assumed constant) and Ω is the atomic volume (also assumed constant). Equilibrium requires that the chemical potential along the surface is constant. For a range of island volumes, shapes which satisfy the constant chemical potential condition are calculated.

Island formation is a mechanism for the reduction of the elastic strain energy which arises in a film due to the constraint of epitaxy. During the early growth stages of some semiconductor heterostructures (e.g. Ge/Si and InGaAs/GaAs), island formation is typical. Another mechanism for strain relaxation is the formation of misfit dislocations at the island-substrate interface. In order to investigate the effect of a dislocation on the free energy of the system and on island morphology, a single misfit dislocation is introduced at the center of the each equilibrium island without concern for the mechanism of dislocation nucleation. The total free energy of the system is calculated to determine whether or not introduction of such a dislocation is energetically favorable. Equilibrium shapes of dislocated islands are then calculated and the resulting values of the chemical potential are compared to those of dislocation-free islands.

493

Mat. Res. Soc. Symp. Proc. Vol. 436 © 1997 Materials Research Society

A CONTINUUM MODEL OF THE INTERFACE

When the influence of island-substrate interface becomes important, it must be taken into consideration by means of some model. Chiu and Gao[2] modeled the interface as a boundary layer across which the surface energy changes abruptly making it energetically unfavorable for the substrate to become exposed. Spencer and Tersoff[3] utilized a thin wetting layer which could not be removed from the substrate. By constraining the wetting layer to be flat, they were able to calculate equilibrium shapes for which the chemical potential is constant only over the island surface.

Here, the interface is taken into account by assuming that there is a narrow transition region through which the strain changes by an amount equal to the mismatch strain ϵ_m. This is similar to Gibbs' notion of a surface transition region as described by Zangwill.[4] This transition region is narrower than any other physical dimensions, and further reduction in its size has no influence on the solutions. Equilibrium can then be satisfied (mathematically) along the entire surface. The island surface seeks out a shape for which the chemical potential is constant and the substrate surface assumes a position within the transition region which matches this value. The fact that the chemical potential is constant across both the island and substrate is an artifact of the model.

Using this model, the stress distribution is determined by a superposition of the mismatch stress σ_{ij}^m, the stress field of a periodic array of dislocations in an infinite medium[5] (σ_{ij}^d) and an equilibrium stress field σ_{ij}^b which is required to satisfy the traction-free boundary condition of the surface $(\sigma_{ij}^b n_j = -\sigma_{ij}^m n_j - \sigma_{ij}^d n_j$ on the surface where n_j is the outward pointing normal). The mismatch stress contains the transition region described above. Within the island, the only nonzero component of the mismatch stress is $\sigma_{xx}^m = M\epsilon_m$ where M is the plane strain film modulus. Below the transition region, within the substrate, $\sigma_{ij}^m = 0$. The nonzero component of the mismatch stress is conveniently represented by $\sigma_{xx}^m = SM\epsilon_m$ where

$$ S = \frac{1}{\pi} \tan^{-1} \left(\frac{y - y_t}{\alpha} \right) + \frac{1}{2}. \tag{2} $$

The interface is defined as the plane $y = y_t$ and α is a parameter which determines the size of the transition region.

NUMERICAL IMPLEMENTATION

The stress distribution σ_{ij} throughout the system is given by the superposition $\sigma_{ij} = \sigma_{ij}^b + \sigma_{ij}^m + \sigma_{ij}^d$. For a given surface shape, the only unknown is σ_{ij}^b. This is calculated via a boundary integral equation developed by Spencer and Meiron[6] The solution procedure is similar to theirs except for some modifications needed to incorporate the transition in mismatch stress and the dislocation field. Only symmetric periodic surface shapes are considered. The integral equation is solved along the free surface by Fourier collocation with 129 equally spaced collocation points per half-period. Once the stress distribution is known, the chemical potential is obtained from (1).

From an assumed initial surface profile, a multi-dimensional secant search is used to find a shape for which the chemical potential is constant. To test the stability of the solutions, sinusoidal perturbations are added to the solutions and the surface is allowed to evolve such that the normal velocity of the surface is $v_n = \bar{\mu} - \mu$ where $\bar{\mu}$ is the average value of the chemical potential along the surface.[7] The evolution always proceeds in the direction of decreasing free energy. If the surface is unstable with respect to a given perturbation,

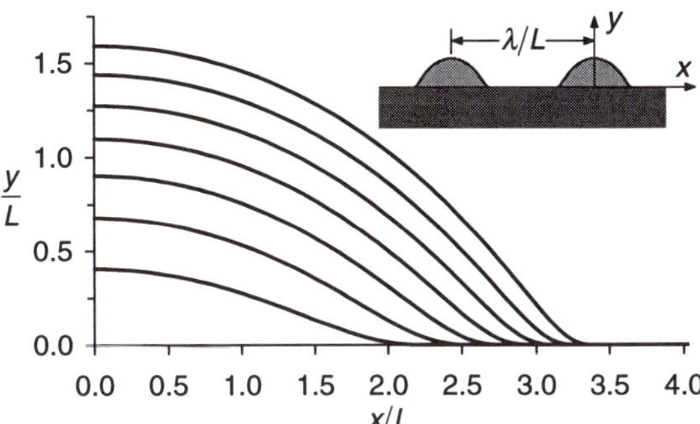

Figure 1. Equilibrium island shapes for wavelength $\lambda/L = 13$ and volumes $V/L^2 = 1, 2, ..., 7$. The characteristic length is $L = \gamma/M\epsilon_m^2$. The shapes are periodic and symmetric about $x/L = 0$. The spacing between islands is at least one island width.

the perturbation will grow and the surface will not return to its original shape. In all cases, the orignal shape is recovered.

EQUILIBRIUM ISLAND PROFILES

A set of equilibrium shapes of dislocation-free islands is shown in Figure 1 for a range of volumes. All lengths are normalized with respect to the characteristic length $L = \gamma/M\epsilon_m^2$. The islands are symmetric about $x/L = 0$ and the spacing between islands is at least one island width. The transition in mismatch stress, from $\sigma_{xx}^m = M\epsilon_m$ to $\sigma_{xx}^m = 0$, is centered at $y/L = 0$. The mismatch stress reaches 99% of its maximum value $M\epsilon_m$ at about 1/5000 of the height of each island. Over most of the surface of the islands, the curvature has a negative value, but in all cases there is an inflection point near the edges of the islands. The inflection point is not due to the presence of the mismatch transition region. In general, it occurs quite far from the end of the transition region. Further reduction in the size of the transition region has no effect on the location of the inflection points nor on the overall island shapes.

The shapes shown are for a wavelength $\lambda/L = 13$ which is large enough for the mechanical interaction between islands to have a small influence on shape. For the largest islands the interaction is not completely negligible and shapes are affected slightly. As the wavelength is decreased, the height-to-width aspect ratio increases for an island of a given volume. Increasing the wavelength further does not change the nature of the solutions.

The island sizes shown are typical of those found during the growth of semiconductor materials.[8] For example, using $\epsilon_m = -0.04$, $M = 10^{11}$ Pa, and $\gamma = 2$ J/m^2 for a Ge/Si heterostructure results in a characteristic length of $L = 12.5$ nm. The widths of the islands in Figure 1 range from 50 to 80 nm and the volumes (per unit depth) range from 160 to 1100 nm^2.

The stress distribution in the islands is nonhomogeneous with σ_{xx} (the tensile stress parallel to the interface) significantly less than the mismatch stress $M\epsilon_m$ over a large portion of the island. At the edges of the islands, stress becomes very large. The component σ_{xx} is plotted as a function of island volume in Figure 2 for three points along the y-axis, namely $y = 0$, $h/2$, and h. The stress at the bottom of the island drops with increasing volume and appears to level off at nearly 60% of the mismatch at a volume of about $V/L^2 = 4$. The stress at the top of the island drops off more dramatically with volume and appears to approach total relaxation as the volume becomes large. The stress at the

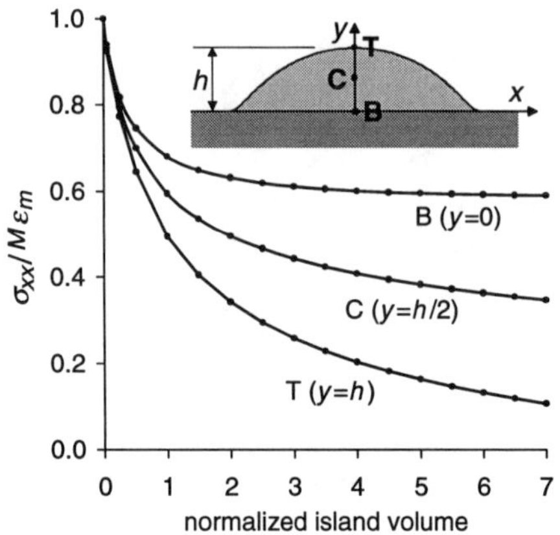

Figure 2. Plots of σ_{xx} normalized by mismatch stress versus normalized island volume V/L^2 at three points along the y-axis: $y = 0$ (point B), $y = h/2$ (point C), and $y = h$ (point T).

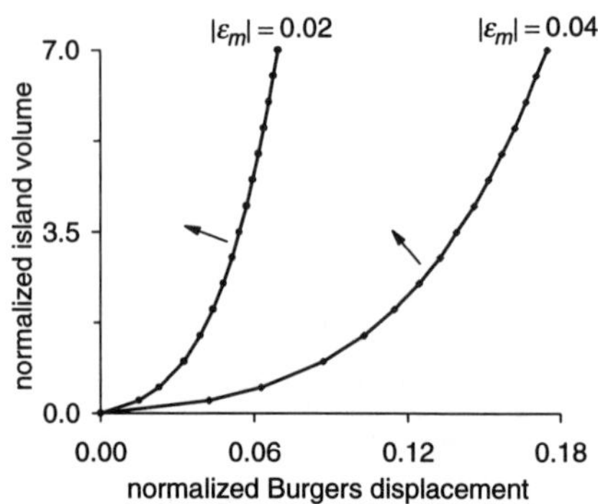

Figure 3. Plots of normalized volume V/L^2 versus normalized Burgers displacement b/L representing systems for which a misfit dislocation causes no net increase in free energy. The arrows point the region in which the total free energy is decreased for a given Burgers displacement, island volume, and mismatch strain. $\nu = 0.3$

middle of the island is roughly the average of that at the top and bottom which is typical of a bending field in an elastic structure.[9,10]

DISLOCATED ISLANDS

Another mechanism for strain relaxation is the formation of misfit dislocations at the island-substrate interface. The large stress concentration at the edge of the islands is a potential site for dislocation nucleation. A nucleation criterion based on this observation has been presented elsewhere.[10] Here, the effect of a misfit dislocation on the free energy of the system is considered without regard for a nucleation mechanism.

A single dislocation, with a Burgers vector parallel to the interface, is introduced at the center of each island in the periodic array of equilibrium islands. The net change in free energy is determined by summing the self-energy of the dislocations (core energy included) and the interaction energy of the dislocations with the stress field due to the elastic strain. Figure 3 is a plot of normalized island volume versus normalized Burgers displacement (b/L) for two values of mismatch strain. The curves represent the combination of island volume and Burgers displacement for which there is no net change in free energy due to the introduction of a misfit dislocation. Above each curve, in the direction of the arrows, the presence of the misfit dislocations results in a *decrease in free energy*. Below each curve, the free energy *increases* upon dislocation introduction. In other words, for a given Burgers displacement, there is a critical island volume above which introduction of a misfit dislocation is energetically favorable. As already noted, this observation does not address the important issue of dislocation nucleation. At the critical volume, dislocations will not nucleate due to a large energy barrier. For a typical system, the critical volumes are extremely small. Consider the example of the last section with $L = 12.5$ nm, $\epsilon_m = 0.04$, and a Burgers displacement of 0.5 nm, which imply a critical volume per unit depth V less than 80 nm^2.

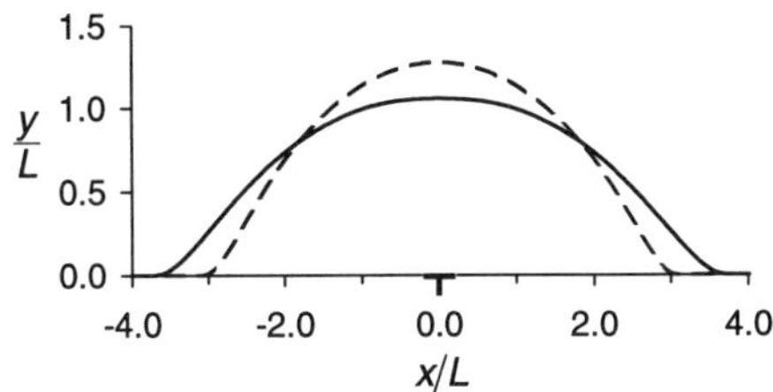

Figure 4. Equilibrium shape for a dislocated island (solid line). The dashed line shows the equilibrium shape of an island of the same volume prior to introduction of a dislocation.

Once a dislocation is introduced into a previously stable island, the stress distribution is altered and the island shape is no longer stable; the constant chemical potential condition is no longer satisfied. Mass diffuses along the surface until equilibrium is again achieved. Figure 4 is a typical example of the relaxation that occurs due to a misfit dislocation. The dashed line represents the surface of the original dislocation-free island and the solid line is the surface of the relaxed dislocated island. A dislocated island generally has a lower aspect ratio than a dislocation-free island of the same volume. This was observed experimentally by LeGoues et al.[11] They found that Ge islands on Si (100) tend to grow in a cyclical manner. The islands grew with an increasing aspect ratio until a dislocation was nucleated, at which time there was an almost immediate decrease in aspect ratio. With further growth, the aspect ratio continued to increase until the next nucleation event.

EQUILIBRIUM CHEMICAL POTENTIAL

The chemical potential of the island surface is equal to the increase in free energy due to the addition of an atomic sized material volume on the surface. A relatively small chemical potential results in a relatively small increase in free energy when an atom is adsorbed on the island surface. A simple coarsening equation for an aggregate of islands can be formulated by using the Gibbs-Thomson relation to relate the chemical potential to the concentration of adatoms on the surface.[12,13] The time rate of change of island volume can then be determined by considering the diffusion of adatoms along the surface. A simple two-dimensional form of the coarsening equation is

$$\dot{V}_i = D\left[\overline{C}/C_p - 1 - (\mu_i - \mu_p)/(kT)\right] \qquad (3)$$

where $\dot{V}_i$ is the volume rate of change of an island with chemical potential μ_i. In (3) $\overline{C}$ is the average adatom concentration on the aggregate surface, C_p is the adatom concentration of a planar surface with chemical potential μ_p, k is Boltzmann's constant, T is the absolute temperature, and D is a positive constant. The detail to note in (3) is that the island volume rate of change is related to the island's chemical potential. As the chemical potential decreases, $\dot{V}_i$ becomes larger. Islands with smaller values of chemical potential coarsen at the expense of islands with larger values.

For the dislocation-free islands, the chemical potential decreases with island volume. The chemical potential of the largest island calculated here is less than 55% of that of a flat film. In light of this and (3), larger islands can be expected to coarsen at the expense of the smaller ones. Also, the chemical potential of a dislocated island is significantly lower than that of a non-dislocated island of the same volume. Upon dislocation introduction the chemical potential decreases by as much as 35%. Consequently, dislocated islands should grow at a higher rate than those without dislocations. Krishnamurthy et al.[8] found this to occur for Ge islands on Si(100). They observed that larger, dislocated islands tended to grow at a higher rate than smaller, non-dislocated ones. The difference in the growth characteristics of dislocated and non-dislocated islands can be attributed to a large decrease in chemical potential upon dislocation nucleation.

CONCLUSIONS

The elastic strain which arises due to the constraint of epitaxy in the presence of a lattice-mismatch can be relaxed by island formation. The top of the island approaches total relaxation as the volume of the island increases. The addition of a misfit dislocation can further relax the elastic strain. Islands larger than a critical volume are energetically susceptible to the formation of misfit dislocations. Upon introduction of a dislocation the island evolves into a shape with a lower height-to-width aspect ratio. In general, a dislocated island has a significantly lower chemical potential than a non-dislocated island of the same volume. The difference in the growth characteristics of dislocated and non-dislocated islands is attributed to this observation.

ACKNOWLEDGEMENTS

The research support of the Office of Naval Research Contract N00014-95-1-0239 and the National Science Foundation MRG Grant DMR-9223683 is gratefully acknowledged.

REFERENCES

1. C. Herring, *Structure and Properties of Solid Surfaces*, edited by R. Gomer and C.S. Smith (Univ of Chicago Press, Chicago 1953), p. 5.
2. C.-h. Chiu and H. Gao, *Mat. Res. Soc. Symp. Proc.* **356**, 33 (1995).
3. B. J. Spencer and J. Tersoff, *Mat. Res. Soc. Symp. Proc.* **399**, in press (1996).
4. A. Zangwill, *Physics at Surfaces*, (Cambridge Univ. Press, New York, 1988), p. 8.
5. J. P. Hirth and J. Lothe, *Theory of Dislocations*, 2nd ed. (John Wiley & Sons, Inc., New York, 1982), pp. 731–735.
6. B. J. Spencer and D. I. Meiron, *Acta Metall. Mater.* **42**, 3629 (1994).
7. F. Jonsdottir, *Modelling Simul. Mater. Sci. Eng.*, **3**, 503 (1995).
8. M. Krishnamurthy, J. S. Drucker, and J. A. Venables, *J. Appl. Phys.* **69**, 6461 (1991).
9. S. P. Timoshenko and J. N. Goodier, *Theory of Elasticity*, 3rd ed. (McGraw-Hill Inc., New York, 1972), pp. 41–50; pp. 141–144.
10. L. B. Freund, H. T. Johnson, and R. V. Kukta, *Mat. Res. Soc. Symp. Proc.* **399**, in press (1996).
11. F. K. LeGoues, M. C. Reuter, J. Tersoff, M. Hammar and R. M. Tromp, *Phys. Rev. Lett.* **73**, 300 (1994).
12. B. K. Chakraverty, *J. Phys. Chem. Solids* **28**, 2401 (1967).
13. C. V. Thompson, *Acta Metall.* **36**, 2929 (1988).

**INDENTATION MODULUS AND HARDNESS IN
HETEROEPITAXIAL $Al_xGa_{1-x}P$ FILMS**

B. ROOS *, C. J. SANTANA **, C. R. ABERNATHY **, K. S. JONES **
*Institute for Semiconductor Physics, 15230 Frankfurt/Oder, Germany,
**University of Florida, Gainesville, FL 32611

ABSTRACT

$Al_xGa_{1-x}P$ layers ($0 \leq x \leq 0.7$), with thicknesses of ≥ 1 μm were grown on Si (100) wafers by metal-organic molecular beam epitaxy (MOMBE) at 450 °C. Transmission electron micrographs of the single crystal films revealed that the microstructure contains stacking faults and microtwins especially near the interface as well as both threading and misfit dislocations. Hardness and elastic modulus were measured using a Nanotest 500 indenter, which can probe the film properties without influence from the substrate.

The hardness H varies linearly according to (11.8 - 2.3x) GPa. The absence of alloy hardening is due to the fact that there is no difference in atomic size of Al and Ga. The indentation modulus $E/(1-v^2)$ decreases monotonically from 136 GPa for GaP to 129 GPa for $Al_{0.7}Ga_{0.3}P$ and bows only slightly (about 2 %) below the straight line of linear interpolation.

INTRODUCTION

Heteroepitaxial growth of III-V semiconductors on Si substrates has attracted much attention because of the potential development of optoelectronic integrated circuits (OEICs). Despite its indirect bandgap, GaP and its alloys are widely used as optically active materials in the red-green region of the visible spectrum. The addition of AlP, which shows complete miscibility with GaP, offers the possibility of emissions in the blue-green visible region. Both GaP and AlP are nearly lattice matched to silicon, making $Al_xGa_{1-x}P$ a suitable system for an effective buffer layer for the growth of polar semiconductors on Si with low defect densities [1]. The difference between the lattice constants of the Si substrate and the films corresponds to a small mismatch of 0.4 to 0.6 % at room temperature and 0.5 to 0.7 % at the typical low growth temperature of 450 °C, depending on the Al content x. Part of the coherency stresses relaxes by introduction of stacking faults, twins and dislocations during deposition of the film. Defects in the $Al_xGa_{1-x}P$ layers have also been attributed to the island growth mode which has been shown to lead to stacking errors [2] and high dislocation densities [3].

Plastic relaxation may proceed by introduction of dislocations and glide. The same processes occur during indentation. Thus, hardness testing is a suitable method to measure the plasticity. To our knowledge, hardness of AlP and of $Al_xGa_{1-x}P$ alloys has not been measured up to now. The remaining stresses in a lattice mismatched film cause tetragonal deformation. The stresses are linked to strains via elastic constants, which have been determined experimentally and theoretically for the binaries GaP and AlP [4]. For pseudobinary alloys like $Al_xGa_{1-x}P$, which show complete miscibility for the group III element, the elastic constants are assumed to vary linearly with composition. However, no experimental data were available.

In this paper we report on nanoindentation experiments in thick $Al_xGa_{1-x}P$ films ($0 \leq x \leq$ 0.7) grown by MOMBE on Si substrates. This technique is capable of measuring simultaneously the hardness and the indentation modulus. The results will be explained with respect to structural properties of $Al_xGa_{1-x}P$ and the technological impact will be discussed.

Mat. Res. Soc. Symp. Proc. Vol. 436 © 1997 Materials Research Society

GROWTH AND STRUCTURAL CHARACTERIZATION OF THE AL$_x$GA$_{1-x}$P FILMS

Al$_x$Ga$_{1-x}$P films ($0 \leq x \leq 0.7$), with thicknesses of ≥ 1 μm were grown on epitaxial Si(100) substrates with on-axis orientation. Epitaxial silicon wafers have been shown to provide a smoother starting surface for the growth. The samples were grown in a Varian Gas Source Gen II on 2 inch diameter substrates which received an ex situ diluted HF chemical treatment in order to remove the native oxide and create a hydrogen passivated surface. Triethylgallium (TEG), dimethylethylamine alane (DMEAA) and phosphine were used as the precursors for growth. The group III sources were transported into the chamber via a hydrogen carrier gas and the phosphine was cracked in a Varian low-pressure cracker set at 950 °C. The growth temperature was 450°C and the growth rates ranged from 1 to 1.5 μm/hr depending on Al composition.

The surface of the Al$_x$Ga$_{1-x}$P layers appeared almost mirror-like. Atomic force microscopy was used to measure the surface roughness of the layers which have average RMS values of (18 ± 4) nm. Transmission electron microscopy (TEM) of the Al$_x$Ga$_{1-x}$P films revealed a highly defected single crystal structure as shown in figure 1. The predominant defects are stacking faults and microtwins especially near the heterointerface where they were also found to annihilate by crossing each other. Many stacking faults extend through the entire thickness of the film and using plan view TEM measurements their densities were calculated to be $\approx 7 \times 10^8/\text{cm}^2$. A typical misfit dislocation grid was not observed at the interface possibly due to the island growth mode of the Al$_x$Ga$_{1-x}$P films on silicon.

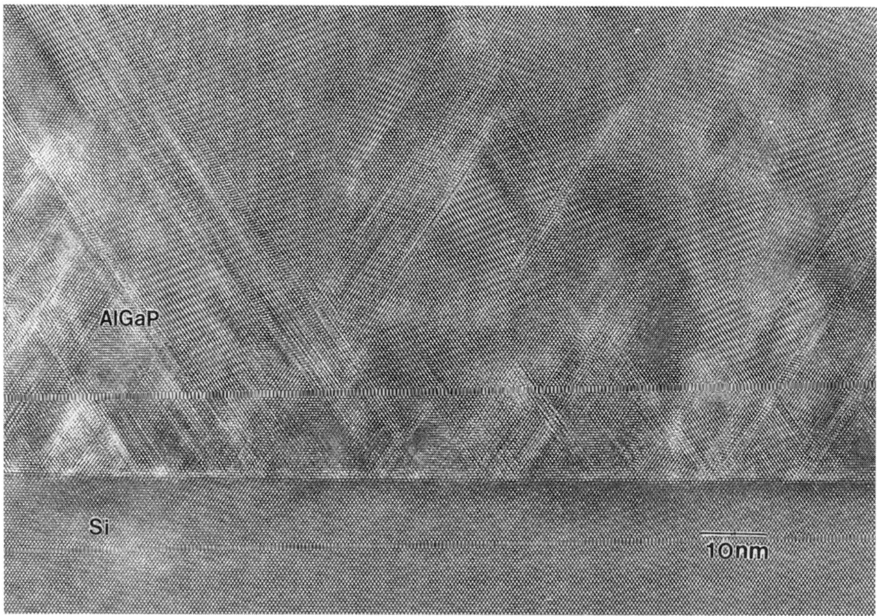

Fig. 1: Cross sectional high resolution TEM micrograph of the interface region of an Al$_{0.5}$Ga$_{0.5}$P/Si sample. Stacking faults and microtwins originate from the interface.

NANOINDENTATION TECHNIQUE

To investigate the plastic and elastic properties of the films indentation experiments have been carried out in a Nanotest 500 nanoindenter described elsewhere [5]. A Berkovich pyramid-shaped (triangular base) diamond tip is pressed into the sample in a load-controlled mode. Load and tip displacement are recorded continuously during loading and unloading thereby producing a load-depth curve. Loading the indenter, the sample is deformed elastically *and* plastically. Unloading the indenter the elastic surface flexure recovers following closely the linear load-depth dependence of Hooke´s law. The intercept of the tangent to the unloading branch in the maximum point with the depth axis yields the plastic depth d_p. According to Doerner and Nix [6] this is the penetration depth under load (= contact depth d_c).

Both the computation of hardness and modulus depend on the contact area (cf. eqs. (1) and (2)). Thus, the tip area function which relates the contact area to the contact depth has to be calibrated for each tip experimentally due to the inevitable tip rounding. We have chosen the following procedure. Based on the assumption that the hardness of fused quartz is independent of the indentatation depth, indentations with widely varying depths were performed. For the deepest ones the area function of the perfect Berkovich indenter, $A(d_c) = 24.5\ d_c^2$ can be used as an initial input for the contact area A. By an iterative fit routine we finally got the area function $A(d_c) = 22.29\ d_c^2 + 1465$ nm d_c (contact depth given in nm) for the tip used here.

For each sample the indentation depth was less than 200 nm, i.e. less than 1/5 of the film thickness due to the small loads (about 10 mN) used in the experiments. Hence, the influence of the substrate on hardness and moduli can be neglected. Considering the anisotropy of the samples, the relative orientation between the tip with threefold symmetry and the sample with fourfold symmetry was kept constant thereby producing an average of all directions. All the results are the mean value of at least 15 measurements made at room temperature.

RESULTS

The hardness H is defined in the usual way - like for microhardness - as the ratio of the load P and the contact area of the indent $A(d_c)$

$$H = P\ /\ A(d_c)\ . \tag{1}$$

Fig. 2 displays the hardness as a function of the fractional aluminum content x. The hardness value for pure GaP is slightly larger than those reported for bulk crystals [7]. Possible reasons will be discussed in the next section. The hardness varies linearly according to H = (11.8 - 2.3x) GPa. The values for x > 0.7 have been obtained by linear extrapolation.

Following Doerner and Nix [6] we analyzed the indentation modulus $E/(1-v^2)$ according to the formula

$$S = \frac{P_{max}}{d_t - d_p} = \frac{2}{\sqrt{\pi}} \left(\frac{1-v^2}{E} - \frac{1-v_i^2}{E_i} \right)^{-1} \sqrt{A(d_c)}. \tag{2}$$

Here S denotes the measured stiffness, i.e. the slope of the tangent to the unloading branch, E and v are Young´s modulus and Poisson´s ratio, respectively, for the specimen, and E_i and v_i are the values for the indenter. The latter have been determined experimentally. Hence, knowing the values for S, $A(d_c)$ and the indenter compliance $(1-v_i^2)/E_i$ one can extract from equ. (2) the indentation modulus (= plate modulus) $E/(1-v^2)$ of the sample to be investigated.

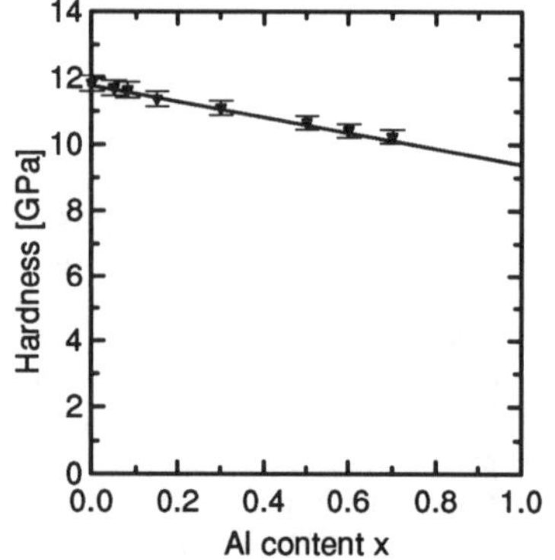

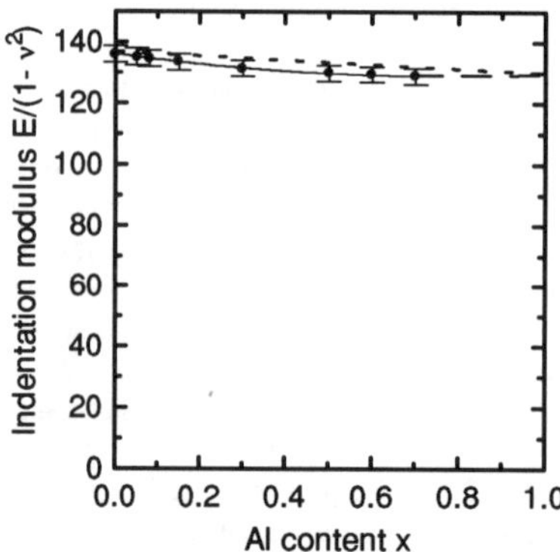

Fig. 2: Hardness of $Al_xGa_{1-x}P$ as a function of the Al content x.

Fig. 3: Indentation modulus of $Al_xGa_{1-x}P$ as a function of the Al content x.

Fig. 3 shows this modulus in dependence of the composition. The full line is the fitting curve from a polynomial fit of the order 2 to the data points. The fitting function $y = 136.2 - 18.7 x + 12.1 x^2$, obtained from the region $x < 0.7$, is extrapolated to $x = 1$ (dashed line), providing a value of 129.6 GPa for pure AlP. The dotted line marks the linear interpolation between the values for pure GaP and pure AlP. The fitting curve shows a small negative bow. Maximum deviation from simple linear interpolation is found for $x = 0.5$. There the measured value is 2.3 % lower than the rule-of-mixtures (ROM) value from linear interpolation.

DISCUSSION AND CONCLUSIONS

<u>Plastic properties - hardness</u>

For reasons of comparison we performed an indentation series in Si (100) using the same parameters as for the indents into GaP. We found a hardness of (11.5 ± 0.2) GPa, in excellent agreement with the result given in ref. [7]. Hence, according to our measurements the hardness of GaP is slightly larger than that of Si. This contradicts other findings and might be due to special properties of the epitaxial layer. First, the epitaxial layer contains a lot of stacking faults. Although the interaction of a dislocation and a stacking fault is weaker than the dislocation-dislocation interaction, the stacking faults may act as obstacles for dislocation movement and therefore increase the hardness. Secondly, the residual stress in the layer is compressive. As well known, a compressive stress shifts hardness to higher values.

The observed ROM behaviour of hardness, i.e. the absence of alloy hardening, resembles the behaviour in the pseudobinary $Al_xGa_{1-x}As$ [8]. According to Cottrell and Fleischer [9] the increment in critical resolved shear stress and therefore also the increment in hardness

$$\Delta H = const \left| \chi \, \frac{1}{a}\frac{da}{dx} - \frac{1}{G}\frac{dG}{dx} \right| x^{1/2} \tag{3}$$

depends linearly on the misfit parameter $\delta = (1/a)(da/dx)$ and on the elasticity parameter $\eta = (1/G)(dG/dx)$. Here G denotes the shear modulus and χ is a weight factor with different values for screw and edge dislocations, respectively. However, δ practically vanishes because the lattice parameter of AlP is only 0.2 % larger than that of GaP. This is due to an anomalous property of the gallium atom. Its atomic size is the same as that of aluminum, since it is the first group-III element with a filled d shell. Furthermore, the parameter η is of the order 1 % , causing an increase in ΔH of less than 1 % which cannot be resolved in our experiments. Hence, there is no solid solution hardening effect for the $Al_xGa_{1-x}P$ alloy.

The hardness is determined by the energy necessary to introduce dislocations and by the interaction energy of these dislocations [10]. The latter one is dominated by the shear modulus which is smaller for AlP when compared to GaP [4]. A measure for the dislocation generation energy (per unit length) is the Peierls energy $U_p = (\sqrt{3}/\sqrt{8})(U_b/b)$. Again, this is smaller for AlP, since the bond breaking energy U_b is smaller and the bond length b is larger than in GaP. Therefore, the hardness of AlP is lower than that of GaP, which agrees well with the value found by extrapolation.

<u>Elastic properties - indentation modulus</u>

Again we start the discussion with a comparison between the indentation moduli of Si and GaP. For Si we find $E/(1-v^2) = 143$ GPa which is about 5 % larger than the value for GaP. This is in reasonable agreement with the ratio of the data for the elastic constants c_{ij} and the bulk modulus B given in ref. [10].

For discussion of the absolute values we have to consider the mechanical anisotropy of the cubic crystals. The indentation modulus $E/(1-v^2)$ is *not* invariant in the {100} planes but is minimum in the <100> directions and maximum in the <110> directions. For GaP it varies between 114 GPa and 144 GPa [11]. The measured value of 136 GPa lies in this interval. This is due to the fact that the penetration of the indenter tip yields an average of all directions, of course. For the III-V compounds the elastic constants c_{ij} determined experimentally scale as $1/b^4$, where b denotes the nearest neighbour distance, i.e. the III-V bond length [10]. The bond length b in AlP is slightly larger than that in GaP, therefore we expect the elastic constants of AlP to be smaller , in correspondence to the extrapolated experimental value.

Considering the composition dependence, the measured slightly negative bow again resembles the behaviour in the pseudobinary $Al_xGa_{1-x}As$ [12]. Below we present arguments for the negative bow in the composition dependence of the elastic properties following the ideas given by Chen et al. [10]. A power law dependence, also $\propto 1/b^4$ has been found for the bulk modulus B in the framework of a valence force-field (VFF) model for zincblende compounds [13]. It should be mentioned here that the indentation modulus is a measure for both the elastic compressibility, expressed in B, and a measure for the shear rigidity, represented by G. However, according to finite element method (FEM) analysis the bulk modulus B is more important for hard materials like semiconductors [14]. Thus, we conclude that the arguments given for B can be applied to the indentation modulus.

First, the lattice constant a and bond length b of $Al_xGa_{1-x}P$ are known to vary linearly with composition, i.e. Vegard´s law holds [15]. The observed negative bow is consistent with the linear interpolation for b due to the $1/b^4$ scaling of B. However, a quantitative assessment is not possible, since there are no models with quantitative predictions to be tested.

Secondly, $Al_xGa_{1-x}P$ is a *disordered* pseudobinary alloy. The Al and Ga atoms are randomly distributed on one face-centered cubic (fcc) sublattice, and the P atoms belong to the other fcc sublattice. In order to get the elastic constants one has to calculate the strain energy and its variation under external loading. Chen et al. used a VFF model to estimate the strain energy and Harrison´s bond orbital method to treat the change of energy of the bonds in the alloy environment [16]. For disordered alloys of the formula $A_xB_{1-x}C$ the smallest structure is a cluster of five atoms centered on a C atom. There are five different types of such clusters.

Obviously, there are more degrees of freedom for elastic displacement of atoms than in the corresponding ordered binaries, thereby decreasing the rigidity of the lattice. A mean field approch can be used since the alloy has an average lattice constant which follows Vegard´s law (virtual crystal approximation, VCA). By calculating the spring constants for the different bonds Chen et al. found that the bulk modulus is very close to, but slighty smaller than the concentration-weighed averaged value for a random alloy, just as seen in Fig. 3.

In conclusion, we have demonstrated the first low-temperature MOMBE growth of high-quality heteroepitaxial $Al_xGa_{1-x}P/Si$ structures. The composition dependence of hardness and elastic indentation modulus of $Al_xGa_{1-x}P$ is similar to that in the $Al_xGa_{1-x}As$ alloy system. The hardness varies linearly with composition whereas the indentation modulus bows slightly below the straight line of linear interpolation. Some arguments for this behaviour have been given, but there is a strong demand for quantitative theoretical studies.

Since there is no solid solution hardening the movement of dislocations is not hindered by addition of Al atoms to GaP. Hence, the $Al_xGa_{1-x}P$ layers aren´t more stable against plastic relaxation by dislocation introduction than GaP. However, the differences in elastic constants offer the possibility of dislocation filtering, i. e. bending of threading dislocations due to inserted strained layers [17].

In order to decrease the high density of stacking faults the island growth mode (Volmer-Weber growth) has to be overcome. Kinetic suppression of islanding due to low-temperature growth appears to be a promising way.

REFERENCES

1   T. George, E. R. Weber, S. Nozaki, A. T. Wu, N. Noto, and M. Umeno, J. Appl. Phys. **67** (1990) 2441

2   F. Ernst and P. Pirouz, J. Mater. Res. **4** (1989) 834

3   Y. Qiu, A. Osinsky, A. A. El-Emawy, E. Littlefield, H. Temkin, and N. Faleev, J. Appl. Phys. **79** (1996) 1164

4   T. Ito, J. Appl. Phys. **77** (1995) 4845

5   B. Roos, H. Richter, and J. Wollweber, in Proc. 6th Autumn Meeting Gettering and Defect Engineering in Semicond. Technol.(GADEST ´95), ed. by H. Richter, M. Kittler, and C. Clays (Scitec Publications, Zürich,1996), Solid State Phenomena **47-48** (1996) 509

6   M. F. Doerner and W. D. Nix, J. Mater. Res. **1** (1986) 601

7   CRC Handbook of Chemistry and Physics, ed. by D.R. Lide (CRC Press, Boca Raton, 1992)

8   K. Hjort, F. Ericson, J.-Å. Schweitz, C. Hallin, E. Janzén, Thin Solid Films **250** (1994) 157

9   R. L. Fleischer, Acta Met. **9** (1961) 996

10 A.-B. Chen, A. Sher, and W. T. Yost, in Semiconductors and Semimetals, edited by R. K. Willardson, A. C. Beer, and E. R. Weber (Academic Press, New York, 1992), Vol. 37, p. 1

11 A. Fischer, H. Kühne, B. Roos, H. Richter, Semicond. Sci. Technol. **9** (1994) 2195

12 M. Krieger, H. Sigg, N. Herres, K. Bachem, K. Köhler, Appl. Phys. Lett. **66** (1995) 682

13 R. M. Martin, Phys. Rev. B **1** (1970) 4005

14 A. K. Bhattacharya and W. D. Nix, Int. J. Solids Structures, **27** (1991) 1047

15 Landoldt-Börnstein, Neue Serie III 22a (Springer, Berlin, 1987)

16 A.-B. Chen, A. Sher, M. A. Berding, Phys. Rev. B **37** (1988) 6285

17 R. J. Dieter, F. Goroncy, J. P. Lay, N. Draida, K. Zieger, W. Kürner, B. Lu, F. Scholz, B. Roos, M. Braun, V. Freese, J. Hilgarth, Proc. 11th Europ. Photovolt. Energy Conf., ed. by L. Guimaraes, W. Palz, C. de Reyff, H. Kiess (Harwood Academic, Chur, 1992), p. 225

ORIGINS OF RESIDUAL STRESS IN Mo AND Ta FILMS: THE ROLE OF IMPURITIES, MICROSTRUCTURAL EVOLUTION, AND PHASE TRANSFORMATIONS

L. J. Parfitt, O. P. Karpenko, Z. U. Rek*, S. M. Yalisove, and J. C. Bilello, The Materials Science and Engineering Department, The University of Michigan, Ann Arbor, MI 48109-2136, * Stanford Synchrotron Radiation Laboratory, Stanford, CA 94025

Abstract

Both the sign and magnitude of residual stress can vary with the thickness of sputter deposited films. The origins of this behavior are not well understood. In this work, we consider the correlation between the residual stress behavior and the depth dependence of impurities in thin (2.5 nm - 150 nm) sputtered Mo and Ta films. We also consider the effects of phase transformations and microstructural changes on the stress behavior. Films were deposited onto Si substrates with native oxide. The residual stress observed in the Mo films varied from highly compressive at 2.5 nm film thickness to ~ 0 at 10 nm thickness. Ta films also exhibited a high compressive stress, which relaxed from highly compressive to tensile between 10 nm and 50 nm film thickness. Impurities in the films may originate from the sputtering targets, the background gases, and the substrate surfaces. Auger Electron Spectroscopy (AES) results showed the presence of O and C contamination near the film/Si interface; these impurities contributed to the compressive stresses in the thinner films. As anticipated, both Mo and Ta films exhibited grain growth as a function of film thickness, which may have contributed to the relaxation in the compressive stress. The Mo films were entirely bcc. The Ta films showed a transformation from the amorphous phase to the β crystalline phase between 2.5 nm and 20 nm film thickness, which contributed to the relaxation in stress observed in that thickness regime.

Introduction

Refractory metals like Mo and Ta are of interest in numerous research and commercial applications due to their high melting points, low coefficients of thermal expansion (CTE), moderate ductility, and relatively good mechanical properties. Films of these metals have been used in transistors[1]; diffusion coatings[2,3]; soft, x-ray mirrors[4]; and multiscalar, multilayer, microlaminates (MMM)[5,6]. In the latter application, the films are used as a strengthening phase and are extremely thin (~ 4 nm.)

Although Mo and Ta offer the potential for being used advantageously in a variety of applications, the residual stress behavior of their films is of concern. Residual stresses may cause changes in the optical, electronic, and mechanical properties of films[4,7,8,9]. In order to further understand the stress behavior of Mo and Ta, we examined the residual stress behavior of very thin sputtered Mo and Ta films. While both metals have been investigated previously, little work has been done in the film thickness regime of < 10 nm. Sputtered films were used because sputtering provides good uniformity and relatively high deposition rates; however, sputtered films exhibit stress behavior that is influenced by a number of parameters. In this study, we focused particularly on the effects of impurities, phase transformations, and microstructure on residual stress behavior as a function of increasing film thickness[8,10].

Experimental Procedure

Mo and Ta films with thicknesses ranging from 2.5 to 150 nm were deposited using dc magnetron sputtering in a cryo-pumped chamber with a "sputter down" geometry, two 2.5 KW power supplies, 99.95% pure Mo and Ta targets, and a base pressure of < 3X10⁻⁶ Torr. In order to remove surface contamination, the targets were sputtered for four minutes prior to each deposition. The Ar sputtering pressure was held at 10 mTorr for all films. C substrates were used for thickness calibration and Si (100) single crystal substrates with native oxide were used for all other analyses.

The films were studied using Double Crystal Diffraction Topography (DCDT) to obtain the residual stress as a function of film thickness[11]. Using DCDT, the curvature of each substrate was measured before and after film deposition. The stress of the film was calculated using the modified

Mat. Res. Soc. Symp. Proc. Vol. 436 © 1997 Materials Research Society

Stoney's equation[12]. In order to understand the mechanisms of the residual stress behavior, the impurity levels, phases, and microstructures of the films were then evaluated.

Film thickness was calibrated using Rutherford Backscattering Spectroscopy (RBS) on a Tandem accelerator with 2 MeV He++ ions. RBS was also used to examine the films for high atomic number (Z) impurities, especially for the sputtering gas Ar.

Further compositional data were obtained for selected films by examining them with Auger Electron Spectroscopy (AES) using a PHI 600 Auger unit with a 3 KeV electron beam. In order to obtain depth information, films were sputtered with an Ar ion beam at an angle of 20 degrees, at a sputtering rate of ~ 2.5 nm/min.

Films with thicknesses ranging from 5 nm to 80 nm were examined to obtain microstructural information using Transmission Electron Microscopy (TEM) on a Philips 420T at 120 KeV. Plan view samples of the films were prepared by backthinning the samples to ~ 150 μm, followed by etching to perforation with an $HF:HNO_3:CH_3COOH$ acid mixture (3:5:3). In addition, one Mo film (10 nm thick) and two Ta films (10 nm and 20 nm thick) were studied in cross section on a JEOL 4000EX High Resolution TEM at 350 KeV. These samples were prepared by cleaving the substrate, polishing to ~50 μm, and ion milling.

Phase information for the films was obtained using several techniques. The films were studied using transmission electron diffraction on the previously described plan view TEM samples, and were also studied using x-ray techniques. The 10 to 80 nm thick Mo and Ta films were studied using low angle Bragg diffraction with a conventional Cu Kα x-ray source.

The 2.5 and 5 nm Mo and Ta films, which have very small diffraction volumes, were studied at the Stanford Synchrotron Radiation Laboratory (SSRL) on wiggler beamline 7-2. These films were studied in the symmetric Grazing Incidence X-ray Scattering (GIXS) geometry, allowing collection of diffraction information from planes nearly perpendicular to the film surface.

The 20 and 80 nm thick Ta films were also studied at SSRL to determine the mixture of crystalline phases in the films. Ta films are known to form both the stable α bcc phase and a metastable tetragonal β phase[13,14]. Due to overlap of the high intensity α Ta and β Ta peaks, we had difficulty ascertaining whether the films were entirely β, or whether they contained a small volume percent of α. Using several geometries, we searched for the α (200) Ta peak, which is of relatively low intensity but is well separated from high intensity β peaks.

Results

The residual stresses of the sputtered Mo and Ta films as a function of thickness are shown in Figures 1 and 2, respectively. In both cases, the residual stresses of the thinnest films were highly compressive, then relaxed with increasing thicknesses. In the case of the Mo films, the stress relaxed to ~ 0 GPa in the 10 nm thick films, whereas for the Ta films the stresses relaxed to ~ 0 GPa between 40 and 50 nm. The thickest Mo films were slightly compressive (~-0.2 GPa), and the thickest Ta films slightly tensile (~0.5 GPa).

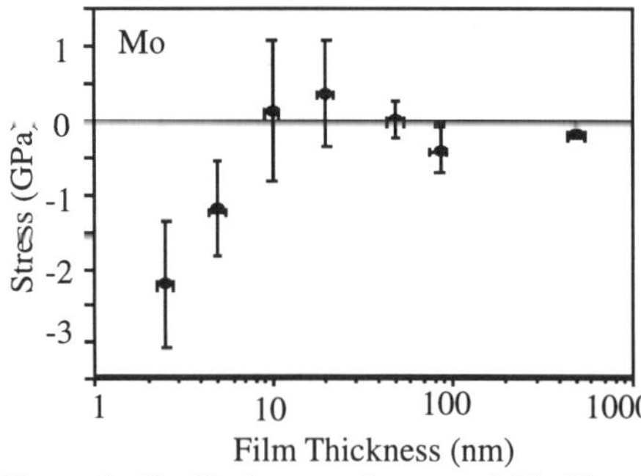

Figure 1: Residual stress of sputtered Mo films showing relaxation of high compressive stress between 2.5 nm and 10 nm film thickness.

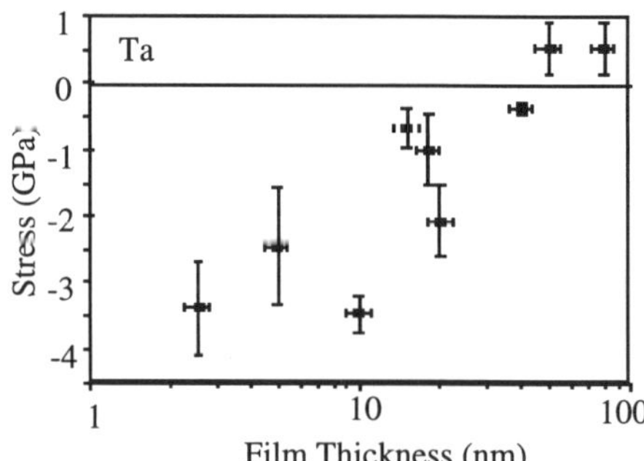

Figure 2: Residual stress of sputtered Ta showing relaxation of high compressive stress between 10 and 50 nm film thickness.

RBS showed that the Mo and Ta films contained no detectable Ar. AES analysis showed both O and C contamination in all Mo and Ta films. The amounts of O and C were highest at the surface-film and film-Si interfaces. In the Mo films, some O contamination existed throughout the 5, 10, and 50 nm thick films. However, the C concentration dropped to zero in the bulk of the 50 nm thick film. In the 5 and 20 nm thick Ta films, O was observed throughout the films, though the C content dropped to zero away from the interfaces. In the 50 nm thick Ta films, both the O and C contents dropped to zero away from the interfaces.

The microstructures of the Ta and Mo films as a function of film thickness are shown in Figures 3 and 4, respectively. In the 5 and 10 nm thick Ta films, no clearly defined grains were observed. In the 20 to 80 nm Ta films, and in all the Mo films studied, the grain sizes were small and increased with increasing film thickness. The cross sectional views of the 10 and 20 nm thick Ta films are shown in Figure 5. In the 10 nm thick film, the structure was largely amorphous with a few regions of crystallinity. In the 20 nm thick film, the structure was almost entirely crystalline.

The GIXS studies of the 2.5 and 5 nm thick Mo films yielded peaks that indexed to the equilibrium bcc structure (Figure 6). TEM analysis (Figure 4) and low angle Bragg diffraction studies also showed that the films had the bcc structure. No rings or peaks corresponding to alternate phases were observed. Thus, all Mo films studied were bcc crystalline, and no phase transformations occurred with increasing film thickness.

The 2.5 nm thick Ta film was determined using GIXS to be amorphous or nano-crystalline; the only observed peak was extremely broad. For the 5 nm thick Ta film, GIXS showed that the film was partially crystalline, with several peaks corresponding to the β Ta phase (Figure 7).The TEM diffraction patterns of the 5 nm Ta film indicated that the film was either amorphous or nano-crystalline (Figure 3).

The TEM diffraction pattern of the 10 nm thick Ta film exhibited several rings; all the rings were broad, with the innermost ring indicating that the film had a partially amorphous structure (Figure 3). The other rings corresponded to the β Ta structure. For the 20 to 80 nm thick Ta films, the TEM and low angle x-ray studies yielded diffraction data corresponding to either β Ta or a mixture of β Ta and α Ta. However, when the films were studied with GIXS, the α (200) peak was not observed in either the 20 nm or 80 nm thick Ta films, suggesting that these films were almost entirely β Ta (Figure 7). Thus, the Ta transformed from amorphous Ta to β Ta between 2.5 nm and 20 nm film thickness, and the 20 to 80 nm thick films were β Ta.

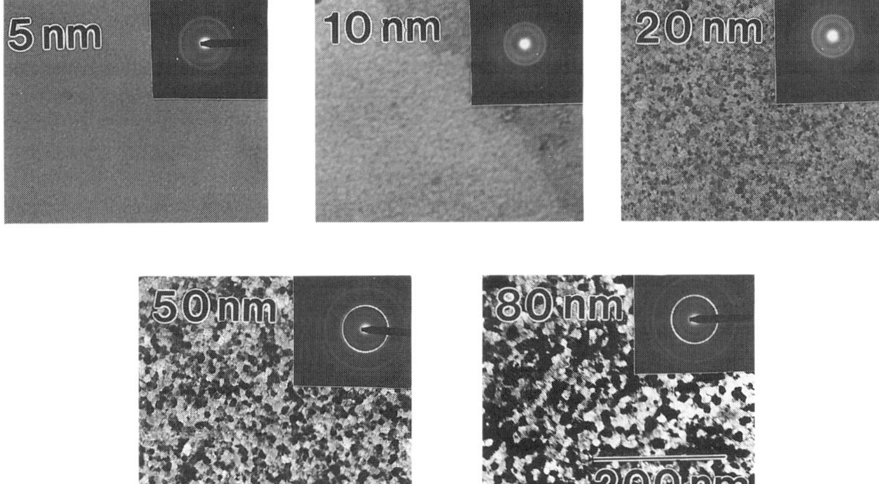

Figure 3: Plan view micrographs and diffraction patterns for Ta films (5 to 80 nm film thickness).

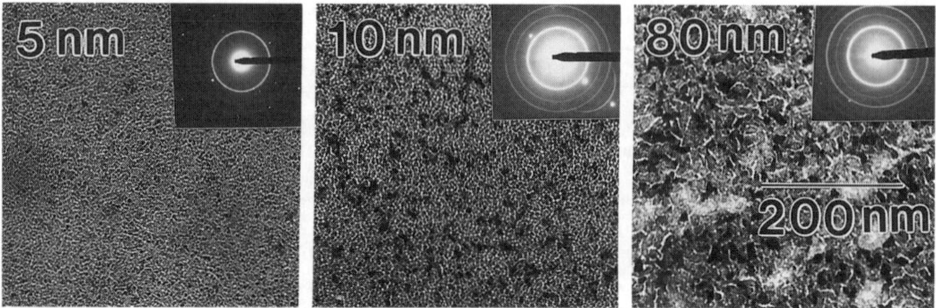

Figure 4: Plan view micrographs and diffraction patterns for Mo (5 to 80 nm film thickness).

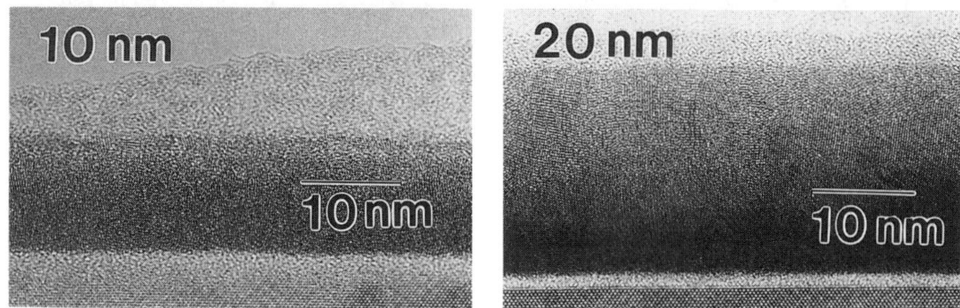

Figure 5: HRTEM cross sections of 10 nm and 20 nm thick Ta films, showing change from largely amorphous to a largely crystalline structure.

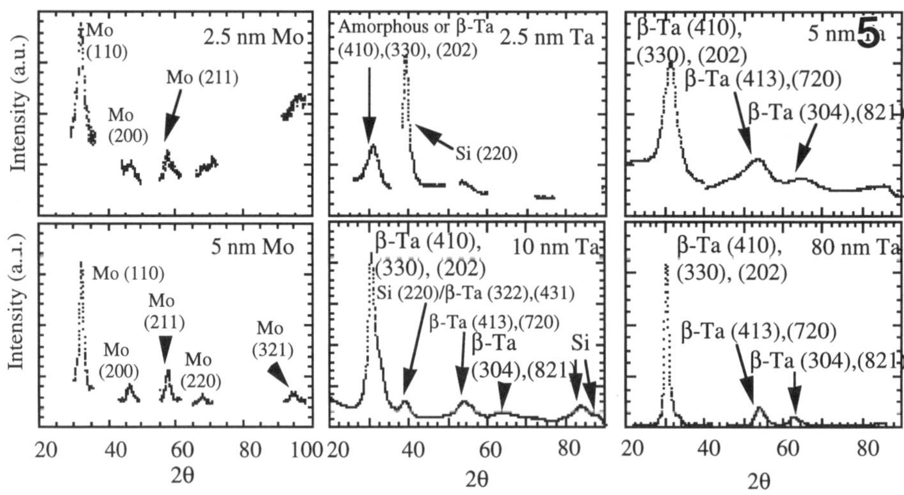

Figure 6: GIXS results from 2.5 nm and 5 nm Mo films, showing bcc structure ($\lambda = 0.124$ nm).

Figure 7: GIXS results from 2.5 nm, 5 nm, 10 nm and 80 nm thick Ta films, suggesting phase transformation from amorphous to β Ta ($\lambda = 0.1278$ nm).

Discussion

Impurities

The residual stress of a film may be affected by the presence of impurities in the film[10,15,16,17]. Ar, O, and C atoms may be incorporated into a film from sputtering gases, background gases, sputtering targets, or the substrates on which the films are deposited. Impurities may diffuse into interstitial sites or form alternate phases[10,18], causing compressive stresses in the films.

In our films, no Ar was observed in any of the films, indicating that the sputtering pressure was above the critical pressure; thus, neither Ar incorporation or ion peening affected the stress behavior of the films[8]. The O and C content of the films as a function of thickness correlated to the level of compressive stress. For both Mo and Ta, the highest compressive residual stresses existed in the thinnest films with high levels of O and C impurities. Impurities originated from the oxide on the Si substrate and were incorporated into the growing film, either by entering into interstitial sites or by forming oxides. As the film thickness increased, the influence of these substrate impurities decreased.

The thicker Mo films contained low levels of O throughout the film, probably originating from impurities in the Mo target. The 50 nm Ta film showed no O in the bulk of the film. The impurity level of the Ta target may have been lower than the Mo target. The difference in O content may explain why the thick Mo films were slightly compressive, and the thick Ta films slightly tensile.

Microstructure

Grain growth in films may also affect the residual stress. The grain size of sputtered films typically increases with increasing thickness. As the film thickens, preferentially oriented grains grow more rapidly and "crowd out" less favorably oriented grains, resulting in grain growth[19]. Provided that this growth occurs only in the direction of film growth, there is little effect on the residual stress. However, if the energetics of the process permit *lateral* growth of favorably oriented grains relative to less favorably oriented grains, grain boundary annihilation occurs. This annihilation of grain boundaries causes a tensile stress, as the film attempts to densify and is restrained by the substrate from doing so[20,21].

Given that the deposition temperature was very low (T/Tm = ~ 0.1), very little mobility was anticipated below the surface of the film during sputtering. Essentially, the film atoms were expected to freeze in place when buried by incoming atoms. However, Chaudhari[20] suggested a mechanism whereby very high compressive stresses could provide a driving force for lateral grain growth. If this mechanism were in effect for these films, lateral grain growth would occur and could contribute to the observed relaxation in stress of the Mo and Ta films.

Phase

Residual stresses may be affected by phase transformations in the films. In very thin films, metastable phases may be stabilized by impurities or may form when the kinetics of the deposition do not permit the film atoms to relax into the most stable form[15,22,23]. If, upon thickening, the film transforms from a less dense to a more dense phase, the phase transformation causes a tensile stress because the film is constrained by the substrate from contracting.

Phase transformations had no apparent effect on the residual stress behavior of the Mo films, since they were bcc throughout. In the Ta films, the films transformed from largely amorphous to β Ta between 10 nm and 20 nm film thickness, which contributed to the stress relaxation observed in that regime.

Conclusions

1. Impurities originating from the native oxide layer of the Si substrates contribute to high compressive stresses in thin Mo and Ta sputtered films.
2. Grain growth may contribute to the relaxation of the high compressive stresses.
3. An amorphous to β Ta transformation in the Ta films causes a relaxation in stress between 10 and 20 nm film thickness. Phase transformations have no effect on the stress behavior of the Mo films, since all Mo films were bcc.

Acknowledgments

We acknowledge the use of the University of Michigan Electron Microscopy and Analysis Laboratory for AES and TEM work. This work was supported by ARO Materials Research Division and ARPA under grant numbers DAAL03-91-0235 and DAAH04-95-1-0120.

References

[1] T. Ohnishi *et al, Appl. Phys. Letters,* **43** (1983) 600-602.

[2] L. A. Clevenger *et al, J. Appl. Phys.,* **73** (1993) 300-308.

[3] K. Holloway and P. M. Fryer, *Appl. Phys. Lett.,* **57** (1990) 1736-1738.

[4] R. R. Kola *et al, Appl. Phys. Letters,* **60** (1992) 3120-3122.

[5] M. Vill *et al, Acta. metall. mater.,* **43** (1995) 427-437.

[6] D. P. Adams *et al, J. Appl. Phys.,* **74** (1993) 1015-1021.

[7] C. Cabral Jr. *et al, Mat. Res. Soc. Proc.,* **308** (1993) 57-75.

[8] J.A. Thorton and D.W.Hoffman, *J.Vac.Sci. Technol.,* **14** (1977) 164-168.

[9] R. R. Kola and G. K. Celler, *Mat. Res. Soc. Proc.,* **226** (1991) 235-240.

[10] J. A. Thornton and D. W. Hoffman, *Thin Sol. Films,* **171**(1989) 5-31.

[11] J. Tao *et al, J. Elect. Mat.,* **20** (1991) 819-825.

[12] G. G. Stoney, *Proc. R. Soc. London A,* **82** (1909) 1729.

[13] M. H. Read and C. Altman, *Appl. Phys.Lett.,* **7** (1965) 51-52.

[14] P. N. Baker, *Thin Sol.Films,* **14** (1972) 3-25.

[15] A. M. Haghiri-Gosnet *et al, J. Vac. Sci. Tech. A,* **7** (1989) 2663-2669.

[16] R. Abermann and R. Koch, *Thin Sol. Films,* **142** (1986) 65-76.

[17] T. Yamaguchi and R. Miyagawa, *Japan. J. Appl. Phys.,* **30** (1991) 2069-2073.

[18] P. Petroff *et al, J. Appl. Phys.,* **44** (1973) 2545-2554.

[19] C. V. Thompson, *Mat. Res. Soc. Proc.,* **343** (1994) 3-12.

[20] P. Chaudhari, *J. Vac. Sci. Tech.,* **9** (1972) 520-522.

[21] M. F. Doerner and W. D. Nix, *CRC Critical Reviews in Solid State and Materials Science,* **14** (1988) 225-268.

[22] S. Sato *et al, Thin Sol. Films,* **86** (1981) 21-30.

[23] M. H. Read and D. H. Hensler, *Thin Sol. Films,* **10** (1972) 123-135.

Stress Determination in Thermally Grown Alumina Scales Using Ruby Luminescence

D. Renusch, B. W. Veal, I. Koshelev, K. Natesan and M. Grimsditch
Argonne National Laboratory, Argonne IL 60439, USA
P. Y. Hou, *Lawrence Berkeley Laboratory, Berkeley CA 94720, USA*

Abstract

By exploiting the strain dependence of the ruby luminescence line, we have measured the strain in alumina scales thermally grown on Fe-Cr-Al alloys. Results are compared, and found to be reasonably consistent with strains determined using x-rays. Oxidation studies were carried out on alloys with compositions Fe - 5 Cr - 28 Al and Fe - 18 Cr - 10 Al (at. %, bal. Fe). Significantly different levels of strain buildup were observed in scales on these alloys. Results on similar alloys containing a "reactive element" (Zr or Hf) in dilute quantity are also presented. Scales on alloys containing a reactive element (RE) can support significantly higher strains than scales on RE-free alloys. With the luminescence technique, strain relief associated with spallation thresholds is readily observed.

I. Introduction

For many systems which operate in high temperature oxidizing environments, thermally grown alumina scales play an essential role in providing protection against environmental attack. However, scale failure can occur because of stresses that develop in the scales. Stresses can occur as a result of growth processes and from thermal expansion mismatch between scale and substrate. For different alloy substrates, sustainable stresses in thermally grown scales might vary significantly. To develop an understanding of the failure mechanisms, it is important to obtain reliable methods for measuring stresses in the scales and for detecting the onset of scale failure [1,2].

We have discovered that stresses can be determined in thermally grown alumina scales by measuring shifts in the "ruby fluorescence", a spectral doublet that occurs near 14000 cm^{-1} in chromium doped aluminas [3]. This fluorescence doublet results from a crystal field excitation in the Cr 3d electrons when Cr atoms are substituted for Al atoms in α-Al$_2$O$_3$. We have also recently learned of similar studies by Clarke , et al [4].

In this paper, we monitor the evolution of the ruby luminescence line from oxide scales grown on Fe - 5 Cr - 28 Al and Fe -18 Cr - 5 Al alloys (at %, bal. Fe), and from scales grown on alloys with similar compositions but containing the reactive elements Zr and Hf, respectively. We find that the amount of strain appearing in scales grown on these alloys differs substantially. For both alloys, the addition of a reactive element dramatically increases the maximum observed strain prior to spallation.

Mat. Res. Soc. Symp. Proc. Vol. 136 © 1997 Materials Research Society

II. Experimental

Oxidation studies were carried out on samples of Fe - 5 Cr - 28 Al [FA71] and Fe - 5 Cr - 28 Al - 0.1 Zr - 0.05 B [FAL], and on samples of Fe - 18 Cr - 10 Al and Fe - 18 Cr - 10 Al - 0.5 Hf (at. %, bal. Fe). All samples were polished with 1 micron alumina polishing grit before oxidation treatments.

The samples were oxidized in air at systematically increasing temperatures (in 50 C or 100 C increments), beginning at 300 C, for durations of one hour. After each oxidation treatment, the ruby fluorescence and Raman spectra were measured at room temperature. Higher temperature anneals were then performed on the same sample; ie, oxide scale growth was cumulative with increasing temperature. Thus samples were thermally cycled between the reaction temperature and ambient as the scale accumulated.

Raman and fluorescence spectra were excited with 50 to 100 mW of 476 nm radiation from a Kr ion laser. The scattered light was analyzed with a triple Jobin-Yvon grating spectrometer and detected with a CCD detector from Princeton Instruments. The Raman spectra were acquired in 500 sec runs at room temperature after the oxidation treatments. A given oxide phase has a characteristic Raman spectrum that can be conveniently used as a "fingerprint" to identify the presence of that phase in the scale. Raman spectroscopy is typically very sensitive to oxides (including Fe_2O_3 and Cr_2O_3) while being completely insensitive to the underlying metal.

The fluorescence radiation from unstrained α-Al_2O_3, doped with Cr^{3+}, appears as a very sharp doublet, detectable at very low levels of Cr doping, with peaks at 14402 and 14432 cm^{-1}. The peak positions are strongly dependent on the state of strain in the sample. (This well known "ruby doublet" has long been used to calibrate pressure in diamond anvil cells.) The peak positions are very weakly dependent on Cr concentration and temperature [4,5]. We find that, for thermally grown scales, the fluorescence doublet is often visible, even when the substrate is nominally Cr - free. Consequently, the fluorescence feature provides a very sensitive and convenient probe of strain in thermally grown scales.

III. Fluorescence Spectroscopy - Strain Analysis

The shift in frequency of the fluorescence doublet is approximately given by

$$\Delta v = \pi_{ij}\, \sigma_{ij} \tag{1}$$

where π_{ij} are the known piezospectroscopic coefficients [4] and σ_{ij} is the stress. Small relative shifts of the two components of the doublet are not considered here. For a strained scale on a substrate, we have

$$\sigma_{11} = \sigma_{22} \text{ and } \sigma_{33} = 0 \tag{2}$$

where σ_{11} and σ_{22} represent the in-plane stress components and σ_{33} is the out-of-plane stress. Since the stress σ_{ij} and strain ε_{kl} are related through

$$\sigma_{ij} = C_{ijkl}\, \varepsilon_{kl}, \tag{3}$$

using the known coefficients C_{ijkl} for sapphire [6] and performing a polycrystalline average, we obtain the ratio of in-plane ($\varepsilon_{in} = \varepsilon_{11} = \varepsilon_{22}$) and out-of-plane ($\varepsilon_{33} = \varepsilon_{out}$) strains. It is interesting to note that the usual in-plane compression ($\varepsilon_{in} < 0$) leads to an out-of-plane expansion ($\varepsilon_{out} > 0$). Combining Eqs., we obtain

$$\varepsilon_{out} = \varepsilon_{33} \approx - 0.48 \, \varepsilon_{in} \qquad (4)$$

and

$$\Delta v = 2810 \, \varepsilon_{in} \qquad (5)$$

where Δv is expressed in cm^{-1}.

IV. Results and Discussion

Since it appears that transient phases of Fe_2O_3 and Cr_2O_3 might significantly influence the early formation of α-phase alumina scales, we have used Raman spectroscopy to study the early stage evolution of these sesquioxides [3]. For the alloys in this study, the Fe_2O_3 signal is first apparent after oxidation at 500 C. Oxidation at higher temperatures increases the Fe_2O_3 signal which reaches a maximum at ~ 750 C, and then decreases and completely disappears by ~ 1000 C. A transient Cr_2O_3 is also visible, between 600 - 800 C on the Fe -18 Cr - 5 Al alloy. When oxidized at temperatures above 1000 C, both the Fe_2O_3 and Cr_2O_3 Raman signals disappeared and were replaced by a Raman signal from α-Al_2O_3.

Fig. 1 shows ruby fluorescence signals from the systematically oxidized Fe - 5 Cr - 28 Al alloy [FA71]. The fluorescence signal is first detectable in the scales after treatment at ~ 750 C. The intensity of this signal grows very rapidly as the sample experiences additional oxidation treatments. The lower spectrum in Fig 1, taken from a fully strain relieved scale, is the same as that from a bulk ruby crystal. Note, however, that the other spectra in Fig 1 show red shifts, of varying amounts, indicative of compressive strain in the scales. There is also substantial broadening of the spectra. Taking the peak position as a measure of the strain induced spectral shift and applying Eq. 5, we obtain the in-plane compressive strain. We make no attempt to interpret the observed broadening because it has contributions from crystal anisotropy, strain inhomogeniety, and perhaps defects in the growing Al_2O_3 crystals.

Fig 2 shows the measured in-plane strain as determined from the fluorescence spectra plotted against the oxidation temperature. The solid circles show strain measurements for Fe - 5 Cr - 28 Al (FA71); the solid triangles are equivalent measurements for the alloy Fe - 5 Cr - 28 Al - 0.1 Zr - 0.05 B (FAL). To ~ 900 C, both alloys show scales that develop increasing strains as oxidation temperature is increased and scale thickness grows. At T ~ 900 C, the FA71 alloy begins to show strain relaxation indicating the onset of failure. With increasing reaction temperature, the scale exhibits catastrophic failure and apparently becomes completely debonded at T > 1000 C. Spalled flakes are visually observable. For the acumulating scale on FAL, however, increasing strain buildup is observed to temperatures of about 1100 C. This is a dramatic manifestation of the reactive element (RE) effect - samples containing a RE develop thermally grown scales that are capable of sustaining substantially larger strains than scales on the RE-free alloys.

The maximum compressive strains obtained from measurements of the FAL scale are surprisingly large, being approximately 2.5 %. The solid line in Fig 2 gives the compressive strain at room temperature from thermal mismatch

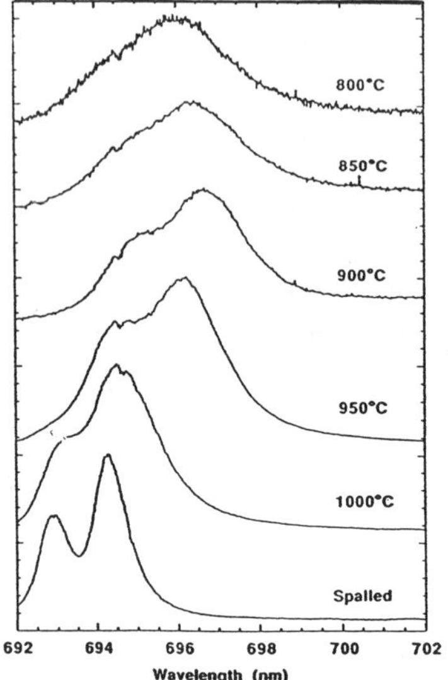

Fig 1. "Ruby" fluorescence spectra from an alumina scale grown on Fe-5Cr-28Al alloy (FAL) after one hour of (cumulative) oxidation at the indicated temperatures. The red shift indicates a compressive strain. The lower spectrum is from a strain relieved scale.

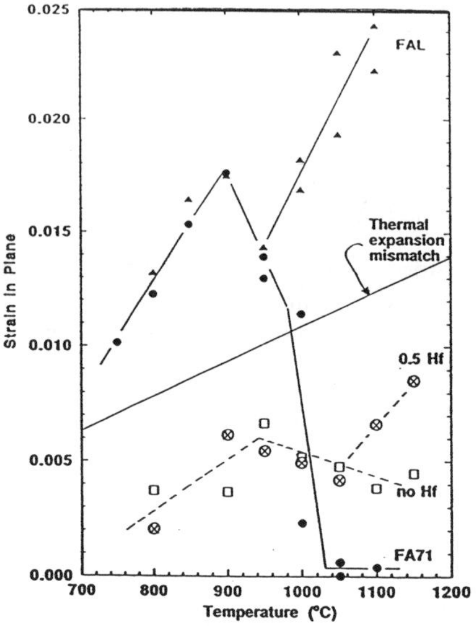

Fig 2. Strain in alumina scales thermally grown on several alloys as a function of exposure temperature. Solid dots FAL; solid triangles FA71; open squares Fe-18Cr-10Al; X's are Fe-18Cr-10Al-0.5Hf. Also shown is the strain calculated from thermal expansion mismatch between alumina and $AlFe_3$.

expected for an Al_2O_3 film on $AlFe_3$, assuming that the film is unstrained at the oxidation temperature. If the thermal expansion coefficient of the Fe-Cr-Al alloys is comparable to that of $AlFe_3$, then, for FA71 and FAL, substantial compressive strain beyond that attributable to thermal mismatch appears in the scales. Apparently a large compressive growth strain, comparable in magnitude to the thermal mismatch strain, develops in these scales.

Another possible explanation for the large strains observed in FAL and FA71 is that our model relating strain to the ruby shift is inaccurate. To test the validity of the model, we have performed a combined x-ray and ruby investigation of oxidized Fe - 18 Cr - 5 Al - 0.5 Hf. For this scale, the strain extracted from the ruby line is plotted in Fig 2 (as X's). A much smaller strain buildup is observed relative to that in FA71 or FAL scales (Fig 2). This reduced strain might be

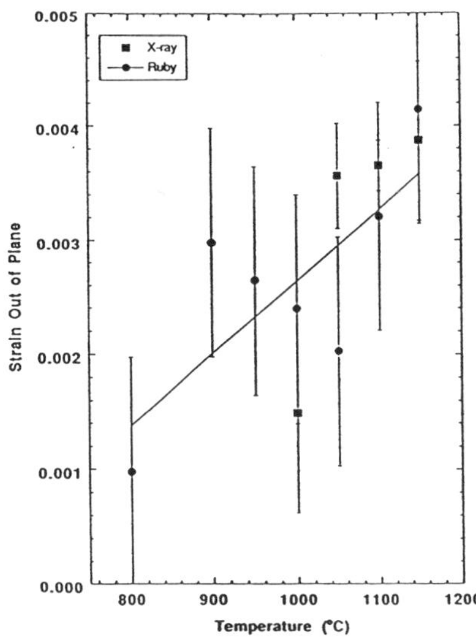

Fig 3. A comparison of out-of-plane strain, as determined by x-ray diffraction and by ruby fluorescence, for the two alloys Fe-18Cr-10Al and Fe-18Cr-10Al-0.5Hf after undergoing a series of oxidation treatments.

related to the peculiar interface morphology known to exist in the Fe - 18 Cr - 5 Al material [7]. Also, significantly less broadening is observed in the fluorescence peaks.

X-ray diffraction measurements were taken using the standard $\theta - 2\theta$ configuration. Because of the geometry of the experiment, only those crystallographic planes parallel to the sample surface were probed; ie, only the out-of-plane strain was measured. Fig 3 shows both the x-ray results and the out-of-plane strain obtained from the fluorescence lineshifts (using Eqs. 4 and 5) for the Fe - 18 Cr - 5 Al - 0.5 Hf scale. Both the x-ray and ruby measurements show a relaxation (expansion) normal to the sample surface as the oxidation temperature is increased. Within the large error bars, agreement between the x-ray and fluorescence results can be considered satisfactory; thereby validating the model used to interpret the fluorescence.

For all of the Fe-Cr-Al alloys, the luminescence line was first detected after heat treating at ~ 750 C, well below the usual formation temperature for the α-phase [eg, see ref. 8]. We believe that the ruby signal provides evidence that α-Al_2O_3 is present in the scales even at these low temperatures [also see ref. 2].

(We have observed that bulk γ-Al_2O_3 shows almost no ruby signal, but the ruby signal is strong after the sample is treated at 1200 C where it transforms to α-Al_2O_3.) A mechanism by which low temperature α-phase growth might occur involves the presence of transient phase Cr_2O_3 or Fe_2O_3 which could serve as templates for the direct formation of α-Al_2O_3. We have noted that, below ~ 900 C, the scales have a significant concentration of Fe_2O_3, and, in some cases, detectable Cr_2O_3. These phases are very similar to α-Al_2O_3; indeed α-Al_2O_3 forms a complete solid solution with Cr_2O_3. Consequently, possible"template phases" exist within the scales. The intensity of the fluorescence signal increases very rapidly with reaction temperature. The rapidly increasing intensity results, in part, from increasing scale thickness, but, more likely from (1) an increased growth rate of α-Al_2O_3 relative to other aluminum oxides and (2) from the conversion of some metastable oxide phase (possibly γ-Al_2O_3) to α-Al_2O_3.

In summary, the growth of alumina scales on several Fe-Cr-Al (alumina former) alloys was studied after samples were oxidized in air. We report room temperature measurements of the strain in thermally grown alumina scales using spectral lineshift measurements of the Cr^{3+} (ruby) fluorescence line and compare with x-ray strain measurements. We observed that, if samples contained a small amount of added "reactive element" (eg, Hf or Zr), the scales could support significantly larger strains. We observed the apparent growth of α-Al_2O_3 at oxidation temperatures as low as 750 C. Presumably, transient Fe_2O_3 or Cr_2O_3 serves as a "template" for the growth of α-Al_2O_3 at these low temperatures. The fluorescence technique provides a sensitive, rapid, and non-destructive method for measuring strains in alumina scales.

Acknowledgements

Work supported by US Department of Energy, Basic Energy Sciences, and Office of Fossil Energy, Advanced Research and Technology Development Materials Program, under Contract No. W-31-109-ENG-38; and by Electric Power Research Institute under contract No. RP 8041-05.

References

1. A. M. Huntz and M. Schütze, Materials at High Temperatures **12**, p. 151 (1994).

2. R. Prescott and M. J. Graham, Oxidation of Metals, **38**, p. 233, (1992).

3. K. Natesan, B. W. Veal, M. Grimsditch, D. Renusch and A. P. Paulikas, Proceedings of Ninth Annual Conference on Fossil Energy Materials, Oak Ridge, TN, May 16-18, 1995; K. Natesan, C. Richier, B. W. Veal, M. Grimsditch, D. Renusch and A. P. Paulikas, Argonne National Laboratory Report ANL/FE-95-02.

4. D. M. Lipkin and D. R. Clarke, Oxidation of Metals (preprint); Q. Ma and D. R. Clarke, J. Am Ceram. Soc., **76**, p. 1433 (1993).

5. A. A. Kaplyanskii, A. K. Przhevuskii and R. B. Rozenbaum, Sov. Phys. Solid State, **10**, p. 1864 (1969).

6. American Inst. of Phys. Handbook, Mc-Graw-Hill, New York, 1963, p. 2-55.

7. P.Y. Hou and J. Stringer, Journal de Physique IV, Colloq C9, supplement to Journal de Physique III, vol. 3, p. 231 (1993).

8. B. A. Pint, J. R. Martin and L. W. Hobbs, Solid State Ionics **78**, p. 99 (1995).

MORPHOLOGICAL INSTABILITY OF A SiC FILM DURING CARBONIZATION

C.-H. CHIU AND L. B. FREUND
Division of Engineering, Brown University, Providence, RI 02912

ABSTRACT

A model is developed to understand the morphological stability of a SiC film on a Si substrate during carbonization where the Si substrate is exposed to a carbon precursor. The morphological stability is determined by considering the surface evolution along a slightly wavy film surface and film-substrate interface. The morphological evolution along the film surface is dominated by surface diffusion and along the interface by a chemical reaction. The kinetic analysis shows the stability is controlled by the film surface energy, the interface energy, the diffusion-reaction process of the carbon precursor, and the strain energy. At small wavelengths of the surface profiles, the two types of surface energy dominate, which results in stable morphology. The diffusion-reaction process dictates the surface stability at large wavelengths. The strain energy may cause the surfaces to become unstable at moderate wavelengths; the instability can be completely suppressed by the diffusion-reaction process and the film surface energy, while it is enhanced by a large value of interface energy.

INTRODUCTION

Silicon carbide is considered to have significant potential for application in high power electronics, high frequency electronics, and other devices because of its excellent material and electronic properties. As a consequence, there has been great interest in growing a heteroepitaxial thin film of SiC on a Si substrate. An important step in growing this system by a CVD method is carbonization where the Si substrate is exposed to a carbon precursor to form a thin layer of SiC, e.g. [1–2]. Formation of defects during carbonization, especially voids on the SiC-Si interface, were reported in the literature [1–6]. Experimental results [1–3] showed that the interface voiding is controlled by the partial pressure of the C-precursor: low partial pressure leads to void formation, while high partial pressure yields a uniform thin film. Mechanisms proposed to explain the interface voiding and its relation with the C-precursor partial pressure include vacancy aggregation [5–6] and island nucleation and coalescence [1–2].

The mechanisms mentioned above include several important factors in the kinetic process of voiding; however, they ignore the effect of the large lattice mismatch (20%) between the SiC film and the Si substrate. Due to the mismatch, the SiC film will be highly stressed during the carbonization process. It is known that large stresses in a film can cause the film surface to be unstable against roughening [7–11] and result in morphological evolution leading to the nucleation of defects such as cusp-cracks and dislocations [12–18].

Accordingly, we propose that it is the surface instability and the accompanying morphological evolution that control the formation of interface voids in the carbonization process. In this paper, we will study the morphological stability issue as a first step toward understanding the voiding mechanism. Of particular interest is how the morphological stability is affected by the partial pressure of the C-precursor.

Mat. Res. Soc. Symp. Proc. Vol. 436 © 1997 Materials Research Society

PROBLEM STATEMENT

Consider a film-substrate system during carbonization. The system is characterized by three phases: a vapor phase g, a SiC film denoted as phase α, and a Si substrate denoted as β. The interface between phases g and α is the film surface Γ_f, and that between α and β is the coherent interface Γ_{int}. The vapor phase is comprised of hydrogen H_2^g and the C-precursor which can be expressed as $C_m H_n^g$ where m and n are integers; the superscript g is used to emphasize the phase. Both the species can diffuse into the SiC film. When $C_m H_n^\alpha$ reaches the film-substrate interface, it reacts with Si to form SiC, resulting in film growth. The penetration of $C_m H_n^\alpha$ and H_2^α into the Si substrate is ignored in this paper. As a result, three chemical reactions are considered here,

$$C_m H_n^\alpha + m\,Si = m\,SiC + \frac{n}{2}\,H_2^\alpha \quad \text{on } \Gamma_{int}, \quad C_m H_n^g = C_m H_n^\alpha \quad \text{and} \quad H_2^g = H_2^\alpha \quad \text{on } \Gamma_f,$$

which are denoted as reactions (I), (II), and (III), respectively. The film and the substrate have a lattice mismatch which induces a biaxial stress T in the film [19] plus other stresses in the structure which depend on the profiles of the two surfaces.

To determine the morphological stability of the system, we first take the film surface Γ_f and the film-substrate interface Γ_{int} to be sinusoidal curves,

$$y = A\cos kx \quad \text{along } \Gamma_f, \qquad y = -H + B\cos kx \quad \text{along } \Gamma_{int} \tag{1}$$

where k is the wavenumber, and H is the average film thickness. We then investigate the morphological evolution of the two surfaces for the case where $kH \ll 1$. If the magnitudes of A and B decrease for any wavenumber k during the evolution, the two surfaces are stable against any perturbation, and they will remain flat in this stage. On the other hand, if $|A|$ and/or $|B|$ increase for some range of k, the two surfaces are unstable and could start to roughen or form islands at the very beginning of the carbonization process.

The morphological evolution along Γ_f is dominated by surface diffusion [7] and along Γ_{int} by reaction (I). For simplicity, reaction (I) is regarded as a sum of the forward and backward processes. In the forward process the reactants combine to form the products, while in the backward process, the products decompose into the reactants. The rate of each process is assumed to obey Boltzmann statistics with the driving force depending on the concentration of the species involved in the process, the strain energy along Γ_{int}, and the interface energy [20]. The dependence of the chemical reaction rate on the strain energy and the interface energy reflects the fact that the two kinds of energy vary with interface migration as reaction (I) proceeds. The interface energy density γ_{int} is given as the sum of γ_β and γ_α; the former is the interface energy involved in the forward process of the chemical reaction, and the latter is that in the backward process.

FIRST-ORDER PERTURBATION ANALYSIS FOR THE SURFACE EVOLUTION

The procedure for deriving the evolution equations for the amplitudes A and B of the two surface profiles is briefly outlined as follows. (1) We solve the elasticity problem of a film coherently bonded to a substrate with a mismatch between them and calculate the strain energy density along the two wavy surfaces by employing the boundary perturbation method [9]. For simplicity, the film and the substrate are taken to be isotropic materials with an identical shear modulus μ and Poisson ratio ν. (2) We then determine the concentration

c_2 of $C_mH_n^\alpha$ and c_3 of H_2^α along Γ_f and Γ_{int} by solving the bulk diffusion problems with the assumption that the variation of c_3 is negligible comparing with that of c_2. The elasticity and the diffusion results are accurate to the first order in Ak and Bk. (3) Substituting the two results into the governing equations for surface diffusion along Γ_f and reaction (I) along Γ_{int} yields the evolution equations for A and B. A detailed discussion about the elasticity, the diffusion, and the morphological evolution problems, as well as the solution procedures for these problems, can be found in [20].

The rates $d\tilde{A}/d\tilde{t}$ and $d\tilde{B}/d\tilde{t}$ for $kH \ll 1$ are found to be

$$\frac{d}{d\tilde{t}}\begin{pmatrix} \tilde{A} \\ \tilde{B} \end{pmatrix} = \begin{pmatrix} 4\tilde{k}^3 - \tilde{k}^4 & -4\tilde{k}^3 \\ m_0 - 4m_1\tilde{k} + 4m_3\tilde{k}^3 - m_3\tilde{k}^4 & -m_0 + 4m_1\tilde{k} - m_2\tilde{k}^2 - 4m_3\tilde{k}^3 \end{pmatrix}\begin{pmatrix} \tilde{A} \\ \tilde{B} \end{pmatrix} \quad (2)$$

where $\tilde{A} = A/L$, $\tilde{B} = B/L$, $\tilde{k} = Lk$, and $\tilde{t} = t/t_L$; L and t_L will be defined later;

$$m_0 = \frac{\hat{c}_2 - 1}{(1 + \theta)^2}\frac{L}{\overline{\overline{H}}}\mathcal{T}_L \qquad\qquad m_1 = \frac{R_1}{1 + \theta}\mathcal{T}_L$$

$$m_2 = \frac{R_1}{1 + \theta}\left(\frac{\gamma_{int}}{\gamma_f} + \frac{\hat{c}_2 - 1}{1 + \theta}\frac{\gamma_\beta}{\gamma_f}\right)\mathcal{T}_L \qquad m_3 = \frac{m\theta c_{min}(\hat{c}_2 + \theta)}{(1 + \theta)^2} \quad (3)$$

Equation (2) shows that the evolution of the two surfaces is coupled. Also, the morphological stability is fully determined by the four parameters m_0, m_1, m_2, and m_3. The stability condition and the meaning of the four parameters will be explored in the next section.

The quantities L and t_L, and those appearing in (3) are defined to be

$$R_1 = \frac{m\Omega w_0}{k_B T_k} \qquad \overline{H} = \frac{D_2}{K^{I+}} \qquad \theta = \frac{K^{I+}}{K^{II-}} \qquad c_{min} = \frac{K^{I-}c_3^{n/2}R_3}{K^{I+}} \qquad t_L = \frac{k_B T_k \gamma_f^3}{\Omega^2 c_s D_s w_0^4}$$

$$R_3 = \exp\left[\frac{m\Omega w_{3d}}{k_B T_k}\right] \qquad \hat{H} = \frac{H}{\overline{H}} \qquad L = \frac{\gamma_f}{w_0} \qquad \hat{c}_2 = \frac{K^{II+}P_4}{K^{II-}c_{min}} \qquad \mathcal{T}_L = \frac{m^2 c_{min} L^2 K^{I+}}{R_1 c_s D_s} \quad (4)$$

where $K^{I\pm}$ are the rates of the forward and backward process of reaction (I) along a flat interface when concentration c_2 and c_3 are unity and when there is no deformation along the interface, $K^{II\pm}$ are the forward and backward rates of reaction (II) at a reference state where $c_2 = 1$, D_2 is the bulk diffusivity of $C_mH_n^\alpha$, $w_0 = (1 - \nu)T^2/4\mu$, $w_{3d} = (1 - \nu)T^2/2\mu(1 + \nu)$ is the strain energy density of a biaxially stressed film, k_B is the Boltzmann constant, T_k is the temperature, γ_f is the film surface energy density, D_s is surface diffusion coefficient, and c_s is the density of moving atoms on the film surface.

MORPHOLOGICAL STABILITY AND THE ENERGETIC FORCES

It follows from eqn (2) that any surface perturbation of Γ_f and Γ_{int} will decay if the real parts of the two eigenvalues, Λ_1 and Λ_2, of the matrix in (2) are negative for all wavenumbers. This stability condition can be rephrased as $\Lambda_1 + \Lambda_2 < 0$ and $\Lambda_1 \cdot \Lambda_2 > 0$, leading to

$$S_1 = m_0 - 4m_2 - 8m_1 + \frac{4m_1^2}{m_2} > 0$$
$$S_2 = m_0 + 4\tilde{k}_0^4 - 4(1 - m_3)\tilde{k}_0^3 + m_2\tilde{k}_0^2 - 4m_1\tilde{k}_0 > 0 \qquad \text{for all } \tilde{k}_0 \quad (5)$$

where $\tilde{k}_0$ is the positive real root of the algebraic equation $d(\Lambda_1 + \Lambda_2)/d\tilde{k} = 0$. The inequality $S_1 > 0$ is equivalent to the auxiliary condition $\Lambda_1 \cdot \Lambda_2 > 0$, and $S_2 > 0$ to $\Lambda_1 + \Lambda_2 < 0$.

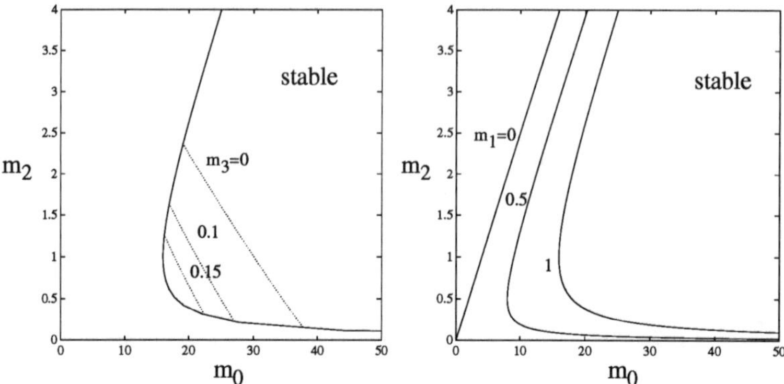

Figure 1: The stability diagram for (a) $m_1 = 1$, and m_3=0, 0.1, and 0.15; and (b) the auxiliary condition $S_1 = 0$ for $m_1 = 0$, 0.5, and 1.

Equation (5) defines a region for the four parameters in which the two surfaces are stable against any roughening. One way to visualize the region is to adopt m_0 and m_2 as the horizontal and vertical axes, respectively, and fix m_1 and m_3, yielding a stability diagram as shown in Figure 1. Part (a) of Figure 1 illustrates the stability diagram for $m_1 = 1$. The solid line corresponds to $S_1 = 0$ which is independent of m_3, and the dashed lines to $S_2 = 0$ for $m_3 = 0, 0.1, 0.15$. The stability region lies in the right-upper corner of the diagram. It is found that increasing m_0 and m_3 will increase morphological stability, while increasing m_2 will reduce it. The effect of m_1 on stability is presented in Part (b) of Figure 1 which depicts $S_1 = 0$ for $m_1 = 0, 0.5$, and 1.0. Evidently, the stability region decreases as m_1 increases. It is also noted that $S_1 = 0$ approaches an asymptotic line $S_a = m_0 - 4m_2 - 8m_1 = 0$ at large m_0 and m_2.

The stability diagram shown in Figure 1 can be used to understand the effect of the partial pressure P_4 of the C-precursor on the morphological stability. First consider the case where P_4 is at the minimum value for film growth. In such a case, reaction (I) is in equilibrium, $\hat{c}_2 = 1$, $m_0 = 0$, and the other three parameters, m_1, m_2, and m_3, are positive. This corresponds to a point on the vertical axis of the stability diagram. One finds when P_4 is at the minimum value, the morphology is unstable. Now suppose the partial pressure P_4 of the C-precursor is raised. It follows from eqn (3) that raising P_4 will increase all the four parameters, though the variation of m_1 can be neglected. Hence, the effect of raising P_4 on stability can be approximately described as moving a point (m_0, m_2) in the right-upper direction on a stability diagram with a fixed m_1. If the point (m_0, m_2) approaches the asymptotic line $S_a = 0$ of $S_1 = 0$, raising P_4 can stabilize the surfaces; if not, higher P_4 would further destabilize the surfaces. This observation suggests the capability of P_4 in surface stabilization can be measured by the value dS_a/dP_4,

$$\frac{dS_a}{dP_4} = \frac{T_{\mathrm{L}} L \gamma_\beta K^{\mathrm{II}+}}{(1+\theta)^2 c_{\min} K^{\mathrm{II}-}} \left[\frac{K^{\mathrm{I}+}}{D_2 \gamma_\beta} - \frac{4m\Omega w_0^2}{k_B T_k \gamma_{\mathrm{f}}^2} \right] \tag{6}$$

One finds the stabilizing capability of P_4 is enhanced by a high characteristic reaction rate $K^{\mathrm{I}+}$, small diffusivity D_2, small strain energy density w_0, large film surface energy density

γ_f, and small interface energy density γ_β. The finding that large interface energy density reduces morphological stability is puzzling. In order to resolve this issue, it is necessary to explain the energetic forces represented by the terms in the first-order evolution result (2) and illustrate how the forces affect the morphological stability as follows.

The morphological evolution of the system is controlled by three energetic forces: the strain energy, the diffusion-reaction process, and the surface energy. The strain energy relaxation, represented by the terms $\pm 4\tilde{k}^3$ and $\pm 4m_1\tilde{k}$ in eqn (2), is the driving force causing the morphology to be unstable. The relaxation is maximum when the rough profiles are symmetric, i.e. $A = -B$, and is zero when the film thickness is uniform, i.e. $A = B$. As a consequence, the symmetric roughening mode tends to be unstable, and the uniform mode is stable. The parameter m_1 is a ratio of the morphology migration rates of the stress-driven surface diffusion and chemical reaction.

The diffusion-reaction process of the C-precursor to form SiC affects the morphological evolution by two ways: a film-thickness effect and a squeezing effect, represented by m_0 and m_3, respectively. The film-thickness effect refers to the variation of film growth rate with the film thickness, which is caused by the fact that c_2 decreases from Γ_f toward Γ_{int}. When the distance between Γ_f and Γ_{int} is slightly larger than the average film thickness, c_2 decreases more and takes a smaller value at Γ_{int}; consequently, the film growth rate will be smaller, reducing the larger thickness. Similarly, a smaller distance will result in higher c_2 and a faster film growth rate, increasing the smaller thickness. This scenario shows that the diffusion-reaction process, in contrast to the strain energy, suppresses the symmetric roughening mode and favors the stable uniform mode. The scenario also explains the result in eqn (6) that a fast reaction rate K^{I+} and a small diffusivity D_2 contribute to surface stabilization. It is noted that the effect is independent of the wavenumber k. Therefore, at a small k, it will dominate the other energetic forces which are all proportional to some power of k.

The squeezing effect, represented by m_3, refers to the phenomenon that when the film surface Γ_f migrates toward the the interface, it induces a higher flux of c_2 toward Γ_{int}, higher c_2 along Γ_{int}, and a faster reaction rate, causing Γ_{int} to move away from Γ_f. Hence, analogous to the film-thickness effect, the squeezing effect tends to stabilize the morphology, as indicated in Part (a) of Figure 1. However, it is also demonstrated in the same figure that this effect cannot fully suppress the instability even if m_3 is large, since the effect prevents the roughening process along the film surface, and has no influence on that along Γ_{int}.

There are two types of surface energy in the system, the film surface energy and the interface energy. The film surface energy is represented by the term $-\tilde{k}^4$ in eqn (2). The energy competes with the strain energy by flattening the film surface and stabilizes the film surface at a large $\tilde{k}$. The interface energy effect is represented by the term $-m_2\tilde{k}^2$ in eqn (2), which also tends to flatten the interface Γ_{int}. However, it is noted that letting $B = 0$ is equivalent to inducing both the uniform and symmetric modes. In this sense, the flattening tendency will reduce the strength of the film-thickness effect in prohibiting the symmetric roughening mode and destabilize the morphology.

CONCLUSIONS

The morphological stability is controlled by the strain energy, the diffusion-reaction process, and the surface energy. At a small wavenumber, the diffusion-reaction process dominates and the two surfaces are stable. At a large wavenumber, the surface energy dominates and stabilizes the morphology. At a moderate wavenumber, the strain energy becomes important and causes the surfaces to be unstable. The strain energy is in competition with the diffusion-reaction process and the film surface energy and is assisted by the interface energy. Hence, at low partial pressure of the C-precursor, the diffusion-reaction effect is small and the surfaces are unstable. Raising the partial pressure enhances both the effects of the diffusion-reaction process and the interface energy. The competition of the two effects determines whether high partial pressure of the C-precursor favors morphological stability.

ACKNOWLEDGEMENTS—The research support of the Office of Naval Research Contract N00014-95-1-0239 is gratefully acknowledged.

REFERENCES

1 A. J. Steckl and J. P. Li, *Thin Solid Films* **216**, 149 (1992).

2 J. P. Li and A. J. Steckl, *J. Electrochem. Soc.* **142**, 634 (1995).

3 T. Yoshinobu, H. Mitsui, Y. Tarui, T. Fuyuki, and H. Matsunami, *J. Appl. Phys* **72**, 2006 (1992).

4 V. Cimalla, K. V. Karagodina, J. Pezoldt, and G. Eichhorn, *Mat. Sci. Eng.* **B29**, 170 (1995).

5 C. J. Mogab and H. J. Leamy, *J. Appl. Phys.* **45**, 1075 (1974).

6 C. C. Chiu and S. B. Desu, *J. Mater. Res.* **8**, 535 (1993).

7 R. J. Asaro and W. A. Tiller, *Metall. Trans.* **3**, 1789 (1972).

8 M. A. Grinfeld, *J. Nonlin. Sci.* **3**, 35 (1993).

9 H. Gao, *Int. J. Solids Structures* **28**, 703 (1991).

10 L. B. Freund and F. Jonsdottir, *J. Mech. Phys. Solids* **41**, 1245 (1993).

11 C.-h. Chiu and H. Gao, *Int. J. Solids Structures* **30**, 2983 (1993).

12 C.-h. Chiu, Ph.D. dissertation, Stanford University, 92–130 (1995).

13 C.-h. Chiu and H. Gao, *Mat. Res. Symp. Proc.* **356**, 33 (1995).

14 C.-h. Chiu and H. Gao, *Mat. Res. Symp. Proc.* **317**, 369 (1994).

15 B. J. Spencer and D. I. Meiron, *Acta Metall. Mater.* **42**, 3629 (1994).

16 W. H. Yang, and D. J. Srolovitz, *J. Mech. Phys. Solids* **42**, 1551 (1994).

17 D. E. Jesson, S. J. Pennycook, J.-M. Baribeau, and C. D. Houghton, *Phys. Rev. Lett.* **71**, 1774 (1993).

18 C. S. Ozkan, W. D. Nix, and H. Gao, *Mat. Res. Soc. Symp.* **399**, in press (1996).

19 W. D. Nix, *Metall. Trans. A* **20A**, 2217 (1989).

20 C.-h. Chiu and L. B. Freund, manuscript in preparation (1996).

STRESS ANALYSIS OF TITANIUM DIOXIDE FILMS BY RAMAN SCATTERING AND X-RAY DIFFRACTION METHODS

Li-jian Meng*, M.P. dos Santos**

*Departamento de Física, Instituto Superior de Engenharia do Porto, Rua de São Tomé, 4200 Porto, Portugal, ljmeng@ci.uminho.pt

**Departamento de Física, Universidade do Minho, 4710 Braga, Portugal

ABSTRACT

Titanium dioxide films have been deposited onto glass substrates by dc reactive magnetron sputtering at different sputtering pressures (2×10^{-3} -- 2×10^{-2} mbar). The films have been characterized by measuring their Raman scattering and X-ray diffraction (XRD). The films stress have been calculated by analyzing their Raman spectra and X-ray diffraction spectra. Two methods give similar results which films prepared at low sputtering pressure have high stress values.

INTRODUCTION

Raman spectroscopy has been extensively applied during the past several years to the characterization of molecular bonding in optical coatings [1-6]. Vibrational spectra obtained by this rapid, nondestructive technique are quite sensitive to perturbations in chemical bonding caused by localized distortions or induced phase transformations. Such structural alterations can affect the ultimate optical response of a thin film or multilayer structure. The use of laser radiation and its unique properties to excite Raman scattering has enhanced the utility of this technique to thin film characterization. Many Raman studies have been done for TiO_2 films [1-15] and recently for nanophase TiO_2 [16-18]. The studies are focused on the structural characterization of TiO_2 films [1-5, 9-13], laser induced damage in the films [8, 15] and the measurement of the stress [6]. Parker *et al.*, have introduced a method to calculate the stoichiometry of nanophase TiO_2 from their Raman spectra [18]. In this paper, Raman scattering effects have been studied for TiO_2 films prepared by dc reactive magnetron sputtering at different sputtering pressures. The stress of the films has been calculated by analyzing the respective peak shift. In addition, the stress of the films has also been calculated by analyzing X-ray diffraction data. The results obtained by these two method have been compared.

EXPERIMENTAL DETAILS

Titanium oxide films were deposited on glass substrates by dc reactive magnetron sputtering in a mixed argon and oxygen atmosphere. The target was titanium metal of 99.6% purity. The substrate was not heated or biased during the deposition processes. For samples N15, N16, N17, N5, N18 and N19, the oxygen partial pressure was kept at 6×10^{-4} mbar, and the total pressures were 2, 4, 6, 8, 10 and 20×10^{-3} mbar respectively. Details of the sample preparation can be found in other paper [19]. Raman scattering experiments were performed with a Coherent Innova 92 Ar+ ion laser and a JOBIN-YVON T64000 triple Raman spectrometer using a 1024 elements Spectraview-2D charge-coupled-device (CCD) as photon detector. The laser beam was focused on the sample surface by a optical input focusing device. A special scattering geometry was used in order to remove the directly reflected light because the incident light was strongly reflected by the films [20]. In all measurements, the entrance and exit slit widths are 100 μm and the laser power on the sample surface is about 70 mW. All the measurements were done at room temperature. All the spectra have been calibrated referred to the TiO$_2$ powder spectra.

EXPERIMENT RESULTS AND DISCUSSION

Figure 1 shows Raman spectra of TiO$_2$ films prepared at different deposition conditions. By comparing these spectra with the spectrum of the anatase phase TiO$_2$ powder, it can be concluded that the films prepared at different conditions consist only of the anatase phase, no other phase can be observed in these films. Although the films prepared at different conditions have only anatase phase structure, their microstructures are different. The film prepared at low total pressure (N15) has a polycrystal structure with a preferred orientation along (101) crystal plane and the one prepared at high total pressure (N19) has an almost amorphous structure [19]. However, these clear variations in the film microstructure do not result in any clear differences in their Raman spectra as shown in Fig.1. In order to study the slight variations in the Raman spectra, we analyze only the strongest Raman peak (143 cm^{-1} Eg mode). The peak was fitted by a Lorentzian distribution:

$$I_L(x,\mu,\Gamma) = \frac{1}{\pi} * \frac{\frac{\Gamma}{2}}{(x-\mu)^2 + (\frac{\Gamma}{2})^2}$$

where I_L is Lorentzian distribution; Γ is the full width at half-maximum and μ is the peak position.

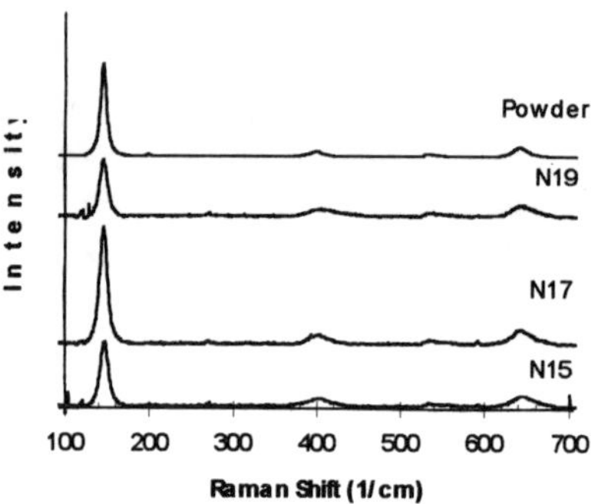

Fig. 1. Raman spectra of films prepared different total pressures. (Exciting wavelength: 514.5 nm; sample surface power: 70mW.)

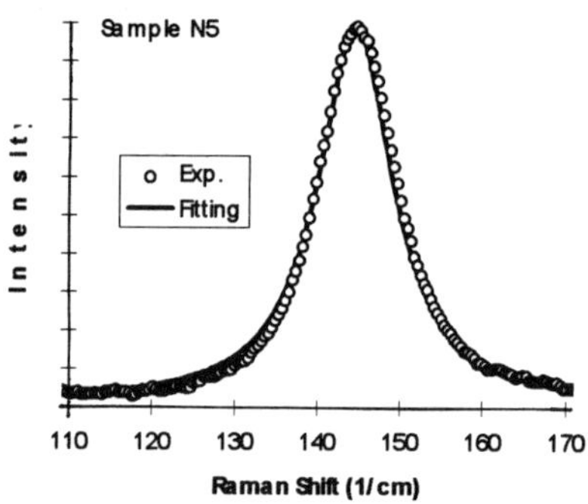

Fig. 2 Raman peak fitting of sample N5.

The fitting result for sample N5 is shown in Fig. 2. The fitting parameters for the films prepared at different total pressures are listed in Table 1. The peak position has been calibrated by taking the TiO$_2$ powder peak value as 143 cm^{-1}. It can be seen from Table 1 that the differences of the Raman parameters are very small for the film prepared at different total pressures. In order to make the results clear, we have studied the Raman spectra of TiO$_2$ films prepared at different total pressures carefully by measuring their spectra many times. The Raman spectra, peak positions and peak widths are shown in Fig. 3 (a), (b) and (c) respectively. As can be seen from Fig. 3 the peak position diverges from the standard value of 143 cm^{-1} both at low and high total pressures (Fig.3 (b)) and the peak width decreases as the total pressure is increased (Fig. 3 (c)).

Table 1. The Raman parameters of the TiO$_2$ films prepared at different total pressures.

Sample Number	N15	N16	N17	N5	N18	N19
Peak Position (cm^{-1})	144.5	144.0	143.2	143.2	143.2	144.0
Peak Width (cm^{-1})	12.0	11.5	11.0	10.5	10.2	10.5

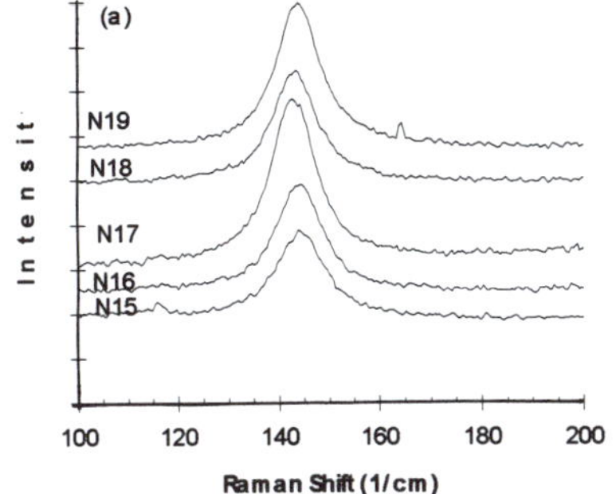

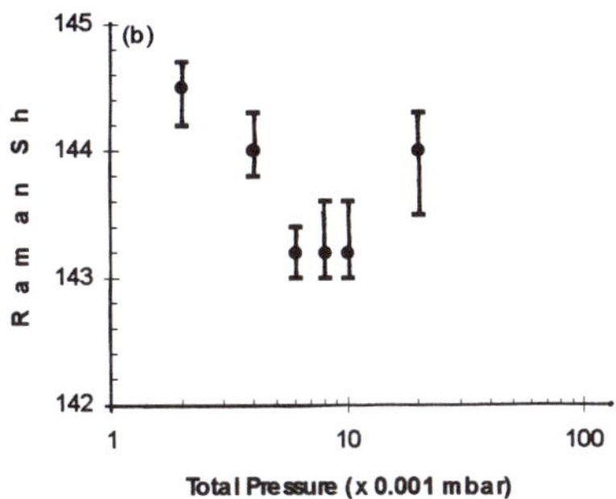

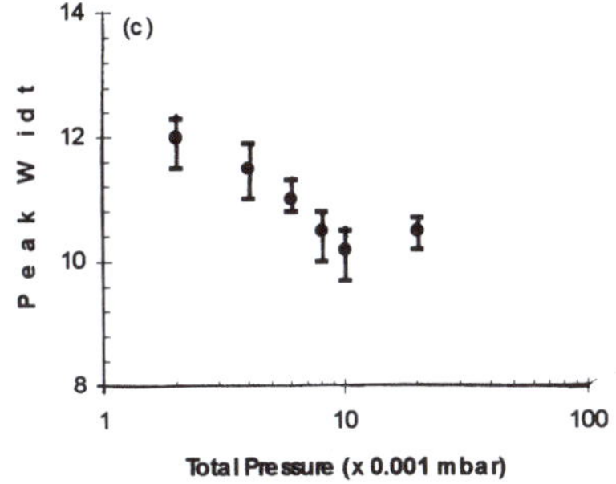

Fig. 3. Raman spectra (a), peak position (b) and peak width (c) of the films prepared at different total pressures.

Calculations of the lattice vibrations at k = 0 of anatase TiO$_2$ have been made [21, 22]. The 143 cm^{-1} (Eg) Raman mode has been shown using the GF-matrix method to be an O-Ti-O bond bending mode. This mode involves the basal planar O atoms moving out of phase with the central Ti atom. The calculations yield a weak force constant owing to the mode's low frequency vibration. It was also implied that the slightly distorted oxygen atoms that lie in the basal plane do not interact to a large extent. These calculations considered only short-range forces and admittedly ignored the long-range Coulomb forces. Parker *et al.*, has indicated that the calculations are in good agreement with Raman data on stoichiometric TiO$_2$ but not on non-stoichiometric TiO$_2$ [17]. Our results are in agreement with the calculations, which means that our TiO$_2$ films are stoichiometric. This conclusion is also in agreement with our Rutherfurd backscattering experiment results [20].

The relative frequency changes of the 143 cm^{-1} mode could be related to the film stress [6, 23, 24]. In general, compressive stress usually induces positive frequency shifts resulting from lattice contraction. From Fig.3 it can be seen that the frequency shifts in our films are positive, that means there is compressive stress in our films and the stress value decreases as the total pressure is increased. The x-ray diffraction results also support this conclusion as shown in Fig. 4.

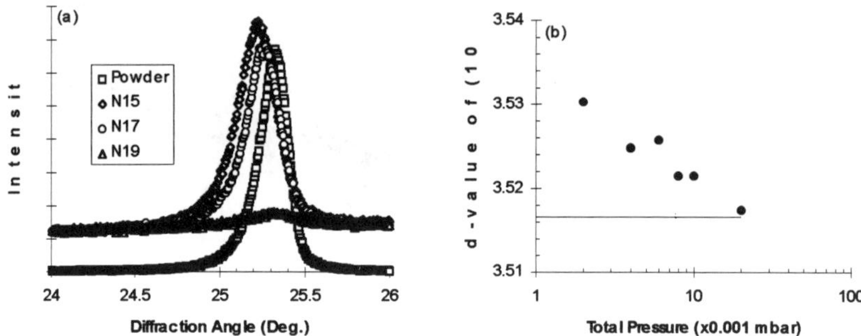

Fig.4 (a) XRD of (101) plane of powder and films prepared at different total pressures, (b) lattice parameter as a function of the sputtering pressure(the line is the value of the powder).

From Fig.4 it can be seen that the diffraction peak shifts to high diffraction angle as the total pressure is increased. The crystal plane distance was calculated from the diffraction angle. The difference between the standard value d_o (3.5166 Å) and the calculated value is shown in Fig.4. The d value is larger than d_o and the value decreases as the total pressure is increased. That means the films prepared at different total pressures have a compressive stress and the stress value decreases as the total pressure is increased. The decrease of the stress results in the narrowing of the Raman peak width as shown in Fig. 3(c).

Exarhos *et al.*, have developed a method to calculate the TiO$_2$ films compressive stress by measuring their Raman spectra[6]. They have given an empirical analytical representation of the pressure *(p)* and temperature *(T)* dependence of the mode frequency determined by temperature-dependent Raman measurements of this material constrained in a diamond anvil cell:

$$\omega_i = \omega_i^0 + a_i p + b_i T + c_i pT$$

where ω_i is the frequency of the mode of interest and ω_i^0, a_i, b_i, and c_i are fitting parameters. For the 143 mode, they give the fitting parameters as follows:

$$\omega_i^0 = 136.8;\ a_i = 0.493;\ b_i = 1.73 \times 10^{-2};\ c_i = -5.50 \times 10^{-4}$$

By using these parameters, we have calculated the compressive stress in our films. The results are shown in Fig. 5.

In addition, the principal residual stress $\sigma_1 + \sigma_2$ can also be estimated from lattice strain ($(d-d_0)/d_0$), Young's modulus E_f and Poisson ratio υ_f of the coating according to the follow equation [25]:

$$\sigma_1 + \sigma_2 = - \frac{E_f}{\nu_f} \frac{d - d_0}{d_0}$$

where d and d_0 are the plane spacing for films and standard sample (TiO_2 powder in our experiment) determined by XRD measurements. For TiO_2 material, the Young's modulus E_f and Poisson ratio υ_f are 1272 kg/cm^2 and 0.4272 respectively [26]. The film stress estimated by this method is also shown in Fig.5. According that formula, the $\sigma_1 + \sigma_2$ should be negative value for the compressive stress. In Fig.5, we only consider its absolute value.

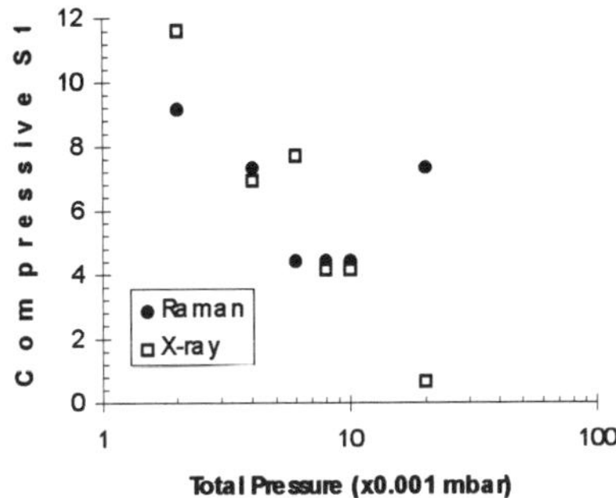

Fig.5. The film compressive stress estimated by Raman scattering (●) and X-ray diffraction (□) methods.

From Fig.5 it can be seen that two methods give similar results of the film stress except for sample N19. The sample N19 has an almost amorphous structure and has a very weak (101) diffraction peak as shown in Fig.4(a). Therefore, its d value was obtained probably with a big error and results in the big divergence. Anyway, Fig.5's results show that the film stress can be estimated by measuring its Raman scattering spectra.

CONCLUSIONS

Raman scattering effect of the films prepared at different deposition conditions have been studied. The results show that the Raman spectra are not sensitive to the variation of the microstructure of TiO_2 films but the variation of the phase.

The film stress has been estimated by measuring their Raman spectra and compared with one obtained by the X-ray diffraction method. The results obtained by these two methods are similar. Therefore, it gives a method to estimate the film stress especially for the films without diffraction peaks.

ACKNOWLEDGMENT

This work is supported by Junta Nacional de Investigação Cientific e Tecnológiga (JNICT). (Contract No. PBIC/C/CTM/1906/95). Li-jian Meng wish to thank Fundação Luso-Americana para o Desenvolvimento for travel support.

REFERENCES

1. W.T. Pawlewicz, G.J. Exarhos and W.E. Conaway, Appl. Opt. **22**, 1837(1983).
2. G.J. Exarhos and W.T. Pawlewiez, Appl. Opt. **23**, 1986(1984).
3. J.C. Tsang, Ph. Avouris and J.R. Kirtley, J. Chem. Phys. **79**, 493(1983).
4. D.M. Friedrich and G.J. Exarhos, Thin Solid Films **154**, 257(1987).
5. C.Y. She, Thin Solid Films **154**, 239(1987).
6. G.J. Exarhos and N.J. Hess, Thin Solid Films **220**, 254(1992).
7. G.J. Exarhos, P.J. Miller and W.M. Risen, J. Chem. Phys. **60**, 4145(1974).
8. L.S. Hsu, R. Solanki, G.J. Collins and C.Y. She, Appl. Phys. lett. **45**, 1065(1984).
9. L.S. Hsu, R. Rujkorakarn, J.R. Sites and C.Y. She, J. Appl. Phys. **59**, 3475(1986).
10. S. Miyake, K. Honda, T. Kohno and Y. Setsuhara, J. Vac. Sci. Technol. **A10**, 3253(1992).
11. C.M. Foster, R.P. Chiarello, H.L.M. Chang, H.You, T.J. Zhang, H.Frase, J.C. Parker and D.J. Lam, J. Appl. Phys. **73**, 2841(1993).
12. A. Turkovic, M. Ivanda, V. Vranesa and A. Drasner, Vacuum **43**, 471(1992).
13. A. Turkovic, M. Ivanda, V. Vranesa and A. Drasner, Thin Solid Films **198**, 199(1991).
14. G.J. Exarhos and M. Aloi, Thin Solid Films **193/194**, 42(1990).
15. R.L. White and G.J. Exarhos, J. Mater Res. **6**, 126(1991).
16. C.A. Melendres, A. Narayanasamy, V.A. Maroni and R.W. Siegel, J. Mater. Res. **4**, 1246(1989).
17. J.C. Parker and R.W. Siegel, J. Mater. Res. **5**, 1246(1990).
18. J.C. Parker and R.W. Siegel, Appl. Phys. Lett. **57**, 943(1990).
19. L.J. Meng and M.P. dos Santos, Thin Solid Films **226**, 22(1993).
20. L.J. Meng, PhD Thesis, University of Miho, Portugal,1994
21. T. Ohsaka, F. Izumi and Y. Fujiki, J. Raman Spectroscopy **6**, 321(1978).
22. T. Ohsaka, S. Yamaoka and O. Shimomura, Solid State Commun. **30**, 345(1979).
23. G.J. Exarhos and N.J. Hess, SPIE **1441**, 190(1990).
24. N.J. Hess and G.J. Exarhos, SPIE **1055**, 194(1989).
25. I.C. Noyan, <u>Residual Stress</u>, Springer, New York, Berlin, Heidelberg, (1987).
26. P. Vretenar, Vacuum **40**, 173(1990).

EFFECT OF THICKNESS AND ANNEALING ON STRESS IN TANTALUM AND TANTALUM NITRIDE THIN FILM HARD COATINGS

RANJANA SAHA, RAMA B. INTURI, JOHN A. BARNARD
Department of Metallurgical and Materials Engineering, The University of Alabama, Tuscaloosa, AL 35487-0202.

ABSTRACT

An understanding of the relationship between stress and the other properties of thin films is extremely useful in the design of hard coatings for long term performance. In our earlier study, sputtered Ta and Ta(N) films were found to exhibit promising hard coating properties. For example, nano hardness as high as 30 GPa was observed in the nitride ($pN_2 = 0.100$ mTorr) films. In this work, we study the variation in the stress in these films with respect to film thickness and annealing. Films in six different thicknesses (50, 250, 350, 500, 750, and 1000 nm) were deposited on oxide coated Si (111) wafers. Stresses in the films in the as-deposited state and as a function of temperature (300°C) were determined using a thin film stress measuring unit.

INTRODUCTION

Virtually all films deposited on substrates are in a state of stress. Depending on the deposition process and the deposition parameters, these stresses can be either compressive or tensile. Extremely high stresses can significantly affect the properties of the films and can even lead to failure. There are two main sources of stress development in thin films : (1) thermal stresses due to the difference in the thermal expansion coefficients between the coating or coatings and the substrate materials, together with the difference between the deposition and application temperatures [1], and (2) intrinsic or growth stresses due to the accumulating effect of the crystallographic flaws that are built into the coating during deposition [2]. In the case of epitaxial films, stresses can arise due to mismatch between the lattice parameters of the film and substrate [3]. Various models have been proposed [4-6] and reviewed [2, 7-9] to explain the origin of internal stresses in thin films.

The effect of deposition parameters on stresses in sputtered thin films has been investigated in detail by Thornton and Hoffman [2, 10-12]. In almost all the materials studied, they observed a transition from compressive to tensile stress with increasing operating pressure. The observed stresses are explained by the atomic peening model proposed by d'Heurle [13] and expanded by Hoffman and Thornton [14]. Stresses in sputtered films are also observed to be a function of substrate temperature, working gas species, deposition rate, angle of incidence (orientation of deposition surface relative to direction of coating flux), apparatus geometry, distance between the substrate and source, substrate bias, and atomic mass of the target material [2].

Stress vs. film thickness has been investigated by many researchers [4, 7, 10-12, 15]. According to their observations, the average stress in thick films is relatively independent of the film thickness when the deposition conditions remained the same. A comprehensive study of 15 different evaporated metal films by Klockholm and Berry [4] shows a leveling of stress values with film thicknesses of ~1000Å or more. They found the stress to be related to μ and T_m, where μ is the shear modulus and T_m is the melting temperature of the metal. Doljack and Hoffman [15] use the grain boundary relaxation model to explain the stresses observed in thin nickel films as a function of thickness.

Thin ceramic coatings are being extensively used in tribological applications because their high defect densities and small grain size leads to a very hard, wear-resistant surface [16]. However, hardness alone is not sufficient when considering the selection of a material for any hard coating application. Other properties such as residual stress may be equally important as they can have a profound effect on the functioning and reliability of the coating. In our earlier study [17], sputtered Ta and Ta(N) films were found to exhibit promising hard coating properties. For example, nano hardness as high as 30 GPa was observed in the nitride ($pN_2 = 0.100$ mTorr) films. In this work, we study the variation in the stress in these films with respect to thickness and

Mat. Res. Soc. Symp. Proc. Vol. 436 © 1997 Materials Research Society

annealing. The structure and mechanical properties, hardness and Young's modulus, before and after annealing were also determined.

EXPERIMENTAL PROCEDURE

A Vac-Tec model 250 batch side sputtering system with six carousel positions was used to deposit the Ta and Ta(N) films. The target was a Ta disk, 4" in diameter and with a purity of 99.9%. The films were deposited on 2" dia. oxidized Si (111) wafers. Pieces of the Si substrate were placed next to the wafer during deposition for thickness, structure, and mechanical properties determination. Base pressure in the chamber prior to sputtering was 4.8 x 10^{-7} Torr or less. Power was set at 100W during sputtering. The Ar gas pressure was maintained at 4 mTorr for the pure Ta films. The Ta(N) films were reactively sputtered in an Ar + N_2 gas mixture. The gas pressure was set at 4 mTorr Ar + 0.026, 0.052, and 0.100 mTorr nitrogen to obtain films with three different nitrogen contents. The deposition times (determined from earlier work [17]) were varied such that films of thicknesses ~ 50, 250, 350, 500, 750, and 1000nm were deposited.

Film thicknesses were measured using a Dektak IIa surface profilometer. The structure of the films in the as-deposited and annealed conditions were determined by x-ray diffraction using the θ-2θ mode. X-ray diffraction was performed on a Rigaku D/Max-2BX XRD system with a thin film attachment using Cu $K\alpha$ radiation. A thin film stress tester (Flexus Stress Measurement System, model FLX 2320) with in situ annealing facilities was used to evaluate stress at room temperature (as-deposited films) and as a function of temperature (annealed films). Stress was determined by measuring the wafer curvature and converting it to film stress by using Stoney's equation. Annealing was carried out in the stress tester itself. The chamber was backfilled with Ar gas and a continuous flow was maintained during the measurements. All the films were heated to 300°C at the rate of 5°C/min. The films were held at this temperature for an hour and then cooled to room temperature (25°C) at the same rate (5°C/min). An extra hour of cooling was provided since it was observed that the films did not attain room temperature at the end of the cooling cycle.

The films were reannealed using the same conditions and the stress-temperature properties were again recorded. The mechanical properties of the as-deposited and annealed films were determined using the Nano Indenter® II (Nano Instruments, Inc., Oak Ridge, TN) mechanical properties microprobe with a berkovich indenter. The indentations were carried out using a constant displacement loading rate in the load segment. The indentation depth varied from 30nm to 275 nm depending on the thickness of the films. Hardness and Young's modulus were determined using the method of Oliver and Pharr [18].

RESULTS AND DISCUSSION

<u>Structure Analysis</u>

Fig. 1 shows the x-ray diffraction profiles of the 250nm thick Ta and Ta(N) films. All the films are highly textured. No substantial change in texture is observed with thickness in any of the films. Ofcourse, the absolute intensity of the peaks increases with increasing thickness. In the case of the pure Ta films, which crystallize in the β tetragonal phase [17], the ratio of the intensities of (004)/(002) diffracting peaks increases with

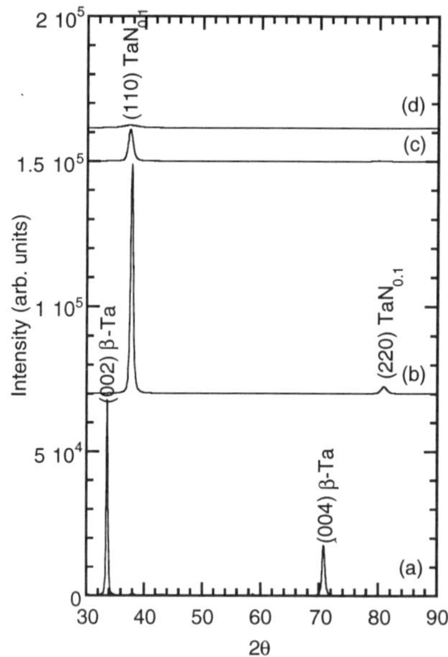

Fig. 1 : X-ray diffraction profile of 250 nm as-deposited films (a) 0 N_2, (b) 0.026 N_2, (c) 0.052 N_2, and (d) 0.100 N_2.

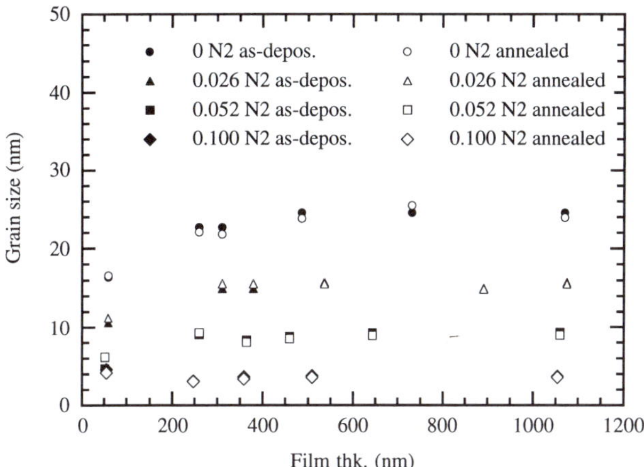

Fig. 2 : Plot of grain size as a function of film thickness for the as deposited and annealed films.

thickness from 17% in the 50nm films to 55% in the 1000nm films. The increase in the intensity ratio with thickness could probably be attributed to the increasing film temperature due to longer deposition times, which gives the (004) orientation a better oppurtunity to evolve. Addition of nitrogen leads to the formation of the $TaN_{0.1}$ phase. The intensity of the peaks decreases with increasing nitrogen (from $pN_2 = 0.026$ mT to 0.100 mT). The grain size, calculated from the width of the diffracting peaks using the Scherrer formula [19], shows little variation with film thickness above 200nm. It is also observed to decrease with increasing nitrogen from ~15nm to ~3.5nm (fig. 2). Annealing at 300°C caused little change in grain size.

Stress measurement

The results of the room temperature stress measurements as a function of film thickness are shown in fig. 3. All films exhibit high compressive stresses. These results are similar to those reported by other researchers [10, 20-22] and is explained by the atomic peening of the depositing film by the impact of high-energy sputtered metal atoms and working gas species at low deposition pressures. In case of the pure Ta films, on heating, the compressive stress increases with temperature. This behavior is exhibited for all thicknesses.

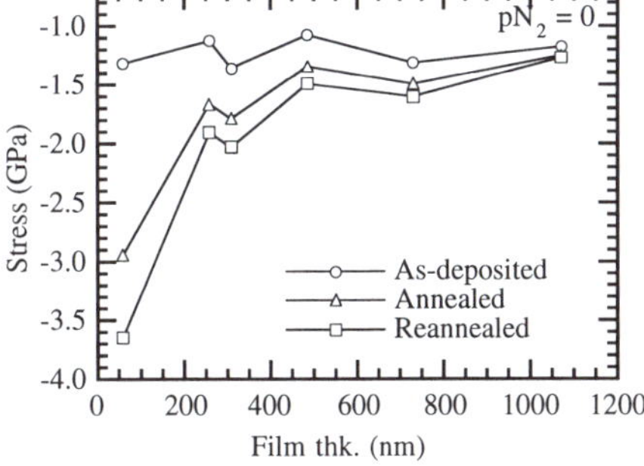

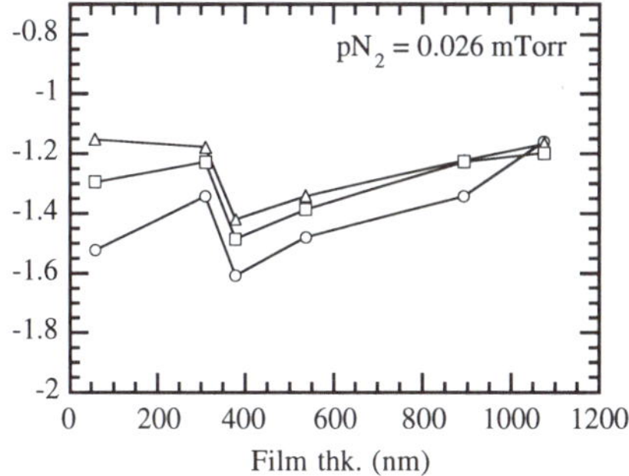

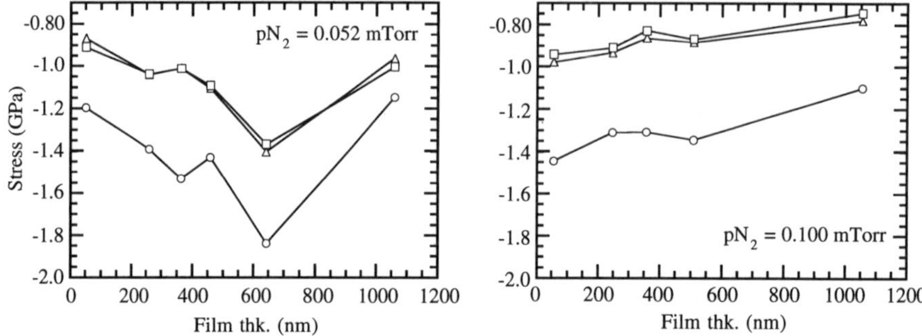

Fig. 3 : Stress as a function of film thickness.

A representative stress-temperature plot of the 250nm thick film is as shown in fig. 4. These results are similar to the results obtained by Mutscheller et. al. [21] and they have attributed this to thermo-elastic deformation in the films. During the one hour isothermal anneal at 300°C, the stresses increased further. This increase decreased with thickness until it was almost negligible in the 1μm film (from -1266.9 MPa in the 50nm film to -27.1 MPa in the 1μm film). A similar trend, i.e. increase in compressive stress, is observed on repeating the annealing cycle, though the extent of increase is less for all the thicknesses as compared to the previous cycle. Cabral et. al [23] have also made similar observations. They demonstrate that 50nm to 200nm thick sputtered β-Ta thin films undergo repeated compressive stress increases when thermally cycled from room temperature to 400°C and back in a purified He ambient. Isothermal annealing of the Ta films at 400°C exhibits a similar behavior. They attribute this effect to the oxygen contamination of the films which occurs due to the low levels of oxygen gettered by the tantalum during the annealing cycle. Thus they conclude that the increase in the oxygen concentration of the tantalum films which occurs upon thermal cycling leads to a repetitive increase in compressive stress leaving the films in an unstable state. This ultimately leads to film failure through cracking when any perturbation of the film occurs and thus relieves the high compressive stress.

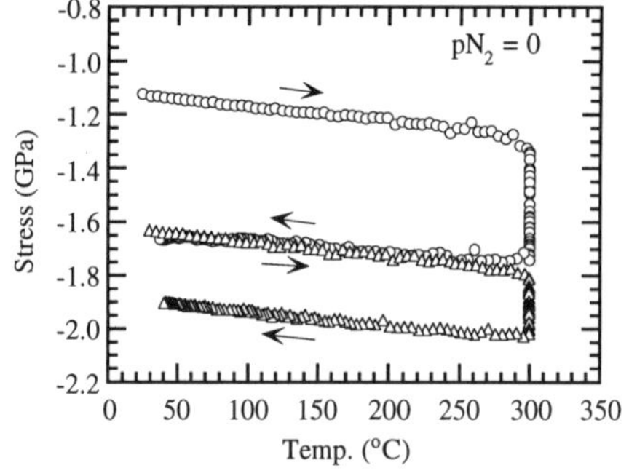

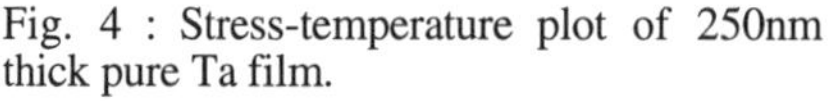

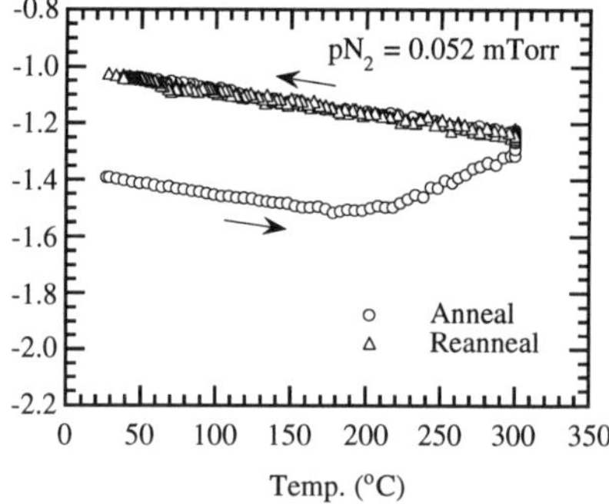

Fig. 4 : Stress-temperature plot of 250nm thick pure Ta film.

Fig. 5: Stress-temperature plot of 250nm thick Ta(N) film with $pN_2 = 0.052$ mTorr.

By contrast, in case of the nitride films, an overall decrease in compressive stress is observed after annealing. Fig. 5 is a representative stress-temperature plot of the behavior exhibited by different thickness and nitrogen content films. On heating, the elastic strain due to the difference in thermal expansion between the nitride film and the silicon substrate results in a linear increase in stress with temperature. On further heating beyond 150°C, stress decreases which could be attributed to densification and defect annihilation in the films. During the isothermal anneal at 300°C, a very slight decrease in compressive stress is observed. On cooling, a linear decrease in stress is observed because the film thermally contracts more than the substrate again due to the difference in their coefficients of thermal expansion. During the second annealing cycle, only elastic behavior was observed in all the films, indicating that the microstructure has been stabilized by the first annealing cycle. In case of the nitride films with pN_2 = 0.026 mTorr, a slight increase in compressive stress is observed after the second annealing cycle, whereas for the films with pN_2 = 0.100 mTorr, a slight decrease is observed.

During annealing, any change in the volume of the film will lead to a change in the state of its stress. If the film is compressively stressed, a volume shrinkage will lead to stress relaxation. For eg., annihilations of excess vacancies, dislocations, and grain boundaries are processes that lead to volume changes due to densification. Also, dilational strains in the film can be produced by phase transformations and composition changes [3]. Stress relaxation due to grain growth when a film is compressively stressed has been explained in detail by Chaudhari [24]. Flinn observed stress relaxation in Al [25] and Cu [26] films due to recrystallization during annealing. However, in our films, X-ray diffraction reveals no apparent change in the structure of the annealed films as compared to the as-deposited films. Neither is a significant difference in grain size observed (fig. 2). Hence grain growth or phase transformation cannot account for the stress relaxation observed in the nitride films. This issue is under further investigation. The hardness of the annealed films (fig. 6) as compared to the as-deposited films show a very slight increase (~ 1 GPa).

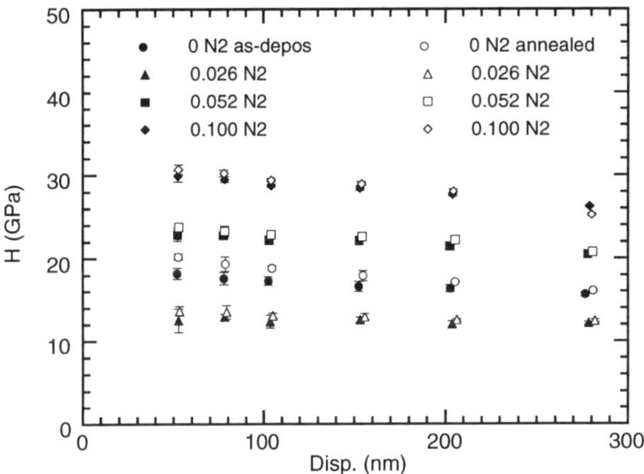

Fig. 6 : Hardness of 1000nm thick Ta and Ta(N) films.

CONCLUSIONS

The tantalum and tantalum nitride thin films deposited by magnetron sputtering exhibit compressive stress in the as-deposited state. On annealing, the stress in the pure Ta films increases with temperature and also when held constant at 300°C. A similar behavior is observed on reannealing. In case of the nitride films, the stress first increases with temperature due to the difference in the thermal expansion coefficients of the film and the substrate. However, in the temperature range of 150 - 200°C, the stress is observed to decrease with temperature. No apparent change in structure or significant change in grain size is observed in the films after annealing.

ACKNOWLEDGMENT

This work is supported by NASA-EPSCoR.

REFERENCES

1. P. S. Alexopoulos and T. C. O'Sullivan, Annu. Rev. Mater. Sci., 20, p. 391 (1990).
2. J. A. Thornton and D. W. Hoffman, Thin Solid Films, 171, p. 5 (1989).
3. W. D. Nix, Met. Trans. A, Vol. 20A, p. 2217 (1989).
4. E. Klockholm and B. S. Berry, J. Electrochem Soc., Vol. 115, No. 8, p. 823 (1968).
5. W. Buckel, J. Vac. Sci. Technol., Vol. 6, No. 4, p.606 (1969).
6. D. S. Campbell in R. Niedermayer and H. Mayers (eds.), Basic Problems in Thin Film Physics, Vandenhoeck and Ruprecht, Gottingen, p. 223 (1966).
7. R. W. Hoffman, in G. Hass and R. E. Thun (eds.), Physics of Thin Films, Vol. 3, Academic Press, New York, p. 211 (1966).
8. Koreo Kinosita, Thin Solid Films, 12, p. 17 (1972).
9. R. Koch, J. Phys. Condens. Matter, 6, p. 9519 (1994).
10. J. A. Thornton and D. W. Hoffman, J. Vac. Sci. Technol., 14, p. 164 (1977).
11. J. A. Thornton and D. W. Hoffman, Thin Solid Films, 45, p. 387 (1977).
12. D. W. Hoffman and J. A. Thornton, J. Vac. Sci. Technol., 17, p. 380 (1980).
13. F. M. d'Heurle, Met. Trans., Vol. 1, p. 725 (1970).
14. D. W. Hoffman and J. A. Thornton, Thin Solid Films, 40, p. 40 (1977).
15. F. A. Doljack and R. W. Hoffman, Thin Solid Films, 12, p. 71 (1972).
16. S. J. Bull ans D. S. Rickerby, Mat. Res. Soc. Symp. Proc., Vol. 188, p. 337 (1990).
17. R. Saha, R. B. Inturi, and J. A. Barnard, Surface and Coatings Technol., (in press) (1995).
18. W. C. Oliver and G. M. Pharr, J. Mater. Res., Vol. 7, No. 6, p. 1564 (1992).
19. B. D. Cullity, Elements of X-ray Diffraction, 2nd edition, Addison-Wesseley Publishing Co., Inc., Reading, Massachussetts (1978), p. 102.
20. D. W. Hoffman and J. A. Thornton, J. Vac. Sci. Technol., Vol. 20, No. 3, p. 355 (1982).
21. A. Mutscheller, L. A. Clevenger, J. M. E. Harper, C. Cabral Jr., and K. Barmak, Mat. Res. Soc. Symp. Proc., Vol. 239, p. 51 (1992).
22. P. R. Stuart, Vacuum, Vol. 19, No. 11, p. 507 (1969).
23. C. Cabral Jr., L. A. Clevenger, and R. G. Schad, Mat. Res. Soc. Symp. Proc., Vol. 308, p. 57 (1993).
24. P. Chaudhari, J. Vac. Sci. Technol., Vol. 9, No. 1, p. 520 (1971).
25. Donald S. Gardner and Paul A. Flinn, J. Appl. Phys., Vol. 67, No. 4, p. 1831 (1990).
26. Paul A. Flinn, J. Mater. Res., Vol. 6, No. 7, p. 1498 (1991).

AUTHOR INDEX

SUBJECT INDEX